MEDICINAL SPICES AND VEGETABLES FROM AFRICA

MEDICINAL SPICES AND VEGETABLES FROM AFRICA

Therapeutic Potential against Metabolic, Inflammatory, Infectious and Systemic Diseases

Edited by

VICTOR KUETE
University of Dschang
Dschang, Cameroon

ACADEMIC PRESS
An imprint of Elsevier
elsevier.com

Academic Press is an imprint of Elsevier
125 London Wall, London EC2Y 5AS, United Kingdom
525 B Street, Suite 1800, San Diego, CA 92101-4495, United States
50 Hampshire Street, 5th Floor, Cambridge, MA 02139, United States
The Boulevard, Langford Lane, Kidlington, Oxford OX5 1GB, United Kingdom

Copyright © 2017 Elsevier Inc. All rights reserved.

No part of this publication may be reproduced or transmitted in any form or by any means, electronic or mechanical, including photocopying, recording, or any information storage and retrieval system, without permission in writing from the publisher. Details on how to seek permission, further information about the Publisher's permissions policies and our arrangements with organizations such as the Copyright Clearance Center and the Copyright Licensing Agency, can be found at our website: www.elsevier.com/permissions.

This book and the individual contributions contained in it are protected under copyright by the Publisher (other than as may be noted herein).

Notices
Knowledge and best practice in this field are constantly changing. As new research and experience broaden our understanding, changes in research methods, professional practices, or medical treatment may become necessary.

Practitioners and researchers must always rely on their own experience and knowledge in evaluating and using any information, methods, compounds, or experiments described herein. In using such information or methods they should be mindful of their own safety and the safety of others, including parties for whom they have a professional responsibility.

To the fullest extent of the law, neither the Publisher nor the authors, contributors, or editors, assume any liability for any injury and/or damage to persons or property as a matter of products liability, negligence or otherwise, or from any use or operation of any methods, products, instructions, or ideas contained in the material herein.

Library of Congress Cataloging-in-Publication Data
A catalog record for this book is available from the Library of Congress

British Library Cataloguing-in-Publication Data
A catalogue record for this book is available from the British Library

ISBN: 978-0-12-809286-6

For information on all Academic Press publications
visit our website at https://www.elsevier.com/

Publisher: Mica Haley
Acquisition Editor: Kristine Jones
Editorial Project Manager: Molly McLaughlin
Production Project Manager: Karen East and Kirsty Halterman
Designer: Alan Studholme

Typeset by Thomson Digital

CONTENTS

List of Contributors   xv
Preface   xix

PART I. DIVERSE DEGENERATIVE DISEASES IN AFRICA   1

1. Diseases in Africa: An Overview   3

J.C.N. Assob, D.S. Nsagha, A.J. Njouendou, D. Zofou, W.F. Sevidzem, A. Ketchaji, A. Chiara, T.E. Asangbeng, K.O. Dzemo, B.M. Samba, A.C. Wenze, E. Malika, V.B. Penlap, V. Kuete

1. General overview of disease burden in Africa   3
2. Infectious diseases   6
3. Neglected tropical diseases   32
4. Noncommunicable diseases   43
5. Malnutrition   57
6. Conclusions   63
References   64

2. Management of Inflammatory and Nociceptive Disorders in Africa   73

G.S. Taïwe, V. Kuete

1. Introduction   73
2. Epidemiology of inflammatory and nociceptive disorders   74
3. Gender and age of patients versus inflammatory and nociceptive disorders in Africa   75
4. Prevention   75
5. Inflammatory pain control policy in the African context   76
6. The inflammatory response   77
7. Pain processing and nociception pathway   84
8. Conventional drugs used for the management of inflammatory and nociception disorders   87
9. Medicinal plants used for the management of inflammatory and nociception disorders   88
10. Conclusions   89
References   90

3. Burden and Health Policy of Cancers in Africa   93

T.J. Makhafola, L.J. McGaw

1. Introduction   93
2. Common cancers in Africa   94

	3. Government control policies for cancer in African countries	100
	4. Conclusions	105
	References	106

4. Metabolic Syndromes and Public Health Policies in Africa   109
E.U. Nwose, P.T. Bwititi, V.M. Oguoma

1. Introduction	109
2. Prevalence of metabolic syndrome in Africa	109
3. Systematic review of public health policies on metabolic syndrome in Africa	117
4. Framework to develop public health policy for medicinal herbs—the case of pepper	121
5. Conclusions	127
References	127

5. Management of Infectious Diseases in Africa   133
R. Seebaluck-Sandoram, F.M. Mahomoodally

1. Introduction	133
2. HIV/AIDS	134
3. Dengue	135
4. Ebola	136
5. Chikungunya	137
6. Cholera	139
7. Cryptococcal meningitis	139
8. Malaria	140
9. Schistosomiasis	141
10. Antimicrobial resistance	142
11. Management of infectious diseases using African biodiversity	143
12. Conclusions	148
References	148

6. Overview of Governmental Support Across Africa Toward the Development and Growth of Herbal Medicine   153
D.O. Ochwang'i, J.A. Oduma

1. Introduction	153
2. Encouragement of use of medicinal plants in health care programs	154
3. Policies for conservation of medicinal plants and local community participation	157
4. Policy for restoring plants harvested in the wild and sustainability of use	158
5. Incentives to collectors and farmers to keep production of medicinal plants sustainable	159
6. Government support on medicinal plant research	159
7. Policies regarding export of medicinal plants	160

		8. Policy for trade in herbal medicine	161
		9. Herbal medicine regulation in Africa	163
		10. Conclusions and recommendations	165
		References	167

7. Preparation, Standardization, and Quality Control of Medicinal Plants in Africa — 171

M.O. Nafiu, A.A. Hamid, H.F. Muritala, S.B. Adeyemi

1. Introduction	171
2. Factors that affect the use of medicinal plant preparations	172
3. Modes of preparation of medicinal plants used in traditional medicine	173
4. Modes of preparation of extracts in research laboratories in Africa	176
5. Parameters for selecting an appropriate extraction method	179
6. Standardization of medicinal plant preparation in Africa and other parts of the world	180
7. Quality control of medicinal preparations	188
8. Conclusions	201
References	201

PART II. THERAPEUTIC POTENTIAL OF AFRICAN MEDICINAL SPICES AND VEGETABLES — 205

8. Antimicrobial Activities of African Medicinal Spices and Vegetables — 207

J.D.D. Tamokou, A.T. Mbaveng, V. Kuete

1. Introduction	207
2. Antimicrobial secondary metabolites and their modes of action	208
3. In vitro screening methods of phytochemicals for antimicrobial activities	213
4. Antimicrobial effects of African medicinal spices and vegetables	219
5. Other antimicrobial spices and vegetables from Africa	228
6. Antimicrobial mode of action of African spices and vegetables	229
7. Conclusions	230
References	231

9. Anti-inflammatory and Anti-nociceptive Activities of African Medicinal Spices and Vegetables — 239

J.P. Dzoyem, L.J. McGaw, V. Kuete, U. Bakowsky

1. Introduction	239
2. Methods used in the screening of antiinflammatory and antinociceptive activity of African spices and vegetables	240
3. Selected African spices with antiinflammatory and antinociceptive activity	247
4. Prominent antiinflammatory active ingredients found in African spices and vegetables	255

		5. Clinical trials	259
		6. Conclusions	259
		References	259

10. Anticancer Activities of African Medicinal Spices and Vegetables — 271
V. Kuete, O. Karaosmanoğlu, H. Sivas

1. Introduction — 271
2. In vitro screening methods of phytochemicals for anticancer activities — 272
3. Anticancer potential of phytochemicals — 277
4. Antiproliferative effects of African medicinal spices — 277
5. Cytotoxicity of other African medicinal vegetables — 286
6. Mode of action of African spices, vegetable's extracts, and derived products — 290
7. Conclusions — 292
References — 292

11. Antiemetic African Medicinal Spices and Vegetables — 299
S. Tchatchouang, V.P. Beng, V. Kuete

1. Introduction — 299
2. Modes of action of antiemetics — 300
3. Antiemetics screening methods — 301
4. In vitro and in vivo antiemetic activities of chemicals — 302
5. African medicinal spices as sources of antiemetics — 304
6. African vegetables as sources of antiemetics — 308
7. Conclusions — 309
References — 310

12. African Medicinal Spices and Vegetables and Their Potential in the Management of Metabolic Syndrome — 315
V. Kuete

1. Introduction — 315
2. In vitro screening methods of phytochemicals against metabolic syndrome — 316
3. Potential of phytochemicals against metabolic syndrome — 317
4. Effects of African medicinal spices on metabolic syndrome — 318
5. Conclusions — 324
References — 324

13. Other Health Benefits of African Medicinal Spices and Vegetables — 329
V. Kuete

1. Introduction — 329
2. African medicinal spices and vegetables in the treatment of age-related and Alzheimer's diseases — 330

3. Oxidative stress and antioxidant effects of African medicinal spices and vegetables	334
4. Medicinal spices and vegetables from Africa used in the treatment of epilepsy	338
5. Prevention of rheumatoid arthritis with African medicinal spices and vegetables	340
6. African spices and vegetables and their potential effects on obesity	341
7. Role of African spices and vegetables in human fertility	344
8. Conclusions	344
References	345

PART III. POPULAR AFRICAN MEDICINAL SPICES AND VEGETABLES, AND THEIR HEALTH EFFECTS — 351

14. *Allium cepa* — 353
V. Kuete

1. Introduction	353
2. Cultivation and distribution of *Allium cepa*	353
3. Chemistry of *Allium cepa*	354
4. Pharmacology of *Allium cepa*	356
5. Patents with *Allium cepa*	359
6. Conclusions	360
References	360

15. *Allium sativum* — 363
V. Kuete

1. Introduction	363
2. Cultivation and distribution of *Allium sativum*	364
3. Chemistry of *Allium sativum*	365
4. Pharmacology of *Allium sativum*	368
5. Toxicity of *Allium sativum*	373
6. Patents with *Allium sativum*	373
7. Conclusions	374
References	374

16. *Canarium schweinfurthii* — 379
V. Kuete

1. Introduction	379
2. Botanical aspects and distribution of *Canarium schweinfurthii*	379
3. Chemistry of *Canarium schweinfurthii*	380
4. Pharmacology of *Canarium schweinfurthii*	381
5. Conclusions	383
References	383

17. Cinnamon Species — 385
A.T. Mbaveng, V. Kuete

1. Introduction — 385
2. Botanical aspects — 385
3. Chemistry of *Cinnamomum cassia* and *Cinnamomum zeylanicum* — 386
4. Pharmacology of *Cinnamomum cassia* and *Cinnamomum zeylanicum* — 387
5. Conclusions — 393
References — 393

18. *Cymbopogon citratus* — 397
O.A. Lawal, A.L. Ogundajo, N.O. Avoseh, I.A. Ogunwande

1. Introduction — 397
2. Chemical constituents of *C. citratus* — 400
3. Pharmacological activities of *C. citratus* — 407
4. Patents — 416
5. Conclusions — 416
References — 416

19. *Curcuma longa* — 425
L.K. Omosa, J.O. Midiwo, V. Kuete

1. Introduction — 425
2. Cultivation and distribution of *Curcuma longa* — 426
3. Chemistry of *Curcuma longa* — 426
4. Pharmacology of *Curcuma longa* — 426
5. Patents with *Curcuma longa* — 432
6. Conclusions — 433
References — 433

20. *Lactuca sativa* — 437
J.A.K. Noumedem, D.E. Djeussi, L. Hritcu, M. Mihasan, V. Kuete

1. Introduction — 437
2. History — 438
3. World and African distribution — 438
4. Chemistry — 438
5. Pharmacological activities — 441
6. Other pharmacological activities — 446
7. Clinical trials — 447
8. Conclusions — 447
References — 447

21. *Mangifera indica* L. (Anacardiaceae) — 451
S. Derese, E.M. Guantai, Y. Souaibou, V. Kuete

1. Introduction — 451
2. Botanical description — 452
3. Origin and distribution — 454
4. Classification and cultivars of *Mangifera indica* — 455
5. Phytochemistry of *Mangifera indica* — 455
6. Pharmacological activity of *Mangifera indica* — 469
7. Clinical trials — 476
8. Toxicity status of *Mangifera indica* — 477
9. Conclusions — 477
References — 478

22. *Moringa oleifera* — 485
V. Kuete

1. Introduction — 485
2. Cultivation and distribution of *Moringa oleifera* — 485
3. Chemistry of *Moringa oleifera* — 486
4. Pharmacology of *Moringa oleifera* — 487
5. Conclusions — 493
References — 493

23. *Myristica fragrans*: A Review — 497
V. Kuete

1. Introduction — 497
2. Botanical aspects and distribution of *Myristica fragrans* — 497
3. Chemistry of *Myristica fragrans* — 498
4. Pharmacology of *Myristica fragrans* — 499
5. Toxicity of Myristica fragrans — 507
6. Patents with *Myristica fragrans* — 508
7. Conclusions — 508
References — 509

24. *Passiflora edulis* — 513
G.S. Taïwe, V. Kuete

1. Introduction — 513
2. Botanical description — 514
3. Propagation — 515
4. Traditional or ethnomedicinal uses — 515

5. Phytochemistry	516
6. In vitro and in vivo pharmacological studies	519
7. Clinical trials	522
8. Safety profile and pharmacovigilance data	522
9. Conclusions	523
References	523

25. *Petroselinum crispum*: A Review — 527
C. Agyare, T. Appiah, Y.D. Boakye, J.A. Apenteng

1. Introduction	527
2. Plant description	528
3. Geographical distribution	529
4. Ethnomedicinal uses	529
5. Pytochemical constituents	530
6. Pharmacological properties	534
7. Toxicity	541
8. Clinical trials	541
9. Patents	542
10. Conclusions	543
References	544

26. *Sesamum indicum* — 549
S.O. Amoo, A.O.M. Okorogbona, C.P. Du Plooy, S.L. Venter

1. Introduction	549
2. Origin of the crop	551
3. Chemical properties of *Sesamum indicum*	551
4. Phytochemical studies	552
5. Pharmacological evaluation of plant extracts	559
6. Production and cultivation of *Sesamum indicum*	572
7. Conclusions	575
References	576

27. African Medicinal Spices of Genus *Piper* — 581
I.A. Oyemitan

1. Introduction	581
2. Taxonomy of the genus *Piper*	582
3. Diversity in the *Piper* species	582
4. Distribution of the genus *Piper*	583
5. Ethnobotanical and ethnosocial importance of the genus *Piper*	584
6. Ethnomedicinal applications of the genus *Piper*	585

7. Economic and commercial prospect of the genus *Piper*	585
8. Bioprospecting and conservation status of the genus *Piper*	586
9. Phytochemical constituents of the genus *Piper*	587
10. Biological and pharmacological effects of the genus *Piper*	589
11. Toxicity profile	592
12. Conclusions	592
References	593

28. *Thymus vulgaris* — 599

V. Kuete

1. Introduction	599
2. Cultivation and distribution of *Thymus vulgaris*	599
3. Chemistry of *Thymus vulgaris*	600
4. Pharmacology of *Thymus vulgaris*	602
5. Patents with *Thymus vulgaris*	606
6. Conclusions	606
References	607

29. *Syzygium aromaticum* — 611

A.T. Mbaveng, V. Kuete

1. Introduction	611
2. Botanical aspect and distribution of *Syzygium aromaticum*	611
3. Chemistry of *Syzygium aromaticum*	612
4. Pharmacology of *Syzygium aromaticum*	613
5. Patents with *Syzygium aromaticum*	621
6. Toxicity of *Syzygium aromaticum*	622
7. Conclusions	622
References	623

30. *Zingiber officinale* — 627

A.T. Mbaveng, V. Kuete

1. Introduction	627
2. Botanical aspect, distribution, and production of *Zingiber officinale*	628
3. Chemistry of *Zingiber officinale*	628
4. Pharmacology of *Zingiber officinale*	630
5. Patents with *Zingiber officinale*	636
6. Conclusions	636
References	637

Index — 641

LIST OF CONTRIBUTORS

Sherif Babatunde Adeyemi
Department of Plant Biology, University of Ilorin, Ilorin, Kwara, Nigeria

Christian Agyare
Department of Pharmaceutics, Kwame Nkrumah University of Science and Technology, Kumasi, Ghana

Ketchaji Alice
Department of Public Health and Hygiene, Faculty of Health Sciences, University of Buea, Buea, Cameroon

Stephen O. Amoo
Agricultural Research Council–Roodeplaat Vegetable and Ornamental Plant Institute, Pretoria, South Africa

John Antwi Apenteng
Department of Pharmaceutical Sciences, Central University College, Accra, Ghana

Theresa Appiah
Department of Pharmaceutics, Kwame Nkrumah University of Science and Technology, Kumasi, Ghana

Tanue Elvis Asangbeng
Department of Medical Laboratory Sciences, University of Buea, Buea, Cameroon

Jules Clement Nguedia Assob
Department of Medical Laboratory Sciences, Faculty of Health Sciences, University of Buea, Buea, Cameroon

Nudewhenu O. Avoseh
Natural Products Research Unit, Department of Chemistry, Faculty of Science, Lagos State University, Lagos, Nigeria

Udo Bakowsky
Department of Pharmaceutical Technology and Biopharmaceutics, Marburg University, Marburg, Germany

Veronique P. Beng
Department of Biochemistry, Faculty of Science, Biotechnology Centre, University of Yaoundé I, Yaoundé, Cameroon

Yaw Duah Boakye
Department of Pharmaceutics, Kwame Nkrumah University of Science and Technology, Kumasi, Ghana

Samba Melvis Bora
Department of Public Health and Hygiene, Faculty of Health Sciences, University of Buea, Buea, Cameroon

Phillip Taderera Bwititi
Faculty of Science, Charles Sturt University, Bathurst, NSW, Australia

Achangwa Chiara
Department of Public Health and Hygiene, Faculty of Health Sciences, University of Buea, Buea, Cameroon

Zofou Denis
Department of Biochemistry and Molecular Biology, Faculty of Science, University of Buea, Buea, Cameroon

Solomon Derese
Department of Chemistry, University of Nairobi, Nairobi, Kenya

Doriane E. Djeussi
Department of Biochemistry, Faculty of Science, University of Dschang, Dschang, Cameroon

Christian Phillipus Du Plooy
Agricultural Research Council–Roodeplaat Vegetable and Ornamental Plant Institute, Pretoria, South Africa

Kibu Odette Dzemo
Department of Medical Laboratory Sciences, University of Buea, Buea, Cameroon

Jean P. Dzoyem
Department of Biochemistry, Faculty of Science, University of Dschang, Dschang, Cameroon; Department of Pharmaceutical Technology and Biopharmaceutics, Marburg University, Marburg, Germany

Eric M. Guantai
Department of Pharmacology and Pharmacognosy, University of Nairobi, Nairobi, Kenya

Abdulmumeen Amao Hamid
Department of Chemistry, University of Ilorin, Ilorin, Kwara, Nigeria

Lucian Hritcu
Department of Biology, Alexandru Ioan Cuza University, Iași, Romania

Oğuzhan Karaosmanoğlu
Department of Biology, Science Faculty, Anadolu University, Eskişehir; Department of Biology, Kamil Özdağ Science Faculty, Karamanoğlu Mehmetbey University, Karaman, Turkey

Victor Kuete
Department of Biochemistry, Faculty of Science, University of Dschang, Dschang, Cameroon

Oladipupo A. Lawal
Natural Products Research Unit, Department of Chemistry, Faculty of Science, Lagos State University, Lagos, Nigeria

Fawzi M. Mahomoodally
Department of Health & Sciences, Faculty of Science, University of Mauritius, Réduit, Mauritius

Tshepiso J. Makhafola
Department of Life and Consumer Sciences, College of Agriculture and Environmental Science, University of South Africa, Pretoria, South Africa

Esembeson Malika
Department of Public Health and Hygiene, Faculty of Health Sciences, University of Buea, Buea, Cameroon

Armelle T. Mbaveng
Department of Biochemistry, Faculty of Science, University of Dschang, Dschang, Cameroon

Lyndy J. McGaw
Phytomedicine Programme, Department of Paraclinical Sciences, Faculty of Veterinary Science, University of Pretoria, Pretoria, South Africa

Jacob O. Midiwo
Department of Chemistry, School of Physical Sciences, University of Nairobi, Nairobi, Kenya

Marius Mihasan
Department of Biology, Alexandru Ioan Cuza University, Iași, Romania

Hamdalat Folake Muritala
Department of Biochemistry, University of Ilorin, Ilorin, Kwara, Nigeria

Mikhail Olugbemiro Nafiu
Department of Biochemistry, University of Ilorin, Ilorin, Kwara, Nigeria

Abdel Jelil Njouendou
Department of Medical Laboratory Sciences, University of Buea, Buea, Cameroon

Jaures A.K. Noumedem
Department of Biochemistry, Faculty of Science, University of Dschang, Dschang, Cameroon

Dickson Shey Nsagha
Department of Public Health and Hygiene, Faculty of Health Sciences, University of Buea, Buea, Cameroon

Ezekiel Uba Nwose
Faculty of Science, Charles Sturt University, Bathurst, NSW, Australia

Dominic O. Ochwang'i
Department of Veterinary Anatomy and Physiology, University of Nairobi, Nairobi, Kenya

Jemimah A. Oduma
Department of Veterinary Anatomy and Physiology, University of Nairobi, Nairobi, Kenya

Akintayo L. Ogundajo
Natural Products Research Unit, Department of Chemistry, Faculty of Science, Lagos State University, Lagos, Nigeria

Isiaka Ajani Ogunwande
Natural Products Research Unit, Department of Chemistry, Faculty of Science, Lagos State University, Lagos, Nigeria

Victor Maduabuchi Oguoma
School of Psychological and Clinical Science, Charles Darwin University, Darwin, NT, Australia

Alfred Oghode Misaiti Okorogbona
Agricultural Research Council–Roodeplaat Vegetable and Ornamental Plant Institute, Pretoria, South Africa

Leonidah Kerubo Omosa
Department of Chemistry, School of Physical Sciences, University of Nairobi, Nairobi, Kenya

Idris Ajayi Oyemitan
Department of Pharmacology, Faculty of Pharmacy, Obafemi Awolowo University, Ile-Ife, Osun, Nigeria; Division of Academic Affairs & Research, Directorate of Research, Innovation & Development, Walter Sisulu University, Mthatha, South Africa

Roumita Seebaluck-Sandoram
Department of Health & Sciences, Faculty of Science, University of Mauritius, Réduit, Mauritius

Wirsiy Frankline Sevidzem
Department of Public Health and Hygiene, Faculty of Health Sciences, University of Buea, Buea, Cameroon

Hülya Sivas
Department of Biology, Science Faculty, Anadolu University, Eskişehir, Turkey

Germain Sotoing Taïwe
Department of Zoology and Animal Physiology, Faculty of Science, University of Buea, Buea, Cameroon

Jean-de-Dieu Tamokou
Department of Biochemistry, Faculty of Science, University of Dschang, Dschang, Cameroon

Serges Tchatchouang
Department of Biochemistry, Faculty of Science, Biotechnology Centre, University of Yaoundé I, Yaoundé, Cameroon

Sonia L. Venter
Agricultural Research Council–Roodeplaat Vegetable and Ornamental Plant Institute, Pretoria, South Africa

Ayima Charlotte Wenze
Department of Public Health and Hygiene, Faculty of Health Sciences, University of Buea, Buea, Cameroon

Souaibou Yaouba
Department of Chemistry, University of Nairobi, Nairobi, Kenya

PREFACE

Medicinal plants constitute a good alternative to conventional medicine, considering the rich biodiversity of the continent. They are widely used in Africa to heal various types of health conditions and it has always been stated that about 80% of the population in the continent use traditional healing as primary health care. This could be true, but these are very irregular consultations that cannot be definitively attributed to a class of persons in the population. For example, a person may appeal to traditional therapy in context of a specific disease and unfavorable financial conditions, but used conventional medicine in case of financial ease or another disease. In Africa, traditional medicine is often stigmatized by many people, especially the middle- and high-income people.

Governments in Africa are also providing few incentives to boost traditional medicine including the use of medicinal plants. This position is understandable insofar as there are few studies on the safety of plants used as well as the methods used to obtain various therapeutic preparations. Meanwhile, African populations are disproportionately affected by infectious and cardiovascular diseases, cancers and other illnesses. This remains a contrast if we take into account the exceptional floral biodiversity of the continent and the number of medicinal plants found in each African country. If African countries are endowed with appropriate infrastructure to demonstrate the efficacy and safety of medicinal plants used in the continent, it would be likely that governments would change point of view and encourage not only the use of these plants, but also the production of improved phytodrugs.

Until this is done, and given the fact that the miseries of the people in health matters will not wait, African researchers should propose acceptable alternatives to conventional medicine, which is very expensive for the majority of the population. One of the most acceptable solutions is to propose the use of functional foods and medicinal herbs, such as spices and vegetables, not only as a palliative to the conventional medicine, but as effective therapeutic alternative. This could be more accepted and popularized worldwide.

Given the scarcity of reference updated books adapted to Africa in this theme, I undertook to edit this document, also having an academic purpose. Hence, the background on diverse degenerative diseases in Africa will be discussed from Chapters 1–7. To place African research globally and for a possible academic use, emphasis was put on the methods used in pharmacological surveys in Part II, dealing with the therapeutic potential of African medicinal spices and vegetables (Chapters 8–13). The reviews of the most popular African functional foods, medicinal spices, and vegetables have been discussed in details in Part III (Chapters 14–30). The topics of this book are of interest to scientists of several fields, including Pharmaceutical Science, Pharmacognosy, Complementary and Alternative Medicine, Ethnomedicine, Pharmacology, Medical and Public

Health Sciences, Medicinal Chemistry, Phytochemistry, and Biochemistry. The highlight of this book is an exhaustive compilation of scientific data related to the Pharmacological survey of African medicinal spices and vegetables by top scholars from several countries worldwide. Finally, I would like to thank Molly McLaughlin, the Editorial Project Manager at 50 Hampshire Street Cambridge, Massachusetts 02139, the technical assistant, and Karen East and Kirsty Halterman, the production managers for their help and fruitful collaboration.

Prof. Dr. Victor Kuete

PART I

Diverse Degenerative Diseases in Africa

1. Diseases in Africa: An Overview — 3
2. Management of Inflammatory and Nociceptive Disorders in Africa — 73
3. Burden and Health Policy of Cancers in Africa — 93
4. Metabolic Syndromes and Public Health Policies in Africa — 109
5. Management of Infectious Diseases in Africa — 133
6. Overview of Governmental Support Across Africa Toward the Development and Growth of Herbal Medicine — 153
7. Preparation, Standardization, and Quality Control of Medicinal Plants in Africa — 171

CHAPTER 1

Diseases in Africa: An Overview

J.C.N. Assob, D.S. Nsagha, A.J. Njouendou, D. Zofou, W.F. Sevidzem, A. Ketchaji,
A. Chiara, T.E. Asangbeng, K.O. Dzemo, B.M. Samba, A.C. Wenze, E. Malika,
V.B. Penlap, V. Kuete

1 GENERAL OVERVIEW OF DISEASE BURDEN IN AFRICA

The notion of disease burden refers to the impact of a health problem as measured by financial cost, mortality, morbidity, or other indicators. It is often quantified in terms of quality-adjusted life years (QALYs) or disability-adjusted life years (DALYs), both of which quantify the number of years lost due to disease (YLDs). One DALY can be thought of as 1 year of healthy life lost, and the overall disease burden can be thought of as a measure of the gap between current health status and the ideal health status (where the individual lives to old age free from disease and disability) (David Kay and Carlos, 2000; Prüss-Üstün et al., 2003; Murray et al., 2013).

People around the world are living longer than ever before, and the population is getting older. The number of people in the world is growing because many countries have made remarkable progress in preventing child deaths. As a result, disease burden is increasingly defined by disability instead of premature mortality. The leading causes of deaths and disabilities have changed from communicable diseases in children to non-communicable diseases in adults; whereas eating too much has overtaken hunger as a leading risk factor for illness (Institute for Health Metrics and Evaluation, 2013). Across the African region, overall progress has been made in decreasing death rates between 1970 and today. Each country has witnessed an increase in its average age of death with some variations; Cape Verde demonstrating the greatest gain (about 28 years) and Chad showing the smallest improvement (1.4 years). Decline in mortality rates largely varied by age, with greatest improvements for young children (Institute for Health Metrics and Evaluation, 2013).

Death rates for children between 1 and 4 years old declined by 65% between 1970 and 2010, whereas mortality rates rose for women and men at different ages (a 39% increase for women aged from 25 to 39 and a 50% increase for men aged from 30 to 34). Premature death and disability caused by some communicable diseases and newborn conditions have decreased, but they remain leading causes of premature death and illness. Between 1990 and 2010, the African region had succeeded in decreasing premature death and disability, also known as healthy years lost, from lower respiratory infections, diarrheal diseases, and protein-energy malnutrition; however, these conditions are still the leading causes of disease burden in the region, especially in lower income countries like Niger and Sierra Leone. Malaria and human immunodeficiency virus (HIV)/acquired immune deficiency syndrome (AIDS) are now the first- and second major causes of premature death and disability (Institute for Health Metrics and Evaluation, 2013).

Noncommunicable diseases are rapidly rising, especially for wealthier countries in the region. Road injuries have taken a growing toll on health in the African region. Healthy years lost from road injuries have increased by 76% between 1990 and 2010, with substantial country variation (ranging from a rise of 9% in Madagascar to 87% in Congo) (Institute for Health Metrics and Evaluation, 2013). To ensure a health system is adequately aligned to a population's true health challenges, policymakers must be able to compare the effects of different diseases that kill people prematurely and cause ill health. The original Global Burden of Disease (GBD) study's creators developed a single measurement, DALYs, to quantify the number of years of life lost as a result of both premature death and disability (Institute for Health Metrics and Evaluation, 2013). One DALY equals one lost year of healthy life. DALYs will be referred to by their acronym, as years of healthy life lost, and years lost due to premature death and disability.

Decision-makers can use DALYs to quickly assess the impact caused by conditions, such as cancer versus depression using a comparable metric. Considering the number of DALYs instead of causes of death alone provides a more accurate picture of the main drivers of poor health. GBD provides high-quality estimates of diseases and injuries that are more rigorous than those published by disease-specific advocates. Beyond providing a comparable and comprehensive picture of causes of premature death and disability, GBD also estimates the disease burden attributable to different risk factors (Institute for Health Metrics and Evaluation, 2013). The GBD approach goes beyond risk-factor prevalence, such as the number of smokers or heavy drinkers in a population. With comparative risk assessment, GBD incorporates both the prevalence of a given risk factor and the relative harm caused by that risk factor. It counts premature death and disability attributable to high blood pressure, tobacco and alcohol use, lack of exercise, air pollution, poor diet, and other risk factors that lead to ill health.

According to the World Health Organization (WHO) figures, the Ebola epidemic in West Africa claimed 11,323 lives (WHO, 2003). Although the virulence and rapid spread of the Ebola virus were major causes of concern, it is important to understand the mortality figures in the broader sub-Saharan African context. Deaths in Africa, for example in 2012, fell largely in all the WHO groups. This was such that:

- Group 1: (death through communicable diseases, and perinatal, maternal, and nutritional causes): 5.9 million deaths amounting to 61.7% of all deaths in sub-Saharan Africa.
- Group 2: death as a result of noncommunicable diseases accounted for 2.7 million deaths or 28.6% of all deaths. This category includes heart disease (293,000 deaths), various forms of cancer (426,000), and diabetes (175,000).

- Group 3: deaths through injury, amounted to 939,000 deaths, or 9.8% of the total. Group 3 causes of death include unintentional injuries, such as road accidents (207,000), and intentional injuries, such as interpersonal violence (132,000) and collective violence (14,000).

Noncommunicable and lifestyle diseases are the top killers in high-income countries, accounting for 67.8% of deaths in 2012. In contrast, many of the top killers in sub-Saharan Africa—lower respiratory tract infections, tuberculosis (TB), diarrheal disease, and malaria—are preventable and treatable, given adequate health care systems and resources. According to WHO's figures, the five top killers in Africa in 2012 were the following: HIV/AIDS, lower respiratory tract infections, diarrheal diseases, malaria, and strokes.

People living in the African region of the WHO confront the world's most dramatic public health crisis, with the hope that over time the region can address the health challenges it faces, given sufficient international support (WHO, 2006a). HIV/AIDS continues to devastate the WHO Africa region, which has 11% of the world's population but 60% of the people with HIV/AIDS. Although HIV/AIDS remains the leading cause of death in adults, more and more people are receiving life-saving treatment. The number of HIV-positive people on antiretroviral medicines increased eightfold, from 100,000 in December 2003 to 810,000 in December 2005 (WHO, 2006a). More than 90% of the estimated 300–500 million malarial cases that occur worldwide every year are in Africans, mainly in children under 5 years, but most countries are moving toward better treatment policies. Of the 42 malaria-endemic countries in the African region, 33 have adopted artemisinin-based combination therapy (ACT)—the most effective antimalarial medicines available today—as first-line treatment (WHO, 2006a). River blindness has been eliminated as a public health problem, and guinea worm control efforts have resulted in a 97% reduction in cases since 1986. Leprosy is close to elimination—which means there is less than one case per 10,000 people in the region (WHO, 2006a).

Most countries are making good progress on preventable childhood illness. Polio is close to eradication, and 37 countries are reaching 60% or more of their children with measles immunization. Overall measles deaths have declined by more than 50% since 1999 (WHO, 2006a). While drawing the world's attention to recent successes, the report offers a candid appraisal of major hurdles, such as the high rate of maternal and newborn mortality overall in the region. Of the 20 countries with the highest maternal mortality ratios worldwide, 19 are in Africa; and the region has the highest neonatal death rate in the world (WHO, 2006a). Then there is the strain on African health systems imposed by the high burden of life-threatening communicable diseases coupled with increasing rates of noncommunicable diseases, such as hypertension and coronary heart disease. Basic sanitation needs remain unmet for many with only 58% of people living in sub-Saharan Africa having access to safe water supplies. Noncommunicable diseases, such as hypertension, heart disease, diabetes, are on the rise, and injuries remain among the top causes of death in the region (WHO, 2006a).

2 INFECTIOUS DISEASES
2.1 HIV/AIDS in Africa

HIV, the virus that causes AIDS, "acquired immunodeficiency syndrome" has become one of the world's most serious health and development challenges. In Canada, and most other western countries, these drugs are readily available and are used by nearly all AIDS patients. These drugs, in addition to effective education and awareness regarding the disease, have helped developed nations get their HIV/AIDS epidemics under control. However, sadly it is not like that everywhere in the world. In Africa particularly, one of the world's poorest places, AIDS remains a wildly virulent disease that is still killing millions and debilitating entire swaths of the population, mainly a result of people not being able to afford treatment. It is a major public health concern and cause of death in many parts of Africa (Michael, 2014). Although the continent is home to about 15.2% of the world's population, sub-Saharan Africa alone accounted for an estimated 69% of all people living with HIV and 70% of all AIDS deaths in 2011 (Apps.who.int, 2016).

Countries in North Africa and the Horn of Africa have significantly lower prevalence rates, as their populations typically engage in fewer high-risk cultural patterns that have been implicated in the virus's spread in sub-Saharan Africa (UNAIDS, 2010). Southern Africa is the most affected region in the continent.

2.1.1 Epidemiology of HIV/AIDS in Africa

There are many reasons why Africa has such a high number of AIDS cases. Between 1999 and 2000 more people died of AIDS in Africa than in all the wars on the continent, as mentioned by the previous UN Secretary General, Kofi Annan (Globalissues.org, 2016). The death toll is expected to have a severe impact on many economies in the region. In some nations, it is already being felt. Life expectancies in some nations are already decreasing rapidly, whereas mortality rates are increasing. Each day, 6000 Africans die from AIDS. Each day, an additional 11,000 are infected (Globalissues.org, 2016). The first cases were reported in 1981 and HIV prevalence rates and the number of people dying from AIDS varied between African countries in 2010 (UNAIDS, 2010): in Somalia and Senegal—HIV prevalence is under 1% of the adult population, whereas in Namibia, Zambia, and Zimbabwe, 10–15% of adults are infected with HIV, in South Africa, the HIV prevalence stands at 17.8%, exceeding 20% in Botswana (24.8%), Lesotho (23.6%), and Swaziland (25.9%). In Cameroon and Gabon, HIV prevalence is 5.3 and 3.6%, respectively, while the prevalence in Nigeria is 4.1%. In East Africa including Uganda, Kenya, and Tanzania the prevalence is above 5%.

2.1.2 Control
2.1.2.1 Prevention
Numerous prevention interventions have been developed to fight HIV, and new tools, such as vaccines, are currently under investigation. Effective prevention strategies include

behavior change programs, condoms, HIV testing, blood supply safety, harm reduction efforts for injecting drug users, and male circumcision. Additionally, recent research has shown that providing HIV treatment to people with HIV significantly reduced the risk of transmission to their negative partners. Preexposure antiretroviral prophylaxis (PrEP) has also been shown to be an effective HIV prevention strategy in individuals at high risk for HIV infection. Experts recommend that prevention be based on "knowing your epidemic," that is, tailoring prevention to the local context and epidemiology, and using a combination of prevention strategies, bringing programs to scale, and sustaining efforts over time.

2.1.2.2 Treatment

Treatments include the use of combination antiretroviral therapy (ART) to attack the virus itself, and medications to prevent and treat opportunistic infections that can occur when the immune system is compromised by HIV. In light of recent research findings, WHO released a guideline in 2015 recommending starting HIV treatment earlier in the course of illness (UNAIDS, 2015). Approximately 76% of all people receiving ART in sub-Saharan Africa are virally suppressed, which means they are likely healthier and less likely to transmit the virus (WHO, 2013).

2.1.2.3 WHO policies

Since 2002, ART programs have slowly improved in Africa. Initially, HIV-infected people had to wait until they were seriously immunocompromised, with a CD4 T-cell count below 200 mm^{-3}, to begin ART. In 2013, WHO released a guideline that recommends daily oral PrEP as a form of prevention for high-risk individuals in combination with other prevention methods (WHO, 2013). The threshold was raised to 350 and then 500, as the importance of earlier initiation of treatment was recognized. Improved tools and strategies followed, as did consensus on treatment guidelines and international funding. The trajectory toward ending AIDS seemed assured, and international goals grew from "3 by 5" (treating 3 million people by 2005), to "15 by 15," to a call from the Joint United Nations Programme on HIV/AIDS for "90-90-90" by 2020: 90% of people living with HIV tested, 90% receiving treatment, and 90% with an undetectable 1-viral load.

2.1.2.4 Challenges

Major progress has been made in Africa's fight against HIV/AIDS but one of the major problems remaining is giving the thousands living with AIDS the drug therapies that can let them live a long life. "In terms of the reality, unfortunately the reality is that while there's been huge progress made in the fight against AIDS, there still remains important efforts to do" (Ellman, 2015).

In addition, the main reason is because many Africans are having unprotected sex, and many people in Africa are not educated about the dangers of AIDS or unprotected sex. Recently the President of South Africa made a comment doubting the existence of AIDS. This is one of the other reasons why people in Africa have high cases of AIDS. Many just don't believe that AIDS exists, and so they don't take proper precautions to avoid contracting the disease. In most of Africa, despite shortages of health workers, initiating ART in a healthy HIV-positive adult is the begining of monthly queues at clinics and pharmacies to see overburdened medical staff. To obtain treatment, such patients face lots of challenges like long walks to health centers or high transportation costs, hours of queuing, and poor, sometimes stigmatizing, consultation. Unlike symptomatic patients, these patients see no short-term benefit from treatment.

Hence, HIV/AIDS is still a challenge in Africa; providing health care, antiretroviral treatment, and support to people with HIV-related illnesses can help to reduce the annual toll of new HIV infections. As guidelines edge closer to universal treatment for HIV-infected people regardless of CD4 count, more people can choose to receive ART before they develop symptoms. Earlier treatment represents an opportunity not only to prevent illness and transmission but also to inform and empower people living with HIV–AIDS while reducing the burden on health care systems.

2.2 Overview of TB in Africa

TB remains one of the world's deadliest communicable diseases; it is currently responsible for more years of healthy life lost (2.5% of all DALYs) than any other infectious disease, like AIDS and malaria (Ibrahim et al., 2015). In 2013, an estimated 9.0 million people developed TB and 1.5 million died from the disease, 360,000 of whom were HIV positive (Corbett et al., 2003). TB is slowly declining each year and it is estimated that 37 million lives were saved between 2000 and 2013 through effective diagnosis and treatment (Corbett et al., 2003). TB is present in all regions of the world and the Global Tuberculosis Report 2014 includes data compiled from 202 countries and territories. The report shows higher global totals for new TB cases and deaths in 2013 than previously, reflecting the use of increased and improved national data (Corbett et al., 2003). A special supplement to the 2014 report highlights the progress that has been made in surveillance of drug resistant TB over the last two decades, and the response at global and national levels in recent years (Corbett et al., 2003).

The year 2015 is a watershed moment in the battle against TB. It marks the deadline for global TB targets set in the context of the millennium development goals (MDGs), and is a year of transitions: from the MDGs to a new era of sustainable development goals (SDGs), and from the Stop TB Strategy to the End TB Strategy. It is also two decades since WHO established a global TB monitoring system; since that time, 20 annual rounds of data collection have been completed (WHO, 2015). A decade ago, the problem of TB in Africa attracted little attention, not even meriting a chapter in the first edition of Disease and Mortality in sub-Saharan Africa. Part of the reason was that incidence of TB was low and falling in most parts of the continent (Organization, 2015). The burden of TB in sub-Saharan Africa is far greater today, due to continuing poverty and political instability in parts of the continent which has inhibited progress in implementing effective TB control measures. However, the principal reason for the resurgence of TB in Africa is not the deterioration of control programs. Rather, it is the link between TB and the HIV and the (HIV/AIDS) (Dye et al., 1998).

2.2.1 Epidemiology

TB is the second-most common cause of death from infectious disease (after those due to HIV/AIDS). In 2004, the incidence rate of TB in the WHO African region was growing at approximately 3% per year, and at 4% per year in eastern and southern Africa (the areas most affected by HIV), faster than on any other continent, and considerably faster than the 1% per year global increase (WHO, 2006c). In Africa, it primarily affects adolescents and young adults. In several African countries, including those with well-organized control programs annual TB case-notification rates have risen more than fivefold since the mid-1980s, reaching more than 400 cases per 100,000 people (Harries et al., 2001). Rates per 100,000 people in different areas of the world were as follows: globally, 178; Africa, 332; the Americas, 36; eastern Mediterranean, 173; Europe, 63; Southeast Asia, 278; and western Pacific, 139 in 2010. About one-third of the population of sub-Saharan Africa is infected with *M. tuberculosis* (Dye et al., 1998). In 2000, an estimated 17 million people in sub-Saharan Africa were infected with both *M. tuberculosis* and HIV—70% of all people coinfected worldwide (Corbett et al., 2003). As more people have become infected and coinfected with HIV, especially in eastern and southern Africa, the incidence of TB has been driven upward; as reflected in estimates derived from population-based surveys and from routine TB surveillance data HIV infection is the most important single predictor of TB incidence across the African continent (Jamison, 2006). Despite the emphasis placed on finding smear-positive cases under DOTS and the new WHO Stop TB Strategy, the proportion of cases reported to be smear positive has fallen in recent years in several African countries with high rates of HIV (Jamison, 2006). Before HIV infection and AIDS emerged and spread in Africa, TB incidence rates were typically under 100 per 100,000 persons per year. HIV has turned a slow decline into a rapid resurgence, especially in eastern and southern Africa. However, the worst HIV epidemics are now almost certainly decelerating, and even turning downward in some countries (UNICEF, 2004). As the interval between HIV infection and the onset of TB is 4–6 years, we can expect TB incidence rates to continue increasing for some years in some African countries.

The increase in HIV prevalence has also been accompanied by a rise in the TB case-fatality rate, and hence the TB death rate in the general population (Gilks et al., 2006). The TB epidemic in Africa is largely fueled by poverty and the simultaneous infection with HIV. Poor people living with HIV are more likely than others to become sick with TB. At least one-third of people living with HIV worldwide in 2014 were infected with

TB bacteria. People living with HIV are 20–30 times more likely to develop active TB disease than people without HIV (Gilks et al., 2006).

HIV and TB form a lethal combination, each speeding the other's progress. In 2014 about 0.4 million people died of HIV-associated TB. Approximately one-third of deaths among HIV-positive people were due to TB in 2014. In 2014 1.2 million new cases of TB were estimated among people who were HIV positive, 74% of whom were living in Africa (WHO Fact sheet number 104, March 2016).

The following have been designated as vulnerable groups at risk for exposure or transmitting TB disease: children, people living with HIV, diabetics, smokers, alcohol and substance users, people who are malnourished or have silicosis, mobile, migrant and refugee populations, and people living and working in poorly ventilated environments, including those living in informal settlements.

2.2.2 Diagnosis

Effective management of TB depends on finding (diagnosing) persons with the disease, initiating appropriate treatment early, and preventing complications and spread of the infection.

Traditionally, TB services have relied on passive case finding where persons with TB symptoms present themselves to the health care facilities. New policies encourage active case finding to look for persons with TB through community- or facility-based interventions. The aim is to screen every person for TB annually. TB diagnosis depends on symptom screening of all patients presenting to a health or workplace facility, contacts of people with laboratory confirmed pulmonary TB disease and key population groups. All those who have symptoms of TB disease must be investigated for the presence of TB infection. Until recently the primary method for rapid diagnosis of TB was through smear microscopy. In 2011, SA introduced *Xpert MTB/RIF* as a replacement for sputum smear microscopy for the diagnosis of pulmonary TB. Unfortunately this test cannot be used for monitoring treatment because it does not distinguish between live and dead bacilli (Crofton et al., 1999).

Culture is more sensitive than smear microscopy, detecting a higher proportion of cases among patients with symptoms. However, it is an expensive and slow diagnostic technique, and results are only available within 4–6 weeks, depending on the bacillary load (Crofton et al., 1999). A sputum culture and drug sensitivity are indicated in smear negative TB suspects (particularly if HIV-infected or sick), TB patients failing treatment or requiring retreatment, and TB patients who fail to convert their sputum from positive to negative at 2 or 3 months or those who convert from negative to positive during the treatment period. Other important testing methods include drug susceptibility testing (DST) to determine whether the isolate is susceptible or resistant to the drugs tested and molecular testing using polymerase chain reaction (PCR) technologies—Gene Xpert (GXP)—useful for rapidly diagnosing TB and screen for rifampicin resistance and line probe assay (LPA) detects resistance to both rifampicin and isoniazid at the same time and reduces time to diagnosis of MDR-TB to 7 days.

Chest x-rays (CXR) are an adjunctive test, which do not show a radiographic pattern specific for TB. However, the presence of infiltrates, lymph nodes, or cavities is highly suggestive of TB (Aisu et al., 1995). They are useful in patients who cannot produce sputum or who have negative Xpert results and are HIV positive, and where extra pulmonary TB (such as pleural effusions and pericardial TB) is suspected.

2.2.3 Treatment
Early initiation of effective treatment use of the correct drugs for the correct length of time reduces individual morbidity and mortality and stops the spread of the disease. TB is a treatable and curable disease (Aisu et al., 1995). Active, drug-susceptible TB disease is treated with a standard 6 month course of four antimicrobial drugs that are provided with information, supervision, and support to the patient by a health worker or trained volunteer. Without such support, treatment adherence can be difficult and the disease can spread. The vast majority of TB cases can be cured when medicines are provided and taken properly (Aisu et al., 1995).

2.2.4 Tuberculosis program
The TB program contributes to the reduction of the TB disease burden in the WHO Africa region through support to all member states to adopt and implement cost-effective TB prevention, treatment, care, and support interventions (WHO, 2015). Countries should continue to intensify efforts to fight HIV which is an important contributor to the TB epidemic. To reach the missing patients and move toward eliminating TB there is need to scale up TB interventions, such as DOTS and TB/HIV collaborative activities among others, especially for the most vulnerable groups (WHO, 2015).

Four major interventions have been proven successful in the control of TB in the African region:
- expansion of the basic package that underpins the Stop TB Strategy (DOTS), resulting in an increased number of countries achieving TB treatment success rates of 85%;
- improved diagnostics, resulting in improved case detection and, detection and treatment of multidrug-resistant and extensively drug-resistant TB;
- implementation of WHO pediatric TB guidelines, leading to improved detection and notification of all forms of TB in children;
- TB/HIV integration, resulting in improved access to HIV testing and treatment for TB patients (WHO, 2015).

2.2.5 WHO response
The WHO End TB Strategy, adopted by the World Health Assembly in May 2014, is a blueprint for countries to end the TB epidemic by driving down TB deaths, incidence, and

eliminating catastrophic costs. It outlines global impact targets to reduce TB deaths by 90% and to cut new cases by 80% between 2015 and 2030, and to ensure that no family is burdened with catastrophic costs due to TB. WHO pursues six core functions in addressing TB.
- Provide global leadership on matters critical to TB.
- Develop evidence-based policies, strategies, and standards for TB prevention, care and control, and monitor their implementation.
- Provide technical support to Member States, catalyze change, and build sustainable capacity.
- Monitor the global TB situation and measure progress in TB care, control, and financing.
- Shape the TB research agenda and stimulate the production, translation, and dissemination of valuable knowledge.
- Facilitate and engage in partnerships for TB action.

Ending the TB epidemic by 2030 is among the health targets of the newly adopted SDGs. WHO has gone one step further and set a 2035 target of 95% reduction in deaths and a 90% decline in TB incidence similar to current levels in low TB incidence countries today.

The strategy outlines three strategic pillars that need to be put in place to effectively end the epidemic:
- Pillar 1: integrated patient-centered care and prevention
- Pillar 2: bold policies and supportive systems
- Pillar 3: intensified research and innovation.

The success of the strategy will depend on countries respecting the following four key principles as they implement the interventions outlined in each pillar:
- government stewardship and accountability, with monitoring and evaluation
- strong coalition with civil society organizations and communities
- protection and promotion of human rights, ethics, and equity
- adaptation of the strategy and targets at country level, with global collaboration.

Concerning TB and HIV, WHO recommends National guidelines that an HIV test should be offered during the diagnostic work-up for TB or soon after the initiation of TB treatment and likewise, TB symptom screening should be done as part of HIV counseling and testing (HCT). On completion of TB treatment all coinfected TB patients must be on ART.

Finally, TB is really a public health issue and whatever the impact of HIV on TB in the next few years, African countries will continue to need vigorous TB control programs that fully implement the new Stop TB Strategy, founded on DOTS. Even with high rates of HIV infection, DOTS implementation is relatively cheap and cost-effective. Other methods for the prevention and treatment of HIV and AIDS will be needed too, but they should be introduced in ways that will be complementary to DOTS. Pragmatic field studies must continue to explore better ways of using the current tools for TB control as new vaccines, drugs, and diagnostics begin to emerge from laboratories.

2.3 Malaria

Malaria is a parasitic disease caused by *Plasmodium* species, which is spread when a female Anopheles mosquito feeds on an infected person's blood, gets the parasites which matures in the mosquito and later deposited via saliva when the mosquito visits an uninfected person for another blood meal. The female Anopheles mosquito is the malaria vector and in the absence of prompt and effective treatment, malaria often causes death with an estimate of 438,000 malarial deaths in 2015 (WHO, 2015). Among the plasmodium parasite species that cause malaria in humans, two of these, *Plasmodium falciparum* and *Plasmodium vivax*, pose the greatest threat. *P. falciparum* is the most prevalent malaria parasite on the African continent and is responsible for most malaria-related deaths globally whereas *P. vivax* has a wider distribution than *P. falciparum*, and predominates in many countries outside of Africa. Globally, an estimated 3.2 billion people in 95 countries and territories are at risk of being infected with malaria and developing the disease, and 1.2 billion are at high risk (>1 in 1000 chance of getting malaria in a year) (Prüss-Üstün et al., 2003). According to the World Malaria Report 2015, there were 214 million cases of malaria globally in 2015 (uncertainty range 149–303 million) and 438,000 malarial deaths (range 236,000–635,000), representing a decrease in malarial cases and deaths of 37% and 60% since 2000, respectively. Between 2000 and 2015, globally, incidences of malaria among populations at risk fell by 37% and malarial death rates among populations at risk fell by 60% among all age groups, and by 65% among children under 5 (WHO, 2015).

2.3.1 Life cycle of malaria parasite

The cycle of malarial infection begins when a female anopheles mosquito, in search for a blood meal, bites a person with malaria and ingests blood that contains reproductive cells of the parasite (WHO, 2015). Once inside the mosquito, the parasite reproduces, develops, and migrates to the mosquito's salivary gland. When the mosquito bites another person, parasites are injected along with the mosquito's saliva. Inside the newly infected person, the parasites move to the liver and multiply again. They typically mature over an average of 1–3 weeks, then leave the liver and invade the person's red blood cells. The parasites multiply yet again inside the red blood cells, eventually causing the infected cells to rupture, releasing parasites to invade other red blood cells. At this point, the cycle can be repeated if a female anopheles mosquito feeds on the person's blood. This causes the diseases to be easily spread from one person to the other with the help of the mosquito as the vector (WHO, 2015).

2.3.2 Overview of the burden of malaria in Africa

An estimated 6.2 million malarial deaths have been averted globally since 2001. Despite this, Africa continues to carry a disproportionately high share of the global malaria burden with 88% of malarial cases and 90% of malarial deaths in 2015 (WHO, 2015). Malaria affects the lives of almost all people living in the area of Africa defined by the southern

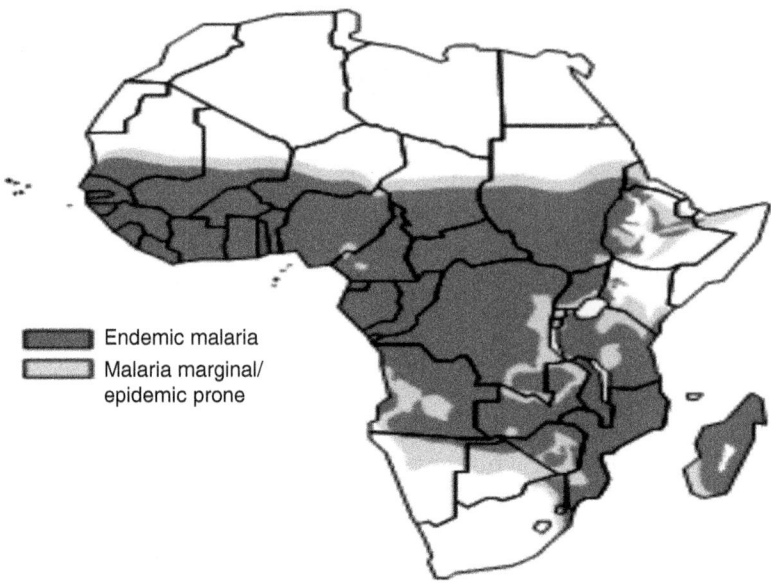

Figure 1.1 *Distribution of endemic malaria on the African Continent. (Mapping Malaria Risk in Africa available at www.mara.org.za).*

fringes of the Sahara Desert. Most people at risk of the disease live in areas of relatively high malaria transmission (Fig. 1.1). Infection is common and occurs with sufficient frequency that some level of immunity develops (WHO, 2015). In this WHO African region, among the estimated 90% of all malarial deaths that occurred in 2015, children under 5 years accounted for more than two-thirds of all deaths (Prüss-Üstün et al., 2003). This is because the majority of infections in Africa are caused by *P. falciparum*, the most dangerous of the four human malaria parasites. It is also because the most effective malaria vector, the mosquito *Anopheles gambiae*, is the most widespread in Africa and the most difficult to control. Malaria remains a major killer of children, particularly in sub-Saharan Africa, taking the life of a child every 2 min. In the African region, the estimated number of malarial deaths in children under 5 fell from 694,000 in 2000 (range: 569,000–901,000) to 292,000 in 2015 (range: 212,000–384,000). Malaria episodes in pregnant women cause anemia, and other complications in the mother and newborn child (WHO). By 2015, more than two-thirds of children under 5 were sleeping under an ITN and the proportion of febrile children receiving ACTs rose from 0% to 13% (WHO) and a lot has been put in place by the various malaria control programs to tackle malaria among populations at risk. Despite incredible progress in fighting malaria in all regions affected by the disease, malaria continues to be one of the main public health problems in the world, especially in the majority of African countries (Prüss-Üstün and Corvalán, 2006; Prüss-Üstün et al., 2003). It particularly affects young children, young adults engaged in economic development activities, pregnant women, and recently those whose systems have been weakened by HIV/AIDS (WHO, 2015).

Looking at the distribution of malaria in Africa as revealed by the WHO 2015 report, in West Africa, about 342 million people in the 17 countries of this subregion are at risk for malaria, with 289 million at high risk (reported incidence >1 per 1000).

In Central Africa, about 158 million people in the 10 countries of this subregion are at some risk for malaria, with 145 million at high risk (Fig. 1.1). Cases are almost exclusively due to *P. falciparum*. All endemic countries in the subregion are in the control phase. In eastern African countries, about 313 million people in the 12 countries of the subregion are at some risk for malaria, with 254 million at high risk (Fig. 1.1). About 25% of the population of Ethiopia and Kenya live in areas that are free of malaria. *P. falciparum* is the predominant species, except in Eritrea and Ethiopia, where *P. vivax* accounts for about 31 and 26% of reported cases, respectively. All countries in the subregion are focused on malaria control (WHO, 2015). In southern African countries, about 21 million people in the 5 countries of this subregion are at some risk for malaria, with 8 million at high risk (Fig. 1.1). About 72%, or 54 million people, live in areas that are free of malaria. Countries in the subregion are focused on malaria control activities.

2.3.3 Vulnerable population

Mostly children, pregnant women, and recently those with immune system have been weakened by diseases like HIV/AIDS (WHO, 2006a; Institute for Health Metrics and Evaluation, 2013).

2.3.4 Malaria prevention and control in Africa

Malaria is preventable and curable if diagnosed early and prompt and effective treatment is used.

Malaria has been well controlled or eliminated in the five northernmost African countries: Algeria, Egypt, Libya, Morocco, and Tunisia. In these countries the disease was caused predominantly by *P. vivax* and transmitted by mosquitoes that were much easier to control than those in Africa south of the Sahara (Prüss-Üstün et al., 2003). Surveillance efforts continue in most of these countries in order to prevent both a reintroduction of malaria parasites to local mosquito populations, and the introduction of other mosquito species that could transmit malaria more efficiently especially in southern Egypt (Prüss-Üstün et al., 2003). Several medications are available for chemoprevention but these medicines are not 100% protective and must be combined with personal protective measures, such as using long lasting insecticidal bed nets (LLINs), wearing long-sleeved shirts and long trousers, using indoor residual spraying (IRS), and also applying insect repellent on any exposed skin, especially when going out at night when mosquitoes are most active (WHO, 2015).

Malaria also continues to prevent many school children from attending school due to illness, diminishing their capacity to realize their full potential (WHO, 2006a). The objectives of the Malaria Program in the African region are to
- reduce morbidity and mortality due to malaria,
- maintain malaria-free areas and expand areas where malaria is controlled, and
- finally reduce the adverse health and socioeconomic consequences due to malaria (WHO, 2015).

2.3.5 The challenge

Though there is a global reduction in incidence of malaria, the rate of reduction is still slow in malaria endemic regions in Africa when compared to other regions; also, surveillance in these regions is not as effective as expected.

2.4 Parasitic infections

A parasite is an organism that lives on or in a host organism and gets its food from or at the expense of its host. There are three main classes of parasites that can cause disease in humans: protozoa, helminths, and ectoparasites (Cdc.gov, 2016). Parasitic infections are common in rural or developing areas of Africa, Asia, and Latin America and are less common in developed area. Parasites usually enter the body through the mouth or skin. Parasites that enter through the mouth are swallowed and can remain in the intestine or burrow through the intestinal wall and invade other organs whereas those that enter through the skin bore directly through the skin or are introduced through the bites of infected vector, like that of female anopheles mosquitoes in the case of malaria transmission (Brooker et al., 2006a,b); walking barefoot on infected soil also favors the entry of some parasites through the soles of the feet and some through the skin when a person swims or bathes in water containing the parasites (Olsen, 2007). It must be noted that rarely, parasites are spread through blood transfusions, in transplanted organs, through injections with a needle previously used by an infected person, or from a pregnant woman to her fetus.

2.4.1 Conclusion and way forward

Enhanced collaboration with international, continental, and regional partners advocates and provides normative guidance and technical assistance for the scale-up of essential interventions in order to reverse the incidence of malaria.

Increased access to services and prevention and treatment interventions, procurement and supply of quality medicines and commodities, diagnostic capacity, routine surveillance, monitoring and evaluation concur to systems strengthening and progress toward national and international targets. Primary health care and community empowerment and involvement are critical for the success of malaria control and progress toward its elimination.

Parasites that infect people include protozoa and worms (helminths, such as hookworms and tapeworms). Protozoa, which reproduce by cell division, can reproduce inside people. Helminths, in contrast, produce eggs or larvae that develop in the environment before they become capable of infecting people. Development in the environment may involve another animal (an intermediate host). A few helminths can develop in the human intestine and cause infection in that person (called autoinfection). Some protozoa (such as those that cause malaria) and some helminths (such as those that cause river blindness) have complex life cycles and are transmitted by insect vectors (Abaver et al., 2011).

2.4.2 The burden of parasitic diseases on Africa
Diseases caused by protozoan and helminthes parasites are among the leading causes of death and disease in tropical and subtropical regions of the world like in Africa (Abaver et al., 2011). Efforts to control these diseases are often difficult partly due to lack of adequate support, except for diseases like malaria. The neglected tropical diseases (NTDs), which have suffered from a lack of attention by the public health community, include parasitic diseases, such as lymphatic filariasis, onchocerciasis, and Guinea worm disease. The NTDs affect more than 1 billion people, one-sixth of the world's population—largely in rural areas of low-income countries (Molyneux et al., 2005). A number of studies revealed that the prevalence of these diseases also correlated with poverty, poor environmental hygiene, and impoverished health services (Nxasana et al., 2014). It is estimated that schistosomiasis, which is prevalent in poor communities without access to safe drinking water and adequate hygiene, had approximately 258 million treated for it in 2014 with 90% being in Africa. Also, according to WHO, the African Trypanosomiasis is a treat to 60 million people throughout Africa. Also, the environment and climate in the tropics favor the existence of some vectors. The infections may occur in places with poor sanitation and unhygienic practices and common in people with a weakened immune system, such as those who have AIDS or who take drugs that suppress the immune system called immunosuppressant. Intestinal parasite infections have enormous consequences on the health of HIV/AIDS patients, especially sub-Sahara Africa which is already overburdened by HIV; infection and chronic diarrhea due to intestinal parasites is also common in this region (Mariam et al., 2009). Hookworm infection occurs in almost half of SSA's poorest people, including 40–50 million school-aged children and 7 million pregnant women in whom it is a leading cause of anemia. Schistosomiasis is the second most prevalent NTD after hookworm (192 million cases), accounting for 93% of the world's number of cases and possibly associated with increased horizontal transmission of HIV/AIDS. Lymphatic filariasis (46–51 million cases) and onchocerciasis (37 million cases) are also widespread in SSA, each disease representing a significant cause of disability and reduction in the region's agricultural productivity (Pappas et al., 2009). Toxoplasmolisis is worldwide and one of the most common human parasites and Africa has a prevalence of about 20–55% (Pappas et al., 2009). This is just a few of the common parasitic infections in Africa as is shown in Table 1.1. They contribute greatly to the disease burden in Africa especially in diseases comorbidities, and therefore remains a major public health problem (Tables 1.2–1.4).

Table 1.1 Common parasitic infections in Africa: protozoans

Common name of organism or disease (causal agent)	Body parts affected	Diagnostic specimen	Prevalence	Source/transmission (reservoir/vector)
Amoebiasis (*Entamoeba histolytica*)	Intestines (mainly large, can go to extraintestinal sites)	Stool (fresh diarrheic stools have amoeba, solid stool has cyst)	Areas with poor sanitation, high population density, and tropics	Fecal–oral transmission of cyst, not amoeba
Giardiasis (*Giardia lamblia*)	Lumen of the small intestine	Stool	Widespread	Ingestion of cysts in fecal contaminated water or food
Malaria (*P. falciparum, P. vivax*)	Red blood cells, liver	Blood film	Tropical—250 million cases/year	Anopheles mosquito, bites at night
Toxoplasmosis (*Toxoplasma gondii*)	Eyes, brain, heart, liver	Blood and PCR	Worldwide: one of the most common human parasites	Ingestion of undercooked pork/lamb/goat meat and in cat feces
Trichomoniasis (*Trichomonas vaginalis*)	Female urogenital tract (males asymptomatic)	Microscopic examination of genital swab	32 million sub-Sahara African	Sexually transmitted infection—only trophozoite form (no cyst)
Sleeping sickness (*Trypanosoma brucei*)	Blood lymph and central nervous systems	Fluid, lymph node aspirates, blood, bone	50,000–70,000 Africans	tsetse fly, day biting fly of the genus Glossina

Table 1.2 Common parasitic infections in Africa: Helminthes (worms)

Common name of organism or disease (causal agent)	Body parts affected	Diagnostic specimen	Prevalence	Source/transmission (reservoir/vector)
Beef tapeworm (*Taenia saginata*)	Intestines	Stool	Worldwide distribution	Ingestion of undercooked beef
Cysticercosis–Pork tapeworm (*Taenia solium*)	Brain, muscle, eye (cysts in conjunctiva/anterior chamber/sub-retinal space)	Stool, blood	Asia, Africa, South America, southern Europe, North America	Ingestion of undercooked pork

Table 1.3 Common parasitic infections in Africa: Flukes

Common name of organism or disease (causal agent)		Body parts affected	Diagnostic specimen	Prevalence	Source/transmission (reservoir/vector)
Liver fluke—Fasciolosis	*Fasciola hepatica, Fasciola gigantica*	Liver, gall bladder	Stool		Freshwater snails
Intestinal schistosomiasis	*Schistosoma mansoni* and *Schistosoma intercalatum*	Intestine, liver, spleen, lungs, skin rarely infects the brain	Stool	All two species found in Africa Africa, Caribbean, South America, Asia, Middle East–83 million people	Skin exposure to water contaminated with infected biomphalaria fresh water snails
Urinary schistosomiasis	*Schistosoma haematobium*	Kidney, bladder, ureters, lungs, skin	Urine	Africa, Middle East	Skin exposure to water contaminated with infected Bulinus sp. snails
Roundworms					
River blindness, Onchocerciasis	*Onchocerca volvulus*	Skin, eye, tissue	Bloodless skin snip	Africa, near cool, fast flowing rivers	Simulium/Black fly, bite during the day
Strongyloidiasis—parasitic pneumonia	*Strongyloides stercoralis*	Intestines, lungs, skin (larva currens)	Stool, blood		Skin penetration
Whipworm	*Trichuris trichiura, Trichuris vulpis*	Large intestine, anus	Stool (eggs)	Common worldwide	Accidental ingestion of eggs in dry grains or soil contaminated with human feces
Elephantiasis lymphatic filariasis	*Wuchereria bancrofti*	Lymphatic system	Thick blood smears stained	Tropical and sub-tropical	Mosquito, bites at night

Source: From analysis of various studies on parasites.

Table 1.4 Comparison of prevalence of schistosomiasis (%) in 1980s and 2010

	Data from 1985–1987	Data in 2010	Percentage difference (%)
Overall	6.28 (5.72–6.89)	7.20 (6.76–7.65)	14.65
Center	10.41 (9.26–11.68)	10.49 (9.80–11.23)	0.77
East	5.70 (4.71–6.88)	2.07 (1.55–2.79)	63.68
West	2.44 (1.89–3.15)	3.97 (3.41–4.64)	62.7

2.4.3 Vulnerable population

Children, those with AIDS, pregnant women, and the elderly are most vulnerable. The degree of vulnerability depends on the type of parasite.

2.4.4 Diagnoses

Diagnoses are done by taking samples of blood, stool, urine, or other infected tissues and examining or sending them to a laboratory for analysis.

2.4.5 Control and preventive measures

Drugs that eliminate parasites are available for most infections. Each parasite requires unique intervention. Research on parasitic infections is targeted at developing a better understanding of the pathogenesis of infections and developing more effective prevention approaches, diagnostics, and treatments for them (Olsen, 2007). There are currently no vaccines to prevent or control the spread of parasitic diseases. Thus, control of these diseases depends heavily on the availability of drugs. Unfortunately, most existing treatments are either not completely effective due to issues of drug resistance, or they are toxic to the body. In a number of cases, even safe and effective drugs are failing as a result of the selection and spread of drug-resistant types or parasites. This is best dramatized by the global spread of drug-resistant *P. falciparum*, the protozoan responsible for the most lethal form of malaria (Olasehinde et al., 2015). Therefore, new treatments are urgently needed.

2.4.6 Conclusion and way forward

The overview of these parasitic diseases in Africa has revealed they are a threat to public health in Africa. Looking at the disease burden in Africa as a result of parasitic infection, it is imperative for more research to be done to see how this can be averted. This means there is a need to support research in the area of diseases that have been given less attention than most of those that fall under the NTDs. This is because of the parasitic infections with unique effect in Africa; cerebral malaria, neurocysticercosis, human African trypanosomiasis, toxoplasmosis, and schistosomiasis are largely preventable conditions, which are rarely seen in resource-equipped settings.

2.5 Fungi infections

Fungi are eukaryotes with a higher level of biological complexity than bacteria. Fungi range from unicellular to differentiating multicellular branching filaments. They reproduce sexually or asexually. Mycoses (fungi infection) vary greatly in their manifestations but tend to be subacute to chronic with indolent, relapsing features. Acute disease, such as that produced by many viruses and bacteria, is uncommon with fungal infections (Ryan and Ray, 2004).

Fungal infections have emerged worldwide as an increasingly frequent cause of opportunistic infections (Hsu et al., 2011). As fungal infections are frequently under recognized and difficult to detect, one of the largest gaps in our understanding of their epidemiology is determining the incidence of disease (Brandt and Park, 2013). The incidence of nosocomial fungal infections has continued to rise over the past two decades in parallel with advances in medical and surgical procedures resulting in considerable morbidity and high mortality rate (Martin et al., 2003; Horn et al., 2009).

Fungal diseases are a global public health problem. Although fungal diseases can affect anyone, including travelers, they pose a serious threat to people who have weakened immune systems, such as those who have cancer or HIV/AIDS. Many people at risk for and suffering from fungal diseases live in resource-limited settings, where diagnosis and treatment of these infections can be challenging. These areas of the world often lack the laboratory infrastructure needed to diagnose fungal diseases, and limited availability of antifungal medications means that some patients who have fungal diseases aren't able to receive the life-saving treatments they need (CDC, 2014). This chapter aims at estimating, from available data, the types, epidemiology, and disease burden caused by fungi.

2.5.1 Epidemiology of fungi infections

Globally, over 300 million people are afflicted with a serious fungal infection and 25 million are at high risk of dying or losing their sight. Estimates for the global burden of fungal diseases are based on population and disease demographics (age, gender, HIV infected, asthma, etc.) (fungal disease frequency. http://www.gaffi.org/why/fungal-disease-frequency/)

Fungal infections take more than 1.3 million lives each year worldwide, nearly as many as TB (Rutgers Biomedical and Health Sciences, 2013). Fungal diseases also appear to be emerging beyond their traditionally described borders for reasons that are not entirely understood. One article in this issue reports the incidence of *Cryptococcus gattii* disease, once believed to be restricted to tropical regions, but which is now found in locations as disparate as Vancouver Island, Canada, and parts of the southeastern United States (Harris et al., 2013). Although this organism is genetically related to *Cryptococcus neoformans*, a cause of meningitis in HIV-infected persons, *C. gattii* is frequently associated with a different spectrum of disease, prominently pneumonia. In another article

Table 1.5 Epidemiology of the most significant invasive fungal infections

Disease (most common species)	Location	Estimated life-threatening infections/year at that location[a]	Mortality rates (% in infected populations)[a]
Aspergillosis (*Aspergillus fumigatus*)	Worldwide	>200,000	30–95
Candidiasis (*Candida albicans*)	Worldwide	>400,000	46–75
Cryptococcosis (*C. neoformans*)	Worldwide	>1,000,000	20–70
Mucormycosis (*Rhizopus oryzae*)	Worldwide	>10,000	30–90
Blastomycosis (*Blastomyces dermatitidis*)	Worldwide	>400,000	20–80
Coccidioidomycosis (*Coccidioides immitis*)	Southwestern United States	~25,000	<1–70
Histoplasmosis (*Histoplasma capsulatum*)	Midwestern United States	~25,000	28–50
Paracoccidioidomycosis (*Paracoccidioides brasiliensis*)	Brazil	~4,000	5–27
Penicilliosis (*Penicillium marneffei*)	Southeast Asia	>8,000	2–75[a]

[a] Most of these figures are estimates based on available data, and the logic behind these estimates can be found in the text and in the Supplementary Materials.

in this issue, Nucci et al. (2013) report an increase in the incidence of community-associated Fusarium spp. infections in a cancer ward in Brazil.

Some fungal diseases are acute and severe (i.e., Cryptococcal meningitis and fungal eye infection (keratitis), other recurrent (i.e., Candida vaginitis or oral candidiasis in AIDS), and other chronic (i.e., chronic pulmonary aspergillosis or fungal hair infection (*Tinea capitis*). The most common life-threatening fungal infections and estimated overall mortality are shown in the subsequent sections.

The statistics presented in Table 1.5 have been largely extrapolated from the few (mostly geographically localized) studies that have been performed and are undermined by the lack of accurate incidence data from many parts of the developing world.

Endemic dimorphicmycoses can occur at many locations throughout the world. However, data for most of those locations are severely limited. For these mycoses, we have estimated the infections per year and the mortality at a specific location, where the most data are available (Brown et al., 2012).

2.5.2 Types of fungi infections

There are a number of different types of infections caused by fungi. Superficial infections are fungal infections that affect the skin or mucous membranes. Superficial fungal infections (e.g., yeast vaginitis, oral thrush, and athlete's foot) affect millions of people worldwide. Although rarely life-threatening, they can have debilitating effects on a person's quality of life and may in some cases spread to other people or become invasive (systemic). Most superficial fungal infections are easily diagnosed and can be treated effectively. The skin, the nails, and/or hair may be infected. Superficial infections of the skin and nails are the most common fungal diseases in humans and affect ~25% (or ~1.7 billion) of the general population worldwide (Havlickova et al., 2008). These infections are caused primarily by dermatophytes, which give rise to well-known conditions, such as athlete's foot (occurs in 1 in 5 adults), ringworm of the scalp (common in young children and thought to affect 200 million individuals worldwide), and infection of the nails (affects ~10% of the general population worldwide, although this incidence increases with age to~50% in individuals 70 years and older) (Thomas et al., 2010). Mucosal infections of the oral and genital tracts are also common, especially vulvo-vaginal candidiasis (or thrush). In fact, 50–75% of women in their childbearing years suffer from at least one episode of vulvo-vaginitis, and 5–8% (~75 million women) have at least four episodes annually (Sobel, 2007). In world regions with limited health care provision, HIV/AIDS adds nearly 10 million cases of oral thrush and 2 million cases of esophageal fungal infections annually. These superficial infections are caused most often by several species of *Candida*, which are the second most numerous agents of fungal infection worldwide.

2.5.3 *Tinea corporis*
Fungal infections of the skin are the most common on the exposed surfaces of the body, namely the face, arms, and shoulders. Tinea or ringworm presents in typical round lesions, which show scaling at the periphery, or in concentric rings. Usually one or a few lesions are seen and only topical treatment is necessary. Multiple, large, or widespread lesions may be seen if a patient delays seeking treatment for a long time or is malnourished or immunosuppressed (van Hees and Naafs, 2001).

2.5.4 *Tinea capitis*
Scalp ringworm is common in children. The fungus grows into the hair follicle and cannot be removed by topical treatment only. Severe pustular forms exist with follicular pustules and nodules and often massive purulent secretion. Lymph nodes in the neck swell and the patient may have a fever and headache (van Hees and Naafs, 2001). The patient should be given following: griseofulvin (500 mg) once daily for 8–12 weeks in adults and griseofulvin (10–15 mg/kg) once daily for 8–12 weeks in children. Then Whitfield's ointment or miconazole should be applied twice daily topically for 4 weeks. The treatment should be continued after 12 weeks if the infection is not cleared completely. Another alternative is to apply ketaconazole (200 mg) twice daily or terbinafine (250 mg) once daily or itraconazole (200 mg) (2 tabs) once daily for 4–8 weeks in adults (van Hees and Naafs, 2001).

2.5.5 Tinea unguium

Fungal infection of the nails is common, especially of the toenails in the elderly, where it generally does not require treatment. There may be a mixed fungal and yeast infection of toenails and/or fingernails. It often occurs in people who frequently wet their hands, such as domestic workers, cleaners, kitchen, and laundry staff (van Hees and Naafs, 2001).

2.5.5.1 Infection of the toenails

Usually this does not require any treatment. Thickened toenails may be softened using Whitfield's ointment or urea (10–40% ointment), and then thinned with a stone or a file. Systemic treatment of infected toenails is sometimes indicated, for example, when there is pain or when the patient is young; griseofulvin (500 mg) should be administered once daily until the affected nails have grown out completely, this may take a year or longer. Recurrences are common; This should be taken into account when deciding whether to use one of the more expensive drugs as listed subsequently for infection of the fingernails.

2.5.5.2 Infection of the fingernails

Griseofulvin (500 mg) should be administered once daily in adults or griseofulvin (10 mg/kg) once daily in children. The treatment should be continued until the affected nails have grown out completely; this may take 4–9 months (van Hees and Naafs, 2001).

2.5.6 Athlete's foot

Athlete's foot is itchy, often macerated whitish scaling lesions and inflammation of the skin in the interdigital spaces of the foot. It is most common between the fourth and fifth toe. The condition is not always caused by fungi but can be caused by bacteria as well. For this reason oral antifungals are often ineffective. The condition is often seen in people wearing rubber boots or rubber/plastic sandshoes (van Hees and Naafs, 2001). Keep the space in-between the toes dry. This may be achieved by drying the skin thoroughly after washing, exposing to air, using betadine scrub, paint, wearing cotton socks, and not wearing shoes that are too tight or hot. Changing socks daily will help prevent reinfection. An imidazole cream or Whitfield's ointment should be applied twice daily until a week after symptoms have cleared. This usually takes a minimum of 4 weeks (van Hees and Naafs, 2001).

2.5.7 Candidiasis

Candidiasis is a yeast infection that is caused by a fungal microorganism, most often the fungus *C. albicans*. Candidiasis is also known as thrush and can cause yeast infections in many areas of the body. These commonly include the mouth (oral thrush), the vagina (vaginal yeast infection, vaginal thrush), and the digestive tract (gastroenteritis)

(Barnett, 2004). Infection of the mouth shows symptoms of oral thrush, yellow, or white patchy lesions of the mouth and tongue. Infection of the vagina shows symptoms, such as thick discharge resembling cottage cheese, odor which is not unpleasant, vaginal itching and irritation, burning with urination, swelling of the vulva, and vaginal tenderness and pain (Hidalgo and Vazquez, 2010). The occurrence of disseminated Candida infections has been surveyed frequently in the United States and in many European countries, and variable reported incidences ranging from 2.4 (Norway) to 29 cases (Iowa, United States) per 100,000 inhabitants have been published (Odds et al., 2007). A median value of 5.9 per 100,000 inhabitants allows an estimate (US and World Population Clocks; http://www.census.gov/main/www/popclock.html) of the annual global incidence of Candida bloodstream infections at ~400,000 cases, with the most occurring in economically developed regions of the world.

2.5.7.1 Pathogenesis of *Candida albicans*

C. albicans hyphae have the capacity to form strong attachments to human epithelial cells. A mediator of this binding may be a surface hyphal wall protein (Hwp1), which is found only on the surface of germ tubes and hyphae. This protein has amino acid sequences similar to those in the substrates of mammalian keratinocyte transaminases, which form crosslinks between squamous epithelial specific proteins. This novel pathogenic strategy makes use of host enzymes to bind the pathogen to epithelial cells (Ryan and Ray, 2004). Other mannoproteins that have similarities to vertebrate integrins may also mediate binding to components of the extracellular matrix (ECM), such as fibronectin, collagen, and laminin. Hyphae also secrete proteinases and phospholipases that are able to digest epithelial cells and probably facilitate invasion (Ryan and Ray, 2004).

2.5.7.2 Clinical manifestations of Candidiasis

Superficial invasion of the mucous membranes by *C. albicans* produces a white, cheesy plaque that is loosely adherent to the mucosal surface. The lesion is usually painless, unless the plaque is torn away and the raw, weeping, invaded surface is exposed. Oral lesions, called thrush, occur on the tongue, palate, and other mucosal surfaces as single or multiple, ragged white patches. A similar infection in the vagina, vaginal candidiasis, produces a thick, curd-like discharge and itching of the vulva. Although most women have at least one episode of vaginal candidiasis in a lifetime, a small proportion suffers chronic and recurrent infections.

Systemic infections occur when fungi get into the bloodstream and generally cause more serious diseases. Systemic fungal infections may be caused either by an opportunistic organism that attacks a person with a weakened immune system, or by an invasive organism that is common in a specific geographic area, such as cocci and histoplasma. Unlike superficial infections, systemic fungal infections can be life-threatening. Invasive fungal infections have an incidence that is much lower than superficial infections, yet invasive diseases are of greater concern because they are associated with unacceptably high mortality rates (van Hees and Naafs, 2001). Many species of fungi are responsible for these invasive infections, which kill about one and a half million people every year. In fact, at least as many, if not more, people die from the top 10 invasive fungal diseases than from TB (WHO, TB; http://www.who.int/mediacentre/factsheets/fs104/en/) or malaria (WHO, Malaria; http://www.who.int/mediacentre/factsheets/fs094/en/). More than 90% of all reported fungal-related deaths result from species that belong to one of four genera: Cryptococcus, Candida, Aspergillus, and Pneumocystis. However, epidemiological data for fungal infections are notoriously poor because fungal infections are often misdiagnosed and coccidioidomycosis (also sometimes called "valley fever") is the only fungal disease that must be reported (CDC, 2014).

2.5.8 Aspergillosis (Aspergillus fumigatus)

Aspergillosis is an infection caused by *Aspergillus*, a common mold (a type of fungus) that lives indoors and outdoors. *Aspergillus* species are ubiquitous molds found in organic matter. Although more than 100 species have been identified, the majority of human illness is caused by *A. fumigatus* and *Aspergillus niger* and, less frequently, by *Aspergillus flavus* and *Aspergillus clavatus* (Harman Eloise M). Most people breathe in *Aspergillus* spores every day without getting sick. However, people with weakened immune systems or lung diseases are at a higher risk of developing health problems due to *Aspergillus*. The types of health problems caused by *Aspergillus* include allergic reactions, lung infections, and infections in other organs (https://www.cdc.gov/fungal/diseases/aspergillosis/). *A. fumigatus* is also a ubiquitous aeroallergen. Globally, millions of susceptible individuals develop pulmonary and nasal allergies to *A. fumigatus* and other airborne fungal particles. Severe asthma is linked to fungal allergy and *A. fumigatus* colonization (or infection) of the airway, principally in adults, but also in children (Menzies et al., 2011). Severe asthma with fungal sensitization (SAFS) is thought to affect between 3.25 million and 13 million adults worldwide and to contribute to the 100,000 people who die from asthma annually (Denning et al., 2013). Allergic bronchopulmonary aspergillosis is a discrete allergic syndrome estimated to affect more than 4 million people with asthma and cystic fibrosis worldwide (Knutsen and Slavin, 2011), whereas allergic fungal rhinosinusitis affects ~1.3% of those with chronic rhinitis, ~12 million people (Denning et al., 2013).

2.5.9 Sporotrichosis

Sporotrichosis is an infection caused by a fungus called *Sporothrix schenckii*, a chronic granulomatous infection. The fungus lives throughout the world in soil, plants, and decaying vegetation. Cutaneous (skin) infection is the most common form of infection and usually occurs after handling contaminated plant material, when the fungus enters the skin through a small cut or scrape: however, outbreaks have been linked to activities that involve handling contaminated vegetation, such as moss, hay, or wood. The initial episode is typically followed by secondary spread with involvement of the draining lymphatics and lymph nodes. *S. schenckii* occurs worldwide in close association with plants. For example, cases have been linked to contact with sphagnum moss, rose thorns, decaying wood, pine straw, prairie grass, and other vegetation. About 75% of cases occur in males, either because of increased exposure or because of an X-linked difference in susceptibility. The incidence is higher among agricultural workers, and sporotrichosis is considered an occupational risk for forest rangers, horticulturists, and workers in similar occupations (van Hees and Naafs, 2001).

The exact incidence of sporotrichosis is unknown, but people at increased risk for sporotrichosis usually have occupational or recreational exposures related to agriculture, horticulture, forestry, or gardening (CDC, 2014). *S. schenckii* can be found throughout the world in soil and plant matter. Peru is suspected to be an area where *S. schenckii* is extremely common in the environment. Outbreaks of sporotrichosis have been documented in the United States, Western Australia, and Brazil (van Hees and Naafs, 2001). The first symptom is usually a small painless nodule (bump) resembling an insect bite. The first nodule may appear any time from 1 to 12 weeks after exposure to the fungus. The nodule can be red, pink, or purple in color, and it usually appears on the finger, hand, or arm where the fungus has entered through a break in the skin. The nodule will eventually become larger in size and may look like an open sore or ulcer that is very slow to heal. Additional bumps or nodules may appear later near the original lesion (CDC, 2014).

Most *Sporothrix* infections only involve the skin. However, the infection can spread to other parts of the body, including the bones, joints, and the central nervous system. Usually, these types of disseminated infections only occur in people with weakened immune systems. In rare cases, a pneumonia-like illness can occur after inhaling *Sporothrix* spores, which can cause symptoms, such as shortness of breath, cough, and fever.

Sporotrichosis is typically diagnosed when your doctor obtains a swab or a biopsy of the infected site and sends the sample to a laboratory for a fungal culture. Serological tests are not always useful in the diagnosis of sporotrichosis due to limitations in sensitivity and specificity (van Hees and Naafs, 2001). Most cases of sporotrichosis only involve the skin and/or subcutaneous tissues and are nonlife-threatening, but the infection requires treatment with prescription antifungal medication for several months. The most common treatment for this type of sporotrichosis is oral itraconazole for 3–6 months. Itraconazole may also be used to treat bone and joint infections, but treatment should continue for at least 12 months (van Hees and Naafs, 2001).

For patients with severe disease, and/or an infection that has spread throughout the body, a lipid formulation of amphotericin B should be used. Itraconazole can be used for step-down therapy once the patient has stabilized. Supersaturated potassium iodide (SSKI) is another treatment option for cutaneous or lymphocutaneous disease. SSKI and azole drugs like itraconazole should not be used during pregnancy. Treatment recommendations may differ for children. There is no vaccine to prevent sporotrichosis. You can reduce your risk of sporotrichosis by wearing protective clothing, such as gloves and long sleeves when handling wires, rose bushes, bales of hay, pine seedlings, or other materials that may cause minor cuts or punctures in the skin. It is also advisable to avoid skin contact with sphagnum moss. Opportunistic infections: the fungi attack people with weakened immune systems. These can be either systemic or superficial infections (http://www.infoplease.com/cig/dangerous-diseases-epidemics/types-infection.html).

2.5.10 Cryptococcus neoformans *infections*

Cryptococci are encapsulated saprophytic yeasts. Two species, transmitted by inhalation, are the principal human pathogens: *C. neoformans* and *C. gattii* (Byrnes et al., 2011). *C. neoformans* var. *grubii* (capsular serotype D) is the most common, and causes 82% of cryptococcal disease worldwide. *C. neoformans* infections are extremely rare among people who have healthy immune systems; however, *C. neoformans* is a major cause of illness in people living with HIV/AIDS, with an estimated 1 million cases of cryptococcal meningitis occurring worldwide each year (Park et al., 2009). Although the widespread availability of ART in developed countries has helped improve the immune systems of many HIV patients so that they don't become vulnerable to infection with *Cryptococcus*, cryptococcal meningitis is still a major problem in resource-limited countries where HIV prevalence is high and access to health care is limited. An estimated 1 million cases of cryptococcal meningitis occur among people with HIV/AIDS worldwide each year, resulting in nearly 625,000 deaths (Park et al., 2009). Most cryptococcal meningitis cases occur in sub-Saharan Africa (Fig. 1.2).

Cryptococcus is now the most common cause of meningitis in adults. Cryptococcal meningitis is therefore one of the leading causes of death in HIV/AIDS patients in sub-Saharan Africa. The immune system of healthy individuals has effective mechanisms for preventing fungal infections, and the current incidence of invasive diseases is largely a result of substantial escalations over the last few decades in immunosuppressive infections, such as HIV/AIDS, and modern immunosuppressive and invasive medical interventions.

For example, the vast majority of patients with cryptococcosis—for which we have comparatively strong epidemiological data, at least from economically developed countries—have quantitative or qualitative defects in cellular immune function, specifically in CD4+ lymphocytes. AIDS is the major risk factor, although individuals who have received immunosuppressive medications, particularly in the setting of solid organ transplantation, are also predisposed. The CDC recently estimated the yearly global burden of cryptococcal meningitis to be nearly 1 million cases, with more than 620,000 deaths in sub-Saharan Africa (Park et al., 2009). Mortality rates in AIDS patients are estimated to range from 15% to 20% in the United States and 55% to 70% in Latin America and sub-Saharan Africa, despite treatment (Park et al., 2009).

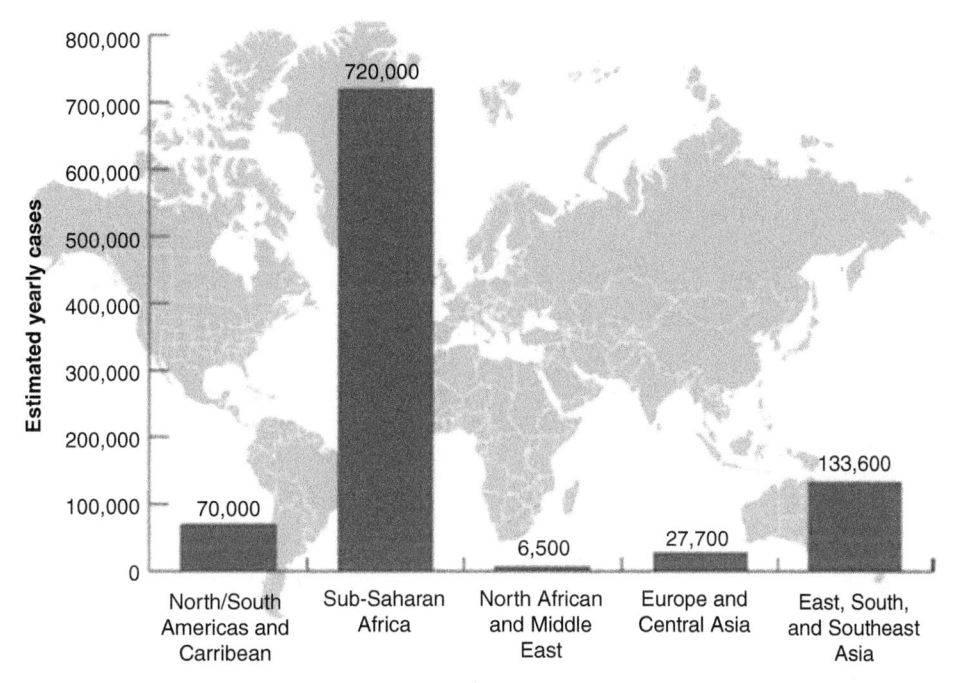

Figure 1.2 *Global burden of HIV-related cryptococcal meningitis. (Adapted from Park, Wannemuehler, Marston, Govender, Pappas and Chiller, 2009. Estimation of the current global burden of cryptococcal meningitis among persons living with HIV/AIDS. AIDS 23(4), 525–530).*

2.5.11 Paracoccidioidomycosis

Paracoccidioidomycosis (PCM) is an acute- to chronic systemic mycosis caused by fungi of the genus *Paracoccidioides*. The genus *Paracoccidioides* belongs to Phylum Ascomicota, Class Euromycetes, Order Onygenales, and Family Ajellomycetaceae (Onygenaceae), the same as *H. capsulatum*, *B. dermatitidis*, *Coccidioides immites*, and *Coccidioides posadasii*, with which it shares the same thermally dimorphic character, infecting forms (arthroconidia and mycelium) and a geographically restricted habitat (Bagagli et al., 2008). A study was published in 2011, confirming that PCM, as a disease, is not limited to the human species. The study described the second case of canine PCM, again in a Dobermann breed dog and once more showing the prevalence of lymph node involvement, with vast histologic, mycological, and molecular evidence (De Farias et al., 2011). Recently, Theodoro et al. (2012) built a map that expresses the predominance (or even exclusiveness) of each species of genus *Paracoccidioides* in South America. The authors used the single nucleotide polymorphisms (SNP) technique as molecular marker along with morphologic data applied to 63 isolates from patients of several countries in South America (10 of which obtained from cutaneous lesions of patients from the endemic region of Botucatu-SP), as well as isolates from armadillos from various geographic areas. This map demonstrated that (until the present) *P. lutzii* is more prevalent in the central areas of Brazil (Theodoro et al., 2012).

PCM, previously named South American blastomycosis, is a systemic endemic granulomatous mycotic disease caused by *P. brasiliensis*, a thermodimorphic fungus. The disease was first described by Adolpho Lutz in the city of São Paulo, Brazil, in 1908 (Colombo et al., 2011). Paracoccidiodomycosis occurs in most Latin America countries, with the endemic zone extending South from Tampico, Mexico, to Buenos Aires, Argentina (Wanke and Londero, 1994). It is considered to be the most important systemic fungal infection affecting countries in South America, with a higher incidence in Brazil (80% of the cases), Colombia, and Venezuela, followed by Argentina, Peru, Ecuador, Uruguay, and Paraguay (Colombo et al., 2011). There is no data as far as Africa is concerned. Paracoccidiocomycosis infection is usually acquired early in life. Both humans and animals are thought to be infected by the respiratory route. Although the clinical forms of the disease are mostly observed in humans, sporadic cases in domestic animals have been reported (Ricci et al., 2004). After being inhaled *P. brasiliensis* usually causes a benign and transient pulmonary infection that may be recognized by a positive paracoccidin intradermal test (Hay et al., 1987). SAFS can be conceptualized as a continuum of fungal sensitization, with asthma at one end and allergic bronchopulmonary aspergillosis at the other. SAFS is caused by many, some alone, some collectively fungi species, such as *A. fumigatus, Penicillium chrysogenum, Cladosporium herbarum, Alternaria alternata, C. albicans, Trichophyton* spp. and probably others (fungi infections—SAFS. http://www.life-worldwide.org/fungal-diseases/safs/). It is diagnosed by the presence of severe asthma, fungal sensitization, and exclusion of allergic bronchopulmonary aspergillosis. Because of the paucity of data and ambiguity in diagnostic criteria, SAFS is currently more of a diagnosis of exclusion than a specific entity. It is found probably worldwide. Severe asthma affects 5–20% of those with asthma, depending on definition and denominator. Of these, 35–50% have SAFS, depending on how extensively they are tested. Six millions of people with SAFS are a conservative estimate (Agarwal, 2011).

2.5.12 Fungal pneumonia: Coccidioidomycosis (valley fever)

Valley fever is an illness caused by the fungus Coccidioides, which lives in soil. People can become infected by inhaling fungal spores. This can cause flu-like symptoms that may last from weeks to months. Valley fever occurs in people who live in or have traveled to areas where Coccidioides is endemic, or native and common in the environment. It is found most often in the southwestern United States (especially Arizona and California) and parts of Mexico, Central America, and South America. An estimated 150,000 more cases go undiagnosed every year. More than 70% of cases occur in Arizona and 25% occur in California. In 2011, more than 20,000 cases were reported in the United States, twice as many cases as TB (Tsang et al., 2010). It results from inhaling the spores (Arthroconidia) of Coccidioides species (*C. immitis* or *C. posadasii*) (Fisher et al., 2002). As the symptoms (fever, cough, headache, rash, muscle aches, or joint pain) are similar to other common illnesses, diagnosis and treatment are often delayed. In a very small proportion of people, the infection can cause chronic pneumonia, spread from the lungs to the rest of the body and cause meningitis (brain or spine infection), or even death. Travelers who have recently visited the region of endemicity or previously infected patients with immunosuppression who experience reactivation of latent infections can develop clinical disease and require medical management outside of the region of coccidioidal endemicity (Galgiani, 1999).

2.5.13 Histoplasmosis

Histoplasmosis is an endemic infection in most of the United States. Disseminated disease is rare but can be fatal if untreated. *H. capsulatum* is a dimorphic fungus found in the temperate zones of the world; it is highly endemic in the Ohio and Mississippi river valleys of the United States (Gurney and Conces, 1996). An estimated 40 million people in the United States have been infected with *H. capsulatum*, with 500,000 new cases occurring each year (Springston, 1998). The mycelial form of *H. capsulatum* is found in the soil, especially in areas contaminated with bird or bat droppings, which provide added nutrients for growth. Infections in endemic areas are typically caused by wind-borne spores emanating from point sources, such as bird roosts, old houses or barns, or activities involving disruption of the soil, such as farming and excavation (Gurney and Conces, 1996).

2.5.14 Prevention and control of fungal infections

As most invasive fungal infections have high mortality rates, reducing the incidence of these diseases often relies on rapid and specific diagnostics, effective antifungal drugs, novel immunotherapeutic strategies, and adherence to infection control and sterility practices. Recently, we have seen examples of successes and failures in this area. In regions with high HIV prevalence, use of novel lateral-flow diagnostic tests for cryptococcal disease has opened the door to systematic screening and point-of care testing in asymptomatic persons with low CD4 cell counts and may result in reduction of deaths caused by this disease (Jarvis et al., 2012). Broader control of fungal exposures in the community can also be improved by awareness, especially education regarding high-risk practices and activities. Outbreaks of histoplasmosis linked to construction and cleaning activities in places contaminated with bird or bat guano have led to production of educational materials describing how risk can be mitigated (Lenhart et al., 2004). Furthermore, recent advances in whole-genome sequencing are being explored to suggest novel vaccine and diagnostic targets for the agent of Valley fever (Whiston et al., 2012). Continued public health efforts toward defining, characterizing, and tracking the emergence of fungal infections can help to focus studies on priority infections and settings. Future translational research is urgently needed to develop novel diagnostics, vaccines, and treatments as more is learned about the pathogenesis of fungal infections and the biology of fungal agents.

3 NEGLECTED TROPICAL DISEASES

3.1 Schistosomiasis in Africa

Schistosomiasis affects more than 200 million individuals, mostly in sub-Saharan Africa, but empirical estimates of the disease burden in this region are unavailable. Infection risk decreased from 2000 onwards, yet estimates suggest that more than 163 million of the sub-Saharan African population was infected in 2012. Mozambique had the highest prevalence of schistosomiasis in school-aged children. Low-risk countries (prevalence among school-aged children lower than 10%) included Burundi, Equatorial Guinea, Eritrea, and Rwanda (Lai et al., 2015). Schistosomiasis is caused by blood flukes (trematodes) of the genus *Schistosoma*. Six species infect humans: *Schistosoma guineensis, S. haematobium, S. intercalatum, Schistosoma japonicum, S. mansoni* and *Schistosoma mekongi*, of which *S. haematobium* and *S. mansoni* are the predominant causes of disease. Schistosomiasis occurs in intestinal and urogenital forms. Transmission of infection begins when human excreta containing parasite eggs reach fresh water bodies and hatched larvae infect susceptible snail hosts. Parasites undergo asexual multiplication in snails and another larval stage, infective to humans, is released into water. People are infected during domestic, occupational, and recreational water contact. The distribution of schistosomiasis is focal, as transmission depends on specific snail hosts and particular human activities, with endemicity continuously changing as a result of environmental alteration, water development schemes, migration, control interventions, and snail host distribution (Rollinson et al., 2013).

Successful schistosomiasis control programs in Egypt, China, Philippines, Brazil, and in some African countries have shown that control of schistosomiasis with progression to its elimination is feasible. In the African region, while most of the programs have been promoted and supported by donor agencies, the majority of the countries now have national comprehensive NTDs control plans developed by the Ministries of Health (Rollinson et al., 2013). Number of people in Africa requiring preventive chemotherapy for schistosomiasis annually and schistosomiasis treatment, by WHO region, 2014 is as follows; number of endemic countries is 43, number of countries requiring preventive chemotherapy is 41, total requiring preventive chemotherapy is 236,590,000, among which 111,430,624 are school age children (Rollinson et al., 2013).

Reports on treatment were received from 23 countries in 2014 compared to 19 in 2013. Angola and the Democratic Republic of the Congo (DRC) reported treatment for the first time. No treatment reports were received from Ethiopia, Guinea, Liberia, Mauritania, Namibia, Sierra Leone, or Zimbabwe. Guinea, Liberia, and Sierra Leone were unable to undertake PC due to the Ebola Virus Disease outbreak that occurred there. Major endemic countries, such as Ghana and Uganda which had not reported in 2013 submitted reports on treatments in 2014. The number of people treated in 2014 in the region was 52,413,796, representing a 52.3% (18 million people) increase over the previous year, and accounting for 85% of the total number of people treated globally. This increase could be explained by the increased supply of praziquantel, mainly from the Merck donation. Nigeria doubled the number of people treated from 3.7 million in 2013 to 7 million in 2014. However, there was a significant drop in the number of people treated in Mali where 553,400 people were reported to be treated in 2014 compared to 4.3 million in 2013. The number of school-age children who received treatment for schistosomiasis in 2014 was 43,725,454, representing 83.4% of the total number of people treated in this region. The number treated in this age group was 70.1% more than in the previous year. In 2014, eight countries—Benin, Burkina Faso, Burundi, Cameroon, Malawi, Mozambique, Niger, and Senegal—reached the target threshold treatment of at least 75% of school-age children. Of these, Burkina Faso accomplished this over a 3-year period and Senegal for over a 2-year period.

3.2 The soil-transmitted helminthes

The soil-transmitted helminthes are a group of parasitic nematode worms causing human infection through contact with parasite eggs or larvae that thrive in the warm and moist soil of the world's tropical and subtropical countries. As adult worms, the soil-transmitted helminthes live for years in the human gastrointestinal tract. More than a billion people are infected with at least one species of helminthes (Polack et al., 2005). Of particular worldwide importance are the roundworms (*Ascaris lumbricoides*), whipworms (*T. trichiura*), and hookworms (*Necator americanus* or *Ancylostoma duodenale*). They are considered together because it is common for a single individual, especially a child living in a less developed country, to be chronically infected with all three worms. Such children have malnutrition, growth stunting, intellectual retardation, and cognitive and educational deficits (Polack et al., 2005; Augusto et al., 2009).

The soil-transmitted helminths are one of the world's most important causes of physical and intellectual growth retardation. Yet, despite their educational, economic, and public health importance, they remain largely neglected by the medical and international community. This neglect stems from three features: first, the people most affected are the world's most impoverished, particularly those who live on less than US $2 per day; second, the infections cause chronic ill health and have insidious clinical presentation; and third, quantification of the effect of soil-transmitted helminth infections on economic development and education is difficult. Over the past 5 years, however, the worldwide community has begun to recognize the importance of these infections after revised estimates showed that their combined disease burden might be as great as those of malaria or TB. Two studies have also highlighted the profound effect of soil-transmitted helminth infection on school performance and attendance and future economic productivity (Bleakley, 2007). Such infections might also increase host susceptibility to other important illnesses, such as malaria, TB, and HIV infection (Fincham et al., 2003). In 2001, the World Health Assembly passed a resolution urging member states to control the morbidity of soil-transmitted helminth infections through large-scale use of antihelmintic drugs for school-aged children in less developed countries. A response to this resolution could establish one of the largest worldwide health initiatives ever undertaken (Horton, 2003). However, such widespread and frequent use of anthelmintics could lead to drug resistance or at least a decline in effectiveness of these front-line drugs in the long-term battle with soil-transmitted helminthes (Albonico et al., 2003).

3.2.1 Epidemiology and burden of disease

Soil-transmitted helminthes infections are widely distributed throughout the tropics and subtropics. Climate is an important determinant of transmission of these infections, with adequate moisture and warm temperature essential for larval development in the soil. Equally important determinants are poverty and inadequate water supplies and sanitation (Crompton, 2001). Under such conditions, soil-transmitted helminthes species are commonly coendemic. There is evidence that individuals with many helminth infections have even heavier infections with soil-transmitted helminths. As morbidity from these infections and the rate of transmission are directly related to the number of worms harbored in the host (Anderson et al., 1992), intensity of infection is the main epidemiological index used to describe soil-transmitted helminth infection. Intensity of infection is measured by the number of eggs per gram of feces, generally by the Kato–Katz fecal thick-smear technique (Katz et al., 1972). For *A. lumbricoides* and *T. trichiura*, the most intense infections are in children aged 5–15 years, with a decline in intensity and frequency in adulthood. Whether such age dependency indicates changes in exposure, acquired immunity, or a combination of both remains controversial. Although heavy hookworm infections also occur in childhood, frequency and intensity commonly remain high in adulthood, even in elderly people (Bethony et al., 2002). Soil-transmitted helminth infections are often referred to as being "overdispersed" in endemic communities, such that most worms are harbored by a few individuals in an endemic area. There is also evidence of familial and household aggregation of infection (Chan et al., 1994a), with the relative contribution of genetics and common household environment debated.

Estimates of annual deaths from soil-transmitted helminth infection vary widely, from 12,000 to as many as 135,000 (Committee, 2002) (Table 1.5). As these infections cause more disability than death, the worldwide burden, as for many NTDs, is typically assessed by DALY. Since the first DALY estimates were provided, there has been much variability in quoted estimates (Committee, 2002) partly because of different emphases on the cognitive and health effects. The lower estimates assume that most hookworm cases do not result in severe anemia or pronounced protein loss by the host, whereas the higher estimates show the long-term results of infection, such as malnutrition and delayed cognitive development, especially in children. For these reasons, school-aged children have been the major targets for antihelmintic treatment, and the scale of disease in this age group was pivotal in leveraging support for school-based control (Bundy et al., in press).

3.2.1.1 Clinical features

The clinical features of soil-transmitted helminth infections can be classified into the acute manifestations associated with larval migration through the skin and viscera, and the acute and chronic manifestations resulting from parasitism of the gastrointestinal tract by adult worms

3.2.1.2 Early larval migration

Migrating soil-transmitted helminth larvae provoke reactions in many of the tissues through which they pass. For example, ascaris larvae that die during migration through the liver can induce eosinophilic granulomas. In the lungs, ascaris larval antigens cause an intense inflammatory response consisting of eosinophilic infiltrates that can be seen on chest radiographs. The resulting verminous pneumonia is commonly accompanied by wheezing, dyspnoea, a nonproductive cough, and fever, with blood-tinged sputum produced during heavy infections. Children are more susceptible to pneumonitis, and the disease is more severe on reinfection. In some regions—such as Saudi Arabia—verminous pneumonia is seasonal and occurs after spring rains. Small numbers of affected children develop status asthmaticus, leading to the idea that *A. lumbricoides* and its zoonotic counterpart, *Toxocara canis*, are occult environmental causes of asthma (Chan et al., 2001). Several cutaneous syndromes result from skin penetrating larvae. Repeated exposure to *N. americanus* and *A. duodenale* hookworm third-stage larvae results in ground itch, a local erythematous and papular rash accompanied by pruritus on the hands and feet (Hotez et al., 2004). By contrast, when zoonotic hookworm third-stage larvae—typically *A. braziliense*—enter the skin, they produce cutaneous larva migrans, which is characterized by the appearance of serpiginous tracks on the feet, buttocks, and abdomen. After skin invasion, hookworm third-stage larvae travel through the vasculature and enter the lungs, although the resulting pneumonitis is not as great as in ascaris infection. Oral ingestion of *A. duodenale* larvae can result in Wakana syndrome, which is characterized by nausea, vomiting, pharyngeal irritation, cough, dyspnoea, and hoarseness (Hotez et al., 2004).

3.2.2 Intestinal parasitism

Generally only soil-transmitted helminth infections of moderate and high intensity in the gastrointestinal tract produce clinical manifestations, with the highest intensity infections most common in children (Chan et al., 1994b). The numerical threshold at which worms cause disease in children has not been established, because it depends on the underlying nutritional status of the host. Each of the major soil-transmitted helminths produces characteristic disease syndromes.

3.2.2.1 Ascariasis

The presence of large numbers of adult ascaris worms in the small intestine can cause abdominal distension and pain. They can also cause lactose intolerance and malabsorption of vitamin A and possibly other nutrients (Taren et al., 1987) which might partly cause the nutritional and growth failure. In young children, adult worms can aggregate in the ileum and cause partial obstruction because the lumen is small. Various grave consequences can ensue, including intussusceptions, volvulus, and complete obstruction (Khuroo et al., 1990), leading to bowel infarction and intestinal perforation. The resulting peritonitis can be fatal, although if the child survives, then wandering adult worms can die and cause a chronic granulomatous peritonitis. Typically, a child with obstruction because of ascaris has a toxic appearance with signs and symptoms of peritonitis. In some cases, a mass can be felt in the right lower quadrant. Adult worms can enter the lumen of the appendix, leading to acute appendicular colic and gangrene of the appendix tip, resulting in a clinical picture indistinguishable from appendicitis. Adult ascaris worms also tend to move in children with high fever, resulting in the emergence of worms from the nasopharynx or anus. Hepatobiliary and pancreatic ascariasis results when adult worms in the duodenum enter and block the ampullary orifice of the common bile duct, leading to biliary colic, cholecystitis, cholangitis, pancreatitis, and hepatic abscess (Khuroo et al., 1990). By contrast with intestinal obstruction, hepatobiliary and pancreatic ascariasis occurs more commonly in adults— especially women—than in children, presumably because the adult biliary tree is large enough to accommodate an adult worm (Khuroo et al., 1990).

3.2.2.2 Trichuriasis

Adult whipworms live preferentially in the caecum, although in heavy infections, whipworms can be seen throughout the colon and rectum. The adult parasite leads both an intracellular and an extracellular existence, with the anterior end embedded in epithelial tunnels within the intestinal mucosa and the posterior end located in the lumen. Inflammation at the site of attachment from large numbers of whipworms results in colitis. Longstanding colitis produces a clinical disorder that resembles inflammatory bowel disease, including chronic abdominal pain and diarrhoea, as well as the sequelae of impaired growth, anemia of chronic disease, and finger clubbing (Bundy and Cooper, 1989). Trichuris dysentery syndrome is an even more serious manifestation of heavy whipworm infection, resulting in chronic dysentery and rectal prolapse. Whipworm infection can also exacerbate colitis caused by infection with *Campylobacter jejuni*.

3.2.2.3 Hookworm infection

In hookworm infection, the appearance of eosinophilia coincides with the development of adult hookworms in the intestine. The major pathology of hookworm infection, however, results from intestinal blood loss as a result of adult parasite invasion and attachment to the mucosa and submucosa of the small intestine (Hotez et al., 2004). Hookworm disease occurs when the blood loss exceeds the nutritional reserves of the host, thus resulting in iron-deficiency anemia. The presence of more than 40 adult worms in the small intestine is estimated to be sufficient to reduce host hemoglobin concentrations below 11 g/dL, although the exact number depends on several factors including the species of hookworm—*A. duodenale* causes more blood loss than *N. americanus*— and the host iron reserves (Hotez et al., 2004). The clinical manifestations of hookworm disease resemble those of iron-deficiency anemia from other causes. The chronic protein loss from heavy hookworm infection can result in hypoproteinaemia and anasarca (Hotez et al., 2004). As children and women of reproductive age have reduced iron reserves, they are at particular risk of hookworm disease. The severe iron-deficiency anemia that can arise from hookworm disease during pregnancy can have adverse results for the mother, the fetus, and the neonate (Christian et al., 2004).

In their definitive host, each adult female whipworm or hookworm produces thousands of eggs per day, and each female ascaris worm produces upward of 200,000 eggs daily. As many soil-transmitted helminth infections present without specific signs and symptoms, the clinician typically needs some index of suspicion, such as local epidemiology or country of origin, to request a fecal examination. In some cases, especially of hookworm infection, persistent eosinophilia is a common presenting finding. Several egg concentration techniques—egg, formalinethyl acetate sedimentation—can detect even light infections (Faust and Russell, 1964). The Kato–Katz fecal-thick smear and the McMaster method are used to measure the intensity of infection by estimating the number of egg counts per gram of feces (Dunn and Keymer, 1986). Ultrasonography and endoscopy are useful for diagnostic imaging of the complications of ascariasis, including intestinal obstruction and hepatobiliary and pancreatic involvement. The treatment goal for soil-transmitted helminth infections is to remove adult worms from the gastrointestinal tract. The drugs most commonly used for the removal of soil-transmitted helminth infections are mebendazole and albendazole. These benzimidazole drugs bind to nematode β-tubulin and inhibit parasite microtubule polymerization, which causes death of adult worms through a process that can take several days. Although both albendazole and mebendazole are deemed broad-spectrum anthelmintic agents, important therapeutic differences affect their use in clinical practice. Both agents are effective against ascaris in a single dose. However, in hookworm, a single dose of mebendazole has a low cure rate and albendazole is more effective (Albonico et al., 2002), Conversely, a single dose of albendazole is not effective in many cases of trichuriasis. For both trichuriasis and hookworm infection, several doses of benzimidazole anthelmintic drugs are commonly needed. Another important difference between the two drugs is that mebendazole is poorly absorbed from the gastrointestinal tract so its therapeutic activity is largely confined to adult worms. Albendazole is better absorbed, especially when ingested with fatty meals, and the drug is metabolized in the liver to a sulfoxide derivative, which has a high volume of distribution in the tissues. For this reason, albendazole is used for the treatment of disorders caused by tissue migrating larvae, such as visceral larva migrans caused by *T. canis*. Systemic toxic effects, such as those on the liver and bone marrow, are rare for the benzimidazole anthelmintic drugs in the doses used to treat soil-transmitted helminth infections. However, transient abdominal pain, diarrhoea, nausea, dizziness, and headache commonly occur because the benzimidazole antiihelmintic drugs are embryotoxic and teratogenic in pregnant rats; there are concerns about their use in children younger than 12 months and during pregnancy. Overall, the experience with these drugs in children younger than 6 years is scarce, although evidence suggests they are probably safe. A review of the use of the benzimidazole antihelmintic drugs in children aged 12–24 months concluded that they can be used "if local circumstances show that relief from ascariasis and trichuriasis is justified." Both pyrantel pamoate and levamisole are regarded as alternative drugs for the treatment of hookworm and ascaris infections, although the former is not effective for the treatment of trichuriasis and they are administered by bodyweight.

The use of antihelmintic drugs nowadays is not restricted to the treatment of symptomatic soil-transmitted helminth infections; the drugs are now used also for large-scale morbidity reduction in endemic communities. Increasing evidence suggests that chronic infection with soil-transmitted helminths results in impaired childhood growth and poor physical fitness and nutritional status. The causal link between chronic infection and impaired childhood development is extrapolated from the recorded improvement in these features after deworming (Bundy et al., 2002). The mechanisms underlying these associations are thought to involve impairment of nutrition, although there is little specific evidence to support this assumption.

Regular treatment with benzimidazole anthelmintic drugs in school-age children reduces and maintains the worm burden below the threshold associated with disease. The benefits of regular deworming in this age group include improvements in iron stores growth and physical fitness, cognitive performance, and school attendance. In younger children, studies have shown improved nutritional indicators, such as reduced wasting, malnutrition, and stunting, and improved appetite (Stoltzfus et al., 1997). Treated children had better scores for motor and language milestones in their early development, although some investigators still find this relation controversial. Relevant to these findings, administration of antihelmintic drugs to children infected with soil-transmitted helminths from 1 year of age is now deemed appropriate. The patients on antihelmintic drugs recommended by WHO have expired, and the drugs can be produced at low cost by generic manufacturers. The cost of drug delivery is also low because after simple training, teachers could be involved in deworming (WHO, 2005). If women in endemic areas are treated once or twice during pregnancy, there are substantial improvements in maternal anemia and birthweight and infant mortality at 6 months (Christian et al., 2004). In areas where hookworm infections are endemic, antihelmintic treatment is recommended during pregnancy except in the first trimester.

An important factor in treatment is reinfection. After community-wide treatment, rates of hookworm infection reach 80% of pretreatment rates within 30–36 months.

A. lumbricoides infection reached 55% of pretreatment rates within 11 months and *T. trichiura* infection reached 44% of pretreatment rates within 17 months (Chan et al., 1994a). Despite reinfection, however, regular treatment to reduce the worm burden consistently could prevent some of the sequelae associated with chronic infection. Drug resistance against the front-line anthelmintics is widespread in nematodes of livestock as a result of frequent treatment of animals kept in close proximity and with little gene flow. If such conditions were replicated in human nematodes, drug resistance would soon arise. Human nematodes have longer reproducing times, are subjected to less frequent treatment (the treatment interval is longer than the parasites' generation time), and the treatment is targeted at certain populations, thereby sparing a circulating pool of sensitive alleles, which should reduce selection pressure. Nevertheless, the effectiveness of drugs must be closely monitored, especially in areas where drug pressure is high, such as regions where mass antihelmintic chemotherapy is also administered for the elimination of lymphatic filariasis. Development of sensitive methods for the early detection of antihelmintic resistance are part of the research agenda, with special attention being given to in vitro tests and molecular biology techniques that could be adapted to field conditions. Because no new anthelmintic drugs are in late-stage development at present, the effectiveness of available products needs to be preserved.

Finally, soil-transmitted helminthic infection in people remains a worldwide public health threat for as long as poverty persists in the developing world. The UN agencies have appropriately recognized the health and educational effect of these infections in children, and have taken steps to distribute antihelmintic drugs in schools and to undertake chemotherapy programs on an unprecedented scale. Large-scale deworming is necessary to reduce the worldwide morbidity of these infections, but without improved water supplies and sanitation this approach cannot be relied on for sustainable reductions in parasite frequency or intensity of infection. The infrastructure that has been established for deworming of children in schools is expected, however, to facilitate introduction of new anthelmintic vaccines and other control tools, and some of the proposed interventions for the integrated control of endemic neglected tropical coinfections, such as lymphatic filariasis, onchocerciasis, schistosomiasis, and trachoma. Such strategies could result in substantial reductions in the worldwide disease burden in the years to come.

3.3 Trachoma

Trachoma is the most common infectious cause of blindness worldwide. It afflicts some of the poorest regions of the globe, predominantly in Africa and Asia. The disease is initiated in early childhood by repeated infection of the ocular surface by *Chlamydia trachomatis*. This triggers recurrent chronic inflammatory episodes, leading to the development of conjunctival scarring. This scar tissue contracts, distorting the eyelids (entropion) causing contact between the eyelashes and the surface of the eye (trichiasis).

This compromises the cornea and blinding opacification often ensues. The WHO is leading a global effort to eliminate blinding trachoma, through the implementation of the SAFE strategy. This involves surgery for trichiasis, antibiotics for infection, facial cleanliness, and environmental improvements to reduce transmission of the organism. Where this program has been fully implemented, it has met with some success. However, there are significant gaps in the evidence base and optimal management remains uncertain.

3.3.1 Clinical features

Clinically, trachoma is subdivided into active (early) and cicatricial (late-stage) disease. Active disease is more commonly found in children and is characterized by a chronic, recurrent follicular conjunctivitis, most prominently of the upper tarsal conjunctiva. Follicles are collections of lymphoid tissue subjacent to the tarsal conjunctival epithelium. Intense cases are characterized by the presence of papillary hypertrophy— engorgement of small vessels with surrounding edema. In more severe cases, there is a pronounced inflammatory thickening of the conjunctiva that obscures the normal deep tarsal blood vessels. During an episode of active disease, the cornea can be affected. There may be minimal symptoms of ocular irritation and a slight watery discharge. The scarring sequelae of trachoma develop in later life, usually from around the third decade, but can present earlier in regions with more severe disease. Recurrent chronic conjunctival inflammation promotes conjunctival scarring, which ranges from a few linear or stellate scars to thick distorting bands of fibrosis with fornix shortening and symblepheron (bands between eyelid and globe). The scar tissue contracts causing in-turning of the eyelids (entropion). Contact between the eyelashes and the eye is called trichiasis. In trachoma, trichiasis commonly results from entropion. However, trichiasis may also arise from misdirection of lashes in a normal position (aberrant lashes) or lashes growing from abnormal positions (metaplastic lashes). Ultimately, blinding corneal opacification can develop. Individuals with entropion and trichiasis frequently experience pain as the lashes scratch the cornea. The clinical features are usually classified using the Simplified WHO Trachoma Grading System (Thylefors et al., 1987). This is reliable and easy to use, yielding useful information on the prevalence of active and cicatricial disease. For research purposes, a more detailed system is sometimes used (Dawson et al., 1981).

3.3.2 Differential diagnosis

Several conditions can produce a chronic follicular conjunctivitis with a similar appearance to active trachoma, including conjunctivitis caused by viruses (e.g., adenovirus) and bacteria (e.g., *Staphylococcus aureus* and Moraxella). Adult inclusion conjunctivitis, infection with genital strains of *C. trachomatis*, is characterized by large opalescent follicles. There are several causes of entropion and trichiasis that should be considered in the differential diagnosis, although most of these are relatively rare in trachoma endemic regions. Cicatricial conjunctivitis can be caused by mucus membrane pemphigoid,

Stevens–Johnson syndrome, systemic sclerosis, chemical injuries, and drugs. In nontrachomatous areas, most cases of entropion are due to involutional changes. Two rare congenital disorders result in lashes touching the eye: epiblepharon (upward riding of skin and orbicularis over the inferior tarsus) and distichiasis.

3.3.3 Epidemiology

Trachoma is probably the third most common cause of blindness worldwide, after cataract and glaucoma. Current estimates indicate that there are 8 million people who are blind or have severe visual impairment from trachoma, 7.6 million unoperated trichiasis cases, and 84 million with active trachoma. During the last two centuries, trachoma retreated from some formerly endemic regions, such as Europe and North America. This change is attributed to general improvements in living standards, rather than specific interventions against the disease (Jones, 1975). Today trachoma is prevalent in large parts of Africa, and in some regions of the Middle East, the Indian Subcontinent, South-east Asia, and South America (Polack et al., 2005). The highest prevalence of trachoma is reported from countries, such as Ethiopia and Sudan where the prevalence of active trachoma in children is often greater than 50% and trichiasis is found in up to 5% of adults (Berhane et al., 2006). For many trachoma endemic countries, the socioeconomic developments that might promote the disappearance of the disease are likely to be very slow in arriving, which in the light demographic trends and in the absence of effective control programs could lead to an increase in the amount of trachoma blindness (Schachter and Dawson, 1990).

3.3.4 Transmission of infection

C. trachomatis, the causative agent in trachoma, is probably transmitted from infected to uninfected individuals within an endemic community by various mechanisms: direct spread from eye to eye during close contact, spread on fingers, indirect spread on fomites (e.g., face cloths) and transmission by eye-seeking flies (Jones, 1975). A combination of these and other modes of transmission probably functions in most environments, with their relative importance varying between different communities and between members of a community. Therefore, a combination of interventions will be necessary to interrupt transmission. The signs of trachoma are strongly related to age. The prevalence of active disease peaks in preschool children and declines to low levels in adulthood (Dawson et al., 1976; West et al., 1991; Dolin et al., 1998). However, this may in part be due to shorter infection/disease episodes with increasing age (Bailey et al., 1999). Where tests have been used to confirm the presence of *C. trachomatis*, the findings have generally paralleled the clinical observations with much of the infection occurring in children (Schachter et al., 1999; Solomon et al., 2003). In contrast to the signs of active disease, the prevalence of trachomatous conjunctival scarring increases with age, reflecting the cumulative nature of the damage. Clinically active trachoma generally occurs with equal prevalence in male and female children. However, in most areas women are more frequently affected by the blinding complications than men (Dolin et al., 1998). About 75% of trichiasis and corneal blindness cases are women, probably due to their greater lifetime exposure to *C. trachomatis* infection through contact with children (Congdon et al., 1993).

3.3.4.1 Risk factors

Various individual and environmental risk factors have been identified, which may facilitate the introduction and transmission of *C. trachomatis* in endemic communities. Migration of people between communities is probably important for maintaining trachoma endemicity through the introduction of new strains of *C. trachomatis* (Burton et al., 2005). Transmission is probably promoted by crowded living conditions (Bailey et al., 1989). Children with active trachoma frequently have infectious ocular and nasal secretions and trachoma is often prevalent in regions where water is scarce, probably because less can be used for face washing (Emerson et al., 2000). Eye-seeking flies are a common feature of life in many trachoma endemic communities and are frequently observed feeding on ocular secretions. There is good evidence that they can act as vectors for *C. trachomatis* transmission in some environments (Emerson et al., 2004). The fly most commonly found in contact with eyes is Musca sorbens, which preferentially breeds in human feces. Lack of latrines has often been associated with increased risk of trachoma, probably due to a larger fly population (Burton et al., 2003). No animal reservoir for *C. trachomatis* has been found in trachoma endemic environments, although there is an association with cattle, which may result in an abundance of flies (De Sole, 1987).

3.3.5 Detection of Chlamydia trachomatis *infection*

The detection of *C. trachomatis* infection is problematic. Operationally, trachoma control programs rely on the clinical signs of disease for diagnosis. However, for research studies, it is often important to know the individual infection status. Various diagnostic tests have been used to detect *C. trachomatis*, but there is no "Gold Standard" test (Schachter et al., 1988; Solomon et al., 2004). The earliest method was Giemsa staining of smears of conjunctival cells to demonstrate the chlamydial inclusion body. This allows assessment of the adequacy of the specimen; it is specific but lacks sensitivity. The sensitivity of microscopy can be increased by direct immunofluorescence with monoclonal antibodies to *C. trachomatis* antigens. *C. trachomatis* can be grown in cell culture from clinical specimens and then detected by microscopy. This approach confirms the viability of the organism; however, it requires stringent conditions and also lacks sensitivity. Enzyme-linked immunoassays are commercially produced which detect chlamydial antigens; however, these have moderate sensitivity and crossreaction with other bacteria is reported, reducing specificity. Nucleic acid amplification tests, such as PCR, are the current favored modality for *C. trachomatis* detection (Solomon et al., 2004). These tests are both highly specific and sensitive, identifying significantly more individuals harboring *C. trachomatis* in endemic populations than previously recognized. However, they are not appropriate for nonresearch use due to expense and complexity. Considerable care needs to be taken in the collection and processing of conjunctival swab specimens to avoid contamination leading to false positive results. Recently quantitative real-time PCR has been used to measure the load of *C. trachomatis* infection in members of trachoma endemic communities to better define the major reservoirs of infection and monitor response to treatment (Burton et al., 2003). Currently, a point-of-care rapid diagnostic test is being developed which may be of use to trachoma control programs in the future.

3.3.6 Trachoma control

The World Health Assembly has resolved to eliminate blinding trachoma by the year 2020. To this end, the Global Alliance for the Elimination of Blinding Trachoma (GET2020) was formed in 1998, including the WHO, trachoma endemic countries and organizations working in the field. Control activities focus on the implementation of the SAFE strategy, surgery for trichiasis, antibiotics for infection, facial cleanliness (hygiene promotion) and environmental improvements, to reduce transmission of the organism. Each of these components tackles the pathway to blindness at different stages.

4 NONCOMMUNICABLE DISEASES

4.1 Cancers in Africa

Cancer is the uncontrolled growth of cells, which can invade and spread to distant sites of the body. Cancer can have severe health consequences, and is an emerging public health problem and a leading cause of death in Africa. According to the International Agency for Research on Cancer (IARC), about 645,000 new cancer cases and 456,000 cancer deaths occurred in 2012 in Africa (GLOBOCAN, 2012). These numbers are projected to nearly double by 2030 simply due to the aging and growth of the population, with the potential to be even higher because of the adoption of behaviors and lifestyles associated with economic development, such as smoking, unhealthy diet, and physical inactivity (Boyle and Levin, 2008). Necessary information related to cancer burden in Africa is available in Chapter 3 (Tables 1.6–1.8).

4.2 Neurological disorders

Neurological disorders are diseases of the central and peripheral nervous system. Structural, biochemical, or electrical abnormalities in the brain, spinal cord, or other nerves can result in a range of symptoms. These disorders include epilepsy, Alzheimer disease and other dementias, cerebrovascular diseases including stroke, migraine and other headache disorders, multiple sclerosis, Parkinson's disease (PD), neuroinfections, brain tumors, traumatic disorders of the nervous system, such as brain trauma, and neurological disorders as a result of malnutrition. Mental disorders, on the other hand, are

Table 1.6 Estimated numbers of new cases and deaths for leading cancer sites in Africa

Males				Females			
Estimated cases		Estimated deaths		Estimated cases		Estimated deaths	
Prostate	39,500	Liver	33,800	Female breast	92,600	Cervix uteri	53,300
Liver	34,600	Prostate	28,000	Cervix uteri	80,400	Breast	50,000
Kaposi sarcoma	22,400	Lung and bronchus	19,400	Liver	16,900	Liver	16,600
Non-Hodgkin lymphoma	21,900	Kaposi sarcoma	19,100	Colon and rectum	15,800	Non-Hodgkin lymphoma	12,700
Lung and bronchus	20,800	Non-Hodgkin lymphoma	18,100	Non-Hodgkin lymphoma	15,300	Colon and rectum	12,300
Colon and rectum	19,000	Esophagus	16,700	Ovary	14,000	Kaposi sarcoma	10,500
Esophagus	17,500	Colon and rectum	14,700	Kaposi sarcoma	12,400	Ovary	10,400
Urinary bladder	16,900	Stomach	11,900	Esophagus	10,400	Esophagus	9,900
Stomach	12,600	Urinary bladder	11,400	Stomach	10,100	Stomach	9,500
Leukemia	11,200	Leukemia	10,600	Leukemia	8,300	Leukemia	7,800
All sites but skin	324,700	All sites but skin	267,200	All sites but skin	390,700	All sites but skin	274,800

Source: GLOBOCAN, 2008.

Table 1.7 Age-adjusted incidence rates for the most common cancers in males and females in Africa, 2008

Site	Africa		Sub-Saharan Africa		Southern Africa		Eastern Africa		Middle Africa		Northern Africa		Western Africa	
	R1	R2	R1	R2	R1	R2	R1	R2	R1	R2	R1	R2	R1	R2
All sites[a]		114.1		115.9		235.9		121.3		88.1		109.2		92
Prostate	1	17.5	1	21.2	1	53.9	3	14.5	2	16.4	4	8.1	1	22.2
Liver	2	11.7	2	13.1	5	13.9	4	7.2	1	18.9	5	7.5	2	16.5
Lung	3	8.4	6	5.9	2	29	9	4.1	7	2.8	1	14.9	7	3.1
Colorectal	4	6.9	5	6.8	4	20.4	6	5.8	5	4.3	6	7	3	5.6
Esophagus	5	6.7	3	8.5	3	22.3	1	14.9		1.5		2		1.4
Urinary bladder	6	6.7	9	3.7	8	7.3	10	3.4		1.5	2	14.5	6	3.9
Non-Hodgkin lymphoma	7	6.3	7	5.5	9	5.7	5	6.2	3	5.4	3	8.4	4	4.8
Kaposi sarcoma	8	6	4	8.1	6	11.5	1	14.9	6	4.1		0.4		1.9
Stomach	9	4.7	8	5		4.1	7	5.6	4	5.3	9	3.9	5	4.5
Leukemia	10	3.2		2.8		3.9		3	8	2.8	7	4.4	9	2.5
Females														
All sites[a]		118.1		124.7		161.1		125.3		96.7		98.9		123.5
Breast	1	28	2	26.3	1	38.1	2	19.3	2	21.3	1	32.7	2	31.8
Cervix uteri	2	25.2	1	31.7	2	26.8	1	34.5	1	23	2	6.6	1	33.7
Liver	3	5.3	3	6.3	7	5.1	9	3.6	3	9.6	8	2.5	3	8.1
Colorectal	4	5	4	4.7	4	8.2	5	4.7	7	3.3	3	5.8	4	4.3
Ovary	5	4.2	6	4		3.8	6	4	6	4.3	5	4.8	5	3.8
Non-Hodgkin lymphoma	6	4.1	7	3.8	9	4.3	8	3.7	4	4.8	4	5	7	3.2
Esophagus	7	3.5	5	4.2	3	11.7	4	6.4		0.8		1.6		1
Stomach	8	3.3	8	3.7		2.2	6	4	5	4.7	9	2.4	6	3.3
Kaposi sarcoma	9	2.8	9	3.6	8	5.1	3	6.8		0.6		0.1		1.2
Corpus uteri	10	2.5	10	2.6	6	6.9		2.4	9	1.9		2.2	10	1.9

Data for Kaposi sarcoma in northern Africa were estimated separately (see data sources). Rates are per 100,000 and age-standardized to the world population.
[a] Excluding nonmelanoma skin cancer; R1: rank; R2: rate.
Source: GLOBOCAN, 2008.

Table 1.8 Age-adjusted death rates for the most common cancers in males and females in Africa, 2008

Site	Africa		Sub-Saharan Africa		Southern Africa		Eastern Africa		Middle Africa		Northern Africa		Western Africa	
	R1	R2	R1	R2	R1	R2	R1	R2	R1	R2	R1	R2	R1	R2
All sites[a]		95.7	1	98.1		172.1		105.4		78.5		89.5		80.1
Prostate	1	12.5	1	15.0	3	19.3	3	11.7	2	13.4	5	6.2	1	18.3
Liver	2	11.7	2	13.2	5	14.0	4	7.3	1	19.2	3	7.4	2	16.5
Lung	3	7.9	5	5.6	1	27.4	8	4.0	7	2.7	1	14.0	7	2.9
Esophagus	4	6.5	3	8.2	2	21.4	1	14.3	9	1.4		2.0	10	1.4
Colorectal	5	5.5	6	5.5	4	15.8	7	4.7	5	3.5	6	5.5	3	4.6
Non-Hodgkinlymphoma	6	5.3	8	4.6	8	4.6	6	5.1	4	4.6	4	6.9	5	4.1
Kaposi sarcoma	7	5.1	4	6.9	6	9.6	2	12.7	5	3.5		0.3	9	1.5
Urinary bladder	8	4.8	9	2.8	7	4.9	10	2.6		1.2	2	9.9	6	3.1
Stomach	9	4.5	7	4.9		3.9	5	5.4	3	5.2	8	3.7	4	4.4
Leukemia	10	3.0	10	2.7		3.6	9	2.8	8	2.6	7	4.1	8	2.3
All sites[a]		86.5		92.8		108.1		95.9		75.6		68.2		91.2
Cervix uteri	1	17.6	1	22.5	2	14.8	1	25.3	1	17.0	4	4.0	1	24.0
Breast	2	16.0	2	15.3	1	19.3	2	11.4	2	13.1	1	17.8	2	18.9
Liver	3	5.5	3	6.6	6	5.0	5	3.8	3	10.6	7	2.5	3	8.3
Colorectal	4	4.0	5	3.8	5	6.1	5	3.8	7	2.7	2	4.5	4	3.5
Non-Hodgkinlymphoma	5	3.5	7	3.2	9	3.5	9	3.1	5	4.1	3	4.1	7	2.7
Ovary	6	3.4	7	3.2	10	2.8	8	3.3	6	3.6	5	3.7	5	3.1
Esophagus	7	3.4	4	4.0	3	11.1	3	6.2		0.8		1.5		1.0
Stomach	8	3.2	6	3.5		2.0	5	3.8	4	4.6	8	2.3	5	3.1
Kaposis arcoma	9	2.3	9	3.1	7	4.4	4	5.8		0.5		0.1	10	1.0
Leukemia	10	2.1	10	1.9		2.0	10	1.7	8	1.8	6	2.8	8	2.0

Data for Kaposi sarcoma in northern Africa were estimated separately (see data sources). Rates are per 100,000 and age-standardized to the world population.
[a]Excluding nonmelanoma skin cancer.
Source: GLOBOCAN, 2008.

"psychiatric illnesses" or diseases which appear primarily as abnormalities of thought, feeling, or behavior, producing either distress or impairment of function.

Very few neurological conditions are curable, and many worsen over time. They produce a range of symptoms and functional limitations that pose daily challenges to individuals and their families. In addition, neurological conditions pose an economic burden to society (Canadian Institute for Health Information, 2007). Current accurate estimates of the numbers of people affected by neurologic disorders are needed to understand the burden of these conditions, to plan research on their causes and treatment, and to assess preventive interventions. Estimating the incidence and prevalence of neurologic disorders can be challenging. For some diseases, there are few published data and for others, available estimates vary greatly (Kurtzke, 1982).

World Health Organization data suggests that neurological disorders are an important and growing cause of morbidity. The magnitude and burden of mental, neurological, and behavioral disorders is huge, affecting more than 450 million people globally. According to the GBD Report, 33% of years lived with disability and 13% of DALYs are due to neurological and psychiatric disorders, which account for four out of the six leading causes of years lived with disability (Mathers et al., 2006). Neurological disorders are responsible of more than 20% of the world's burden of disease (BOD) (Bower et al., 2007). Neurological disorders contributed to 92 million DALY in 2005 and were projected to 103 million in 2030 (Bower et al., 2007). The burden of these neurological diseases is higher in developing countries that constitute about 85% of the world's population (Bower et al., 2007).

Unfortunately, the burden of these disorders in developing countries remains largely unrecognized. Moreover, the burden imposed by such chronic neurological conditions in general can be expected to be particularly devastating in poor populations. Primary manifestations of the impact on the poor—including the loss of gainful employment, with the attendant loss of family income; the requirement for caregiving, with further potential loss of wages; the cost of medications; and the need for other medical services—can be expected to be particularly devastating among those with limited resources. In addition to health costs, those suffering from these conditions are also frequently victims of human rights violations, stigmatization, and discrimination. Stigmatization and discrimination further limit patients' access to treatment. These disorders, therefore require special attention in developing countries (Chandra et al., 2006).

4.2.1 Causes and risk factors of neurological disorders

Although the brain and spinal cord are surrounded by tough membranes, enclosed in the bones of the skull and spinal vertebrae, and chemically isolated by the so-called blood–brain barrier, they are very susceptible if compromised. Nerves tend to lie deep under the skin but can still become exposed to damage. Individual neurons, and the neural networks and nerves into which they form, are susceptible to electrochemical and structural disruption. Neuroregeneration may occur in the peripheral nervous system and thus overcome or work around injuries to some extent; it is thought to be rare in the brain and spinal cord. The specific causes of neurological problems vary, but can include genetic disorders, congenital abnormalities or disorders, infections, lifestyle or environmental health problems including malnutrition, and brain injury, spinal cord injury or nerve injury, prenatal, perinatal, and neonatal factors, poverty, and trauma. The problem may start in another body system that interacts with the nervous system. For example, cerebrovascular disorders involve brain injury due to problems with the blood vessels (cardiovascular system) supplying the brain; autoimmune disorders involve damage caused by the body's own immune system; lysosomal storage diseases, such as Niemann–Pick disease can lead to neurological deterioration. A review of the neurological disorders seen at the pediatric neurology clinic of the University of Nigeria Teaching Hospital in Enugu revealed that perinatal problems, such as birth asphyxia, severe neonatal jaundice, and infections, were the most common etiological factors identified (Izuora and Iloeje, 1989). Studies from Côte d'Ivoire, Nigeria, and Zimbabwe found hypertension to be the main risk factor for both ischemic and hemorrhagic strokes (Matenga, 1997). Hypertension is an increasingly important public health problem in African countries, where it may affect up to 10% of the population and contributes to coronary heart disease, as well as to hemorrhagic and thrombotic strokes. Few studies from sub-Saharan Africa document the impact of the many infections on neurological morbidity and mortality. Bacterial meningitis particularly that due to pneumococcal and meningococcal organisms is still common. The neurological outcomes of these readily treatable infections depend on the availability of and access to health services, which differ tremendously across the region. The situation is made worse where populations are displaced by conflicts. Epidemics of meningococcal disease are particularly frequent in the "meningococcal belt," which extends from Guinea to the Sudan and northern Uganda and where, every 5–10 years, up to 100,000 cases occur. Viral encephalitis is on the increase, particularly where the prevalence of HIV infection is high as well. The HIV enters the nervous system within hours of an individual becoming infected. Acute inflammatory demyelinating polyneuropathy (Guillain–Barré syndrome), which can cause paralysis leading to death from respiratory failure.

4.2.2 Consequences of neurological disorders

Neurological disorders in sub-Saharan Africa impose a significant burden on the family and community, as well as on the affected individual. Some disorders, such as epilepsy, are well recognized but are not socially and culturally accepted. The enormous stigma attached to epilepsy often leads to the patients being denied access to proper care. Lack of care then leads to severe complications, social isolation, and early death. Other disorders, such as dementia, paraplegia, and stroke, put undue stress on the caring family because institutional or community care support is limited or nonexistent. This stress leads to further neglect of the afflicted person and eventually his or her premature death.

Protein–calorie and micronutrient malnutrition contribute to impaired cognitive development, which compromises the future productivity of a nation's workforce (Silberberg and Katabira, 2006).

4.2.3 Parkinson's disease

PD is the most common movement disorder besides essential tremor and the second most common neurodegenerative disease (Tanner and Aston, 2000). PD is neuropathologically characterized by nigrostriatal cell loss and presence of intracellular a-synuclein-positive inclusions called Lewy bodies. PD is one of the most common age-related neurodegenerative disorders, second in frequency only to Alzheimer's disease. In the United States, at least half a million people are diagnosed as having PD, and the frequency of PD is predicted to triple over the next 50 years as the average age of the population increases (Goldman and Tanner, 1998). PD is a slowly progressive neurodegenerative disorder with no identifiable cause. The definition of PD does not include neurological signs suggesting more extensive injury of the motor or sensory pathways extending beyond the pigmental brain stem nuclei. These signs are suggestive of other neurodegenerative disorders, often termed atypical Parkinsonism. These include multiple system atrophy, progressive supranuclear palsy, striatonigral degeneration, and other less common conditions. Originally understood as a dopamine-deficit disorder, increasing evidence suggests PD to be a multisystem brain disease in which various nondopaminergic transmitter systems are affected at motor onset and become more prominent during the course of the disease (Perry et al., 1991). Epidemiology has played an important role not only in health care and planning but also as a tool for the investigation of the cause of PD. Analytic epidemiology seeks to identify risk factors that could lead to clues for causative agents of the disease. By providing a profile of disease parameters, such as prevalence, incidence, and mortality, descriptive epidemiology of PD is useful in the etiological investigation of PD.

4.2.3.1 Prevalence, incidence, and distribution of PD

Several studies have been carried out to determine the prevalence, incidence, and distribution of this disease in different regions. Prevalence quantifies the proportion of the total number of current subjects with PD in a population at a given time. Approximately 1–2% of the population over 65 years suffers from PD. This figure increases from 3% to 5% in people 85 years and older (Fahn, 2003). As PD is mainly an illness of later life, it is more common in developed countries where people live longer. Most community-based prevalence studies across Europe found crude prevalence rates between 100 and 200 per 100,000 inhabitants (von Campenhausen et al., 2005). Crude prevalence of PD has also been reported to vary from 15 (per 100,000 population) in China to 657 in Argentiṅa in door-to-door surveys (Wang, 1991) and to vary from 100 to 250 in North America and Europe, respectively.

Incidence is a better estimate frequency, and it quantifies the number of new subjects with PD occurring in a given time period for a population of individuals at risk. It is relatively unaffected by factors affecting disease, such as mortality. However, as the clinical manifestations of PD may be preceded by a long latent stage and have a slow clinical progression, accurate measurements of the incidence of PD are relatively difficult.

Both prevalence and incidence of PD vary greatly across age groups. PD is less common before 50 years of age and increases steadily with age thereafter up to the ninth decade. The decline among the most elderly seen in some studies probably results from the very few people in this age group and may also reflect diagnostic and ascertaining difficulties. A recent study showed that the prevalence in Yonago City, Japan, increased from 80.6 (per 100,000 population) in 1980 to 117.9 in 1992, but that the age- and sex-adjusted prevalence decreased from 103.9 per 100,000 to 99.5 (Kusumi et al., 1996). There was no significant difference in incidence between 1980 and 1992, although the age-adjusted incidence in those under 55 years of age in 1992 was lower than in those under 55 in 1980. This study suggests that the increased prevalence might be mainly due to the aging of the population. Although gender-specific differences reveal more variability than association with increasing age, PD appears to be slightly more common in men than in women in most studies, usually ranging from a 1.2:1 ratio up to a 1.5:1 ratio. The prevalence and incidence of PD vary in different countries, partly reflecting variations in racial composition of the population surveyed. Generally, white people in Europe and North America have a higher prevalence, around 100–350 per 100,000 populations. Asians in Japan and China and black Africans have lower rates, around one-fifth to one-tenth of those in whites. However, age-adjusted PD prevalence was not significantly different in whites and blacks in a door-to-door screening conducted in Mississippi, USA (Schoenberg et al., 1985). Meanwhile, two studies reported that PD incidence in African-American men and women (Mayeux et al., 1995) and in Asian-American men (Morens et al., 1996) was similar to rates for Americans of European origin.

4.2.3.2 Risk factors for Parkinson's disease

Identifying the risk factors provide the clues to the causative agents of PD. Although the cause or causes of PD remain obscure, a number of factors have been associated with increased or decreased risk of PD. Demographic factors, such as age, gender, and racial origin are associated with an increased risk of PD. Family history has been implicated as a significant risk factor for PD in several large epidemiological studies, and the estimated prevalence of positive family history ranges from 5% to 40% (Tanner et al., 1997). All familial PD is not necessarily genetic, for families share the same environment. Several studies suggested that environmental factors play an important role in the cause of PD. Head injury, emotional stress, and premorbid personality have been linked to PD in numerous reports. A number of studies reported that different lifestyle elements, such as rural living, farming activity, or well-water drinking may act as risk factors for PD.

4.2.3.3 Signs and symptoms
When PD becomes clinically overt, tremor, rigidity, bradykinesia, and postural instability are considered to be the cardinal signs of the disease. The course of the disease is chronic and progressive, and may be complicated by a wide range of motor and nonmotor features, many of which contribute to increased disability as well as diminished quality of life in patients and caregivers (Schrag et al., 2000).

4.2.3.4 Mortality
Although the mortality rate represents a unique population-based statistic and has been used to examine both the time trends and the geographical distribution of PD, Phillips and colleagues found only 37% of patients had PD coded as the underlying cause of death in all diagnosed during life as having PD. The reason is that PD is not a primary or direct cause of death. Generally, mortality rates for PD increased in the older age groups but decreased for younger ages.

4.2.4 Epilepsy
Epilepsy is a neurological disorder that affects people in every country throughout the world. Epilepsy is also one of the oldest conditions known to mankind. It is characterized by a tendency to recurrent seizures and is defined by two or more unprovoked seizures. The clinical manifestations of seizures will therefore vary and depend on where in the brain the disturbance first starts and how far it spreads. Transient symptoms can occur, such as loss of awareness or consciousness and disturbances of movement, sensation (including vision, hearing, and taste), mood, or mental function (Shnayder et al., 2011). Worldwide, there are an estimated at least 65 million people living with epilepsy (Ngugi et al., 2010) Reported estimates of epilepsy occurrence vary substantially among populations studied, but, in sum, indicate that in developed countries, the annual incidence of epilepsy is nearly 50 per 100,000 population, whereas the prevalence approximates 700 per 100,000 (Hirtz et al., 2007). In low- and middle-income countries, estimates of the corresponding rates are generally higher (Hauser, 1995; Ngugi et al., 2010) Throughout the world, therefore epilepsy imposes a substantial public health burden. Epilepsy is an important health problem in developing countries, where its prevalence can be up to 57 per 1000 population. The prevalence of epilepsy is particularly high in Latin America and in several African countries, notably Liberia, Nigeria, and the United Republic of Tanzania. Parasitic infections, particularly neurocysticercosis, are important etiological factors for epilepsy in many of these countries. Other reasons for the high prevalence include intracranial infections of bacterial or viral origin, perinatal brain damage, head injuries, toxic agents, and hereditary factors. Many of these factors are, however, preventable or modifiable, and the introduction

of appropriate measures to achieve this could lead to a substantial decrease in the incidence of epilepsy in developing countries (Senanayake and Román, 1993). Better understanding of the epidemiology of epilepsy is a prerequisite for improving epilepsy care.

4.2.4.1 Risk factors

A reported risk factor for idiopathic (presumed genetic) epilepsy is family history of epilepsy. Reported risk factors for symptomatic epilepsy include prenatal or perinatal causes (obstetric complications, prematurity, low birthweight, neonatal asphyxia). Data suggest that the effect of obstetric complications or neonatal asphyxia may have been overemphasized. Prematurity, low birthweight, and neonatal seizures may be independent risk factors as well as markers of underlying disease. Other causes include traumatic brain injuries, central nervous system infections, cerebrovascular disease, brain tumors, and neurodegenerative diseases (Casetta et al., 2002; Leone et al., 2002).

4.2.4.2 Signs and symptoms of epilepsy

Epilepsy is caused by abnormal activity in brain cells; seizures can affect any process your brain coordinates. Seizure signs and symptoms may include temporary confusion, a staring spell, uncontrollable jerking movements of the arms and legs, loss of consciousness or awareness, and psychic symptoms. Symptoms vary depending on type of seizure.

4.2.4.3 Burden of epilepsy

The BOD estimates for epilepsy include epilepsy and status epileptics. Mathers and others (2004) estimate the DALYs for epilepsy as 6,223,000, with slightly higher rates for males (3,301,000) than for females (2,922,000). Many risk factors for epilepsy are linked with a lower level of economic development; thus, the burden is highest in South Asia followed by sub-Saharan Africa. A notable observation is the reportedly low burden in the Middle East and North Africa, despite the fact that parts of that region are relatively underdeveloped. Epilepsy imposes a large economic burden on patients and their families. It also imposes a hidden burden associated with stigmatization and discrimination against patients and even their families in the community, workplace, school, and home. Social isolation, emotional distress, dependence on family, poor employment opportunities, and personal injury add to the suffering of people with epilepsy.

4.2.5 Stroke

Stroke is a form of cardiovascular disease affecting the blood supply to the brain. Also referred to as cerebrovascular disease or apoplexy, strokes actually represent a group of diseases. In order to function properly, nerve cells within the brain must have a continuous supply of blood, oxygen, and glucose (blood sugar). If this supply is impaired, parts of the brain may stop functioning temporarily. If the impairment is severe, or lasts long enough, brain cells die and permanent damage follows. Because the movement and functioning of various parts of the body are controlled by these cells, they are affected also. The symptoms experienced by the patient will depend on which part of the brain is affected. Stroke is a leading cause of death and disability in sub-Saharan Africa. To date, most data on mortality have been hospital-based, although the majority of stroke deaths in the region are thought to occur at home (Kahn and Tollman, 1999).

4.2.5.1 Signs and symptoms of a stroke

There are two main types of stroke: ischemic, due to lack of blood flow, and hemorrhagic, due to bleeding. They result in part of the brain not functioning properly. Signs and symptoms of a stroke may include an inability to move or feel on one side of the body, problems understanding or speaking, feeling like the world is spinning, or loss of vision to one side among others. Signs and symptoms often appear soon after the stroke has occurred. If symptoms last less than 1 or 2 h it is known as a transient ischemic attack (TIA). Hemorrhagic strokes may also be associated with a severe headache. The symptoms of a stroke can be permanent. Long-term complications may include pneumonia or loss of bladder control.

4.2.5.2 Risk factors for stroke

The most important modifiable risk factors for stroke are high blood pressure and atrial fibrillation [although magnitude of this effect is small, the evidence from the Medical Research Council trials is that 833 patients have to be treated for 1 year to prevent one stroke (Greenberg, 1985; Thomson, 2009)]. Other modifiable risk factors include high blood cholesterol levels, diabetes mellitus, cigarette smoking (Hankey, 1999; Wannamethee et al., 1995) (active and passive), heavy alcohol consumption (Reynolds et al., 2003) and drug use (Sloan et al., 1991), lack of physical activity, obesity, processed red meat consumption (Larsson et al., 2011), and unhealthy diet. Alcohol use could predispose to ischemic stroke, and intracerebral and subarachnoid hemorrhage via multiple mechanisms (e.g., via hypertension, atrial fibrillation, rebound thrombocytosis, and platelet aggregation and clotting disturbances) (Gorelick, 1987). Drugs, most commonly amphetamines and cocaine, can induce stroke through damage to the blood vessels in the brain and/or acute hypertension (Westover et al., 2007; Fauci et al., 2012).

4.2.5.3 Burden of stroke

The BOD estimates for stroke include subarachnoid hemorrhage, intracerebral hemorrhage, cerebral infarction, and sequelae of cerebrovascular disease. The estimated DALYs for cerebrovascular disease are 72,024,000, with the burden being almost similar for females (36,542,000) and males (35,482,000) (Mathers et al., 2006). The burden is highest in East Asia and the Pacific, followed by South Asia and by Europe and Central Asia. The burden in sub-Saharan Africa is higher than in the

Middle East and North Africa, which may suggest an etiology for stroke other than atherosclerotic disease. Health experts anticipate that the number of stroke cases will increase, particularly in developing countries, because of aging populations and increased exposure to major risk factors. Corresponding to this increase in the number of stroke cases will be an increase in the number of people with disabilities surviving after stroke. In South Africa, stroke accounts for 8–10% of all reported deaths and 7.5% of deaths among people of prime working age, between 25 and 64 years old (Kahn and Tollman, 1999). A prospective community survey in rural South Africa reported that stroke accounted for 25% of all noncommunicable diseases, including in many younger individuals. Stroke was responsible for 5.5% of all deaths and 10.3% in those aged 35–64 years. Stroke ranked second as the cause of death in those aged 35–64 years, first in those aged 55–74 years (11% of all deaths), and second among those aged 75 and older (6% of all deaths) (Kahn and Tollman, 1999). In a rural hospital in Zambia, stroke accounted for 9% of admissions, but used 14% of the intensive care unit's bed days (Birbeck and Munsat, 2002). The mortality following stroke was 50%, far higher than in wealthy countries, reflecting the lack of resources for early recognition and access to treatment.

4.2.6 Neurological infections

This section shall discuss some infections associated with neurological disorders. The AIDS is caused by a retrovirus known as the HIV, which attacks and impairs the body's natural defense system against disease and infection. HIV is a slow-acting virus that may take years to produce illness in a person. During this period, an HIV-infected person's defense system is impaired, and other viruses, bacteria, and parasites take advantage of this "opportunity" to further weaken the body and cause various illnesses, such as pneumonia, TB, and mycosis. When a person starts having such opportunistic infections, he or she has AIDS. Neurological complications occur in 39–70% of patients with AIDS and significantly impact functional capacity, quality of life, and survival (Rajabiun, 2001) (Table 1.9).

The worldwide use of highly active antiretroviral therapy (HAART) has played an important role in changing the incidence of neurological complications in AIDS patients. Recent studies have shown that HAART has produced both quantitative and qualitative changes in the pattern of HIV neuropathology: an overall decrease in the incidence of some cerebral opportunistic infections, such as toxoplasmosis and cytomegalovirus encephalitis, for which successful treatment is available, whereas other uncommon types and new variants of brain infections, such as varicella zoster encephalitis, herpes simplex virus encephalitis, or HIV encephalitis, are being reported more frequently as ART promotes some immune recovery and increases survival (Gray et al., 2003). Some viral infections associated with neurological disorders include viral encephalitis, poliomyelitis, and rabies (Tables 1.6–1.10).

Table 1.9 Neurological diseases in the HIV-infected individual

Type of condition	Examples
Primary HIV-related syndromes HIV-associated cognitive–motor complex	HIV-associated myelopathy
	HIV-associated polyneuropathy
	HIV-associated myopathy
Opportunistic conditions	Toxoplasma encephalitis
	Cryptococcal meningitis
	Cytomegalovirus encephalitis/polyradiculitis
	Progressive multifocal leukoencephalopathy
	Primary central nervous system lymphoma
Inflammatory conditions	Acquired demyelinating neuropathies
	Aseptic meningitis
Treatment-associated conditions	Zidovudine-induced myopathy
	Nucleoside analog-induced neuropathy

Table 1.10 Number of undernourished and prevalence (%) of undernourishment

Place		PREVALENCE (%)		
	1990–92 (No.)	1990–2 (%)	2014–16 (No.)	2014–16 (%)
World	1,010.6	18.6	794.6	10.9
Developed regions	20.0	<5	14.7	<5
Developing regions	990.7	23.3	779.9	12.9
Africa	181.7	27.6	232.5	20.0
Sub-Saharan Africa	175.7	33.2	220.0	23.2
Asia	741.9	23.6	511.7	12.1
Eastern Asia	295.4	23.2	145.1	9.6
South-eastern Asia	137.5	30.6	60.5	9.6
Southern Asia	291.2	23.9	281.4	15.7
Latin America and Caribbean	66.1	14.7	34.3	5.5
Oceana	1.0	15.7	1.4	14.2

Source: FAO, 2015. WFP. The State of Food Insecurity in the World.

4.2.6.1 Bacterial infections

TB is a leading infectious disease and one of the causes of morbidity and mortality worldwide. TB is one of the most common opportunistic infections among HIV/AIDS patients which has pernicious effects. Among extrapulmonary cases, the most common sites involved are the lymph nodes and the pleura, but the sites of TB associated with neurological disorders (meninges, brain, and vertebrae) also constitute an important group. Meningeal TB has a high case-fatality rate, and neurological sequelae are common among survivors. Cerebral tuberculoma usually presents as a space-occupying lesion with focal signs depending on the location in the brain.

The key steps in diminishing the global burden of neurological disorder associated with TB are to promote the following: investment in full implementation of the Stop TB Strategy and International Standards for Tuberculosis Care; full immunization coverage so that all neonates are protected by BCG from risk of disseminated and severe TB; and better understanding of the epidemiology of TB disease associated with neurological disorder through improved surveillance in countries with high TB prevalence.

Bacterial meningitis is a very common cause of morbidity, mortality, and neurological complications in both children and adults, especially in children. It has an annual incidence of 4–6 cases per 100,000 adults (defined as patients older than 16 years of age), and *Streptococcus pneumoniae* and *Neisseria meningitidis* are responsible for 80% of all cases (Van de Beek et al., 2006). In developing countries, overall case-fatality rates of 33–44% have been reported, rising to over 60% in adult groups (Wright and Ford, 1995). Bacterial meningitis can occur in epidemics that can have a serious impact on large populations. This disease has the highest burden in sub-Saharan Africa. However, the burden has substantially reduced because of the introduction of vaccines. Other diseases associated with neurological disorders are tetanus, leprosy neuropathy.

4.2.6.2 Parasitic diseases

Cysticercosis is the parasitic disease caused by the larvae of the pork tapeworm (*T. solium*) that most frequently affects the CNS and is one of the major health problems of developing countries in Africa, Asia, and Latin America. In addition, because of high immigration rates from endemic to nonendemic areas and tourism, neurocysticercosis is now commonly seen in countries that were previously free of the disease. Despite the advances in diagnosis and therapy, neurocysticercosis remains endemic in most low-income countries, where it represents one of the most common causes of acquired epilepsy (Medina et al., 1990). Almost 50,000 deaths attributable to neurocysticercosis occur every year. Many more patients survive but are left with irreversible brain damage, with all the social and economic consequences that this implies (Gemmel et al., 1982). Neurocysticercosis is one of a few conditions included in a list of potentially eradicable infectious diseases of public health importance (Satcher et al., 1993). The control strategy that seems promising at the moment is a combination of different available tools in order to interrupt or reduce the cycle of direct person-to-person transmission: mass human chemotherapy to eliminate the tapeworm stage, enforced meat inspection and control, improvement of pig husbandry and inspection, treatment of infected animals, surveillance, identification and treatment of individuals who are direct sources of contagion (human carriers of adult tapeworm) and their close contacts, combined with hygiene education and better sanitation.

Malaria remains a serious public health problem in the tropics, mostly in Africa. There exist four Plasmodium species that affect humans; of these, only *P. falciparum* can sequester in capillaries of the CNS and cause cerebral malaria. In cerebral malaria, neuroimaging studies may demonstrate brain swelling, cerebral infarcts, or small hemorrhages in severe cases (Mung'Ala-Odera et al., 2004). In addition to the aforementioned parasitic diseases, other diseases causing neurological disorders include toxoplasmosis, African trypanosomiasis (Sleeping sickness), American trypanosomiasis (Chargas disease), schitosomiasis and Hydatidosis.

Neurological disorders are currently estimated to affect as many as a billion people worldwide. These disorders are found among all age groups and in all geographical regions. As a result of demographic transition from predominantly youthful populations to older and aging ones, neurological disorders, such as Alzheimer and other dementias and PD increase with increase in age. As a consequence, many low-income countries especially Africa face the double burden of a continuing high level of infections including some that result in neurological disorders (e.g., HIV and malaria)—and increases in noncommunicable diseases. The number of people with neurological disorders is estimated to increase considerably in years to come. The stigma often associated with neurological disorders adds to the social and economic burden.

5 MALNUTRITION

Malnutrition literally means "bad nutrition" and technically includes both over- and undernutrition. The World Food Programme (WFP) defines malnutrition as "a state in which the physical function of an individual is impaired to the point where the body can no longer maintain adequate bodily performance process, such as growth, pregnancy, lactation, physical work and resisting and recovering from disease (World Food Programme, 2000). In this chapter, undernutrition is also referred to as malnutrition. Inadequate diet and disease, in turn, are closely linked to the general standard of living, the environmental conditions, and whether a population is able to meet its basic needs, such as food, housing, and health care. Malnutrition is thus a health outcome as well as a risk factor for disease and exacerbated malnutrition and it can increase the risk both of morbidity and mortality.

There were 793 million undernourished people in the world in 2015 [13% of the total population (FAO, 2015)]. This is a reduction of 216 million people since 1990 when 23% were undernourished (Joint FAO/WHO, 2010). In 2012 it was estimated that another billion people had a lack of vitamins and minerals. In 2013, protein-energy malnutrition was estimated to have resulted in 469,000 deaths—down from 510,000 deaths in 1990 (Naghavi et al., 2015). Other nutritional deficiencies, which include iodine deficiency and iron deficiency anemia, result in another 84,000 deaths (Lozano et al., 2013). In 2010, malnutrition was the cause of 1.4% of all DALYs (Murray et al., 2013). About one-third of deaths in children are believed to be due to undernutrition, although the deaths are rarely labeled as such contributing to more than half of deaths in children worldwide; child malnutrition was associated with 54% of deaths in children in developing countries in 2001

(Blössner et al., 2005). The WHO says that malnutrition is by far the largest contributor to child mortality globally, currently present in 45% of all cases. Underweight births and interuterine growth restrictions are responsible for about 2.2 million child deaths annually in the world. Deficiencies in vitamin A or zinc cause 1 million deaths each year. WHO adds that malnutrition during childhood usually results in worse health and lower educational achievements during adulthood. Malnourished children tend to become adults who have smaller babies. While malnutrition used to be seen as something which complicated diseases such as measles, pneumonia, and diarrhea, it often works the other way round—malnutrition can cause diseases to occur. The United Nations Food and Agriculture Organization estimates that about 795 million people of the 7.3 billion people in the world, or 1 in 9, were suffering from chronic undernourishment in 2014–2016. Almost all the hungry people, 780 million, live in developing countries, representing 12.9%, or 1 in 8, of the population of developing counties. There are 11 million undernourished people in developed countries (Table 1.10).

5.1 Malnutrition in women and children

Substantial global progress has been made in reducing child deaths since 1990. The number of under-5 deaths worldwide has declined from 12.7 (12.6, 13.0) million in 1990 to 5.9 (5.7, 6.4) million in 2015—16,000 every day compared with 35,000 in 1990. Malnourished children, particularly those with severe acute malnutrition, have a higher risk of death from common childhood illness such as diarrhoea, pneumonia, and malaria. Nutrition-related factors contribute to about 45% of deaths in children under 5 years of age.

Children who are severely malnourished typically experience slow behavioral development, even mental retardation may occur. Even when treated, undernutrition may have long-term effects on children, with impairments in mental function and digestive problems persisting—in some cases for the rest of their lives. The nutritional status of women and children is particularly important, because the effects of malnutrition on women and their off-springs can be propagated to future generations. A malnourished mother is likely to give birth to a low birth-weight (LBW) baby susceptible to disease and preterm death, which only further undermines the economic development of the family and society, and continues the cycle of poverty and malnutrition. Although child malnutrition declined globally during the 1990s, with the prevalence of underweight children falling from 27% to 22%, national levels of malnutrition still vary considerably (0% in Australia; 49% in Afghanistan) (De Onis and Blössner, 2003). In Africa, the number of underweight children increased between 1990 and 2000 (from 26 to 32 million), and 25% of all children under 5 years old were underweight, which signals that little changed from a decade earlier. The projection for 2005 was that the prevalence of child malnutrition will continue to decline in all regions but Africa, which is dominated by the trend in sub-Saharan Africa (De Onis et al., 2004). Women are a critical link, biologically and socially, in the well-being of households and communities, and they are often more vulnerable than men to malnutrition. Although women, being smaller, will need less dietary energy, they require the same amount or more of many nutrients, so they must eat a higher proportion of nutrient-rich foods. Pregnant women need an additional 300 kcal per day; this increases to 500 kcal daily while breastfeeding. Malnutrition puts women at greater risk of complications and death during pregnancy and childbirth. Pregnancy-related factors are the leading cause of death for young women of 15–19 years old. These adolescents, whose bodies have not finished growing and who are often nutrient deficient themselves, face a 20–200% greater risk of dying than mothers aged 20–24. Malnutrition also threatens their babies.

More than half of the annual 12 million deaths of children under 5 are related to malnutrition, often due to the mother's poor nutrition during pregnancy. Evidence shows that infant mortality rates for children of very young mothers are higher—sometimes twice as high—than for children born to older mothers. As children are one of the most vulnerable segments of the population, their health status is usually a good indicator of the health of a community. In particular, they are usually the first victims of micronutrient deficiencies. Each year up to 500,000 children become partially or totally blind due to vitamin A deficiency. It also increases susceptibility to disease, retards growth and development, and is associated with increased death rates from measles, diarrhoea, and respiratory diseases. Iodine deficiency is the single most important cause of preventable brain damage in children, and also increases the incidence of miscarriages, stillbirths, and maternal deaths. There are more than 16 million cretins and nearly 49.5 million people suffering from brain damage caused by iodine deficiency. The good news is that micronutrient deficiencies are easily prevented and corrected with proper diet, fortified foods, and supplements. Worldwide, 70% of salt is now iodized. This has reduced the number of babies born cretins each year by more than half since 1990—to less than 55,000. Vitamin A supplementation saved the lives of at least 300,000 young children in developing countries in 1997 alone.

5.2 Causes of malunitrition

Poor nutrition arises from multifactor and interrelated circumstances and determinants. The immediate causes of malnutrition include inadequate dietary intake, poor nutritional status, poor water and sanitation with their related diseases and disease perpetuating nutrient loss. Intermediate causes include household food insecurity through agricultural production and income, inadequate care for children and women, unhealthy household environment and lack of accessible health and education services. Underlying these causes are longer-term, more complicated determinants such as poverty as a major factor, along with gender inequalities, and larger political, economic, social, and cultural environments which affect institutions and leadership from the community to national level (Golden, 2009; Shrimpton et al., 2001).

5.2.1 Health and infectious diseases

Health is one of the immediate causes of malnutrition. The ability to access primary health care including issues such as distance, affordability, and quality of care can have implications on nutritional outcomes. Food supply, underlying health, and health care interact in important ways, and their combined effect is synergistic. Malnutrition is one of the primary causes of immunodeficiency worldwide, particularly for infants. There is a strong relationship between malnutrition and infection and infant mortality, with poor nutrition leaving children underweight, weakened, and susceptible to infections (Calder and Jackson, 2000). Not only does malnutrition put a child at risk to infections but infections also contribute to the symptoms of malnutrition, causing a vicious cycle.

5.2.2 Water and sanitation

Poor water and sanitation causes a wide variety of infections. Poor water and sanitation has been associated with increased risk of infections in children (Daniels et al., 1990; Huttly et al., 1990) and increased malnutrition (Adair and Guilkey, 1997). Conversely, improved water and sanitation has been associated with lower risk of malnutrition (Huttly et al., 1990). In sub-Saharan Africa, the BOD attributable to malnutrition is 32.7% and to poor water and sanitation 10.1% (Murray and Lopez, 1997).

5.3 Agriculture

Throughout Africa, massive challenges persist, with 60% of rural populations living on less than $1.25 a day (Heinemann et al., 2011) and a huge proportion of the population resides in landlocked, resource-scarce countries (Collier, 2007). Tropical Africa is essentially stuck in a poverty trap and thus experiences low levels of economic growth. Many places are too poor to grow mainly due to high transport costs, small markets, and low-productivity agriculture hence leading to high disease and malnutrition burden. (Millennium Assessment, 2005). In many countries within Africa, agriculture remains the backbone of the rural economy. Increasing agricultural outputs impact economic growth by enhancing farm productivity and food availability (Diao et al., 2007), while providing an economic and employment buffer during times of crisis (FAO, 2009). Food production in Africa, to feed Africa, is still an issue. Many in Africa live in rural areas trapped in a combination of low-productivity agriculture, poor health, and undernutrition.

Climate change is increasingly viewed as a current and future cause of hunger and poverty. Increasing drought, flooding, and changing climatic patterns requiring a shift in crops and farming practices that may not be easily accomplished are three key issues. Another key issue is the future of industrialization and higher standards of living, as the principal cause of climate change appears to be carbon dioxide produced by high energy use with industrialization and higher standards of living (World Hunger and Poverty Facts and Statistics 2015 http://www.worldhunger.org/articles/Learn/world%20hunger%20facts%202002.htm).

5.4 Consequences of malnutrition

5.4.1 Health consequences

Our objective in this section is to discuss the various effects of malnutrition and how it affects individuals and community. Either alone or in association with infectious diseases or other causes of malnutrition, protein energy malnutrition (PEM) is one of the main consequences of malnutrition. At the individual level, the term protein–energy malnutrition is a generic name used in the medical literature to group the whole range of mild to severe clinical and biochemical signs present in children as a consequence of deficient intake and/or utilization of foods of animal origin. Kwashiorkor and marasmus are the names given to the two extreme clinical varieties of the syndrome (Franco et al., 1999).

5.4.1.1 Marasmus

This disease is caused by combined protein and energy deficiency (protein energy malnutrition. Available at http://emedicine.medscape.com/article/1104623-overview#showal). It is commonly seen when adequate quantity of food is not available. It occurs when infants below 1 year are weaned from breast milk, and their diet is replaced with less nutritive food. Characteristics of a marasmic patient include massive reduction in body weight, wasting of body tissues, and reduction in brain weight, ribs become visible through the skin, sunken eyes, and brain development is affected and this may result in mental retardation.

5.4.1.2 Kwashiorkor

Deficiency of protein in the presence of energy results from insufficient protein supply and occurs in children between 1 and 3 years. Characteristics of kwashiorkor include irritability, cracked and scaly skin, stunted growth, thinning of the body, bulging eyes, edema (foot, hands), muscle wasting, and hair discoloration (http://emedicine.medscape.com/article/1104623-clinical#b4).

In 2013, 52 million children under age 5 (10% of the global population) were wasted, meaning that, due to acute malnourishment, they had low weight for their height. Other 165 million children in the world, a quarter of the world's under-5 population, were too short for their age, or stunted, which can impact the child's physical and mental development (http://www.thelancet.com/series/maternal-and-child-nutrition).

On the other hand, overnutrition can lead to obesity. This is one of the predisposing factors of many noncommunicable diseases such as cardiovascular diseases in the world today.

5.4.2 Social consequences

The consequences of stunting on education are also dramatic. Various studies show that child stunting is likely to impact brain development and impair motor skills. According to UNICEF, stunting in early life is linked to 0.7 grade loss in schooling, a 7-month delay in starting school and between 22% and 45% reduction in lifetime earnings. Stunted children become less educated adults, thus making malnutrition a long-term and intergenerational problem (http://www.unicef.org/eu/files/EU-UNICEF_Africa.pdf).

5.4.3 Economic consequences

Malnutrition also slows economic growth and perpetuates poverty (World Bank, 2006). Mortality and morbidity associated with malnutrition represent a direct loss in human capital and productivity for the economy. At a microeconomic level, it is calculated that 1% loss in adult height as a result of childhood stunting equals to a 1.4% loss in productivity of the individual. Malnutrition affects the brain development leading to low cognitive function and reduced school attainment in early childhood. In fact, the education gap will lower academic excellence and skill level of workforce thus leading to a substantial delay in the development of countries affected by malnutrition. Undernutrition in early childhood also makes an individual more prone to noncommunicable diseases later in life, including diabetes and heart disease, significantly increasing health costs in resource constrained health systems. Therefore, improving nutrition is essential to eradicate poverty and accelerate the economic growth of low- and middle-income countries. There is much evidence that improved nutrition drives stronger economic growth.

5.5 Treatments for malnutrition

The type of malnutrition treatment recommended depends mainly on its severity, and whether the patient has an underlying condition/illness which is a contributory factor. If so, that underlying illness/condition needs to be treated or addressed. NICE (National Institute for Health and Clinical Excellence), UK, has guidelines for malnutrition treatment. They state that the needs and preferences of the patient need to be taken into account. The patient, along with health care professionals, should be able to make informed decisions about care and treatment (http://www.medicalnewstoday.com/articles/179316.php?page=3).

NICE guidelines say that individuals who are receiving nutritional support, as well as their caregivers, should be fully informed about their treatment, should be given tailored information, and should be given the opportunity to discuss diagnosis, treatment options, and relevant physical, psychological, and social issues.

A good health care professional discusses good eating and drinking habits with the patient and provides advice regarding healthy food choices. The aim is to make sure the patient is receiving a healthy, nutritious diet. The doctor or dietitian works with the patient to make sure enough calories are being consumed from carbohydrates, proteins, fats, and diary, as well as vitamins and minerals. If the patient cannot get their nutritional requirements from the food they eat, oral supplements may be needed. An additional 250–600 kcal may be advised (malnutrition: treatment and prevention http://www.medicalnewstoday.com/articles/179316.php?page=3).

There are two main types of artificial nutritional support, mainly for patients with severe malnutrition (http://www.medicalnewstoday.com/articles/179316.php?page=3).
- *Enteral nutrition (tube feeding)*— a tube is placed in the nose, the stomach, or small intestine. If it goes through the nose it is called a nasogastric tube or nasoenteral tube. If the tube goes through the skin into the stomach it is called a gastrostomy or percutaneous endoscopic gastrostomy (PEG) tube. One that goes into the small intestine is called a jejunostomy or percutaneous endoscopic jejunostomy (PEJ) tube.
- *Parenteral feeding*—a sterile liquid is fed directly into the bloodstream (intravenously). Some patients may not be able to take nourishment directly into their stomach or small intestine.

5.6 Global response: sustainable development Goal 3

The SDGs adopted by the United Nations in 2015 aimed to ensure healthy lives and promote well-being for all children. The SDG goal 3 target 3.2 is to end preventable deaths of newborns and under-5 children by 2030. Target 3.2 is closely linked with target 3.1, to reduce the global maternal mortality ratio to less than 70 per 100,000 live births, and target 2.2 on ending all forms of malnutrition, as malnutrition is a frequent cause of death for under-5 children. These have been translated into the new Global Strategy for Women's, Children's and Adolescent's Health (Global Strategy), which calls for ending preventable child deaths while addressing the emerging child health priorities. To achieve the SDG targets, the global community has set goals and targets for tackling the unfinished child survival agenda to achieve under-5 mortality of 25 or less per 1000 live births by 2030. This has been translated into several global initiatives:
- "ending preventable maternal mortality" and "every newborn action plan" to promote universal coverage of high-quality maternal and newborn care;
- the "global action plan for the prevention and treatment of pneumonia and diarrhoea";
- a "comprehensive implementation plan on maternal, and infant and young child nutrition" to reduce undernutrition and obesity;
- the Global Technical Strategy for Malaria to reduce global malarial case incidence and mortality by 2030; and
- the Global Vaccine Action Plan to prevent childhood diseases through vaccination.

Malnutrition is a serious health problem in our society today. It mostly affects children and women, especially pregnant women, and contributes to the increase in death rate in children less than 5 years of age. Malnutrition predisposes individuals to infectious diseases and noncommunicable diseases and on the other hand, infectious diseases predispose individuals to malnutrition. Diseases caused by malnutrition are treatable and preventable mainly by improving on the healthy diet consumption and intake of diet supplements.

6 CONCLUSIONS

In Africa, many of the top killers are HIV/AIDS, lower respiratory tract infections, TB, diarrheal disease, and malaria which are preventable and treatable, if adequate health care systems and resources are available. These communicable diseases are a fertile ground to noncommunicable and lifestyle diseases which are continuously affecting the quality of life of people. Global efforts put in place by governments, the WHO, the World Bank to fight these ailments have to be sustained and reinforced. Researchers in Africa are called upon to work harder to provide appropriate solutions to their problems.

REFERENCES

Abaver, D., Nwobegahay, J., Goon, D., Iweriebor, B., Anye, D., 2011. Prevalence of intestinal parasitic infections among HIV/AIDS patients from two health institutions in Abuja, Nigeria. Afr. Health Sci. 11 (3), 24–27.

Adair, L.S., Guilkey, D.K., 1997. Age-specific determinants of stunting in Filipino children. J. Nutr. 127 (2), 314–320.

Agarwal, R., 2011. Severe asthma with fungal sensitization. Curr. Allergy Asthma Rep. 11 (5), 403–413.

Aisu, T., Raviglione, M.C., van Praag, E., Eriki, P., Narain, J.P., Barugahare, L., et al., 1995. Preventive chemotherapy for HIV-associated tuberculosis in Uganda: an operational assessment at a voluntary counselling and testing centre. AIDS 9 (3), 267.

Albonico, M., Bickle, Q., Ramsan, M., Montresor, A., Savioli, L., Taylor, M., 2003. Efficacy of mebendazole and levamisole alone or in combination against intestinal nematode infections after repeated targeted mebendazole treatment in Zanzibar. Bull. World Health Organ. 81 (5), 343–352.

Albonico, M., Ramsan, M., Wright, V., Jape, K., Haji, H., Taylor, M., et al., 2002. Soil-transmitted nematode infections and mebendazole treatment in Mafia Island school children. Ann. Trop. Med. Parasitol. 96 (7), 717–726.

Anderson, R.M., May, R.M., Anderson, B., 1992. Infectious diseases of humans: dynamics and controlVol. 28Wiley Online Library.

Apps.who.int., 2016. GHO. By category. Number of deaths due to HIV/AIDS—Estimates by country. WHO. Available from http://apps.who.int/gho/data/node.main.623?lang=en.

Augusto, G., Nalá, R., Casmo, V., Sabonete, A., Mapaco, L., Monteiro, J., 2009. Geographic distribution and prevalence of schistosomiasis and soil-transmitted helminths among school children in Mozambique. Am. J. Trop. Med. Hyg. 81 (5), 799–803.

Bagagli, E., Theodoro, R.C., Bosco, S.M., McEwen, J.G., 2008. *Paracoccidioides brasiliensis*: phylogenetic and ecological aspects. Mycopathologia 165 (4–5), 197–207.

Bailey, R., Duong, T., Carpenter, R., Whittle, H., Mabey, D., 1999. The duration of human ocular *Chlamydia trachomatis* infection is age dependent. Epidemiol. Infect. 123 (03), 479–486.

Bailey, R., Osmond, C., Mabey, D., Whittle, H., Ward, M., 1989. Analysis of the household distribution of trachoma in a Gambian village using a Monte Carlo simulation procedure. Int. J. Epidemiol. 18 (4), 944–951.

Barnett, J.A., 2004. A history of research on yeasts 8 taxonomy. Yeast 21 (14), 1141–1193.

Berhane Y., Worku A., Bejiga A., 2006. National survey on blindness, low vision and trachoma in Ethiopia. Federal Ministry of Health of Ethiopia.

Bethony, J., Chen, J., Lin, S., Xiao, S., Zhan, B., Li, S., et al., 2002. Emerging patterns of hookworm infection: influence of aging on the intensity of Necator infection in Hainan Province, People's Republic of China. Clin. Infect. Dis. 35 (11), 1336–1344.

Birbeck, G.L., Munsat, T., 2002. Neurologic services in sub-Saharan Africa: a case study among Zambian primary healthcare workers. J. Neurol. Sci. 200 (1), 75–78.

Bleakley, H., 2007. Disease and development: evidence from hookworm eradication in the American South. Q. J. Econ. 122 (1), 73.

Blössner, M., De Onis, M., Prüss-Üstün, A., Campbell-Lendrum, D., Corvalán, C., Woodward, A., 2005. Quantifying the Health Impact at National and Local Levels. WHO, Geneva.

Bower, J., Asmera, J., Zebenigus, M., Sandroni, P., Bower, S., Zenebe, G., 2007. The burden of inpatient neurologic disease in two Ethiopian hospitals. Neurology 68 (5), 338–342.

Boyle, P., Levin, B., 2008. World cancer report 2008. IARC Press, International Agency for Research on Cancer. Available from http://www.iarc.fr/en/publications/pdfs-online/wcr/2008/

Brandt, M.E., Park, B.J., 2013. Think fungus—prevention and control of fungal infections. Emerg. Infect. Dis. 19 (10), 1688.

Brooker, S., Clements, A.C., Bundy, D.A., 2006a. Global epidemiology, ecology and control of soil-transmitted helminth infections. Adv. Parasitol. 62, 221–261.

Brooker, S., Clements, A.C., Hotez, P.J., Hay, S.I., Tatem, A.J., Bundy, D.A., Snow, R.W., 2006b. The co-distribution of *Plasmodium falciparum* and hookworm among African school children. Malaria J. 5 (1), 1.

Brown, G.D., Denning, D.W., Gow, N.A., Levitz, S.M., Netea, M.G., White, T.C., 2012. Hidden killers: human fungal infections. Sci. Transl. Med. 4 (165), 165rv113.

Bundy, D., Cooper, E., 1989. Human *Trichuris and trichuriasis*. Adv. Parasitol. 28, 107–173.

Bundy, D.A.P., Michael, E., Guyatt, H., 2002. Epidemiology and control of nematode infection and disease in humans. In: Lee, D.L. (Ed.), The biology of nematodes. Taylor and Francis, London, pp. 599–617.

Bundy, D.A., Shaeffer, S., Jukes, M., et al., (in press) School-based health and nutrition programs. In: Jamison, D., Breman, J., Meacham, A., et, al., (Eds.), Disease control priorities in developing countries, second ed. World Bank, World Health Organization, Fogarty International Center of the National Institutes of Health, US Department of Health and Human Services, New York.

Burton, M.J., Holland, M.J., Faal, N., Aryee, E.A., Alexander, N.D., Bah, M., et al., 2003. Which members of a community need antibiotics to control trachoma? Conjunctival *Chlamydia trachomatis* infection load in Gambian villages. Invest. Ophthal. Vis. Sci. 44 (10), 4215–4222.

Burton, M.J., Holland, M.J., Makalo, P., Aryee, E.A., Alexander, N.D., Sillah, A., et al., 2005. Re-emergence of *Chlamydia trachomatis* infection after mass antibiotic treatment of a trachoma-endemic Gambian community: a longitudinal study. Lancet 365 (9467), 1321–1328.

Byrnes, III, E.J., Bartlett, K.H., Perfect, J.R., Heitman, J., 2011. *Cryptococcus gattii*: an emerging fungal pathogen infecting humans and animals. Microbes Infect. 13, 895–907.

Calder, P.C., Jackson, A.A., 2000. Undernutrition, infection and immune function. Nutr. Res. Rev. 13 (01), 3–29.

Canadian Institute for Health Information, 2007. The Burden of Neurological Diseases. Disorders and Injuries in Canada. CIHI, Ottawa.

Casetta, I., Monetti, V.C., Malagu, S., Paolino, E., Govoni, V., Fainardi, E., et al., 2002. Risk factors for cryptogenic and idiopathic partial epilepsy: a community-based case-control study in Copparo, Italy. Neuroepidemiology 21 (5), 252–154.

Cdc.gov., 2016. CDC—Parasites—About Parasites.

Centers For Disease Control and Prevention. 2014. Sporotrichosis. Available at : http://www.cdc.gov/fungal/diseases/sporotrichosis/

Chan, P.W., Anuar, A.K., Fong, M.Y., Debruyne, J.A., Ibrahim, J., 2001. Toxocara seroprevalence and childhood asthma among Malaysian children. Pediatr. Int. 43 (4), 350–353.

Chan, L., Bundy, D., Kan, S.P., 1994a. Aggregation and predisposition to *Ascaris lumbricoides* and *Trichuris trichiura* at the familial level. Trans. R. Soc. Trop. Med. Hyg. 88 (1), 46–48.

Chan, M., Medley, G., Jamison, D., Bundy, D., 1994b. The evaluation of potential global morbidity attributable to intestinal nematode infections. Parasitology 109 (03), 373–387.

Chandra, V., Pandav, R., Laxminarayan, R., Tanner, C., Manyam, B., Rajkumar, S. et al., 2006. Neurological disorders. In : Jamison, D.T., Breman, J.G., Measham, A.R., et al., (Eds). Disease Control Priorities in Developing Countries, second ed. The International Bank for Reconstruction and Development/The World Bank, Washington, DC; Oxford University Press, New York. pp. 627–644 (Chapter 32).

Christian, P., Khatry, S.K., West, K.P., 2004. Antenatal anthelmintic treatment, birthweight, and infant survival in rural Nepal. Lancet 364 (9438), 981–983.

Collier, P., 2007. Poverty reduction in Africa. Proc. Natl. Acad. Sci. 104 (43), 16763–16768.

Colombo, A.L., Tobón, A., Restrepo, A., Queiroz-Telles, F., Nucci, M., 2011. Epidemiology of endemic systemic fungal infections in Latin America. Med. Mycol. 49 (8), 785–798.

Committee W.E., 2002. Prevention and control of schistosomiasis and soil-transmitted helminthiasis, World Health Organization technical report series 912 i, p. 1–67.

Congdon, N., West, S., Vitale, S., Katala, S., Mmbaga, B., 1993. Exposure to children and risk of active trachoma in Tanzanian women. Am. J. Epidemiol. 137 (3), 366–372.

Corbett, E.L., Watt, C.J., Walker, N., Maher, D., Williams, B.G., Raviglione, M.C., Dye, C., 2003. The growing burden of tuberculosis: global trends and interactions with the HIV epidemic. Arch. Int. Med. 163 (9), 1009–1021.

Crofton, J., Horne, N., Miller, F., 1999. Crofton's Clinical Tuberculosis, third edition. Macmillan Education, Oxford, OX43PP. 200 pages.

Crompton, D., 2001. Ascaris and ascariasis. Adv. Parasitol. 48, 285–375.

Daniels, D.L., Cousens, S.N., Makoae, L., Feachem, R.G., 1990. A case-control study of the impact of improved sanitation on diarrhoea morbidity in Lesotho. Bull. World Health Organ. 68 (4), 455.

David Kay, A.P., Carlos, C., 2000. Methodology for Assessment of Environmental Burden of Disease. Buffalo, New York.

Dawson, C., Daghfous, T., Messadi, M., Hoshiwara, I., Schachter, J., 1976. Severe endemic trachoma in Tunisia. Br. J. Ophthalmol. 60 (4), 245–252.

Dawson, C.R., Jones, B.R., Tarizzo, M.L., 1981. Guide to Trachoma Control in Programmes for the Prevention of Blindness. Geneva, Switzerland.

De Farias, M.R., Condas, L.A.Z., Ribeiro, M.G., Bosco, S.d.M.G., Muro, M.D., Werner, J., et al., 2011. Paracoccidioidomycosis in a dog: case report of generalized lymphadenomegaly. Mycopathologia 172 (2), 147–152.

De Onis, M., Blössner, M., 2003. The World Health Organization global database on child growth and malnutrition: methodology and applications. Int. J. Epidemiol. 32 (4), 518–526.

De Onis, M., Blössner, M., Borghi, E., Frongillo, E.A., Morris, R., 2004. Estimates of global prevalence of childhood underweight in 1990 and 2015. JAMA 291 (21), 2600–2606.

De Sole, G., 1987. Impact of cattle on the prevalence and severity of trachoma. Br. J. Ophthalmol. 71 (11), 873–876.

Denning, D.W., Pleuvry, A., Cole, D.C., 2013. Global burden of allergic bronchopulmonary aspergillosis with asthma and its complication chronic pulmonary aspergillosis in adults. Med. Mycol. 51 (4), 361–370.

Diao, X., Hazell, P.B., Resnick, D., Thurlow, J., 2007. The Role of Agriculture in Development: Implications for Sub-Saharan Africa vol. 153. Research report/International Food Policy Research Institute.

Dolin, P., Faal, H., Johnson, G., Ajewole, J., Mohamed, A., Lee, P.S., 1998. Trachoma in the Gambia. Br. J. Ophthalmol. 82 (8), 930–933.

Dunn, A., Keymer, A., 1986. Factors affecting the reliability of the McMaster technique. J. Helminthol. 60 (04), 260–262.

Dye, C., Garnett, G.P., Sleeman, K., Williams, B.G., 1998. Prospects for worldwide tuberculosis control under the WHO DOTS strategy. Lancet 352 (9144), 1886–1891.

Ellman, T., 2015. Demedicalizing AIDS prevention and treatment in Africa. New Engl. J. Med. 372 (4), 303–305.

Emerson, P.M., Cairncross, S., Bailey, R.L., Mabey, D., 2000. Review of the evidence base for the 'F'and 'E'components of the SAFE strategy for trachoma control. Trop. Med. Int. Health 5 (8), 515–527.

Emerson, P.M., Lindsay, S.W., Alexander, N., Bah, M., Dibba, S.-M., Faal, H.B., Walraven, G.E., 2004. Role of flies and provision of latrines in trachoma control: cluster-randomised controlled trial. Lancet 363 (9415), 1093–1098.

Fahn, S., 2003. Description of Parkinson's disease as a clinical syndrome. Ann. NY Acad. Sci. 991 (1), 1–14.

FAO, 2009. The state of food insecurity in the world: Economic crises-impacts and lessons learned: FAO of the United Nations Rome.

FAO, I. 2015. WFP. The State of Food Insecurity in the World.

Fauci, W., Braunwald, E., Kasper, D., Hauser, S., Longo, D., Jameson, J., 2012. Harrison Principle of Internal Medicine, Eighteenth edition McGraw-Hill, New York.

Faust, E.C., Russell, P.F., 1964. Craig and Faust's clinical parasitology. Acad. Med. 39 (9), 867.

Fincham, J.E., Markus, M., Adams, V., 2003. Could control of soil-transmitted helminthic infection influence the HIV/AIDS pandemic. Acta Trop. 86 (2), 315–333.

Fisher, M., Koenig, G., White, T., Taylor, J., 2002. Molecular and phenotypic description of *Coccidioides posadasii* sp. nov., previously recognized as the non-California population of Coccidioides immitis. Mycologia 94 (1), 73–84.

Franco, V.H.M., Hotta, J.K.S., Jorge, S., Dos Santos, J., 1999. Plasma fatty acids in children with grade III protein-energy malnutrition in its different clinical forms: marasmus, marasmic kwashiorkor, and kwashiorkor. J. Trop. Pediatr. 45 (2), 71–75.

Galgiani, J.N., 1999. Coccidioidomycosis: a regional disease of national importance: rethinking approaches for control. Ann. Int. Med. 130 (4_Part_1), 293–300.

Gemmel, M., Matyas, Z., Pawlawski, Z., Larralde, C., 1982. Guidelines for surveillance prevention and control of taeniasis/cysticercosis. VPH/83.49. World Health Organization, Geneva, Switzerland, pp. 207.

Gilks, C.F., Crowley, S., Ekpini, R., Gove, S., Perriens, J., Souteyrand, Y., et al., 2006. The WHO public-health approach to antiretroviral treatment against HIV in resource-limited settings. Lancet 368 (9534), 505–510.

Globalissues.org. 2016. AIDS in Africa—Global Issues. Available from http://www.globalissues.org/article/90/aids-in-africa

Golden, M., 2009. Proposed nutrient requirements of moderately malnourished populations of children. Food Nutr. Bull. 30 (3), S267–S342.

Goldman, S., Tanner, C., 1998. Etiology of Parkinson's disease. In: Jankovic, J., Tolosa, E. (Eds.), Parkinson's Disease and Movement Disorders. Williams and Wilkins, London, UK, pp. 133–158.

Gorelick, P.B., 1987. Alcohol and stroke. Stroke 18 (1), 268–271.

Gray, F., Chrétien, F., Vallat-Decouvelaere, A.V., Scaravilli, F., 2003. The changing pattern of HIV neuropathology in the HAART era. J. Neuropathol. Exp. Neurol. 62 (5), 429–440.

Greenberg, G., 1985. MRC trial of treatment of mild hypertension-principal results. Br. Med. J. 291 (6488), 97–104.

Gurney, J.W., Conces, D., 1996. Pulmonary histoplasmosis. Radiology 199 (2), 297–306.

Hankey, G.J., 1999. Smoking and risk of stroke. Eur. J. Cardiovasc. Risk 6 (4), 207–211.

Harries, A., Hargreaves, N., Gausi, F., Kwanjana, J., Salaniponi, F., 2001. High early death rate in tuberculosis patients in Malawi. Int. J. Tuberculosis Lung Dis. 5 (11), 1000–1005.

Harris, J.R., Lockhart, S.R., Sondermeyer, G., Vugia, D.J., Crist, M.B., D'Angelo, M.T., Smelser, C., 2013. *Cryptococcus gattii* infections in multiple states outside the US Pacific Northwest. Emerg. Infect. Dis. 19 (10), 1621–1627.

Hauser, W., 1995. Recent developments in the epidemiology of epilepsy. Acta Neurol. Scand. 92 (s162), 17–21.

Havlickova, B., Czaika, V.A., Friedrich, M., 2008. Epi demiological trends in skin mycoes worldwide. Mycoses 4, 2–15.

Hay, R., Rose, P., Jones, T., 1987. Paracoccidioidin sensitization in Guyana—a preliminary skin test survey in hospitalized patients and laboratory workers. Trans. R. Soc. Trop. Med. Hyg. 81 (1), 46–48.

Heinemann, E., Prato, B., Shepherd, A., 2011. Rural Poverty Report 2011. International Fund for Agricultural Development (IFAD), Rome.

Hidalgo, J.A., Vazquez, J.A., 2010. Candidiasis.emedicine.com, updated January, 11.

Hirtz, D., Thurman, D., Gwinn-Hardy, K., Mohamed, M., Chaudhuri, A., Zalutsky, R., 2007. How common are the "common" neurologic disorders? Neurology 68 (5), 326–337.

Horn, D.L., Neofytos, D., Anaissie, E.J., Fishman, J.A., Steinbach, W.J., Olyaei, A.J., Webster, K.M., 2009. Epidemiology and outcomes of candidemia in 2019 patients: data from the prospective antifungal therapy alliance registry. Clin. Infect. Dis. 48 (12), 1695–1703.

Horton, J., 2003. Global anthelmintic chemotherapy programmes: learning from history. Trends Parasitol. 19 (9), 405–409.

Hotez, P.J., Brooker, S., Bethony, J.M., Bottazzi, M.E., Loukas, A., Xiao, S., 2004. Hookworm infection. New Engl. J. Med. 351 (8), 799–807.

Hsu, J.L., Ruoss, S.J., Bower, N.D., Lin, M., Holodniy, M., Stevens, D.A., 2011. Diagnosing invasive fungal disease in critically ill patients. Crit. Rev. Microbiol. 37 (4), 277–312.

Huttly, S.R., Blum, D., Kirkwood, B.R., Emeh, R.N., Okeke, N., Ajala, M., Feachem, R.G., 1990. The Imo State (Nigeria) drinking water supply and sanitation project, 2. Impact on dracunculiasis, diarrhoea and nutritional status. Trans. R. Soc. Trop. Med. Hyg. 84 (2), 316–321.

Ibrahim, S.a., Hamisu, I., Lawal, U., 2015. Spatial pattern of tuberculosis prevalence in Nigeria: a comparative analysis of spatial autocorrelation indices. Am. J. Geogr. Inform. Syst. 4 (3), 87–94.

Institute for Health Metrics, and Evaluation, 2013. The global burden of disease: generating evidence, guiding policy: IHME Seattle.

Izuora, G., Iloeje, S., 1989. A review of neurological disorders seen at the Paediatric Neurology Clinic of the University of Nigeria Teaching Hospital, Enugu. Ann. Trop. Paediatr. 9 (4), 185–190.

Jamison, D.T., 2006. Disease and Mortality in Sub-Saharan Africa. World Bank Publications, https://openknowledge.worldbank.org/handle/10986/7050.

Jarvis, J.N., Govender, N., Chiller, T., Park, B.J., Longley, N., Meintjies, G., Harrison, T.S., 2012. Cryptococcal antigen screening and preemptive therapy in patients initiating antiretroviral therapy in resource-limited settings: a proposed algorithm for clinical implementation. J. Int. Assoc. Phys. AIDS Care (JIAPAC) 11 (6), 374–379, 1545109712459077.

Joint FAO/WHO Expert Committee on Food Additives. Meeting, 2010. Evaluation of Certain Food Additives: Seventy-First Report of the Joint FAO/WHO Expert Committee on Food Additivesvol. 71World Health Organization, Geneva.

Jones, B., 1975. The prevention of blindness from trachoma. Trans. Ophthalmol. Soc. UK 95 (1), 16.

Kahn, K., Tollman, S.M., 1999. Stroke in rural South Africa—contributing to the little known about a big problem. S. Afr. Med. J. 89 (1), 63–65.

Katz, N., CHAVES, A., Pellegrino, J., 1972. A simple device for quantitative stool thick-smear technique in *Schistosomiasis mansoni*. Rev. Inst. Med. Trop. São Paulo 14 (6), 397–400.

Khuroo, M.S., Zargar, S.A., Mahajan, R., 1990. Hepatobiliary and pancreatic ascariasis in India. Lancet 335 (8704), 1503–1506.

Knutsen, A.P., Slavin, R.G., 2011. Allergic bronchopulmonary aspergillosis in asthma and cystic fibrosis. Clin. Dev. Immunol. 2011:843763.

Kurtzke, J.F., 1982. The current neurologic burden of illness and injury in the United States. Neurology 32 (11), 1207.

Kusumi, M., Nakashima, K., Harada, H., Nakayama, H., Takahashi, K., 1996. Epidemiology of Parkinson's disease in Yonago City, Japan: comparison with a study carried out 12 years ago. Neuroepidemiology 15 (4), 201–207.

Lai, Y.-S., Biedermann, P., Ekpo, U.F., Garba, A., Mathieu, E., Midzi, N., et al., 2015. Spatial distribution of schistosomiasis and treatment needs in sub-Saharan Africa: a systematic review and geostatistical analysis. Lancet Infect. Dis. 15 (8), 927–940.

Larsson, S.C., Virtamo, J., Wolk, A., 2011. Red meat consumption and risk of stroke in Swedish men. Am. J. Clin. Nutr. 94 (2), 417–421.

Lenhart, S., Schafer, M., Singal, M., Hajjeh, R., 2004. Histoplasmosis: protecting workers at risk [cited 2013 July 17].

Leone, M., Bottacchi, E., Beghi, E., Morgando, E., Mutani, R., Cremo, R., et al., 2002. Risk factors for a first generalized tonic-clonic seizure in adult life. Neurol. Sci. 23 (3), 99–106.

Lozano, R., Naghavi, M., Foreman, K., Lim, S., Shibuya, K., Aboyans, V., et al., 2013. Global and regional mortality from 235 causes of death for 20 age groups in 1990 and 2010 a systematic analysis for the Global Burden of Disease Study 2010. Lancet 380 (9859), 2095–2128.

Mariam, Z.T., Abebe, G., Mulu, A., 2009. Opportunistic and other intestinal parasitic infections in AIDS patients, HIV seropositive healthy carriers and HIV seronegative individuals in southwest Ethiopia. East Afr. J. Public Health 5 (3), 169–173.

Martin, G.S., Mannino, D.M., Eaton, S., Moss, M., 2003. The epidemiology of sepsis in the United States from 1979 through 2000. New Engl. J. Med. 348 (16), 1546–1554.

Matenga, J., 1997. Stroke incidence rates among black residents of Harare—a prospective community-based study. S. Afr. Med. J. 87 (5), 606–609.

Mathers, C., Fat, D.M., Boerma, J.T., Ebrary, I. 2004. World Health Organization the global burden of disease: 2004 update. Geneva, Switzerland.

Mathers, C.D., Lopez, A.D., Murray, C.J.L., 2006. The burden of disease and mortality by condition: data, methods and results for 2001. In: Lopez, A.D., Mathers, C.D., Ezzati, M., Murray, C.J.L., Jamison, D.T. (Eds.), Global Burden of Disease and Risk Factors. Oxford University Press, New York, pp. 45–240.

Mayeux, R., Marder, K., Cote, L.J., Denaro, J., Hemenegildo, N., Mejia, H., et al., 1995. The frequency of idiopathic Parkinson's disease by age, ethnic group, and sex in northern Manhattan, 1988-1993. Am. J. Epidemiol. 142 (8), 820–827.

Medina, M.T., Rosas, E., Rubio-Donnadieu, F., Sotelo, J., 1990. Neurocysticercosis as the main cause of late-onset epilepsy in Mexico. Arch. Intern. Med. 150 (2), 325–327.

Menzies, D., Holmes, L., McCumesky, G., Prys-Picard, C., Niven, R., 2011. Aspergillus sensitization is associated with airflow limitation and bronchiectasis in severe asthma. Allergy 66 (5), 679–685.

Michael, D., 2014. AIDS In Africa: Devastating An Entire Generation. http://www.faze.ca/aids-in-africa-devastating-an-entire-generation/

Millennium Assessment U., 2005. Investing in development: a practical plan to achieve the millennium development goals, Report to UN Secretary General. J. Sachs, United Nations, 94. Available from http://www.unmillenniumproject.org/documents/MainReportComplete-lowres.pdf.

Molyneux, D.H., Hotez, P.J., Fenwick, A., 2005. Rapid-impact interventions: how a policy of integrated control for Africa's neglected tropical diseases could benefit the poor. PLoS Med. 2 (11), e336.

Morens, D., Davis, J., Grandinetti, A., Ross, G., Popper, J., White, L., 1996. Epidemiologic observations on Parkinson's disease Incidence and mortality in a prospective study of middle-aged men. Neurology 46 (4), 1044–1050.

Mung'Ala-Odera, V., Snow, R.W., Newton, C.R., 2004. The burden of the neurocognitive impairment associated with *Plasmodium falciparum* malaria in sub-Saharan Africa. Am. J. Trop. Med. Hyg. 71 (Suppl. 2), 64–70.

Murray, C.J., Lopez, A.D., 1997. Global mortality, disability, and the contribution of risk factors: Global Burden of Disease Study. Lancet 349 (9063), 1436–1442.

Murray, C.J., Vos, T., Lozano, R., Naghavi, M., Flaxman, A.D., Michaud, C., et al., 2013. Disability-adjusted life years (DALYs) for 291 diseases and injuries in 21 regions, 1990–2010: a systematic analysis for the Global Burden of Disease Study 2010. Lancet 380 (9859), 2197–2223.

Naghavi, M., Wang, H., Lozano, R., Davis, A., Liang, X., Zhou, M., Abd-Allah, F., 2015. Global, regional, and national age-sex specific all-cause and cause-specific mortality for 240 causes of death, 1990–2013: a systematic analysis for the Global Burden of Disease Study 2013. Lancet 385 (9963), 117–171.

Ngugi, A.K., Bottomley, C., Kleinschmidt, I., Sander, J.W., Newton, C.R., 2010. Estimation of the burden of active and life-time epilepsy: a meta-analytic approach. Epilepsia 51 (5), 883–890.

Nucci, M., Varon, A.G., Garnica, M., Akiti, T., Barreiros, G., Trope, B.M., Nouér, S.A., 2013. Increased incidence of invasive fusariosis with cutaneous portal of entry. Brazil. Emerg. Infect. Dis. 19 (10), 1567–1572.

Nxasana, N., Baba, K., Bhat, V., Vasaikar, S., 2014. Prevalence of intestinal parasites in primary school children of mthatha, eastern Cape Province. S. Afr. Ann. Med. Health Sci. Res. 3 (3), 511–516.

Odds, F.C., Hanson, M.F., Davidson, A.D., Jacobsen, M.D., Wright, P., Whyte, J.A., et al., 2007. One year prospective survey of Candida bloodstream infections in Scotland. J. Med. Microbiol. 56 (Pt 8), 1066–1075.

Olasehinde, G., Ojurongbe, D., Akinjogunla, O., Egwari, L., Adeyeba, A.O., 2015. Prevalence of malaria and predisposing factors to antimalarial drug resistance in southwestern Nigeria. Res. J. Parasitol. 10 (3), 92–101.

Olsen, A., 2007. Efficacy and safety of drug combinations in the treatment of schistosomiasis, soil-transmitted helminthiasis, lymphatic filariasis and onchocerciasis. Trans. R. Soc. Trop. Med. Hyg. 101 (8), 747–758.

Organization, W.H., 2015. Global tuberculosis report.

Pappas, G., Roussos, N., Falagas, M.E., 2009. Toxoplasmosis snapshots: global status of Toxoplasma gondii seroprevalence and implications for pregnancy and congenital toxoplasmosis. Int. J. Parasitol. 39 (12), 1385–1394.

Park, B.J., Wannemuehler, K.A., Marston, B.J., Govender, N., Pappas, P.G., Chiller, T.M., 2009. Estimation of the current global burden of cryptococcal meningitis among persons living with HIV/AIDS. AIDS 23 (4), 525–530.

Perry, E., McKeith, I., Thompson, P., Marshall, E., Kerwin, J., Jabeen, S., et al., 1991. Topography, extent, and clinical relevance of neurochemical deficits in dementia of Lewy body type, Parkinson's disease, and Alzheimer's disease. Ann. NY Acad. Sci. 640 (1), 197–202.

Polack, S., Brooker, S., Kuper, H., Mariotti, S., Mabey, D., Foster, A., 2005. Mapping the global distribution of trachoma. Bull. World Health Organ. 83 (12), 913–919.

Prüss-Üstün, A., Corvalán, C., World Health Organization, 2006. Preventing Disease Through Healthy Environments: Towards an Estimate of the Environmental Burden of Disease. World Health Organization, Geneva, Switzerland.

Prüss-Üstün, A., Mathers, C., Corvalán, C., Woodward, A., 2003. Introduction and methods: assessing the environmental burden of disease at national and local levels. World Health Organization, Geneva (WHO Environmental Burden of Disease series, No. 1).

Rajabiun, S., 2001. HIV/AIDS: A Guide for Nutrition, Care, and Support: Food and Nutrition Technical Assistance Project, Academy for Educational Development, Washington, D.C.

Reynolds, K., Lewis, B., Nolen, J.D.L., Kinney, G.L., Sathya, B., He, J., 2003. Alcohol consumption and risk of stroke: a meta-analysis. JAMA 289 (5), 579–588.

Ricci, G., Mota, F.T., Wakamatsu, A., Serafim, R., Borra, R.C., Franco M, 2004. Canine paracoccidioidomycosis. Med. Mycol. 42 (4), 379–383.

Rollinson, D., Knopp, S., Levitz, S., Stothard, J.R., Tchuenté, L.-A.T., Garba, A., et al., 2013. Time to set the agenda for schistosomiasis elimination. Acta Trop. 128 (2), 423–440.

Rutgers Biomedical and Health Sciences, 2013. Attacking fungal infection, one of world's major killers. ScienceDaily. 23 December 2013.

Ryan, K.J., Ray, C.G. (Eds.), 2004. Sherris Medical Microbiology. fourth ed. McGraw Hill, New York, pp. 362–368.

Satcher, D., Foege, W.H., Watson, Jr., W.C., Hopkins, D.R., Ortiz, T.G., Walters, C.C., et al., 1993. Recommendations of the International Task Force for Disease Eradication. Morbid. Mort. Wkly Rep. 42, i-38.

Schachter, J., Dawson, C.R., 1990. The epidemiology of trachoma predicts more blindness in the future. Scand. J. Infect. Dis. Suppl. 69, 55–62.

Schachter, J., Dawson, C., Sheppard, J., Courtright, P., Said, M., Zaki, S., Lorincz, A., 1988. Nonculture methods for diagnosing chlamydial infection in patients with trachoma: a clue to the pathogenesis of the disease? J. Infect. Dis. 158 (6), 1347–1352.

Schachter, J., West, S.K., Mabey, D., Dawson, C.R., Bobo, L., Bailey, R., Sallam, S., 1999. Azithromycin in control of trachoma. Lancet 354 (9179), 630–635.

Schoenberg, B.S., Anderson, D.W., Haerer, A.F., 1985. Prevalence of Parkinson's disease in the biracial population of Copiah County, Mississippi. Neurology 35 (6), 841–845.

Schrag, A., Jahanshahi, M., Quinn, N., 2000. What contributes to quality of life in patients with Parkinson's disease? J. Neurol. Neurosurg. Psychiatr. 69 (3), 308–312.

Senanayake, N., Román, G.C., 1993. Epidemiology of epilepsy in developing countries. Bull. World Health Organ. 71 (2), 247.

Shnayder, N., Dmitrenko, D., Sadykova, A., Sharavii, L., Shulmin, A., Shapovalova, E., et al., 2011. Epidemiological studies on epilepsy in Siberia. Med. Health Sci. J. 6, 35–42.

Shrimpton, R., Victora, C.G., de Onis, M., Lima, R.C., Blössner, M., Clugston, G., 2001. Worldwide timing of growth faltering: implications for nutritional interventions. Pediatrics 107 (5), e75.

Silberberg, D., Katabira, E., 2006. Neurological disorders. In: Jamison, D.T., Feachem, R.G., Makgoba, M.W. et al., (Eds.), Disease and Mortality in Sub-Saharan Africa, second ed. The International Bank for Reconstruction and Development/The World Bank, Washington, DC.

Sloan, M.A., Kittner, S.J., Rigamonti, D., Price, T.R., 1991. Occurrence of stroke associated with use/abuse of drugs. Neurology 41 (9), 1358.

Sobel, J.D., 2007. Vulvovaginal candidosis. Lancet 369 (9577), 1961–1971.

Solomon, A.W., Holland, M.J., Burton, M.J., West, S.K., Alexander, N.D., Aguirre, A., Johnson, G.J., 2003. Strategies for control of trachoma: observational study with quantitative PCR. Lancet 362 (9379), 198–204.

Solomon, A.W., Peeling, R.W., Foster, A., Mabey, D.C., 2004. Diagnosis and assessment of trachoma. Clin. Microbiol. Rev. 17 (4), 982–1011.

Springston, J., 1998. The birds. Occup. Health Saf (Waco, TX) 67 (5), 86–89.

Stoltzfus, R.J., Dreyfuss, M.L., Chwaya, H.M., Albonico, M., 1997. Hookworm control as a strategy to prevent iron deficiency. Nutr. Rev. 55 (6), 223–232.

Tanner, C.M., Aston, D.A., 2000. Epidemiology of Parkinson's disease and akinetic syndromes. Curr. Opin. Neurol. 13 (4), 427–430.

Tanner, C., Hubble, J., Chan, P., 1997. Epidemiology and genetics of Parkinson's disease. Mov. Disord., 137–152.

Taren, D.L., Nesheim, M., Crompton, D., Holland, C.V., Barbeau, I., Rivera, G., Tucker, K., 1987. Contributions of ascariasis to poor nutritional status in children from Chiriqui Province, Republic of Panama. Parasitology 95 (03), 603–613.

Theodoro, R.C., de Melo Teixeira, M., Felipe, M.S.S., dos Santos Paduan, K., Ribolla, P.M., San-Blas, G., Bagagli, E., 2012. Genus Paracoccidioides: species recognition and biogeographic aspects. PLoS One 7 (5), e37694.

Thomas, J., Jacobson, G., Narkowicz, C., Peterson, G., Burnet, H., Sharpe, C., 2010. Toenail onychomycosis: an important global disease burden. J. Clin. Pharm. Ther. 35 (5), 497–519.

Thomson, R., 2009. Evidence based implementation of complex interventions. Br. Med. J. 339, b3124.

Thylefors, B., Dawson, C.R., Jones, B.R., West, S., Taylor, H.R., 1987. A simple system for the assessment of trachoma and its complications. Bull. World Health Organ. 65 (4), 477.

Tsang, C.A., Anderson, S.M., Imholte, S.B., Erhart, L.M., Chen, S., Park, B.J., Sunenshine, R.H., 2010. Enhanced surveillance of coccidioidomycosis, Arizona, USA, 2007–2008. Emerg. Infect. Dis. 16 (11), 1738–1746.

UNAIDS, J., 2010. Global Report: UNAIDS Report on the Global AIDS Epidemic 2010. UNAIDS, Geneva.

UNAIDS, 2015. How AIDS Changed Everything—MDG6: 15 Years 15 Lessons of Hope from the AIDS Response. UNAIDS, Geneva.

United Nations Children's Fund, 2004. Joint United Nations Programme on HIV/AIDS, World Health Organization, Médecins Sans Frontières. Sources and prices of selected medicines and diagnostics for people living with HIV/AIDS. UNICEF, UNAIDS, WHO, MSF, Geneva.

Van de Beek, D., de Gans, J., Tunkel, A.R., Wijdicks, E.F., 2006. Community-acquired bacterial meningitis in adults. New Engl. J. Med. 354 (1), 44–53.

van Hees, C., Naafs, B., 2001. Common Skin Diseases in Africa: An Illustrated Guide. van Hees, http://www.telemedicine.itg.be/telemedicine/uploads/skin.pdf.

von Campenhausen, S., Bornschein, B., Wick, R., Bötzel, K., Sampaio, C., Poewe, W., et al., 2005. Prevalence and incidence of Parkinson's disease in Europe. Eur. Neuropsychopharmacol. 15 (4), 473–490.

Wang, Y., 1991. [The incidence and prevalence of Parkinson's disease in the People's Republic of China]. Zhonghua Liu Xing Bing Xue Za Zhi 12 (6), 363–365.

Wanke, B., Londero, A.T., 1994. Epidemiology and Paracoccidioidosis infection. In: Franco, M. (Ed.), Paracoccidioidomycoses. CRC Press, Boca Raton, pp. 109–120.

Wannamethee, S.G., Shaper, A.G., Whincup, P.H., Walker, M., 1995. Smoking cessation and the risk of stroke in middle-aged men. JAMA 274 (2), 155–160.

West, S.K., Munoz, B., Turner, V.M., Mmbaga, B., Taylor, H.R., 1991. The epidemiology of trachoma in central Tanzania. Int. J. Epidemiol. 20 (4), 1088–1092.

Westover, A.N., McBride, S., Haley, R.W., 2007. Stroke in young adults who abuse amphetamines or cocaine: a population-based study of hospitalized patients. Arch. Gen. Psychiatr. 64 (4), 495–502.

Whiston, E., Wise, H.Z., Sharpton, T.J., Jui, G., Cole, G.T., Taylor, J.W., 2012. Comparative transcriptomics of the saprobic and parasitic growth phases in Coccidioides spp. PLoS One 7 (7), e41034.

WHO, 2003. The World Health Report 2003: Shaping the Future. World Health Organization, Geneva, http://www.who.int/whr/2003/en/whr03_en.pdf?ua=1.

WHO, 2005. Deworming for health and development: report of the Third Global Meeting of the Partners for Parasite Control.

WHO, 2006a. The African Regional Health Report: The Health of the People [Monografia en Internet]. WHO Regional office for Africa.

WHO, 2006c. WHO Report 2006: Global Tuberculosis Control: Surveillance, Planning, Financing. World Health Organization, Geneva.

WHO, 2013. Global update on HIV treatment 2013 results, impact and opportunities.

WHO, 2015. Global Tuberculosis Report 2014. World Health Organization, Geneva.

World Bank, 2006. Repositioning Nutrition as Central to Development: A Strategy for Large-Scale Action. World Bank, Washington DC.

World Food Programme, 2000. Food and Nutrition Handbook. World Food Programme, Rome.

Wright, J.P., Ford, H.L., 1995. Bacterial meningitis in developing countries. Trop. Doctor 25 (1), 5–8.

CHAPTER 2

Management of Inflammatory and Nociceptive Disorders in Africa

G.S. Taïwe, V. Kuete

1 INTRODUCTION

Inflammation is a localized protective reaction of cells/tissues of the body to allergic or chemical irritation, injury, and/or infections. The symptoms of inflammation are characterized by pain, heat, redness, swelling, and loss of function that result from dilation of the blood vessels leading to an increased blood supply and from increased intercellular spaces resulting in the movement of leukocytes, protein, and fluids into the inflamed regions (Parham, 2000). Pain is a complex interaction of sensory, emotional, and behavioral factors. There are no pain pathways, only nociception pathways. WHO study found that people who live with chronic pain are 4 times more likely to suffer from inflammation, depression or anxiety. Moreover the damaged primary afferent fibers appear to have altered excitability and conduction patterns during the initiation and maintenance phases of pathological pain. In addition, these altered patterns reflect changes in multiple ion channels, such as their density or operating characteristics. Differential alterations of specific ion channels due to the damaged afferent fibers may be important determinants of primary afferent discharge and conduction (Gold et al., 2003). The physical and psychological effects of chronic pain influence the course of disease (Gureje et al., 1998).

As the production, distribution, and dispensation of controlled medicines are under exclusive government control, governments must also put in place an effective system of distribution in order to provide health care providers and pharmacies with a continuous and adequate supply of the medications. Inflammation and pain treatment medications are not evenly distributed worldwide. Yet, many governments, as a result of resource limitations or lack of political will, have failed to put in place effective supply systems for controlled medicines (Cleary et al., 2013). More fundamentally, there is a lack of a common understanding of opiate pain relief needs and accessibility. Besides a probable basis in international law, the provision of adequate pain relief has some foundation in domestic law through the vehicles of national constitutions, domestic legislation, and the law of negligence. Many of the world's nations have written constitutions that enumerate the right of their citizens to receive adequate health care. None expressly articulate a right to pain relief. A 2006 African Palliative Care Association survey found that, while drug control agencies in Kenya, Tanzania, and Ethiopia believed the regulatory system worked well, morphine consumption in each of these countries was far below the estimated need and the palliative care providers surveyed the identified myriads of problems with the regulatory system (Harding et al., 2007; Cleary et al., 2013).

Although the precise mechanisms of action of many herbal drugs are yet to be determined, some of them have been shown to exert antiinflammatory and/or antinociceptive effects in a variety of cells in the human and animal bodies. There is increasing evidence to indicate that both peripheral and central nervous system cells play a prominent role in the chronic inflammatory responses in the body system and antiinflammatory herbal medicine and its constituents are being proved to be a potent protector against various proinflammatory mediators in diseases and nociceptive disorders.

This chapter provides the state-of-the-art of the inflammatory and nociceptive disorders in Africa, gives a brief description of pathophysiology of these diseases, and outlines the common obstacles to the provision of pain treatment and palliative care. It also emphasizes on statistics in various African countries and government's control policies, describes the mechanisms of action of conventional drugs and family of compounds extracted from the medicinal plants used for the management of inflammatory and/or nociceptive disorders.

2 EPIDEMIOLOGY OF INFLAMMATORY AND NOCICEPTIVE DISORDERS

Epidemiological studies have shown wide variation in their estimates of inflammation and pain in the community ranging from 13% to 68%. Most of these studies are cross-sectional; that is, they provide information at only one point of time. However, a recent longitudinal study reported pain prevalence in many African countries over a 5-year period. There was no significant difference in pain reports between men and women at the baseline (Andersson, 1999; Harding et al., 2007; Cleary et al., 2013). In Europe, chronic pain affects about 17–45% of the population (Andersson, 1999; Eliot et al., 2002; Harding et al., 2007; Cleary et al., 2013) and it is believed that the same range or even more may be found in developing countries where population cannot afford manufactured drugs. Moreover, inflammatory disorders, chronic pain is often resistant to existing therapy. So, there is a great need to search for new and better drugs (Wang and Wang, 2003).

The proportion of the population reporting chronic pain or inflammation significantly increased with age. The overall prevalence of pain increased from 63.5% of the population at baseline to 73.6% at follow-up; that is, an increase of 8.1% over 4 years. There was a larger increase in prevalence among women than men and the increase was highest in the youngest age group (17–34 years). Of those who had chronic pain at baseline, the pain persisted in 78.5% and was resolved in 13.5% (Andersson et al., 1993; Wang and Wang, 2003). This study reinforces previous work that has shown chronic pain to be a common and persistent problem in the community. In this study, health factors appeared to be better predictors of onset or recovery from chronic pain than sociodemographic factors (Andersson et al., 1993).

3 GENDER AND AGE OF PATIENTS VERSUS INFLAMMATORY AND NOCICEPTIVE DISORDERS IN AFRICA

There is no simple relationship between gender and pain. Not only does the pattern vary across different conditions, but also across different age groups (Andersson et al., 1993). Epidemiological data on pain in children is still scarce. Although some conditions associated with pain have been studied (e.g., juvenile rheumatoid arthritis, malaria, microbial infection), data on pain is usually absent. Cancer in children is thankfully less common than in adults. The common types of cancer are also different (e.g., leukemia) and are associated with less pain than the common cancers found in adults (Andersson et al., 1993). Evidence concerning the prevalence of pain in older people must be interpreted with caution, as few studies have specifically addressed this problem. Acute pain probably has a similar incidence across all age groups, but chronic pain or inflammation increases with age (up to about 45 years of age). Several painful conditions which are more likely to occur in older people (e.g., pain caused by HIV infection, cancer, or postherpetic neuralgia) are relatively uncommon, and so tend to be underestimated in general surveys. The most common problem is joint pains and this reflects an increasing pathological load with advancing age (Andersson et al., 1993).

4 PREVENTION

Primary prevention is intended to prevent a disease or symptom from occurring. Secondary prevention is aimed at early detection so that treatment begins before it becomes chronic (Pauwels et al., 2001). In the context of pain, strategies to prevent acute back pain from becoming chronic, or antiviral treatment of acute herpes zoster infection to prevent postherpetic neuralgia would be good examples. Tertiary prevention seeks not to prevent disease or symptoms, but to minimize disability and handicap arising from it. Illness behavior and psychological morbidity can develop following the onset of chronic pain and tertiary prevention would be aimed at reducing these sequelae. Causative agents include physical trauma (such as surgery, lifting heavy weights, and accidents) and infective agents (including herpes zoster). Host factors can modify the impact of these physical factors (Langford and Thompson, 2005). Potential host influences, such as psychological, immunological, physiological, and anatomical factors, are also important. Environmental factors comprising the social context in which the patient lives and works are a significant influence. Environmental factors relating to work, such as compensation for work injury or work dissatisfaction may be relevant (Main et al., 2008).

The use of natural products in treatment of inflammatory and nociceptive disorders results from their influence on different stages of inflammation and pain.

Some acute pain guidelines with many information on acute pain management can be used in a variety of ways in South Africa and other African countries. It is primarily intended to serve as a guide to good practice and to promote a consistent and cohesive approach to care. The document is intended to be realistic and practical. It is a basis for developing and improving acute pain management, stimulating learning among medical teams, promoting effective interdisciplinary team working, and measuring quality in acute pain management (Andersson et al., 1993). This document must be considered as an aid to any health care professional managing acute pain, rather than a "recommended" regimen. It remains the prerogative of the practitioner to evaluate the patient and to adapt any of the suggestions to the circumstances surrounding that particular patient. It is hoped that using the information provided in this publication will be meaningfully beneficial to both medical professional and the patient. The use of the principles provided will ensure that the right of all patients to adequate and effective pain management will be fulfilled (Ready and Edwards, 1992).

5 INFLAMMATORY PAIN CONTROL POLICY IN THE AFRICAN CONTEXT

Sub-Saharan Africa faces a very high burden of incurable terminal disease. During 2007 there were 22.5 million people living with HIV infection; 1.7 million adults and children became infected with HIV; and 1.6 million died of AIDS. The burden of cancer is just beginning to be understood and receive attention with clinical research and policy. The burden of other nonmalignant diseases is unknown, although heart failure is recognized as a leading cause of death in Southern Africa. Palliative care can relieve the suffering of patients and families affected by these diseases and a strong body of evidence has demonstrated this. The pain, symptoms, insight, and anxiety associated with HIV can also be controlled under palliative care. For this reason, policy and legal frameworks have sought to remove this unnecessary suffering and to promote cheap and effective palliative care as both a public health issue and as a human right. Medical ethics requires that we "do no harm." Leaving a patient in pain in the era when pain relief and symptom control is affordable, is breaking our ethical code (Merriman and Harding, 2010). Since the modern palliative care speciality was developed and grew in the 1960s across Western Europe, Australia, and North America it has (to a greater or lesser degree) become integrated into the health system. The palliative care needs of a range of patients and families can be met (although still only a proportion of those who could benefit actually get access). Therefore it is difficult to imagine, and impossible to justify, a country where there is no medicine for severe pain. This, however, is the case for many countries in Africa today. While there are centers of excellence providing care, education, and advocacy, they are few and far between (Merriman and Harding, 2010). To reach the poorest and those suffering in an appropriate and ethical manner, a public health policy needs to be accepted and delivered in every country. The introduction of this essential component of health care was a particular challenge in an African country such as Uganda in 1993. Uganda, which lacks economic resources, has few doctors (1:19,000 population), and nurses (1:5,000 population) and where 57% of the population never sees a health worker. Traditional healers are the first port of call in illness, and there is 1 traditional healer to 450 population. These traditional healers are holistic in their approach and culturally more acceptable to those in the village than Western medicine which can be expensive. Curative options are usually few for people who present with cancer—firstly they present very late, as they spend some time with the traditional healer, fear hospitals, and have very little money. There is also only one source of radiotherapy and chemotherapy and that is in Kampala, far away from most Ugandan dwellings, and cannot meet the needs of all patients (Merriman and Harding, 2010).

6 THE INFLAMMATORY RESPONSE

6.1 Biology and physiology of inflammation

Inflammation has been described as the basis of many pathologies of human disease. It is a complex process initiated by several factors ranging from bacterial infection and chemical injury to environmental pollution that result in cell injury of death (O'Byrne and Dalgleish, 2001). While there is a tendency to consider the inflammatory response as a reaction that is harmful to the body, a more balanced view is that it is actually a protective and restorative response in which the body attempts to either rid itself of chemical toxins or foreign invaders, or to repair itself following an injury. It is when inflammation becomes excessive or uncontrolled that we may begin to see delayed healing or chronic inflammatory conditions. When tissue injury occurs, numerous substances are released by the injured tissues, which cause changes to the surrounding uninjured tissues. Some of the tissue products that cause the inflammatory reaction include: histamine (which increases permeability, causes contraction of smooth muscle, and constriction of the bronchioles), serotonin, lipid mediators (prostaglandins, leukotrienes and lipoxins, and platelet-activator factor), bradykinin, products of the complement system, products of the blood clotting system, and substances released by the sensitized lymphocytes (lymphokines) (Fantone and Ward, 1999). These substances are the messengers of the inflammation process, and have been viewed as areas of therapeutic intervention. Collectively they are called autocoids. Autocoids are substances released from the cells in response to various stimuli to elicit normal physiological responses locally. An imbalance in the synthesis and release of the autocoids contributes significantly to pathological conditions such as inflammation, allergy, hypersensitivity, and ischemia-reperfusion (Fantone and Ward, 1999; Levine and Reichling, 1999). The proposed general classification of autocoids is: biogenic amines—histamine, serotonin; biogenic peptides—kinins (bradykinin, kallidin), angiotensin; small peptides—cytokines [e.g., interleukins (ILs), chemokines, lymphokines, interferon (IFN), tumor necrosis factor (TNF)] are small soluble proteins with low molecular weight, their function is to act as chemical messengers for regulation of innate and acquired immunity and stimulate hematopoiesis, they are produced in just about all cells involved with immunity, but in particular the T-helper cells; membrane derived—leukotrienes, prostaglandins, thromboxane A2, platelet-activating factor, prostacyclin, lipoxins, and hepoxylins; endothelial derived—nitric oxide.

Inflammation is characterized by the following events: vasodilatation of the local blood vessels; increased capillary permeability (which causes an increase in interstitial fluid); clotting of the interstitial fluid (caused by fibrinogen); migration of monocytes and granulocytes; swelling of the tissues. As you can read from the preceding description the inflammatory process is connected to the vasculature. You may recall the cardinal features of inflammation: tumor, calor, rubor, and dolor (Fantone and Ward, 1999; Levine and Reichling, 1999). Acute inflammation is the early (almost immediate) response to injury. It is nonspecific and may be evoked by any injury short of one that is immediately fatal. It is usually of short duration and typically occurs before the immune response becomes established and is aimed primarily at removing the injurious agent and limiting the extent of tissue damage. Acute infections usually are self-limiting and rapidly controlled by the host defenses. In contrast, chronic inflammation is self-perpetuating and may last for weeks, months, or even years. It may develop during a recurrent or progressive acute inflammatory process or from low-grade, smoldering responses that fail to evoke an acute response (Levine and Reichling, 1999).

Tissue macrophages are the first line of defense, followed by neutrophil invasion, the second line of defense, and subsequent neutrophilia. A third line of defense occurs with a secondary macrophage invasion into the injured area. The fourth line of defense is the increased production of granulocytes and monocytes by the bone marrow. Activated macrophages produce and release numerous growth factors in the inflamed tissue (Fantone and Ward, 1999). The dominant factors (cytokines) released by the macrophages include the following: TNF; interleukin-1 (IL-1); granulocyte–monocyte colony-stimulating factor (GM-CSF); granulocyte colony-stimulating factor (G-CSF); and monocytes colony-stimulating factor (M-CSF). These factors cause an increase in production of granulocytes and monocytes by the bone marrow. These factors also provide a powerful feedback mechanism to help remove the cause of the inflammation (Levine and Reichling, 1999).

6.2 Innate and acquired immunity

You should recall, from the immune system module, the distinction between innate immunity and acquired immunity. Innate immunity consists of the following: phagocytosis by the tissue macrophage system; stomach acid and digestive enzymes; resistance of the skin; presence of certain substances in the blood (lysozymes, polypeptides, complement, and natural killer lymphocytes). Acquired (adaptive) immunity forms antibodies and activated lymphocytes that attack and destroy specific organisms and toxins (Fantone and Ward, 1999; Levine and Reichling, 1999; O'Byrne and Dalgleish, 2001)

6.3 Interactions of cellular and humoral immunity as defense against invaders

The inflammatory response is a localized, nonspecific response to infection. Infected or injured cells release chemical alarm signals, most notably histamine and prostaglandins. These chemicals promote the dilation of local blood vessels, which increases the flow of blood to the site of infection or injury and causes the area to become red and warm. They also increase the permeability of capillaries in the area, producing the edema (tissue swelling) so often associated with infection. The more permeable capillaries allow phagocytes (monocytes and neutrophils) to migrate from the blood to the extracellular fluid, where they can attack bacteria. Neutrophils arrive first, spilling out chemicals that kill the bacteria in the vicinity (as well as tissue cells and themselves); the pus associated with some infections is a mixture of dead or dying pathogens, tissue cells, and neutrophils. Monocytes follow, become macrophages, and engulf pathogens and the remains of the dead cells (Watkins and Maier, 1999). Some neurological diseases like—nociception disorders can influence cytokines production. TNF-α, IL-1, and IL-6 pathway is associated with altered pain perception; hyperalgesia induced by TNF-α via stimulating release of IL-1; hyperalgesia induced by peripheral inflammation is associated with IL-1 overexpression; spinal cord glia and glially derived proinflammatory cytokines suggested to be powerful modulators of pain; interleukin-1 β mediated induction of cyclooxygenase-2 (COX-2) in neurons of the central nervous system contributes to inflammatory pain hypersensitivity; bradykinin B2 receptors are suggested to be involved with the acute phase of the inflammatory and pain response; TNF-α expression is suggested to be upregulated in Schwann cells influencing central in painful neuropathies (Watkins et al., 1995; Watkins and Maier, 1999; Samad et al., 2001; Rutowski and DeLeo, 2002; Wieseler-Frank et al., 2005). T-lymphocyte stimulation through the antigen receptor causes early activation of a tyrosine kinase (Samelson et al., 1986; Patel et al., 1987; Trevillyan et al., 1990) and the generation of phosphatidylinositol (PI) biphosphate (PIP2)-derived second messengers, namely inositol triphosphare (IP3) and diacyl glycerol (DAG), via activation of phospholipase C (Koretzky et al., 1990; Trevillyan et al., 1990; Ledbetter et al., 1991). Several cellular substrates are phosphorylated through the activation of protein kinase PTK, this eventually lead to the expression of IL-1. It is now understood that the proliferative signal is generated by members of a family of PTKs that catalyze the phosphorylation of cellular substrates, which in turn leads to T-cell proliferation (Rudd, 1990).

Regardless of the source, when an antigen is present in the body, the production of two types of lymphocytes is enhanced: the B lymphocytes (B cells) and T lymphocytes

(T cells). Like macrophages, these lymphocytes circulate throughout the blood and lymph system and are concentrated in the spleen and lymph nodes. Because lymphocytes recognize specific antigens, they are said to display specificity. B and T cells are able to recognize specific antigens when they have the correct antigen receptor on their cell membrane (Yamanashi et al., 1991). A single B or T cell bears over 100,000 identical copies of the same receptor for one specific antigen. The antigen receptors that are present on B cells, called antibodies, are proteins produced in response to the presence of antigens. The antigen receptors on T cells are called T-cell receptors. B-lymphocyte activation, like T-cell activation, is accompanied by phosphorylation of tyrosine on particular B-cell proteins (Campbell and Sefton, 1990; Gold et al., 1990; Lane et al., 1991; Yamanashi et al., 1991).

6.4 Triggers of immune response and inflammatory disorders

Even though we have a single immune system, it is diversified into two subsystems so we can combat the multitude of infectious agents we encounter in our lifetimes. This diversification is a result of the differing approaches that B and T cells have to ridding the body of infectious agents once they are found. B cells provide a response called humoral immunity, while T cells provide a cell-mediated immunity. There are a host of stimuli that can activate the immune response, and therefore inflammation. The specific response generated by the immune system is triggered by proteins and carbohydrates on the outer membranes of bacterial cells or cells that have been infected by a virus. Molecules that are foreign to the host and stimulate the immune system to react are called antigens (Sitkovsky et al., 2004). The following is a list of some of the triggers: oxidative stress (reactive oxygen species, especially oxidized lipids); radiation; psychological stress; injury; food and environmental allergens; viral infections; nutrient deficient/poor diet (e.g., diet high in refined sugar, SAD diet); intestinal hyperpermeability (leaky gut); pathogen-associated molecular patterns (PAMPs)—PAMPs are molecules that trigger an immune response by activating toll-like receptors. Toll-like receptors are transmembrane proteins expressed by cells of the innate immune system and the antigen presenting cell. Once activated, the toll-like receptors signal the pathways of the inflammatory response, namely nuclear factor kappa B (NF-kB) (Tracey et al., 2008).

6.5 Proinflammatory compounds regulated by nuclear factor kappa B and their physiological effects

6.5.1 *Tumor necrosis factor*

Tumor necrosis factors (TNF) refer to a group of cytokines that can cause cell death (apoptosis). The first two members of the cytokines family to be identified were tumor necrosis factor, formerly known as TNFα or TNF alpha, is the best-known member of this class (Tracey et al., 2008). TNF is a monocyte-derived cytotoxin that has been implicated in tumor regression, septic shock, and cachexia. TNF-α is a pleiotrophic inflammatory cytokine that serves as a mediator in various pathologies. Some examples are cancer, multiple sclerosis, rheumatoid arthritis, diabetes, and AIDS (Feghali and Wright, 1997).

The protein is synthesized as a prohormone with an unusually long and atypical signal sequence, which is absent from the mature secreted cytokine. A short hydrophobic stretch of amino acids serves to anchor the prohormone in lipid bilayers (Tracey et al., 2008). Both the mature protein and a partially processed form of the hormone can be secreted after cleavage of the propeptide. Lymphotoxin-alpha, formerly known as tumor necrosis factor-beta (TNF-β), is a cytokine that is inhibited by interleukin 10. TNF-β acts on a plethora of different cells. It induces the synthesis of GM-CSF, G-CSF, IL-1, and prostaglandin-E2 in fibroblasts. It also promotes the proliferation of fibroblasts and is involved in wound healing (Tracey et al., 2008).

6.5.2 Colony-stimulating factors

Colony-stimulating factors (CSFs) are secreted glycoproteins that bind to receptor proteins on the surfaces of hemopoietic stem cells, thereby activating intracellular signaling pathways that can cause the cells to proliferate and differentiate into a specific kind of blood cell (usually white blood cells). They may be synthesized and administered exogenously (Isner and Asahara, 1999). However, such molecules can at a later stage be detected, since they differ slightly from the endogenous ones in, for example, features of posttranslational modification. G-CSF stimulates the bone marrow to produce granulocytes. GM-CSF stimulates stem cells to produce granulocytes and monocytes (Isner and Asahara, 1999).

6.5.3 Interleukins

ILs are a group of cytokines (secreted proteins and signal molecules) that were first found to be expressed by white blood cells (leukocytes). The function of the immune system depends in a large part on ILs, and rare deficiencies of a number of them have been described, all featuring autoimmune diseases or immune deficiency (Kaplan et al., 1996). During an infection macrophages may release cytokines, such as IL1, that travel to the hypothalamus and induce a change in the thermostat setting (Kaplan et al., 1996). IL-1 is produced by macrophages, monocytes, dendritic cells, and fibroblasts, as well as other cells. IL-1 is released into the local environment as a part of the inflammatory reaction. IL-1 causes the endothelial cells to secrete chemokines such as MCP-1 and upregulates the expression of vascular adhesive molecules such as E-selectin, ICAM-1, and VCAM-1. Interleukin-2 (IL-2) was initially identified as a T cell growth factor. IL-2 can stimulate the growth and (differentiation of B cells, natural killer cells, monocytes, macrophages, and oligodendrocytes. Interleukin-6 (IL-6) plays a role in acute-phase reactions, hematopoiesis, bone metabolism, and cancer progression. Interleukin-12 (IL-12) is produced by macrophages and B cells and has been shown to have multiple effects on T cells and natural killer cells (Yudkin et al., 2000).

6.5.4 Interferon

IFNs are naturally occurring glycoproteins involved in nonspecific immune responses. They are a group of signaling proteins made and released by host cells in response to the presence of several pathogens, such as viruses, bacteria, parasites, and also tumor cells (Whyte, 2007). In a typical scenario, a virus-infected cell will release IFNs causing nearby cells to heighten their antiviral defenses. IFNs do just as their name states they "interfere" with viral growth. IFN-β (IFN-β is known to upregulate and downregulate a wide variety of genes, most of which are involved in the antiviral immune response (Whyte, 2007).

6.5.5 Chemokines

Chemokines (Greek; *kinos*, movement) are a family of small cytokines, or signaling proteins secreted by cells (Zlotnik and Yoshie, 2000). Their name is derived from their ability to induce directed chemotaxis in nearby responsive cells; they are chemotactic cytokines. Interleukin-8 (IL-8) is a chemoattractant for neutrophils, basophils, eosinophils, and T cells. IL-8 related chemotactic cytokines activates neutrophils and basophils (Zlotnik and Yoshie, 2000).

6.5.6 Adhesion molecules

Cell adhesion molecules (CAMs) are proteins located on the cell surface involved in binding with other cells or with the extracellular matrix (ECM) in the process called cell adhesion (Aplin et al., 1998). In essence, CAMs help cells stick to each other and to their surroundings. Intercellular adhesion molecule 1 (ICAM-1) also known as CD54 (cluster of differentiation 54) is a protein that in humans is encoded by the ICAM1 gene. E-selectin is expressed on inflamed endothelial cells by cytokine activity and is referred to as endothelial leukocyte adhesion molecule (Aplin et al., 1998). Vascular cell adhesion molecule (V-CAM) mediates adhesion of lymphocytes, monocytes, eosinophils, and basophils to the vascular endothelium. V-CAM is upregulated by a variety of cytokines. V-CAM-1 plays a role in the formation of atherosclerosis (Aplin et al., 1998).

6.5.7 Enzymes

Enzymes are natural proteins produced in tiny quantities by all living organisms (bacteria, plants, and animals) and function as highly selective biochemical catalysts in converting one molecule into another. Enzymes are essential to life because they speed up metabolic reactions to a very great extent, but do not undergo any change themselves. COX-2 is an enzyme that catalyzes the production of proinflammatory prostaglandins. Inducible nitric oxide synthase (iNOS) plays a part in the oxidative burst from the macrophages. Recall that the burst participates in antimicrobial and antitumor activity under normal circumstances (Aplin et al., 1998).

6.6 Acute-phase response and acute-phase proteins

The acute-phase response is the answer of the organism to disturbances of its homeostasis due to infection, tissue injury, neoplastic growth, or immunological disorders. The acute-phase response is thought to be beneficial to the injured organism with the aim of restoring the disturbed physiological homeostasis (Carpintero et al., 2005). Acute-phase proteins are a class of proteins whose plasma concentrations increase (positive acute-phase proteins) or decrease (negative acute-phase proteins) in response to inflammation. This response is called the acute-phase reaction (also called acute-phase response). Acute-phase proteins mostly comprised glycoproteins that are found in the serum. They are mainly synthesized by the liver cells and released into the bloodstream in response to tissue injury from a variety of sources including trauma, acute infections, chronic infections, thermal injury, and malignancy. IL-1, IL-6, and TNF are the cytokines thought to stimulate their production. The acute-phase response is general and nonspecific. A research study investigating the role of protein malnutrition and acute-phase reactants indicated that a protein deficient diet by itself induces IL-6 production (Gruys et al., 2005). There are numerous acute-phase proteins. Some of the acute-phase proteins is the C-reactive protein (CRP) which is increased in connective tissue disorders, bacterial infections, and neoplastic disease. CRP is an independent risk factor of the development of type 2 diabetes. CRP is synthesized in the liver in response to IL-1. CRP binds to membrane phospholipids in microbial membranes. CRP functions as an opsonin (a molecule that promotes phagocytosis by binding to a microbe). A research study on sleep loss and inflammatory markers revealed both acute total and short-term partial sleep deprivation resulted in elevated high-sensitivity CRP concentrations. The conclusion was that failure to obtain adequate amounts of healthy sleep promotes low-level systemic inflammation (Volanakis, 2001). Alpha-1-acid glycoprotein is an acute-phase (acute-phase protein) plasma alpha-globulin glycoprotein and is modulated by two polymorphic genes. It is synthesized primarily in hepatocytes and has a normal plasma concentration between 0.6–1.2 mg/mL (1–3% plasma protein). Alpha-1 acid glycoprotein, promotes fibroblasts and interacts with collagen. Many others acute-phase proteins are alpha-1 antitrypsin (protease inhibitor); haptoglobins (hemoglobin scavenger); ceruloplasmin (copper transport protein); serum amyloid A (cholesterol scavenger); fibrinogen (clotting formation of fibrin matrix repair); ferritin an iron transport protein (increased in inflammation, malignancy, and liver disease); and complement components (Volanakis, 2001).

In summary, the physiological response to tissue injury (trauma, infection, surgery, radiation, oxidative stress) involves both local and systemic reactions. The normal inflammatory response maintains homeostasis and allows for tissue healing (Aplin et al., 1998; Yudkin et al., 2000; Zlotnik and Yoshie, 2000; Volanakis, 2001; Carpintero et al., 2005; Gruys et al., 2005; Whyte, 2007). Chronic inflammation leads to multiple organ dysfunctions. From a functional medicine perspective, you need to keep in mind the body systems involved in acute and chronic inflammation, which include; the sympathetic nervous system, the immune system, the endocrine system, the gastrointestinal system, and the vascular endothelium.

7 PAIN PROCESSING AND NOCICEPTION PATHWAY

7.1 Biology and physiology of nociceptors

One of the cardinal features of inflammatory states is that normally innocuous stimuli produce pain. Pain is defined by the International Association for the Study of Pain (IASP) as an unpleasant sensory and emotional experience associated with actual or potential tissue damage, or described in terms of such damage (Merskey and Bogduk, 1994). It is universally recognized that factors in three domains contribute to the human experience of chronic pain: biological (nociceptive and neuropathic), psychological, and social (environmental). In fact, all three domains almost certainly exert their influence at various levels of the nervous system, with much overlapping. A stimulus of intensity sufficient to threaten tissue damage activates nociceptors, which are specialized nerve endings (Merskey and Bogduk, 1994). A major advance has been the discovery of the transduction process of noxious temperature (heat and cold) and chemical stimuli to generate electrical energy that conducts along the axon to the spinal cord. The transduction process for the remaining category of noxious stimuli associated with pressure will likely be elucidated soon. Existing knowledge of temperature and chemical mechanisms includes the key role of transient receptor potential vanilloid (TRPV), acid-sensing ion channel (ASIC) and P2X receptor sites, which have become exciting new targets to block nociception at a peripheral level. The cell bodies of these nociceptors (first-order neurons) are located outside the spinal cord in the dorsal root ganglia and extend their trigeminal neuralgia in an idiopathic paroxysmal recurrent pain in the distribution of one or more branches of the trigeminal (fifth cranial) nerve (MacFarlane et al., 1997).

Acute pain is defined as pain of recent onset and probably limited duration. It usually has an identifiable temporal and causal relationship to injury or disease. Chronic pain commonly persists beyond the tie of healing of an injury and frequently there may not be any clearly identifiable cause (Ready and Edwards, 1992). There are many ways to classify pain; for example, by duration, etiology, or intensity (Table 2.1). As understanding of the cellular mechanisms of pain has increased, proposals have been advanced to classify pain according to the predominant pathophysiological mechanism thought to be involved.

7.2 Pathophysiology of nociception

Pain is thought to be caused by vascular compression of the trigeminal ganglion or its branches, but bony abnormalities or otherwise inapparent multiple sclerosis could also be contributors. Rarely, a space occupying lesion (e.g., tumor) in the cerebellopontine

Table 2.1 Possible ways of classifying pain

Using duration	Using probable mechanism
• Acute pain	• Tissue damage
• Subacute pain	• Inflammation
• Chronic	• Central sensitization of nociceptors
	• Nerve-damage-triggered neuroplasticity changes
Using etiology	Using probable mechanism
• Cancerous	• Glia-derived neural sensitization
• Ischemic	• Loss of inhibition
• Postoperative	• Brain neuroplasticity changes
	• "Crosstalk" between sympathetic and sensory neurons
Using intensity	Using type of injured tissues
• Mild	• Nociception
• Moderate	• Neuropathic
• Severe	• Visceral
	• Somatic

angle can be a cause of trigeminal neuralgia, particularly if there is loss of sensation in trigeminal territory—sometimes called atypical trigeminal neuralgia (Patten, 1996).

Compression of the peripheral branches of the trigeminal nerve can also occur intraorally or in the region of the chin as the result of trauma, metastatic tumor, or injury during alveolar or mandibular bone excision during tooth extraction. About 5% of people with trigeminal neuralgia have other family members with the disorder, which suggests a possible genetic cause in some cases (Patten, 1996).

Injury to the nerve root renders axons and axotomized neurons in the Gasserian ganglion hyperexcitable. A discharge from the Gasserian ganglion is then thought to spread to neighboring neurons, triggering them to fire in turn. Therefore, although trigeminal neuralgia may have an initially inapparent peripheral origin, the clinical syndrome results from abnormal discharges within clusters of central neurons in the trigeminal nucleus and/or abnormal central processing of afferent neural impulses (Schaible and Richter, 2004).

7.3 Nociception in uninjured skin

The nerve fibers (axons) within a compound nerve include both afferent nerves and efferent (motor and autonomic) nerves. The speed at which an individual nerve fiber conducts action potentials is related to the diameter of the fiber. In the larger myelinated fibers, the conduction velocity in meters per second is to a first approximation 6 times the axon diameter given in microns (Boveja and Widhany, 2005). The histogram of the distribution of conduction velocities has four peaks: the slowest conducting fibers are unmyelinated and designated C; the faster conducting myelinated fibers are designated

Aδ, Aβ, and Aα (Fein, 2012; Chowdhury, 2013). Hence, to allow for this possibility, the designation used here is that the signal from nociceptors is carried by unmyelinated C-fibers and myelinated A-fibers conducting in the A (δ-β) conduction velocity range. It should be kept in mind that the reverse is not true, not all C-fibers and A (δ-β) fibers are nociceptors. The C and A (δ-β) fibers also carry signals for nonnoxious innocuous mechanical, warm and cold stimuli (Fein, 2012). Nociceptors respond to noxious cold, noxious heat, and high threshold mechanical stimuli as well as a variety of chemical mediators. However, not every nociceptor responds to each of the noxious stimuli. The apparent lack of a response to a noxious stimulus may result because the stimulus intensity is insufficient. Additionally, application of a high intensity stimulus of one modality may alter the response properties of the nociceptor to other modalities (Fein, 2012).

7.4 Acute pain sensations, nociception activities, and injury response

Acute pain is one of the activators of the complex neurohumoral and immune response to injury, and both peripheral and central injury responses have a major influence on acute pain mechanisms. Thus acute pain and injury of various types are inevitably interrelated and if severe and prolonged, the injury response becomes counterproductive and can have adverse effects on outcome (Clarke et al., 2013).

Although published data relate to the *combination* of surgery or trauma and the associated acute pain, some data have been obtained with experimental pain in the absence of injury. Electrical stimulation of the abdominal wall results in a painful experience (visual analog scale 8/10) and an associated hormonal/metabolic response, which includes increased cortisol, catecholamines, and glucagon, and a decrease in insulin sensitivity (Clarke et al., 2013). Although acute pain is only one of the important triggers of the injury response, as the magnitude and duration of the response is related to the magnitude and duration of the stimulus, effective pain relief can have a significant impact on the injury response (Clark et al., 2007). Release of proinflammatory cytokines may contribute to postoperative ileus, but the impact of modulating this response on overall patient outcome requires further evaluation.

7.5 Hyperalgesia and allodynia

Hyperalgesia and allodynia are frequent symptoms of disease and may be useful adaptations to protect vulnerable tissues. Enhanced sensitivity for pain may, however, persist long after the initial cause for pain has disappeared, then pain is no longer a symptom but rather a disease in its own right. In an early definition hyperalgesia was considered "a state of increased intensity of pain sensation induced by either noxious or ordinarily nonnoxious stimulation of peripheral tissue." Allodynia is a pain in response to a nonnociceptive stimulus (Sandkühler, 2009). The proper function of the nociceptive system enables and enforces protective behavioral responses such as withdrawal or avoidance to acutely painful stimuli. In case of an injury, the vulnerability of the affected tissue increases. The nociceptive system adapts to this enhanced vulnerability by locally lowering the nociceptive thresholds and by facilitation of nocifensive responses, thereby adequate tissue protection is ensured. The behavioral correlates of these adaptations are allodynia and hyperalgesia. Thus neither hyperalgesia nor allodynia is per se pathological or a sign of an inadequate response but may rather be an appropriate shift in pain threshold to prevent further tissue damage. Painful syndromes are typical for a large number of diseases and pain intensity if often used by the patients and their health care professionals to evaluate the progression of the disease or the success of the therapy (Sandkühler, 2009). Hyperalgesia is prevalent in many disease states. The hyperalgesia that occurs in inflammation (e.g., rheumatoid arthritis) is comparable to primary hyperalgesia and is thought to result from nociceptor sensitization. The hyperalgesia to light touch that occurs in neuropathic pain (e.g., shingles) is comparable to secondary hyperalgesia and is thought to be the result of central sensitization (MacFarlane et al., 1997).

8 CONVENTIONAL DRUGS USED FOR THE MANAGEMENT OF INFLAMMATORY AND NOCICEPTION DISORDERS

In multiple African countries, access to even the simplest pain-relieving or inflammation medication is limited. With nearly 1.1 billion inhabitants living in more than 50 countries, Africa is the world's poorest and socioeconomically most underdeveloped continent (UNAIDS, 2013). Despite some advances for individual states, many African countries have very low opioid consumption and, overall, the continent has the lowest consumption per capita of any in the world. Approximately 826 million people in Africa are affected by the human immunodeficiency virus (HIV) and acquired immune deficiency syndrome (AIDS) pandemic, resulting in a high prevalence of HIV/AIDS and resultant mortality (UNAIDS, 2013). In 2010, global mean consumption of morphine in milligrams per capita was 5.99; the majority of African countries consumed less than 1 mg per capita with the exception of South Africa where the average morphine consumption was 10.93 mg per capita. In South Africa, palliative care services are firmly established in hospital settings (Clark et al., 2007). WHO has promoted several initiatives and palliative care projects for patients with cancer and HIV/AIDS in Botswana, Ethiopia, Tanzania, Uganda, and Zimbabwe; however, the development of modern palliative care practices in Africa remains scattered and poor (Clark et al., 2007; Sandkühler, 2009; UNAIDS, 2013). From 2000 to 2002, 29 African countries, including Sudan, Chad, Mali, and Nigeria, reported no morphine utilization, and these countries had no known palliative care activity during this time period (Clark et al., 2007). Low levels of opioid utilization across the continent and limited statistical data for medical opioid use indicate that cultural and social barriers exist with regard to opioid use in pain relief (Clark et al., 2007; Sandkühler, 2009; UNAIDS, 2013). Some significant movements to improve morphine availability have been made in various African countries in recent years. For example, in Uganda, free morphine therapy for patients with cancer and HIV/AIDS has allowed health care professionals to prescribe morphine within their clinical practices (Jagwe and Merriman, 2007). However, opioid distribution remains a challenge in many of the developing countries, such as Zimbabwe, because of barriers with regard to drug availability, policymaking, and education.

Nonsteroidal antiinflammatory drugs (NSAIDs) such as aspirin, paracetamol, ibuprofen, naproxen, and indomethacin are extensively used as analgesics and antiinflammatory agents and produce their therapeutic effects through the inhibition of prostaglandin synthesis (Gilman and Goodman, 1990). Aspirin and other NSAIDs block the formation of colon cancer in experimental animals, and there is epidemiological evidence that chronic NSAID usage decreases the incidence of colorectal cancer in humans (Gupta and DuBois, 1999). Although diclofenac is not registered for veterinary use in South Africa, a wide variety of other NSAID drugs are widely used, including phenyl-butazone (used in cattle, especially dairy cows, horses, pigs, and dogs), flunixin meglumine (horses, cattle, pigs, dogs, and cats), eltenac (horses), carprofen (dogs and cats), meloxicam (dogs and cats), and vedaprofen (cattle, horses, pigs, and dogs). NSAIDs have antiinflammatory, analgesic, and antipyretic activity and are particularly used for painful musculoskeletal conditions and treatment of tissue inflammation in all animal species and humans. Corticosteroids are also used in Africa for the control of inflammation and chronic pain (Gupta and DuBois, 1999; Clark et al., 2007).

9 MEDICINAL PLANTS USED FOR THE MANAGEMENT OF INFLAMMATORY AND NOCICEPTION DISORDERS

Approximately 80% of the population in Africa still relies on traditional healing practices, especially medicinal plants for their different health problems. The modulations of the various functions of inflammatory cells in the body and pain sensitization are regulated by pharmacological physiological constituents present in herbal products. In Africa the use of plants to treat many diseases is widely practiced. The bioactive principles in these plant species have been linked to secondary metabolites such as phenolic compounds (curcumins, flavonoids, and tannins), saponins, terpenoids, and alkaloids. Biological and therapeutic properties attributed to these plant metabolites include antioxidant, antiinflammatory, antimicrobial, anticancer, and antinociceptive activities. The mechanisms of action of many phenolic compounds such as flavonoids, alkaloids, saponins, tannins, and curcumins are thought to be via their free radical scavenging activities or the inhibition of proinflammatory enzymes such as cyclooxygenases and lipoxygenases in the inflammatory cascades (Kamatou et al., 2008).

The immune system is a highly complex, intricately regulated group of cells whose integrated function is essential to defend the body from diseases. Cells of the immune system may interact in a cell–cell manner and may also respond to intercellular messages through the transfer of hormones, cytokines, and autacoids elaborated by various cells (Kulmatycki and Jamal, 2005). In the central nervous system, the cells that modulate inflammatory effects are the neurons, microglia, astrocytes, and endothelial cells while in the peripheral sites, the function of lymphocytes (T cells, B cells, macrophages, NK cells) and leukocytes (basophils, monocytes, neutrophils, eosinophils), mast cells, and platelets are involved (Licinio and Wong, 1999). These are described briefly later to expose the biochemical markers that constituent of phytomedicines might inhibit inflammatory processes. The antiinflammatory properties of several phytomedicines origin, that contain substances like phytoestrogens, flavonoids, and its derivatives, phytosterol, tocopherol, ascorbic acid, curcumin, genistein, and others can be the inhibitors of the molecular targets of proinflammatory mediators in inflammatory responses. There are other plants that contain alkaloids, tannin, saponins, anthraquinones, triterpenoids, and other constituents which have been reported to possess a diverse range of bioactivities including anticancer, immunostimulatory, antibacterial, antimalarial, and antituberculosis activities bearing in mind that some of the causative organisms and factors responsible for the initiating and promoting inflammation could be remove or neutralized to suppress the expression of proinflammatory agents (Iwalewa et al., 2007). Previous results obtained on study of medicinal plants' efficacy demonstrate that the central and peripheral effects of flavonoids, alkaloids, saponins, and tannins might partially or wholly be due to the stimulation of peripheric opioid receptors through the action of the nitric oxide–cyclic GMP-ATP-sensitive K^+(NO/cGMP/ATP)-channel pathway and/or facilitation of the GABAergic transmission. Some of the results confirm the popular use of medicinal plants as an antinociceptive and contribute to the pharmacological knowledge of the studied species because it was shown that the aqueous extract and the alkaloid fraction produced dose-related antinociception in models of chemical and thermal nociception through mechanisms that involve an interaction with opioidergic pathway (Taïwe et al., 2011).

10 CONCLUSIONS

The purposes of this review were to provide the state-of-the-art of the inflammatory and nociceptive disorders in Africa, to give a brief description of pathophysiology of these diseases, and to outline the common obstacles to the provision of pain treatment and palliative care. It also emphasizes on statistics in various African countries and government's control policies, describes the mechanisms of action of conventional drugs and family of compounds extracted from the medicinal plants used for the management of inflammatory and/or nociceptive disorders. Untreated or undertreated inflammation significantly decreases a patient's quality of life by causing sleep disorders, nociception, depression, impaired activity, mood alterations, abnormal appetite, inability to focus, and poor hygiene. Opioids, NSAIDs and drugs derived from the natural products are a popular choice for the treatment of intractable painful conditions, but barriers to effective pain assessment and management exist in both developed and Africa. Huge disparities exist in conventional antiinflammatory and antinociceptive drugs production versus actual needs, as well as in the distribution in developed countries versus developing countries. Imbalances between conventional drugs consumption and availability persist despite international efforts in recent years. Health care professionals and patients are often concerned about the side effects of conventional drugs, particularly the potential for tolerance and addiction. Comprehensive guidelines for goal-directed and patient-friendly chronic therapy will potentially enhance the outlook for future chronic pain management. The improvement of pain education in undergraduate and postgraduate training will benefit patients and clinicians in Africa. The promise of new medications, along with the utilization of multimodal approaches, has the potential to provide effective pain relief to future generations of sufferers.

REFERENCES

Andersson, G.B., 1999. Epidemiological features of chronic low-back pain. Lancet 354, 581–585.
Andersson, H.I., Ejlertsson, G., Leden, I., Rosenberg, C., 1993. Chronic pain in a geographically defined general population: studies of differences in age, gender, social class, and pain localization. Clin. J. Pain 9, 174–182.
Aplin, A.E., Howe, A., Alahari, S.K., Juliano, R.L., 1998. Signal transduction and signal modulation by cell adhesion receptors: the role of integrins, cadherins, immunoglobulin-cell adhesion molecules, and selectins. Pharmacol. Rev. 50 (2), 197–264.
Boveja, B., Widhany, A., 2005. U.S. Patent Application No. 11/035,374.
Campbell, M.A., Sefton, C.M., 1990. Protein tyrosine phosphorylation is induced in murine B lymphocytes in response to stimulation with anti-immunoglobulin. EMBO J. 9, 2125–2131.
Carpintero, R., Pineiro, M., Andres, M., Iturralde, M., Alava, M.A., Heegaard, P.M., Jobert, J.L., Madec, F., Lampreave, F., 2005. The concentration of apolipoprotein AI decreases during experimentally induced acute-phase processes in pigs. Infect. Immun. 73 (5), 3184–3187.
Chowdhury, E.A., 2013. Study of the Efficacy of the Distribution of F-Latency (DFL) in the Diagnosis of Cervical Spondylosis. Dissertation, University of Dhaka Bangladesh.
Clarke, Kathy, W., Cynthia, M.T. (Eds.), 2013. Veterinary Anaesthesia. Elsevier Health Sciences.
Clark, D., Wright, M., Hunt, J., Lynch, T., 2007. Hospice and palliative care development in Africa: a multi-method review of services and experiences. J. Pain Symptom Manage. 33 (6), 698–710.
Cleary, J., Powell, R.A., Munene, G., Mwangi-Powell, F.N., Luyirika, E., Kiyange, F., Merriman, A., Scholten, W., Radbruch, L., Torode, J., Cherny, N.I., 2013. Formulary availability and regulatory barriers to accessibility of opioids for cancer pain in Africa: a report from the Global Opioid Policy Initiative (GOPI). Ann. Oncol. 24 (Suppl. 11), xi14–xi23.
Eliot, A.M., Smith, B.H., Hannaford, P.C., Smith, W.C., Chambers, W., 2002. The course of chronic pain in the community: results of a 4-year follow-up study. Pain 99, 299–307.
Fantone, J.C., Ward, P.A., 1999. Inflammation. In: Rubin, E., Farber, J.L. (Eds.), Pathology. Lippincott Williams & Wilkins, Philadelphia, pp. 37–75.
Feghali, C.A., Wright, T.M., 1997. Cytokines in acute and chronic inflammation. Front. Biosci. 2 (1), d12–d26.
Fein, A., 2012. Nociceptors and the perception of pain. University of Connecticut Health Center, 4, 61–67.
Gilman, A.G., Goodman, L.S., 1990. The Pharmacological Basis of Therapeutics. In: Rall, T.W., Nies, A.S. Taylor, P. (Eds.). pp. 1146–1164.
Gold, M.R., Law, D.A., DeFranco, A.L., 1990. Stimulation of protein tyrosine phosphorylation by the B-lymphocyte antigen receptor. Nature 345, 810–813.

Gold, M.S., Weinreich, D., Kim, C.S., Wang, R., Treanor, J., Porreca, F., Lai, J., 2003. Redistribution of Na(V)1.8 in uninjured axons enables neuropathic pain. J. Neurosci. 23, 158–166.

Gruys, E., Toussaint, M.J.M., Niewold, T.A., Koopmans, S.J., 2005. Acute phase reaction and acute phase proteins. J. Zhejiang Univ. Sci. B 6 (11), 1045–1056.

Gupta, R.A., DuBois, R.N., 1999. Aspirin, NSAIDs, and colon cancer prevention. Mechanism? Gastroenterology 114, 1095–1098.

Gureje, O., Von Korff, M., Simon, G.E., Gater, R., 1998. Persistent pain and well-being: a World Health Organization study in primary care. JAMA 80, 147–151.

Harding, R., Powell, R.A., Kiyange, F., Downing, J., Mwangi-Powell, F., 2007. Pain Relieving Drugs in 12 African PEPFAR Countries: Mapping Current Providers, Identifying Current Challenges and Enabling Expansion of Pain Control Provision in the Management of HIV/AIDS. African Palliative Care Association, Kampala.

Isner, J.M., Asahara, T., 1999. Angiogenesis and vasculogenesis as therapeutic strategies for postnatal neovascularization. J. Clin. Investig. 103 (9), 1231–1236.

Iwalewa, E.O., McGaw, L.J., Naidoo, V., Eloff, J.N., 2007. Inflammation: the foundation of diseases and disorders. A review of phytomedicines of South African origin used to treat pain and inflammatory conditions. Afr. J. Biotechnol. 6 (25), 2868–2885.

Jagwe, J.J., Merriman, A.A., 2007. Uganda: delivering analgesia in rural Africa: opioid availability and nurse prescribing. J. Pain Symptom Manage. 33, 547–551.

Joint United Nations Programme on HIV/AIDS (UNAIDS), 2013. UNAIDS report on the global AIDS epidemic. Joint United Nations Programme on HIV/AIDS (UNAIDS), Geneva, Switzerland.

Kamatou, G.P., Makunga, N.P., Ramogola, W.P., Viljoen, A.M., 2008. South African Salvia species: a review of biological activities and phytochemistry. J. Ethnopharmacol. 119 (3), 664–672.

Kaplan, M.H., Sun, Y.L., Hoey, T., Grusby, M.J., 1996. Impaired IL-12 responses and enhanced development of Th2 cells in Stat4-deficient mice. Nature 382 (6587), 174–177.

Koretzky, G.A., Picus, I., Thomas, M.L., Weiss, A., 1990. Tyrosine phosphatase CD45 is essential for coupling T-cell antigen receptor to the phosphatidyl inositol pathway. Nature 346, 66–68.

Kulmatycki, K.M., Jamal, F., 2005. Drug disease interactions: role of inflammatory mediators in disease and variability in drug response. J. Pharm. Pharm. Sci. 8 (3), 602–625.

Lane, P.J.L., Ledbetter, J.A., McConnell, F.N., Draves, K., Deans, J., Schieven, G.L., Clark, E.A., 1991. The role of tyrosine phosphorylation in signal transduction through surface Ig in human B cells: inhibition of tyrosine phosphorylation prevents intracellular calcium release. J. Immunol. 146, 715–722.

Langford, R., Thompson, J.D., 2005. Mosby's handbook of diseases. Elsevier Health Sciences.

Ledbetter, J.A., Schieven, G.L., Uckun, G.M., Imboden, J.B., 1991. CD45 cross-linking regulates phospholipase C activation and tyrosine phosphorylation of specific substrates in CD3/Ti stimulated T cells. J. Immunol. 146, 1577–1583.

Levine, J.D., Reichling, D.B., 1999. Peripheral mechanisms of inflammatory pain. In: Wall, P.D., Melzack, R.D. (Eds.), Textbook of Pain. Churchill Livingstone, London, pp. 59–84.

Licinio, J., Wong, M.L., 1999. The role of inflammatory mediators in the biology of major depression: central nervous system cytokines modulate the biological substrate of depressive symptoms, regulate stress-responsive systems, and contribute to neurotoxicity and neuroprotection. Mol. Psychiatr. 4 (4), 317–327.

MacFarlane, B.V., Wright, A., O'Callaghan, J., Benson, H.A.E., 1997. Chronic neuropathic pain and its control by drugs. Pharmacol. Ther. 75 (1), 1–19.

Main, C.J., Sullivan, M.J., Watson, P.J. (Eds.), 2008. Pain Management: Practical Applications of the Biopsychosocial Perspective in Clinical and Occupational Settings. Elsevier Health Sciences.

Merriman, A., Harding, R., 2010. Pain control in the African context: the Ugandan introduction of affordable morphine to relieve suffering at the end of life. Philos. Ethics Humanit. Med. 5 (1), 1.

Merskey, H., Bogduk, N., 1994. Classification of Chronic Pain, IASP Task Force on Taxonomy. IASP Press, Seattle.

O'Byrne, K.J., Dalgleish, A.G., 2001. Chronic immune activation and inflammation as the cause of malignancy. Br. J. Cancer 85, 473–483.

Parham, P. (Ed.), 2000. The immune system. Elements of the immune systems and their roles in defense. Garland Publishing, New York, pp. 1–31 (Chapter 1).

Patel, M.D., Samelson, L.E., Klausner, R.D., 1987. Multiple kinases and signal transduction: phosphorylation of the T cell antigen receptor complex. J. Biol. Chem. 262, 5831–5838.
Patten, J.P., 1996. The brain stem. Neurological Differential Diagnosis. Springer, Berlin Heidelberg, pp. 162–177.
Pauwels, R.A., Buist, A.S., Calverley, P.M., Jenkins, C.R., Hurd, S.S., 2001. Global strategy for the diagnosis, management, and prevention of chronic obstructive pulmonary disease. Am. J. Respir. Crit. Care Med. 163 (5), 1256–1276.
Ready, L.B., Edwards, W.T., 1992. Management of Acute Pain: A Practial Guide. Taskforce on Acute Pain. IASP Publications, Seattle.
Rudd, C.E., 1990. CD4, CD8 & the TCR-CD3 complex: a novel class of proteintyrosine kinase receptor. Immunol. Today 11, 400–406.
Rutowski, M.D., DeLeo, J.A., 2002. The role of cytokines in the initiation and maintenance of chronic pain. Drug News Perspect. 15, 626–632.
Samad, T.A., Moore, K.A., Sapirstein, A., Billet, S., Allchrone, A., Poole, S., Bonventre, J.V., Woolf, C.J., 2001. Interleukin-1 β-mediated induction of Cox-2 in the CNS contributes to inflammatory pain hypersensitivity. Nature 410, 471–475.
Samelson, L.E., Patel, M.D., Weissman, A.M., Harford, J.B., Klausner, R.D., 1986. Antigen activation of murine T cells induces tyrosine phosphorylation of a polypeptide associated with the T cell antigen receptor. Cell 46, 1083–1090.
Sandkühler, J., 2009. Models and mechanisms of hyperalgesia and allodynia. Physiol. Rev. 89 (2), 707–758.
Schaible, H.G., Richter, F., 2004. Pathophysiology of pain. Langenbecks Arch. Surg. 389 (4), 237–243.
Sitkovsky, M.V., Lukashev, D., Apasov, S., Kojima, H., Koshiba, M., Caldwell, C., Thiel, M., 2004. Physiological control of immune response and inflammatory tissue damage by hypoxia-inducible factors and adenosine a2a receptors. Annu. Rev. Immunol. 22, 657–682.
Taïwe, G.S., Ngo Bum, E., Dimo, T., Talla, E., Weiss, N., Sidiki, N., Amadou, D., Moto, O.F.C., Dzeufiet, P.D., De Waard, M., 2011. Antipyretic and antinociceptive effects of *Nauclea latifolia* and possible mechanisms of action. Pharm. Biol. 49, 15–25.
Tracey, D., Klareskog, L., Sasso, E.H., Salfeld, J.G., Tak, P.P., 2008. Tumor necrosis factor antagonist mechanisms of action: a comprehensive review. Pharmacol. Ther. 117 (2), 244–279.
Trevillyan, J.M., Lu, Y., Atluru, D., Phillips, C.A., Bjorndahl, J.M., 1990. Differential inhibition of T cell receptor signal transduction and early activation events by a selective inhibitor of protein-tyrosine kinase. J. Immunol. 145, 3223–3230.
Volanakis, J.E., 2001. Human C-reactive protein: expression, structure, and function. Mol. Immunol. 38 (2), 189–197.
Wang, L.X., Wang, Z.J., 2003. Animal and cellular models of chronic pain. Adv. Drug Deliv. Rev. 55, 949–965.
Watkins, L.R., Maier, S.F., 1999. Implications of immune-to-brain communication for sickness and pain. Proc. Natl. Acad. Sci. 96, 7710–7713.
Watkins, L.R., Maier, S.F., Goehler, L.E., 1995. Immune activation: the role of pro-inflammatory cytokines in inflammation, illness responses and pathological pain states. Pain 63, 289–302.
Whyte, S.K., 2007. The innate immune response of finfish–a review of current knowledge. Fish Shellfish Immunol. 23 (6), 1127–1151.
Wieseler-Frank, J., Maier, S.F., Watkins, L.R., 2005. Immune-to-brain communication dynamically modulates pain: physiological and pathological consequences. Brain Behav. Immun. 19, 104–111.
Yamanashi, Y., Kakiuchi, T., Milizuguchi, J., Yamamoto, T., Toyoshima, K., 1991. Association of B cell antigen receptor with protein tyrosine kinase Lyn. Science 251, 192–194.
Yudkin, J.S., Kumari, M., Humphries, S.E., Mohamed-Ali, V., 2000. Inflammation, obesity, stress and coronary heart disease: is interleukin-6 the link? Atherosclerosis 148 (2), 209–214.
Zlotnik, A., Yoshie, O., 2000. Chemokines: a new classification system and their role in immunity. Immunity 12 (2), 121–127.

CHAPTER 3

Burden and Health Policy of Cancers in Africa

T.J. Makhafola, L.J. McGaw

1 INTRODUCTION

Cancer is one of the leading causes of morbidity and mortality worldwide. External factors, such as tobacco, an unhealthy diet, and infectious organisms, as well as internal factors including genetic mutations, hormones, and immune conditions may cause cancer. Currently, cancer is the second leading cause of death in high-income countries, following cardiovascular diseases, and the third leading cause of death in low- to middle-income countries, following cardiovascular diseases and infectious and parasitic diseases (American Cancer Society, 2015). Estimates for total cancer deaths up to 2012 were 8.2 million with 2.9 million in economically developed countries and 5.3 million in economically developing countries (American Cancer Society, 2015). One in seven deaths globally is due to cancer, which is more than deaths caused by AIDS, tuberculosis, and malaria combined (American Cancer Society, 2015). It is estimated that 715,000 new cancer cases and 542,000 cancer deaths occurred in 2008 (Boyle and Levin, 2008; Ferlay et al., 2010). With 847,000 new cases and 591,000 cancer deaths occurring in 2012, the International Agency for Research on Cancer (IARC) projects that cancer incidence and mortality will double to 1.28 million new cases and 970,000 deaths per year by 2030 (Parkin et al., 2008, 2015).

Until recently, cancer was predominantly depicted as a disease of more economically developed countries but is now recognized as a global problem (Sylla and Wild, 2012; Parkin et al., 2015). Cancer has emerged as a substantial public health problem in Africa. Aging and growth of populations combined with increased prevalence of risk factors associated with economic transition contribute to the rise of cancer burden in Africa. The change to lifestyle behaviors that are known to increase cancer risk such as smoking, poor diet, physical inactivity, and reproductive changes including fewer pregnancies have further increased the cancer burden in Africa (Jemal et al., 2010; Torre et al., 2015). Infections leading to cancer are more prevalent in developing than in developed countries, with about threefold greater incidence of cancer attributable to infections in the latter countries (American Cancer Society, 2015). Approximately 16% of all incidences of cancers worldwide are caused by infections (Cancer Genome Atlas Network, 2012). The future burden of cancers in developing countries will most likely expand owing to expected increases in life expectancy and growth of the population (Thun et al., 2010).

Despite this growing cancer burden, Africa is the continent least prepared to cope with the cancer endemic (both incidence and mortality). Cancer continues to receive a relatively low public health priority in Africa. Much attention is diverted to other pressing public health problems including communicable diseases such as human immunodeficiency virus (HIV)/AIDS, malaria, and tuberculosis which, in contrast with cancer, are on the decline. This increase in cancer cases and cancer deaths and its inherent consequences will only extend the delay suffered by Africa in its economic and social development (Sylla and Wild, 2012). The late diagnosis of cancers and lack of access to cancer therapies in Africa increase the burden of cancer. Other factors contributing to increased burden of cancers in Africa include low socioeconomic status and deficiently equipped healthcare systems and infrastructure in most African countries resulting in inadequate access to preventive and cancer therapies. Additionally, the lack of awareness among policy makers, the general public, and international private or public health agencies concerning the magnitude of the current and future cancer burden and its economic impact contributes to the burden of cancer in Africa (Sylla and Wild, 2012; Jemal et al., 2012)

In this chapter, several review articles and research papers analyzing data extracted from GLOBOCAN 2008 and GLOBOCAN 2012 were studied to provide the overall burden of cancers including the most common cancers in Africa. GLOBOCAN is a database of the IARC which presents the estimates of incidence of, and mortality from, all cancers as well as the 27 major types in 184 countries or territories worldwide for 2008 and 2012 (globocan.iarc.fr).

2 COMMON CANCERS IN AFRICA

Few exact data are available to calculate the extent of the public health problem posed by cancer in Africa, as GLOBOCAN data for African countries are mostly estimates (Stefan et al., 2013). Accurate data should be located in national cancer registries which are able to then guide national policies, but these are often not available. In 2009, the International Association of Cancer Registries reported the existence of only 15 national registries in Africa, with a further 60 being only hospital or city registries. Worth noting is that the activity of the registries may often be suspended owing to staff or funding shortages. Data up to 2002 were reported to the IARC incidence database by only 12 African countries (Stefan et al., 2013).

Cancer incidence and mortality patterns vary remarkably across regions within Africa because of the substantial regional differences in economic development and social, cultural, and other environmental factors, including major known risk factors for cancers. The African regions include the following: northern Africa, eastern Africa, middle Africa, southern Africa, western Africa, and sub-Saharan Africa (Jemal et al., 2012).

Table 3.1 Estimated numbers of new cases and deaths of leading cancer sites in Africa for 2008

Males		Females	
Cancer type	Estimated cases	Cancer type	Estimated cases
Prostate	39,500	Female breast	92,600
Liver	34,600	Cervix uteri	80,400
Kaposi sarcoma	22,400	Liver	16,900
NHL	21,900	Colon and rectum	15,800
Lung and bronchus	20,800	NHL	15,300
Colon and rectum	19,000	Ovary	14,000
Esophagus	17,500	Kaposi sarcoma	12,400
Urinary bladder	16,900	Esophagus	10,400
Stomach	12,600	Stomach	10,100
Leukemia	11,200	Leukemia	8,300
All sites but skin	342,700	All sites but skin	390,700
Cancer type	Estimated deaths	Cancer type	Estimated deaths
Liver	33,800	Cervix uteri	53,300
Prostate	28,000	Female breast	50,000
Lung and bronchus	19,400	Liver	16,600
Kaposi sarcoma	19,100	NHL	12,700
NHL	18,100	Colon and rectum	12,300
Esophagus	16,700	Kaposi sarcoma	10,500
Colon and rectum	14,700	Ovary	10,400
Stomach	11,900	Esophagus	9,900
Urinary bladder	11,400	Stomach	9,500
Leukemia	10,600	Leukemia	7,800
All sites but skin	267,200	All sites but skin	274,800

NHL, non-Hodgkin lymphoma.
(*Source:* GLOBOCAN, 2008, sourced from Ferlay, Shin, Bray, Forman, Mathers, Parkin, 2010. GLOBOCAN 2008—Cancer Incidence and Mortality Worldwide: IARC CancerBase No 10. International Agency for Research on Cancer, Lyon).

Cancers of the breast, cervix, liver, prostate, non-Hodgkin lymphoma (NHL), colon and rectum, Kaposi sarcoma, esophagus, lung, stomach, and bladder are the most common cancers in Africa (Parkin et al., 2008). Some of these cancers, for example, liver, stomach, and esophagus are predominant in Africa and are difficult to treat, even in countries with state-of-the-art diagnosis and therapy (Sylla and Wild, 2012). Overall, 715,000 new cancer cases and 542,000 deaths occurred in 2008 and an increased 847,000 new cases and 591,000 deaths occurred in 2012 in Africa (Jemal et al., 2012; Parkin et al., 2015). The GLOBOCAN data show a rise in the number of cancer incidents and cancer deaths from 2008 to 2012. Estimated numbers of new cases and deaths of leading cancer sites in Africa for 2008 (Ferlay et al., 2010) and 2012 (Ferlay et al., 2015) are shown in Tables 3.1 and 3.2, respectively.

Table 3.2 Estimated numbers of new cases and deaths of cancer sites in Africa for 2012

Males		Females	
Cancer type	Estimated cases	Cancer type	Estimated cases
Oral cavity	10,200	Oral cavity	7,000
Nasopharynx	5,300	Nasopharynx	3,000
Other pharynx	3,400	Other pharynx	1,900
Esophagus	16,100	Esophagus	11,500
Stomach	13,200	Stomach	10,600
Colorectum	21,200	Colorectum	19,900
Liver	38,700	Liver	20,000
Gall bladder	1,600	Gall bladder	3,000
Pancreas	6,600	Pancreas	5,500
Larynx	7,600	Larynx	1,100
Lung	21,800	Lung	8,600
Melanoma	2,800	Melanoma	3,800
Kaposi sarcoma	23,800	Kaposi sarcoma	13,700
Prostate	59,500	Breast	133,900
Testis	1,500	Cervix uteri	99,000
Kidney	5,100	Corpus uteri	11,400
Bladder	17,700	Ovary	17,800
Brain, CNS	7,600	Kidney	4,900
Thyroid	2,900	Bladder	6,800
Hodgkin	4,700	Brain, CNS	6,300
NHL	21,100	Thyroid	9,100
Multiple myeloma	3,000	Hodgkin	3,200
Leukemia	13,300	NHL	15,700
All sites but skin	362,000	Multiple myeloma	2,900
		Leukemia	10,700
		All sites but skin	484,900
Cancer type	Estimated deaths	Cancer type	Estimated deaths
Oral cavity	6,100	Oral cavity	4,300
Nasopharynx	3,500	Nasopharynx	2,100
Other pharynx	2,600	Other pharynx	1,500
Esophagus	14,700	Esophagus	10,500
Stomach	12,000	Stomach	9,800
Colorectum	15,100	Colorectum	14,300
Liver	37,000	Liver	19,000
Gall bladder	1,500	Gall bladder	2,800
Pancreas	6,400	Pancreas	5,300
Larynx	4,300	Larynx	600
Lung	19,400	Lung	7,700
Melanoma	1,500	Melanoma	2,100
Kaposi sarcoma	16,300	Kaposi sarcoma	9,200
Prostate	42,800	Breast	63,200
Testis	900	Cervix uteri	60,100
Kidney	4,200	Corpus uteri	4,000

Table 3.2 Estimated numbers of new cases and deaths of cancer sites in Africa for 2012 (*cont.*)

Cancer type	Estimated deaths	Cancer type	Estimated deaths
Bladder	9,400	Ovary	13,100
Brain, CNS	5,400	Kidney	4,000
Thyroid	1,500	Bladder	3,900
Hodgkin	2,800	Brain, CNS	4,600
NHL	15,000	Thyroid	4,000
Multiple myeloma	2,600	Hodgkin	2,000
Leukemia	11,600	NHL	11,400
All sites but skin	277,800	Multiple myeloma	2,500
		Leukemia	9,500
		All sites but skin	313,300

(*Source*: GLOBOCAN, 2012, sourced from Ferlay, J., Soerjomataram, I., Dikshit, R., Eser, S., Mathers, C., Rebelo, M., Parkin, D.M., Forman, D., Bray, F., 2015. Cancer incidence and mortality worldwide: Sources, methods and major patterns in GLOBOCAN 2012. *Int. J. Cancer* 136, E359–E386).

2.1 Cervical cancer

Cervical cancer is the second most frequently diagnosed cancer in Africa (99,000 cases) according to the 2012 estimates with a particularly high incidence in sub-Saharan Africa (Parkin et al., 2015). It accounted for 21% of the total newly diagnosed cancers in females in 2008 (Jemal et al., 2011). The highest rates are found in eastern and southern Africa. On average, approximately 80,000 women are diagnosed with cervical cancer and more than 60,000 die from the disease. It is a relatively rare disease in countries that have instituted and maintained national screening programs (Hicks et al., 2006). Treatment of cervical cancer in Africa is hampered by the lack of diagnostic and treatment facilities, lack of health care infrastructure, and poor pathology services (Denny and Anorlu, 2012). In developing countries, patients often present with advanced stage disease, which is associated with a poor prognosis and high mortality rate. For example, in 2007, 197 women were diagnosed with cervical cancer in Sudan and 141 of them had advanced stage disease (Hicks et al., 2006; Ibrahim et al., 2011).

Cervical cancer is a preventable and curable disease. It is preventable by vaccination and curable if identified at an early stage. The incidence of cervical cancer varies considerably by region in Africa. Based on the 2012 GLOBOCAN projections (Parkin et al., 2015), incidence and death rates in East Africa and West Africa were five times as high as the rates in North Africa. Some countries in East Africa, including Zambia, Malawi, Mozambique, and Tanzania have among the highest cervical cancer rates worldwide. This is attributed to high prevalence of human papillomavirus (HPV), which causes cancer and the lack of preventive/screening services for early detection of the disease (Ferlay et al., 2010). Low levels of awareness and poor knowledge of cervical cancer coupled with unavailability and inaccessibility of cervical cancer screening services are responsible for only a very small number of women being screened in sub-Saharan Africa (Anorlu, 2008).

2.2 Breast cancer

Breast cancer is the most common cancer in women both in developed and developing countries. It was the most commonly diagnosed cancer and the second leading cause of cancer death among women in 2008 (92,600 cases, 50,000 deaths) and 2012 (63,100 deaths) in Africa. Southern African women have the highest breast cancer incidence rates of all African regions, in part because of a higher prevalence of reproductive risk factors for breast cancer. According to 2012 estimates, breast cancer is the most commonly diagnosed cancer in Africa, with the highest rates seen in Egypt, Algeria, Nigeria, and Republic of South Africa. It is the leading cause of death from cancer. Although the reasons for the increasing importance of breast cancer are speculative, they most likely include increases in the prevalence of risk factors such as early menarche, late child bearing, having fewer children, obesity, and increased awareness and detection, which are associated with urbanization and economic development (Jemal et al., 2012; Parkin et al., 2015).

2.3 Liver cancer

Liver cancer (hepatocellular carcinoma) is a rapidly lethal cancer, with life expectancy after diagnosis rarely exceeding 3 years in developed countries and even less in developing countries. In the middle and western African countries, liver cancer was the most commonly diagnosed cancer as estimated in 2008. It was the second most commonly diagnosed cancer and the leading cause of cancer death in men and the third most common cancer as well as the third leading cause of cancer death in women (Scheff, 2007; Jemal et al., 2012). Given the poor prognosis of liver cancer, the number of new cases (58,500) and deaths (56,000) estimated in 2012 are rather similar, and in terms of both indicators, liver cancer ranked as the fourth most frequent cancer on the African continent in 2012 (Parkin et al., 2015). The predominant risk factors for the development of liver cancer in Africa include chronic infection by hepatitis B virus (HBV) and infection by hepatitis C virus (HCV). Chronic infections with HBV and HCV in sub-Saharan regions and northern Africa are major causes of liver cancer accounting for approximately 65–80% of the total cases. Additionally, consumption of staple foods such as maize and ground nuts with contaminated aflatoxins, known hepatocarcinogens produced by molds, is another contributing factor to the liver cancer burden in sub-Saharan regions (Scheff, 2007; Franceschi and Raza, 2009).

2.4 Lung cancer

Lung cancer killed approximately 1,590,000 persons in 2012 and currently is the leading cause of cancer death worldwide. There is large variation in mortality rates across the world in both males and females (Ismali et al., 2015). In 2008, lung cancer was the leading cause of cancer death among men in southern Africa and northern Africa and the fourth leading cause of death among women in southern Africa. The incidence and mortality rates in southern Africa in both men and women are twice as high as the second highest rates in northern Africa (Jemal et al., 2012). Lung cancer was still the most commonly diagnosed cancer among males in most countries in northern Africa, including Tunisia, Libya, Morocco, and Algeria (Parkin et al., 2015). About 30,300 new lung cancer cases and 27,000 deaths were estimated to have occurred in 2012 in Africa, with men accounting for over 70% of the total cases and deaths. There was over a 30-fold difference in incidence and mortality rates between countries in both males and females, with the lowest rates found in the western Africa and middle Africa and the highest rates in southern and northern Africa. Smoking accounts for 65% of lung cancer cases in African countries, particularly South Africa, where tobacco use is similar to that in developed countries (Sitas et al., 2004).

2.5 Prostate cancer

Prostate cancer (PCa) was the second most common cause of cancer and the sixth leading cause of cancer death among men worldwide with an estimated 899,000 new cases and 258,000 new deaths in 2008. PCa is common among men in southern Africa and western Africa, including South Africa, Nigeria, and Cameroon (Ferlay et al., 2010; Jemal et al., 2012). Even though it is the most frequently diagnosed cancer in men, PCa lies in fourth position in North Africa (after lung, liver, and bladder). Almost 60,000 new cases were estimated in 2012. It is the third most common neoplasm overall (after breast and cervix) in Africa as a whole (Parkin et al., 2015).

2.6 Colorectal cancer

This cancer is the fifth most common malignancy with 41,000 new cases and around 29,000 deaths in 2012, and a slight preponderance of cases in men (Parkin et al., 2015).

2.7 Esophageal cancer

This was the leading cause of cancer death among both men and women in East Africa and among men in South Africa. About 27,900 new cancer cases and 26,600 deaths from esophageal cancer occurred in 2008. About 27,500 new cancer cases and 25,200 deaths from esophageal cancer were estimated to have occurred in Africa in 2012, with 89% of these in sub-Saharan Africa (Jemal et al., 2012; Parkin et al., 2015).

2.8 Bladder cancer

Bladder cancer (BCa) is the 11th most commonly diagnosed cancer and the 14th leading cause of cancer deaths worldwide, with an estimated 382,700 new cases and 150,300 deaths in 2008 (Ferlay et al., 2010). Its incidence and mortality rates among men in northern Africa are twice as high as those in southern Africa, which has the second highest regional rates (Jemal et al., 2012). The incidence of BCa in Africa was much lower in 2012; however, it is the fourth most common cancer of men in North Africa. Incidence and mortality rates among men in northern Africa are twice as high as those in southern Africa (Parkin et al., 2015). Egyptian men have by far the highest incidence rates of BCa in Africa and worldwide (Parkin, 2008a). Incidence and mortality rates among men in northern Africa are twice as high as those in southern Africa, which has the second highest regional rate (Parkin et al., 2015).

2.9 Kaposi sarcoma

Kaposi sarcoma is an HIV associated cancer caused by human herpes virus-8 (Ziegler et al., 2003). The incidence rates for Kaposi sarcoma increased by several folds in East Africa and other parts of sub-Saharan Africa during the 1990s, consistent with the HIV/AIDS epidemic in these regions. An estimated 37,500 cases of KS (23,800 cases in males and 13,700 cases in females) were diagnosed in Africa in 2012. The area of highest incidence is in East Africa, where in six countries it is the most common cancer in males (Parkin et al., 2015).

2.10 Non–Hodgkin lymphoma

The incidence rates of NHL in Africa are higher than the world average. An estimated 37,200 new cases and 30,900 deaths from NHL occurred in 2008. These estimates increased to 36,700 new cases and 26,400 deaths in 2012 (Jemal et al., 2012, Parkin et al., 2015).

2.11 Nasopharyngeal cancer

The incidence rates of nasopharyngeal cancer are twice as high in men as in women. The highest incidence rates were in the Republic of South Africa and in the countries of the Maghreb (Morocco, Algeria, and Tunisia) as estimated in 2008. An estimated 8700 incidences of cancer cases and 5500 cancer deaths occurred in 2008 (Jemal et al., 2012).

3 GOVERNMENT CONTROL POLICIES FOR CANCER IN AFRICAN COUNTRIES

3.1 Cancer control policies: requirements

A well-planned and well-managed national cancer control program decreases incidences of cancer and improves the life of patients with cancer, regardless of a country's constraints in terms of resources. A national cancer control program is a public health program that is specifically designed to reduce the number of cancer cases and deaths and to improve quality of life of people suffering from cancer. This is achieved via the systematic and equitable implementation of evidence-based strategies for prevention, early detection, diagnosis, treatment, and palliative care, making the best use of available resources (WHO, 2002). To be most efficient, a comprehensive national cancer program evaluates different ways of controlling disease and implementing measures that are most cost-effective and beneficial for the largest proportion of the population. Emphasis is placed on preventing cancers or detecting cases early enough to allow a cure, and providing relief to patients with advanced disease.

A sound approach to a cancer control program incorporates prevention, early detection, and effective treatment, which includes palliative care (American Cancer Society, 2015). National cancer control policies and programs should raise awareness of cancer, help to limit exposure to cancer risk factors, provide information and support for more healthy lifestyles, and increase the proportion of cancers detected early if they are to be successful. Many types of cancers can be prevented, including those caused by tobacco use and excessive alcohol consumption. In 2010, nearly 1.5 million of the probable 8 million cancer deaths in the world were caused by tobacco smoking. The World Cancer Research Fund also estimated that between one-fifth and one-quarter of cancers globally are related to being overweight or obese, physical inactivity, and poor nutrition, and thus could also be prevented (World Cancer Research Fund, 2007; American Cancer Society, 2015).

Cancers related to infectious agents, such as liver, stomach, and cervical cancers, are potentially preventable (Jemal et al., 2010). Many cancers related to infectious agents, for example, HPV, HBV, HCV, HIV, and *Helicobacter pylori*, could be avoided through behavioral changes, infection control procedures, vaccinations, or treatment of infections. Many skin cancer cases could be prevented by protecting skin from excessive sun exposure and avoiding indoor tanning (American Cancer Society, 2015). Screening can also result in early detection of cancer before symptoms appear, which often results in more effective treatment and better outcomes. Diagnosis of cancer is the first step in managing the disease, and appropriate treatment protocols must be applied following staging to determine treatment options and prognosis. Cancer treatment may include a combination of surgery, chemotherapy, radiotherapy, hormone therapy, immune therapy, and targeted therapy, or any of these alone. Where patients are diagnosed with advanced-stage disease, which is often the case in low-resource countries, pain relief and palliative care may be the only effective options. Oral pain medications ranging from aspirin to opiates, depending on the needs of an individual patient, may be employed at this stage.

The WHO has developed guidelines and policies for establishing an effective national cancer control program according to the capacity and economic development of a country (WHO, 2002, 2009). In developing countries, this includes raising awareness of the rising burden of cancer, reducing the prevalence of major risk factors such as tobacco use, obesity, and infectious agents, the application of low-technology, cost-effective approaches to prevention or early detection of cervical cancer, and improving the availability of palliative care (Jemal et al., 2010). Several developing countries have developed national cancer control programs, although these programs are inadequately funded owing to limited resources and other competing public health programs (Ngoma, 2006). International public health agencies and other donors can play a significant part in maintaining existing cancer control programs or implementing new programs to halt the growing incidence of cancer in economically developing countries (Boyle et al., 2008). Developing a cancer control program should include establishing a cancer registry to assess the cancer burden and identify priorities, as well as to evaluate the effectiveness of the program (Jemal et al., 2010).

3.2 Establishing and maintaining cancer control programs in Africa

The WHO has published comprehensive guidelines for cancer control policies on a regional and national level according to national economic development (WHO, 2002). The organization recommends a stepwise approach for the establishment of cancer control programs in Africa via the implementation of one or two key priorities in a demonstration project (American Cancer Society, 2011). Priority actions in countries with low resources may begin with, for example, palliative care as an entry point to a more comprehensive approach (WHO, 2002). The WHO provides further useful advice for establishing the other components of a cancer control program, namely prevention, early diagnosis, and treatment (WHO, 2002).

African countries need to make the political commitment to invest in the programs adequately by providing the required staff together with a dedicated budget, assisted where possible by donors and international public agencies (American Cancer Society, 2011). The integration of cancer control programs with established disease control platforms is sensible as risk factors or routes of transmission may often coincide, such as in the case of certain infectious diseases which have a strong probability of leading to cancer. For example, unsafe sex is a risk factor for HIV and HPV, which is commonly associated with cervical cancer, so ongoing HIV prevention programs could be expanded to include aspects of cervical cancer prevention policies.

A reliable, population-based cancer registry is an essential component of a cancer control program as this is required to assess the burden of cancer, to set priorities as well as to implement and evaluate cancer control programs (Parkin, 2008b; Valsecchi and Steliarova-Foucher, 2008). In 2006 it was estimated that only 11% of the African population was covered by a population-based cancer registry (Parkin, 2006). Additionally, many cancer registries in Africa do not meet the criteria of the IARC for high-quality incidence data in terms of completeness, validity, and timelines (Parkin et al., 2001). A high-quality cancer registry enables identification of causes and risk factors for cancer in Africa, allowing the development of cancer prevention measures that take into account culture, diet, and other environmental factors. Africa therefore needs to concentrate on establishing and maintaining population-based cancer registration systems to allow the implementation of effective, evidence-based cancer control programs (American Cancer Society, 2011).

Table 3.3 Countries in Africa with operational policies, strategies, or action plans for cancer in 2013

Countries with plans	Countries with no plans	Data not received
Algeria, Benin, Cameroon, Congo (Republic of), Cote d'Ivoire, Djibouti, Egypt, Gabon, Eritrea, Ghana, Guinea, Kenya, Madagascar, Maldives, Mauritania, Morocco, Mozambique, Niger, Nigeria, Rwanda, South Africa, Sudan, Togo, Tunisia, Zambia, Zimbabwe	Botswana, Burkina Faso, Burundi, Central African Republic, Comoros, Equatorial Guinea, Gambia, Guinea-Bissau, Lesotho, Liberia, Libya, Malawi, Mali, Namibia, Sao Tome and Principe, Senegal, Seychelles, Somalia, Swaziland, Uganda	Angola, Cabo Verde, Chad, Congo (Democratic Republic of), Ethiopia, Mauritius, Sierra Leone, South Sudan, United Republic of Tanzania

(*Source*: Data sourced from WHO, 2013. Global Health Observatory data repository: NCD country capacity survey 2013. Available from: http://apps.who.int/gho/data/; Stefan, Elzawawy, Khaled, Ntaganda, Asiimwe, Addai, Wiafe, Adewole, Developing cancer control programs in Africa: examples from five countries. 2013. Lancet Oncol. 14, 189–195).

3.3 Country-specific cancer control policies in Africa

In African countries, there is a wide range of national cancer control plans, where they exist. These range from development of basic facilities needed for cancer care in countries like Rwanda, to existing well-conceived plans in more developed economies such as South Africa and Nigeria (Stefan et al., 2013). However, a major constraint to cancer care in African countries with relatively more advanced cancer control policies is the limited political will to fund such plans, and heightened awareness of the increasing burden of cancer is needed to promote further investment in this area (Stefan et al., 2013).

A worldwide assessment of the status of national efforts to combat noncommunicable diseases including cancer (WHO, 2011, 2013) was undertaken in 2010. Less than half of the respondent countries in Africa had operational policies, strategies, or plans to control cancer—not necessarily formal cancer control policies (Table 3.3). An alarming statistic emerging from the survey was that only 17% of national programs against noncommunicable diseases were funded in Africa. Although an in-depth analysis of cancer control programs or strategies in all African countries where they exist is beyond the scope of this chapter, a few cases will be expanded upon to provide an indication of the range of control programs supported.

A recent publication (Stefan et al., 2013) described in detail the obstacles encountered in the establishment and implementation of national cancer control plans in three relatively populous countries in Africa with the largest economies (Nigeria, South Africa, and Egypt) as well as Ghana as a medium-sized country and Rwanda representing a relatively small country. Nigeria is the most densely populated country in Africa with a population of over 170 million, and almost 102,000 new cases of cancer occur annually according to GLOBOCAN data (Ferlay et al., 2010). There is no national population-based cancer registry but data located in smaller population-based and hospital-based registries suggest that the incidence of cancer is increasing. Nigeria also carries a heavy burden of infectious diseases and so is in the unfortunate position of facing the "double burden" of communicable and noncommunicable diseases (Stefan et al., 2013). The country instituted a national cancer control plan in 2008 (Federal Ministry of Health, 2008) and this identifies priority cancers with goals to control each one. However, control measures for risk factors are not defined, and the process by which the goals are to be achieved as well as the necessary funding and personnel are lacking (Stefan et al., 2013).

There is a growing awareness of cancer and other noncommunicable diseases in South Africa, and GLOBOCAN estimated the equivalent of 101,000 new cancer cases in a population of 50 million in 2008 (Ferlay et al., 2010). A national cancer registry was established in 1986 but lack of funding after 2004 contributed to decline in maintenance of this registry, and only in 2011 did the Department of Health institute compulsory cancer registration (Stefan et al., 2013). Death notifications in 2009 suggested that cancer caused 6.3% of all deaths, making this the third most common cause of mortality, following tuberculosis and then influenza grouped together with pneumonia (Statistics South Africa, 2011). A national cancer control program was proposed in 1993, and was finally adopted in 1999 with later updates and revisions, but this is still awaiting adequate funding, as the scourge of AIDS and associated illnesses such as tuberculosis in South Africa demand a substantial share of the health budget (Stefan et al., 2013; National Department of Health, 2014).

Egypt has a population of approximately 83 million, and cancer and cardiovascular diseases dominate the noncommunicable diseases burden (Stefan et al., 2013). It has a network of population-based registries contributing to the national cancer registry, and has almost 70,000 new cancer cases per year (Ferlay et al., 2010). Egypt has more facilities than any other African country for the treatment of cancer (Stefan et al., 2013), but its cancer control strategy may still be improved. Recommendations by the WHO need to be taken into account in preparing a structured national cancer control program as this is lacking, and optimum allocation of available resources is needed (Stefan et al., 2013; WHO, 2002).

In contrast to Nigeria, South Africa, and Egypt, Ghana (with a population of 24 million) has no systematic national cancer program and only a rudimentary national cancer registry (Wiredu and Armah, 2006). Accurate estimation of the cancer burden is therefore not possible, and there is thus no basis upon which provision of cancer care can be matched to what is needed to allow efficient use of resources and equitable access (Stefan et al., 2013). Roughly 16,000 cases of cancer are estimated to occur annually in Ghana (Ferlay et al., 2010), with cancer being the fourth most common cause of death in the country. In the past few years, cancer care has improved a great deal in Ghana, but many patients with cancer still present at late stages (Clegg-Lamptey et al., 2009). There is a shortage of qualified workers and technological resources, so cost-effective measures over the entire spectrum of cancer management are needed (Stefan et al., 2013).

Rwanda has a population of over 11 million people and 6,600 new cases of cancer occur every year, with 5,300 people dying from the disease annually (Ferlay et al., 2010). Rwanda has no formal cancer control program according to the WHO guidelines, but plans are afoot to formalize several existing initiatives into a national cancer program (Stefan et al., 2013). The Ministry of Health in Rwanda launched an initiative to establish a national cancer registry in 2010, and a national cancer protocol was created and adopted in 2012 for treatment of patients with cancer (Stefan et al., 2013). Funding for cancer management remains a major issue as does development of necessary skills, and these are issues shared with many low-income countries.

In many African countries, as in other predominantly low-income, developing countries around the world, there are several common obstacles to the establishment of effective cancer control policies. The first is the lack of comprehensive and accurate national cancer registries, but steps are being taken to address this and create such registries or to support existing ones (Stefan et al., 2013). Particularly in countries with relatively small, agricultural economies, a second constraint is the scarcity of resources available to combat cancer. The development and retention of skills is a crucial issue for poorer countries, which often rely on international collaboration for the training of doctors and nurses specializing in oncology (Stefan et al., 2013). Those countries in Africa with larger economies, such as South Africa and Nigeria, may have national cancer control plans, but funding is again an issue as health budgets are largely spent on priorities such as infectious diseases, including HIV, tuberculosis, and malaria, which substantially affect the young, economically active population group. To deliver comprehensive cancer control in Africa, relevant programs must integrate with existing platforms targeted toward combating HIV/AIDS, tuberculosis, and malaria (Lingwood et al., 2008).

4 CONCLUSIONS

The African continent is experiencing a heightened risk of cancers previously associated with a western lifestyle following a rise in smoking, physical inactivity, and poor diet in developing countries. This, together with a substantial number of cancers associated with infectious diseases, is leading to an inexorable increase in the burden of cancer in Africa. Many governmental and nongovernmental organizations, as well as other associations, institutions, and organizations, contribute to the fight against cancer. However, it is necessary to focus and efficiently organize such anticancer efforts. This can be done most effectively through development of a national cancer control policy, adapted to conditions in each country, as recommended by the WHO. Comprehensive guidelines have been published by the WHO, although disappointingly few African countries have current strategies aimed at dealing with cancer. It is encouraging, however, that awareness is growing and several countries have in recent years made efforts or initiated cancer strategies. With the multitude of challenges facing countries in Africa, it is not surprising that public health funds are limited in terms of fighting diseases like cancer, but it is clear that more needs to be done to raise awareness and promote prevention measures as part of efforts to contain the growing burden of cancer in the African continent.

REFERENCES

American Cancer Society, 2011. Cancer in Africa, American Cancer Society, American Cancer Society, Atlanta.
American Cancer Society, 2015. Global Cancer Facts & Figures, third ed. American Cancer Society, Atlanta.
Anorlu, R.I., 2008. Cervical cancer: the sub-Saharan African perspective. Reprod. Health Matters 16, 41–49.
Boyle, P., Anderson, B.O., Andersson, L.C., et al., 2008. Need for global action for cancer control. Ann. Oncol. 19, 1519–1521.
Boyle, P., Levin, B. (Eds.), 2008. 2008. World Cancer Report. World Health Organization, International Agency for Research on Cancer, Lyon, France.
Cancer Genome Atlas Network, 2012. In: Comprehensive molecular portraits of human breast tumours. Nature 490, 61–70.
Clegg-Lamptey, J., Dakubo, J., Attobra, Y.N., 2009. Why do breast cancer patients report late or abscond during treatment in Ghana? A pilot study. Ghana Med. J. 43, 127–131.
Denny, L., Anorlu, R., 2012. Cervical cancer in Africa. Cancer Epidemiol. Biomarkers Prev. 21, 1434–1438.
Ferlay, J., Soerjomataram, I., Dikshit, R., Eser, S., Mathers, C., Rebelo, M., Parkin, D.M., Forman, D., Bray, F., 2015. Cancer incidence and mortality worldwide: Sources, methods and major patterns in GLOBOCAN 2012. Int. J. Cancer 136, E359–E386.
Federal Ministry of Health, 2008. Nigeria Cancer Control Plan 2008—2013. Federal Ministry of Health, Abuja.
Ferlay, J., Shin, H.R., Bray, F., Forman, D., Mathers, C., Parkin, D.M., 2010. GLOBOCAN 2008—Cancer Incidence and Mortality Worldwide: IARC CancerBase No 10. International Agency for Research on Cancer, Lyon.
Franceschi, S., Raza, S.A., 2009. Epidemiology and prevention of hepatocellular carcinoma. Cancer Lett. 286, 5–8.
Hicks, M.L., Yap, O.S., Mathews, R., Parham, G., 2006. Disparities in cervical cancer screening, treatment and outcomes. Ethnicity Dis. 16, S3.
Ibrahim, A., Rasch, V., Pukkala, E., Aro, A.A., 2011. Predictors of cervical cancer being at an advanced stage at diagnosis in Sudan. Int. J. Women's Health 3, 385–389.
Ismali, F., Torre, L.A., Jemal, A., 2015. Global trends of lung cancer mortality and smoking prevalence. Transl. Lung Cancer Res. 4, 327–338.
Jemal, A., Bray, F., Center, M.M., Ferlay, J., Ward, E., Forman, D., 2011. Global cancer statistics. CA: Cancer J. Clin. 61, 69–90.
Jemal, A., Bray, F., Forman, D., O'Brien, M., Ferlay, J., Center, M., Parkin, D.M., 2012. Cancer burden in Africa and opportunities for prevention. Cancer 118, 4372–4384.
Jemal, A., Center, M.M., DeSantis, C., Ward, E.M., 2010. Global patterns of cancer incidence and mortality rates and trends. Cancer Epidemiol. Biomarkers Prev. 19, 1893–1907.
Lingwood, R.J., Boyle, P., Milburn, A., Ngoma, T., Arbuthnott, J., McCaffrey, R., 2008. The challenge of cancer control in Africa. Nat. Rev. Cancer 8, 398–403.
National Department of Health, 2014. National Plan for the Prevention and Control of Cancer in South Africa 2015–2020. Department of Health, South Africa.
Ngoma, T., 2006. World Health Organization cancer priorities in developing countries. Ann. Oncol. 17 (Suppl. 8), viii9-14.
Parkin, D.M., 2006. The evolution of the population-based cancer registry. Nat. Rev. Cancer 6, 603–612.
Parkin, D.M., 2008a. The global burden of urinary bladder cancer. Scand. J. Urol. (Suppl. 218), 12–20.
Parkin, D.M., 2008b. The role of cancer registries in cancer control. Int. J. Clin. Oncol. 13, 102–111.

Parkin, D.M., Bray, F., Ferlay, J., Jemal, A., 2015. Cancer in Africa 2012. Cancer Epidemol. Biomarkers Prev. 23, 953–966.

Parkin, D.M., Sitas, F., Chirenje, M., Stein, L., Abratt, R., Wabinga, H., 2008. Part I: cancer in indigenous Africans—burden, distribution, and trends. Lancet Oncol. 9, 683–692.

Parkin, D.M., Wabinga, H., Nambooze, S., 2001. Completeness in an African cancer registry. Cancer Causes Control 12, 147–152.

Scheff, N., 2007. Liver cancer in Africa. Med. J. Ther. Afr. 1 (No 3), 236–243.

Sitas, F., Urban, M., Bradshaw, D., Kielkowski, D., Bah, S., Peto, R., 2004. Tobacco attributable deaths in South Africa. Tob. Control 13, 396–399.

Statistics South Africa, 2011. Mortality and causes of death in South Africa, 2009: findings from death notification. Statistical Release P0309.3. Statistics South Africa, Pretoria.

Stefan, D.C., Elzawawy, A.M., Khaled, H.M., Ntaganda, F., Asiimwe, A., Addai, B.W., Wiafe, S., Adewole, I.F., 2013. Developing cancer control programs in Africa: examples from five countries. Lancet Oncol. 14, 189–195.

Sylla, B.S., Wild, C.P., 2012. Cancer burden in Africa in 2030: invest today and save tomorrow. Afr. J. Cancer 4, 1–2.

Thun, M.J., DeLancey, J.O., Center, M.M., Jemal, A., Ward, E.M., 2010. The global burden of cancer: priorities for prevention. Carcinogenesis 31, 100–110.

Torre, L.A., Bray, F., Siegel, R.L., Ferlay, J., Lortet-Tieulent, J., Jemal, A., 2015. Global Cancer Statistics, 2012. CA Cancer J. Clin. 65, 8–108.

Valsecchi, M.G., Steliarova-Foucher, E., 2008. Cancer registration in developing countries: luxury or necessity? Lancet Oncol. 9, 159–167.

World Health Organization, 2002. National Cancer Control Programmes: Policies and Managerial Guidelines, second ed. World Health Organization, Geneva.

World Health Organization, 2009. 2008-2013 Action Plan for the Global Strategy for the Prevention and Control of Noncommunicable Diseases: Prevent and Control Cardiovascular Diseases, Cancers, Chronic Respiratory Diseases and Diabetes. World Health Organization, Geneva.

World Health Organization, 2011. Global Status Report on Non-Communicable Diseases 2010. World Health Organization, Geneva, Available from: http://who.int/nmh/publications/ncd_report2010/en/index.html (accessed 1 April, 2016).

World Health Organization, 2013. Global Health Observatory data repository: NCD country capacity survey 2013. Available from: http://apps.who.int/gho/data/

Wiredu, E.K., Armah, H.B., 2006. Cancer mortality patterns in Ghana: a 10-year review of autopsies and hospital mortality. BMC Public Health 6, 159.

World Cancer Research Fund/American Institute for Cancer Research, 2007. Food, Nutrition, Physical Activity, and the Prevention of Cancer: A Global Perspective. American Institute for Cancer Research, Washington, DC.

Ziegler, J., Newton, R., Bourboulia, D., 2003. Risk factors for Kaposi sarcoma: a case-control study of HIV-seronegative people in Uganda. Int. J. Cancer 103, 233–240.

CHAPTER 4

Metabolic Syndromes and Public Health Policies in Africa

E.U. Nwose, P.T. Bwititi, V.M. Oguoma

1 INTRODUCTION

It has been known for some time that public health care in Africa; especially in Sub-Saharan Africa (SSA) is under pressure, which includes scarcity of resources that perhaps undermines health care providers (Streefland, 2005). Until recently, noncommunicable diseases such as metabolic syndrome (MetS) were thought to be rare in Africa but now realized to be prevalent enough to require instituting a public health policy (Akpalu et al., 2011; Belfki et al., 2013). Among others, MetS is associated with malnutrition, which is a risk factor (Kimani-Murage, 2013; Zeba et al., 2012) and it is also linked to depressive symptoms (Hamer et al., 2011). Medicinal spices are commonly consumed in Africa and what is yet to be evaluated is whether African countries have public health policy for MetS or guidelines for use of these spices, in the context of medical nutritional therapy to manage some of the chronic noncommunicable diseases (CNCDs).

This chapter is divided into three parts leading to concluding recommendations and the parts include a cursory narrative review of prevalence of MetS in Africa; a systematic review of literature for public health policies on MetS in Africa; and case narratives on biochemical bases of medicinal spices with regards to medical nutritional therapy.

2 PREVALENCE OF METABOLIC SYNDROME IN AFRICA

2.1 Overview

MetS is the presence of three or more components: central obesity, hypertension, hypertriglyceridemia, impaired fasting blood glucose level and low level of high density lipoprotein (HDL) (Belachew, 2015; Tran et al., 2011). Generally, there is an increase in the prevalence of CNCDs and MetS throughout the world and this increase and the associated burden are prominent in developing countries (Belachew, 2015; BeLue et al., 2009; Tran et al., 2011). The major risk factors for MetS and CNCDs encompass smoking, stress inclusive of stress-related problems of poverty, urbanization, adoption of westernized lifestyle, which includes unhealthy diet and reduced physical activity. These are on rise in most low- to middle-income communities and such factors superimposed on overnutrition and high prevalence of undernutrition, which are sometimes observed in some developing communities, increase the load of diseases (Belachew, 2015; BeLue et al., 2009).

In SSA, infectious diseases account for the majority of mortality (69%), while CNCDs account for 25% of deaths but elevated mortality from CNCDs is now being seen and it is predicted that in 2030, the prevalence of CNCDs and perhaps the associated morbidity and mortality will be more than that of infectious diseases (Belachew, 2015; BeLue et al., 2009). A problem with analyzing the epidemiology of MetS is that there are variations across regions and this depends on a number of factors including the definition or criteria that is used to determine the syndrome, age, sex, ethnic origin, and lifestyle of the population (Bello-Rodriguez et al., 2013; Ben Ali et al., 2014; Cameron et al., 2004). To this end, MetS definition has been modified over the years and there are now various criteria, which compounds the difficulties in comparing data from various communities (Bello-Rodriguez et al., 2013). Furthermore, it is also paramount to be mindful that most of the data comes from hospital-based studies, namely, patients with MetS-related illnesses such as diabetes, hypertension, and dyslipidemia. A study by Oguoma and Nwose shows that for over 12 years of clinical and public health research on MetS in Nigeria, 9 studies (28.1%) utilized the World Health Organization (WHO) criteria, 19 (59.4%) used the ATP III, while 10 (31.3%) studies employed the International Diabetes Federation (IDF) criteria (Oguoma et al., 2016). Another report also shows that in Nigerian studies of cardiometabolic syndrome (CMS), 20 (62.5%) were hospital-based involving diabetic, hypertensive, HIV, asthmatic, and thyroid disorder patients and the remaining 12 (37.5%) were population-based studies (Oguoma et al., 2015). It is acknowledged that these studies give different results thus, may also not account for the whole Sub-Saharan region. Africa is a large continent with different ethnic groups who have varied cultural, socioeconomic, and possibly genetic factors, which possibly influence pattern of diseases (Oguoma et al., 2016; Okafor, 2012). For these reasons, this section is not only about figures because it is clear from studies that the prevalence of MetS and noncommunicable diseases (NCDs) is high (Tables 4.1 and 4.2) and is increasing worldwide but

Table 4.1 Prevalence (%) of MetS using the IDF criteria

Country	Population	Age group (years)	All sexes	Female	Male	References
Kenya	Urban	≥18	34.6	40.2	29	Kaduka et al. (2012)
Seychelles	General	25–64	a	34.5	25.1	Kelliny et al. (2008)
Ethiopia	Urban	Adults	17.9	24	14	Tran et al. (2011)
South Africa	Rural	>15	23.3	a	a	Motala et al. (2011)
Algeria	Urban	12–18	a	0.5	1.3	Benmohammed et al. (2015)
Nigeria	Semiurban	25–64	18	a	a	Ulasi et al. (2010)
Nigeria	Rural	25–64	10	a	a	Ulasi et al. (2010)

MetS, Metabolic syndrome.
[a] Not indicated.

Table 4.2 Prevalence (%) of MetS using the NCEP-ATP III criteria

Country	Population	Age group (years)	All sexes	Female	Male	References
Ethiopia	Urban	Adults	12.5	16.6	10	Tran et al. (2011)
Seychelles	General			32.2	24	Kelliny et al. (2008)
Cameron	Urban	25–55	17.7			Assah et al. (2011)
South Africa[a]	Urban	Adults		57	30	Duki and Naidoo (2016)
South Africa	Rural	>15	18.5			Motala et al. (2011)
Tunisia	General	35–70		35.9		Ben Ali et al. (2014)
Tunisia	Urban	13–18	0.4	Same as males		Harrabi et al. (2009)
Angola	Urban	>20	17.7	25.9	8.5	Magalhaes et al. (2014)

[a] Indian community.

also attempts to explain the reasons for the increase in the African context hence data on some of the components of MetS such as obesity will be mentioned.

2.2 Prevalence

In a study of MetS among working adults in Ethiopia conducted in Addis Ababa on bank employees and teachers, the data shows that using the modified Third Report of the National Cholesterol Education Program Adult Treatment Panel (NCEP-ATP III) criteria, the prevalence was 12.5% overall (10.0% in men and 16.6% among women) and using the IDF criteria, the prevalence was 17.9% overall (14.0% of men and 24.0% of women). The study also showed that regardless of criteria, the prevalence increased with age in both sexes but the prevalence was highest between the age of 45–54 years with 40.5% of men and 53.7% of women categorized as having MetS. The study also noted that while some participants did not meet the MetS criteria of ATP III and IDF, there were many participants with one or two components of MetS and such people are at risk of developing the syndrome later. Using the ATP III criteria, 20.4% of women and 18.6% of men had two MetS components in the study and the IDF criteria showed 46.5% of women and 34.7% of men having central obesity plus one additional component, thus were at risk of developing MetS (Tran et al., 2011). A Cameroon study on adults aged 25–55 years in Yaoundé and Bamenda reported a higher frequency of MetS in middle class urban dwellers than rural dwellers (17.7% vs. 3.5%) using the NCEP-ATP III criteria (Assah et al., 2011).

A rural South African black adult community study aged >15 years showed an age-adjusted crude frequency of MetS of 22.1%, with a higher prevalence in women (25.0%) than in men (10.5%). Peak prevalence in the sample was in the ≥ 65 year age-group in women (44.2%) and 45–54 year age-group in men (25.0%). Further, the crude prevalence appeared higher using the Joint Interim Statement (JIS) criteria (26.5%) than with the

IDF (23.3%) or the NCEP-ATP III criteria (18.5%) (Motala et al., 2011). Another South African study in Cape Town (urban), on mainly black adults aged 25–74 years, observed the overall MetS components to be higher in women compared with men and the components include; central obesity and low-density lipoprotein (LDL) cholesterol in women, while in men raised blood pressure was the most frequent. MetS was found to be highly prevalent in this urban Black population, with the age-standardized prevalence of 31.7% according to the JIS criteria, the prevalence was higher in women in comparison with men (Peer et al., 2015).

A Ghanaian study in urban and rural communities to ascertain the sociodemographic associations of obesity, which is a component of MetS, in adults of 25 years or over, reported the overall crude prevalence of overweight and obesity to be 23.4 and 14.1%, respectively. The rates of overweight (27.1% vs. 17.5%) and obesity (20.2% vs. 4.6%) were higher in females than in males. In addition, the study also reported that obesity increased with age of up to 65 years and that there were more overweight and obese individuals in the urban high-class residents compared with low-class residents and also in urban than in rural subjects (Amoah, 2003).

In Tunisian women aged 35–70 years, the overall prevalence of MetS using NCEP-ATP III criteria was 35.9% and higher in postmenopausal (45.7% vs. 25.6%) than in premenopausal women. This study also showed that the prevalence of MetS increased with age and that the urban area was significantly associated with an increased risk of MetS. The study further showed that the risk of MetS increased among widowed and divorced women compared to married women (Ben Ali et al., 2014). A survey on a sample representative of the general population aged 25–64 years in the Seychelles also reported the prevalence of MetS to increase with age and using the NCEP-ATP III, WHO and IDF criteria, the prevalence was, respectively, 24.0, 25.0, and 25.1% in men and 32.2, 24.6, 35.4% in women (Kelliny et al., 2008). In a household cross-sectional survey of adults aged ≥18 years in Nairobi, Kenya, the prevalence of MetS according to IDF was 34.6% and it was more in women (40.2%) than in men (29%) and the most observed components were raised blood pressure, higher waist circumference and low HDL cholesterol. This study also noted that the major factors associated with MetS were advancing age, socioeconomic status, and education (Kaduka et al., 2012).

Using the IDF criteria in Benin, West Africa a study on rural, semirural, and urban adults aged 25–60 years reported a positive rural–urban gradient, that is, rural to semirural to urban for the overall prevalence of MetS (4.1, 6.4, and 11%, respectively). In men, a positive rural–urban gradient (rural to semirural to urban) prevalence was 0, 4.7, and 5.0%, respectively, while in women it was 8.2, 8.2, and 17.0%, respectively (Ntandou et al., 2009). A study on CMS in adults aged 25–64 years in semiurban and rural communities in 36 states in Nigeria showed the overall prevalence of CMS (both sexes combined) to be 18.0% in the semiurban community against 10.0% in the rural community, using the IDF criteria (Ulasi et al., 2010). In a study of cardiovascular (CV) risk factors adults aged 20 years or older who were employees of a university in Luanda, Angola, reported the overall crude prevalence of MetS to be 17.6% according to the NCEP-ATP III and 27.8% using the JIS criteria. In the same study, the JIS criteria, gave an overall crude prevalence of 27.8% with men accounting for 17.0% and females 37.7%. Both criteria showed prevalence to be higher in women than in men and also the prevalence increased with age although women showed higher prevalence in all age-groups from 30 years and older. Prevalence of smoking was higher in men and that of overweight, obesity, and low-HDL-C levels was higher in women, compared with the other sex group. Both sexes showed similar prevalence of hypertension, diabetes, hypercholesterolemia, hypertriglycerides and high LDL-C and looking at socioeconomic class and educational levels, no significant relationship of these factors with the MetS were observed in both sexes (Magalhaes et al., 2014).

Obesity and overweight are components of MetS and the prevalence of childhood overweight is increasing world-wide and this is more so in Africa and Asia. Between 2000 and 2013, the prevalence of overweight in children aged less than 5 years increased from 11% to 19% in some countries in Southern Africa and in 2013, there were about 11 million overweight children in Africa (WHO, 2014). The few studies on MetS in African children show a low prevalence in comparison with adults and this is not surprising since as mentioned the prevalence of MetS increases with age, physical inactivity, smoking, and stress among others and these are associated more with adults than children. In Constantine (urban) in Algeria, the prevalence for MetS in adolescents, 12–18 years was 1.3% for boys and 0.5% for girls using the IDF criteria (Benmohammed et al., 2015) while in urban Sousse in Algeria, 13- to 18-year-olds had a prevalence of MetS of 0.4% using the modified ATP III definition and there was no statistical difference between the sexes (Harrabi et al., 2009). Tables 4.1 and 4.2 indicate summary of prevalence of MetS in some African countries based on IDF and ATPIII, respectively

2.3 Burden and complications

African countries such as those in SSA are faced with increasing prevalence of CNCDs. This is superimposed on the prevalence of infectious diseases and strategies are required to address both problems. Diabetes is a component of MetS and studies show that the increasing prevalence of diabetes most likely interferes with tuberculosis control program by perhaps increasing the number of susceptible individuals in communities where tuberculosis is endemic, thus making management difficult. HIV infection and the antiretroviral therapy (ART) of HIV infection are associated with an increase in prevalence of MetS in developing countries (Belachew, 2015). The life expectancy for HIV-infected people is now long due to the effective and improved accessibility of ART. However, extended life expectancy coupled with sometimes poor diet and/or increasing sedentary lifestyles predispose HIV-infected persons to developing CNCDs such as diabetes mellitus, CV disease, and cancers (Ali et al., 2014).

Studies are showing that HIV-infected persons are increasingly at risk of developing metabolic diseases such as diabetes, dyslipidemias and overweight/obesity. With the projected increase in global prevalence of diabetes from 382 million in 2013, to 592 million in 2035, studies therefore need to identify effective DM prevention strategies in HIV-infected persons. It is not clear if HIV infection itself increases the risk of developing dysglycemia, that is, prediabetes or diabetes but it has been shown that dyslipidemia, a component of MetS occurs in ART-naive HIV-infected patients and increased triglyceride (TG) and reduced total LDL and HDL cholesterol subfractions are seen in advanced HIV disease. It must be pointed out that most of these studies were conducted in high income communities and often in populations where obesity and overweight are common and literature from low- and middle-income countries is scarce. Studies in ART-naive HIV-infected persons in Cameron, Tanzania, and South Africa have shown normal LDL levels but dyslipidemias (low HDL and high TG) have been seen in patients with long ART exposure as reported in a few South American studies. The pathophysiology of dyslipidemias in HIV-infected persons is complex and is possibly associated with the HIV infection, ART, inflammation, genetic factors, age, obesity, lipodystrophy, DM, and liver disease and these relationships are not clearly understood. Studies show that HIV-infected persons on ART gain weight and those with a higher BMI, a component of MetS at ART initiation are at greater risk of becoming obese (Ali et al., 2014).

The WHO reports that the human, social, and economic consequences of NCDs are felt by all countries especially the poor and vulnerable communities hence decreasing the burden of NCDs is a priority for sustainable development. Globally, NCDs accounted for 38 million (68%) of the world's deaths in 2012 and more than 40% of these (16 million) were under 70 years. About 75% of all NCD deaths (28 million) and the majority of premature deaths (82%) occur in low- and middle-income countries. It is estimated that during 2011–25, the cumulative economic losses due to NCDs under "business as usual" scenario in low- and middle-income countries to be USD 7 trillion and this is more than the annual USD 11.2 billion cost of effecting high-impact interventions to reduce the NCDs burden. The annual number of deaths due to infectious diseases is projected to go down but the annual number of NCD deaths is expected to rise by 2030 but sadly funding for surveillance, monitoring, and evaluation is low in Africa (WHO, 2014).

A study that looked at the components of MetS, found the age standardized prevalence of raised blood pressure compared to other parts of the world, to be highest in Africa, at 30% for adults combined and equally divided between sexes. The association between obesity, poor health outcomes, and all-cause mortality is known; obesity increases the chances of getting diabetes, hypertension, coronary heart disease and stroke, certain cancers, obstructive sleep apnea, and osteoarthritis and negatively affects reproductive performance. Unfortunately, obesity is on the rise globally and in 2014, 39% of adults aged 18 years and above (38% of males and 40% of females) were overweight and prevalence of obesity nearly doubled between 1980 and 2014, worldwide. Women are more likely to be obese than men and in Africa, women have roughly double the obesity prevalence of men (WHO, 2014).

2.4 Factors possibly affecting prevalence

Cocurrent with understanding the prevalence and the other epidemiological data associated with MetS, it is important to understand the factors and the limitations that affect the data. As mentioned earlier, a problem with MetS is that there are various definitions and there appears to be no consensus about the application of any particular criteria. The criteria have different cut-off values and thus yield different outcomes (Tran et al., 2011). For instance, a study showed that in an Indian adult community in Phoenix, Durban in South Africa, the prevalence of MetS was 30 and 45%, in males while in females 57 and 60% using the NCEP III and JIS criteria, respectively (Duki and Naidoo, 2016). Such data shows that the criteria and the sex of the samples influence prevalence. The WHO definition of obesity is sometimes used in defining obesity and MetS as there is no specific definition for African populations as well as other populations. This definition uses a higher cut point for waist circumference for men than for women and therefore it may not be suitable in some African populations, for example, Cameron where the mean waist circumference of women is greater than that of men. This possibly leads to the reported higher prevalence of abdominal obesity in women compared to men (Fezeu et al., 2006).

SSA is reported to be going through epidemiological transitions involving urbanization-influenced lifestyle changes (BeLue et al., 2009). Studies have over the years reported and predicted that the standard of living in most African countries is rising hence an adequate food supply in most communities and unhealthy behavior involving food consumption and energy expenditure are now being observed and some of these are determinants of obesity (Fezeu et al., 2006). Other authors have also cited unhealthy habits and lifestyle as well as risk factors that include smoking, alcohol ingestion, obesity, diabetes, hypertension, and high serum cholesterol levels to have raised the incidence of NCDs especially CVD (Kadiri and Salako, 1997).

Researchers predicted a surge in life expectancy in developing countries and attributed this to: decline in deaths in infancy; childhood and adolescence; improved and effective public response to perinatal, infectious, and nutritional-deficiency disorders; improved economies and female literacy in some areas. These factors increase longevity and also increase the exposure to risk factors of, for example, CVD. Nutrition transition involving accessibility to food has resulted in increased consumption among low-income countries. Together with that, the globalization of food production and marketing contributes to the increased consumption of energy-foods poor in dietary fiber and several micronutrients. Also implicated is tobacco consumption that is rising in some developing countries that contrasts with a decline in developed countries (Reddy and Yusuf, 1998). Studies show that urbanization and economic development have also resulted in the emergence of a nutritional transition mentioned and this is characterized by a shift to a higher caloric-content diet and/or reduction in physical activity. Modern societies appears to be living on a diet high in saturated fats, sugar, and refined foods but low in fiber and on a lifestyle characterized by lower levels of activity and these changes are reflected in nutritional outcomes, such as changes in average stature, body composition, and morbidity (Popkin and Gordon-Larsen, 2004). The irony is that there is a rapid increase in overweight occurring across most low- and middle-income countries, even in the midst of global food and economic crises (Popkin and Slining, 2013).

Cultural and social factors are implicated as contributory to MetS in Ghanaian women, specifically in obese and overweight groups and this has been reported to be more prominent in certain tribes and also among people with tertiary education. Such studies also noted that urban compared to rural dwellers were more prone to obesity due to westernization and transition in dietary habits since it was the educated people that afford fast foods available as a result of globalization and westernization and this is reinforced by media perception of what constitutes good life. Also noted in this study was that Ghanaians appear to take less exercise and irregularly since previously people walked long distances to school or work. It was also pointed out that Ghanaians generally associate fatness with beauty in women and success in men and women and that generally, men in Ghana prefer fuller to thin women (Amoah, 2003) and such observations are common in most African countries. Equally, common is the notion that obese/weight are perhaps a rooted status symbol in some developing countries and it is difficult to change that health belief (Fezeu et al., 2006) hence education on weight reduction may not be very effective.

During the early part of the HIV epidemic, the major concern was weight loss and in some communities there was stigma associated with weight-loss as this was perceived to be due to AIDS. Studies now show that in high and low-income countries overweight/obesity is prominent as wasting in HIV-infected individuals. Studies also show that HIV-infected persons on ART can gain weight and those with a high BMI at ART initiation are at greater risk of becoming obese (Ali et al., 2014) as mentioned. In Addis Ababa, Ethiopia it has been shown that the prevalence of MetS was 14% in men and 20% in women but this was worse in people with HIV/AIDS with prevalence of MetS, 25% in the ART group compared to 22% in ART naive group (Belachew, 2015).

Perhaps a contribution of high prevalence of MetS in women is the report in Tunisia that showed that the risk of MetS increased among widowed and divorced women compared to married women and that could be explained by the fact that married women are more likely to engage in positive health behaviors than widowed and divorced women (Ben Ali et al., 2014). Similar studies need to be carried out in men to ascertain the association of marital status and MetS.

Studies in Benin, West Africa show that the main occupation and transport means were major contributors to physical activity and most women were involved only in small trade and household chores that are not physically challenging but accounts for high obesity prevalence in women by comparison with men. Other factors that the authors articulated were diet quality and physical activity that was higher in rural and semiurban compared to urban subjects thus also accounting for differences in prevalence of MetS in those groups. The study also found that physical activity was lower in cities compared with semirural and urban dwellers, especially in women, who were particularly prone to obesity (Ntandou et al., 2009).

WHO (2014) has listed a number of factors, which are applicable to all communities and not just to low-income countries, with regard to raised blood pressure, which is a component of MetS: eating food containing too much salt and fat; not eating enough fruits and vegetables; overweight and obesity; harmful use of alcohol; physical inactivity; aging; genetic factors; psychological stress; socioeconomic determinants; inadequate access to health care. These factors are applicable to all countries and not only to developing countries. Other reports have also noted that the prevalence of diabetes has been increasing globally and has particularly accelerated in low- and middle-income countries due to modifiable risk factors particularly physical inactivity, overweight, and obesity (Finucane et al., 2011).

3 SYSTEMATIC REVIEW OF PUBLIC HEALTH POLICIES ON METABOLIC SYNDROME IN AFRICA

The MetS is a diet-related cluster of NCDs such as diabetes, hypertension, obesity, and lipid disorders and the increasing prevalence of MetS and its components in Africa have reached epidemic levels as mentioned. Although calls for global action for prevention of this modifiable ill condition have been resonating for years, progress in national health policy development in African countries, especially toward diet-related NCDs have been largely underaccessed (Lachat et al., 2013).

In Africa, traditional medicinal herbs are used in management of various diseases including diabetes and its complications (Ezuruike and Prieto, 2014) and several herbs have been validated clinically as hypotensive and antihypertensive therapeutic agents (Tabassum and Ahmad, 2011) and evidence of this within the West African region was brought into global purview in the 1980s (Bever, 1980). These herbs are administered in various forms: chewed raw or cooked; used to bath; and blended and sieved for a drinkable liquid extract. Quite considerable in some African countries is that inequitable access to health care services; ineffective public health policies; inadequate implementation of health policies; and high disease burden exacerbated by convergence of communicable and NCDs in part, have sustained the widespread utilization of herbalists as first line of treatment of NCDs (Onwujekwe, 2005). For instance, an estimated 80% of Africans use herbal medicine of some sort and there is enhanced investment and research in ethnomedicinal values of indigenous herbs (Tilburt and Kaptchuk, 2008). Given the increased use of herbs in treatment and management of diseases such as diabetes and hypertension, there is still lack of information about the clinical/public health use or in the monitoring of these plants, despite widespread use (Ezuruike and Prieto, 2014).

There is a large body of evidence, which indicates that a high intake of fruit and vegetables is associated with low risk of MetS (Esmaillzadeh et al., 2006; Kim and Jo, 2011) and WHO estimates that low fruit and vegetable intake contributes to about 2.7 million deaths a year and ranks it the sixth risk factor for mortality in the world (Ruel et al., 2005). Effective prevention of MetS can be achieved through prioritization of these simple and pragmatic preventive measures into national strategic action plan and policy. Public health policy focuses on how societies prioritize their health issues, make decisions, and act to shape health care systems. The concern with MetS and other CNCDs in some African countries is the lack of a clear policy and plan for effective prevention strategies of this complex cluster.

Even though few countries have policies on NCDs, they do not strategize to tackle MetS as a sole entity. Based on the 2014 WHO NCDs country profiles, it is noteworthy that the majority of African countries do not have an operational NCD unit/branch or department in their Ministries of Health (World Health Organization, 2014). In other words, there is predominant unavailability of operational national policy, strategy, or action plan that integrates several NCDs and shared risk factors. The same trend applies to operational policy, strategy, or action plan to reduce unhealthy diet and/or promote healthy diets.

A review of the WHO's NCD country profiles in the African region shows that 11/14 (78.6%) countries in West Africa, 5/10 (50%) East Africa, 1/8 (12.5%) Southern Africa, and 4/6 (66.7%) North Africa lacks an operational NCD unit or department within their Ministry of Health (Tables 4.3–4.7). This is evidence that governments and their health systems are unprepared to challenge the increasing prevalence of individual risk factors of MetS such as diabetes and hypertension. Furthermore, it is alarming

Table 4.3 Noncommunicable diseases profile in West African countries (World Health Organization, 2014)

PH operation	Benin	B'Faso	Gambia	Ghana	Guinea	G'Bissau	Liberia	Mali	M'tania	Niger	Nigeria	Senegal	STP	Togo
NCD unit[a]	No	Yes	No	No	Yes	No	No	No	No	No	No	No	No	Yes
Multisectoral[b]	No	No	No	Yes	Yes	No	No	No	No	No	No	No	No	Yes
Policy[c] on diet	No	No	Yes	Yes	Yes	No	No	No	Yes	No	No	No	Yes	Yes

B' Faso, Burkina Faso; G' Bissau, Guinea Bissau; M'tania, Mauritania; NCD, noncommunicable disease; PH, public health.
[a] NCD unit/branch or department in Health Ministry.
[b] Multisectoral national policy, strategy, or action plan.
[c] Policy, strategy, or action plan on healthy diets.

Table 4.4 Noncommunicable diseases profile in East African countries (World Health Organization, 2014)

PH operation	Burundi	Kenya	Uganda	Rwanda	Djibouti	Eritrea	Somalia	Madagascar	Comoros	Seychelles
NCD unit[a]	No	No	Yes	Yes	No	Yes	No	Yes	No	Yes
Multisectoral[b]	No	No	No	No	No	Yes	No	Yes	No	No
Policy[c] on diet	NR	No	No	No	No	Yes	No	Yes	No	No

NCD, Noncommunicable disease; NR, data unavailable; PH, public health.
[a] NCD unit/branch or department in Health Ministry.
[b] Multisectoral national policy, strategy, or action plan.
[c] Policy, strategy, or action plan on healthy diets.

Table 4.5 Noncommunicable diseases profile in Southern African countries (World Health Organization, 2014)

PH operation	Botswana	Lesotho	Malawi	Mozambique	Namibia	Swaziland	Zambia	Zimbabwe
NCD unit[a]	Yes	No	Yes	Yes	Yes	Yes	Yes	Yes
Multisectoral[b]	No	No	No	Yes	No	No	No	No
Policy[c] on diet	No	No	No	Yes	No	No	Yes	No

NCD, Noncommunicable disease; PH, public health.
[a] NCD unit/branch or department in Health Ministry.
[b] Multisectoral national policy, strategy, or action plan.
[c] Policy, strategy, or action plan on healthy diets.

Table 4.6 Noncommunicable diseases profile in North African countries (World Health Organization, 2014)

Public health operation	Algeria	Egypt	Libya	Morocco	Sudan	Tunisia
NCD unit[a]	Yes	No	Yes	No	No	No
Multisectoral[b]	No	No	No	No	No	No
Policy[c] on diets	Yes	No	No	No	Yes	No

NCD, Noncommunicable disease; PH, public health.
[a] NCD unit/branch or department in Health Ministry.
[b] Multisectoral national policy, strategy, or action plan.
[c] Policy, strategy, or action plan on healthy diets.

Table 4.7 Dietary contents[a] of various foods relevant to oxidative medicine

	Burrito[b]		Chimichanga[b]		Sauce[c] (cooked pepper)		Uncooked pepper	
	Normal	+Cheese	Normal	+Cheese	Green	Red	Green	Red
Vitamin E (mg)		NS			2	2.2	4.1	4.1
β-Carotene (mg)					2.0	1.5	4.0	3.2
Vitamin C (mg)	5	7	1	6	408	180	1455	864
Cholesterol (mg)	162	336	30	168	0	0	0	0
Total fat (g)	49	49	60	59	<1	4	1	3
Total fiber (g)		NS			11.4	4.2	9.0	9.0
Calories (kcal)	1272	1248	1398	1212	120	126	240	240

NS, not shown in data set.
[a] 600 g pepper/day at the rate of 200 g/meal 3 times daily.
[b] Fast food meal.
[c] Equivalence of conventional pepper soup.

to note that over 80% of countries in Africa do not possess any form of multisectoral national policy or action plan that integrate several NCDs and shared risk factors. Tables 4.3–4.7, show that 11/14 countries in West Africa, 8/10 in East Africa, 7/8 in Southern Africa, and 6/6 in North Africa have no national policy for preventing NCDs and its associated risk factors.

It appears that the West African region has the weakest operational policy and strategy to challenge NCDs by, for example, reduction of unhealthy diet and/or promotion of healthy diets. Studies emphasize that most policies lack concrete plans, mechanisms, and incentives to promote multistakeholder and cross-sector collaboration. Further, unambiguous and prioritized strategies are required in Africa to harness the growing epidemics of MetS and it would be of immense value if these policy documents are available to share, promote engagement with the stakeholders, and to stimulate accountability and leadership in the fight against the burden of NCDs (Lachat et al., 2013). Development of such policies would therefore need to be innovative and less reliant on what exists in the developed countries. This is pertinent since developed countries perhaps lack incentive in the use of medicinal spices and where there is accessible in developed countries, conventional health care and urbanization may have eradicated the naturally growing herbal plants to research on, that is, compared to Africa and other low-midincome countries.

Further, it is imperative to recognize the teaching–research–practice nexus, which involves the ministry or public health department proposing a policy that is backed up by sufficient evidence. The need to analyze public-health decision-making from an ethical perspective has been highlighted (Hyder et al., 2008) and this calls for research or scientific articles written for Public Health practitioners as target audience. Besides the review of NCDs profile in West African countries (World Health Organization, 2014) presented in Table 4.3, attempt has been made to briefly and systemically review journal articles on public health policy for MetS in Africa. Using *PubMed* (including the *Journal of Public Health Policy*) as search engine, entry of "public health policy" yielded 1374 articles. Inputting "Africa" as additional term reduced the number, which became abysmal when MetS and its components were added. While there is evidence of significance of medical nutritional therapy for MetS management, there is little or no literature indicating studies that are specifically on "medicinal herbs in MetS" or literature on substantial methodical report to drive initiation of a public health policy for the use of herbal use.

4 FRAMEWORK TO DEVELOP PUBLIC HEALTH POLICY FOR MEDICINAL HERBS—THE CASE OF PEPPER

4.1 Pepper as a typical medicinal spice from Africa

Pepper soup is consumed by the English-speaking countries of Western Africa: Ghana, Liberia, Nigeria, and Sierra Leone. This soup does not necessarily contain more pepper than other African soups but pepper is the major ingredient, besides salt and water, hence, it is called pepper soup. Other ingredients are added according to choice and taste.

Outside Africa the African herbs and spices are not commonly found in supermarkets except in stores specializing in African Foods but there are substitutes or similar herbs world-wide and some of these herbs and spices are available in major supermarkets. While many Westerners consider pepper as a spice (Govindarajan et al., 1987), this is not the case with pepper soup in African countries where it is consumed hence in the preparation of this soup, other spices are added (Edet, 2003). The recipe and procedure for preparing the soup is available (Edet, 2003) and is passed down generations as well as available in most African cook books.

Pepper soup can be eaten as appetizer/entrée in two or three course meals as supper or in combination with macronutrients such as boiled potatoes, pounded yam, or steamed rice. Frying may contribute antinutritional factors that could counteract the nutritional values, for example, frying will destroy vitamins E (Andersen, 1995), which is otherwise preserved in boiling, which is used in preparing the soup.

4.2 Antioxidant potential of pepper soup for oxidative medicine

The nutritional value of pepper soup is dependent on the ingredients used in the preparation. The fish and/or meat that can be added make it a protein-rich meal. The spices and pepper provides calories and micronutrients as well as the flavors that make it an appetizer (Antonious et al., 2006; Barua et al., 2008). Dwelling specifically on pepper being the major ingredient, the nutritional value depends on age or developmental stage of the pepper (Govindarajan et al., 1987; Sun et al., 2007). It is a source of many amino acids including those necessary for synthesis of reduced glutathione (GSH) in the erythrocyte (Sekura and Meister, 1974; USDA, 2008). It contains no cholesterol and its fibers have the capacity to lower cholesterol and reduce oxidative stress (Luqman and Rizvi, 2006). Herbal cum nutritional therapies include *n*-acetylcysteine (NAC), carotenoid, GSH, vitamin C, and vitamin E (Bradley et al., 2007; Head, 2006) and it is known that pepper contains carotenoid as well as vitamins C and E (Antonious et al., 2006; Barua et al., 2008; USDA, 2008). NAC is a nonessential micronutrient and it is the more stable form of cysteine and a precursor for GSH synthesis. However, what is infamous is that it is contained in pepper (The Nutros Team, 2010) and further, it is reported that pepper contains sulfur in its seed (Barua et al., 2008), which is an essential element in GSH structure. Pepper also contains capsaicin (Barua et al., 2008; Dairam et al., 2008), which induces relaxation of the blood vessels in humans (Gupta et al., 2007).

4.3 Oxidative medicine value of pepper being utilized in ethnomedicinal practice

Normal metabolic processes inherently generate free radicals and when a free radical reacts with a molecule, a new radical is formed. The new radical reacts with another molecule to produce another radical and this continues exponentially until the radicals react with a chain-breaking antioxidant resulting in the formation of a stable product. However, the chain-breaking antioxidants are often depleted in conditions, such as hyperglycemia or increased energy need, where glucose metabolism is increased (Gagnier et al., 2007; Guarrera et al., 2008).

GSH is a frontline antioxidant in hyperglycemia toxicity and is abundantly available in cell membranes including those of the erythrocyte (Engel and Vemulpad, 2009). There are several ways by which it acts as a chain-breaking antioxidant and one involves breaking the chain of exponential lipid peroxidation process. Upon lipid (L) peroxidation, vitamin E (tocopherol: TOH) donates one electron (H• radical) to the resultant lipid peroxyl radical (LOO•) to form lipid peroxide and tocopheroxyl (TO•) radicals. The former is acted upon by GSH, in the presence of glutathione peroxidase, to form a stable lipid-OH (Fig. 4.1). Another pathway in which GSH is involved as an antioxidant is in regeneration of other antioxidants (Fig. 4.1) (Nwose, 2009, 2010; Nwose et al., 2008a). Vitamins C and E depend strongly on GSH to prevent the prooxidant activities of their oxidized forms/radicals that would otherwise lead to oxidative stress and feed-forward to affect blood flow and testicular dysfunction amongst other complications. The TO• that is formed can be reduced by GSH (Ho and Chan, 1992; Nwose et al., 2008a). It is important that health practitioners including dietitians and oxidative stress researchers consider therapeutic value of meals and Table 4.7 shows distribution of some dietary contents in various foods (Nwose, 2010).

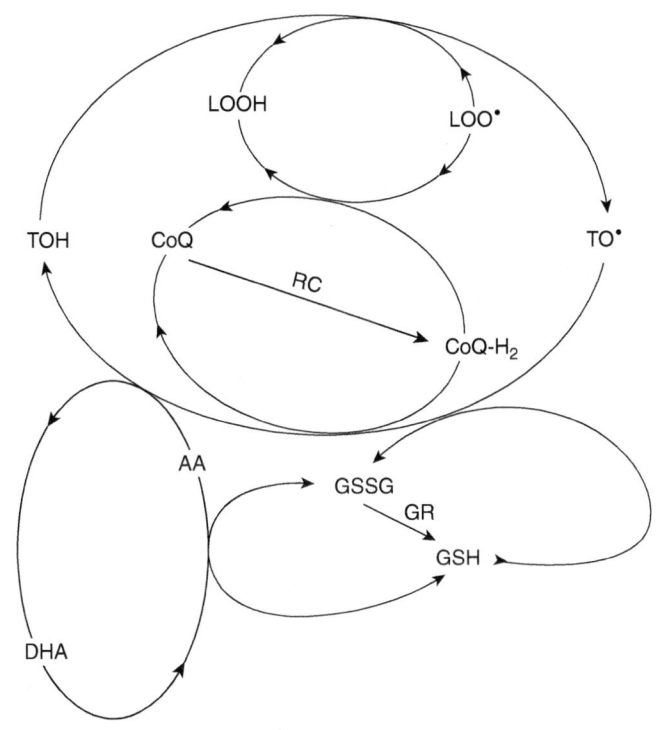

Figure 4.1 *Glutathione dependence by other antioxidants to avoid vascular complications (Nwose et al., 2008a).* AA, Ascorbic acid; CoQ, oxidized coenzyme-Q_{10}; CoQ-H_2, coenzyme-Q_{10}; CVD, cardiovascular disease; DHA, dehydro-ascorbic acid (free radical); GR, glutathione reductase; GSH, reduced glutathione; GSSG, oxidized glutathione; LOO·, lipid peroxyl radical; LOOH, lipid peroxide; OS, oxidative stress; RC, mitochondrial respiratory chain; TOH, tocopherol (vitamin e); TO·, tocopheroxyl radical.

The data in Table 4.7 show that consumption of 200 g pepper/meal 3 times a day provides
- When taken as pepper soup/sauce or raw: the necessary antioxidants and calories and a little less than 25% the recommended dietary indication (RDI), except vitamin C and there is also no cholesterol.
- When taken in combination with other macronutrients such as beef and cheese: calories far greater than 25% RDI for a normal healthy person are consumed and there is increased total fat content and the cholesterol content is greater than 25% RDI of 300 mg.

On the premise of RDI, vitamin C of 40 mg and 30 mg for 76 kg men and 62 kg women, respectively (Olson and Hodges, 1987) or the additional 300–500 kcal/day requirement of ante- or postnatal woman (Porth, 2007), Table 4.7 shows that consumption of 200 g pepper/meal 3 times a day provides more than 25% these RDI requirements. Therefore, there are cohorts of the general population who would derive discrete clinical benefit at the same time avoiding cholesterol. It is possible that the antioxidant potential of pepper can ameliorate several disease states including MetS as illustrated by these selected examples.

4.3.1 Maintenance of blood flow/viscosity

Fig. 4.2 illustrates that inadequate recycling of vitamins C and/or E by GSH could lead to oxidative stress and CV complications, which is the hallmark of MetS. Fig. 4.2 also shows that maintaining the endogenous level of GSH by recycling depends on glutathione reductase (GR). Nevertheless, the body requires replenishment of GSH through endogenous synthesis or from nutrition and this is where nutritional therapy as source

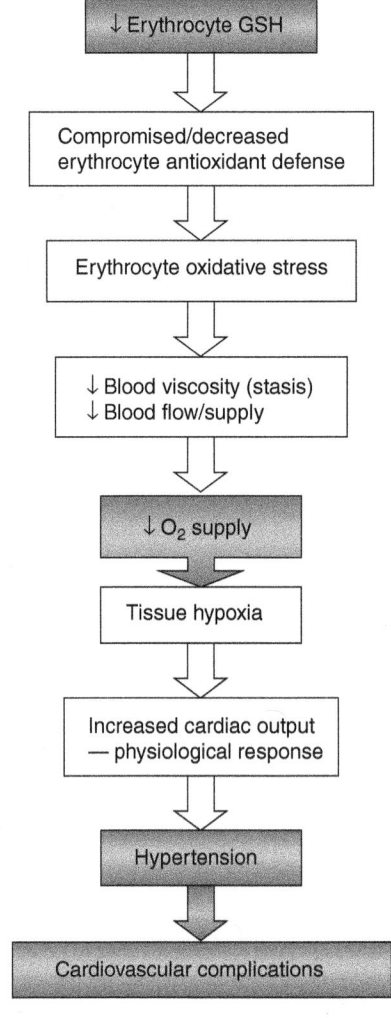

Figure 4.2 *Causality of glutathione deficiency in cardiovascular (CV) complications (Nwose, 2010).*

of NAC to augment GSH synthesis comes into perspective (Ristoff and Larsson, 2007; The Nutros Team, 2010; USDA, 2008).

It is pertinent to note that the antioxidant activity depends on the type of oxidant (Young and Woodside, 2001). For instance, increased glucose metabolism, which may occur in hyperglycemia, a component of MetS can lead to the depletion of GSH level when the maximum GR potential to regenerate GSH is exceeded (Taniyama and Griendling, 2003). When this occurs, regeneration of ascorbic acid and vitamin E is reduced and the unrecycled dehydroascorbic acid and tocopheroxyl radical go into exponential free radical reactions that culminate in oxidative stress and in turn lead to vascular complications (Nwose et al., 2008a,b; Stocker and Keaney, 2004). The point here is that the level of GSH is a factor in the occurrence of stasis and reduced blood flow en route to CV complications (Fig. 4.2) and pepper as a medicinal spice has some value that requires methodical articulation for the purpose of public health policy.

To avoid the untoward effect of overwhelmed GR potential, the GSH-regeneration process is inherently backed up in the physiological system by γ-glutamyl cysteine ligase and GSH-synthetase enzymes and both enzymes function to synthesize GSH. Nevertheless, the activities of these enzymes are subject to availability of cysteine as substrate, which is where NAC comes as a vital supplement for GSH, that is, to supply cysteine (Stocker and Keaney, 2004; Taniyama and Griendling, 2003). Studies report capsaicin and NAC to be alternative therapies for peripheral vascular diseases since they ameliorate the effects of reactive oxygen species (Ristoff and Larsson, 2007; Sun et al., 2007). What is being emphasized here is the medicinal/nutritional value of pepper as a spicy plant that is rich in capsaicin and NAC.

4.3.2 Diabetes management
In diabetes, hyperglycemia-induced oxidative stress overwhelms the inherent potential to maintain GSH level (Taniyama and Griendling, 2003) and oxidative stress induction of whole blood viscosity as a vasculopathy has been illustrated (Nwose, 2007; Nwose et al., 2008b). NAC, vitamin C, and vitamin E among others have been identified as nutritional ingredients necessary for diabetes management (Bradley et al., 2007) and this identifies the effects of nutritional ingredients in the management of oxidative damage in DM and that pepper soup is one option of oxidative medicine to consider for diabetes management.

4.3.3 Adoptability in policy formulation
Central to the objective of this hypothesis is that if medicinal spices, such as pepper are intended to manage MetS, cholesterol-rich or fat-riddled foods necessarily need to be limited. In the context of drug–disease interaction, these foods would counteract the pharmacodynamic effects of pepper hence research to drive initiation of public health policy must necessarily consider this factor. As previously published (Nwose, 2010), it is

pertinent to acknowledge that ascertaining the clinical benefits of foods remains a challenge compared to pharmaceutical agents because while pharmaceutical products are measured and purified single active ingredients food is mostly a medley of unmeasured and unpurified agents. While the medicinal benefit of several medicinal herbs such as pepper are known, the challenge is defining the benefits; establish how to scientifically validate the benefits as well as the policy or procedure to monitor. It has been suggested that studies should focus on validating the antioxidant capacity of any nutrition and testing the effects on markers of oxidation (Tapsell et al., 2006).

4.3.4 Clinical diagnostic tests indicating and optimizing pepper soup as a therapy

There would also be the need to review clinical laboratory monitoring to optimize the outcomes of dietary evaluations. Clinical algorithms have been suggested to guide the use of dietary therapy using laboratory measures (Bradley et al., 2007; Gordon et al., 2008) but the suggestions do not include identifiable antioxidants. That is, the crucial factor of oxidative stress is being overlooked in the algorithms. For instance, the position statement and recommendations from American Diabetes Association is that deficiencies of antioxidant vitamins are difficult to ascertain (Franz et al., 2003, 2008), which is perhaps not clear. This position statement needs to be reviewed and dietitians reminded that validated methods for in vitro diagnostic methods for vitamins C and E are available. Below are suggestion on how laboratory evaluations can be used to optimize outcomes of antioxidant nutrition therapy, including pepper soup and there are occasions when results for the parameters provide a basis to make one of two decisions.

- Concomitant normal levels may indicate antioxidant/oxidant balance. In this situation, it is recommended not to prescribe antioxidant vitamin supplement, but nutrition that aims to maintain normalcy.
- Deficiency of vitamins C and/or vitamin E may be indicative of failed regeneration system. In this case, it has to be remembered that the vitamins may be in prooxidant forms and requiring regeneration. In such condition, other antioxidants such as GSH or other sulfur-containing amino acids should be considered (Atmaca, 2004). If the probability is true, the follow-up evaluation tests would present improved levels of vitamins C and E.

This is when and where pepper soup comes as an option and since pepper is used in other food preparations, it is important to appreciate the following three rationales:

- Intravenous NAC has been reported to show side effects but not the case with the dietary form (Dodd et al., 2008). Thus, one practical rationale of pepper soup is oral administration of NAC in a lifestyle manner and thus avoiding the supplement that has side effects.
- The process of preparation is a major factor, while cooking may destroy much of the vitamin C content, the vitamin E is preserved. The nutritional NAC boosts GSH level, which sustains the latter to recycle and maintain the endogenous level of vitamin

C. This is in contrast to frying or other forms of cooking that involve oil where vitamins E is significantly destroyed (Steinhart and Rathjen, 2003), whilst the added oil enhances postprandial lipemia and associated vascular pathologies (Delgado-Lista et al., 2008; Jackson et al., 2007).

- When appetizing junk foods need to be avoided, pepper soup provides an appetizing spicy alternative. Furthermore, breakfast cereals may be rich in fiber and could reduce cholesterols but cereals are often taken with milk, which is fat-riddled if unprocessed or may be of lesser quality if processed (Ma et al., 2008; Rocquelin et al., 1998; Rudloff and Lonnerdal, 1992; Wigertz et al., 1996). Thus, another rationale for pepper soup lies on the premise of being prepared and taken with no cholesterol-containing ingredients.

Indeed, research has shown that treatment with pepper prevents lipid peroxidation (Anandakumar et al., 2008); and what this article adds is that pepper soup may provide sufficient natural antioxidants against lipid peroxidation and needs to be investigated to develop a framework for incorporation into clinical practice. In a nutshell, pepper soup is a traditional food consumed in West Africa and prepared mainly with pepper, salt, and water, plus other optional ingredients. It provides calories and micronutrients including antioxidants but it is yet to be scholarly looked into as antioxidant nutrition therapy. It is possible that diagnostic tests can indicate this cheap-and-easy-to-prepare meal to maintain level of endogenous antioxidants and manage blood flow-related diseases including diabetes. The potential of pepper as a medicinal spice or pepper soup as an alternative medical nutrition therapy requires methodical evaluation in animal and human case studies and in clinical trials to enable initiation of a public health policy for its use.

5 CONCLUSIONS

Research into ethnopharmacology as well as medical nutritional therapy has come a long way. Yet, it is neither assessed in clinical practice in terms of therapeutic monitoring by laboratory methods, nor are there public health policies or procedures to enable guided adoption by medical practitioners during treatment. In MetS and its components as well as other metabolic diseases that are on the increase, there is implication for the role of antioxidants as alternative diagnostic and therapeutic tools. A major importance of this chapter is the clinical point of view that monitoring by laboratory methods that are readily applicable and available in clinical practice should be part of validation studies to drive policy development.

REFERENCES

Akpalu, J., Akpalu, A., Ofei, F., 2011. The metabolic syndrome among patients with cardiovascular disease in Accra, Ghana. Ghana Med. J. 45 (4), 161–166.

Ali, M.K., Magee, M.J., Dave, J.A., Ofotokun, I., Tungsiripat, M., Jones, T.K., et al., 2014. HIV and metabolic, body, and bone disorders. JAIDS 67, S27–S39.

Amoah, A., 2003. Sociodemographic variations in obesity among Ghananian adults. Public Health Nutr. 6 (8), 751–757.

Anandakumar, P., Jagan, S., Kamaraj, S., Ramakrishnan, G., Titto, A.A., Devaki, T., 2008. Beneficial influence of capsaicin on lipid peroxidation, membrane-bound enzymes and glycoprotein profile during experimental lung carcinogenesis. J. Pharm. Pharmacol. 60 (6), 803–808.

Andersen, G.D., 1995. Vitamin E, vol 13. Dynamic Chiropractic. Available from http://www.dynamicchiropractic.com/mpacms/dc/article.php?id=40148

Antonious, G.F., Kochhar, T.S., Jarret, R.L., Snyder, J.C., 2006. Antioxidants in hot pepper: variation among accessions. J. Environ. Sci. Health B 41 (7), 1237–1243.

Assah, F., Ekelund, U., Brage, S., Mbanya, C., Wareham, N., 2011. Urbanization, physical activity, and metabolic health in Sub-Saharan Africa. Diab. Care 34 (2), 491–496.

Atmaca, G., 2004. Antioxidant effects of sulfur-containing amino acids. Yonsei Med. J. 45 (5), 776–788.

Barua, A.G., Hazarika, S., Pathak, J.S., Kalita, C., 2008. Spectroscopic investigation of the seeds of chilli (*Capsicum annuum* L.). Int. J. Food Sci. Nutr. 59 (7–8), 671–678.

Belachew, T., 2015. Are we ready for the rising silent epidemic of metabolic syndrome and chronic noncommunicable disease in Ethiopia? Ethiop. J. Health Sci. 25 (1), 1–2.

Belfki, H., Ben Ali, S., Aounallah-Skhiri, H., Traissac, P., Bougatef, S., Maire, B., et al., 2013. Prevalence and determinants of the metabolic syndrome among Tunisian adults: results of the Transition and Health Impact in North Africa (TAHINA) project. Public Health Nutr. 16 (4), 582–590.

Bello-Rodriguez, B.M., Sanchez-Cruz, G., Delgado-Bustillo, F., Asiama, G., 2013. The relationship between metabolic syndrome and target organ damage in Ghanaian with stage-2 hypertension. Ghana Med. J. 47 (4), 189–196.

BeLue, R., Okoror, T., Iwelunmor, J., Taylor, K., Degboe, A., Agyemang, C., Ogedegbe, G., 2009. An overview of cardiovascular risk factor burden in Sub-Saharan African countries: a socio-cultural perspective. Global. Health 5, 10.

Ben Ali, S., Belfki-Benali, H., Aounallah-Skhiri, H., Traissac, P., Maire, B., Delpeuch, F., et al., 2014. Menopause and metabolic syndrome in Tunisian women. Biomed Res. Int. 2014, 457131.

Benmohammed, K., Valensi, P., Benlatreche, M., Nguyen, M.T., Benmohammed, F., Paries, J., et al., 2015. Anthropometric markers for detection of the metabolic syndrome in adolescents. Diab. Metab. 41 (2), 138–144.

Bever, B.O., 1980. Oral hypoglycaemic plants in West Africa. J. Ethnopharmacol. 2 (2), 119–127.

Bradley, R., Oberg, E.B., Calabrese, C., Standish, L.J., 2007. Algorithm for complementary and alternative medicine practice and research in type 2 diabetes. J. Altern. Complement. Med. 13 (1), 159–175.

Cameron, A.J., Shaw, J.E., Zimmet, P.Z., 2004. The metabolic syndrome: prevalence in worldwide populations. Endocrinol. Metab. Clin. North Am. 33 (2), 351–375.

Dairam, A., Fogel, R., Daya, S., Limson, J.L., 2008. Antioxidant and iron-binding properties of curcumin, capsaicin, and S-allylcysteine reduce oxidative stress in rat brain homogenate. J. Agric. Food Chem. 56 (9), 3350–3356.

Delgado-Lista, J., Lopez-Miranda, J., Cortes, B., Perez-Martinez, P., Lozano, A., Gomez-Luna, R., et al., 2008. Chronic dietary fat intake modifies the postprandial response of hemostatic markers to a single fatty test meal. Am. J. Clin. Nutr. 87 (2), 317–322.

Dodd, S., Dean, O., Copolov, D.L., Malhi, G.S., Berk, M., 2008. N-acetylcysteine for antioxidant therapy: pharmacology and clinical utility. Expert Opin. Biol. Ther. 8 (12), 1955–1962.

Duki, Y., Naidoo, D.P., 2016. Relationship of Body anthropometry with cardiovascular risk factors in a random community sample: The Phoenix Lifestyle Project. Metab. Syndr. Relat. Disord. 14 (2), 102–107.

Edet, L., 2003. Pepper soup. Available from: http://www.onlinenigeria.com/links/Recipesadv.asp?blurb=407

Engel, R.M., Vemulpad, S., 2009. Progression to chronic obstructive pulmonary disease (COPD): could it be prevented by manual therapy and exercise during the 'at risk' stage (stage 0)? Med. Hypotheses 72 (3), 288–290.

Esmaillzadeh, A., Kimiagar, M., Mehrabi, Y., Azadbakht, L., Hu, F.B., Willett, W.C., 2006. Fruit and vegetable intakes, C-reactive protein, and the metabolic syndrome. Am. J. Clin. Nutr. 84 (6), 1489–1497.

Ezuruike, U.F., Prieto, J.M., 2014. The use of plants in the traditional management of diabetes in Nigeria: pharmacological and toxicological considerations. J. Ethnopharmacol. 155 (2), 857–924.

Fezeu, L., Minkoulou, E., Balkau, B., Kengne, A.P., Awah, P., Unwin, N., et al., 2006. Association between socioeconomic status and adiposity in urban Cameroon. Int. J. Epidemiol. 35 (1), 105–111.

Finucane, M., Stevens, G., Cowan, M., Danaei, G., Lin, K., Paciorek, C., et al., 2011. National, regional, and global trends in body mass index since 1980 systematic analysis of health examination surveys and epidemiological studies with 960 country-years and 9.1 million participants. Lancet 377 (9765), 557–567.

Franz, M., Bantle, J., Beebe, C., Brunzell, J., Chiasson, J., Garg, A., et al., 2003. Evidence-based nutrition principles and recommendations for the treatment and prevention of diabetes and related complications. Diab. Care 26 (90001), S51–S61.

Franz, M.J., Boucher, J.L., Green-Pastors, J., Powers, M.A., 2008. Evidence-based nutrition practice guidelines for diabetes and scope and standards of practice. J. Am. Diet Assoc. 108 (Suppl. 4), S52–S58.

Gagnier, J.J., van Tulder, M.W., Berman, B., Bombardier, C., 2007. Herbal medicine for low back pain: a Cochrane review. Spine 32 (1), 82–92.

Gordon, L., Morrison, E., McGrowder, D., Young, R., Fraser, Y., Zamora, E., et al., 2008. Effect of exercise therapy on lipid profile and oxidative stress indicators in patients with type 2 diabetes. BMC Complement. Altern. Med. 8 (1), 21.

Govindarajan, V.S., Rajalakshmi, D., Chand, N., 1987. Capsicum—production, technology, chemistry, and quality Part IV. Evaluation of quality. Crit. Rev. Food Sci. Nutr. 25 (3), 185–282.

Guarrera, P., Lucchese, F., Medori, S., 2008. Ethnophytotherapeutical research in the high Molise region (Central-Southern Italy). J. Ethnobiol. Ethnomed. 4 (1), 7.

Gupta, S., Lozano-Cuenca, J., Villalón, C., de Vries, R., Garrelds, I., Avezaat, C., et al., 2007. Pharmacological characterisation of capsaicin-induced relaxations in human and porcine isolated arteries. Naunyn Schmiedebergs Arch. Pharmacol. 375 (1), 29–38.

Hamer, M., Malan, N.T., Harvey, B.H., Malan, L., 2011. Depressive symptoms and sub-clinical atherosclerosis in Africans: role of metabolic syndrome, inflammation and sympathoadrenal function. Physiol. Behav. 104 (5), 744–748.

Harrabi, I., Bouaouina, M., Maatoug, J., Gaha, R., Ghannem, H., 2009. Prevalence of the metabolic syndrome among urban schoolchildren in Sousse, Tunisia. Int. J. Cardiol. 135 (1), 130–131.

Head, K.A., 2006. Peripheral neuropathy: pathogenic mechanisms and alternative therapies. Altern. Med. Rev. 11 (4), 294–329.

Ho, C.T., Chan, A.C., 1992. Regeneration of vitamin E in rat polymorphonuclear leucocytes. FEBS Lett. 306 (2–3), 269–272.

Hyder, A.A., Merritt, M., Ali, J., Tran, N.T., Subramaniam, K., Akhtar, T., 2008. Integrating ethics, health policy and health systems in low- and middle-income countries: case studies from Malaysia and Pakistan. Bull. World Health Org. 86 (8), 606–611.

Jackson, K.G., Armah, C.K., Minihane, A.M., 2007. Meal fatty acids and postprandial vascular reactivity. Biochem. Soc. Trans. 35 (Pt. 3), 451–453.

Kadiri, S., Salako, B.L., 1997. Cardiovascular risk factors in middle aged Nigerians. East Afr. Med. J. 74 (5), 303–306.

Kaduka, L.U., Kombe, Y., Kenya, E., Kuria, E., Bore, J.K., Bukania, Z.N., Mwangi, M., 2012. Prevalence of metabolic syndrome among an urban population in Kenya. Diab. Care 35 (4), 887–893.

Kelliny, C., William, J., Riesen, W., Paccaud, F., Bovet, P., 2008. Metabolic syndrome according to different definitions in a rapidly developing country of the African region. Cardiovasc. Diabetol. 7, 27.

Kim, J., Jo, I., 2011. Grains, vegetables, and fish dietary pattern is inversely associated with the risk of metabolic syndrome in South Korean adults. J. Am. Diet Assoc. 111 (8), 1141–1149.

Kimani-Murage, E.W., 2013. Exploring the paradox: double burden of malnutrition in rural South Africa. Glob. Health Action 6, 19249.

Lachat, C., Otchere, S., Roberfroid, D., Abdulai, A., Seret, F.M., Milesevic, J., et al., 2013. Diet and physical activity for the prevention of noncommunicable diseases in low- and middle-income countries: a systematic policy review. PLoS Med. 10 (6), e1001465.

Luqman, S., Rizvi, S.I., 2006. Protection of lipid peroxidation and carbonyl formation in proteins by capsaicin in human erythrocytes subjected to oxidative stress. Phytother. Res. 20 (4), 303–306.

Ma, Y., Li, W., Olendzki, B.C., Pagoto, S.L., Merriam, P.A., Chiriboga, D.E., et al., 2008. Dietary quality 1 year after diagnosis of coronary heart disease. J. Am. Diet Assoc. 108 (2), 240–246.

Magalhaes, P., Capingana, D.P., Mill, J.G., 2014. Prevalence of the metabolic syndrome and determination of optimal cut-off values of waist circumference in university employees from Angola. Cardiovasc. J. Afr. 25 (1), 27–33.

Motala, A., Esterhuizen, T., Pirie, F., Omar, M., 2011. The prevalence of metabolic syndrome and determination of the optimal waist circumference cutoff points in a rural South African community. Diab. Care 34 (4), 1032–1037.

Ntandou, G., Delisle, H., Agueh, V., Fayomi, B., 2009. Abdominal obesity explains the positive rural–urban gradient in the prevalence of the metabolic syndrome in Benin, West Africa. Nutr. Res. 29 (3), 180–189.

Nwose, E.U., 2007. Evaluation of Erythrocyte Oxidative Stress in Diabetic Macrovascular Complications [Dissertation]. Charles Sturt University, Australia.

Nwose, E.U., 2009. Laboratory evaluations to optimize outcomes of antioxidant nutrition therapy in diabetes management. North Am. J. Med. Sci. 1 (3), 137–143.

Nwose, E.U., 2010. Oxidative Stress Concepts in Clinical Practice—Towards Optimizing Medical Nutrition Therapy With Laboratory Methods, LAP Lambert Academic Publishing, Germany.

Nwose, E.U., Jelinek, H.F., Richards, R.S., Kerr, P.G., 2008a. The 'vitamin E regeneration system' (VERS) and an algorithm to justify antioxidant supplementation in diabetes—a hypothesis. Med. Hypotheses 70 (5), 1002–1008.

Nwose, E.U., Richards, R.S., Kerr, P.G., Tinley, R., Jelinek, H.F., 2008b. Oxidative damage indices for the assessment of subclinical diabetic macrovascular complications. Br. J. Biomed. Sci. 65 (3), 136–141.

Oguoma, V.M., Nwose, E.U., Richards, R.S., 2015. Prevalence of cardio-metabolic syndrome in Nigeria: a systematic review. Public Health 129 (5), 413–423.

Oguoma, V.M., Nwose, E.U., Ulasi, I.I., Akintunde, A.A., Chukwukelu, E.E., Araoye, M.A., et al., 2016. Maximum accuracy obesity indices for screening metabolic syndrome in Nigeria: a consolidated analysis of four cross-sectional studies. Diab. Metab. Syndr. 10 (3), 121–127.

Okafor, C., 2012. The metabolic syndrome in Africa: current trends. Indian J. Endocrinol. Metab. 16, 56–66.

Olson, J.A., Hodges, R.E., 1987. Recommended dietary intakes (RDI) of vitamin C in humans. Am. J. Clin. Nutr. 45 (4), 693–703.

Onwujekwe, O., 2005. Inequities in healthcare seeking in the treatment of communicable endemic diseases in Southeast Nigeria. Soc. Sci. Med. 61 (2), 455–463.

Peer, N., Lombard, C., Steyn, K., Levitt, N., 2015. High prevalence of metabolic syndrome in the Black population of Cape Town: the Cardiovascular Risk in Black South Africans (CRIBSA) study. Eur. J. Prev. Cardiol. 22 (8), 1036–1042.

Popkin, B.M., Gordon-Larsen, P., 2004. The nutrition transition: worldwide obesity dynamics and their determinants. Int. J. Obes. Relat. Metab. Disord. 28, S2–S9.

Popkin, B.M., Slining, M.M., 2013. New dynamics in global obesity facing low- and middle-income countries. Obes. Rev. 14, 11–20.

Porth, C.M., 2007. Stress and Adaptation Essentials of Pathophysiology—Concepts of Altered Health States. Lippincott Williams & Wilkins, Sydney, pp. 151–164.

Reddy, K., Yusuf, S., 1998. Emerging epidemic of cardiovascular disease in developing countries. Circulation 97, 596–601.

Ristoff, E., Larsson, A., 2007. Inborn errors in the metabolism of glutathione. Orphanet J. Rare Dis. 2 (1), 16.

Rocquelin, G., Tapsoba, S., Mbemba, F., Gallon, G., Picq, C., 1998. Lipid content and fatty acid composition in foods commonly consumed by nursing Congolese women: incidences on their essential fatty acid intakes and breast milk fatty acids. Int. J. Food Sci. Nutr. 49 (5), 343–352.

Rudloff, S., Lonnerdal, B., 1992. Solubility and digestibility of milk proteins in infant formulas exposed to different heat treatments. J. Pediatr. Gastroenterol. Nutr. 15 (1), 25–33.

Ruel, M.T., Minot, N., Smith, L., 2005. Patterns and Determinants of Fruit and Vegetable Consumption in Sub-Saharan Africa: A Multicountry Comparison. WHO, Geneva.

Sekura, R., Meister, A., 1974. Glutathione turnover in the kidney; considerations relating to the gamma-glutamyl cycle and the transport of amino acids. Proc. Natl. Acad. Sci. USA 71 (8), 2969–2972.

Steinhart, H., Rathjen, T., 2003. Dependence of tocopherol stability on different cooking procedures of food. Int. J. Vitam. Nutr. Res. 73 (2), 144–151.

Stocker, R., Keaney, J.F., 2004. Role of oxidative modifications in atherosclerosis. Physiol. Rev. 84 (4), 1381–1478.

Streefland, P., 2005. Public health care under pressure in Sub-Saharan Africa. Health Policy 71 (3), 375–382.

Sun, T., Xu, Z., Wu, C.T., Janes, M., Prinyawiwatkul, W., No, H.K., 2007. Antioxidant activities of different colored sweet bell peppers (*Capsicum annuum* L.). J. Food Sci. 72 (2), S98–S102.

Tabassum, N., Ahmad, F., 2011. Role of natural herbs in the treatment of hypertension. Pharmacogn. Rev. 5 (9), 30–40.

Taniyama, Y., Griendling, K.K., 2003. Reactive oxygen species in the vasculature: molecular and cellular mechanisms. Hypertension 42 (6), 1075–1081.

Tapsell, L.C., Hemphill, I., Cobiac, L., Patch, C.S., Sullivan, D.R., Fenech, M., et al., 2006. Health benefits of herbs and spices: the past, the present, the future. Med. J. Aust. 185 (4 Suppl.), S4–S24.

The Nutros Team, 2010. NAC. In: The Natural Supplement Review. Available from: http://www.nutros.com/nsr-02025.html

Tilburt, J.C., Kaptchuk, T.J., 2008. Herbal medicine research and global health: an ethical analysis. Bull. World Health Org. 86 (8), 594–599.

Tran, A., Gelaye, B., Girma, B., Lemma, S., Berhane, Y., Bekele, T., et al., 2011. Prevalence of metabolic syndrome among working adults in Ethiopia. Int. J. Hypertension 2011, 1–8.

Ulasi, I.I., Ijoma, C.K., Onodugo, O.D., 2010. A community-based study of hypertension and cardio-metabolic syndrome in semi-urban and rural communities in Nigeria. BMC Health Serv. Res. 10, 71.

USDA, 2008. National Nutrient Database for Standard Reference Release 21. Available from: www.nal.usda.gov/fnic/foodcomp/cgi-bin/list_nut_edit.pl

WHO, 2014. Global Status Report on Noncommunicable Diseases 2014, WHO Switzerland.

Wigertz, K., Hansen, I., Hoier-Madsen, M., Holm, J., Jagerstad, M., 1996. Effect of milk processing on the concentration of folate-binding protein (FBP), folate-binding capacity and retention of 5-methyltetrahydrofolate. Int. J. Food Sci. Nutr. 47 (4), 315–322.

World Health Organization, 2014. Noncommunicable Diseases and Mental Health: Noncommunicable Diseases Country Profiles 2014. World Health Organization, Geneva.

Young, I.S., Woodside, J.V., 2001. Antioxidants in health and disease. J. Clin. Pathol. 54 (3), 176–186.

Zeba, A.N., Delisle, H.F., Renier, G., Savadogo, B., Baya, B., 2012. The double burden of malnutrition and cardiometabolic risk widens the gender and socio-economic health gap: a study among adults in Burkina Faso (West Africa). Public Health Nutr. 15 (12), 2210–2219.

CHAPTER 5

Management of Infectious Diseases in Africa

R. Seebaluck-Sandoram, F.M. Mahomoodally

1 INTRODUCTION

Infectious diseases are regarded as the leading cause of deaths among children, adolescents, and adults worldwide (CSIS, 2016). Common infectious diseases with high mortality and morbidity include human immune-deficiency virus (HIV)/acquired immune deficiency syndrome (AIDS), malaria, diarrheal diseases, respiratory infections, tuberculosis, and neglected tropical diseases (NTDs) which accounts for approximately 13.4 million of deaths annually. It has been found that most of these deaths occur in low- and middle-income regions of the globe where preventive or curative interventions are inaccessible to the needy populations (CSIS, 2016). NTDs comprise 17 viral, bacterial, and parasitic diseases that occur mainly in tropical regions (WHO, 2015b) and the most prevalent NTDs are buruli ulcer, chagas disease, dengue, chikungunya, dracunculiasis (guinea-worm disease), echinococcosis, endemic treponematoses (yaws), foodborne trematodiases, and human African trypanosomiasis (sleeping sickness) (WHO, 2016f). Africa is depicted as the continent with the largest infectious disease burden and has poor public health infrastructure and access to modern drugs among all regions in the world (Falarin et al., 2014). Transmission of infectious diseases in sub-Saharan Africa is attributed to various environmental and socioeconomic factors. Global warming and deforestation have further impacted on the tropical climate of Africa making it prone to breeding area for emerging pathogens. Extreme weather conditions such as floods and drought can provide new breeding sites for vectors of malaria, dengue fever, chikungunya, and other diseases (Laino, 1999). Socioeconomic factors such as poverty, poor sanitation, and hygiene as well as inaccessibility of potable water for domestic use have influenced the transmission of infectious diseases in sub-Saharan Africa. The unavailability of potable water has forced people to use natural water bodies such as lakes, rivers, ponds, and other water sources contaminated with parasites (Adenowo et al., 2015). Other factors such as human migrations, wild animal migrations, trade, and farming systems have also influenced the spread of infections in Africa (Rweyemamu et al., 2006).

Infectious diseases are caused by various pathogenic microorganisms such as bacteria, fungus, parasites, and virus and diseases can disseminate directly or indirectly from one person to another (WHO, 2016e). Infections can be transmitted via different pathways namely airborne, foodborne, waterborne, vectors, direct contact, and waste disposal (Rweyemamu et al., 2006).

The chapter endeavors to address the different strategies used for the management of common infectious diseases such as HIV/AIDS, dengue, ebola, chikungunya, cholera, cryptococcal meningitis (CM), malaria, and schistomiasis in Africa. Prophylactic measures such as health education, vector control, sanitation and hygienic conditions, screening programs, and disease surveillance are among the recommended policies for the elimination of most infectious diseases. Moreover, we have reviewed both conventional and traditional approaches that are in common use to alleviate signs and symptoms of various ailments including diarrhoeal infections, malarial, dengue and chikungunya fever, skin infections, as well as venereal diseases.

2 HIV/AIDS

HIV/AIDS poses a serious threat to humanity worldwide (Abdu et al., 2016). This disease was diagnosed in 1981 and approximately 25 million people have died since the identification of the virus. More than 100 million people have been infected by HIV (Abdu et al., 2016) whereas approximately 36.9 million people were living with HIV at the end of 2014 (WHO, 2015a). Sub-Saharan Africa is considered as the most affected region with 25.8 million people living with HIV in 2014 (WHO, 2015a). The majority of people affected with HIV in sub-Saharan Africa are women which accounts to 60% of the population. Thus, it is the principal cause of premature deaths in Africa mainly among reproductive young people and women. Nigeria is the third country after India and South Africa to be greatly affected by HIV/AIDS pandemic with approximately 10% of the cases globally (Abdu et al., 2016).

The disease AIDS is caused by HIV and the transmission of this disease occurs through direct contact of mucous membrane or the blood stream with a bodily fluid containing HIV including blood, semen, vaginal fluid, preseminal fluid, and breast milk (Abdu et al., 2016). There are various risk factors of contracting HIV such as having unprotected sex, having another sexually transmitted infection (syphilis, herpes, chlamydia, gonorrhea, bacterial vaginosis), sharing contaminated syringes, needles, and drug solutions, receiving unsafe injections, blood transfusion, medical procedures that comprise unsterile cutting or piercing and subjected to accidental needle stick injuries (WHO, 2015a).

World Bank programs have provided more than $2 billion catering to HIV prevention, treatment, care, and support in more than 30 sub-Saharan African countries and five regional programs since 2000. Since 2006, this program has been successful in providing HIV counselling and testing among 4.3 million people, finance for above 65,000 civil society projects, and antennal care to 3 million pregnant women. Moreover, AIDS projects were approved in Nigeria and Botswana with a budget of $225 and $50 million, respectively (UNDP, 2011). Africa is among the leading countries that are increasing access to antiretroviral therapy with 7.6 million people obtaining antiretroviral therapy across the continent in 2012 which accounted for 7.5 million people in sub-Saharan Africa. The number of people receiving treatment in eastern and southern Africa has doubled between 2006 and 2012. Western and Central African countries are increasing the provision of treatment at a slower pace (UNAIDS, 2013).

Antiretroviral therapy has proved successful in alleviating AIDS. Advantage of this therapy includes the following:
- Saving lives of about 5.5 million people in low- and middle-income countries mainly sub-Saharan Africa from the peak in 1995 to 2012.
- Decreasing the risk of HIV transmission by up to 96%.
- Declining the risk of tuberculosis infection in population having HIV by 65%.
- Increasing productivity of the human resource as people living with HIV can return to work earlier when they obtain treatment and reducing suffering among affected families.
- Helping financially and encouraging development (UNAIDS, 2013).

African governments are encouraging global efforts to assemble funds for antiretroviral therapy. Domestic support involves approximately half of all expenditures on HIV treatment and care across sub-Saharan Africa. Countries such as Angola, Botswana, and South Africa spend more than 80% of funds from public and private sectors on antiretroviral therapy. Some countries are searching innovative approaches to expand funding sources and generate renewable sources of funding for HIV programs. Examples include Cape Verde and Ivory Coast which are taxing tobacco and alcohol to create resources for their AIDS response (UNAIDS, 2013).

3 DENGUE

Africa has been reported as the second most region (16%) to be affected by dengue after Southeast or Southcentral Asia (CDC, 2015b). However, not enough updated surveillance data are available on this disease in Africa (WHO, 2009). In 2010, approximately 390 million people were affected by this infection throughout the tropics. From 1960 to 2010, 20 laboratories confirmed the outbreaks in 15 countries in Africa which were mainly from eastern Africa (Amarasinghe et al., 2011). Dengue, also termed yellow fever, is a viral disease and transmission of virus to humans occurs through the bites of infected female *Aedesaegypti* mosquitoes (WHO, 2009). Other dengue vectors present in Africa include *Aedes albopictus*, *Australopithecus africanus*, and *Aedes luteocephalus* (Were, 2012). *A. aegypti* feeds during the day and the peak biting time is in early morning and in the evening before dusk (WHO, 2009). Dengue virus (DENV) was first isolated in 1960s in Nigeria. Four DENV serotypes have been isolated in Africa. However, DENV-2 is the cause of most epidemics followed by DENV-1 (Amarasinghe et al., 2011).

Dengue affects infants, children, and adults and the fever can reach up to 40°C. The symptoms of this disease include severe headache, pain behind the eyes, muscle and joint pains, swollen glands or rash, nausea, and vomiting (Amarasinghe et al., 2011). The duration of the symptoms is about 2–7 days. According to WHO (2009), severe dengue is defined by the following conditions: plasma leakage that may cause shock and/or fluid accumulation, with or without respiratory distress and/or severe bleeding, and/or severe organ impairment (WHO, 2009).

Dengue fever has no specific treatment. In case of severe dengue, it is important to maintain the patient's body fluid volume (WHO, 2016d). The first dengue vaccine namely Dengvaxia (CYD-TDV) was registered in late 2015 and early 2016 for use in several countries among people of 9–45 years old living in endemic regions. The development of further tetravalent live-attenuated vaccines is in progress and is in phase II clinical trials whereas three other vaccine candidates (based on subunit, DNA, and purified inactivated virus platforms) are at earlier stages of clinical development (WHO, 2016d). Technical advice and guidance are provided by WHO to countries and private partners for vaccine research and evaluation. Dengue vaccine will be reviewed by the Strategic Advisory Group of Experts (SAGE) on immunization and recommendations are expected in April 2016 (WHO, 2016d).

The diagnosis of dengue is difficult due to scarcity of diagnostic infrastructure (CDC, 2015b). DENV transmission can be decreased or prevented by controlling mosquito vectors or interrupting human-vector contact (WHO, 2009). Preventive strategies in Africa emphasize on vector control although the main objective of such efforts is usually the prevention of malaria (Were, 2016). Breeding sites and vector populations are destroyed with insecticides throughout all areas. Insecticide-impregnated bed nets are provided in many regions and personal protection is available for travelers such as insect repellants and information to promote awareness on dengue disease (Were, 2012).

4 EBOLA

Ebola virus disease (EVD), previously known as Ebola hemorrhagic fever, is a severe, often lethal disease in humans and nonhuman primates (including monkeys, gorillas, and chimpanzees) (DMHS, 2014). In 2015, WHO reported 27,541 cases which included 11,235 deaths from 10 countries affected by West African epidemic of EVD that started in December 2013 (ECDC, 2015). Major causes of death are due to disruption of vascular endothelium, disseminated intravascular coagulation (DIC), and fibrinolysis and multiorgan hemorrhage (Tan et al., 2015). Ebola outbreaks have occurred in various African countries such as Democratic Republic of the Congo (DRC), Gabon, South Sudan, Ivory Coast, Uganda, Republic of the Congo (ROC), and South Africa (CDC, 2015a). This disease is caused by the genus Ebolavirus. Five subspecies of the Ebolavirus have been identified and four of them are responsible for causing the infection in humans namely Ebola virus (*Zaire ebolavirus*); Sudan virus (*Sudan ebolavirus*); Taï Forest virus (*Taï Forest ebolavirus*, formerly *Côte d'Ivoire ebolavirus*); and Bundibugyo virus (*Bundibugyo ebolavirus*). The fifth, Reston virus (*Reston ebolavirus*), has caused disease in nonhuman primates (DMHS, 2014). The infection is transmitted to humans through close contact with the blood, secretions, organs, or other bodily fluids of infected animals (Were, 2012). EVD propagates in the human population through human to human transmission by direct contact with the blood, secretions, or other infected body fluids and with surfaces of materials which have been infected with these fluids (WHO, 2016a). Ebola does not disseminate through air or by water (CDC, 2015a). No evidence is available on the transmission of this infection by mosquitoes or other insects. Some signs and symptoms of EVD are fever, severe headache, fatigue, muscular pain, weakness, diarrhea, vomiting, stomach ache, and unexplained bleeding or bruising (CDC, 2015a).

Diagnosis of this disease is conducted by antibody-capture enzyme-linked immunosorbent assay (ELISA), antigen-capture detection tests, serum neutralization test, reverse transcriptase polymerase chain reaction (RT-PCR) assay, electron microscopy, and virus isolation by cell culture (WHO, 2016a). Currently, there is no proven treatment available for this infection. Treatment of EVD includes supportive care-rehydration with oral or intravenous fluids. Nevertheless, there are various treatments that are being evaluated including blood products, immune therapies, and drug therapies (WHO, 2016a). Vaccines against ebolavirus comprising virus vectors such as adenovirus type 5, human parainfluenza virus type 3, vesicular stomatitis virus; virus-like particles with VP40 (viral matrix protein), NP (nucleoprotein) and GP (glycoprotein) as well as recombinant ebolavirus have been evaluated in animal models. However, further studies are required to assess the efficacy of these vaccines in humans (To et al., 2015).

5 CHIKUNGUNYA

Chikungunya fever is a severe febrile disease caused by an arthropodborne alphavirus, Chikungunya virus (CHIKV). Chikungunya is transmitted to humans via the bite of infected *Aedesa aegypti* and *Aedesa albopictus* mosquitoes (Nunes et al., 2015; Staples et al., 2009). CHIKV was first identified as a human pathogen during 1950s in Africa (Staples et al., 2009). The virus circulates in an enzootic cycle in Africa between forest-dwelling mosquitoes and nonhuman primates (Coffey et al., 2014). This infection has been reported in 25 countries of the African continent (CDC, 2016). The large outbreaks were recorded in 2005-06 from Réunion Island, Mauritius, Mayotte, and various Indian states with more than 250,000 suspected cases (ECDC, 2014). The signs and symptoms of this disease include fever, joint pain, headache, nausea, fatigue, and rash (WHO, 2016b). Some of the clinical signs of this disease are similar to dengue and can be misdiagnosed in regions where dengue is popular. There is no definite antiviral remedy; however treatment involves alleviating the symptoms by using antipyretics for joint pain, optimal analgesics, and fluids (WHO, 2016b). Acetaminophen or paracetamol can be used during onset of fever. Narcotics or nonsteroidal antiinflammatory drug (NSAIDS) are used in severe cases (CDC, 2014a). NSAIDS, corticosteroids, or physiotherapy can provide relief from persistent joint pain. Aspirin or other NSAIDS (such as ibuprofen, naproxen, toradol) are not advised for patients suffering from dengue fever until they have been afebrile for ≥48 h and show no signs of severe dengue (CDC, 2014a). Dissemination of CHIKV occurs in three distinct stages namely intradermal stage, blood stage, and finally to target organs such as liver, muscle, joints, and brain (Caglioti et al., 2013). Diagnosis of CHIKV can be conducted through serological tests such as ELISA, indirect immunofluorescence assays (IFA), hemoagglutination inhibition (HI), and microneutralization (MNt) which confirm the presence of IgM and IgG antichikungunya antibodies (Caglioti et al., 2013). The level of IgM antibody is high 3–5 weeks after the onset of infection and persists for about 2 months. Both serological and virological assays (RT-PCR) should be conducted for samples collected during the first week after the onset of symptoms (WHO, 2008, 2016b).

According to WHO (2008), vector control is the only public health strategy to prevent and control the outbreaks and guidelines for prevention and control of chikungunya fever involve key components including integrated vector management (IVM). The guidelines comprise different activities that should be conducted at household level as well as at institutional level such as school, universities, hospitals, and other establishments (WHO, 2008). At household level, natural and artificial water-filled container habitats of mosquitoes should be eliminated by spraying insecticides on the surfaces in and around containers. Clothes that minimize skin exposure to the day-biting vectors are recommended. Insect repellents that contain DEET (*N,N*-diethyl-3-methylbenzamide), IR3535 (3-[*N*-acetyl-*N*-butyl]-aminopropionic acid ethyl ester), or icaridin (1-piperidinecarboxylic acid, 2-(2-hydroxyethyl)-1-methylpropylester) can be applied on exposed skin (WHO, 2008, 2016b). Pyrethroid-based aerosols can be sprayed in bedrooms during peak biting time of mosquitoes that is early morning or late afternoon. Insecticide-treated mosquito nets can be used during daytime sleep especially for young children, sick, or old people. Mosquito coils, electric mat, and vaporizers can also be used to provide protection. Water in birdbaths and plant pots or drip trays should be changed regularly. Larvivorous fish such a gambusia and guppy can be introduced in ornamental water tanks and garden pools to reduce vector population (WHO, 2008). At school level, health education on all aspects of chikungunya fever should be taught among school children and they should also be asked to practice the different activities for eradication of mosquito breeding areas. At community level, people are encouraged to keep the surroundings clean and improve basic sanitation in domestic and public utility regions (WHO, 2008).

6 CHOLERA

Cholera is caused by *Vibrio cholerae*, which is a major pathogen responsible for outbreaks of life-threatening diarrheal disease worldwide, mainly in developing countries (Thompson et al., 2011). In 2014, 190,549 cases of cholera were reported to WHO which included 55% from Africa and 15% from the Americas. However, the true number of cholera cases is known to be much higher. Cholera represents an estimated burden of 1.4–4.0 million cases, and 21,000 to 143,000 deaths per year worldwide (WHO, 2016g). The different phases of cholera are asymptomatic, mild, or severe, about 1 in 20 infected people has severe illness characterized by profuse watery diarrhea, vomiting, and leg cramps. Rapid loss of body fluids can lead to dehydration, electrolyte disturbances, and hypovolemic shock. Death can occur within hours if treatment is not provided (Gaffga et al., 2007). The mode of transmission of infection is mainly by the fecal–oral route and some of the sources of infection include consumption of water that has been contaminated during storage or usage, contaminated foods, vegetables that have been fertilized with human excreta and fish especially shellfish taken from contaminated water and consumed raw or insufficiently cooked (HDHD, 2001).

In 2001, the National department of Health Directorate (NDHD) of Pretoria emphasized on prevention and control of cholera. Preventive measures described the different sources of contamination and ways to evade infection. Focus was drawn toward good sanitation which can reduce the risk of cholera and other intestinal pathogens. Adults, infants, and children should practice basic hygiene including thorough hand wash following contact with excreta. Public awareness included campaigning for sensitizing communities through intensive health education, encouraging people to use purified or boiled water for consumption, ensuring proper disposal of human waste without contaminating water sources, controlling flies, and consuming food that has been properly cooked. Street food vendors or restaurants are advised to follow good hygienic conditions through health education activities (HDHD, 2001). Treatment of cholera includes administration of oral rehydration salts. Patients with severe dehydration necessitate administration of intravenous fluids and suitable antibiotics to reduce the duration of diarrheal condition. Oral cholera vaccines namely Dukoral and Shanchol are currently being used in mass vaccination campaigns supported by WHO (2016c).

7 CRYPTOCOCCAL MENINGITIS

CM is fatal brain infection caused by soil-dwelling fungus *Cryptococcus*. About 1 million new cases of CM occur each year with 625,000 mortality worldwide (CDC, 2014b). CM is the main cause of adult meningitis in sub-Saharan Africa and 20% of AIDS-related deaths are recorded due to this disease. CM is a severe infection among people with HIV/AIDS. The increase in antiretroviral therapy programs in African centers has not reduced the incidence of CM (Muzoora et al., 2012). *Cryptococcus* can also infect the skin, lungs, or other parts of the body. Symptoms of this infection include headache, fever, neck pain, sensitivity to light, nausea and vomiting, altered mental status which range from confusion to coma and can lead to death (CDC, 2014b). This infection can be diagnosed by lumbar puncture (LP) and computed tomography (CT) scan of the brain (Arthington-skaggs et al., 2007). LP may help to reduce symptoms such as headache, altered level of consciousness, and sixth nerve palsies which are result of increased intracranial pressure (HDHD, 2001). Cryptoccocal antigen can be detected in serum or cerebrospinal fluid and this technique is rapid with sensitivity greater than 90% (Opota et al., 2014). Cost-effective strategy to prevent deaths due to CM includes "targeted screening" of HIV-infected patients for the presence of cryptococcal antigen which is a chemical marker for this disease. A patient with positive test is advised to take oral fluconazole to fight against early stage of the infection (Arthington-skaggs et al., 2007). Dipstick test is an inexpensive and rapid method to detect crytococcal antigen using small sample of serum. However, these facilities are inaccessible in most district and provincial laboratories in sub-Saharan Africa. Treatment of CM involves intravenous administration of amphotericin B for duration of 2 weeks (Muzoora et al., 2012). It has been observed that necessary medications including amphotericin B and flucytosine that have saved patients with advanced disease are lacking in various regions of sub-Saharan Africa. Many centers in Africa depend on fluconazole donation program for treatment of CM (Muzoora et al., 2012). Moreover, sertraline is considered as a safe, inexpensive, and accessible oral drug which has been found to inhibit *Crytococcus neoformans* in in vitro and in vivo assays (Jarvis and Harrison, 2016). The Centre for Disease Control and Prevention (CDC) is providing assistance to countries in sub-Saharan Africa and Southeast Asia for the screening program as well as cryptococcal treatment (CDC, 2014b).

8 MALARIA

Malarial infection is caused by parasites of the genus *Plasmodium*, with *Plasmodium falciparum* and *Plasmodium vivax* being the most prevalent that are able to infect humans (Stanisic and Good, 2015). In 2015, 214 million of malarial cases were reported. There was 37% global decrease in incidences of malaria between 2000 and 2015 while the decrease in global malaria mortality rates was 60% between the same periods (WHO, 2016l). In 2015, 90% of deaths were recorded due to malaria in sub-Saharan Africa. In 2013, malaria was responsible for 78% of deaths among children under 5 years old. The symptoms of malaria appear between 10 and 15 days after the mosquito bite and include fever, headache, chills, and vomiting (WHO, 2016k). Malaria is diagnosed using microscopy, PCR, and rapid diagnostic tests (RDTs). Microscopy technique can detect approximately 50 parasites/μL of whole blood, detection limit of PCR is 0.002 parasites/μL whereas that of RDTs is lower than 200 parasites/μL (Doctor et al., 2016). WHO recommends chloroquine against *P. vivax* malaria where the drug remains effective and artemisin-based combination therapies against uncomplicated malaria caused by *P. falciparum* (WHO, 2016j). Complicated malarial cases are treated with injectable artesunate (intramuscular or intravenous) and the treatment includes a complete course of artemisin-based combination therapy when the patient has the ability to take oral medications (WHO, 2016j). There is a long history of research into the development and implementation of different strategies which aim to both prevent and treat malarial infection. These include vector control, chemoprevention [e.g., Intermittent Preventive Treatment (IPT) for vulnerable populations such as pregnant women and infants], and the prompt diagnosis and treatment of confirmed cases with appropriate antimalaria drugs (Stanisic and Good, 2015). Control of vector using insecticide-treated bed nets and indoor residual spraying has proved to be efficient means to reduce malarial infection. Spraying of houses and structures with insecticide can effectively kill mosquitoes after they have fed. This process should be carried out for a period greater than 6 months. These techniques have eliminated or decreased malaria epidemic considerably in many countries worldwide between 1940s and 1960s. Consequently, DDT was used in public-health and malaria control campaigns which were conducted between 1955 and 1965. Spraying with multiple insecticides has effectively reduced resistance to DDT and was found useful in the control of *Anopheles funestus*, *Anopheles gambiae*, and *Anopheles melas* in Equatorial Guinea (Church and Smith, 2016). Significant progress is being made in developing a malaria vaccine which is the most effective long-term strategy for preventing malaria (Kpanake et al., 2016).

9 SCHISTOSOMIASIS

Schistosomiasis, a common NTD also termed as bilharziasis or bilharzia (Gray et al., 2011), is ranked second among the most prevalent parasitic disease in various countries of sub-Saharan Africa (Adenowo et al., 2015). The global estimation of schistosomiasis is 207 million and sub-Saharan Africa accounts for 93% (192 million). The top five countries in Africa with high prevalence of schistosomiasis are Nigeria (29 million) followed by United Republic of Tanzania (19 million), Ghana and Democratic Republic of Congo (15 million), and Mozambique (13 million) (Adenowo et al., 2015). According to WHO, more than 61.6 million people were treated for schistosomiasis in 2014 and 258 million people necessitated preventive treatment for this illness in the same year (WHO, 2016h). The different factors responsible for this disease in sub-Saharan Africa include climate change and global warming, proximity to water bodies, irrigation, and dam construction whereas socioeconomic factors involve occupational activities and poverty (Adenowo et al., 2015). *Schistosoma haematobium*, *Schistosoma masoni*, and *Schistosoma intercalatum* are the main causes of infections in sub-Saharan Africa (Gray et al., 2011). This disease develops into three distinct phases namely acute, chronic, and advanced stage. Symptoms of acute stage include fever, myalgia, fatigue, malaise, nonproductive cough, and diarrhea and right upper quadrant pain. Chronic and advanced phases are associated with gastrointestinal and liver disease as well as genitourinary disease. The most severe complication of this infection is neuroshistosomiasis and the signs and symptoms are intracranial pressure, myelopathy, and radiculopathy. Schistomiasis can be diagnosed through laboratory investigations which include stool/urine examination for the presence of schistosome eggs, full blood count, coagulation profile, urea, electrolytes and liver function, serology, rectal or bladder biopsy and radiography such as chest radiograph, abdominal ultrasound, and pelvic ultrasound (Gray et al., 2011).

Praziquantel is an oral anthelmintic drug used for the treatment of schistomiasis for adults and children (aged 4 and above) at dosage of 40 or 60 mg/kg (Stothard et al., 2013). In 2010, 34.8 million people from 30 countries were treated with praziquantel and the drug was accessible through yearly donation of 250 million tablets (Adenowo et al., 2015). However, praziquantel can be rendered ineffective due to drug resistance. The drug showed low cure rates of 18–36% in the *S. mansoni* outbreak in northern Senegal. Oxamniquine, an aminoethyltetrahydroquinolone derivative is efficient only against *S. mansoni* mainly during invasive stages and adult worms; male worms are more sensitive to the drug compared to female worms (Inobaya et al., 2014). Schistomiasis control program includes implementation of mass drug administration in many countries such as Uganda, Sierra Leone, Burkina Faso, Brazil, Mali, Niger, China, and Philippines (Inobaya et al., 2014). This disease can be controlled by implementing key strategies such as provision of potable water, adequate sanitation, hygiene education, and snail control in various countries of sub-Saharan Africa (Jember, 2014).

10 ANTIMICROBIAL RESISTANCE

Resistant pathogens continue to pose increasing therapeutic challenges (Bush, 2014) and the key area of policy concern is the growth of resistance in bacteria to antibiotics (Fowler et al., 2014). The effective life span of an antimicrobial agent lies in the gap between the development of the new drug and the establishment of widespread resistance. According to Fowler et al. (2014), the advancement of resistance has prompted the market to produce novel antibiotics. However, there have been no new classes of antibiotics discovered since the late 1980s while antiviral and antifungal agents are in development (Fowler et al., 2014). The development and proliferation of methicillin-resistant *S. aureus* (MRSA) emanated in the 1960s persists as a major clinical and epidemiological threat in hospital environments (Martinez et al., 2014). Currently, even the third generation cephalosporins have proved ineffective to ward off infections such as gonorrhea which consequently leads in increased rates of illness and complications such as infertility, adverse pregnancy outcomes, and neonatal blindness (WHO, 2016). Urinary tract infections caused by *E. coli* have shown resistance against fluoroquinolones. Globally, 6% of new tuberculosis (TB) cases and 20% of previously treated TB cases are estimated to have multidrug-resistant (MDR)-TB, with substantial differences in the frequency of MDR-TB among countries. In 2012, there were approximately 450,000 new cases of MDR-TB in the world. Extensively drug-resistant TB (XDR-TB, defined as MDR-TB plus resistance to any fluoroquinolone and any second-line injectable drug) has been identified in 92 countries, in all regions of the world. Resistance is an emerging concern for treatment of HIV infection, after the rapid expansion in access to antiretroviral drugs in recent years (WHO, 2016i). European data state that the mortality rate for septicemia, blood stream infections caused by MDR *E. coli* is twice (30%) compared to susceptible *E. coli* (15%) (Fowler et al., 2014). The relatively rapid increase in carbapenem-resistant Enterobacteriaceae (CRE) has challenged the infectious disease community (Bush, 2014). In the United States, carbapenem resistance has been indicated for 4% of *E. coli* and more than 10% of *Klebsiella pneumoniae* isolates associated with some device-related infections. The resistance can be transferred by mobile plasmids between different strains or even species and genera of bacteria (Fowler et al., 2014). CDC estimated that 2 million infections arise each year due to resistant bacteria with a death of 23,000 people and similar rates of disease burden have been stated for Europe (Rex et al., 2014). The pharmaceutical arsenal available to control antibiotic-resistant bacteria is limited (Martinez et al., 2014). The emergence of multiple drug-resistant strains in human pathogenic organisms has further necessitated the search for novel antimicrobial agents from natural sources (Pandey and Kumar, 2013).

11 MANAGEMENT OF INFECTIOUS DISEASES USING AFRICAN BIODIVERSITY

World Health Organization (WHO) advocates that 70–95% of the global population in developing countries depend on plants for their primary health care (Mabona and Vuuren, 2013). Interestingly, 11% from the 252 drugs considered as basic and essential by the WHO are solely of plant origin and a significant amount are synthetic drugs obtained from natural precursors (Rates, 2001). Examples include arecolin, antithelmintic drug from *Areca catechu*, quinine and quinidine, antimalarial drugs from *Cinchona* spp., hemsleyadin, bacillary dysentery drug from *Helmsleya amabilis* Diels. African traditional medicine is the oldest medicinal system which is often culturally termed as the Cradle of Mankind (Vuuren, 2008).

Traditional herbal medicines are used in Africa against various infectious diseases (Table 5.1) including HIV/AIDS and HIV-related problems such as dermatological disorders, nausea, depression, insomnia, and weakness (Mills et al., 2005). *Hypoxis hemerocallidea* (common name: African potato-Hypoxidaceae) and *Sutherlandia frutescens* (Fabaceae) are two major African herbal compounds used for the treatment of HIV/AIDS in sub-Saharan Africa. The South African Ministry of Health supported the use of these two herbal medicines for HIV management. The different members of the South

Table 5.1 Ethnomedicinal uses of medicinal plants against infectious diseases in Africa

Plant species	Family	Country	Part used	Disease/condition	Preparation	References
Acalypha hispida Burm. f.	Euphorbiacea	NI	L, F	Leprosy, gonorrhea	NI	Onocha et al. (2010)
Acalypha integrifolia Willd. subsp. *Integrifolia* var. *integrifolia*	Euphorbiaceae	Mauritius	L	Intestinal worms, skin infections	Leaf decoction is consumed	Gurib-Fakim and Gueho (1996)
Acalypha wilkesiana Müll. Arg.	Euphorbiaceae	Mauritius	L	Dysentery	Leaf decoction in combination with that of *Psidium cattleianum* is drunk	Gurib-Fakim and Gueho (1996)
Alchornea cordifolia (Schumach.) Muell.	Euphorbiaceae	Democratic Republic of Congo	L	Intestinal parasites, diarrhoea, cough, bronchitis, dysentery	NA	Memvanga et al. (2015)
Allium sativum L.	Alliceae	Southern Africa	T	Tuberculosis	NA	Green et al. (2010)
Azadirachta indica A. Juss.	Meliaceae	Mauritius	L, B	Scabies, malarial fever	Leaves are applied on infected skin. Decoction of bark is used against malarial fever	Gurib-Fakim (2007)
Cardiospermum halicacabum L.	Sapindaceae	Mauritius	L	Scabies, eczema	Leaf decoction of CH in combination with the leaves of *Paederia foetida* and *Calendula officinalis* is used to wash infected body parts	Gurib-Fakim (2007)
Cassia occidentalis L.	Caesalpiniaceae	Democratic Republic of Congo	L	Gonorrhea, fever, stomach pain, anemia	NA	Memvanga et al. (2015)

Clerodendrum rotundifolium Oliv.	Lamiaceae	Uganda	L	Malaria, deworming, and stomach pain, against intestinal parasites	NA	Adia et al., 2016
Cissampelos mucronata A.Rich.	Menispermaceae	Democratic Republic of Congo	R	Conjunctivitis, cough, sexually transmitted diseases, fever	NA	Memvanga et al. (2015)
Ficus carica L.	Moraceae	Namibia	L, B	Malaria, gonorrhoe, diarrhoea	Leaf maceration and bark decoction is consumed	Chinsembu and Hedimbi (2010)
Harungana madagascariensis Lam. ex Poir.	Guttiferaces	Gabon	B, L	Hepatitis	NA	Tchouya et al. (2015)
Kigelia africana (Lam.) Benth.	Bignoniaceae	Namibia	S, L	Eczema, psoriasis, leprosy, herpes simplex	Decoction of crushed stem bark and leaves is used to wash or rub onto infected skin	Chinsembu and Hedimbi (2010)
Lippia javanica (Burm.f.) Spreng	Verbaneceae	Southern Africa	L	Scabies and lice	Leaf infusion is used	Green et al. (2010)
			L, S	Coughs, fever and bronchitis	Infusion of leaf and stem is taken as tea	Green et al. (2010)
Maerua edulis (Gilg&Bened.) De Wolf.	Capparidaceae	Tanzania	R	Venereal diseases such as gonorrhea and syphillis	Peeled roots infusion is consumed	Maregesi et al. (2007)
Mangifera indica L.	Anacardiaceae	Mauritius	BR	Skin infection, acne	Resin of bark is applied on skin infections and acne	Gurib-Fakim and Gueho (1996)

(*Continued*)

Table 5.1 Ethnomedicinal uses of medicinal plants against infectious diseases in Africa (cont.)

Plant species	Family	Country	Part used	Disease/condition	Preparation	References
Microglossa pyrifolia (Lam.) O. Ktze	Asteraceae	Uganda	L	Malaria, abdominal disorders, cough, chest pain, syphilis, and skin allergy	NA	Adia et al., 2016
Momordica charantia L.	Cucurbitaceae	Mauritius	NA	Cholera	NA	Gurib-Fakim and Gueho (1996)
Momordica foetida Schumach	Cucurbitaceae	Uganda	L	Malaria, flu, and against worms	NA	Adia et al., 2016
Morinda citrifolia L.	Rubiaceae	Mauritius	L	Diarrhoea	NA	Gurib-Fakim and Gueho (1996)
Musanga cecropioides R.Br. ex Tedlie	Moraceae	Gabon	B, L	Tuberculosis	NA	Tchouya et al. (2015)
Ozoroa reticulate (Bak.f.) R.A.	Anacardiaceae	Tanzania	R or SB	Cholera, dysentery	Decoction is taken	Maregesi et al. (2007)
Persea americana Mill.	Lauraceae	Gabon	B, L	Venereal	NA	Tchouya et al. (2015)
Zanthoxylum chalybeum Engl.	Rutaceae	Uganda	SB	Malaria	NA	Adia et al., 2016
Ziziphus mucronata Willd.	Rhamnaceae	Namibia	B	Gonorrhea, chlamydia, diarrhea, dysentery	Crushed bark is boiled in water and the decoction is consumed	Chinsembu and Hedimbi (2010)

CH, *Cardiospermum halicacabum*; NA, not available; S, stem, B, bark; L, leaves, R, root, SB, stem bark, T, tuber.

African Development Community (SADC) also encouraged their use. *H. hemerocallidea* is used as an immunostimulant for patients with HIV/AIDS. Main chemical constituent of the plant includes hypoxoside, a nonlignan glycoside. Sterols (β-sitosterol, stigmasterol) and their glycosides as well as stanols namely stigmastanol are also present in the plant (Mills et al., 2005). *S. frutescens* is traditionally used against fever, wounds, cancer, diabetes, kidney and liver problem, rheumatism, and stomach ailments. This species also alleviate various symptoms and conditions such as depression and stress, skin and inflammatory diseases, influenza, hemorrhoids, urinary tract infections, back pain, and gonorrhea (Aboyade et al., 2014). The plant contains triterpenoids, saponins, γ-aminobutyric acid (GABA), pinitol, and 1-canavanine (2-amino-4-guanidinooxybutyric acid) which have been patented for the treatment of various side effects of HIV/AIDS. 1-canavanine is a structural analog of 1-arginine and has been reported to possess anticancer activity and antiviral activity against influenza and retroviruses (Aboyade et al., 2014). *S. frutescens* is available in different dosage forms including tablets, capsules, and gel for topical application, creams, liquid extracts, and ointments. *Sutherlandia* tablets have been reported to be effective in the treatment of muscle wasting in patients with HIV/AIDS (Aboyade et al., 2014). Clinicians in South Africa and Australia reported that *Sutherlandia* improved CD4 counts and reduced virus load in AIDS patients (Sutherlandia, 2016).

Artemisinin derivatives, artesunate and artemether isolated from *Artemisia annua* (Compositae) are well-known drugs against malaria. Artemisinin derivatives are being considered for the treatment of schistomiasis (Inobaya et al., 2014). However, in 1980s these drugs were discovered to cure schistomiasis caused by *S. japonicum*. Artemisinin is effective against younger stages of schistosome but less potent against adult stage. In vivo study using mice showed that artemether was more effective against female worms than male worms. Artemisinin-based combination therapies such as artemether with lumefantine showed 100% cure rate against *P. falciparum* and *S. mansoni* infections (Inobaya et al., 2014). The roots of *Cryptolepis sanguinolenta* are used against malaria in Ghana and are also available as phytopharmaceutical product, Phytolaria (Wright, 2010). Cryptolepine, an indoloquinoline alkaloid, is the main component of *C. sanguinolenta* and inhibited chloroquinine-sensitive and chloroquinine-resistant *P. falciparum* but showed toxicity in mice at 20 mg/kg. Synthetic halogenated derivatives suppressed parasitaemia (90%) in mice infected by *Plasmodium berghei* and did not show any toxic effects (Wright, 2010). Spermine alkaloids isolated from *Albizia gummifera*, species of Kenya revealed strong antiplasmodial activities against chloroquine-sensitive *P. falciparum* strain NF54 (IC_{50}: 0.18–0.24 μM) and chloroquinine-resistant strain ENT30 (IC_{50}: 1.43–1.79 μM). Tazopsine and sinococuline, morphinan alkaloids obtained from Malagasy *Strychnopsisthouarsii* species prevented malaria by targeting early liver stages of *Plasmodium*. Both alkaloids inhibited *Plasmodium yoelii* liver stage parasites in vitro with IC_{50}: 3.1 and 4.5 μM, respectively (Wright, 2010). Xanthones namely demthycalabaxanthone, calothwaitesixanthone, and 6-deoxy-gamma-mangostin from *Calophyllum caledonicum* and *Garcinia veillardii* (Clusiaceae) species have been found active against chloroquine-resistant strains of *P. falciparum* (IC$_{50}$: 1.0 μg/mL) (Gurib-Fakim, 2006).

Annickia affinis and *Anickia chlorantha* (Annonaceae) species from West and Central Africa are used as traditional medicines for the treatment of malaria, tuberculosis, hepatitis, HIV/AIDS, typhoid fever, yellow fever, syphilis, and other infectious diseases (Olivier et al., 2015). Seed extract of *Garcinia kola* (Clusiaceae or Guttiferae) inhibited Ebola virus in cell culture at nontoxic concentrations. The extract was also found active against various viruses including influenza virus (Muanya, 2016).

12 CONCLUSIONS

Infectious diseases such as HIV/AIDS, tuberculosis, malaria, and dengue remain the leading causes of mortality and morbidity worldwide. The chapter has summarized the management of prevalent infectious diseases including HIV/AIDS, dengue, ebola, chikungunya, cholera, CM, malaria, and schistomiasis in African region. Prevention and control are among the key strategies for managing incurable and treatable diseases and priority approach involves vector control, improves sanitation, health education, mass screening programs as well as accessibility of treatment to low-income countries in Africa. The African continent has a rich and diverse flora and many native, indigenous and endemic plant species have been used in folk medicine for the treatment of various infectious diseases. However, clinical evaluation of the different species should be conducted for the safety approval of their therapeutic uses.

REFERENCES

Abdu, M., Umar, A., Faisal, B.H., Tajuddin, S.H., Suria, B.I., Yakasai, M.G., 2016. Effectiveness of HIV/AIDS educational intervention in increasing knowledge, attitude and practices for primary school teachers in some part of Africa. HIV AIDS Rev. 15, 17–25.

Aboyade, O.M., Styger, G., Gibson, D., Hughes, G., 2014. *Sutherlandia frutescens*: the meeting of science and traditional knowledge. J. Altern. Complement. Med. 20 (2), 71–76.

Adenowo, A.F., Oyinloye, B.E., Ogunyinka, B.I., Kappo, A.P., 2015. Impact of human schistosomiasis in sub-Saharan Africa. Braz. J. Infect. Dis. 19 (2), 196–205.

Adia, M.M., Emami, S.N., Byamukama, R., Faye, I., Borg-Karlson, A.-K., 2016. Antiplasmodial activity and phytochemical analysis of extracts from selected Ugandan medicinal plants. J. Ethnopharmacol. 186, 14–19.

Amarasinghe, A., Kuritsky, J.N., Letson, G.W., Margolis, H.S., 2011. Dengue virus infection in Africa. Emerg. Infect. Dis. 17 (8), 1349–1354.

Arthington-skaggs, B., Bicanic, T., Cotton, M., Chiller, T., Govender, N., Harrison, T., Karstaedt, A., Maartens, G., Varavia, E., Venter, F., Vismer, H., 2007. Guidelines for the prevention, diagnosis and management of cryptococcal meningitis and disseminated cryptococcosis in HIV-infected patients. S. Afr. J. HIV Med. 8, 25–35.

Bush, K., 2014. Introduction to antimicrobial therapeutics reviews: infectious diseases of current and emerging concern. Ann. NY Acad. Sci. 1323, 5–6.

Caglioti, C., Lalle, E., Castilleti, C., Carletti, F., Bordi, M.R., 2013. Chikungunya virus infection: an overview. New Microbiol. 36, 211–227.

Centers for Disease Control and Prevention (CDC), 2014a. Chikungunya for healthcare providers. Available from http://www.cdc.gov/chikungunya/pdfs/CHIKV_Clinicians.pdf

Centers for Disease Control and Prevention (CDC), 2014b. Cryptococcal meningitis. Available from http://www.cdc.gov/fungal/global/cryptococcal-meningitis.html

Centers for Disease Control and Prevention (CDC), 2015a. Available from http://www.cdc.gov/vhf/ebola/pdf/ebola-factsheet.pdf

Centers for Disease Control and Prevention (CDC), 2015b. Unveiling the burden of Dengue in Africa. http://blogs.cdc.gov/publichealthmatters/2015/07/unveiling-the-burden-of-dengue-in-africa/

Centers for Disease Control and Prevention (CDC), 2016. Chikungunya virus. Available from http://www.cdc.gov/chikungunya/geo/

Chinsembu, K.C., Hedimbi, M., 2010. An ethnobotanical survey of plants used to manage HIV/AIDS opportunistic infections in Katima Mulilo, Caprivi region, Namibia. J. Ethnobiol. Ethnomed. 6, 25.

Church, K.E.M., Smith, R.J., 2016. Comparing malaria surveillance with periodic spraying in the presence of insecticide-resistant mosquitoes: should we spray regularly or based on human infections? Math. Biosci. 276, 145–163.

Coffey, L.L., Failloux, A.-B., Weaver, S.C., 2014. Chikungunya virus-vector interactions. Viruses 6, 4628–4663.

Center for Strategic & International Studies (CSIS), 2016. Infectious diseases a persistent threat. Available from http://www.smartglobalhealth.org/issues/entry/infectious-diseases

Directorate Medical & Health Service (DMHS), 2014. Clinical case management guidelines of ebola virus disease (EVD). Available from http://www.vbch.dnh.nic.in/pdf/ebola/ClinicalCaseManagement-Guidelines.pdf

Doctor, S.M., Liu, Y., Whitesell, A., Thwai, K.L., Taylor, S.M., Janko, M., Emch, M., Kashamuka, M., Muwonga, J., Tshefu, A., Meshnick, S.R., 2016. Diag. Microbiol. Infect. Dis. 85, 16–18.

European Centre for Disease Prevention and Control (ECDC), 2014. Chikungunya outbreak in Caribbean region. Available from http://ecdc.europa.eu/en/publications/Publications/chikungunya-caribbean-june-2014-risk-assessment.pdf

European Centre for Disease Prevention and Control (ECDC), 2015. Available from http://ecdc.europa.eu/en/publications/Publications/Ebola-west-africa-12th-update.pdf

Falarin, O.A., Happi, A.I., Happi, C.T., 2014. Empowering African genomics for infectious disease control. Genome Biol. 15 (11), 515.

Fowler, T., Walker, D., Davies, S.C., 2014. The risk/benefit of predicting a post-antibiotic era: is the alarm working? Ann. NY Acad. Sci. 1323 (1), 1–10.

Gaffga, N.H., Tauxe, R.V., Mintz, E.D., 2007. Cholera: a new homeland in Africa? Am. J. Trop. Med. Hyg. 77 (4), 705–713.

Gray, J.D., Ross, A.G., Li, Y.-S., McManus, D., 2011. Diagnosis and management of schistosomiasis. Br. Med. J. 342, d2651, Available from http://www.ncbi.nlm.nih.gov/pubmed/21586478.

Green, E., Samie, A., Obi, C.L., Bessong, P.O., Ndip, R.N., 2010. Inhibitory properties of selected South African medicinal plants against *Mycobacterium tuberculosis*. J. Ethnopharmacol. 130, 151–157.

Gurib-Fakim, A., 2006. Medicinal plants: traditions of yesterday and drugs for tomorrow. Mol. Aspect Med. 27, 1–93.

Gurib-Fakim, A., 2007. Plantes Médicinal de Maurice et d'ailleurs. Caractère Ltée, Mauritius, 63 and 205.

Gurib-Fakim, A., and Gueho, J., 1996. Plantes Médicinal de Maurice, TOME 2. Edition de L'Ocean Indien, Mauritius, 71–77.

Inobaya, M.T., Olveda, R.M., Chau, T.N., Olveda, D.U., Allen, G.P., 2014. Prevention and control of schistosomiasis: a current perspective. J. Res. Rep. Trop. Med. 5, 65–75.

Jarvis, J.N., Harrison, T.S., 2016. Forgotten but not gone: HIV-associated cryptoccocal meningitis. Lancet, Available from http://dx.doi.org/10.1016/S1473-3099(16)00128-6.

Jember, T.H., 2014. Challenges of schistomiasis prevention and control in Ethiopia: literature review and current status. J. Parasitol. Vector Biol. 6 (6), 80–86.

Joint United Nations Programme on HIV/AIDS (UNAIDS), 2013. Access to antiretroviral therapy in Africa. Malaria. Available from http://www.unaids.org/sites/default/files/media_asset/20131219_AccessARTAfricaStatusReportProgresstowards2015Targets_en_0.pdf.

Kpanake, L., Sorum, P.C., Mullet, E., 2016. The potential acceptability of infant vaccination against malaria: a mapping of parental positions in Togo. Vaccine 34, 408–412.

Laino, C., 1999. Africa, the infectious continent. Available from http://www.nbcnews.com/id/3072106/ns/us_news-only/t/africa-infectious-continent/

Mabona, U., Vuuren, S.F., 2013. Southern African medicinal plants used to treat skin diseases. S. Afr. J. Bot. 87, 175–193.

Maregesi, S.M., Ngassapa, O.D., Peters, L., Vlietinck, A.J., 2007. Ethnopharmacological survey of the Bunda district, Tanzania: plants used to treat infectious diseases. J. Ethnopharmacol. 113, 457–470.

Martinez, M.A., Mattana, C.M., Satorres, S.E., Sosa, A., Fusco, M.R., Laciar, A.L., Alcaraz, L.E., 2014. Screening phytochemical and antibacterial activity of three San Luis native species belonging at the Fabaceae family. Pharmacol. Online Arch. 3, 1–6.

Memvanga, P.B., Tona, G.L., Mesia, G.K., Lusakibanza, M.M., Cimanga, R.K., 2015. Antimicrobial activity of medicinal plants from the Democratic Republic of Congo: a review. J. Ethnopharmacol. 169, 76–98.

Mills, E., Cooper, C., Seely, D., Kanfer, I., 2005. African herbal medicines in the treatment of HIV: *Hypoxis* and *Sutherlandia*. An overview of evidence and pharmacology. Nutr. J. 4, 19.

Muanya, C., 2016. Nigeria: panel probes herbal cure claims for ebola virus, others. Available from http://allafrica.com/stories/201408071329.html

Muzoora, C.K., Kabanda, T., Ortu, G., Sentamu, J., Hearn, P., Mwesigye, J., Longley, N., Jarvis, J.N., Jaffar, S., Harrison, T.S., 2012. Short course amphotericin B with high dose fluconazole for HIV-associated cryptococcal meningitis. J. Infect. 64, 76–81.

National Department of Health Directorate (NDHD), 2001. Guidelines for cholera control. Available from http://www0.sun.ac.za/ruralhealth/ukwandahome/rudasaresources2009/DOH/Guidelines for Cholera Control.pdf

Nunes, M.R., Faria, N.R., Mota de Vasconcelos, J., Golding, N., Kraemer, M.U., Freitas de Oliveira, L., Azevedo, R., Andrade da Silva, D.E., Pinto da Silva, E.V., Patroca da Silva, S., Carvalho, V.L., Coelho, G.E., Cruz, A.C., Rodrigues, S.G., Vianez, J.L., Nunes, B.T., Cardoso, J.F., Tesh, R.B., Hay, S.I., Pybus, O.G., Vasconcelos, P.F., 2015. Emergence and potential for spread of Chikungunya virus in Brazil. BMC Med. 13, 102.

Olivier, D.K., Vuuren, S.F., Moteetee, A.N., 2015. *Annickia affinis* and *A. chlorantha* (Enantiachlorantha)—A review of two closely related medicinal plants from tropical Africa. J. Ethnopharmacol. 176, 438–462.

Onocha, P.A., Oloyede, G.K., Afolabi, Q.O., 2010. Phytochemical investigation, cytotoxicity and free radical scavenging activities of non-polar fractions of *Acalypha hispida* (leaves and twigs). Exp. Clin. Sci. Int. Online J. Adv. Sci. 10, 1–8.

Opota, O., Desgraz, B., Kenfak, A., Jaton, K., Cavassini, M., Greub, G., Prod'hom, G., Giulieri, S., 2014. *Cryptoccocus neoformans* meningitis with negative cryptococcal antigen: evaluation of a new immuno chromatographic detection assay. New Microbes New Infect. Dis. 4, 1–4.

Pandey, A.K., Kumar, S., 2013. Perspective on plant products as antimicrobials agents: a review. Pharmacologia 4 (7), 469–480.

Rates, S.M.K., 2001. Plants as source of drugs. Toxicon 39 (5), 603–613.

Rex, J.H., Goldberger, M., Barry, I., Eisenstein, I., Harney, C., 2014. The evolution of the regulatory framework for antibacterial agents. Ann. NY Acad. Sci. 1323, 11–21.

Rweyemamu, M., Otim-Nape, W., Serwadda, D., 2006. Foresight. Infectious diseases: preparing for the future. Africa Office of Science and Innovation, London. Available online.

Stanisic, D., Good, M.F., 2015. Whole organism blood stage vaccines against malaria. Vaccine 33, 7469–7475.

Staples, J.E., Breiman, R.F., Powers, A.M., 2009. Chikungunya fever: an epidemiological review of a re-emerging infectious disease. Clin. Infect. Dis. 49, 942–948.

Stothard, J.R., Sousa-Feigueiredo, J.C., Betson, M., Bustinduy, A., Reinhard-Rupp, J., 2013. Schistomiasis in Africa infants and preschool children: let them now be treated! Trends Parasitol. 29 (4), 197–205.

Sutherlandia, 2016. Sutherlandia and HIV/AIDS. Available from http://www.sutherlandia.org/aids.html

Tan, D.-X., Korkmaz, A., Reiter, R.J., Manchester, L.C., 2015. Ebola virus disease: potential use of melatonin as a treatment. J. Pineal Res. 57, 381–384.

Tchouya, G.R., Souza, A., Tchouankeu, J.C., Yala, J.-F., Boukandou, M., Foundikou, H., Obiang, G.D., Boyom, F.F., Mabika, R.M., Memkem, E.Z., Ndinteh, D.T., Lebibi, J., 2015. Ethnopharmacological surveys and pharmacological studies of plants used in traditional medicine in the treatment of HIV/AIDS opportunistic diseases in Gabon. J. Ethnopharmacol. 163, 306–316.

Thompson, C.C., Freitas, F.S., Marin, M.A., Fonseca, E.L., Okeke, I.N., Vicente, A.C., 2011. *Vibrio cholera* 01 lineages driving cholera outbreaks during seventh cholera pandemic in Ghana. Infect. Genet. Evol. 11, 1951–1956.

To, K.K., Chan, J.F., Tsang, A.K., Cheng, V.C., Yuen, K.-Y., 2015. Ebola virus disease: a highly fatal infectious disease re-emerging in West Africa. Microbes Infect. 17, 84–97.

United Nations Development Programme (UNDP), 2011. HIV/AIDS Regional Update—Africa. Available from http://web.worldbank.org/WBSITE/EXTERNAL/COUNTRIES/AFRICAEXT/EXTAFRHEANUTPOP/EXTAFRREGTOPHIVAIDS/0,contentMDK:20415756~menuPK:1830800~pagePK:34004173~piPK:34003707~theSitePK:717148,00.html

Vuuren, S.F., 2008. Antimicrobial activity of South African medicinal plants. J. Ethnopharmacol. 119, 462–472.

Were, F., 2012. The dengue situation in Africa. Paediatr. Int. Child Health 32 (S1), 18–21.

World Health Organization (WHO), 2008. Guidelines on clinical management of Chikungunya fever. Available fromhttp://www.wpro.who.int/mvp/topics/ntd/Clinical_Mgnt_Chikungunya_WHO_SEARO.pdf

World Health Organization (WHO), 2009. Dengue guidelines for diagnosis, treatment, prevention and control. Available from http://www.who.int/tdr/publications/documents/dengue-diagnosis.pdf

World Health Organization (WHO), 2015a. HIV/AIDS. Available from http://www.who.int/mediacentre/factsheets/fs360/en/

World Health Organization (WHO), 2015b. Neglected tropical diseases. Available from http://www.afro.who.int/en/neglected-tropical-diseases.html

World Health Organization (WHO), 2016a. Ebola virus disease. Available from http://www.who.int/mediacentre/factsheets/fs103/en/

World Health Organization (WHO), 2016b. Chikungunya. Available from http://www.who.int/mediacentre/factsheets/fs327/en/

World Health Organization (WHO), 2016c. Cholera. Available from http://www.who.int/mediacentre/factsheets/fs107/en/

World Health Organization (WHO), 2016d. Dengue and severe dengue. Available from http://www.who.int/mediacentre/factsheets/fs117/en/

World Health Organization (WHO), 2016e. Infectious diseases. Available from http://www.who.int/topics/infectious_diseases/en/

World Health Organization (WHO), 2016f. Neglected tropical diseases. Available from http://www.who.int/neglected_diseases/diseases/en/

World Health Organization (WHO), 2016g. Number of reported cholera cases. Available from http://www.who.int/gho/epidemic_diseases/cholera/cases_text/en/

World Health Organization (WHO), 2016h. Schitosomiasis. Available from http://www.who.int/mediacentre/factsheets/fs115/en/

World Health Organization (WHO), 2016i. WHO's first global report on antibiotic resistance reveals serious, worldwide threat to public health. Available from http://www.who.int/mediacentre/news/releases/2014/amr-report/en/

World Health Organization (WHO), 2016j. Overview of malaria. Available from http://www.who.int/malaria/areas/treatment/overview/en/

World Health Organization (WHO), 2016k. Malaria. Available from http://www.who.int/topics/malaria/en/

World Health Organization (WHO), 2016l. Malaria. Available from http://www.who.int/mediacentre/factsheets/fs094/en/

Wright, C.W., 2010. Recent developments in research on terrestrial plants used for the treatment of malaria. Nat. Prod. Rep. 27, 961–968.

CHAPTER 6

Overview of Governmental Support Across Africa Toward the Development and Growth of Herbal Medicine

D.O. Ochwang'i, J.A. Oduma

1 INTRODUCTION

Africa, with a population of 1.1 billion by 2013 (Gudmastad, 2013) (now it's 1.18 billion), accounts for about 15% of the world's population. Being a young and developing continent with the fastest growing population, poses a serious challenge to her health care; a basic need of her people. There are two systems of health care in Africa: one is traditional and prescientific; the other modern, scientific and Western in derivation. The two exist side by side, yet remain functionally unrelated in any intentional sense: with the traditional, ethnic system being disregarded by the government-supported modern system, although it is the dominant mode of health care for over three quarters of the population of the developing world. The structure of this dualism with particular reference to Africa should be more widely recognized and used in national health care planning. Many governments do themselves a grave disservice by officially ignoring traditional medicine and not considering its partial or full incorporation in health care planning. Traditional medicine (TM), variously known as ethnomedicine, folk medicine, native healing, or complementary and alternative medicine (CAM), is the oldest form of health care system that has stood the test of time. It is an ancient and culture-bound method of healing that humans have used to cope and deal with various diseases that have threatened their existence and survival. Hence, TM is broad and diverse (Abdullahi, 2011). According to World Health Organization, TM is "the sum total of the knowledge, skills and practices based on the theories, beliefs and experiences indigenous to different cultures, whether explicable or not, used in the maintenance of health, as well as in the prevention, diagnosis, improvement or treatment of physical and mental illnesses" (WHO, 2000b:1). Traditional healer, on the other hand, is "a person who is recognized by the community where he or she lives as someone competent to provide health care by using plant, animal and mineral substances and other methods based on social, cultural and religious practices" (WHO, 2000a:11). The African Regional Director of WHO has outlined a few guidelines on the responsibilities of all African nations for the realistic development of TAM, in order to sustain our health agenda and perpetuate our culture (Elujoba et al., 2005).

Africa is a rich source of medicinal plants. Perhaps, the best known species is *Phytolacca dodecandra*. Extracts of the plant, commonly known as *endod*, are used as an effective molluscicide to control schistosomiasis (Lemma, 1991). Other notable examples are *Catharanthus roseus*, which yields antitumor agents such as vinblastine and vincristine; and *Ricinus communis*, which yields the laxative—castor oil. In Botswana, Lesotho, Namibia, and South Africa, *Harpagophytum procumbens* is produced as a crude drug for export. Similarly, *Hibiscus sabdariffa* is exported from Sudan and Egypt. Other exports are *Pausinystalia yohimbe* from Cameroon, Nigeria, and Rwanda, which yields *yohimbine*; and *Rauvolfia vomitoria*, from Madagascar, Mozambique, and Zaire, which is exploited to yield reserpine and ajmaline.

The World Bank in its document *Medicinal Plants; A Growing Role in Development* (Srivastava et al., 1995) suggests that any national strategy for medicinal plant development needs to consider the following pointers: Is the use of medicinal plants encouraged in health care programs? Are there policies for conserving medicinal plants and incentives to encourage local community participation? Is there a policy for restoring plants harvested in the wild? Are there incentives for collectors and farmers to keep the production of medicinal plants sustainable? Does the government support research into these plants? What are the policies regarding the export of medicinal plants? Are only raw materials exported? And is "in-country" processing (which may further help the trade in medicinal plants) being promoted?

2 ENCOURAGEMENT OF USE OF MEDICINAL PLANTS IN HEALTH CARE PROGRAMS

Prior to the introduction of the cosmopolitan medicine, TM used to be the dominant medical system available to millions of people in Africa in both rural and urban communities. Indeed, it was the only source of medical care for a greater proportion of the population (Romero-Daza, 2002). There are strong indications that traditional health care systems are still in use by the majority of the people not only in Africa but across the world. In Africa, the healers are variously addressed as *Babalawo*, *Adahunse*, or *Oniseegun* among the Yoruba speaking people of Nigeria; *Abia ibok* among the Ibibio community of Nigeria; *Dibia* among the Igbo of Nigeria; *Boka* among the Hausa speaking people of Nigeria; and *Sangoma* or *Nyanga* among South Africans and *Mitishamba* in Kenya (Cook, 2009). In indigenous African communities, the traditional doctors are well known for treating patient holistically. The traditional doctors usually attempt to reconnect the social and emotional equilibrium of patients based on community rules and relationships unlike medical doctors who only treat diseases in patients (Hillenbrand, 2006). In countries like Ghana, Mali, Zambia, and Nigeria, the first line of treatment for 60% of children with high fever resulting from malaria is the use of herbal medicine (WHO, 2002b).

Diversity, flexibility, easy accessibility, broad continuing acceptance in developing countries and increasing popularity in developed countries, relative low cost, low levels of technological input, relative low side effects and growing economic importance are some of the positive features of TM (WHO, 2002a). These remedies have gained broad acceptability as they were reportedly reliable. It is therefore imperative to document and preserve such important information as a cultural heritage and to form a basis for scientific validation for development of better therapeutic drugs (Ochwang'i et al., 2014). Furthermore, there is a critical need to mainstream TM into public health care to achieve the objective of improved access to health care facilities. According to WHO some of the major policy challenges include safety, efficacy, quality, and rational use of TM. In some countries traditional herbal medicine systems have been professionalized since last millennia and have been integrated into the national health programs. For instance in India, there is a Central Council for Indian Medicine and there are national institutes for each of the six systems of medicine. The education system is well developed with over 300 university level programs across the country. Around 600,000 licensed practitioners are registered under the Indian Medicine Practitioners Act and there are over 9,000 licensed TCAM industries in the country (AYUSH, 2009). There is a central research authority with research and development programs on several aspects and the education and practice are regulated under the Indian Medicine Central Council Act 1970. The recent term "alternative medicine" generally used in developed countries indicates a practice which does not fall under the realm of conventional medicine and the word "complementary medicine" refers to a medical practice which is used along with conventional medicine but has not been integrated into the formal health system. In such a perspective what is considered complementary or alternative in one country may be a mainstream practice in another. The scientific status of health care makes society more and more dependent on professionalized practices, hindering health professionals to permeate and listen to local knowledge in primary health care (PHC) (Tesser and Barros, 2008). Around 80% of the population continues to use TM in Africa, Asia and Latin America and many governments in these regions have incorporated TM practices to help meet their PHC needs. In industrialized countries, almost half the population now regularly use some form of TCAM (United States, 42%; Australia, 48%; France, 49%; Canada, 70%), and considerable use exists in many developing countries (China, 40%; India, 70%; Chile, 71%; Colombia, 40%; up to 80% in African countries) (Bodeker and Kronenberg, 2002; WHO, 2007).

During the last decade the Government of Tanzania put forth legislation to address national health needs, traditional knowledge, and the resource base for TM (e.g., practitioners, biodiversity). This is one of the few countries in Africa that is encouraging TM growth. TM is the most common form of health care, and that the HIV pandemic has highlighted the need to work across health sectors. New legislation has facilitated this need. In Tanzania TM is experiencing a renaissance in being formally recognized, integrated into mainstream health care, formal establishment of practitioners, and gaining the interests of different sectors. More studies on bioactivity, safety, domestication, and sustainability of use of medicinal plants are needed. Development of TM can also, other than making a significant contribution to health care and livelihoods, provide income possibilities (Stangeland et al., 2008) (Tables 6.1 and 6.2).

High per capita distribution of TCAM practitioners in developing countries is an important reason for the widespread use of TCAM. WHO report cites example of Uganda, Tanzania, and Zambia where the ratio between healers to population is 1:200 to 1:400 while the ratio of allopathic practitioners to population is 1:20,000. In India, according to government sources, for the 65% of population TM is the only available source of health care (WHO, 2002a). It is also an affordable source of health in many countries. Another reason is that it is firmly embedded in the belief systems and can be termed "culturally compatible." In developed countries higher income and higher education are guiding factors of patient preference for TM. Due to difficulties in accessing modern

Table 6.1 Usage of TCAM in developing countries

Developing country	Usage of CAM (%)
Uganda	60
Tanzania	60
Rwanda	70
India	70
Benin	80
Ethiopia	90

Source: WHO, 2002. WHO Traditional Medicine Strategy 2002–2005, World Health Organization, Geneva; Abdullahi, A.A., 2011. Trends and challenges of traditional medicine in Africa, Afr. J. Tradit. Complement. Altern. Med. 8(5), 115–123.

Table 6.2 Sample ratio of THPs compared with the ratio of medical doctors to the population

Countries	Ratio of traditional practitioners to population	Ratio of medical doctors/ practitioners to population
Kenya, Urbam (Mathare)	1:833	1:987
Rural (Kilungu)	1:143–345	1:70,000
Zimbabwe	1:600	1:6,250
Swaziland	1:100	1:10,000
Nigeria (Benin City)	1:100	1:16,400
National average	No data	1:15740
South Africa (Venda area)	1:700–1,200	1:17,400
Ghana	1:200	1:20,000
Uganda	1:700	1:25,000
Tanzania	1:400	1:33,000
Mozambique	1:200	1:50,000

Source: Chatora, R., 2003. An overview of the traditional medicine situation in the African Region. African Health Monitor 4 (1), 4–7; Abdullahi, A.A., 2011. Trends and challenges of traditional medicine in Africa, Afr. J. Tradit. Complement. Altern. Med. 8(5), 115–123.

health care, ethnic minorities in developed societies who are disadvantaged both economically and socially, use TCAM as a first health care choice, making it noncomplementary (Bodeker and Kronenberg, 2002). Currently in many countries there exists a contradiction between personal choices and public policies with respect to health care involving TMs. While on the one hand the public at large actively integrate various health systems for a variety of reasons such as access, cost, efficacy, convenience, ethical and moral reasons, the policy and institutional mechanisms are slow to address various issues related to such integration (Van der Geest et al., 1990).

According to a regional overview in the WHO *African region* (AFR) only 50% of the population has access to essential health care while 80% continue to rely on African traditional medicines (ATRM) including herbal medicine, spiritual therapies, and manual therapies. TM is largely transmitted as oral knowledge and around 4000 species are used in ATRM which is predominantly (90%) plant based. Currently over half (56%) of the countries in the region have formulated TM policies and majority have established national departments in the health ministry and developed strategies for promotion. In 2001 a regional strategy for promoting ATRM was adopted following which at the World Health Assembly 2003 a resolution on ATRM was also passed. An ATRM day is observed for advocacy on August 31st each year among member countries and African summit of heads of state in 2001 declared 2001–2010 as ATRM decade. These initiatives have been supported by WHO and African summit of heads of state, especially for research and integration in the management of HIV/AIDS, tuberculosis, malaria, and other infectious diseases. Self-regulatory bodies such as healers associations have been established in many countries (Kasilo et al. 2005).

3 POLICIES FOR CONSERVATION OF MEDICINAL PLANTS AND LOCAL COMMUNITY PARTICIPATION

This great surge of public interest in the use of plants as medicines has been based on the assumption that the plants will be available on a continuing basis. However, no concerted effort has been made to ensure this, in the face of the threats posed by increasing demand, a vastly increasing human population and extensive destruction of plant-rich habitats such as the tropical forests, wetlands, Mediterranean ecosystems, and parts of the arid zone. Today many medicinal plants face extinction or severe genetic loss, but detailed information is lacking. For most of the endangered medicinal plant species no conservation action has beencle material of them in gene banks. The WHO's Traditional Medicine Strategy of 2002–05 highlights some urgent needs such as national policy and regulation for safety, efficacy and quality, access, and rational use of TMs. Only 66 countries out of 213 WHO member states have TM policies, while around 43 states have some kind of legislation and 20 member states are in the process of establishing some regulatory policies (Bodeker and Kronenberg, 2002). Key elements suggested in a national policy are definition of TCAM, definition of governments' role in developing TCAM, provision of safety and quality assurance for therapies and products, legislation related to TCAM providers, provision of education and training, promotion of proper/rational use, provision of capacity building for human resources including allocation of financial resources, provision of coverage by public health insurance, and consideration of intellectual property right issues. For most countries especially in Africa, data related to pattern, modalities and outcomes of TM usage are not recorded. In many developed countries though there are strict regulations for usage of TM, there is a growing trend of TM use in the guise of "health supplements" and spas. Increased adoption of national policies would facilitate creation of internationally accepted norms and standards for research into safety and efficacy of TCAM, rational use, sustainable usage of natural resources, and protection and equitable use of knowledge of TM. There are several nongovernment organizations, civil society groups, and other self-regulated associations advocating TCAM. Some of the local approaches to preserve and promote TM include documentation, building of databases, assessment through community-based approaches, self-help approaches through home/community herbal gardens, community health workers training on various aspects, organizing and training local healers, interventions and research initiatives on specific conditions (such as malaria, HIV, anemia), orientation of conventional health professionals on TCAM, consumer watch, livelihood promotion through economic activities, conservation and sustainable use of resources, multistakeholder participation, and evolving guidelines on benefit sharing and knowledge protection.

4 POLICY FOR RESTORING PLANTS HARVESTED IN THE WILD AND SUSTAINABILITY OF USE

Wild harvesting is the collection of plant material such as the *herba* (plant above ground), *flos* (flowers), *folia* (leaves), *lignum* (wood), or *radix* (roots) from wild sources. In many traditions of medicine, wild harvested material is considered to have higher therapeutic benefits, and therefore, for example in China, commands higher prices. TCAM is highly dependent on biodiversity and there is increasing demand for plants, animal and mineral resources. This has led to a situation of endangering many important medicinal plants. There is still no countrywise estimation of medicinal plant diversity, data on cultivated and wild sources, and trade data in terms of domestic and export demand. There is also insufficient data on agrotechnology of medicinal plants. WHO has prepared guidelines on good agricultural practices but the implementation of this has also been low. The government should regulate the collection of medicinal plants from the wild and should prohibit the collection from the wild of threatened medicinal plants except for propagation purposes. Such legislation however does not exist across Africa.

5 INCENTIVES TO COLLECTORS AND FARMERS TO KEEP PRODUCTION OF MEDICINAL PLANTS SUSTAINABLE

The best way to provide the plant material needed for medicine is to cultivate the plants. This is far better than collecting the plant material from the wild since it does not deplete wild stocks. Furthermore, in many cases, the declining habitats of native plants can no longer supply the expanding market for medicinal plant products. In the case of rare, endangered or overexploited plants, cultivation is the only way to provide material without further endangering the survival of those species. Cultivation also has pharmacological advantages over wild-collecting. Wild-collected plants normally vary in quality and composition, due to environmental and genetic variabilities. African governments should therefore develop a tradition of cultivating these medicinal plants and incentives may be needed to encourage cultivation, especially in parts of the tropics where many medicinal plants are slow-growing trees. Indeed there should be an effort to educate the masses about sustainability of this important resource.

6 GOVERNMENT SUPPORT ON MEDICINAL PLANT RESEARCH

A large number of present modern drugs are derived from traditional medical knowledge. Experience of drugs like Artemisia and St. John's wort has boosted confidence among pharmaceuticals to establish the efficacy of other extensively used TCAM therapies (Patwardhan, 2005). However, recent reviews have shown that clinical trials in TCAM have been scanty and inadequately designed (WHO, 2002a). The low level of research has slowed development of national standards and integration efforts. Epidemiological and public health mapping exercises are neglected aspects in the TCAM field. They are important to study population-based effects of TCAM use as well as creating data on the presence and quality of service by TCAM providers, especially in areas where there is limited access to conventional health care (Bodeker and Kronenberg, 2002). There are also insufficient contributions from social sciences to TCAM and most studies consider cultural knowledge as a stumbling block for health sector development.

Each country should identify and support one or more institutions to plan, coordinate, and implement ethnobotanical surveys. So far, most ethnobotanical surveys have been carried out by individuals, rather than by institutions. If the useful information of traditional peoples is to be documented before it is too late, ethnobotanical activities must be broadened and accelerated. To do this, the primary responsibility should move from the individual researchers to selected institutions, which can then provide the support, encouragement, coordination, and implementation that are needed. The selected institution(s) should implement a nationwide program of surveys on the use of plants for medicinal purposes in traditional societies; however this is not happening across all the African countries.

Under the present health care reform of the Federal Government of Nigeria, TM is purportedly recognized as an important component of health care delivery system especially at the primary care level. The Federal Government of Nigeria has established the Nigeria Natural Medicine Development Agency (NNMDA) to study, collate, document, develop, preserve, and promote Nigerian TM products and practices and to also fast-track the integration of the TM into the mainstream of modern health care system in line with happenings in China and India (The Sun news online, 2010). Ebomoyi (2009) found out that Nigerian medical students have reservation for the integration of TM into the mainstream of health care provision in the country. This is an indication that not much is being done in medical schools to encourage the teaching of TM as they keep unfolding in some parts of the world.

Some centers have been formed in Africa to help in carrying out clinical trials, production of standardized drugs, and regulatory work. The centers include:

- Center for scientific research into plant medicine, Ghana. They have helped to make sure that drug production is carried out to provide well-formulated, suitable, standardized, and safe preparations from plants for clinical evaluation, utilization, and monitoring in a clinical setting.
- Center for research on pharmacopoeia and TM in Rwanda produces drugs which are used in curing different diseases.
- The "village chemist" in development of pharmacognosy, Obafemi Awolowo University, Ile-Ife, Nigeria. Manufactures standardized and efficacious phytomedicine for managing different likely infections associated with people living with HIV/AIDS, like antithrush, antifever, antidysentery and antidiarrhea, anticough, and anti-infective against skin pathogens and diseases.
- Swaziland center for research in medicinal and indigenous food plants, University of Swaziland, Swaziland. They analyze medicinal plants collected by rural people familiar with that traditional medical system and study ethnobotanical information on medicinal plants administered by TMPs.
- Department of traditional medicine, Bamako, Mali. They keep ethnobotanical information on medicinal plants in rural areas of Mali and research on their plants to validate claims.
- Centre for Traditional Medicine and Drug Research (CTMDR) in Kenya Medical Research Institute (KEMRI). To identify and develop effective traditional/alternative medicines and drugs for use against human diseases in partnership with relevant institutions and government ministries. In addition, the program is also intended to provide information on quality of selected drugs in the Kenyan market.

7 POLICIES REGARDING EXPORT OF MEDICINAL PLANTS

Medicinal plants are important for pharmacological research and drug development, not only when plant constituents are used directly as therapeutic agents, but also as starting materials for the synthesis of drugs or as models for pharmacologically active compounds. Regulation of exploitation and exportation is therefore essential, together with international cooperation and coordination for their conservation so as to ensure their availability for the future (Jayasuriya, 1990). The government should control trade in medicinal plants and their products and it should be fully controlled, but in most countries it may only be practical to control trade in and out of the country, making use of customs facilities. If a government wishes to regulate trade in a species not on CITES (Convention on International Trade in Endangered Species), the best means may be by requiring export permits. More than 35,000 plant species are being used in various human cultures around the world for medical purposes and many of them are subjected to uncontrolled local and external trade. So far, natural products from fewer than 40 tropical species have been incorporated into modern medicine and only a fraction of the tropical flora has been thoroughly analyzed for their pharmacological activity. Therefore, the annual extinction rate of an estimated 3000 plant species is a matter of great concern as it could imply the loss of a potential drug against an incurable condition, such as dementia, cancer, influenza, or AIDS.

In Africa, Cunningham (1996) found that formal export trade of medicinal plants was only limited to a number of species that were traded in significant volumes. Cunningham cites Cameroon which exports four species to Europe including *Prunus africana* (a bark extract of which is used in the treatment of benign prostatic hypertrophy), all of which is exported to France; and *Pausinystalia johimbe*, 65% of which is exported to Holland, 18% to Germany and the rest to Belgium, Luxembourg, and France. Strong and working policies should be implemented by governments across Africa to protect overexploitation especially of endangered species.

8 POLICY FOR TRADE IN HERBAL MEDICINE

Medicinal plants play a critical role in the health care provision of much of the world's population. Whether they are used to make a decoction in rural Africa, to extract an alkaloid in Switzerland, or as a health food supplement in the United States, demand is increasing. Demand for medicinal plants for pharmaceutical companies is significant with an estimated 25% of prescription drugs in the United States containing plant extracts or active principles prepared from higher plants (Farnsworth and Soejarto, 1985). There are no reliable data on the number of plant species that are currently traded in high volume; indeed, such a list is badly needed. A study undertaken by WWF found that in several African countries, wild harvesting for local requirements was not detrimental to plant survival as the quantity collected tended to be small and also most of the material collected came from the more common varieties (Hamilton, 1990). What is of major concern is the fact that a major part of wild harvested material is now traded commercially. As the prices paid to the gatherers tend to be very low, commercial plant gatherers often "mine" the natural resources rather than manage them, as their main objective is to generate an income.

Most countries including the ones in the African continent have little or no regulations controlling the collection of material from the wild. India is one exception and has banned the export of several wild species in the form of raw material although the export of finished products containing the material is allowed. Despite this, an estimated 95% of medicinal plants collected in India are gathered from the wild and the process of collection is said to be destructive (Vinay, 1996). Equally, a major part of the high range Himalayan plants are wild harvested and many of these are close to extinction from over-harvesting or unskillful harvesting, for example, *Nardostachys jatamansi*, *Aconitum* spp. Major part of wild harvested material is sourced from developing countries, unfortunately most of this countries have no policy to guard against overexploitation. National trade in medicinal plants can involve hundreds of species. This trade would be undertaken at regional medicinal plant markets where hundreds of plant species are traded. For example in urban markets in Kwa Zulu, Natal, more than 400 species were being traded out of a total 1000 that were used medicinally in the area (Cunningham 1996). There is an informal trade of medicinal plants across national borders but within the same continent. This trade tends to consist of fewer numbers of species, although many of these are threatened. For example in Africa, *Warburgia salutaris* and *Siphonochilus aethiopicus* are two species in high demand and very scarce supply. In Asia, *Nardostachys grandiflora* and *Valeriana jatamansi* are examples of species which are threatened, but both are still traded from Nepal to India (Cunningham, 1996).

Although there are only a relatively small number of species that are traded in any significant volume, the fact that so few species (50–100) (Lange, 1996) are produced entirely under cultivation is a matter of great concern. Examples of major cultivated species are: *C. roseus*, *Chamomilla recutita*, *Cinchona* spp., *Digitalis lanata*, *Digitalis purpurea*, *Duboisia* spp., *Mentha piperita*, *Papaver somniferum* and *Plantago ovata* (Lewington, 1993). There exist a number of reasons as to why the trade in wild harvested material has been left to reach such a critical point; the legislation that exists to control harvesting and trade of medicinal plants is inadequate and ineffective in its current form in most African countries; new policies and easier mechanisms to control the trade are needed. There is lack of awareness among many of the end users, as to the extent to which wild harvested materials are used; indeed, it is only during the last 5–10 years that wild harvesting has become a subject of concern. Moreover there is an attempt to control the market, where the traders will give virtually no information on the extent of wild-harvesting. Actually the low price of wild-harvested material has made the procurements of alternative sources of raw material (via cultivation) financially unattractive.

The African plant species *P. africana* (Hook.f.) Kalkman makes for an interesting and important historical case study (Bodeker et al., 2014). This is due to its popularity in the European pharmaceutical market, the lack of access and benefit sharing arrangements that have characterized its exploitation by European companies, and its susceptibility to overharvesting as a habitat-specific, slowly maturing, and destructively harvested species

(Schippmann et al., 2002; Bodeker, 2003). In Ethiopia, a market survey undertaken in Jimma, Bonga, Gambella, and Addis Ababa in 1998 reported on some of the common medicinal plants and their trade routes within the country and to external markets indicating that there are no exports particularly for medicinal purposes (Desissa, 2001). Marshall (1998) also reported that Ethiopia had no legal export and import of products for medicinal use but plants of medicinal importance are exported to Djibouti and other countries as agricultural products.

In some African countries the traditional health care systems enjoy official status and in these countries a more formal trade exists (e.g., Botswana, Mozambique). The interest in plant-derived allopathic drug in Africa has risen recently following the international trend by some pharmaceutical companies of investing in new drug discovery program in Africa. Botswana, Namibia, Mozambique, Zimbabwe, and Malawi import medicinal plants from South Africa. The South African medicinal plant market trade in raw material alone worth around R270 million annually. The international demand for African medicinal plants is growing and major importers are European countries like Germany, France, Italy, Spain, and United Kingdom. Egypt and Sudan are the major exporters to Europe. Competition in the international trade for Ethiopia's major export products, such as coffee, is building fast and there is no absolute certainty to maintain current export earnings indefinitely. Such unstable situations can only be improved by a rational and coordinated policy of research and development at a national level toward identifying other additional candidate commodities. The position and contribution by the medicinal, spice and herb species is obvious.

Annually, TM trade contributes not less than R2.9 billion to the South African economy (Mander et al., 2007). Even in countries where TM is not legally recognized, TM trade has contributed very significantly to the revenue accrued by those countries. For instance, in 1996, about 2.8 million traditional Chinese medicine consultations were reported in Australia. This represented an annual turnover of about 84 million Australian dollars (WHO, 2000a) (Table 6.3).

9 HERBAL MEDICINE REGULATION IN AFRICA

Introduction of Western medicine and culture gave rise to "cultural-ideological clash" which had hitherto created an unequal power-relation that practically undermined and stigmatized the traditional health care system in Africa because of the overriding power of the Western medicine. This became manifested in South Africa during the Apartheid regime. In some extreme cases, TM was outrightly banned. For instance, the South African Medical Association outlawed traditional medical system in South Africa in 1953. In addition, the Witchcraft Suppression Act of 1957 and the Witchcraft Suppression Amendment Act of 1970 also declared TM unconstitutional thereby disallowing the practitioners from doing their business in South Africa (Hassim et al., n.d.).

Table 6.3 Some phytomedicines in the international market

Plant species	Action	Active constituents	Countries
Ancistrocladus abbreviatus	Anti-HIV	Michellamine B	Cameroon and Ghana
Corynanthe pachyceras	Male stimulant	Corynanthidine, corynanthine, yohimbine	Ghana
Tamarindus indica	Insecticides	Pectins	Egypt
R. vomitoria	Tranquilizer and antihypertensive	Reserpine, yohimbine	Nigeria, DR Congo, Rwanda, Mozambique
Cinchona succirubra	Antimalarial	Quinine	West African countries
Syzygium aromaticum	Dental remedy	Eugenol, terpenoids	East Africa countries, Madagascar
Agave sisalana	Corticosteroids and oral contraceptives	Hecogenin	Tanzania
Physostigma venenosum	Opthalmia	Physostigimine (eserine)	Calabar (Nigeria), Ghana, Cote D'ivoire
P. africana	Prostate gland hypertrophy	Sterols, triterpenes, *n*-docosanol	Cameroon, Kenya, Madagascar
C. roseus	Antileukemia and Hodgkin's disease	Triterpenoids, tannins and alkaloids	Madagascar
Zingiber officinale (Ginger)	Spice, carminative and medicinal products	Gingerol	Nigeria
Chrysanthemum cinerariifolium	Insecticides	Pyrethrins	Ghana, Kenya, Rwanda, Tanzania, South Africa

Source: Adapted from Okigbo, R.N., Mmeka, E.C., 2006. An appraisal of phytomedicine in Africa. KMITL Sci. Technol. J. 4 (2), 83–94.

Indeed, local efforts were initiated to challenge the condemnation and stigmatization of TM in some African communities during and after colonialism. Erinosho (1998, 2006) reported that the first protest against the marginalization of TM in Nigeria is dated back to 1922 when a group of native healers insisted that their medicine be legally recognized. In postindependence Africa, concerted efforts have been made to recognize TM as important aspect of health care delivery system in Africa. For instance, in Nigeria, the Federal Government through the Ministry of Health encouraged and authorized the University of Ibadan in 1966 to conduct research into the medicinal properties of local herbs with a view to standardize and regulate TM (WHO, 2001). In 1980s, policies were put in place to accredit and register native healers and regulate their practice.

The Division of Traditional Medicine, a collaborating center of WHO and recognized by the Organization of African Unity, has started the industrial exploitation of medicinal plants, carrying out activities such as: a survey of practitioners; identification of natural areas of growth of medicinal plants in Mali; botanical, chemical, and pharmacological studies; development of improved TMs; improvement of quality control; and training in TM. Since 1974, associations of traditional therapists have been established (Keita, 1993).

In South Africa, the trade in crude indigenous herbal products is completely unregulated. However, once a health-related claim is made for a finished product, it has to go through the full drug evaluation procedure in the Medicines Control Council (MCC) before marketing. Specific regulations for registration and control of new "traditional" herbal medicines do not exist. At present, there is no possibility for an abridged application procedure, and there is neither a list of therapeutic indication claims suitable for treatment with TMs, nor a national herbal medicines formulary of a pharmacopoeia (Gericke, 1995). However, TMs are included in the drug policy section of the government's Reconstruction and Development Programme. Ethiopia has policies and strategies that support the development and utilization of plant resources in a sustainable manner. The policies are reflected under various sectors including environmental protection, development of the natural resources, and diversification of the domestic and export commodities. Medicinal plants fit in the development activities that support public efforts in meeting livelihood requirements. There are few institutions concerned with the medicinal plants and assisted through government budgetary support. The Ethiopian Health and Nutrition Institute receives annual budget of about ETB 1.1 million while a department at IBC concerned with medicinal plant conservation gets ETB 100,000 per annum. The recent ongoing support made through the project funded by the World Bank namely the conservation and sustainable use of medicinal plants project has an annual budget of ETB 5.9 million per year during the project life. Such a support indicates the importance Ethiopia has given to the sector. The health sector strategy of Ethiopia declares that structural, functional TM into the official health care system is advantageous for improving the health coverage in the country (Ministry of Health, 1995). However, suitable institutional mechanisms and detailed implementation strategies and action plans has to be put in place.

The WHO has helped most developing countries of the world by utilizing several expert committees' policy decisions and resolutions in providing guidelines that will aid countries to develop and utilize their indigenous medicines for their national health agenda (WHO, 1978).

10 CONCLUSIONS AND RECOMMENDATIONS

Following the growing demand for TM and the contributions of the medicine to the overall health delivery system particularly in Africa, some authors have suggested that traditional medical system be integrated into the mainstream of health care services to improve accessibility to health care (Erinosho, 1998, 2005, 2006; Obute, 2005; Odebiyi, 1990; Okigbo and Mmeka, 2006). There are certain problems and challenges to be overcome in order to fully achieve the objective of regulation, standardization, and integration of TM in Africa. The ethnocentric and medicocentric tendencies of the Western hegemonic mentality that are usually paraded by most stakeholders in modern medicine remains a very serious challenge. It is a general belief in medical circle that TM defies scientific procedures in terms of objectivity, measurement, codification, and classification. The acceptance of Western religion, education, urbanization, and globalization phenomena in Africa is affecting the use of TM. These challenges notwithstanding, there is increasing evidence that TM would continue to hold sway in both rural and urban communities of Africa even when modern health care facilities are available (Bamidele et al. 2009) to meet wide range of health care needs. Therefore the various governments in Africa should support unwaveringly this system of health provision. Osowole et al. (2005) argue that more than 50% of the traditional healers they studied referred patients at least once to modern health facilities for further treatment. Indeed, traditional healers can provide a lead to scientific breakthrough in modern medicine. There is also an urgent need for appropriate legal frameworks to checkmate the quacks and charlatans in the practice of TM, though such legal frameworks are beginning to unfold in some African countries. It is only when these are genuinely pursued that the objective of regulation, rebranding, and standardization of TM products as well as the proposed cooperation between traditional and modern medicines can be accomplished for the benefit of millions of people who depend on TM in Africa. WHO has proffered a memorandum that will help African member states in institutionalizing African traditional medicine into their health system. This has been adopted by some countries in Africa within their health agenda (Elujoba et al., 2005). The African Regional Director of WHO has placed a big challenge before the different African research centers on traditional medicine that must be able to cure priority diseases in Africa (e.g., malaria, HIV/AIDS, sickle-cell anemia, diabetes and hypertension) (Elujoba et al., 2005). Different steps in the integration of TM into the health scheme (WHO Guidelines and Memorandum) WHO (WHO, 1978) has provided guidelines for institutionalization of TM into the health scheme. The steps include:

1. *Political Recognition*: The government and heads of state should be aware and help in the development of TM; this has already been achieved when the African Summit of Heads of State declared 2001–10 as "Decade of African Traditional Medicine." Research on TM for the treatment of priority disease to be given more importance.
2. *Development of policy, legal and regulatory framework*: Government should formulate national policies, legal frameworks, and registration. WHO has provided guidelines for the assessment of herbal medicine. There is need to establish regional regulatory mechanisms for regulating herbal medicine as reported by Calixto (2000), national expert committees, national programs and training programs for health (Elujoba et al., 2005).

3. *Promoting scientific research on TM and collaboration work*: Scientific research should be conducted on safety, efficacy, and quality of TM as proposed by WHO (Akerele, 1993). Countries of Africa are doing a lot on this, conducting researches to validate claims made on quality, safety, efficacy of TM used for the management of priority diseases like malaria, HIV/AIDS, sickle-cell anemia, diabetes, and hypertension (Elujoba et al., 2005). Collaboration of traditional medicine practitioners with others in the scientific community is very crucial for the supply of initial information on the plants to the scientists; it is achieved through staff exchange and training, sharing of expensive equipment and joint publications (Makhubu, 2006). This has been achieved in Africa with countries like Burkina Faso, Madagascar, Mali, and Tanzania making partnership arrangements between traditional herbal practitioners and the private sector for the integration of TM (Elujoba et al., 2005).
4. *Ensuring that intellectual property rights are protected*: Intellectual property rights are a priority item on the agenda of member states to protect indigenous knowledge about TM and legislation should be made on this (Calixto, 2000).
5. *Disseminating appropriate information to the general public on the use of TM*: Appropriate information should be given to the general public to empower them with knowledge and skills for the proper use of TM (WHO, 1978). This is achieved through organization of seminars to raise awareness as recorded by Makhubu (2006).
6. *Providing a good economic environment*: The government should ensure that a good economic, political, and regulatory environment is established for local production by traditional herbal practitioners as well as develop industries that can produce standardized remedies to increase access (WHO, 1978). Provide funding for their smooth operations.

Collaborative work could be achieved through staff exchange and training and funding for capital building; the government should help in funding researches on phytomedicine; the private sector as well as nongovernmental agencies should also help finance researches; organization of seminars to raise awareness to the general public on the benefits of medicinal plants, and also remove the perception that scientists are out to harness their knowledge for money making; abandoning outdated legislation (such as Witchcraft Act, 1901) and passing new legislation to protect indigenous traditional knowledge and for the integration of TM into the health scheme (Makhubu, 2006). The integration or harmonization of phytomedicine should be developed in such a way to work hand in hand with orthodox medicine with minimum threat to each other.

REFERENCES

Abdullahi, A.A., 2011. Trends and challenges of traditional medicine in Africa. Afr. J. Tradit. Complement. Altern. Med. 8 (5 Suppl.), 115–123.

Akerele, O., 1993. Summary of WHO guideline for the assessment of herbal medicines. Herbalgram 28, 13–17.

AYUSH, 2009. AYUSH Department, Government of India. Available from: http://indianmedicine.nic.in/

Bamidele, J.O., Adebimpe, W.O., Oladele, E.A., 2009. Knowledge, attitude and use of alternative medical therapy amongst urban residents of Osun State Southwestern Nigeria. Afr. J. Tradit. Complement. Altern. Med. 6 (3), 281–288.

Bodeker, G., 2003. Traditional medical knowledge, intellectual property rights & benefit sharing. Cardozo J. Int. Comp. Law 11, 785.

Bodeker, G., Kronenberg, F., 2002. A public health agenda for traditional, complementary and alternative medicine. Am. J. Public Health 92 (10), 1582–1591.

Bodeker, G., van't Klooster, C., Weisbord, E., 2014. *Prunus africana* (Hook.f.) Kalkman: the overexploitation of a medicinal plant species and its legal context. J. Altern. Complement. Med. 20 (11), 810–822.

Calixto, J.B., 2000. Efficacy, safety, quality control, marketing and regulatory guidelines for herbal medicines (phytotherapeutic agents). Braz. J. Med. Biol. Res. 33 (2), 179–189.

Cook, C.T., 2009. Sangomas: problem or solution for South Africa's health care system. J. Natil. Med. Assoc. 101 (3), 261–265.

Cunningham, A.B., 1996. Medicinal plant trade, conservation and the MPSG (Medicinal Plant Specialist Group). Medicinal Plant Conservation 2, 2–3.

Desissa, D., 2001. A preliminary economic valuation of medicinal plants in Ethiopia trade, volume and price. In: Zewdu, M., Demissie, A. (Eds.), Conservation and sustainable use of medicinal plants in Ethiopia. Proceedings of the National Workshop on Biodiversity Conservation and Sustainable Use of Medicinal plants in Ethiopia, Institute of Biodiversity Conservation and research, Addis Ababa, pp. 176–187.

Ebomoyi, E.W., 2009. Genomics in traditional African healing and strategies to integrate traditional healers into western-type health care services—a retrospective study. Researcher 1 (6), 69–79.

Elujoba, A.A., Odeleye, O.M., Ogunyemi, C.M., 2005. Traditional medicine development for medical and dental primary health care delivery system in Africa. African J. Tradit. Complement. Altern. Med. 2 (1), 46–61.

Erinosho, O.A., 1998. Health Sociology for Universities, Colleges and Health Related Institutions. Sam Bookman, Ibadan.

Erinosho, O.A., 2006. Health Sociology for Universities, Colleges and Health Related Institutions. Reprint. Bulwark Consult, Ibadan, Abuja.

Farnsworth, N.R., Soejarto, D.D., 1985. Potential consequence of plant extinction in the United States on the current and future availability of prescription drugs. Econ. Bot. 39 (3), 231–240.

Gericke, N., 1995. The Regulation and Control of Traditional Herbal Medicines. An International Overview With Recommendations for the Development of a South African Approach. Traditional Medicines Programme at the University of Cape Town. Working draft document, December 1995. Unpublished.

Gudmastad, E., 2013. 2013 World Population Data Sheet. Population Reference Bureau. Available from: www.prb.org

Hamilton, A., 1990. La conservation des plantes médicinales et autres végétaux utiles. Bull. Soc. Ind. Mulhouse 819, 19–24.

Hassim, A., Heywood, M., Berger, J., n.d. Health and Democracy. Available from: http://www.alp.org.za

Hillenbrand, E., 2006. Improving traditional–conventional medicine collaboration: perspectives from Cameroonian traditional practitioners. Nordic J. Afr. Stud. 15 (1), 1–15.

Jayasuriya, D.C., 1990. A Review of Legislation Concerning Medicinal Plants. Unpublished report.

Kasilo, O.M.J., Alley, E.S., Wambebe, C., Chatora, R., 2005. Regional overview: Africa Region. In: Bodeker, G., Ong, C.K., Grundy, C., Burford, G., Shein, K. (Eds.), WHO Global Atlas on Traditional, Complementary and Alternative Medicine. WHO Centre for Health and Development, Kobe, pp. 3–12.

Keita, A., 1993. Pharmacopoeia in Mali. Situation of Research and Perspectives of Development. Lecture held at the Morris Arboretum Symposium, Philadelphia, 1921, April 1993. Unpublished.

Lange, D., 1996. Untersuchungen zum Heilpflanzenhandel in Deutschland. Bundesamt fur Naturschutz, Bonn, Germany.

Lemma, A., 1991. The Potentials and challenges of endod, the Ethiopian soapberry plant for control of schistosomiasis. In: Science in Africa: Achievements and Prospects, American Association for the Advancement of Sciences (AAAS), Washington, DC, USA.

Lewington, A., 1993. A Review of the Importation of Medicinal Plants and Plant Extracts into Europe. TRAFFIC International, Cambridge, UK, 37 p.

Makhubu, L., 2006. Traditional medicine: Swaziland. Afr. J. Tradit. Complement. Altern. Med. 5 (2), 63–71.

Mander, M., Ntuli, L., Diederichs, N., Mavundla, K., 2007. Economics of the Traditional Medicine Trade in South Africa. Available from: http://www.hst.org.za/uploads/files/chap13_07

Marshall, N.T., 1998. Searching for a cure, conservation of medicinal wildlife resources in East and Southern Africa. TRAFFIV network report.

Ministry of Health, 1995. Health sector strategy. Ministry Health, Addis Ababa, Ethiopia.

Obute, G.C., 2005. Ethnomedicinal Plant Resources of South Eastern Nigeria. Available from: http://www.siu.edu/~ebl/leaflets/-obute.htm

Ochwang'i, D.O., Kimwele, C.N., Oduma, J.A., Gathumbi, P.K., Mbaria, J.M., Kiama, S.G., 2014. Medicinal plants used in treatment and management of cancer in Kakamega County, Kenya. J. Ethnopharmacol. 151 (3), 1040–1055.

Odebiyi, A.I., 1990. Western trained nurses assessment of the different categories of traditional healers in south western Nigeria. Int. J. Nurs. Stud. 27 (4), 333–342.

Okigbo, R.N., Mmeka, E.C., 2006. An appraisal of phytomedicine in Africa. KMITL Sci. Technol. J. 6 (2), 83–94.

Osowole, O., Ajaiyeoba, E., Bolaji, Akinboye, D., Fawole, O., Gbotosho, G., Ogbole, O., Ashidi, J., Abiodun, O., Falade, C., Sama, W., Oladepo, O., Itiola, O., Oduola, A., 2005. A survey of treatment practices for febrile illnesses among traditional healers in the Nigerian Middle Belt Zone. Afr. J. Tradit. Complement. Altern. Med. 2 (3), 337–344, 281–288.

Patwardhan, B., 2005. Traditional Medicine: A Novel Approach for Available, Accessible and Affordable Health Care, A Paper Submitted for Regional Consultation on Development of Traditional Medicine in the South-East Asia Region, Korea, World Health Organization.

Romero-Daza, N., 2002. Traditional medicine in Africa. Ann. Am. Acad. Pol. Soc. Sci. 583, 173–176.

Schippmann, U., Leaman, D.L., Cunningham, A.B., 2002. Impact of cultivation and gathering of medicinal plants on biodiversity: global trends and issues. FAO, Rome.

Srivastava, J., Lambert, J., Vietmeyer, N., 1995. Medicinal Plants: A Growing Role in Development. Agricultural and Natural Resources Department, The World Bank, Washington, DC.

Stangeland, T., Dhillion, S.S., Reksten, H., 2008. Recognition and development of traditional medicine in Tanzania. J. Ethnopharmacol. 117 (2), 290–299.

Tesser, C.D., Barros, N.F., 2008. Medicalização social e medicina alternativa e complementar: pluralização terapêutica do Sistema Único de Saúde. Rev. Saude Publica, v.42, n.5, pp. 914–920, 2008.

The Sun News, 2010. Lagos. Available from: http://www.thesunnewsonline.com

Van der Geest, S., Speckmann, J.D., Streefland, P.F., 1990. Primary healthcare from a multi-level perspective: towards a research agenda. Soc. Sci. Med. 30 (9), 1025–1034.

Vinay, T., 1996. Camp Workshop: Plants Under Threat—New List Forged, Medicinal Plant Conservation, Volume 2. Newsletter of the IUCN Species Survival Commission, Bonn, Germany, Bundesamt für Naturschutz.

WHO, 1978. Alma Ata Declaration Primary Health Care. Health For All Series No 1.

WHO, 2000a. Traditional and Modern Medicine: Harmonising the Two Approaches. Western Pacific Region, World Health Organization, Geneva.

WHO, 2000b. General Guidelines for Methodologies on Research and Evaluation of Traditional Medicine, World Health Organization, Geneva.

WHO, 2001. Legal Status of Traditional Medicine and Complementary/Alternative Medicine: A Worldwide Review, World Health Organization, Geneva.

WHO, 2002a. Traditional Medicine Strategy 2002–2005, World Health Organization.

WHO, 2002b. Traditional Medicine—Growing Needs and Potential, World Health Organization, Geneva.

WHO, 2007. WHO Guidelines for Assessing Quality of Herbal Medicines with Reference to Contaminants and Residues, World Health Organization, Geneva.

CHAPTER 7

Preparation, Standardization, and Quality Control of Medicinal Plants in Africa

M.O. Nafiu, A.A. Hamid, H.F. Muritala, S.B. Adeyemi

1 INTRODUCTION

Traditional medicine is the sum total of knowledge, skills, and practices based on the theories, beliefs, and experiences indigenous to different cultures that are used to maintain health, as well as to prevent, diagnose, improve, or treat physical and mental illnesses (WHO, 2003). Traditional medicine that has been adopted by other populations (outside its indigenous culture) is often termed complementary or alternative medicine (CAM) (WHO, 2003; Gurib-Fakim, 2006). The World Health Organization (WHO) reported that 80% of the emerging world's population relies on traditional medicine for therapy. During the past decades, the developed world has also witnessed an ascending trend in the utilization of CAM, particularly herbal remedies (Chintamunnee and Mahomoodally, 2012). Herbal medicines include herbs, herbal materials, herbal preparations, and finished herbal products that contain parts of plants or other plant materials as active ingredients. Although 90% of the population in Ethiopia use herbal remedies for their primary health care, surveys carried out in developed countries like Germany and Canada tend to show that at least 70% of their population have tried CAM at least once (Gurib-Fakim, 2006; Chintamunnee and Mahomoodally, 2012). It is likely that the profound knowledge of herbal remedies in traditional cultures, developed through trial and error over many centuries, along with the most important cures was carefully passed on verbally from one generation to another. Indeed, modern allopathic medicine has its roots in this ancient medicine, and it is likely that many important new remedies will be developed and commercialized in the future from the African biodiversity, as it has been till now, by following the leads provided by traditional knowledge and experiences (Chintamunnee and Mahomoodally, 2012; Gurib-Fakim, 2006; Nunkoo and Mahomoodally, 2012; Shohawon and Mahomoodally, 2013). The extensive use of traditional medicine in Africa, composed mainly of medicinal plants, has been argued to be linked to cultural and economic reasons. This is why the WHO encourages African member states to promote and integrate traditional medical practices in their health system (WHO, 2003). Plants typically contain mixtures of different phytochemicals, also known as secondary metabolites that may act individually, additively, or in synergy to improve health. Indeed, medicinal plants, unlike pharmacological drugs, commonly have several chemicals working together catalytically and synergistically to produce a combined effect that surpasses the total activity of the individual constituents.

Plant-derived substances have recently become of great interest owing to their versatile applications. Medicinal plants are the richest bioresource of drugs of traditional systems of medicine, modern medicines, nutraceuticals, food supplements, folk medicines, pharmaceutical intermediates, and chemical entities for synthetic drugs. WHO encourages, recommends, and promotes traditional/herbal remedies in national health care programs because these drugs are easily available at low cost, safe, and people have faith in them (Pandey and Tripathi, 2014).

2 FACTORS THAT AFFECT THE USE OF MEDICINAL PLANT PREPARATIONS

All medicines, whether synthetic or of plant origin, should fulfill the basic requirements of being safe and effective (EMEA, 2005; WHO, 1990, 2002). The term herbal drugs denote plants or plant parts that have been converted into phytopharmaceuticals by means of simple processes involving harvesting, drying, and storage (EMEA, 1998). Hence they are capable of variation. This variability is also caused by differences in growth, geographical location, and time of harvesting. The use of medicinal herbs is affected by the summarized factors as follows.

2.1 Inadequate standardization and lack of quality specifications

Medicinal plant preparation is administered for its holistic value. Each herbal ingredient of the preparation has different chemical constituents with complex molecular structures. This medicinal herb preparation is a source of polypharmacy within itself. Hence, standardization of medicinal plant preparation or its ingredients becomes a highly complex issue.

2.2 Lack of scientific data

Works on medicinal herb preparation claim the use of scientific data in support of the medicinal activity, and their safety and efficacy are assumed. Hence, there is need to incorporate certain parameters of the ethnomedicinal evaluation of herbs on modern lines (Wani, 2007).

2.3 Ineffective in acute medical care

Medicinal plant materials are not very effective to treat any acute illness; as most of the medicines are prepared to work at molecular level of physiology, the drug takes its time to deliver the results.

2.4 Drug interactions

More patients are taking herbal preparation either as health supplement or for treatment of ailments, before consulting a physician. Medicinal plant preparation presents greater risks of adverse effects and interactions than any other complement medication therapy. This occurs in elderly patients with chronic cardiovascular disease as many herbs alter bleeding time (garlic, ginger, ginseng) and interact with conventional cardiac medications (digoxin, diuretics, antiarrhythmic) (Seow and Thong, 2005).

3 MODES OF PREPARATION OF MEDICINAL PLANTS USED IN TRADITIONAL MEDICINE

In traditional herbal medicine systems, herbal remedies are prepared in several rather standardized ways which usually vary based upon the plant utilized, and sometimes, what condition is being treated. Some of these methods include infusions (hot teas), decoctions (boiled teas), tinctures (alcohol and water extracts), and macerations (cold soaking) which are detailed more fully herein. In indigenous Indian medicine systems, medicine men or shamans generally use these same methods in addition to others. Others include preparing plants in hot baths (in which the patient is soaked in it or bathed with it), inhalation of powdered plants (like snuff), steam inhalation of various aromatic plants boiled in hot water, and even aromatherapy (Taylor, 2005). Some of the methods used are fully discussed herein.

3.1 Concoctions

These are aqueous preparations of different plant parts soaked or boiled in water for particular period of time. If boiled, it is done for 15–20 min, but if soaked, it is soaked for 3 days before consumption to ensure thorough extraction. Concoction can also be prepared using plant parts as soup popularly called "aseje" in Yoruba traditional system of medicine. These soups can be prepared by adding several other materials such as dried chameleon, snail, and dried toads among others. The soup is prepared for a day's consumption.

3.2 Decoctions

Decoctions are aqueous preparations of plant parts boiled in water for 15–20 min until the water volume is halved. Decoctions are prepared by breaking the plant parts into small pieces before soaking in a given amount of water in an earthenware container (Palayok). The water should cover the plant parts, then the container is covered, and the preparation is boiled until the water volume is halved. The boiled preparation is strained, cooled, and refrigerated which can be kept for 2 or 3 days.

3.2.1 Strong decoctions

Depending on the type of plant material used, strong decoctions are prepared in two general ways. The first involves boiling the mixture for a longer time. This is usually indicated when working with larger woody pieces of bark. Longer boiling time, up to 2 h or more, is sometimes necessary to break down, soften, and extract the larger pieces. Alternatively, when smaller woody pieces are used yet a stronger remedy is required, the decoction is prepared as above (boiling for 20 min), then it is allowed to sit/soak overnight before straining out the herb. When straining, it is ensured that the cut herb pieces are pressed in the strainer to get as much moisture/decoction out of the herb pieces (Taylor, 2005).

3.2.2 Dried decoctions

Dried decoctions were developed in Japan in the 1950s and have become a major method of providing herbs in Japan, Taiwan, the United States, and Europe. The dried decoctions are produced by making very large batches of the herb formulas as decoctions (in large tanks), and then draining the liquid from the dregs. The liquid is then evaporated (using heat and vacuum) to form a syrup. The syrup is then put into a spray drier along with a powder carrier (usually starch or the dried, powdered, herb dregs), and the remaining water is evaporated, leaving a dry powder (Dharmananda, 1997).

3.3 Infusions

As in preparing tea, infusions are prepared using dried or fresh herbs. The plant materials are soaked in hot water and allowed to stand for 10 min. Then strain the tea and drink it hot or ice cold. It's just for a day's use.

3.4 Pills

This preparation is also called honey pills. Pills can be made by mixing thoroughly the dried and powdered herbal materials with equal quantity of honey cooked to bright red syrup. The mixture is allowed to cool off after which it would be rolled to desired tubular strands and cut into small pieces. The pieces are air dried in clean place and bottled neatly. If honey is not available, concentrated syrup of cane sugar can be substituted.

3.5 Powder

In Nigeria, this preparation is called "Agunmu" by traditional people of Yoruba land. It is the most common and easiest way of preparing herbal materials. With mortar and pestle, the well dried plant materials are crushed and ground into a fine uniform powder. The powder is stored in clean bottles, and should be as fine as possible to ensure faster solubility. Although the local healers of Oromo, Gambi district of Southwest Ethiopia employed several methods of preparation of traditional medicines from plants, powdering and pounding were the most frequently used methods of traditional medicine preparation (Abera, 2014).

3.6 Tinctures

This preparation is also called alcoholic decoction. To prepare tinctures, the herbal materials either fresh or dried are placed in 40–60% proof alcohol. It is prepared by placing one part of herb to five parts of distilled spirit and is kept in an airtight container. The mixture is shaken or stirred at least once daily. Alcoholic decoction extracts preserve the essential ingredients for the longest possible time. The extract is strained and stored in an airtight glass jar. Dosages are usually five to twenty drops taken directly or added to water. Also, this can be prepared by boiling the alcohol alongside the desired herbs in water. Then the solution would be poured into container and tightly sealed. After 2 weeks of usage, the residue of the solution can be used to prepare ointment.

3.7 Tablets

The herbs of interest are powdered and thoroughly mixed. If small sized tablets with high drug concentration are desired, a portion of the dried drug material may be decocted into a thick concentrated solution and then mixed with the other powdered materials. In making the tablet, a sufficient amount of starch or rice paste is added to the mixture and is forcefully mixed and kneaded by hands. Small globular tablets are made of the kneaded paste-like material with the aid of improvised tablet making devices which may be constructed from wood or metal.

3.8 Syrup

This preparation is applicable for children and infants. This is prepared by dissolving certain amount of cane sugar in certain volume of boiling water until the sugar thoroughly dissolves in the water. The desired herb is placed in water, boiled, and decanted. The resulting extract (generally every 1 mL of the decoction fluid contains 1 g of the concentrated drug) is added to the prepared cane sugar syrup in 1:1 proportion. The decoction should be treated with sufficient amount of fungicide like benzoic acid for long storage if the syrup is not to be added.

3.9 Poultice

Poultice, sometimes referred to as paste, is prepared by grinding and crushing the desired plant materials either dried or fresh (preferable) with a little water, oil, or honey. The resulting paste is then spread on a square of a clean cloth or banana trunk and applied or tied to the affected area. However the crushed plant can also be boiled to achieve a pulp.

3.10 Compresses

These are usually milder than poultice. An infusion or decoction is used to soak a clean cloth or a banana trunk and placed on the affected areas.

3.11 Juices

Juices are prepared by pounding fresh plant materials and filtering through a fine piece of cloth or by squeezing the plant parts to extract the juice.

4 MODES OF PREPARATION OF EXTRACTS IN RESEARCH LABORATORIES IN AFRICA

Extraction, as the term is used pharmaceutically, involves the separation of medicinally active portions of plant or animal tissues from the inactive or inert components by using selective solvents in standard extraction procedures. The products so obtained from plants are relatively impure liquids, semisolids, or powders intended only for oral or external use. These include classes of preparations known as decoctions, infusions, fluid extracts, tinctures, pilular (semisolid) extracts, and powdered extracts. Such preparations have been popularly called galenicals, named after Galen, the second century Greek physician. The purposes of standardized extraction procedures for crude drugs are to attain the therapeutically desired portion and to eliminate the inert material by treatment with a selective solvent known as menstruum. The extract thus obtained may be ready for use as a medicinal agent in the form of tinctures and fluid extracts; it may be further processed to be incorporated in any dosage form such as tablets or capsules, or it may be fractionated to isolate individual chemical entities such as ajmalicine, hyoscine, and vincristine, which are modern drugs. Thus, standardization of extraction procedures contributes significantly to the final quality of the herbal drug (Handa et al., 2008).

4.1 Plant tissue homogenization

Plant tissue homogenization in solvent has been widely used by researchers. Dried or wet, fresh plant parts are grinded in a blender to fine particles, put in a certain quantity of solvent, and shaken vigorously for 5–10 min or left for 24 h after which the extract is filtered. The filtrate then may be dried under reduced pressure and redissolved in the solvent to determine the concentration. Some researchers, however, centrifuged the filtrate for clarification of the extract (Das et al., 2010).

4.2 Counter-current extraction

In counter-current extraction (CCE), wet raw material is pulverized using toothed disc disintegrators to produce a fine slurry. In this process, the material to be extracted is moved in one direction (generally in the form of a fine slurry) within a cylindrical extractor where it comes in contact with extraction solvent. The further the starting material moves, the more concentrated the extract becomes. Complete extraction is thus possible when the quantities of solvent and material and their flow rates are optimized. The process is highly efficient, requiring little time and posing no risk from high temperature. Finally, sufficiently concentrated extract comes out at one end of the extractor while the marc (practically free of visible solvent) falls out from the other end (Handa et al., 2008).

This extraction process has significant advantages:
- A unit quantity of the plant material can be extracted with much smaller volume of solvent as compared to other methods like maceration, decoction, and percolation.
- CCE is commonly done at room temperature, which spares the thermolabile constituents from exposure to heat; this is employed in most other techniques.
- As the pulverization of the drug is done under wet conditions, the heat generated during comminution is neutralized by water. This again spares the thermolabile constituents from exposure to heat.
- The extraction procedure has been rated as more efficient and effective than continuous hot extraction.

4.3 Sonication

The procedure involves the use of ultrasound with frequencies ranging from 20 to 2000 kHz; this increases the permeability of cell walls and produces cavitation. Although the process is useful in some cases, like extraction of rauwolfia root, its large-scale application is limited due to the higher costs. One disadvantage of the procedure is the occasional but known deleterious effect of ultrasound energy (more than 20 kHz) on the active constituents of medicinal plants through formation of free radicals and consequently undesirable changes in the drug molecules (Cowan, 1999).

4.4 Percolation

This is the procedure used most frequently to extract active ingredients in the preparation of tinctures and fluid extracts. A percolator (a narrow, cone-shaped vessel open at both ends) is generally used. The solid ingredients are moistened with an appropriate amount of the specified menstruum and allowed to stand for approximately 4 h in a well-closed container, after which the mass is packed and the top of the percolator is closed. Additional menstruum is added to form a shallow layer above the mass, and the mixture is allowed to macerate in the closed percolator for 24 h. The outlet of the percolator is then opened and the liquid contained therein is allowed to drip slowly. Additional menstruum is added as required, until the percolate measures about three quarters of the required volume of the finished product. The marc is then pressed and the expressed liquid is added to the percolate. Sufficient menstruum is added to produce the required volume, and the mixed liquid is clarified by filtration or by standing followed by decanting (Handa et al., 2008).

4.5 Serial exhaustive extraction

It is another common method of extraction which involves successive extraction with solvents of increasing polarity from a nonpolar (hexane) to a more polar solvent (methanol) to ensure that a wide polarity range of compound could be extracted. Some researchers employ soxhlet extraction of dried plant material using organic solvent. This method cannot be used for thermolabile compounds as prolonged heating may lead to degradation of compounds (Das et al., 2010).

4.6 Soxhlet extraction

Soxhlet extraction is only required where the desired compound has a limited solubility in a solvent, and the impurity is insoluble in that solvent. If the desired compound has a high solubility in a solvent then a simple filtration can be used to separate the compound from the insoluble substance. The advantage of this system is that instead of many portions of warm solvent being passed through the sample, just one batch of solvent is recycled. This method cannot be used for thermolabile compounds as prolonged heating may lead to degradation of compounds (Sutar et al., 2010).

4.7 Maceration

In this process, the whole or coarsely powdered crude drug is placed in a stoppered container with the solvent and allowed to stand at room temperature for a period of at least 3 days with frequent agitation until the soluble matter has dissolved. The mixture is then strained, the marc (the damp solid material) is pressed, and the combined liquids are clarified by filtration or decantation after standing (Handa et al., 2008). This method is best suitable for use in the case of the thermolabile drugs.

4.8 Digestion

This is a kind of maceration in which gentle heat is applied during the maceration extraction process. It is used when moderately elevated temperature is not objectionable and the solvent efficiency of the menstruum is increased thereby (Bimakr, 2010).

4.9 Decoction

This method is used for the extraction of the water soluble and heat stable constituents from crude drug by boiling it in water for 15 min, cooling, straining, and passing sufficient cold water through the drug to produce the required volume (Bimakr, 2010).

4.10 Infusion

It is a dilute solution of the readily soluble components of the crude drugs. Fresh infusions are prepared by macerating the solids for a short period of time with either cold or boiling water (Bimakr, 2010).

5 PARAMETERS FOR SELECTING AN APPROPRIATE EXTRACTION METHOD

According to Handa et al. (2008), the following must be considered for the selection of appropriate method for effective and efficient extraction.

1. Authentication of plant material should be done before performing extraction. Any foreign matter should be completely eliminated.
2. Use the right plant part and, for quality control purposes, record the age of plant and the time, season, and place of collection.
3. Conditions used for drying the plant material largely depend on the nature of its chemical constituents. Hot or cold blowing air flow for drying is generally preferred. If a crude drug with high moisture content is to be used for extraction, suitable weight corrections should be incorporated.
4. Grinding methods should be specified and techniques that generate heat should be avoided as much as possible.
5. Powdered plant material should be passed through suitable sieves to get the required particles of uniform size.
6. Nature of constituents:
    a. If the therapeutic value lies in nonpolar constituents, a nonpolar solvent may be used. For example, lupeol is the active constituent of *Crataeva nurvala* and, for its extraction, hexane is generally used. Likewise, for plants like *Bacopa monnieri* and *Centella asiatica*, the active constituents are glycosides and hence a polar solvent like aqueous methanol may be used.
    b. If the constituents are thermolabile, extraction methods like cold maceration, percolation, and CCE are preferred. For thermostable constituents, Soxhlet extraction (if nonaqueous solvents are used) and decoction (if water is the menstruum) are useful.
    c. Suitable precautions should be taken when dealing with constituents that degrade while being kept in organic solvents, for example, flavonoids and phenyl propanoids.
    d. In case of hot extraction, higher than required temperature should be avoided. Some glycosides are likely to break upon continuous exposure to higher temperature.
    e. Standardization of time of extraction is important, as (1) insufficient time means incomplete extraction and (2) if the extraction time is longer, unwanted constituents may also be extracted. For example, if tea is boiled for too long, tannins are extracted which impart astringency to the final preparation.
    f. The number of extractions required for complete extraction is as important as the duration of each extraction.
    g. The quality of water or menstruum used should be specified and controlled.

h. Concentration and drying procedures should ensure the safety and stability of the active constituents. Drying under reduced pressure (e.g., using a Rotavapor) is widely used. Lyophilization, although expensive, is increasingly employed.
i. The design and material of fabrication of the extractor are also to be taken into consideration.
j. Analytical parameters of the final extract, such as thin layer chromatography (TLC) and high performance liquid chromatography (HPLC) fingerprints, should be documented to monitor the quality of different batches of the extracts.

6 STANDARDIZATION OF MEDICINAL PLANT PREPARATION IN AFRICA AND OTHER PARTS OF THE WORLD

Standardization is defined as the process of implementing and developing technical standards for selection in making appropriate choices for ratification coupled with consistent decisions for maintaining obtained *standards*. This view includes the case of "spontaneous standardization processes," to produce de facto standards (Pandey and Tripathi, 2014). Standardization of herbal medicinal plant is the process of prescribing a set of standards or inherent characteristics, constant parameters, definitive qualitative and quantitative values that carry an assurance of quality, efficacy, safety, and reproducibility. It is the process of developing and agreeing upon technical standards (Kunle et al., 2012).

Standardization should involve the compilation of complete data on medicinal herbs such as the season in which it is harvested, the ripeness, and the taste, smell, appearance, drying, storage, processing, and fingerprinting which needs much larger spectrum of five or more active marker constituents (Kunle et al., 2012). Standardization is also considered as the way to deal with the regulations framed by regulatory authorities that require drug measurability and the active ingredients to be stated on product labels (Page, 2001).

6.1 The need for standardization

According to WHO (2002), over 80% of the world's population especially in the developing world relies on medicinal plants as sources of medicines for their primary health care. Traditional system of medicine which depends mainly on medicinal plants is rich in ethnomedical knowledge of the uses of medicinal plants in the treatment of infectious conditions (Iwu and Anyawu, 1982). These medicinal plants employed in traditional medicine represent potential sources of cheap and effective standardized herbal medicines (phytomedicine) and leads to the discovery of novel molecules for the development of new chemotherapeutic agents (Farnsworth and Morris, 1976) Several infectious diseases including malaria, diarrhea, dysentery, gonorrhea, and fungal infections have been successfully managed in traditional medical practice employing medicinal plants (Sofowora, 1993).

Inconsistent and variable biological effects of herbal medicinal plants are the main discouraging issues in natural products. Reproducible efficacy and safety of the phytopharmaceuticals are based on reproducible quality. Therefore, phytopharmaceuticals could only be considered as a rational drug if these are standardized and their pharmaceutical quality is approved, while their composition needs to be well documented in order to obtain reproducible results in pharmacological, toxicological, and clinical studies. In other words, to gain public trust and bring medicinal plant product into mainstream of today's health care system, the researchers, manufacturers, and the regulatory agencies must apply rigorous scientific methodologies to ensure the quality and consistency of the traditional medicinal plant products (Wani, 2007). However, due to the complex nature and inherent variability of the constituents of plant-based drugs, it is difficult to establish quality control parameter though modern analytical techniques are expected to help in circumventing this problem (Sofowora, 1993).

Furthermore, the constituents responsible for the claimed therapeutic effects are frequently unknown or only partly explained. This is further complicated by the use of combination of herbal ingredients as being used in traditional practice. It is common to have as many as five different herbal ingredients in one product. Thus batch to batch variation starts from the collection of raw material itself in the absence of any reference standard for identification. These variations multiply during storage and further processing. Hence for herbal drugs and products, standardization should encompass the entire field of study from cultivation of medicinal plant to its clinical application. Plant materials and herbal remedies derived from them represent substantial portion of global market and in this respect internationally recognized guidelines for their quality assessment and quality control are necessary. Hence, the WHO, taking cognizance of these problems and/or complications, has published guidelines to ensure the reliability and repeatability of research on medicinal plant preparations in Africa and other parts of the world, and these guidelines must also be followed in the commercial production and therapeutic application of phytopharmaceuticals (EMEA, 1998, 2005; WHO, 1991, 1996, 2000)

Standardized medicinal plant extracts are of two main types:
- An active constituent extract where there is a known and accepted active biochemical principle. It is like a drug and may have undesirable side effects that are normally absent in the herb (Tierra, 2012).
- A marker extract where the active biochemical principle is unknown and a characteristic compound is used as a marker. As no single active components are known, the entire extract is treated as active (Tierra, 2012).

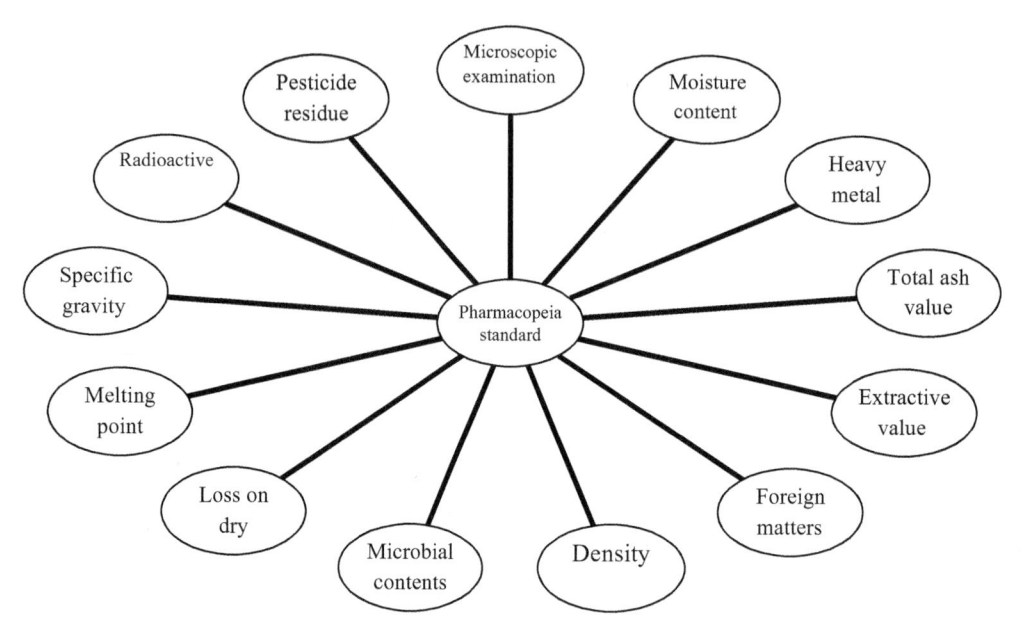

Figure 7.1 *Standardization parameter for plant drugs. (Pandey and Tripathi, 2014. Concept of standarization extraction and prephytochemical screening strategies for herbal drug. J. Pharmacognosy Phytochem. 2(5) 117.)*

6.2 Steps involved in the standardization of medicinal plant preparations

The standardization of medicinal plant preparations in Africa and other parts of the world includes the following steps (Fig. 7.1) (Evans, 1996; Hylands, 2002; Indian Pharmacopoeia, 1996).

- Macro and microscopic examination: for identification of right variety and search of adulterants.
- Foreign organic matter: this involves removal of matter other than source plant to get the drug in pure form.
- Ash values: these are criteria to judge the identity and purity of crude drug—total ash, sulfated ash, water soluble ash, acid insoluble ash, etc.
- Moisture content: checking moisture content helps reduce errors in the estimation of the actual weight of drug material. Low moisture suggests better stability against degradation of product.
- Extractive values: these are indicative weights of the extractable chemical constituents of crude drug under different solvents' environment.
- Crude fiber: this helps to determine the woody material component, and it is a criterion for judging purity.

- Qualitative chemical evaluation: This covers identification and characterization of crude drug with respect to phytochemical constituent. It employs different analytical techniques to detect and isolate the active constituents. Phytochemical screening techniques involve botanical identification, extraction with suitable solvents, purification, and characterization of the active constituents of pharmaceutical importance.
- Chromatographic examination: This includes identification of crude drug based on the use of major chemical constituents as markers.
- Quantitative chemical evaluation: to estimate the amount of the major classes of constituents.
- Toxicological studies: this helps to determine the pesticide residues, potentially toxic elements, safety studies in animals like LD_{50} and microbial assay to establish the absence or presence of potentially harmful microorganisms.

The aforementioned processes involve a wide array of scientific investigations, which include physical, chemical, and biological evaluation employing various analytical methods and tools. The specific aims of such investigation in assuring herbal quality are as varied as the processes employed.

6.3 Techniques in standardization of medicinal plant preparations

The following are some of the required techniques for the standardization of medicinal plant preparations as reported by Kunle et al. (2012).

6.3.1 Physical evaluation
Each monograph contains detailed botanical, macroscopic, and microscopic descriptions of the physical characteristics of each plant that can be used to ensure both identity and purity. Each description is accompanied by detailed illustrations and photographic images which provide visual documentation of accurately identified material.

6.3.2 Microscopic evaluation
Full and accurate characterization of plant material requires a thorough physical examination. Microscopic analyses of plants are invaluable for assuring the identity of the material and as an initial screening test for impurities.

6.3.3 Chemical evaluation
This covers screening, isolation, identification, and purification of the chemical components. Chemical analysis of the drug is done to assess the potency of vegetable material in terms of its active principles. The chemical screening or tests may include color reaction test, which help to determine the identity of the drug substance and possible adulteration.

6.3.4 Biological evaluation

Pharmacological activity of certain drugs has been applied to evaluate and standardize them. The assays on living animal and on their intact or isolated organs can indicate the strength of the drug or their preparations. These assays are known as biological assays or bioassay.

6.3.5 Purity determination

Each monograph includes standards for purity and other qualitative indices already mentioned.

6.3.6 Analytical methods

Critical to compliance with any monograph standard is the need for appropriate analytical methods for determining identity, quality, and relative potency. There are a plethora of analytical methods available. However, it is often difficult to know which is the most appropriate to use, but critical among known analytical tools in monograph standardization is chromatography.

6.3.7 Chromatography

Chromatography is the science which studies the separation of molecules based on differences in their structure and/or composition. In general, chromatography involves moving a preparation of the materials to be separated, "the "test preparation," over a stationary support. The molecules in the test preparation will have different interactions with the stationary support leading to separation of similar molecules. Test molecules which display tighter interactions with the support will tend to move more slowly through the support than those molecules with weaker interactions. In this way, different types of molecules can be separated from each other as they move over the support material. Chromatographic separations can be carried out using a variety of supports, including immobilized silica on glass plates (thin layer chromatography), very sensitive high performance thin layer chromatography (HPTLC), volatile gases [gas chromatography(GC)], paper (paper chromatography), and liquids which may incorporate hydrophilic, insoluble molecules [liquid chromatography (LC)].

HPTLC is a valuable quality assessment tool for the evaluation of botanical materials. It allows for the analysis of a broad number of compounds both efficiently and cost effectively. Additionally, numerous samples can be run in a single analysis thereby dramatically reducing analytical time. With HPTLC, the same analysis can be viewed collectively in different wavelengths of light thereby providing a more complete profile of the plant than is typically observed with more specific type of analysis.

6.3.8 Quantitative analysis

The most appropriate quantitative analytical method with accompanying chromatograms is desirable. The primary goal of the methods is to provide validated methods to be used to quantify the compounds most correlated with pharmacological activity or qualitative markers (Wani, 2007).

6.3.9 Control of starting material

Control of the starting materials is essential to ensure reproducible quality of herbal medicinal products (De Smet, 2004; Gaedcke and Steinhoff, 2003; WHO, 2002). The following points are to be considered in the control of starting materials:
- authentication and reproducibility of herbal ingredients,
- plant populations where collections are made (wild or cultivated medicinal plants).

The problems associated with unregulated herbal products highlight the major public health issues that can arise when their herbal ingredients have not been authenticated correctly. Herbal ingredients must be accurately identified by macroscopic and microscopic comparisons with authentic material or accurate descriptions of authentic herbs (Houghton, 1998). It is essential that herbal ingredients are referred to by their binomial Latin names of genus and species; only permitted synonyms should be used. Even when correctly authenticated, it is important to realize that different batches of the same herbal ingredient may differ in quality due to a number of factors such as:
- Inter- or intraspecies variation: the variation in constituents is mostly genetically controlled and may be related to the country of origin.
- Environmental factors: the quality of an herbal ingredient can be affected by environmental factors like climate, altitude, and other conditions under which it was cultivated.
- Time of harvesting: For some herbs the optimum time of harvesting should be specified as it is known that the concentrations of constituents in a plant can vary during the growing cycle or even during the course of a day.
- Plant part used: active constituents usually vary between plant parts and it is not uncommon for a herbal ingredient to be adulterated with parts of the plant not normally utilized. In addition, plant material that has been previously subjected to extraction and is therefore "exhausted" is sometimes used as adulterants to increase the weight of a batch of herbal ingredient.
- Postharvesting factors: storage conditions and processing treatments can greatly affect the quality of a herbal ingredient. Inappropriate storage after harvesting can result in microbial contamination, and processes such as drying may result in a loss of thermolabile active constituents.

6.3.10 Adulteration/substitution

There are instances when herbal remedies have been adulterated with other plant materials and conventional medicines. Reports of herbal products devoid of known active constituents have reinforced the need for adequate quality control of herbal remedies.

6.3.11 Identity and purity

In order to try to ensure the quality of licensed herbal medicines, it is essential not only to establish the botanical identity of a herbal ingredient but also to ensure batch-to-batch reproducibility. Thus, in addition to macroscopic and microscopic evaluation, identity tests are necessary. Such tests include simple chemical tests, for example, color or precipitation and chromatographic tests. Thin-layer chromatography is commonly used for identification purposes but for herbal ingredients containing volatile oils, a gas–liquid chromatographic test may be used. Although the aim of such tests may be to confirm the presence of active principles, it is frequently the case that the nature of the active principle has not been established. In such instances chemical and chromatographic tests help to provide batch-to-batch comparability and the chromatogram may be used as a "fingerprint" for the herbal ingredient by demonstrating the profile of some common plant constituents such as flavonoids, alkaloids, and terpenes.

To prove identity and purity, criteria such as type of preparation, sensory properties, physical constants, adulteration, contaminants, moisture, ash content, and solvent residues have to be checked. Identity can be achieved by macro- and microscopical examinations. Voucher specimens are reliable reference sources. Outbreaks of diseases among plants may result in changes to the physical appearance of the plant and lead to incorrect identification (De Smet, 1999).

Assaying for those herbal ingredients with known active principles is another method of ensuring product's identity and purity. An assay should be established in order to set the criterion for the minimum accepted percentage of active substances. Such assays should, wherever possible, be specific for individual chemical substances, and high-pressure LC and gas–liquid chromatography are the methods of choice. Where such assays have not been established, then nonspecific classical methods such as titration or colorimetric assays may be used to determine the total content of a group of closely related compounds. Purity is closely linked with the safe use of drugs and deals with factors such as values, contaminants (e.g., foreign matter in the form of other herbs), and heavy metals. However, due to the application of improved analytical methods, modern purity evaluation also includes microbial contamination, aflatoxins, radioactivity, and pesticide residues. Analytical methods such as photometric analysis, TLC, HPLC, and GC can be employed to establish the constant composition of herbal preparations (WHO, 1996, 1998b).

A special form of assay is the determination of essential oils by steam distillation. When the active constituents (e.g., sennosides in Senna) or markers (e.g., alky amides in Echinacea) are known, a vast array of modern chemical analytical methods such as ultraviolet/visible spectroscopy (UV/VIS), TLC, HPLC, GC, mass spectrometry (MS), or a combination of GC and MS (GCMS), can be employed (Watson, 1999).

6.3.12 Good agricultural/manufacturing practices

Quality control and the standardization of herbal medicines also involve several other steps like source and quality of raw materials, good agricultural practices (GAP), and good manufacturing practices. These practices play a pivotal role in guaranteeing the quality and stability of herbal preparations (EMEA, 2002; WHO, 2000, 2003; Blumenthal et al., 1998). The quality of a plant product is determined by the prevailing conditions during growth, and accepted GAP can control this. These include seed selection, growth conditions, fertilizers application, harvesting, drying, and storage. In fact, GAP procedures are integral part of quality control. Factors such as the use of fresh plants, age and part of plant collected, period, time and method of collection, temperature of processing, exposure to light, availability of water, nutrients, drying, packing, transportation of raw material and storage, can greatly affect the quality, and hence the therapeutic value of herbal medicines. Apart from these criteria, factors such as the method of extraction, contamination with microorganisms, heavy metals, and pesticides can alter the quality, safety, and efficacy of herbal drugs. Using cultivated plants under controlled conditions instead of those collected from the wild can minimize most of these factors (Blumenthal et al., 1998; Eskinazi et al., 1999). Sometimes, the active principles are destroyed by enzymatic processes that continue for long periods from collection to marketing, resulting in a variation of composition. Thus, proper standardization and quality control of both the raw material and the herbal preparations should be conducted.

6.3.13 Contaminants of herbal ingredients

Herbal ingredients of high quality should be free from insects, animal matter, and excreta. It is usually not possible to remove completely all contaminants; hence specifications should be set in order to limit them:

- Ash values: incineration of a herbal ingredient produces ash which constitutes inorganic matter. Treatment of the ash with hydrochloric acid results in acid-insoluble ash which consists mainly of silica and may be used to act as a measure of soil present. Limits may be set for ash and acid-insoluble ash of herbal ingredients.
- Foreign organic matter: it is not possible to collect a herbal ingredient without small amounts of related parts of plant or other plants. Standards should be set to limit the percentage of such unwanted plant contaminants.
- Microbial contamination: aerobic bacteria and fungi are normally present in plant material and may increase due to faulty growing, harvesting, storage, or processing. Herbal ingredients, particularly those with high starch content, may be prone to increased microbial growth. Pathogenic organisms including *Enterobacter, Enterococcus, Clostridium, Pseudomonas, Shigella,* and *Streptococcus* have been shown to contaminate herbal ingredients. It is essential that limits be set for microbial contamination and the European Pharmacopoeia now gives nonmandatory guidance on acceptable limits (Barnes et al., 2007).

- Pesticides: herbal ingredients, particularly those grown as cultivated crops, may be contaminated by DDT (dichlorodiphenyltrichloroethane) or other chlorinated hydrocarbons, organophosphates, carbamates, or polychlorinated biphenyls. Limit tests are necessary for acceptable levels of pesticide contamination of herbal ingredients. The European Pharmacopoeia includes details of test methods together with mandatory limits for 34 potential pesticide residues (Barnes et al., 2007).
- Fumigants: ethylene oxide, methyl bromide, and phosphine have been used to control pests which contaminate herbal ingredients. The use of ethylene oxide as a fumigant with herbal drugs is no longer permitted in Europe (Barnes et al., 2007).
- Toxic metals: lead, cadmium, mercury, thallium, and arsenic have been shown to be contaminants of some herbal ingredients. Limit tests for such toxic metals are essential for herbal ingredients.
- Radioactive contamination: there are many sources of ionization radiation, including radionuclides, occurring in the environment. Hence, a certain degree of exposure is inevitable (AOAC, 2005; WHO, 2000).
- Other contaminants: as standards increase for the quality of herbal ingredients it is possible that tests to limit other contaminants such as endotoxins and mycotoxins will be utilized to ensure high quality for medicinal purposes (Barnes et al., 2007).

7 QUALITY CONTROL OF MEDICINAL PREPARATIONS

Quality control systems are important for the production of high-quality herbal products. Lack of quality control may lead to problems due to unidentified problems in the production process that can lead to inferior or inconsistent products or even accidents due to intoxication or allergic reactions. The rising popularity of herbal products as food and feed supplements and as phytotherapeutic drugs has also given rise to many reports describing adverse health effects and variable quality, efficacy, and contents of herbal products (Taylor, 2005). Quality control for the efficacy and safety of herbal products is essential. The quality control of phytopharmaceuticals may be defined as the status of a drug, which is determined either by identity, purity, content, and other chemical, physical, or biological properties, or by the manufacturing process. Compared with synthetic drugs, the criteria and the approach for herbal drugs are much more complex (Bandaranayake, 2006). Quality control is an important tool in the production of high-quality herbal products, whereas lack of control can result in inferior products that may lead to health problems in the consumers (Groot and Van der Roest, 2006)

Phytopharmaceuticals are always mixtures of many constituents and are therefore very variable and difficult to characterize. The active principle(s) in phytopharmaceuticals are not always known. The quality criteria for herbal drugs are based on a clear scientific definition of the raw material. Depending on the type of preparation, sensory properties, physical constants, moisture, ash content, solvent residues, and adulterations have to be checked to prove identity and purity. Microbiological contamination and foreign materials, such as heavy metals, pesticide residues, aflatoxins, and radioactivity, also need to be tested. To prove the constant composition of herbal preparations, appropriate analytical methods have to be applied and different concepts have to be used in order to establish relevant criteria for uniformity (Bandaranayake, 2006).

7.1 Challenges and factors affecting the quality control of herbal drugs
7.1.1 Identity of plant material
Quality control of botanicals starts right from identification of plant. According to WHO general guidelines for methodologies on research and evaluation of traditional medicines, first step in assuring quality, safety, and efficacy of traditional medicines is correct identification. Plant can be named in four different ways: the common English name, the transliterated name, the Latinized pharmaceutical name, and the scientific name. When binomial names are not used misidentification can occur. For example, the scientific name of the Chinese herb that is variously transliterated as "dong quai," "dong guai," and "tang kuei" is *Angelica polymorpha* (formerly *sinensis*). The common English name "Angelica" and Latinized name "*Radix angelica*" could refer either to this species, which is used in Australia, or to the European species *Angelica archangelica*, depending upon the country of the origin (Shinde, 2004).

According to Raina (2003), miscomprehension of problems in procurement of authentic plant materials is due to the following:
- Collection of wildly growing plants from forests and wastelands.
- Traders or suppliers generally have limited knowledge of medicinal plants.
- Illiterates and lower class people who are not fully aware of the identity of the drugs always do collections.
- Nonhomogeneity of plant material due to collection from wild sources and different geographical locations.

Chemical analysis is so far the best method for standardization, for detecting contamination, and for plant identification and authentication of medicinal plants. Molecular biology techniques can also applied to authentication of medicinal plants as complementary techniques (Shinde et al., 2007).

7.1.2 Variations in botanicals
Consistency in composition and biologic activity are essential requirements for the safe and effective use of therapeutic agents. However, botanical preparations rarely meet this standard, as a result of problems in identifying plants, genetic variability, variable growing conditions, differences in harvesting procedures and processing of extracts, and above all, the lack of information about active pharmacologic principles (Donald and Arthur, 2002). Environmental conditions such as sunlight, rainfall, altitude, temperature, soil, storage conditions as well as different harvesting procedures, time and

method of collection, manufacturing processes such as selecting, drying, purifying, extracting, and genetic variability can create substantial variability in product quality and in the concentration of plant chemicals within different products. Ecological conditions like insect feeding, microbial infections may affect secondary metabolites and in turn chemical composition of the plant. Also different parts of same plant (e.g., roots, stems, and leaves) contain different concentrations of chemical constituents. At the same time diurnal variations (e.g., paclitaxel, opium alkaloids) and seasonal changes also account for variability in herbal medicines. The therapeutic or toxic components of plant vary depending on the part of the plant used as well as stages of ripeness (Anna and Stephen, 1997). Products from different manufacturers vary considerably and it is not possible to control all the factors that affect the chemical composition of plants (Michael, 1999).

7.1.3 Adulteration and contamination of botanicals

Adulteration of botanical preparations is another important issue. Due to overexploitation of certain plants, habitat loss, and fragmentation of the forest, many medicinal plants have reached the level of the endangered or rare species. These and many other factors (like cost of raw material) cause problem for availability of genuine drug, which encourages the adulteration of plant by substitution with inferior commercial varieties, artificially manufactured substances, exhausted drugs or cheaper plant or by another vegetative part (Mukherjee, 2002). Several reports suggest that many herbal products contain undisclosed pharmaceuticals and heavy metals (Robert et al., 2004).

The intentional use of pharmaceutical adulterant is possible. Agrochemicals are used to protect the plant from infections and insects, which occur as contaminant in the crude plant material. Moreover mechanism of action, pharmacokinetics, and drug–drug interactions of many herbs are still in infancy. At the same time growing number of reports about fatal or adverse effects of herbal preparations intensifies need for national regulation and registration of herbal medicines and establishment of safety monitoring. Clinicians should not prescribe or recommend herbal remedies without well-established efficacy as if they were medications that had been proved effective by rigorous study. However, these products continue to have great appeal to patients, and this reality cannot be ignored. Thus, it is imperative to ask patients. This also signifies real need for quality control of botanicals (De Smet, 2002).

Patwardhan (2000) summarized the need for quality control and standardization of herbal product as follows:
- When traditional medicines were developed technology and concept of standardization were quite different.
- During past thousand years dynamic process of evolution may have changed the identity of plant material.

- Due to commercialization, supply of genuine raw material has become a challenge. Properties of botanicals may have undergone change due to time and environmental factors.

All these factors adversely affect the quality of crude drug and, consequently, formulations manufactured by using them as ingredient (Shinde et al., 2008).

7.1.4 Parameters for quality control of herbal drugs
7.1.4.1 Microscopic evaluation

Quality control of herbal drugs has been traditionally based on the appearance and today microscopic evaluation is indispensable in the initial identification of herbs, as well as, in identifying small fragments of crude or powdered herbs, and detection of foreign matter and adulterants. A primary visual evaluation, which seldom needs more than a simple magnifying lens, can be used to ensure that the plant is of the required species, and that the right part of the plant is being used. At other times, microscopic analysis is needed to determine the correct species and/or that the correct part of the species is present. For instance, pollen morphology may be used in the case of flowers to identify the species, and the presence of certain microscopic structures such as leaf stomata can be used to identify the plant part used. Although this may seem obvious, it is of prime importance, especially when different parts of the same plant are to be used for different treatments. Stinging nettle (*Urtica urens*) is a classic example where the aerial parts are used to treat rheumatism, while the roots are applied for benign prostate hyperplasia (AOAC, 2005).

7.1.4.2 Foreign matter

Herbal drugs should be made from the stated part of the plant and be devoid of other parts of the same plant or other plants. They should be entirely free from molds or insect, including excreta and visible contaminants such as sand and stones, poisonous and harmful foreign matter, and chemical residues. Animal matters such as insects and "invisible" microbial contaminants, which can produce toxins, are also among the potential contaminants of herbal medicines (EMEA, 2002; WHO, 2003). Macroscopic examination can easily be employed to determine the presence of foreign matter, although, microscopy is indispensable in certain special cases (e.g., starch deliberately added to "dilute" the plant material). Furthermore, when foreign matter consists, for example, of a chemical residue, TLC is often needed to detect the contaminants (AOAC, 2005; WHO, 1998b).

7.1.4.3 Ash content

To determine ash content, the plant material is burnt and the residual ash is measured as total and acid-insoluble ash. Total ash is the measure of the total amount of material left after burning and includes ash derived from the part of the plant itself and acid-insoluble ash. The latter is the residue obtained after boiling the total ash with dilute hydrochloric acid, and burning the remaining insoluble matter. The second procedure measures the amount of silica present, especially in the form of sand and siliceous earth (AOAC, 2005).

7.1.4.4 Heavy metals

Contamination by toxic metals can either be accidental or intentional. Contamination by heavy metals such as mercury, lead, copper, cadmium, and arsenic in herbal remedies can be attributed to many causes, including environmental pollution, and can pose clinically relevant dangers for the health of the user and should therefore be limited (AOAC, 2005; WHO, 1998a). The potential intake of the toxic metal can be estimated on the basis of the level of its presence in the product and the recommended or estimated dosage of the product. This potential exposure can then be put into a toxicological perspective by comparison with the so-called provisional tolerable weekly intake (PTWI) values for toxic metals, which have been established by the Food and Agriculture Organization of the World Health Organization (FAO-WHO) (De Smet, 1999). A simple, straightforward determination of heavy metals can be found in many pharmacopoeias and is based on color reactions with special reagents such as thioacetamide or diethyldithiocarbamate, and the amount present is estimated by comparison with a standard (WHO, 1988). Instrumental analyses have to be employed when the metals are present in trace quantities, in admixture, or when the analyses have to be quantitative. Generally, the main methods commonly used are atomic absorption spectrophotometry (AAS), inductively coupled plasma (ICP), and neutron activation analysis (NAA) (Watson, 1999).

7.1.4.5 Microbial contaminants and aflatoxins

Medicinal plants may be associated with a broad variety of microbial contaminants, represented by bacteria, fungi, and viruses. Inevitably, this microbiological background depends on several environmental factors and exerts an important impact on the overall quality of herbal products and preparations. Risk assessment of the microbial load of medicinal plants has therefore become an important subject in the establishment of modern hazard analysis and critical control point (HACCP) schemes. Herbal drugs normally carry a number of bacteria and molds, often originating in the soil. Poor methods of harvesting, cleaning, drying, handling, and storage may also cause additional contamination, as may be the case with *Escherichia coli* or *Salmonella* spp. while a large range of bacteria and fungi are from naturally occurring microflora, aerobic spore-forming bacteria that frequently predominate (WHO, 2000).

Laboratory procedures investigating microbial contaminations are laid down in the well-known pharmacopoeias, as well as, in the WHO guidelines (WHO, 1998b, 2000). Limit values can also be found in the sources mentioned. Generally, a complete procedure consists of determining the total aerobic microbial count, the total fungal count, and the total Entero-bacteriaceae count, together with tests for the presence of *E. coli, Staphylococcus aureus, Shigella*, and *Pseudomonas aeruginosa* and *Salmonella* spp.

The European Pharmacopoeia also specifies that *E. coli* and *Salmonella* spp. should be absent from herbal preparations. Materials of vegetable origin tend to show much higher levels of microbial contamination than synthetic products and the requirements for microbial contamination in the European Pharmacopoeia allow higher levels of microbial contamination in herbal remedies than in synthetic pharmaceuticals. The allowed contamination level may also depend on the method of processing of the drug. For example, higher contamination levels are permitted if the final herbal preparation involves boiling with water (WHO, 2000).

The presence of fungi should be carefully investigated and/or monitored, since some common species produce toxins, especially aflatoxins. Aflatoxins in herbal drugs can be dangerous to health even if they are absorbed in minute amounts (WHO, 2000). Aflatoxin-producing fungi sometimes build up during storage (De Smet et al., 1992). Procedures for the determination of aflatoxin contamination in herbal drugs are published by the WHO (2000). After a thorough clean-up procedure, TLC is used for confirmation. In addition to the risk of bacterial and viral contamination, herbal remedies may also be contaminated with microbial toxins, and as such, bacterial endotoxins and mycotoxins, at times may also be an issue (De Smet et al., 1992).

There is evidence that medicinal plants from some countries may be contaminated with toxigenic fungi (*Aspergillus, Fusarium*). Certain plant constituents are susceptible to chemical transformation by contaminating microorganisms. Withering leads to enhanced enzymatic activity, transforming some of the constituents to other metabolites not initially found in the herb. These newly formed constituent(s) along with the molds such as *Penicillium nigricans* and *P. jensi* may then have adverse effects (De Smet et al., 1992).

7.1.4.6 Pesticide residues

Herbal drugs are liable to contain pesticide residues, which accumulate from agricultural practices, such as spraying, treatment of soils during cultivation, and administering of fumigants during storage. However, it may be desirable to test herbal drugs for broad groups in general, rather than for individual pesticides. Many pesticides contain chlorine in the molecule, which, for example, can be measured by analysis of total organic chlorine. In an analogous way, insecticides containing phosphate can be detected by measuring total organic phosphorus. Samples of herbal material are extracted by a standard procedure, impurities are removed by partition and/or adsorption, and individual pesticides are measured by GC, MS, or GC-MS. Some simple procedures have been published by the WHO and the European Pharmacopoeia has laid down general limits for pesticide residues in medicine (AOAC, 2005; WHO, 1998b, 2000).

7.1.4.7 Radioactive contamination

The WHO, in close cooperation with several other international organizations, has developed guidelines in the event of a widespread contamination by radionuclides resulting from major nuclear accidents. These publications emphasize that the health risk, in general, due to radioactive contamination from naturally occurring radio nuclides is not a real concern, but those arising from major nuclear accidents such as the nuclear accident in Chernobyl and Fukushima may be serious and depend on the specific radionuclide, the level of contamination, and the quantity of the contaminant consumed. Taking into account the quantity of herbal medicine normally consumed by an individual is unlikely to be a health risk. Therefore, at present, no limits are proposed for radioactive contamination (AOAC, 2005; De Smet et al., 1992; WHO, 2000).

7.1.4.8 Analytical methods

Published monographs in a pharmacopoeia are the most practical approach for quality control of herbal drugs and there are many available (EMEA, 2005; WHO, 1998b). When pharmacopoeia monographs are unavailable, development and validation of analytical procedures have to be carried out by the manufacturer. The best strategy is to follow closely the pharmacopoeia definitions of identity, purity, and content or assay. Valuable sources for general analytical procedures are included in the pharmacopoeias, in guidelines published by the WHO (2000). The plant or plant extract can be evaluated by various biological methods to determine pharmacological activity, potency, and toxicity.

A simple chromatographic technique such as TLC may provide valuable additional information to establish the identity of the plant material. This is especially important for those species that contain different active constituents. Qualitative and quantitative information can be gathered concerning the presence or absence of metabolites or breakdown of products (AOAC, 2005). TLC fingerprinting is of key importance for herbal drugs made of essential oils, resins, and gums, which are complex mixtures of constituents that no longer have any organic structure. It is a powerful and relatively rapid solution to distinguish between chemical classes, where macroscopy and microscopy may fail. Chromatograms of essential oils, for example, are widely published in the scientific literature, and can be of invaluable help in identification. The instruments for UV-visible determinations are easy to operate, and validation procedures are straightforward but at the same time precise. Although measurements are made rapidly, sample preparation can be time consuming and works well only for less complex samples, and those compounds with absorbance in the UV-visible region. HPLC is the preferred method for quantitative analysis of more complex mixtures. Although the separation of volatile components such as essential and fatty oils can be achieved with HPLC, it is best performed by GC or GC-MS. The quantitative determination of constituents has been made easy by recent developments in analytical instrumentation. Recent advances in the isolation, purification, and structure elucidation of naturally occurring metabolites have made it possible to establish appropriate strategies for the determination and analysis of quality and the process of standardization of herbal preparations (WHO, 2002).

Classification of plants and organisms by their chemical constituents is referred to as chemotaxonomy. TLC, HPLC, GC, quantitative TLC (QTLC), and HPTLC can determine the homogeneity of a plant extract. Overpressured layer chromatography (OPLC), infrared and UV-Visible spectrometry, MS, GC, LC used alone, or in combinations such as GC-MS and LC-MS, and nuclear magnetic resonance (NMR), electrophoretic techniques, especially by hyphenated chromatographic techniques, are powerful tools, often used for standardization and to control the quality of both the raw material and the finished product. The results from these sophisticated techniques provide a chemical fingerprint as to the nature of chemicals or impurities present in the plant or extract (WHO, 2002). Based on the concept of photo equivalence, the chromatographic fingerprints of herbal medicines can be used to address the issue of quality control. Methods based on information theory, similarity estimation, chemical pattern recognition, spectral correlative chromatograms (SCC), multivariate resolution, the combination of chromatographic fingerprints, and chemometric evaluation for evaluating fingerprints are all powerful tools for quality control of herbal products.

7.1.4.9 Validation

The validation of herbal products is a major public health concern both in developed and resource-poor countries, where fakers selling adulterated herbal medicines are common. In this regard, there is no control by the government agencies, despite the existence of certain guidelines in some individual countries and those outlined by the WHO. If the herbal products are marketed as therapeutic agents, and irrespective of whether the products really have any positive effects to cure and reduce the severity of the disease, it is necessary to ensure scientific validation and periodic monitoring of the quality and efficacy by drug control administrators. It is feasible that the introduction of scientific validation would control the production of impure or adulterated herbal products and would eventually ensure their rational use. This could also lead to the regulation of the industry so that only qualified physicians and health providers are allowed to prescribe the medication (Bandaranayake, 2006).

Several of the principal pharmacopoeias contain monographs outlining standards for herbal drugs. The major advantage of an official monograph published in a pharmacopoeia is that standards are defined and available, and that the analytical procedures used are fully validated. This is of major importance, since validation can be a rather time-consuming process. By definition, validation is the process of proving that an analytical method is acceptable for its intended purpose for pharmaceutical methods. Guidelines from the United States Pharmacopoeia (USPC, 1994–2001), the International Conference on Harmonization (ICH), and the US Food and Drug Administration (FDA) provide a framework for performing such validations. Generally, validation investigations must include studies on specificity, linearity, accuracy, precision, range, detection, and quantitative limits, depending on whether the analytical method used is qualitative or quantitative (De Smet, 1997). Also, of utmost importance is the availability of standards. For macroscopic and microscopic procedures in general this means that reliable reference samples of the plant must be available. A defined botanical source (e.g., voucher specimens) will normally solve this problem. Standards for chromatographic procedures are less easy to obtain. Characteristic plant constituents, either active or markers, are seldom available commercially. Sometimes an LC-MS approach can be referred to as a mode of characterization. Going one step further, after isolation of such a compound, elucidations to prove its definite structure will not be easy. The method often employed is to use readily available compounds that behave similarly in the chosen chromatographic systems, and to calculate retention values and/or times toward these compounds as a standard. Qualitative chemical examination is designed to detect and isolate the active ingredients. TLC and HPLC are the main analytical techniques commonly used. In cases when active ingredients are not known or too complex, the quality of plant extracts can be assessed by a "fingerprint" chromatogram (De Smet, 1997).

7.1.4.10 Labeling of herbal products

The quality of consumer information about the product is as important as the finished herbal product. Warnings on the packet or label will help to reduce the risk of inappropriate uses and adverse reactions (De Smet, 1997). The primary source of information on herbal products is the product label. Currently, there is no organization or government body that certifies herb or a supplement as being labeled correctly. It has been found that herbal remedy labels often cannot be trusted to reveal what is in the container. Studies of herbal products have shown that consumers have less than a 50% chance of actually getting what is listed on the label, and published analyses of herbal supplements have found significant differences between what is listed on the label and what is in the bottle. The word "standardized" on a product label is no guarantee of higher product quality, as there is no legal definition of the word "standardized." Consumers are often left on their own to decide what is safe and effective for them and the lack of consistent labeling on herbal products can be a source of consumer frustration. Certain information such as "the product has been manufactured according to Pharmacopoeia standards," listing of active ingredients and amounts, directions such as serving quantity (dosage), and frequency of intake of the drug, must be in the label (Bandaranayake, 2006).

7.2 Approaches in quality control of herbal medicine

The specific quality control of the tested samples usually was achieved through the identification and determination. Due to uncertainty and complexity, there is great difficulty in establishing a specific method of quality control of HMs. The techniques of authentication are not powerful enough to identify all the ingredients in a HM, target setting is too general to determine the active ingredient, and the components that are regarded as markers are often not active. Hereby there are some challenges in developing specific and objective quality standards of herb medicines. Thus, adulterated and poor quality drugs appeared in markets. To eliminate these quality problems, on one hand we must actively foster the concept of pharmaceutical production of good quality and the concept of legal system, and on the other hand, the scientific and technologically advanced quality control standards should be established as soon as possible.

7.2.1 Fingerprint approach
By definition, a chromatographic fingerprint of an HM is, in practice, a chromatographic pattern of the extract of some common chemical components of pharmacologically active or chemically characterized (Liang et al., 2011). Specifically, fingerprints of herbal medicine refer to the profiles which can illustrate the specific properties of the analyte including raw materials, slices, semifinished products, and finished products after appropriate processing, and be obtained by certain analysis techniques. The research of finger printing of herbal medicines is really an interdisciplinary and comprehensive research, which is based on the chemical composition of traditional Chinese medicine system. It needs crossover of herbal medicine, separation science, analytical science, and bioinformatics to provide a platform for the quality control of traditional herbal medicines. Those features make fingerprint analysis especially suitable for research on HMs which bear characteristics of a complex mixture of chemical compounds. It can evaluate the integrative and holistic properties of herbal medicines by comparing the similarity and correlation of the analytes among the whole producing process, such as manufacture, processing, and storage of raw materials for preparation, intermediate products, finished products, and distribution products. The fingerprint analysis have been internationally accepted as one of the efficient methods to control the quality of herbal medicines (Liang et al., 2011).

Fingerprinting, generally, was divided into chemical and biological fingerprint patterns. Chemical fingerprint is used to analyze the chemical constituents in HMs, consisting of chromatographic fingerprints, such as TLC, HPLC, GC, CE and spectral fingerprint, for instance, UV, IR, MS, X-ray, and so on, and their hyphenated techniques. The biological fingerprints mainly refer to genomics fingerprints. As genetic composition is unique for each individual, DNA methods for HMs' identification are less affected by age, physiological conditions, environmental factors, harvest, storage, and processing methods. Genomic fingerprint has been used widely for the differentiation of plant individual, genus, homogeneity analysis, and detection of adulterants (Cheng et al., 1997; Wang et al., 2007; Zhang et al., 2006). However, as for herbal instances processed or extractions of plants, DNA fingerprinting techniques usually cannot do anything.

7.2.2 Metabolomics approach

Over the past several decades, the chromatographic fingerprint technique has been used as a more accurate approach for controlling the quality of herbal medicines or their products due to the systemic characterization of compositions of samples and focusing on identifying and assessing the stability of the plants. The fingerprint analysis technique has been introduced and accepted by the WHO (1991), FDA (2000), and European Medicines Agency (EMEA) (EMEA, 2005) as a strategy for assessing consistency between batches of botanical drugs. In 2004, the State Food and Drug Administration (SFDA) of China also required that all injections made from herbal medicines or their raw materials should be standardized by chromatographic fingerprinting (State Food and Drug Administration of China, 2000). However, a validated fingerprint method has not been accepted in general quality control standards of herbal medicines until now. One of the main difficulties is the lack of an analytical method for scientifically evaluating the complex chromatograms of herbal medicines (Sticher, 1993). The technique uses the traditional concept of pharmaceutical analysis and combines it with an analysis of chromatographic peaks. However, the current fingerprint analysis technique does not correct for chromatographic shifts among different runs or from different experimental conditions and cannot compare the fingerprints of TCMs from different species of herbs, grown at different locations, from different harvesting seasons, or extracted and processed using different methods. Therefore, an effective quality control method for TCMs or herbal medicines should be developed to further develop their standardization and modernization.

A metabolomics approach is uniquely suited for developing a new and effective method for the quality control of the processing and manufacturing of TCMs. The central goal of this method is to ensure that representative bioactive components in one preparation are consistent, that the content of each bioactive component meets the stipulated standards, and that concentration proportions of mass constituents are within a reasonable range (Yongyu et al., 2011). Metabolomics, or metabonomics, which aims to identify and quantify the full complement of low molecular weight (<1000 Da), soluble metabolites in actively metabolizing tissues (Fiehn, 2002; Rochfort, 2005) has garnered extensive awareness and interest in the research community, especially in botanical studies for the quality control of plants, metabolic phenotyping (Dai et al., 2010), and the pharmacological effects of TCMs (Liang et al., 2011).

The metabolomics strategy enables the determination of the global metabolite profile of preparations and simultaneously quantifies the key ingredients to ascertain treatment efficacy and minimize probable side effects, which will greatly enhance pharmacological evaluation. This approach will help to perfect the current quality control method used for complicated herbal materials or medicines, improve the quality of TCMs, and ensure the efficacy of the products used in the clinic (Li et al., 2013).

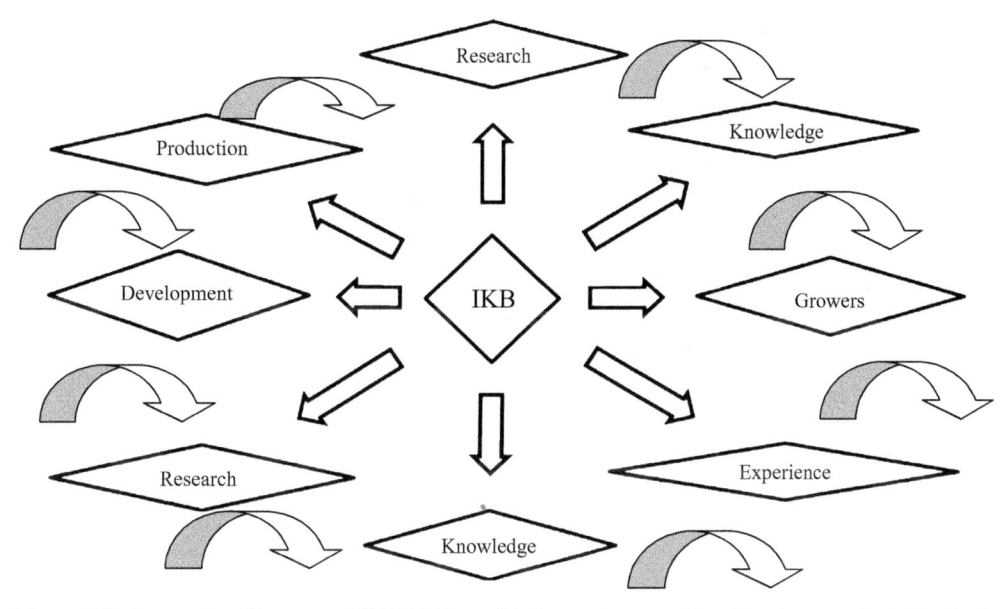

Figure 7.2 *Integral quality control (IQC) linking all information together.* (Bandaranayake, 2006. Quality control, screening, toxicity, and regulation of herbal drugs. Modern Phytomedicine: Turning Medicinal Plants into Drugs. Available from: http://doi.org/10.1002/9783527609987.ch2)

7.2.3 Integral quality control

Quality control is an important tool in the production of high-quality herbal products, whereas lack of control can result in inferior products that may lead to health problems in the consumers (Groot and Van der Roest, 2006). Specific knowledge is linked to the different chains of the production, but a central point that connects all these segments is necessary. This point is formed by the integral quality control (IQC) that collects the information from each point and makes the information available for the whole system, as is illustrated in the knowledge mill (Fig. 7.2). To develop such a system, an integral quality control system has to be built and managed. Managing an integral quality system implies (1) a build-up in phases (product, process, service); (2) starting with basic supplies (baseline measurement); (3) filling-in specific supplies (bottlenecks); (4) registration, control, and assurance (procedures); (5) documentation in a quality manual (binder).

An IQC system has many benefits, such as providing insight into company management, improvement in the efficiency and effectiveness of the production process, provision of products with constant quality, economic methods for production, and demonstrable activities. Moreover, quality control provides a guarantee for the customer, means cost reduction for the supplier and, maybe most importantly, leads to identification and traceability of errors. Although quality can be defined in many ways, all measurements share the common assumption that "quality is defined by the customer." The most used definition is "fitness for use" (Juran and Godfrey, 1999). Other

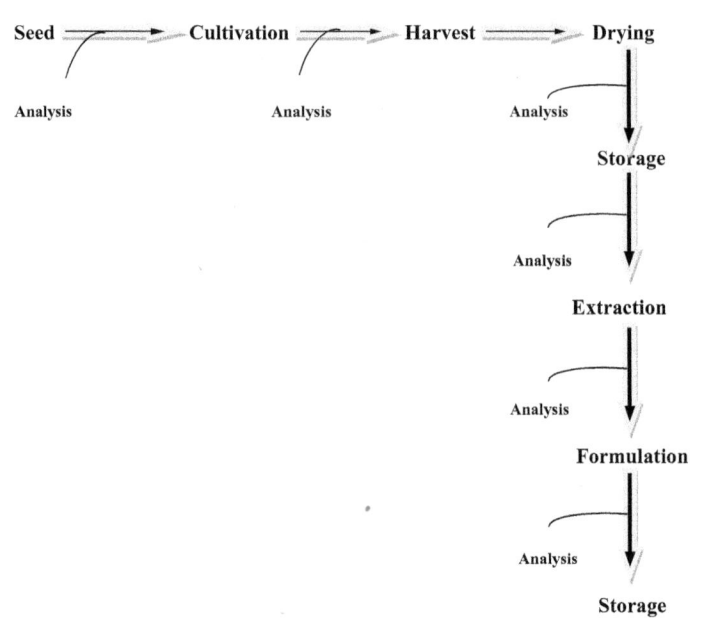

```
Seed ⇌→ Cultivation ⇌→ Harvest ⇌→ Drying
  |              |              |         |
Analysis     Analysis       Analysis      ↓
                                       Storage
                                          |
                                       Analysis
                                          ↓
                                       Extraction
                                          |
                                       Analysis
                                          ↓
                                       Formulation
                                          |
                                       Analysis
                                          ↓
                                       Storage
```

Figure 7.3 *Quality control points in the production chain.* (Bandaranayake, 2006. Quality control, screening, toxicity, and regulation of herbal drugs. Modern Phytomedicine: Turning Medicinal Plants into Drugs. Available from: http://doi.org/10.1002/9783527609987.ch2)

quality experts emphasize particular aspects, such as W. Edwards Deming who calls for "continuous improvement," whereas Philip Crosby uses the phrase "conformance to requirements." In contrast, the Japanese quality expert, Kaoru Ishikawa, combines several aspects and thinks in terms of a product that is "most economical, most useful, and always satisfactory to the consumer." Quality, which has a direct impact on product performance and hence on customer satisfaction, thus begins with customer needs and ends with customer satisfaction (Cortada, 1993). Quality refers to the product in terms of technical specifications and to the organization of the production process and the continuity of service. The know-how and control of the production process and the coordination of all links are essential for good quality. Monitoring should be done at selected steps in production process (Fig. 7.3).

Quality control at the farm level is concerned with sampling, specifications, and testing, whereas quality control at the supply-chain level is concerned with organization, documentation, and release. These procedures serve to ensure that the necessary and relevant tests are actually done and that materials are not released for use and products not released for sale until the product quality has been judged to be satisfactory. The basic requirements of quality control for herbal production are that:
- adequate facilities, trained personnel, and approved procedures are available for sampling, inspecting, and testing of starting materials, packaging materials, and intermediate, bulk and finished products;

- samples of starting materials, packaging materials, and intermediate, bulk and finished products are taken by personnel and by methods approved by quality control;
- test methods are validated;
- records, made manually and/or by recording instruments, demonstrate that all the required sampling, inspecting, and testing procedures were actually done and any deviations are fully recorded and investigated;
- the finished products contain active ingredients complying with the qualitative and quantitative compositions of the Marketing Authorization are of the required purity, and are enclosed within proper containers and correctly labeled;
- records of results from inspection, material testing, and intermediate, bulk and finished products are formally assessed against specification, including a review and evaluation of relevant production documentation and an assessment of deviations from specified procedures;
- no batch of product is released for sale or supply unless certified by a qualified person that all material is in accordance with the requirements of the Marketing Authorization;
- sufficient reference samples of starting materials and products are retained to permit future examination of the product if necessary, and the product is retained in the final pack unless exceptionally large packs are produced (www.pharmacos.eudra.org).

8 CONCLUSIONS

Although herbal products have become increasingly popular throughout the world, one of the impediments in its acceptance is the lack of standard quality control profile. The quality of herbal medicine has implications in efficacy and safety. Africa with its rich heritage of traditional medicine has not benefited optimally from this natural endowment due to the lack of standardized mode of preparation of medicinal herb unlike China and India that have developed and integrated the herbal medicine into their health care system and this has been flourishing in many countries. The review is thus a wakeup call to the African scientists and traditional herbal practitioners to cooperate and synergize to come up with a unified system that will enable maximal benefit from our natural flora. Government should also provide enabling environment, adequate funding, and appropriate legislation to enhance the effort.

REFERENCES

Abera, B., 2014. Medicinal plants used in traditional medicine by Oromo people, Ghimbi District, Southwest Ethiopia. J. Ethnobiol. Ethnomed. 10, 40, Available from: http://doi.org/doi: 10.1186/1746-4269-10-40.

Anna, K.D., Stephen, P., 1997. Safety issues in herbal medicine: implications for the health professions. Med. J. Aust. 166, 538–541.

AOAC, 2005. Official Methods of Analysis of AOAC International, eighteenth ed. AOAC International, Gaithersburg, MD.

Bandaranayake, W.M., 2006. Quality control, screening, toxicity, and regulation of herbal drugs. Modern Phytomedicine: Turning Medicinal Plants into Drugs. Available from: http://doi.org/10.1002/9783527609987.ch2

Barnes, J., Anderson, L.A., Phillipson, J., 2007. Herbal Medicine, third ed. Pharmaceutical Press, London.

Bimakr, M., 2010. Comparison of different extraction methods for the extraction of major bioactive flavonoid compounds from spearmin. Food. Bioprod. Process. 4, 1–6.

Blumenthal, M., Brusse, W.R., Goldberg, A., Gruenwald, J., Hall, T., Riggins, C.W., Rister, R., 1998. Complete German Commission E Monographs. Therapeutic Guide to Herbal Medicines. The American Botanical Council, Austin, TX.

Cheng, K.T., Chang, H.C., Su, C.H.F.H., 1997. Identification of dried rhizomes of coptis species using random amplified polymorphic DNA. Bot. Bull. Acad. Sinica 38, 241–244.

Chintamunnee, V., Mahomoodally, M., 2012. Herbal medicine commonly used against infectious diseases in the tropical island of Mauritius. J. Herbal Med. 2, 113–125.

Cortada, J.W., 1993. TQM for Sales and Marketing Management. McGraw-Hill, New York.

Cowan, M.M., 1999. Plant products as antimicrobial agents. Clin. Microbiol. Rev. 12 (4), 564–582.

Dai, H., Xiao, C., Liu, H., Tang, H., 2010. Combined NMR and LC-MS analysis reveals the metabonomic changes in salvia miltiorrhiza bunge induced by water depletion. J. Proteome Res. 9 (3), 1460–1475.

Das, K., Tiwari, R.K.S., Shrivastava, D.K., 2010. Techniques for evaluation of medicinal plant products as antimicrobial agent: Current methods and future trends. J. Med. Plants Res. 4 (2), 104–111, Available from: http://doi.org/10.5897/JMPR09.030.

De Smet, P., 1999. Overview of herbal quality control. Drug Inform. J. 33, 717–724.

De Smet, P., 2002. Herbal remedies. N. Engl. J. Med. 347 (25), 2046–2056.

De Smet, P., 2004. Health risks of herbal remedies: an update. Clin. Pharmacol. Ther. 76 (1), 1–17.

De Smet, P.A.G., 1997. Adverse Effects of Herbal Drugs. In: De Smet, P.A.G.M., Keller, K., Hansel, R., Chandler, R.F. (Ed.), vol. 3. Springer-Verlag, Heidelberg.

De Smet, P.A.G.M., Keller, K. Hansel, R., Chandler, R.F., 1992. Adverse Effects of Herbal Drugs. vol. 1. Springer-Verlag, Heidelberg.

Dharmananda, S., 1997. The Methods of Preparation of Herb Formulas: Decoctions, Dried Decoctions, Powders, Pills, Tablets, and Tinctures. Institute for Traditional Medicine, Portland, Oregon. Available from: http://www.itmonline.org/arts/methprep.htm

Donald, M.M., Arthur, P., 2002. Botanical medicines—the need for new regulations. N. Engl. J. Med. 347 (25), 2073–2076.

EMEA, 1998. Guidelines on Quality of Herbal Medicinal Products. European Agency for the Evaluation of Medicinal Products, London. EMEA/adhocHMPWG/114/98. http://www.eudra.org/emea.html

EMEA, 2002. Points to Consider on Good Agricultural and Collection Practice for Starting Materials of Herbal Origin, London. Available from: http://doi.org/EMEA/HMPWP/31/99

EMEA, 2005. Guidelines on Quality of Herbal Medicinal Products/Traditional Medicinal Products. European Agency for the Evaluation of Medicinal Products, London. Available from: http://doi.org/EMEA/CVMP/814OO Review.

Eskinazi, D., Blumenthal, M., Farnsworth, N., Riggins, R., 1999. Botanical Medicine: Efficacy, Quality, Assurance, and Regulation. Mary Ann Liebert, New York.

Evans, W., 1996. Techniques in Microscopy: Quantitative Microscopy. A Textbook of Pharmacognosy, fourteenth ed. WB Saunders Company Ltd, London.

Farnsworth, N.R., Morris, R., 1976. Higher plants—the sleeping giant of drug development. Am. J. Pharm. Sci. Support Public Health 148 (2), 46–52.

FDA, 2000. Guidance for Industry—Botanical Drug Products (Draft Guidance). Rockville, MD, USA.

Fiehn, O., 2002. Metabolomics—the link between genotypes and phenotypes. Plant Mol. Biol. 48 (1–2), 155–177.

Gaedcke, F., Steinhoff, B., 2003. Quality assurance of herbal medicinal products. medpharm GmbH Scientific Publishers, Stuttgart.

Groot, M.J., Van der Roest, J., 2006. Quality control in the production chain of herbal products. In: Bogers, R.J., Craker, L.E., Lange, D. (Eds.), Medicinal and Aromatic Plants. Springer, USA, pp. 253–260.

Gurib-Fakim, A., 2006. Medicinal plants: traditions of yesterday and drugs of tomorrow. Mol. Aspects Med. 27 (1), 1–93.

Handa, S.S., Khanuja, S.P.S., Longo, G., Rakesh, D.D., 2008. Extraction techniques of medicinal plants. Extraction Technologies for Medicinal and Aromatic Plants, International Centre for Science and High Technology, Trieste 1–10. Available from: http://doi.org/http://dx.doi.org/10.1024/0301-1526.37. S71.3

Houghton, P., 1998. Establishing identification criteria for botanicals. Drug Inform. J. 32, 461–469.

Hylands, P., 2002. New Approaches for the Quality Control and Standardization of Plant Delivered Pharmaceutical and Neuraceutical Products. Oxford Natural Products Publications, Oxford.

Indian Pharmacopoeia, 1996. IP, vol II. Published by the Controller of Publications, Ministry of Health and Family Welfare, Government of India, New Delhi. A47–A48, A50–A54, A96, A99.

Iwu, M.M., Anyawu, B., 1982. Phytotherapeutic profile of Nigerian herbs. 1. Anti-inflammatory and antiarthritic agents. J. Ethnopharmacol. 63, 263–274.

Juran, J.M., Godfrey, A.B., 1999. Juran's Quality Handbook, Digital ed McGraw-Hill Companies, Inc., USA.

Kunle, O.F., Omoregie, H., Ochogu, P., 2012. Standardization of herbal medicines—A review, 101–112. Available from: http://doi.org/10.5897/IJBC11.163

Li, X., Chen, H., Jia, W., Xie, G., 2013. A Metabolomics-Based Strategy for the Quality Control of Traditional Chinese Medicine: Shengmai Injection as a Case Study. *Evid. Based Complement. Alternat. Med.* 2013, 836179.

Liang, X., Chen, X., Liang, Q., 2011. Metabonomic study of Chinese medicine Shuanglong formula as an effective treatment for myocardial infarction in rats. J. Proteome Res. 10 (2), 790–799.

Michael, D., 1999. Herbal medicine: a practical guide to safety and quality assurance. West J. Med. 171, 172–175.

Mukherjee, P., 2002. Quality control of herbal drugs: an approach to evaluation of botanicals. Bus. Horizons 1, 113–119.

Nunkoo, H., Mahomoodally, M., 2012. Ethnopharmacological survey of native remedies commonly used against infectious diseases in the tropical island of Mauritius. J. Ethnopharmacol. 143 (2), 548–564.

Page, L., 2001. Whole herbs or standardized plant constituents? Total Health 23 (4), 20.

Pandey, A., Tripathi, S., 2014. Concept of standardization extraction and prephytochemical screening strategies for herbal drug. J. Pharmacognosy Phytochem. 2 (5), 117.

Patwardhan, B., 2000. Ayurveda: the designer medicine: a review of ethnopharmacology and bioprospective research. Indian Drugs 37 (5), 213–227.

Raina, M., 2003. Quality control of herbal and herbo-mineral formulations. Indian J. Nat. Prod. 19 (1), 11–15.

Robert, B.S., Stefanos, N.K., Janet, P., Michael, J.B., David, M.E., Roger, B.D., Russell, S., 2004. Heavy metal content of Ayurvedic herbal medicine products. J. Am. Med. Assoc. 292 (23), 2868–2873.

Rochfort, S., 2005. Metabolomics reviewed: a new "omics" platform technology for systems biology and implications for natural products research. J. Nat. Prod. 68 (1), 1813–1820.

Seow, W.K., Thong, K., 2005. Erosive effects of common beverages on extracted premolar teeth. Aust. Dent. J. 50 (3), 173–178.

Shinde, V., 2004. Exploration of Molecular Markers in Quality Control of Herbal Medicines. Bharati Vidyapeeth University Erandwane, Pune, Maharashtra, India.

Shinde, V., Dhalwal, K., Mahadik, K., 2008. Issues related to pharmacognosy. Pharmacognosy Rev. 2 (3), 1–5.

Shinde, V.M., Dhalwal, K., 2007. Pharmacognosy: the changing scenario. Pharmacognosy Rev. 1 (1), 1–6.

Shohawon, S., Mahomoodally, M., 2013. Complementary and alternative medicine use among Mauritian women. Complement Ther. Clin. Pract. 19 (1), 36–43.

Sofowora, A., 1993. Medicinal Plants and Traditional Medicine in Africa. Spectrum Books Ltd (Pub.), Ibadan.

State Food and Drug Administration of China, 2000. Note for Studying Fingerprint of Traditional Chinese Medicine Injections (Draft). Shanghai, China.

Sticher, O., 1993. Quality of Ginkgo preparations. Planta Med. 59 (1), 2–11.

Sutar, N., Garai, R., Sharma, U.S., Sharma, U., 2010. Anthelmintic activity of Platycladus orientalis leaves extract. Int. J. Parasitol. Res. 2 (2), 1–3.

Taylor, L., 2005. The Healing Power of Rainforest Herbs: A Guide to Understanding and Using Herbal Medicinals. Square One Publishers, Inc., New York.

Tierra, M., 2012. Why Standadized Herbal Extract? An Herbalist Perspective. Available from: https://www.planetherbs.com/phytotherapy/why-standardized-herbal-extracts.html

Wang, C.Z., Li, P., Ding, J.Y., Peng, X., Yuan, C.S., 2007. Simultaneous identification of Bulbus Fritillariae cirrhosae using PCR-RFLP analysis. Phytomedicine 14, 628–632, 0944-7113.
Wani, M.S., 2007. Herbal medicine and its standarization. Pharma Info 5 (6), 1–6.
Watson, D., 1999. Pharmaceutical Analysis. Churchill Livingstone, Edinburgh.
WHO, 1988. The International Pharmacopeia, Quality Specifications for Pharmaceutical Substances, Excipients, and Dosage Forms, third ed. WHO, Geneva.
WHO, 1990. The Use of Essential Drugs. Eighth Report of the WHO Expert Committee. WHO, Geneva.
WHO, 1991. Guidelines for the Assessment of Herbal Medicines. Programme on Traditional Medicines. WHO, Geneva.
WHO, 1996. Guidelines for the Assessment of Herbal Medicines. WHO Technical Report Series. WHO, Geneva.
WHO, 1998a. Basic Tests for Drugs, Pharmaceutical Substances, Medicinal Plant Materials and Dosage Forms. WHO, Geneva.
WHO, 1998b. Guidelines for the Appropriate Use of Herbal Medicines. WHO Regional Publications, Western Pacific Series. Manila.
WHO, 2000. The World Health Organisation (WHO) Recommended Classification of Pesticides by Hazard and Guidelines to Classification. WHO, Geneva.
WHO, 2002. Traditional Medicine. Growing Needs and Potential. WHO Policy Perspective on Medicine. WHO, Geneva.
WHO, 2003. WHO Guidelines on Good Agricultural and Collection Practices (GACP). WHO, Geneva.
Yongyu, Z., Shujun, S., Jianye, D., 2011. Quality Control Method for Herbal Medicine—Chemical Fingerprint Analysis. Quality Control of Herbal Medicines and Related Areas, Prof. Yukihiro Shoyama, editor. ISBN: 978-953-307-682-9, InTech: http://www.intechopen.com/books/quality-control-of-herbal-medicines-and-related-areas/quality-controlmethod-for-herbal-medicine-chemical-fingerprint-analysis
Zhang, X., Xu, Q., Xiao, H., Liang, X., Huang, L., Liu, 2006. J. World Sci. Tech./Modern Trad. Chin. Med. Mat. Med. 8, 33.

PART II

Therapeutic Potential of African Medicinal Spices and Vegetables

8. Antimicrobial Activities of African Medicinal Spices and Vegetables 207
9. Antiinflammatory and Antinociceptive Activities of African Medicinal Spices and Vegetables 239
10. Anticancer Activities of African Medicinal Spices and Vegetables 271
11. Antiemetic African Medicinal Spices and Vegetables 299
12. African Medicinal Spices and Vegetables and Their Potential in the Management of Metabolic Syndrome 315
13. Other Health Benefits of African Medicinal Spices and Vegetables 329

CHAPTER 8

Antimicrobial Activities of African Medicinal Spices and Vegetables

J.D.D. Tamokou, A.T. Mbaveng, V. Kuete

1 INTRODUCTION

In the 21st century, infectious diseases continue to ravage the human population, and they account for approximately half of the mortality rates in tropical countries. These alarming statistics indicate their devastating nature. Unfortunately, the worldwide dissemination of multidrug-resistant bacteria has severely reduced the efficacy of antibacterial agents, thus increasing therapeutic failures (Pages et al., 2009). According to WHO, at least 39 new pathogens have been identified since 1967, including human immunodeficiency virus (HIV), Ebola, and Marburg hemorrhagic fevers (Kuete, 2010). In addition, "centuries-old threats" like influenza, malaria, and tuberculosis continue to thrive due to a combination of biological mutations, rising resistance to antibiotics, and weak health systems (Kuete, 2010). In the last 5 years, WHO has verified more than 1100 epidemic events worldwide (Kuete, 2010). Infectious diseases cause about 70% of deaths in children in developing countries and more than a third of those deaths occur in neonates (Kuete, 2010). More than 80% of tuberculosis cases occur in Asia and Africa (Zager and McNerney, 2008). Medicinal plants constitute an arsenal of chemicals that could be exploited by human to prevent microbial invasion. They have been a major source for drug development. Plant extracts and products have been used in the treatment of bacterial, fungal, and viral infections since ancient time (Bruneton, 1999; Cowan, 1999). In an excellent review of medicinal plant as antimicrobial agents (Cowan, 1999), it was estimated that at least 12,000 active compounds have been isolated from plants, representing less than 10% of the total. Several recent reviews have highlighted the underutilized potential of plant species and natural products as sources of antimicrobial drugs (McGaw et al., 2008). Plant-derived antimicrobial compounds belong to exceptionally wide diversity of classes, such as alkaloids, terpenoids, peptides, and phenolics (Cowan, 1999). Various plant secondary metabolites from African flora have been found active against both drug-sensitive and drug-resistant pathogenic bacteria. Some of them include extracts from *Bersama engleriana* (Kuete et al., 2008a); *Dorstenia barteri* (Kuete et al., 2008b, 2010a; Mbaveng et al., 2008); *Ficus cordata, Ficus polita,* and *Ficus ovata* (Kuete et al., 2009a, 2011b); *Irvingia africana*; *Tridesmostemon omphalocarpoides* (Kuete et al., 2006) etc., as well as many spices and vegetables which will be further discussed in this chapter. Some prominent bioactive compounds from African medicinal plants include laurentixanthone B (**1**; xanthone), plumbagin (**2**; naphthoquinone), 4-hydroxylonchocarpin (**3**; flavonoid) and MAB3 (**4**; coumarin) (Kuete et al., 2011b), isobavachalcone (**5**; flavonoid), and diospyrone (**6**; naphthoquinone) (Kuete et al., 2010b), and crassiflorone (**7**; naphthoquinone) (Kuete et al., 2009b). In this chapter, we will review the antimicrobial potency of African medicinal spices and vegetables against drug-sensitive and drug-resistant phenotypes together with the screening methods, cutoff points, as well as other important aspects of antimicrobial drug discovery from higher plants.

2 ANTIMICROBIAL SECONDARY METABOLITES AND THEIR MODES OF ACTION

Plant-derived compounds of therapeutic values mostly belong to various classes of secondary metabolites. They have a wide range of activity, according to the species, the topography, and climate of the country of origin, and may contain different categories of active principles (Arruda et al., 2011; Assob et al., 2011). Variations in the chemical composition modify their antimicrobial activities. Among 109 new antibacterial drugs approved in the period 1981–2006, 69% originated from natural products, and 21% of antifungal drugs were natural derivatives or compounds mimicking natural products (Osman et al., 2012). Useful antimicrobial phytochemicals belong to three main groups including phenolics, terpenoids, and alkaloids (Cowan, 1999). The main categories of antimicrobial botanicals and their modes of action will be discussed in this section (Table 8.1).

2.1 Alkaloids

Naturally occurring alkaloids are heterocyclic nitrogenous compounds; they usually display antimicrobial effects (Omulokoli et al., 1997). Good antibacterial and antifungal activities of alkaloids from *Epinetrum villosum* with minimal inhibitory concentration (MIC) of 31 μg/mL have been reported (Otshudi et al., 2005). The alkaloids from *Sida acuta* were also found to exert good in vitro antibacterial activity against

Table 8.1 Mode of action of phytochemical groups against microorganisms (Cowan, 1999)

Class	Mechanism of action
Alkaloids	Intercalate into cell wall and DNA
Terpenes	Membrane disruption
Phenols polyphenols	Substrate deprivation, metal ion complexation, membrane disruption, bind to adhesins, complex with cell wall, enzymes inhibition
Flavonoids	Inactivate enzymes, complex with cell wall, bind to adhesins
Quinones	Bind to adhesins, complex with cell wall, enzymes inhibition
Tannins	Bind to proteins, enzyme inhibition, substrate deprivation, complex with cell wall, metal ion complexation
Coumarins	Interaction with eucaryotic DNA

several pathogenic bacteria with MIC values that ranged from 16 to 400 µg/mL (Karou et al., 2005). Cosmoline, a bisbenzylisoquinoline alkaloid isolated from the root bark of *E. villosum* also showed good antimicrobial activity against several microorganisms (Otshudi et al., 2005). Recently, one new carbazole alkaloid, clausamine H, isolated from *Clausena anisata* leaves together with three known carbazoles, ekeberginine, girinimbine, and murrayamine A showed antibacterial properties against Gram-positive and Gram-negative bacteria (Tatsimo et al., 2015). Berberine (**8**) (Fig. 8.1), an isoquinoline alkaloid, was isolated from roots and stem bark of *Berberis* species, used in traditional medicine against bacterial, fungal, protozoal, and viral infections (Kim et al., 2002). It is active on several microorganisms, targeting RNA polymerase, gyrase, and topoisomerase IV as well as nucleic acids (Yi et al., 2007). Solamargine, a glycoalkaloid from the berries of *Solanum khasianum*, and other alkaloids may be useful against HIV infection (Sethi, 1983).

2.2 Terpenoids

Terpenes, also referred to as isoprenoids and their derivatives containing additional elements, usually oxygen, are called terpenoids. The antibacterial activities of some monoterpenes (C10), sesquiterpenes (C15), diterpenes (C20), triterpenes (C30), and their derivatives were recently reviewed (Kuete, 2013). However, the antimicrobial activity of terpenoids is rather poor compared to that of phenolics and alkaloids, though some compounds of this group were reported active against bacteria and fungi. Two known sesquiterpene lactones, vernolide and vernodalol isolated from *Vernonia amygdalina* displayed good antimicrobial activity against several bacterial and fungal species (Erasto et al., 2006). Two clerodane diterpenoids isolated from Cameroonian plant, *Microglossa angolensis* namely 6β-(2-methylbut-2(Z)-enoyl)-3α,4α,15,16-bis-epoxy-8β,10βH-ent-cleroda-13(16),14-dien-20,12-olide, and 10β-hydroxy-6-oxo-3α,4α,15,16-bis-epoxy-8βH-cleroda-13(16),14-dien-20,12-olide as well as spinasterol also displayed antifungal and antibacterial activities in a MIC range of 1.56–100 µg/mL (Tamokou et al., 2009a). The mechanism of action of terpenoids is not fully understood, but is speculated to involve membrane disruption by the lipophilic compounds (Termentzi et al., 2011).

2.3 Phenolics

Phenolics are a large group of aromatic compounds, consisting of flavones, flavanoids, and flavanols containing one carbonyl group, quinones with two carbonyl groups, tannins, polymeric phenolic substances, and coumarins, phenolic compounds with fused benzene and pyrone groups (Cowan, 1999; Kuete, 2013). This group of compounds has good antimicrobial effects and serves as plant defense mechanisms against pathogenic microorganisms. Simple phenols and phenolic acid are bioactive phytochemicals consisting of a single substituted phenolic ring. The inhibition of microorganisms by phenolic compounds may be due to iron deprivation or hydrogen bonding with vital proteins, such as microbial enzymes (Scalbert, 1991). Phenolics are vulnerable to polymerization

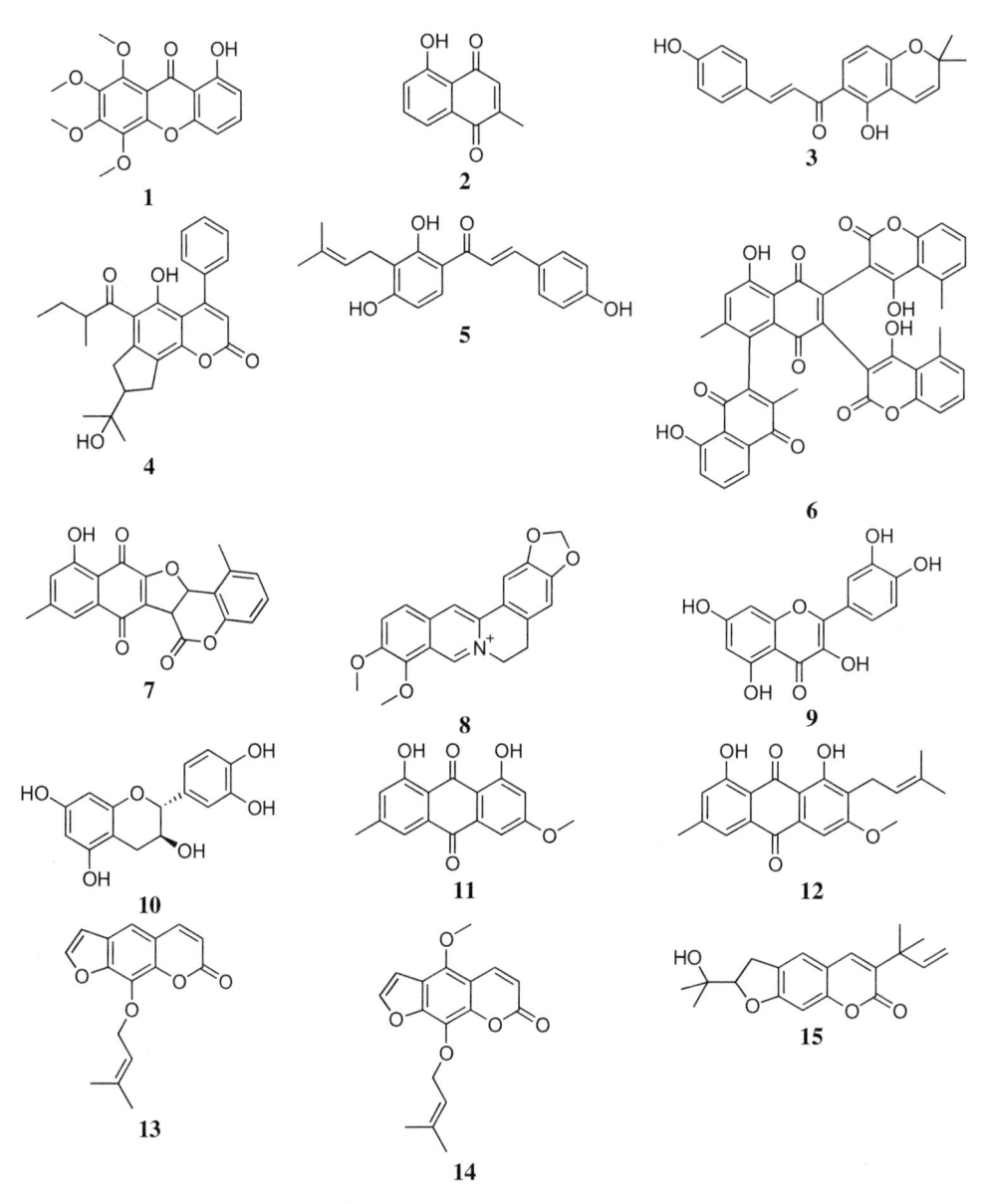

Figure 8.1 *Chemical structures of selected plant antimicrobials.* Laurentixanthone B (**1**); plumbagin (**2**); 4-hydroxylonchocarpin (**3**); MAB3 (**4**); isobavachalcone (**5**); diospyrone (**6**); crassiflorone (**7**) berberine (**8**); quercetin (**9**); catechin (**10**); physcion (**11**); vismiaquinone C (**12**); imperatorin (**13**); phellopterin (**14**); chalepin (**15**).

in air through oxidization reactions. Therefore, an important factor governing their toxicity is their polymerization size. Oxidation and condensation may result in the microbial growth inhibitory effect as well as detoxification (Field and Lettinga, 1992).

2.3.1 Flavonoids

The basic structural feature of flavonoid compounds is the 2-phenylbenzopyrane or flavane nucleus, consisting of two benzene rings linked through a heterocyclic pyrane ring. In total, there are 14 classes of flavonoids, differentiated on the basis of the chemical nature and position of substituents on the different rings. They are synthesized by plants in response to microbial infections and are often found effective in vitro as antimicrobial substance against a wide array of microorganisms (Cowan, 1999). Numerous studies have documented the effectiveness of flavonoids, such as swertifrancheside (Pengsuparp et al., 1995), glycyrrhizin (Watanbe et al., 1996), and chrysin (Critchfield et al., 1996) against HIV. More than one study has found that flavone derivatives are inhibitory to respiratory syncytial virus (Barnard et al., 1993). Galangin (3,5,7-trihydroxyflavone), derived from the perennial herb *Helichrysum aureonitens*, seems to be a particularly useful compound, since it has shown activity against a wide range of Gram-positive bacteria as well as fungi (Afolayan and Meyer, 1997) and viruses, in particular herpes simplex virus 1 and coxsackie B virus type 1 (Meyer et al., 1997). More recently, new flavonoids: isogancaonin C, bolusanthin III, and bolunsanthin IV with antibacterial and antioxidant activities have been isolated from *Bolusanthus speciosus* (Erasto et al., 2004). The antimicrobial properties of quercetin (**9**), apigenin-7-O-β-D-glucuronopyranoside, quercetin 3-O-β-D-galactopyranoside, and quercetin 3-O-α-L-rhamnopyranosyl (1→6) β-D-glucopyranoside have been reported against bacteria (*Enterobacter aerogenes*, *Escherichia coli*, *Klebsiella pneumoniae*, and *Staphylococcus aureus*) and fungi (*Candida parapsilosis*, *Candida albicans*, and *Cryptococcus neoformans*) (Djouossi et al., 2015). Catechins (basic structure **10**; Fig. 8.1), the most reduced form of the C3 unit in flavonoid compounds, inhibited in vitro *Vibrio cholerae* O1, *Streptococcus mutans*, and *Shigella* sp. (Cowan, 1999). Catechins were found to inactivate cholera toxin in *V. cholerae* and inhibit the bacterial glucosyltransferases in *S. mutans*, possibly due to complexing activities (Cowan, 1999). Different species *Hypericum*, used in traditional medicine, contained several compounds including hyperenone A, hypercalin B, and hyperphorin, responsible for their antibacterial activities against resistant trains of *S. aureus* and *Mycobacterium tuberculosis* (Osman et al., 2012; Shiu et al., 2012). In particular, hyperenone A inhibited the ATP-dependent MurE ligase of *M. tuberculosis*, a crucial enzyme in the cytoplasmic steps of peptidoglycan biosynthesis. The antibacterial properties of flavonoids are thought to come from their ability to form complexes with both extracellular and soluble proteins, as well as with bacterial membranes (Cowan, 1999). Many research groups have sought to elucidate the antibacterial mechanisms of action of selected flavonoids; the activity of compound **9** has been at least partially attributed to the inhibition of DNA gyrase, whereas sophoraflavone G and (−)-epigallocatechin gallate inhibited the cytoplasmic membrane function; licochalcones A and C inhibited energy metabolism (Savoia, 2012).

2.3.2 Quinones

Quinones are characteristically highly reactive, colored compounds with two ketone substitutions in aromatic ring. These are another significant group of secondary metabolites with potential antimicrobial properties. They provide a source of stable free radicals and complex irreversibly with nucleophilic amino acids in microbial proteins leading to the loss of their functions (Saleem et al., 2010). Anthraquinones in particular, had a large spectrum of antibacterial (including antimycobacterial) activities, based on the inactivation and loss of function of bacterial proteins, such as adhesins, cell wall polypeptides, and membrane-bound enzymes (Kurek et al., 2011), consequently leading to the death of the pathogens. Kazmi et al. (1994) have reported an anthraquinone from *Cassia italica*, a Pakistani tree, which was bacteriostatic against *Bacillus anthracis*, *Corynebacterium pseudodiphthericum*, and *Pseudomonas aeruginosa* and bactericidal for *Pseudomonas pseudomallei*. Physcion (**11**), an anthraquinone isolated from *Vismia rubescens* showed antimicrobial activity against yeasts, Gram-negative and Gram-positive bacteria with MIC values ranging from 12.50 to 100 µg/mL (Tamokou et al., 2009b). Geranyloxy-6-methyl-1,8-dihydroxyanthraquinone and vismiaquinone C (**12**) isolated from *Vismia laurentii* displayed antifungal and antibacterial activities (Kuete et al., 2007b).

2.3.3 Tannins

Tannins are polymeric phenolic substances possessing astringent property. These compounds are soluble in water, alcohol, and acetone and give precipitates with proteins (Basri and Fan, 2005). Tannins are either hydrolyzable or condensed. Hydrolyzable tannins are based on gallic acid; condensed tannins, often called proanthocyanidins are based on flavonoid monomers, flavones derivatives, or quinine units. Hydrolyzable and condensed tannins, derived from flavanols, exert antimicrobial activity via antiperoxidation properties, inhibiting particularly the growth of uropathogenic *E. coli* (Okuda, 2005). Condensed tannins have been determined to bind cell walls of ruminal bacteria, preventing growth and protease activity (Jones et al., 1994). Tannins in plants inhibit insect growth and disrupt digestive events in ruminal animals (Cowan, 1999).

2.3.4 Coumarins

Coumarins are phenolic substances made of fused benzene and α-pyrone rings. They have a characteristic odor and several of them have antimicrobial properties. For example, one known coumarin, scopoletin was isolated as antitubercular constituents of the whole plant *Fatoua pilosa* (Garcia et al., 2012). Hydroxycinnamic acids, related to coumarins, seem to be inhibitory to Gram-positive bacteria (Fernandez et al., 1996). Also, phytoalexins, which are hydroxylated derivatives of coumarins, are produced in carrots in response to fungal infections and can be presumed to have antifungal activity (Hoult and Paya, 1996). Compounds, such as 7-methoxy-6(2′-oxo-3′-methyl butyl) coumarin, 7-[(*E*)-7-hydroxy-3,7-dimethylocta-2,5dienyloxyl]-coumarin, imperatorin

(13), phellopterin (14), and chalepin (15) isolated from a Cameroonian medicinal plant, *C. anisata* showed antibacterial activities (MIC and MBC ranging from 16 to 256 μg/mL) against multidrug-resistant (MDR) enteropathogenic bacteria including the clinical MDR isolates of toxigenic *V. cholerae* and *Shigella* sp. (Tatsimo et al., 2015).

3 IN VITRO SCREENING METHODS OF PHYTOCHEMICALS FOR ANTIMICROBIAL ACTIVITIES

3.1 Selection of extraction solvent in biological activity screenings

Targeting biologically active compounds from plant material depends on the type of solvent used in the extraction procedure. The choice of solvent is guided by what is intended with the extract. Since the end product will contain traces of residual solvent, the solvent should be nontoxic and should not interfere with the bioassay (Ncube et al., 2008). However, residual solvent should be eliminated as well as possible in oven or at reasonable laboratory temperature, not above 40°C. The affinity of various plant metabolites to common extraction solvents has been established. Polyphenolic compounds, such as flavonols and most other reported bioactive compounds are generally soluble in polar solvents, such as methanol (Ncube et al., 2008). It was shown in several reports that antimicrobials are not water soluble and thus organic solvent extracts have been found to be more potent (Ncube et al., 2008; Parekh et al., 2006). For example, hydrosoluble flavonoids (mostly anthocyanins) have no antimicrobial significance meanwhile water soluble phenolics are only important as antioxidant molecules (Nang et al., 2007; Ncube et al., 2008; Yamaji et al., 2005). In contrast, water-soluble molecules, including polysaccharides and polypeptides (fabatin and various lectins) are commonly more effective as inhibitors of pathogen adsorption and have no real impact as antimicrobial agents (Cowan, 1999). Solvents currently used for the investigation of the antimicrobial activity of plants include methanol, ethanol, and water as well as dichloromethane and acetone (Ncube et al., 2008). Acetone, although not a commonly used solvent, has been used by a number of authors and has been recommended by Masoko and Eloff as a good extractant of antimicrobial phytochemicals (Masoko and Eloff, 2006). Compounds extracted by various solvent are summarized in Table 8.2.

Table 8.2 Solvents used for active component extraction (Cowan, 1999; Ncube et al., 2008; Sandjo et al., 2014)

Solvent	Compounds extracted
Chloroform	Terpenoids, flavonoids
Acetone	Phenols, flavonols
Ether	Terpenoids, fatty acid, coumarins, alkaloids
Ethanol	Terpenoids, polyphenols, polyacetylenes, flavonoids, tannins, alkaloids
Methanol	Terpenoids, saponins, phenones, polyphenols, flavonoids, flavones, quassinoids, lactones, anthocyanins, tannins, alkaloids, xanthoxyllines
Water	Terpenoids, polypeptides, saponins, lectins, anthocyanins, tannins, starches

3.2 Phytochemical screenings of plant extracts

It is important to have an idea on the gross phytochemical composition of plant extracts to use for pharmacological screenings. This allows a preliminary understanding on the nature possible bioactifs compounds. However, this is only qualitative and speculative, taking in account the fact that within different classes of secondary metabolites, all compounds do not have biological activities. In addition, the activity will depend on the amount of active compounds in the crude extract as well as in possible interactions with other constituents. Several methods have been described to evaluate the phytochemical composition of plant extracts (Bruneton, 1999; Harbone, 1973; Kuete, 2013; Tiwari et al., 2011). The most currently used are described in following sections.

3.2.1 Terpenoids and steroids tests

Assays for terpenoids and steroids in plant extract can be screened by Salkowski and Liebermann–Burchard tests.

1. *Salkowski's test.* The plant extract is dissolved in chloroform and filtered. The filtrate is then treated with few drops of concentrated sulfuric acid, shaken, and allowed to stand. The appearance of golden yellow color indicates the presence of triterpenes.
2. *Liebermann–Burchard test.* The plant extract is dissolved in chloroform and filtered. The filtrate is then treated with few drops of acetic anhydride, boiled, and cooled. Then, a concentrated sulfuric acid is added. The formation of brown ring at the junction indicates the presence of phytosterols.

Steroids and terpenoids are also available as glycoside, and carbohydrate tests can be used to detect the presence of sugar ring.

3.2.2 Detection of diterpenes

The presence of diterpenes in plant extract can be monitored using *copper acetate test*. Hence, the extract is dissolved in water and treated with 3–4 drops of copper acetate solution. Formation of emerald green color indicates the presence of diterpenes.

3.2.3 Detection of carbohydrates

To detect the presence of carbohydrates, the plant extract is dissolved in distilled water and filtered.

1. *Molisch test.* The filtrate is treated with two drops of alcoholic alpha-naphthol solution in a test tube. The formation of the violet ring at the junction indicates the presence of carbohydrates.
2. *Benedict test.* The filtrate is treated with Benedict's reagent and heated gently. Orange red precipitate indicates the presence of reducing sugars.
3. *Fehling test.* Filtrate is hydrolyzed with diluted HCl, neutralized with alkali, and heated with Fehling's A and B solutions. The formation of red precipitate indicates the presence of reducing sugars.

3.2.4 Anthraquinone glycosides
It is possible to screen both free and bound forms of anthraquinones in plant extracts.
1. *Bontrager test for free anthraquinone.* The extract is dissolved in hot water for aqueous extract and alcohol extract; both are put in water bath to steam for 5 min, then the solution is filtered hot and the filtrate is extracted with chloroform layer, then taken off. This layer is washed with water and shaken with ammonia solution. Appearance of red color in ammonia upper phase indicates the presence of anthraquinone.
2. *Bound anthraquinone.* Hydrolyze the extract with diluted HCl, and then proceed to test the presence of glycosides. Treat another portion of the extract with ferric chloride solution and immerse in boiling water for about 5 min. Then cool and extract the mixture with equal volumes of benzene. The benzene layer is then separated and treated with ammonia solution. Formation of rose-pink color in the ammoniacal layer indicates the presence of anthranol glycosides.

3.2.5 Detection of saponins
1. *Froth test.* Dilute the extract with distilled water (about 20 mL) and shake in a graduated cylinder for 15 min. Formation of 1-cm layer of foam indicates the presence of saponins.
2. *Foam test.* Dilute about 500 mg of extract with water (about 2 mL) and shake. If foam produced persists for 10 min it indicates the presence of saponins.

3.2.6 Detection of phenols
The ferric chloride test can be used as follows to detect phenols: crude extract is treated with 3–4 drops of ferric chloride solution. Formation of bluish black color indicates the presence of phenols.

3.2.7 Detection of tannins
In the gelatin test, 1% gelatin solution containing sodium chloride is added to crude extract. The formation of white precipitate indicates the presence of tannins.

3.2.8 Detection of flavonoids
1. *Alkaline reagent test.* The crude extract is treated with few drops of sodium hydroxide solution. The formation of intense yellow color, which becomes colorless on addition of dilute acid, indicates the presence of flavonoids.
2. *Lead acetate test.* The crude extract is treated with few drops of lead acetate solution. The formation of yellow color precipitate indicates the presence of flavonoids.
3. *Ferric chloride test.* The crude extract is treated with 0.2 mL of 10% ferric chloride. The mixture is shaken and the appearance of wooly brownish precipitate indicates the presence of flavonoids

3.2.9 Cardiac glycosides

In Kedde's test, a diluted extract is mixed volume–volume with 2% solution of 3,5-dinitrobenzoic acid in methanol and 5.7% aqueous sodium hydroxide (v/v). The appearance of brownish precipitate indicates the presence of cardenolide.

3.2.10 Detection of alkaloids

Various assays can be used to detect the presence of alkaloids in plant extracts; first of all, the extract is dissolved in dilute hydrochloric acid and filtered.

1. *Mayer's test.* The filtrate is treated with Mayer's reagent (potassium mercuric iodide). The formation of a yellow colored precipitate indicates the presence of alkaloids.
2. *Wagner's test.* The filtrate is treated with Wagner's reagent. The formation of brown/reddish precipitate indicates the presence of alkaloids.
3. *Dragendorff's test.* The filtrate is treated with Dragendorff's reagent (solution of potassium bismuth iodide). The formation of red precipitate indicates the presence of alkaloids.
4. *Hager's test.* The filtrate is treated with Hager's reagent (saturated picric acid solution). The presence of alkaloids is confirmed by the formation of yellow colored precipitate.

3.3 Susceptibility testing methods

Numbers of assay systems and organisms have been used to screen plant extracts and their constituents for antimicrobial activities. Test systems should ideally be simple, rapid, reproducible, inexpensive, and maximize sample throughput (Hostettmann et al., 1997). Agar dilution method has been standardized for quantitative determination of antibiotics (Klancnik et al., 2010). Broth dilution methods for inhibitory determination are also recommended by the Clinical and Laboratory Standards Institute (CLSI), using different principles to assess microbial growth or its inhibition (CLSI, 2003; Klancnik et al., 2010). Colorimetric methods could represent an alternative approach, using tetrazolium salts as indicators, since bacteria convert them to colored formazan derivatives that can be quantified (Eloff, 1998; Grare et al., 2008; Johnson et al., 1985). Colorimetric methods are good indicators of bacterial growth; some of them include XTT (3′-{1-[(phenylamino)-carbonyl]-3,4-tetrazolium}-bis(4-methoxy-6-nitro)benzenesulfonic acid hydrate), TTC (2,3,5-triphenyl tetrazolium chloride), and resazurin (Carson et al., 1995; Mann and Markham, 1998; Rahman et al., 2004). However, difficulties arising because of autofluorescence, salt reduction, and the antioxidant properties of plant products, could make them less suitable indicators for MIC assay in certain cases. Microbial growth or its inhibition can also be measured by a number of ways, such as viable counts, direct microscopic counts, turbidity measurement, bioluminescence, and fluorimetry.

Microbroth dilution method. The microbroth dilution method seems to be more appropriate when investigating the activity of botanicals. In fact, this method has several advantages compared to the agar diffusion method. The microbroth dilution method is quantitative, allows the use of small quantities of compounds or plant extracts as well as culture media, and is well adapted for drug intended to systemic use (Chung et al., 1995). Colorimetric microbroth techniques using various reagents, such as tetrazolium salts (Babula et al., 2009; Eloff, 1998), or color indicator (Kuete et al., 2007a) allow easy MIC detection and increase the credibility of this method. Its main disadvantage comes from the fact that it is not well appropriate for hydrophobic samples, such as terpenoids. The microbroth dilution is also less expensive and less cumbersome than the macrodilution method and it yields reproducible results. Besides, the use of microplates allows large amounts of data to be generated quickly. Bacterial growth could be assessed either visually by grading turbidity or spectrophotometrically by measuring optical density. However, on one hand, visual assessment of bacterial growth lacks objectivity and precision; also, the accuracy of spectrophotometric readings may be hampered by (1) additives or antibacterial compounds that affect the spectral characteristics of growth media, (2) the aggregation of bacteria, or (3) bacterial pigments (Eloff, 1998). Bioluminescence or fluorimetry measurements are sensitive but demand expensive equipment and can require extensive laboratory work prior to each assay. Consequently, few conventional methods of antimicrobial susceptibility testing are both precise and reproducible when applied to bacteria that have specific growth requirements or a slow rate of growth. In this context, colorimetric methods could represent an alternative approach (Grare et al., 2008). For the antimycobacterial tests of plant-derived substances, a number of bioassay systems have been used including agar diffusion and dilution assays and radiorespirometry (using a BACTEC 460 instrument) (McGaw et al., 2008).

Diffusion methods. Diffusion methods are attractive because of their simplicity and their low cost, but they are, like all agar-based methods, labor- and time-intensive. They can also be adapted for MIC determination, using dilution of sample. However, higher MIC values are obtained when the disk diffusion method are used, making it not reliable for screening the antimicrobial activity of plant extracts (Klancnik et al., 2010). In agar disk or hole diffusion method, the absence of an inhibition zone did not necessarily mean the compound is inactive, especially for less polar compounds, which diffuse more slowly into the culture medium (Klancnik et al., 2010; Moreno et al., 2006). Hence, the diffusion assay is not suitable to natural antimicrobial compounds that are scarcely soluble or insoluble in water and thus their hydrophobic nature prevents uniform diffusion through the agar media (Mann and Markham, 1998). Agar diffusion is more suitable for samples intended to external use, such as treatment of dermatophytes but should be avoided for those to be used systemically.

Bioautography. This is a variation of the agar diffusion method where the analyte is adsorbed onto a thin layer chromatography (TLC) plate. Bioautography is also employed as a preliminary phytochemical screening technique, by bioassay-guided fractionation, to detect active components (Ncube et al., 2008; Nostro et al., 2000; Schmourlo et al., 2005). The bioautography assay may represent a useful tool for purification of antimicrobial substances if tests are performed using chromatograms. Bioautography allows easy localization of activity even in complex matrix as those derived from natural products (Hamburger and Cordell, 1987; Valgas et al., 2007). Chromatogram comparison developed under identical conditions and visualized with the use of suitable chromogen reagent can provide useful information about nature of active compounds. An advantage of bioautographic method is the possibility of using mobile phases containing solvents of low volatility as *n*-butanol, in case of its complete removal before carrying out tests. However, too acid or too alkali solvents remain on TLC plate after long drying time, inhibiting possible bacterial growth (Hamburger and Cordell, 1987; Valgas et al., 2007).

3.4 Cutoff points for interpretation of antimicrobial activity of phytochemicals

In microbial susceptibility testing, statistical analyses are not necessary to show the significance of recorded MIC values. This is normally due to the fact that MIC values are determined following twofold dilutions, with differences between consecutive concentrations being too high to make credible the obtention of different results. In a triplicated experiment, MIC should be taken as the same value obtained thrice, or a value obtained in two of the three assays rather than an average of three assays. However, when diffusion methods are concerned, the significance of diameters of zone of inhibition of samples could necessitate statistical analyses. Though, statistical analyses are not necessary as far as MIC are concerned, antimicrobial cutoff points have been defined by several authors to enable better understanding of the potential of phytochemicals. Fabry et al. (1998) considered extracts having MIC values below 8 mg/mL to have noteworthy antimicrobials. According to Simões et al. (2009), phytochemicals are routinely classified as antimicrobials on the basis of susceptibility tests that produce MIC in the range of 100–1000 μg/mL. MIC value of 100 μg/mL was used as a criterion for antimicrobial activity classification in accordance with some authors who consider a MIC value between 100 and 200 μg/mL as positive for plant extracts (Jimenez-Arellanes et al., 2003). More elaborated cutoff points for antimicrobial activities were lately defined by Kuete and Efferth (Kuete, 2010; Kuete and Efferth, 2010) as follows: (1) for crude extracts—significant activity (MIC < 100 μg/mL), moderate activity (100 < CMI ≤ 625 μg/mL), or weak activity (CMI > 625 μg/mL); (2) for pure compounds—significant activity (MIC < 10 μg/mL), moderate activity (10 < MIC ≤ 100 μg/mL), and low activity (MIC > 100 μg/mL).

One of the major limitations of the cutoff points method is that scientists have not made any difference between toxic plants and other medicinal plants or edible plants having antimicrobial potential. In this chapter, we propose new cutoff points for the antimicrobial activity of edible plant extracts or extract from edible parts of plants as follows, as far as the dilution series include 1 μg/mL (e.g., 1, 2, 4, 8, 16 μg/mL, … etc.):
- Highly active: MIC below 100 μg/mL
- Significantly active: 100 ≤ MIC ≤ 512 μg/mL

- Moderately active: 512 < MIC ≤ 2048 µg/mL
- Low activity: MIC > 2048 µg/mL
- Considered not active: MIC > 10 mg/mL

Such values can moderately vary according to the dilution scale used; therefore the set points of 512 and 2048 µg/mL can be changed without any significant effect with 625 and 2500 µg/mL, respectively. Values between these set points can also be considered according to the selected dilution scale. It is worth noting that plant extracts are merely devoid of any antibacterial activity; therefore a high limit concentration of 10 mg/mL has been defined in this chapter as the value above, which the activity cannot be taken into consideration. For compounds from medicinal plants, we agree with many literature reviews to define the cutoff points, with slight modifications when the dilutions are similar to the aforementioned as follows:

- Highly active: MIC below 1 µg/mL (or 2.5 µM)
- Significantly active: 1 ≤ MIC ≤ 10 µg/mL (or 2.5 ≤ MIC < 25 µM)
- Moderately active: 10 < MIC ≤ 100 µg/mL (or 25 ≤ MIC < 250 µM)
- Low activity: 100 < MIC ≤ 1000 µg/mL (or 250 ≤ MIC < 2500 µM)
- Considered not active: MIC > 1000 µg/mL (or >2500 µM)

The following cutoff points can also be modified as discussed earlier, with regard to the selected dilution series.

4 ANTIMICROBIAL EFFECTS OF AFRICAN MEDICINAL SPICES AND VEGETABLES

Spice plants and vegetables as well as their essential oils are important sources of antimicrobial agents in addition to their ability to stimulate the digestive system. The antimicrobial effectiveness of mustard, clove, cinnamon, and their essential oils were reported for the first time around 1880s (Rahman et al., 2011). Spice and vegetables have been used traditionally as coloring agents, flavoring agents, preservatives, food additives and as well as antiparasitic, anthelmintic, analgesic, expectorant, sedative, antiseptic, and antidiabetic substances in many parts of the world (Lee et al., 2004; Rahman et al., 2011). In addition, they possess biological activities, such as antioxidant (Miura et al., 2002) and hypocholesterolemic (Craig, 1999). Several African medicinal spices and vegetables were screened in the past 10 years against drug-sensitive as well as drug-resistant microorganisms. In this section, we will bring together, family by family, the available knowledge as retrieved from scientific publications on spices and vegetables used in Africa.

4.1 Annonaceae

Annonaceae are flowering plants consisting of trees, shrubs, or rarely lianas having 2106 accepted species and more than 130 genera. Several genera produce edible fruits, most notably *Annona*, *Anonidium*, *Asimina*, *Rollinia*, and *Uvaria*. The family is concentrated in the tropics, and about 900 species are neotropical, 450 are Afrotropical, and the other species are Indomalayan. Some of the spices in this family are used as food ingredients having medicinal properties.

Anonidium mannii (oliv) Engl. et Diels. is a culinary spice also used in folk medicine in case of spider and snakebites, bronchitis, dysentery, gastroenteritis, syphilis, diarrhea, malaria. The methanol extract from leaves showed selective antibacterial activities against various strains of Gram-negative bacteria including *E. coli*, *E. aerogenes*, *K. pneumoniae*, *Providencia stuartii*, *Enterobacter cloacae*, and *P. aeruginosa* with MIC values varying from 512 to 1024 µg/mL (Djeussi et al., 2013).

Xylopia aethiopica (Dunal) A. Rich. is a native to the lowland rainforest and moist fringe forests in the savanna zones of Africa. It is a culinary spice used in traditional medicine to treat wounds and skin infections, fever, tapeworm, stomach ache, dysentery, and stomach ulcer (Irvine, 1961; Thomas, 1965). The methanol extract from the seeds showed high antibacterial activity against *E. coli* ATCC and *K. pneumoniae* KP63 (MIC of 64 µg/mL), good to moderate activity with MIC ranging from 256 to 1024 µg/mL against other strains of *E. coli* as well as *E. aerogenes*, *P. sturatii*, and *K. pneumoniae* (Fankam et al., 2011). The antimicrobial activities of the essential oil from the seed were also reported against Gram-positive bacteria, such as *S. aureus*, *Bacillus subtilis*, *Bacillus cereus*, *Streptococcus faecalis* as well as against the fungal strain *Aspergillus flavus* (Tatsadjieu et al., 2003).

4.2 Apiaceae

The Apiaceae or Umbelliferae, commonly known as the celery, carrot, or parsley family, are a family of mostly aromatic plants with hollow stems. Many plants of this family are condiments or vegetables with some of them having medicinal properties.

Apium graveolens L. or celery is a cultivated plant, commonly used as a vegetable throughout Africa. It has medicinal properties, such as antiinflammatory, antihypertensive and is also used to treat bronchitis, hepatitis, gastrointestinal infections, and asthma (Singh and Handa, 1995). The methanol extract from celery displayed antibacterial activity against various strains of *K. pneumoniae*, *P. stuartii*, *E. coli*, *E. aerogenes* and *E. cloacae* (MIC of 512–1024 µg/mL); however, extract was found not active against *P. aeruginosa* strain (Tankeo et al., 2014).

Petroselinum crispum Ngn. is a condiment also used medicinally as diuretic, emmenagogue, stimulating, antimicrobial, and antioxidant (Wong and Kitts, 2006). This plant is a well-known spice and vegetable cultivated in almost all African countries. The antimicrobial activities toward *B. subtilis* and *E. coli* of different extracts were reported meanwhile bacterial cell damage resulting in significant greater growth inhibition of the two bacteria, corresponded to ferrous sequestering activity of methanol-derived stem extracts (Wong and Kitts, 2006). The pharmacological properties of this plant are largely discussed in Chapter 25.

4.3 Asteraceae

Asteraceae or Compositae is an exceedingly large and widespread family of flowering plants with more than 23,600 currently accepted species, spread across 1,620 genera and 13 subfamilies. Many of them are spices used in traditional medicine to cure microbial infections.

Echinops giganteus A. Rich. is not only used as culinary spice but also as traditional medicine to treat cancers, heart, and gastric troubles (Tene et al., 2004; Kuete et al., 2011c). The methanol extract of the roots of *E. giganteus* was shown to be highly active against *K. pneumoniae* K24 (MIC of 32 µg/mL); significantly active against *E. coli* MC4100, *E. aerogenes* EA294, and *K. pneumoniae* KP63 (MIC varied from 256 to 512 µg/mL); and moderately active with MIC value of 1024 µg/mL against various other strains of these bacterial species (Fankam et al., 2011). The methanol extract of *E. giganteus* roots also exhibited significant activity with MIC values of 32 and 16 µg/mL, respectively against *M. tuberculosis* $H_{37}Ra$ and $H_{37}Rv$ (Tekwu et al., 2012).

Lactuca sativa Linn., popularly known as lettuce is a worldwide leaf vegetable and is cultivated in almost all African countries. The medicinal potential of this vegetable including its antimicrobial activities are discussed in Chapter 20.

4.4 Brassicaceae

Brassicaceae or the cabbage family is an economically important family of flowering plants, with 372 genera and 4060 species. The family contains well-known species, such as *Brassica oleracea*, *Brassica rapa*, *Brassica napus*, *Raphanus sativus*, *Armoracia rusticana*, *Arabidopsis thaliana*, and many others. The antibacterial activities of ethanol extracts of *B. oleracea* L. var. italica was reported active against *S. aureus*, *B. cereus*, and *P. aeruginosa* while the antifungal activity of the plant was also shown on *Saccharomyces cerevisiae*, *Torulopsis etchellsii*, *Hansenula mrakii*, and *Pichia membranifaciens* (Kyung and Fleming, 1997). Touani et al. (2014) have investigated the antimicrobial potency of *B. oleracea* L. var. *italica* and *B. oleracea* L. var. *butyris* against a panel of sensitive and MDR bacteria. The investigators showed that the methanol extracts of both vegetables were not active against *P. aeruginosa* strains but were moderately and selectively active against strains of *E. coli*, *E. aerogenes*, *K. pneumoniae*, and *P. stuartii*.

4.5 Gnetaceae

Gnetum africanum known as *Eru*, is a Gnetaceae plant and is traditionally considered to be a wild vegetable. *Eru* has many common names and is grown in various countries across Africa, including: Cameroon (Eru, okok, m'fumbua, or fumbua), Angola (KoKo), Nigeria (ukase or afang), Gabon (KoKo), Central African Republic (KoKo), Congo (KoKo), and the Democratic Republic of Congo (m'fumbua or fumbua). It is used as antidote to some forms of poison and snakebite, as antiseptic, antipains, diuretic, fungicide, and as wounds antiseptic (Ekop, 2007). Tankeo et al. (2014) have reported the exceptional antibacterial activity of the methanol extract of the leaves of *G. africanum* against *E. coli* AG100A and *K. pneumoniae* K24 with a MIC value of 64 µg/mL. This extract also displayed good antibacterial activity against various other strains of *E. coli*, *E. aerogenes*, *E. cloacae*, *K. pneumoniae*, *P. stuartii*, and *P. aeruginosa* with MIC values ranging from 128 to 1024 µg/mL (Tankeo et al., 2014). The antibacterial activities of ethanol extracts of the leaves against *S. aureus*, *E. coli*, and *Salmonella typhi* were also reported (Obiukwu and Nwanekwu, 2010).

4.6 Lamiaceae

The Lamiaceae or Labiatae are a family of flowering plants with a cosmopolitan distribution containing about 236 genera and has been stated to contain 6900–7200 species. The largest genera are *Salvia* (900), *Scutellaria* (360), *Stachys* (300), *Plectranthus* (300), *Hyptis* (280), *Teucrium* (250), *Vitex* (250), *Thymus* (220), and *Nepeta* (200). Many of them are used as spices and vegetables. The well-known thyme (*Thymus vulgaris*; Chapter 28) belongs to Lamiaceae.

Ocimum basilicum Linn. also belonging to Lamiaceae, is a famous culinary herb in Africa. It is medicinally used in case of diarrhea as well as hypoglycemic, antiinflammatory, antibacterial, antioxidant, immune-stimulatory, and antiviral (Charvat et al., 2006; Uma, 2001). Another famous culinary herb in Africa, *Ocimum gratissimum* L. has been used in traditional medicine as pulmonary antiseptic, antifungal, antitussive, antispasmodic, and antibacterial (Ngassoum et al., 2003; Okigbo and Ogbonnaya, 2006). Methanol extracts from both condiments were shown to have good to moderate antibacterial activity against sensitive and drug-resistant Gram-negative bacteria, including various strains of *E. coli*, *E. aerogenes*, *E. cloacae*, *K. pneumoniae*, *P. stuartii*, and *P. aeruginosa* with MIC values varying from 128 to 1024 µg/mL (Tankeo et al., 2014). The essential oil of *O. gratissimum* was also found active against *B. cereus*, *B. subtilis*, *S. aureus*, *S. faecalis* as well as *E. coli* (Ngassoum et al., 2003).

4.7 Liliaceae

Liliaceae comprises 50 genera and approximately 600 species of flowering plants with many of them been used as spices and vegetables. *Allium cepa* (Chapter 14) and *Allium sativum* (Chapter 15) belong to this family. *Allium porrum* commonly known as leek, is a vegetable belonging, along with onion and garlic, to the genus *Allium*. *A. porrum* is largely used as soup ingredient and vegetable throughout Africa, and is known to possess anthelmintic, antiasthmatic, anticholesterolemic, antiseptic, antispasmodic, cholagogue, diaphoretic, diuretic, expectorant, febrifuge, stimulant, stings, stomachic, tonic, vasodilator, antibacterial, antioxidant, cytotoxic, insecticidal, fungicidal properties (Akroum et al., 2009; Auger et al., 2002). Akroum et al. (2009) have reported the antimicrobial activities of the methanol extract against *S. aureus*, *B. subtilis*, *B. cereus*, *E. coli* as well as the inhibitory effects of the ethanol extract against *C. neoformans*. Tankeo et al. (2014)

also showed that the methanol extract of the plant was selectively active against Gram-negative bacterial strains, such as *E. coli* AG102 and AG100ATet, *E. aerogenes* EA289 and EA294, *K. pneumoniae* K24, and *E. cloacae* BM67 with MIC values ranging from 512 to 1024 μg/mL.

4.8 Moraceae

Moraceae are monoecious or dioecious trees, shrubs, lianas, or rarely herbs comprising 40 genera and 1000 species, nearly all with milky sap. *Dorstenia psilurus* Welwitch belonging to Moraceae is a culinary spice used in African folk medicine in case of snakebite, rheumatism, head and stomach ache, hypertension, and cancer (Abegaz et al., 2000; Dimo et al., 2001; Kuete et al., 2011c). The methanol extract from the roots of this plant showed antibacterial activity against a panel of Gram-negative bacteria including MDR phenotypes. The reported species included *E. coli* ATCC 10536 (MIC of 128 μg/mL), *E. coli* AG100A, AG100Atet, AG102, MC4100, W3110, *E. aerogenes* CM64, and *K. pneumoniae* KP55 and KP63 (MIC of 512 μg/mL), *E. aerogenes* ATCC 13048 and EA289, *E. cloacae* BM47 and BM67, *K. pneumoniae* ATCC 11296, K2, and K24, *P. stuartii* ATCC 29916 and PS299645 (MIC of 1024 μg/mL) (Voukeng et al., 2012).

4.9 Pedaliaceae

Pedaliaceae is a flowering plant family well-known as pedalium or sesame family. Within this family, Seukep et al. (2013) reported the antibacterial activity of the methanol extract from the beans of *Sesamum indicum* and the stem bark and leaves of *Sesamum radiatum* Schum et Thom. The investigators showed that *S. indicum* had significant inhibitory effects against *E. aerogenes* EA27 and *K. pneumoniae* K2 and K24 (MIC of 256 μg/mL) while *S. radiatum* also displayed MIC value of 256 μg/mL against *P. stuartii* ATCC 29914. Authors also showed that both plants were active against many other drug-sensitive and drug-resistant strains of *E. coli*, *E. aerogenes*, *E. cloacae*, *K. pneumoniae*, and *P. stuartii* with MIC values ranging from 512 to 1024 μg/mL. Tane et al. (2005) also showed that essential oil of *S. radiatum* was active against *Citrobacter* sp., *E. coli* ATCC 25922, *K. pneumoniae*, *Proteus mirabilis*, *P. aeruginosa*, *S. typhi*, and *S. aureus*.

4.10 Piperaceae

The Piperaceae or pepper family contains roughly 3600 species in 13 genera, mainly distributed into two main genera: *Piper* (2000 species) and *Peperomia* (1600 species). The most well-known species is *Piper nigrum*, which yields most peppercorns that are used as spices, including black pepper, although its relatives in the family include many other spices. The antimicrobial activity of some *Piper* sp. have been reported.

Piper capense Lin.f (Piperaceae) is a culinary spice also used in folk medicine as sleep-inducing remedy and as anthelmintic (Kokoworo, 1976; Van Wyk and Gericke, 2000). The fruits of *P. capense* showed antibacterial activity against various strains of Gram-negative bacteria, including *E. coli* ATCC 10536, *E. aerogenes* EA294, *K. pneumoniae* KP63, and *P. stuartii* NAE16 (MIC of 256 μg/mL), *E. aerogenes* EA27, *E. cloacae* BM67, and *K. pneumoniae* K2 (MIC of 512 μg/mL), *E. coli* AG100ATet, AG102, MC4100, W3110, *E. aerogenes* EA3 and EA27, *K. pneumoniae* ATCC 11296, KP55, K24, and *P. stuartii* ATCC 29916 (MIC of 1024 μg/mL) (Fankam et al., 2011).

Piper guineense (Schum and Thonn) is another culinary spice used in traditional medicine to manage cough, bronchitis, rheumatism, insecticidal, anemia, carminative, stomach ache, and cancer (Konning et al., 2004). The methanol extract of the seeds also displayed antibacterial activity against sensitive and MDR strains of *E. coli*, *E. aerogenes*, *K. pneumoniae*, *P. stuartii*, and *E. cloacae* with MIC values ranging from 512 to 1024 μg/mL (Voukeng et al., 2012).

P. nigrum L., a popular culinary spice, is also used traditionally as medicine for cardiovascular diseases; intoxication; inflammation; bacterial, fungal, and parasitic infections; respiratory diseases; and asthma (Nisar et al., 2012). Noumedem et al. (2013a) have reported the antibacterial activity of the methanol extract of the seeds of this plant against a panel of Gram-negative bacteria, with very high inhibitory effects against the nosocomial agent, *P. aeruginosa* (MIC of 32 μg/mL) and significant effects with MIC ranging from 128 to 1024 μg/mL on several other strains of *E. coli*, *E. aerogenes*, *K. pneumoniae*, *P. stuartii*, and *E. cloacae* as well as against the MDR strains of *P. aeruginosa* PA124. The antimicrobial effects of the ethanol extract of the seeds was also reported against Gram-positive (*B. subtilis*, *S. aureus*, and *Staphylococcus epidermidis*) as well as against Gram-negative bacteria (*E. coli* and *P. aeruginosa*) (Erturk, 2006; Nisar et al., 2012). The antifungal effect of this plant was also reported (Umit and Akgun, 2009).

4.11 Solanaceae

The Solanaceae or nightshades are an economically important family consisting of herbs, shrubs, or trees with about 85 genera and 2800 species that are frequently lianous or creeping. The family has worldwide distribution, including a number of important agricultural crops, medicinal plants, spices, weeds, and ornamentals. Many members of the family contain potent alkaloids, and some are highly toxic, but many cultures eat nightshades, in some cases as staple foods. The most economically important genus of the family found in Africa include *Solanum*, which contains *Solanum tuberosum* (potato), *Solanum lycopersicum* (tomato), and *Solanum melongena* (eggplant or aubergine) as well as *Capsicum* sp.

Capsicum annuum is a source of popular sweet peppers and hot chilis with numerous varieties cultivated all around the world. It is used medicinally as antioxidant, antibacterial, and antiviral (Lampe, 2003). The methanol extract of *C. annuum* fruits displayed selectively good to moderate antibacterial activities against various strains of *E. coli*, *E. aerogenes*, *E. cloacae*, *P. stuartii*, and *P. aeruginosa* with MIC values ranging from 256 to 1024 μg/mL (Tankeo et al., 2014). *Capsicum frutescens* is a species of chili pepper that is sometimes considered to be part of the species *C. annuum*. Medicinally, it is used as laxative (N'Guessan et al., 2009) and stimulant (Noumedem et al., 2013b), as well as to treat wound, male virility (Hamisy et al., 2000), rheumatism (N'Guessan et al., 2009), toothache pain, cough, asthma, sore throat, stomach ache, and seasickness (Noumedem et al., 2013b). The methanol extract of the fruits of *C. frutescens* was shown to have selective antibacterial activity against various strains of *E. coli*, *E. aerogenes*, *E. cloacae*, *K. pneumoniae*, and *P. stuartii* with MIC values generally ranging from 256 to 1024 µg/mL (Noumedem et al., 2013b).

S. melongena L.Var inerme D.C Hiern. is used in nutrition and commonly consumed as a vegetable; medicinally, it is used in case of inflammation, to lower blood cholesterol levels (Guimarães et al., 2000), as antihemorrhoidal and hypotensive, as antidote to poisonous mushrooms, to heal burns, abscesses, cold sores, intestinal infections, hemorrhages, piles, and toothache, as an astringent (Sudheesh et al., 1999), cardiac debility, neuralgia, ulcers of nose, cholera, bronchitis, and asthma (Noumedem et al., 2013a). The methanol extract from the fruits of this plant was shown to have antibacterial activities against various strains of *E. coli* (MIC of 512–1024 µg/mL), *E. aerogenes* (MIC of 128–1024 µg/mL), *K. pneumoniae* (MIC of 256–1024 µg/mL), *P. stuartii* (MIC of 128–1024 µg/mL), and *E. cloacae* BM67 (MIC of 512 µg/mL) (Noumedem et al., 2013a).

S. tuberosum is a starchy tuberous crop containing vitamins and minerals, as well as an assortment of phytochemicals, such as carotenoids and natural phenols. Chlorogenic acid constitutes up to 90% of the potato tuber natural phenols. Other studies found that potatoes contain 4-O-caffeoylquinic acid (cryptochlorogenic acid), 5-O-caffeoylquinic acid (neochlorogenic acid), 3,4-dicaffeoylquinic acid, and 3,5-dicaffeoylquinic acid. Potatoes also contain toxic compounds known as glycoalkaloids, of which the most prevalent are solanine and chaconine. Anthocyanins extract from *S. tuberosum* L. var Vitelotte was shown to have antimicrobial activity on a variety of microorganisms, including Gram-positive bacteria: *S. aureus* ATCC 13709 and *Enterococcus faecalis* ATCC 14428 (MIC of 15.6 µg/mL); Gram-negative bacteria: *P. mirabilis* ATCC 7002, *E. aerogenes* ATCC 13048, *S. typhi* ATCC 19430, *E. cloacae* ATCC 10699, *Proteus vulgaris* ATCC 12454 (MIC of 31.3 µg/mL), *P. aeruginosa* ATCC 27853 (MIC of 62.5 µg/mL), and *K. pneumoniae* ATCC 27736 (MIC of 125 µg/mL), pathogenic yeast *C. albicans* (MIC of 500 µg/mL); and two filamentous phytopathogenic fungi: *Botrytis cinerea* (MIC of 500 µg/mL) and *Rhizoctonia solani* (MIC of 250 µg/mL) (Bontempo et al., 2013).

4.12 Tiliaceae

Culinary plants of this family, such as *Triumfetta cordifolia* A. Rich. and *Corchorus olitorius* displayed antimicrobial activities. *T. cordifolia* located in tropical Africa, is a shrub of about 5-m high, used in Cameroon as foodstuff as well as in traditional medicine to manage several ailments including microbial infections. *C. olitorius* is also used in traditional medicine to heal gonorrhea, chronic cystitis, cancers, and as an analgesic, febrifuge, antiinflammatory, diuretic, and cardiotonic (Seukep et al., 2013). A flavonoid glucoside identified as *trans*-tiliroside isolated from the leaves of *T. cordifolia* was active against *E. coli*, *P. aeruginosa*, *S. typhi*, *B. cereus*, *S. aureus*, and *S. faecalis* (Sandjo et al., 2009). *C. olitorius* also displayed antibacterial effects on a panel of Gram-negative bacteria, such as *E. coli*, *E. aerogenes*, *K. pneumoniae*, *E. cloacae*, *P. stuartii*, and *P. aeruginosa* (Seukep et al., 2013). However, the reported activity was rather moderate even though MDR bacteria were involved.

4.13 Rutaceae

Rutaceae, commonly known as citrus family, is a family of flowering plants with approximatively 160 genera also having flowering species. The most economically important genera in the family are *Citrus*, which includes the orange (*Citrus sinensis*), lemon (*Citrus limon*), grapefruit (*Citrus paradisi*), and lime (mostly *Citrus aurantifolia*) as well as *Zanthoxylum* or *Fagara* and *Agathosma*. Spices of the genus *Fagara* have been found to have antimicrobial activities.

Fagara leprieurii (Guill and Perr) Engl. is used traditionally in case of gastritis, gingivitis, bilharzia, diarrhea, cancer, ulcer, kidney ache, sterility, gonorrhea as well as laxative and other infectious diseases (Ngane et al., 2000; Kuete et al., 2011c). *Fraxinus xanthoxyloides* is also used as antiseptic and laxative (Ngane et al., 2000). Ngane et al. (2000) have evaluated the antifungal properties of aqueous-ethanol 90% extracts of leaves, roots, and stem barks of *F. leprieurii* as well as *F. xanthoxyloides* against nine fungi. They found that these extracts, to varying extents, inhibited the in vitro growth of *C. albicans*, *C. neoformans*, and filamentous fungi, *Microsporum gypseum*, *Trichophyton mentagrophytes*, *Trichophyton rubrum*, *B. cinerea*, *Aspergillus fumigatus*, *A. flavus*, and *Scopulariopsis brevicaulis*. They also observed that only the extracts obtained from the roots and stem barks of *F. xanthoxyloides* had antifungal activity on the studied microorganisms, with MIC values varying from 0.5 to 1 mg/mL for the roots and from 0.125 to 1 mg/mL for the stem bark.

4.14 Zingiberaceae

Zingiberaceae commonly known as ginger family, is a family of flowering plants comprising more than 1300 species divided into about 52 genera of aromatic perennial herbs with creeping horizontal or tuberous rhizomes, distributed throughout tropical Africa, Asia, and the America. Numbers of plants of this family showed significant antimicrobial activities.

Aframomum citratum (Pereira) Schum is used traditionally to treat bacterial infections, malaria, cancers and as an aphrodisiac (Titanji et al., 2008; Kuete et al., 2011c). The methanol extract from the bark of this plant was found significantly or moderately active against various strains of *E. coli* (MIC: 512–1024 µg/mL), *E. aerogenes* (MIC: 512–1024 µg/mL on the majority of the strains), *E. cloacae* (MIC: 1024 µg/mL), *K. pneumoniae* (MIC: 512–1024 µg/mL), *P. stuartii* (MIC: 256–1024 µg/mL), and *P. aeruginosa* (MIC: 512–1024 µg/mL) (Fankam et al., 2011).

Aframomum kayserianum K. Schum is used in traditional medicine as an antimumps, vermifuge and in case of dysmenorrheas (Tane et al., 2005). Methanol extract from the fruits of this plant was found to be highly active against the resistant *E. aerogenes* EA289 strain (MIC of 64 μg/mL) and also very active against *E. coli* ATCC 10536, AG100, AG100 Atet, and W3110, *E. aerogenes* EA298, *K. pneumoniae* K24, and *P. stuartii* ATCC 29914 strains with a MIC value of 256 μg/mL (Seukep et al., 2013). Moderate antibacterial activities were also recorded with this extract against many other strains of the previously mentioned species; meanwhile no effect was even found on some of them, suggesting that complete healing effect with the plant could not be expected in all bacterial infections (Seukep et al., 2013). The active constituent of the plant was identified as aframodial, found to be effective against various pathogens, such as *S. cerevisiae*, *Schizosaccharomyces pombe*, *Hansenula anomala*, *Candida utilis*, *Sclerotinia libertiana*, *Penicillium crustosum*, *Mucor mucedo*, *Rhizopus chinensis*, *Aspergillus niger*, *S. aureus*, *B. subtilis*, *E. coli*, and *P. aeruginosa* (Ayafor et al., 1994).

Aframomum melegueta (Roscoe) K. Schum. has been used in traditional medicine in case of malaria, dysentery, dysmenorrhea, fertility, rubella, leprosy, cancers, and as carminative (Konning et al., 2004; Titanji et al., 2008). A decoction of the leaves is used for rheumatism and as an antiemetic agent and a decoction of the fruits for dysenteric conditions (Konning et al., 2004). The seed's methanol extract of *A. melegueta* was found active against Gram-positive bacteria (*S. aureus* and *B. subtilis*), Gram-negative bacteria (*E. coli* and *P. aeruginosa*), and fungi (*C. albicans* and *A. niger*) based on diffusion assay with inhibition zones ranging from 5.7 to 15.4 mm (Konning et al., 2004). Voukeng et al. (2012) also reported the moderate activity of the methanol extract against sensitive and MDR strains of *E. coli*, *E. aerogenes*, *E. cloacae*, *K. pneumoniae*, and *P. stuartii* with MIC values generally ranging from 512 to 1024 μg/mL. However, the investigators found that this extract was not active against *P. aeruginosa* PA01 and PA124 strains.

Djeussi et al. have reported the antibacterial activities of the methanol extracts of the rhizomes of *Aframomum alboviolaceum* and *Aframomum polyanthum* on various strains of *E. coli*, *E. aerogenes*, *E. cloacae*, *K. pneumoniae*, and *P. stuartii*. MIC value as low as 32 μg/mL was obtained with *A. polyanthum* extract against *E. aerogenes* EA294 strain (Djeussi et al., 2013).

Zingiber officinale is used traditionally for several health conditions including analgesic, antioxidant, sedative, antipyretic, insecticidal, cancer, microbial, and respiratory tract infections, rheumatic diseases, vomiting, nausea, and convulsion (Ukeh et al., 2009; Tankeo et al., 2014). Tankeo et al. (2014) reported that the methanol extract from the rhizomes of this medicinal spice was not active against *P. aeruginosa* but displayed moderate activity against sensitive and MDR strains of *E. coli*, *E. aerogenes*, *E. cloacae*, *K. pneumoniae*, and *P. stuartii* with MIC values generally ranging from 512 to 1024 μg/mL. The pharmacological potency of this plant is discussed in detail in Chapter 30.

5 OTHER ANTIMICROBIAL SPICES AND VEGETABLES FROM AFRICA

In the family Bombacaceae, the methanol extract from the fruits of *Adansonia digitata* used as febrifuge, antidysentery, antioxidant, analgesic, antidiarrheal, immunostimulant, hepatoprotector, antismall pox, antirubella, showed antibacterial activity against *E. coli* ATCC 10536, *E. aerogenes* EA298, *K. penumoniae* K2, and *P. stuartii* PS2636 with MIC values ranging from 256 to 1024 µg/mL (Djeussi et al., 2013; Seukep et al., 2013). Nevertheless, the fruit extract was not active on many other bacterial strains belonging to the previously mentioned species contrary to the leaf extract that showed a broader antibacterial spectrum.

The Cucurbitaceae spicy plant *Telfairia occidentalis*, used in folk medicine as aphrodisiac, and to treat male sexual disorders (Tajuddin et al., 2004; Sharma et al., 2011), inflammation, and microbial infections (Moleyar and Narasimham, 1992; Pinto et al., 2009; Palombo, 2011), showed antibacterial activity against various microorganisms including various Gram-negative strains of *E. coli*, *E. aerogenes*, *K. pneumoniae*, *P. aeruginosa*, and *P. stuartii* (Noumedem et al., 2013a).

The methanol extract of the Mimosaceae plant *Dichrostachys glomerata* (Forsk) Chuov was found active on a panel of sensitive and MDR strains of Gram-negative strains of *E. coli* (ATCC8739, ATCC 10536, AG100, AG100A, AG100ATet, AG102, MC4100 W3110), *E. aerogenes* (ATCC 13048, CM64, EA27, EA289, EA298, EA294), *K. pneumoniae* (ATCC 11296, KP55, KP63, K24, K2), *E. cloacae* (ECCI69, BM47, BM67), *P. aeruginosa* (PA01, PA124), and *P. stuartii* (ATCC 29916, NEA16, PS2636, PS299645) with significant to moderate MIC values ranging from 128 to 1024 µg/mL (Fankam et al., 2011). The antibacterial activity of *D. glomerata* was also reported against Gram-positive strains of *S. aureus* and *Listeria monocytogenes*, and Gram-negative strains *E. coli* and *Salmonella typhymurium* (Tamokou et al., 2013).

In the family Ramineae, the methanol extract of *Imperata cylindrica* Beauv. var. koenigii Durand et Schinz used in folk medicine as diuretic, antiinflammatory, dysentery, urinary tract infections, and cancer (Krishnaiah et al., 2009; Kuete et al., 2011c) displayed antibacterial activities against *E. coli*, *E. aerogenes*, *E. cloace*, *K. pneumoniae*, and *P. stuartii* (Voukeng et al., 2012).

The Periplocaceae plant *Mondia whitei* (Hook F) Skell as well as the Olacaceae, *Olax subscorpioidea* Oliv. fruits rather showed poor antibacterial effects (Fankam et al., 2011).

Other medicinal spices and vegetables from Africa, including plant of the family Lauraceae spices, such as *Cinnamomum zeylanicum* Blume and *Cinnamomum cassia* (Chapter 17); Liliaceae with *A. cepa* (Chapter 14), *A. sativum* (Chapter 15); Myrtaceae with *Syzygium aromaticum* (Chapter 29); and Lamiaceae with *T. vulgaris* L. (Chapter 28) also showed antimicrobial effects.

6 ANTIMICROBIAL MODE OF ACTION OF AFRICAN SPICES AND VEGETABLES

6.1 Antibiotic-potentiating extracts

The association of natural products, such as plant extracts and antibiotics constitutes an alternative in the fight against MDR bacteria. Fankam et al. have investigated the synergistic potential of the methanol extract of the roots of the medicinal spice *D. glomerata* and several antibiotic currently used clinically, against eight MDR bacteria, *E. coli* AG100, AG100$_{TET}$, *K. pneumoniae* KP55, *E. aerogenes* EA3, EA27, EA289, CM64 in addition to *P. aeruginosa* PA124. They found that significant synergistic effects were obtained with this extract when it was combined with several antibiotics, especially on 25% of the tested bacteria for cloxacillin (CLX) and ampicillin (AMP), 50% (4/8) for kanamycin (KAN), 62.5% (5/8) for chloramphenicol (CHL), cefepime (FEP), streptomycin (STR), ciprofloxacin (CIP), 75% (6/8) for erythromycin (ERY), and 87.5% (7/8) for norfloxacin (NOR) and tetracycline (TET). An eightfold increase in MIC values was recorded at MIC/2 of the extract with CHL, TET, STR, CIP, NOR. At MIC/5 of the crude extract, synergistic effects were noted on 50% of the eight tested MDR bacteria in the case of STR and CIP, 62.5% in the case of ERY, and 75% in the case of CHL, TET, and NOR (Fankam et al., 2011). The methanol extracts of the roots of *D. psilurus* also showed synergistic effects with antibiotics inhibiting bacterial cell wall synthesis (AMP and CEF) on *K. pneumoniae* KP63 (Voukeng et al., 2012). Noumedem et al. (2013a) also reported the synergistic effect of the methanol extract of the seeds of *P. nigrum* with antibiotics, such as TET, DOX, CIP, NOR, STR, KAN, CHL, ERY but not with beta-lactams (AMP, FEP, and CLX). The synergistic effects of the extract of the leaves of *G. africanum* was obtained when it was combined with antibiotics, such as TET, DOX, CIP, NOR, STR, KAN, CHL, ERY but not with beta-lactams (Tankeo et al., 2014).

6.2 Role of bacterial efflux pumps

Multidrug efflux systems display the ability to transport a variety of structurally unrelated drugs from a cell and consequently are capable of conferring resistance to a diverse range of chemotherapeutic agents (Paulsen et al., 1996; Kuete et al., 2011a). Multidrug efflux systems which use the proton motive force to drive drug transport have been largely described. These proteins are likely to operate as multidrug/proton antiporters and have been identified in both prokaryotes and eukaryotes (Paulsen et al., 1996). Such proton-dependent multidrug efflux proteins belong to three distinct families or superfamilies of transport proteins: the major facilitator superfamily (MFS), the small multidrug resistance (SMR) family, and the resistance/nodulation/cell division (RND) family (Paulsen et al., 1996). Tripartite efflux systems, mainly those clinically described as AcrAB-TolC in Enterobacteriaceae or MexAB-OprM in *P. aeruginosa*, are associated with a major human health problem as they play a central role in multidrug resistance of pathogenic Gram-negative bacteria (Blot et al., 2007; Papadopoulos et al., 2008; Pietras et al., 2008). Phenylalanine arginine β-naphthylamide (PAβN) has been reported as a potent inhibitor of the RND efflux systems and is especially active on AcrAB-TolC and MexAB-OprM (Lomovskaya and Bostian, 2006; Pages et al., 2009; Pietras et al., 2008). Several resistance genes encoding multidrug efflux pumps which can lead to reduced susceptibilities to hydrophilic fluoroquinolones, as well as to other unrelated compounds, have been identified in Gram-positive bacteria (Paulsen et al., 1996). Two of these, NorA and Bmr, have been shown to be inhibited by reserpine (Brenwald et al., 1997). To determine the role of efflux pumps, scientists have been using PAβN or reserpine in combination with plant extracts to improve their activity against MDR bacterial species. It has been demonstrated that significant increase of the antibacterial activity was obtained when spices extracts such *D. glomerata* extract were combined with PAβN against resistant strains of *E. coli*, *K. pneumoniae*, and *P. stuartii* (Fankam et al., 2011). In the presence of PAβN, significant increase of the activity of the methanol extract of the seeds of *P. nigrum* was noted against 13/14 MDR bacteria (Noumedem et al., 2013a).

7 CONCLUSIONS

This chapter was aimed to report the antimicrobial potential of spices and vegetable found in Africa, in order to encourage their consumption as another way of improving human health condition worldwide. We first reviewed the methods used for antimicrobial screening and highlighted the advantages of broth microdilution technique. The preliminary phytochemical screening methods for plant secondary metabolites were also compiled. Herein, we also reviewed the cutoff points to define the significance of antimicrobial agents and proposed new ones for extract from edible parts of plants, with regard to their innocuity compared to other medicinal plants. We also discussed the antimicrobial potency of African medicinal spices and vegetables belonging to Annonaceae, Apiaceae, Brassicaceae, Gnetaceae, Lamiaceae, Liliaceae, Moraceae, Pedaliaceae, Piperaceae, Solanaceae, Tiliaceae, Rutaceae, Zingiberaceae, etc. However, some commonly used spices and vegetables described further in this book were not discussed in this chapter. The role of some African spices and vegetables as antibiotic-potentiating agents was also discussed as well as their possibility to fight bacterial multidrug resistance. Finally, it can be concluded that African medicinal spices and vegetable have the potential to combat microbial infections. This chapter aims to encourage their consumption in the continent to improve health condition of African population.

REFERENCES

Abegaz, B.M., Ngadjui, B.T., Dongo, E., Bezabiha, M.T., 2000. Chemistry of the genus *Dorstenia*. Curr. Org. Chem. 4 (10), 1079–1090.

Afolayan, A.J., Meyer, J.J., 1997. The antimicrobial activity of 3,5,7-trihydroxyflavone isolated from the shoots of *Helichrysum aureonitens*. J. Ethnopharmacol. 57 (3), 177–181.

Akroum, S., Satta, D., Lalaoui, K., 2009. Antimicrobial, antioxidant, cytotoxic activities and phytochemical screening of some Algerian plants. Eur. J. Sci. Res. 31, 289–295.

Arruda, A.L., Vieira, C.J., Sousa, D.G., Oliveira, R.F., Castilho, R.O., 2011. *Jacaranda cuspidifolia* Mart. (Bignoniaceae) as an antibacterial agent. J. Med. Food 14 (12), 1604–1608.

Assob, J.C., Kamga, H.L., Nsagha, D.S., Njunda, A.L., Nde, P.F., Asongalem, E.A., 2011. Antimicrobial and toxicological activities of five medicinal plant species from Cameroon traditional medicine. BMC Complement. Altern. Med. 11, 70.

Auger, J., Dugravot, S., Naudin, A., Abo-Ghalia, A., Pierre, D., Thibout, E., 2002. Utilisation des composés allelochimiques des Allium en tant qu'insecticides. IOBC WPRS Bull. 25, 1–13.

Ayafor, J.F., Tchuendem, M.H., Nyasse, A.B., Tillequin, F., Anke, H., 1994. Aframodial and other bioactive diterpenoids from *Aframomum* species. Pure Appl. Chem. 66, 2327–2330.

Babula, P., Adam, V., Kizek, R., Sladký, Z., Havel, L., 2009. Naphthoquinones as allelochemical triggers of programmed cell death. Environ. Exp. Bot. 65 (2–3), 330–337.

Barnard, D.L., Huffman, J.H., Meyerson, L.R., Sidwell, R.W., 1993. Mode of inhibition of respiratory syncytial virus by a plant flavonoid, SP-303. Chemotherapy 39 (3), 212–217.

Basri, D.F., Fan, S.H., 2005. The potential of aqueous and acetone extracts of galls of *Quercus infectoria* as antibacterial agents. Indian J. Pharmacol. 37 (1), 26–29.

Blot, S., Depuydt, P., Vandewoude, K., De Bacquer, D., 2007. Measuring the impact of multidrug resistance in nosocomial infection. Curr. Opin. Infect. Dis. 20 (4), 391–396.

Bontempo, P., Carafa, V., Grassi, R., Basile, A., Tenore, G.C., Formisano, C., 2013. Antioxidant, antimicrobial and anti-proliferative activities of *Solanum tuberosum* L. var. Vitelotte. Food Chem. Toxicol. 55, 304–312.

Brenwald, N.P., Gill, M.J., Wise, R., 1997. The effect of reserpine, an inhibitor of multi-drug efflux pumps, on the in-vitro susceptibilities of fluoroquinolone-resistant strains of *Streptococcus pneumoniae* to norfloxacin. J. Antimicrob. Chemother. 40 (3), 458–460.

Bruneton, J., 1999. Pharmacognosie: phytochimie, Plantes Medicinales, third ed. Tec & Doc, Paris.

Carson, C.F., Hammer, K.A., Riley, T.V., 1995. Broth microdilution method for determination the susceptibility of *Escherichia coli* and *Staphylococcus aureus* to the essential oil of *Melaleuca alternifolia* (tea tree oil). Microbios 82, 181–185.

Charvat, T.T., Lee, D.J., Robinson, W.E., Chamberlin, A.R., 2006. Design, synthesis, and biological evaluation of chicoric acid analogs as inhibitors of HIV-1 integrase. Bioorg. Med. Chem. 14 (13), 4552–4567.

Chung, G.A., Aktar, Z., Jackson, S., Duncan, K., 1995. High-throughput screen for detecting antimycobacterial agents. Antimicrob. Agents Chemother. 39 (10), 2235–2238.

CLSI, 2003. Methods for Dilution Antimicrobial Susceptibility Tests for Bacteria That Grow Aerobically; Approved Standard M7-A6, sixth ed. National Committee for Clinical Laboratory Standards, Wayne, PA, USA.

Cowan, M.M., 1999. Plant products as antimicrobial agents. Clin. Microbiol. Rev. 12 (4), 564–582.

Craig, W.J., 1999. Health-promoting properties of common herbs. Am. J. Clin. Nutr. 70, 491–499.

Critchfield, J.W., Butera, S.T., Folks, T.M., 1996. Inhibition of HIV activation in latently infected cells by flavonoid compounds. AIDS Res. Hum. Retroviruses 12 (1), 39–46.

Dimo, T., Rakotonirina, A., Tan, P., Dongo, E., Dongmo, A., Kamtchouing, P., 2001. Antihypertensive effects of *Dorstenia psilurus* extract in fructose-fed hyperinsulinemic, hypertensive rats. Phytomedicine 8, 101–106.

Djeussi, D.E., Noumedem, J.A., Seukep, J.A., Fankam, A.G., Voukeng, I.K., Tankeo, S.B., 2013. Antibacterial activities of selected edible plants extracts against multidrug-resistant Gram-negative bacteria. BMC Complement. Altern. Med. 13 (1), 164.

Djouossi, M.G., Tamokou, J.D., Ngnokam, D., Kuiate, J.R., Tapondjou, L.A., Harakat, D., 2015. Antimicrobial and antioxidant flavonoids from the leaves of *Oncoba spinosa* Forssk. (Salicaceae). BMC Complement. Altern. Med. 15, 134.

Ekop, A.S., 2007. Determination of chemical composition of *Gnetum africanum* (AFANG) seeds. Pak. J. Nutr. 6, 40–43.

Eloff, J.N., 1998. A sensitive and quick microplate method to determine the minimal inhibitory concentration of plant extracts for bacteria. Planta Med. 64 (8), 711–713.

Erasto, P., Bojase-Moleta, G., Majinda, R.R., 2004. Antimicrobial and antioxidant flavonoids from the root wood of *Bolusanthus speciosus*. Phytochemistry 65 (7), 875–880.

Erasto, P., Grierson, D.S., Afolayan, A.J., 2006. Bioactive sesquiterpene lactones from the leaves of *Vernonia amygdalina*. J. Ethnopharmacol. 106 (1), 117–120.

Erturk, O., 2006. Antibacterial and antifungal activity of ethanolic extracts from eleven spice plants. Biologia 61 (3), 275–278.

Fabry, W., Okemo, P., Ansorg, R., 1998. Antibacterial activity of East African medicinal plants. J. Ethnopharmacol. 60, 79–84.

Fankam, A.G., Kuete, V., Voukeng, I.K., Kuiate, J.R., Pages, J.M., 2011. Antibacterial activities of selected Cameroonian spices and their synergistic effects with antibiotics against multidrug-resistant phenotypes. BMC Complement. Altern. Med. 11, 104.

Fernandez, M.A., Garcia, M.D., Saenz, M.T., 1996. Antibacterial activity of the phenolic acids fractions of *Scrophularia frutescens* and *Scrophularia sambucifolia*. J. Ethnopharmacol. 53 (1), 11–14.

Field, J.A., Lettinga, G., 1992. Toxicity of tannic compounds to microorganisms. Plant polyphenols: synthesis, properties, significance. Basic Life Sci. 59, 673–692.

Garcia, A., Bocanegra-Garcia, V., Palma-Nicolas, J.P., Rivera, G., 2012. Recent advances in antitubercular natural products. Eur. J. Med. Chem. 49, 1–23.

Grare, M., Fontanay, S., Cornil, C., Finance, C., Duval, R.E., 2008. Tetrazolium salts for MIC determination in microplates: why? Which salt to select? How? J. Microbiol. Methods 75 (1), 156–159.

Guimarães, P.R., Galvão, A.M.P., Batista, C.M., Azevedo, G.S., Oliveira, R.D., Lamounier, R.P., 2000. Eggplant (*Solanum melongena*) infusion has a modest and transitory effect on hypercholesterolemic subjects. Braz. J. Med. Biol. Res. 33 (9), 1027–1036.

Hamburger, M.O., Cordell, G.A., 1987. A direct bioautographic TLC assay for compounds possessing antibacterial activity. J. Nat. Prod. 50 (1), 19–22.

Hamisy, W.C., Mwaseba, D., Zilihona, I.E., Mwihomeke, S.T., 2000. Status and Domestication Potential of Medicinal Plants in the Uluguru Mountain Area, Tanzania. Wildlife Conservation Society of Tanzania (WCST), p. 55.

Harbone, J. (Ed.), 1973. Phytochemical Methods: A Guide to Modern Techniques of Plant Analysis. Chapman & Hall, London.

Hostettmann, K., Wolfender, J.L., Rodriguez, S., 1997. Rapid detection and subsequent isolation of bioactive constituents of crude plant extracts. Planta Med. 63 (1), 2–10.

Hoult, J.R., Paya, M., 1996. Pharmacological and biochemical actions of simple coumarins: natural products with therapeutic potential. Gen. Pharmacol. 27 (4), 713–722.

Irvine, R., 1961. Woody Plant of Ghana. Oxford University Press, London.

Jimenez-Arellanes, A., Meckes, M., Ramirez, R., Torres, J., Luna-Herrera, J., 2003. Activity against multidrug-resistant *Mycobacterium tuberculosis* in Mexican plants used to treat respiratory diseases. Phytother. Res. 17 (8), 903–908.

Johnson, T.L., Forbes, B.A., O'Connor-Scarlet, M., Machinski, A., McClatchey, K.D., 1985. Rapid method of MIC determinations utilizing tetrazolium reduction. Am. J. Clin. Pathol. 83 (3), 374–378.

Jones, G.A., McAllister, T.A., Muir, A.D., Cheng, K.J., 1994. Effects of sainfoin (*Onobrychis viciifolia* Scop.) condensed tannins on growth and proteolysis by four strains of ruminal bacteria. Appl. Environ. Microbiol. 60 (4), 1374–1378.

Karou, D., Dicko, M.H., Simpore, J., Traore, A.S., 2005. Antioxidant and antibacterial activities of polyphenols from ethnomedicinal plants of Burkina Faso. Afr. J. Biotechnol. 4, 823–828.

Kazmi, M.H., Malik, A., Hameed, S., Akhtar, N., Noor Ali, S., 1994. An anthraquinone derivative from *Cassia italica*. Phytochemistry 36 (3), 761–763.

Kim, S.H., Lee, S.J., Lee, J.H., Sun, W.S., Kim, J.H., 2002. Antimicrobial activity of 9-O-acyl- and 9-O-alkylberberrubine derivatives. Planta Med. 68 (3), 277–281.

Klancnik, A., Piskernik, S., Jersek, B., Mozina, S.S., 2010. Evaluation of diffusion and dilution methods to determine the antibacterial activity of plant extracts. J. Microbiol. Methods 81 (2), 121–126.

Kokowaro, J., 1976. Medicinal Plants of East Africa. East African Literature Bureau, Kampala, Uganda.

Konning, G.H., Agyare, C., Ennison, B., 2004. Antimicrobial activity of some medicinal plants from Ghana. Fitoterapia 75 (1), 65–67.

Krishnaiah, D., Devi, T., Bono, A., Sarbatly, R., 2009. Studies on phytochemical constituents of six Malaysian medicinal plants. J. Med. Plant Res. 3 (2), 67–72.

Kuete, V., 2010. Potential of Cameroonian plants and derived products against microbial infections: a review. Planta Med. 76 (14), 1479–1491.

Kuete, V. (Ed.), 2013. Medicinal Plant Research in Africa: Pharmacology and Chemistry. Elsevier, London, England.

Kuete, V., Alibert-Franco, S., Eyong, K.O., Ngameni, B., Folefoc, G.N., Nguemeving, J.R., 2011a. Antibacterial activity of some natural products against bacteria expressing a multidrug-resistant phenotype. Int. J. Antimicrob. Agents 37 (2), 156–161.

Kuete, V., Efferth, T., 2010. Cameroonian medicinal plants: pharmacology and derived natural products. Front. Pharmacol. 1, 123.

Kuete, V., Kamga, J., Sandjo, L.P., Ngameni, B., Poumale, H.M., Ambassa, P., 2011b. Antimicrobial activities of the methanol extract, fractions and compounds from *Ficus polita* Vahl. (Moraceae). BMC Complement. Altern. Med. 11, 6.

Kuete, V., Krusche, B., Youns, M., Voukeng, I., Fankam, A.G., Tankeo, S., 2011c. Cytotoxicity of some Cameroonian spices and selected medicinal plant extracts. J. Ethnopharmacol. 134 (3), 803–812.

Kuete, V., Mbaveng, A.T., Tsaffack, M., Beng, V.P., Etoa, F.X., Nkengfack, A.E., 2008a. Antitumor, antioxidant and antimicrobial activities of *Bersama engleriana* (Melianthaceae). J. Ethnopharmacol. 115 (3), 494–501.

Kuete, V., Metuno, R., Ngameni, B., Tsafack, A.M., Ngandeu, F., Fotso, G.W., 2007a. Antimicrobial activity of the methanolic extracts and compounds from *Treculia obovoidea* (Moraceae). J. Ethnopharmacol. 112 (3), 531–536.

Kuete, V., Nana, F., Ngameni, B., Mbaveng, A.T., Keumedjio, F., Ngadjui, B.T., 2009a. Antimicrobial activity of the crude extract, fractions and compounds from stem bark of *Ficus ovata* (Moraceae). J. Ethnopharmacol. 124 (3), 556–561.

Kuete, V., Ngameni, B., Mbaveng, A.T., Ngadjui, B., Meyer, J.J., Lall, N., 2010a. Evaluation of flavonoids from *Dorstenia barteri* for their antimycobacterial, antigonorrheal and anti-reverse transcriptase activities. Acta Trop. 116 (1), 100–104.

Kuete, V., Ngameni, B., Simo, C.C., Tankeu, R.K., Ngadjui, B.T., Meyer, J.J., 2008b. Antimicrobial activity of the crude extracts and compounds from *Ficus chlamydocarpa* and *Ficus cordata* (Moraceae). J. Ethnopharmacol. 120 (1), 17–24.

Kuete, V., Ngameni, B., Tangmouo, J.G., Bolla, J.M., Alibert-Franco, S., Ngadjui, B.T., 2010b. Efflux pumps are involved in the defense of Gram-negative bacteria against the natural products isobavachalcone and diospyrone. Antimicrob. Agents Chemother. 54 (5), 1749–1752.

Kuete, V., Nguemeving, J.R., Beng, V.P., Azebaze, A.G., Etoa, F.X., Meyer, M., 2007b. Antimicrobial activity of the methanolic extracts and compounds from *Vismia laurentii* De Wild (Guttiferae). J. Ethnopharmacol. 109 (3), 372–379.

Kuete, V., Tangmouo, J.G., Marion Meyer, J.J., Lall, N., 2009b. Diospyrone, crassiflorone and plumbagin: three antimycobacterial and antigonorrhoeal naphthoquinones from two *Diospyros* spp. Int. J. Antimicrob. Agents 34 (4), 322–325.

Kuete, V., Tangmouo, J.G., Penlap Beng, V., Ngounou, F.N., Lontsi, D., 2006. Antimicrobial activity of the methanolic extract from the stem bark of *Tridesmostemon omphalocarpoides* (Sapotaceae). J. Ethnopharmacol. 104 (1–2), 5–11.

Kurek, A., Grudniak, A.M., Kraczkiewicz-Dowjat, A., Wolska, K.I., 2011. New antibacterial therapeutics and strategies. Pol. J. Microbiol. 60, 3–12.

Kyung, K., Fleming, H., 1997. Antimicrobial activity of sulfur compounds derived from cabbage. J. Food Prot. 60 (1), 67–71.

Lampe, J.W., 2003. Spicing up a vegetarian diet: chemopreventive effects of phytochemicals. Am. J. Clin. Nutr. 78 (3 Suppl.), 579S–583S.

Lee, K.W., Everts, H., Beynen, A.C., 2004. Essential oils in broiler nutrition. Int. J. Poult. Sci. 3, 738–752.

Lomovskaya, O., Bostian, K.A., 2006. Practical applications and feasibility of efflux pump inhibitors in the clinic—a vision for applied use. Biochem. Pharmacol. 71 (7), 910–918.

Mann, C.M., Markham, J.L., 1998. A new method for determining the minimum inhibitory concentration of essential oils. J. Appl. Microbiol. 84 (4), 538–544.

Masoko, P., Eloff, J.N., 2006. Bioautography indicates the multiplicity of antifungal compounds from twenty-four Southern African *Combretum* species (Combretaceae). Afr. J. Biotechnol. 5 (18), 1625–1647.

Mbaveng, A.T., Ngameni, B., Kuete, V., Simo, I.K., Ambassa, P., Roy, R., 2008. Antimicrobial activity of the crude extracts and five flavonoids from the twigs of *Dorstenia barteri* (Moraceae). J. Ethnopharmacol. 116 (3), 483–489.

McGaw, L.J., Lall, N., Meyer, J.J., Eloff, J.N., 2008. The potential of South African plants against *Mycobacterium* infections. J. Ethnopharmacol. 119 (3), 482–500.

Meyer, J.J., Afolayan, A.J., Taylor, M.B., Erasmus, D., 1997. Antiviral activity of galangin isolated from the aerial parts of *Helichrysum aureonitens*. J. Ethnopharmacol. 56 (2), 165–169.

Miura, K., Kikuzaki, H., Nakatani, N., 2002. Antioxidant activity of chemical components from sage (*Salvia officinalis* L.) and thyme (*Thymus vulgaris* L.) measured by the oil stability index method. J. Agric. Food Chem. 50 (7), 1845–1851.

Moleyar, V., Narasimham, P., 1992. Antibacterial activity of essential oil components. Int. J. Food Microbiol. 16 (4), 337–342.

Moreno, S., Scheyer, T., Romano, C.S., Vojnov, A.A., 2006. Antioxidant and antimicrobial activities of rosemary extracts linked to their polyphenol composition. Free Radic. Res. 40 (2), 223–231.

N'Guessan, K., Kadja, B., Zirihi, G.N., Traoré, D., Aké-Assi, L., 2009. Screening phytochimique de quelques plantes médicinales ivoiriennes utilisées en pays Krobou (Agboville, Côte-d'Ivoire). Sci. Nat. 6 (1), 1–15.

Nang, H.L.L., May, C.Y., Ngan, M.A., Hock, C.C., 2007. Extraction and identification of water soluble compounds in palm pressed fiber by SC-CO_2 and GC-MS. Am. J. Environ. Sci. 3 (2), 54–59.

Ncube, N.S., Afolayan, A.J., Okoh, A.I., 2008. Assessment techniques of antimicrobial properties of natural compounds of plant origin: current methods and future trends. Afr. J. Biotechnol. 7 (12), 1797–1806.

Ngane, A.N., Biyiti, L., Zollo, P.H., Bouchet, P., 2000. Evaluation of antifungal activity of extracts of two Cameroonian rutaceae: *Zanthoxylum leprieurii* Guill. et Perr. and *Zanthoxylum xanthoxyloides* Waterm. J. Ethnopharmacol. 70 (3), 335–342.

Ngassoum, M.B., Essia-Ngang, J.J., Tatsadjieu, L.N., Jirovetz, L., Buchbauer, G., Adjoudji, O., 2003. Antimicrobial study of essential oils of *Ocimum gratissimum* leaves and *Zanthoxylum xanthoxyloides* fruits from Cameroon. Fitoterapia 74 (3), 284–287.

Nisar, A., Fazal, H., Abbasi, B.H., Farooq, S., Ali, M., Khan, M.A., 2012. Biological role of *Piper nigrum* L. (black pepper): a review. Asian Pac. J. Trop. Biomed. 2, 1–10.

Nostro, A., Germano, M.P., D'Angelo, V., Marino, A., Cannatelli, M.A., 2000. Extraction methods and bioautography for evaluation of medicinal plant antimicrobial activity. Lett. Appl. Microbiol. 30 (5), 379–384.

Noumedem, J.A., Mihasan, M., Kuiate, J.R., Stefan, M., Cojocaru, D., Dzoyem, J.P., 2013a. In vitro antibacterial and antibiotic-potentiation activities of four edible plants against multidrug-resistant Gram-negative species. BMC Complement. Altern. Med. 13, 190.

Noumedem, J.A., Mihasan, M., Lacmata, S.T., Stefan, M., Kuiate, J.R., Kuete, V., 2013b. Antibacterial activities of the methanol extracts of ten Cameroonian vegetables against Gram-negative multidrug-resistant bacteria. BMC Complement. Altern. Med. 13, 26.

Obiukwu, C.E., Nwanekwu, K.E., 2010. Evaluation of the antimicrobial potentials of 35 medicinal plants from Nigeria. Int. Sci. Res. J. 2, 48–51.

Okigbo, R.N., Ogbonnaya, U.O., 2006. Antifungal effects of two tropical plant leaf extracts (*Ocimum gratissimum* and *Aframomum melegueta*) on postharvest yam (*Dioscorea* spp.) rot. Afr. J. Biotechnol. 5, 727–731.

Okuda, T., 2005. Systematics and health effects of chemically distinct tannins in medicinal plants. Phytochemistry 66 (17), 2012–2031.

Omulokoli, E., Khan, B., Chhabra, S.C., 1997. Antiplasmodial activity of four Kenyan medicinal plants. J. Ethnopharmacol. 56 (2), 133–137.

Osman, K., Evangelopoulos, D., Basavannacharya, C., Gupta, A., McHugh, T.D., Bhakta, S., 2012. An antibacterial from *Hypericum acmosepalum* inhibits ATP-dependent MurE ligase from *Mycobacterium tuberculosis*. Int. J. Antimicrob. Agents 39 (2), 124–129.

Otshudi, A.L., Apers, S., Pieters, L., Claeys, M., Pannecouque, C., De Clercq, E., 2005. Biologically active bisbenzylisoquinoline alkaloids from the root bark of *Epinetrum villosum*. J. Ethnopharmacol. 102 (1), 89–94.

Pages, J.M., Lavigne, J.P., Leflon-Guibout, V., Marcon, E., Bert, F., Noussair, L., 2009. Efflux pump, the masked side of beta-lactam resistance in *Klebsiella pneumoniae* clinical isolates. PLoS One 4 (3), e4817.

Palombo, E., 2011. Traditional medicinal plant extracts and natural products with activity against oral bacteria: potential application in the prevention and treatment of oral diseases. Evid. Based Complement. Alternat. Med. 2011, 680354.

Papadopoulos, C.J., Carson, C.F., Chang, B.J., Riley, T.V., 2008. Role of the MexAB-OprM efflux pump of *Pseudomonas aeruginosa* in tolerance to tea tree (*Melaleuca alternifolia*) oil and its monoterpene components terpinen-4-ol, 1,8-cineole, and alpha-terpineol. Appl. Environ. Microbiol. 74 (6), 1932–1935.

Parekh, J., Karathia, N., Chanda, S., 2006. Screening of some traditionally used medicinal plants for potential antibacterial activity. Indian J. Pharm. Sci. 68 (6), 832–834.

Paulsen, I.T., Brown, M.H., Skurray, R.A., 1996. Proton-dependent multidrug efflux systems. Microbiol. Rev. 60 (4), 575–608.

Pengsuparp, T., Cai, L., Constant, H., Fong, H.H., Lin, L.Z., Kinghorn, A.D., 1995. Mechanistic evaluation of new plant-derived compounds that inhibit HIV-1 reverse transcriptase. J. Nat. Prod. 58 (7), 1024–1031.

Pietras, Z., Bavro, V.N., Furnham, N., Pellegrini-Calace, M., Milner-White, E.J., Luisi, B.F., 2008. Structure and mechanism of drug efflux machinery in Gram negative bacteria. Curr. Drug Targets 9 (9), 719–728.

Pinto, E., Vale-Silva, L., Cavaleiro, C., Salgueiro, L., 2009. Antifungal activity of the clove essential oil from *Syzygium aromaticum* on *Candida*, *Aspergillus* and dermatophyte species. J. Med. Microbiol. 58, 1454–1462.

Rahman, M., Kuhn, I., Olsson-Liljequist, B., Mollby, R., 2004. Evaluation of a scanner-assisted colorimetric MIC method for susceptibility testing of gram-negative fermentative bacteria. Appl. Environ. Microbiol. 70 (4), 2398–2403.

Rahman, S., Parvez, A.K., Islam, R., Khan, M.H., 2011. Antibacterial activity of natural spices on multiple drug resistant *Escherichia coli* isolated from drinking water, Bangladesh. Ann. Clin. Microbiol. Antimicrob. 10, 10.

Saleem, M., Nazir, M., Ali, M.S., Hussain, H., Lee, Y.S., Riaz, N., 2010. Antimicrobial natural products: an update on future antibiotic drug candidates. Nat. Prod. Rep. 27 (2), 238–254.

Sandjo, L.P., Kuete, V., Tchangna, R.S., Efferth, T., Ngadjui, B.T., 2014. Cytotoxic benzophenanthridine and furoquinoline alkaloids from *Zanthoxylum buesgenii* (Rutaceae). Chem. Cent. J. 8 (1), 61.

Sandjo, L.P., Simo, I.K., Kuete, V., Hannewald, P., Yemloul, M., Rincheval, V., 2009. Triumfettosterol Id and triumfettosaponin, a new (fatty acyl)-substituted steroid and a triterpenoid 'dimer' bis(β-D-glucopyranosyl) ester from the leaves of wild *Triumfetta cordifolia* A. Rich. (Tiliaceae). Helv. Chim. Acta 92 (9), 1748–1759.

Savoia, D., 2012. Plant-derived antimicrobial compounds: alternatives to antibiotics. Future Microbiol. 7 (8), 979–990.

Scalbert, A., 1991. Antimicrobial properties of tannins. Phytochemistry 30, 3875–3883.

Schmourlo, G., Mendonca-Filho, R.R., Alviano, C.S., Costa, S.S., 2005. Screening of antifungal agents using ethanol precipitation and bioautography of medicinal and food plants. J. Ethnopharmacol. 96 (3), 563–568.

Sethi, M.L., 1983. Enzyme inhibition VI: inhibition of reverse transcriptase activity by protoberberine alkaloids and structure-activity relationships. J. Pharm. Sci. 72 (5), 538–541.

Seukep, J.A., Fankam, A.G., Djeussi, D.E., Voukeng, I.K., Tankeo, S.B., Noumdem, J.A., 2013. Antibacterial activities of the methanol extracts of seven Cameroonian dietary plants against bacteria expressing MDR phenotypes. Springerplus 2, 363.

Sharma, A., Kumar, M., Kaur, S., 2011. Modulatory effects of *Syzygium aromaticum* (L.) Merr. & Perry and *Cinnamomum tamala* Nees & Ebrem. on toxicity induced by chromium trioxide. Phytopharmacology 1 (4), 71–81.

Shiu, W.K., Rahman, M.M., Curry, J., Stapleton, P., Zloh, M., Malkinson, J.P., 2012. Antibacterial acylphloroglucinols from *Hypericum olympicum*. J. Nat. Prod. 75 (3), 336–343.

Simões, M., Bennett, R., Rosa, E., 2009. Understanding antimicrobial activities of phytochemicals against multidrug resistant bacteria and biofilms. Nat. Prod. Rep. 26, 746–757.

Singh, A., Handa, S.S., 1995. Hepatoprotective activity of *Apium graveolens* and *Hygrophila auriculata* against paracetamol and thioacetamide intoxication in rats. J. Ethnopharmacol. 49 (3), 119–126.

Sudheesh, S., Sandhya, C., Sarah Koshy, A., Vijayalakshmi, N.R., 1999. Antioxidant activity of flavonoids from *Solanum melongena*. Phytother. Res. 13 (5), 393–396.

Tajuddin, Ahmad, S., Latif, A., Qasmi, I., 2004. Effect of 50% ethanolic extract of *Syzygium aromaticum* (L.) Merr. & Perry. (clove) on sexual behaviour of normal male rats. BMC Complement. Altern. Med. 4 (1), 17.

Tamokou, J.D., Chouna, J.R., Fischer-Fodor, E., Chereches, G., Barbos, O., Damian, G., Benedec, D., Duma, M., Nkeng-Efouet, A.P., Wabo, K.H., Kuiate, J.R., Mot, A., Dumitrescu, S.R., 2013. Anticancer and antimicrobial activities of some antioxidant-rich Cameroonian medicinal plants. PLoS One 8 (2), e0055880.

Tamokou, J.D., Kuiate, J.R., Tene, M., Tane, P., 2009a. Antimicrobial clerodane diterpenoids from *Microglossa angolensis* Oliv. et Hiern. Indian J. Pharmacol. 41 (2), 60–63.

Tamokou, J.D., Tala, F.M., Wabo, K.H., Kuiate, J.R., Tane, P., 2009b. Antimicrobial activities of methanol extract and compounds from stem bark of *Vismia rubescens*. J. Ethnopharmacol. 124, 571–575.

Tane, P., Tatsimo, S., Ayimele, G., Connolly, J., 2005. Bioactive metabolites from *Aframomum* species. 11th NAPRECA Symposium Book of Proceedings. NAPRECA, Antananarivo, Madagascar, pp. 214–223.

Tankeo, S.B., Lacmata, S.T., Noumedem, J.A., Dzoyem, J.P., Kuiate, J.R., Kuete, V., 2014. Antibacterial and antibiotic-potentiation activities of some Cameroonian food plants against multi-drug resistant gram-negative bacteria. Chin. J. Integr. Med. 20 (7), 546–554.

Tatsadjieu, L., Essia Ngang, J., Ngassoum, M., Etoa, F., 2003. Antibacterial and antifungal activity of *Xylopia aethiopica*, *Monodora myristica*, *Zanthoxylum xanthoxyloides* and *Zanthoxylum leprieurii* from Cameroon. Fitoterapia 74, 469–472.

Tatsimo, N.S.J., Tamokou, J.D.D., Lamshöft, M., Mouafo, T.F., Lannang, M.A., Sarkar, P., 2015. LC-MS guided isolation of antibacterial and cytotoxic constituents from *Clausena anisata*. Med. Chem. Res. 24 (4), 1468–1479.

Tekwu, E.M., Askun, T., Kuete, V., Nkengfack, A.E., Nyasse, B., Etoa, F.X., 2012. Antibacterial activity of selected Cameroonian dietary spices ethno-medically used against strains of *Mycobacterium tuberculosis*. J. Ethnopharmacol. 142 (2), 374–382.

Tene, M., Tane, P., Sondengam, B.L., Connolly, J.D., 2004. Lignans from the roots of *Echinops giganteus*. Phytochemistry 65 (14), 2101–2105.

Termentzi, A., Fokialakis, N., Skaltsounis, A.L., 2011. Natural resins and bioactive natural products thereof as potential antimicrobial agents. Curr. Pharm. Des. 17 (13), 1267–1290.

Thomas, S., 1965. Chemical basis of drug action. Drug Plants of Africa. Publications of University of Pennsylvania, Philadelphia, Pennsylvania.

Titanji, V.P., Zofou, D., Ngemenya, M.N., 2008. The antimalarial potential of medicinal plants used for the treatment of malaria in Cameroonian folk medicine. Afr. J. Tradit. Complement. Altern. Med. 5 (3), 302–321.

Tiwari, P., Kumar, B., Kaur, M., Kaur, G., Kaur, H., 2011. Phytochemical screening and extraction: a review. Int. Phramaceut. Sci. 1 (1), 98–106.

Touani, F.K., Seukep, A.J., Djeussi, D.E., Fankam, A.G., Noumedem, J.A., Kuete, V., 2014. Antibiotic-potentiation activities of four Cameroonian dietary plants against multidrug-resistant Gram-negative bacteria expressing efflux pumps. BMC Complement. Altern. Med. 14, 258.

Ukeh, D.A., Birkett, M.A., Pickett, J.A., Bowman, A.S., Luntz, A.J., 2009. Repellent activity of alligator pepper, *Aframomum melegueta*, and ginger, *Zingiber officinale*, against the maize weevil, *Sitophilus zeamais*. Phytochemistry 70 (6), 751–758.

Uma, D.P., 2001. Radioprotective, anticarcinogenic and antioxidant properties of the Indian holy basil, *Ocimum sanctum* (Tulasi). Indian J. Expl. Biol. 39, 185–190.

Umit, A.K.I., Akgun, K.O., 2009. Antifungal activity of aqueous extracts of spices against bean rust (*Uromyces appendiculatus*). Allelopathy J. 24, 207–213.

Valgas, C., de Souza, S.M., Smânia, E.F.A., Smânia, Jr., A., 2007. Screening methods to determine antibacterial activity of natural products. Braz. J. Microbiol. 38, 369–380.

Van Wyk, B., Gericke, N., 2000. Peoples Plants: A Guide to Useful Plants in Southern Africa. Brizza Publications, Pretoria, South Africa.

Voukeng, I.K., Kuete, V., Dzoyem, J.P., Fankam, A.G., Noumedem, J.A., Kuiate, J.R., 2012. Antibacterial and antibiotic-potentiation activities of the methanol extract of some cameroonian spices against Gram-negative multi-drug resistant phenotypes. BMC Res. Notes 5, 299.

Watanbe, H., Miyaji, C., Makino, M., Abo, T., 1996. Therapeutic effects of glycyrrhizin in mice infected with LP-BM5 murine retrovirus and mechanisms involved in the prevention of disease progression. Biotherapy 9 (4), 209–220.

Wong, P.Y.Y., Kitts, D.D., 2006. Studies on the dual antioxidant and antibacterial properties of parsley (*Petroselinum crispum*) and cilantro (*Coriandrum sativum*) extracts. Food Chem. 97 (3), 505–515.

Yamaji, K., Ishimoto, H., Usui, N., Mori, S., 2005. Organic acids and water soluble phenolics produced by *Paxillus* species. Mycorrhiza 15 (1), 17–23.

Yi, Z.B., Yan, Y., Liang, Y.Z., Bao, Z., 2007. Evaluation of the antimicrobial mode of berberine by LC/ESI-MS combined with principal component analysis. J. Pharm. Biomed. Anal. 44 (1), 301–304.

Zager, E.M., McNerney, R., 2008. Multidrug-resistant tuberculosis. BMC Infect. Dis. 8, 10.

CHAPTER 9

Anti-inflammatory and Anti-nociceptive Activities of African Medicinal Spices and Vegetables

J.P. Dzoyem, L.J. McGaw, V. Kuete, U. Bakowsky

1 INTRODUCTION

Inflammation is a complex process initiated by several factors related to physical or chemical noxious stimuli or microbiological toxins. The inflammatory response is intended to inactivate or destroy invading organisms, remove irritants, and set the stage for tissue repair. In addition, inflammation is considered to be one of the major causes for the development of various diseases including cancer, cardiovascular disease, diabetes, obesity, osteoporosis, rheumatoid arthritis, inflammatory bowel disease, asthma, and central nervous system-related diseases, such as Alzheimer's disease, depression, and Parkinson's disease (Laveti et al., 2013). Inflammation is generally characterized by certain regular events, such as redness, swelling, heat, and pain, sometimes leading to exudation and loss of function. The process of inflammation involves several events and inflammatory mediators which are potent chemical substances found in the body tissues, such as prostaglandins, leukotrienes, prostacyclins, lymphokines, and chemokines like interferon-α (IFN-α), IFN-γ, interleukin (IL)-1, IL-8, histamine, 5-hydroxytryptamine (5-HT), and tumor necrosis factor-α (TNF-α) (Serhan and Savill, 2005). According to the modern concept, inflammation is a healthy process resulting from some disturbance or disease. To overcome this problem various types of safe and effective antiinflammatory agents are available, such as aspirin and other nonsteroidal antiinflammatory drugs, with many other drugs under development. Although steroidal antiinflammatory drugs and NSAIDs are currently used to treat acute inflammation, these drugs have not been entirely successful in curing chronic inflammatory disorders, and most cause an increased risk of blood clotting, resulting in heart attacks and strokes. Reports suggest that almost 90% of the drugs used against inflammation produce drug-related toxicities, iatrogenic reactions, and adverse effects complicating the treatment process (Lanas, 2009). Therefore, a shift in the area of antiinflammatory treatment has been observed from the use of synthetics to natural therapy. The development of potent antiinflammatory drugs from natural products has long been under consideration. Natural resources form a rich basis for the discovery of new drugs because of their chemical diversity. Plant medicines, including those particularly used for their aromatic properties (aromatic plants), such as spices and vegetables, are of great importance in primary health care in many developing countries.

Spices are distinguished from herbs, which are leafy, green plant parts used for flavoring purposes. Herbs, such as basil or oregano may be used fresh; spices, however, are almost always dried. The use of aromatic plants depends on the content and composition of their active compounds, which are preferentially located in roots, stems, leaves, flowers, or seeds. Most of these compounds can be characterized as secondary metabolites, like terpenoids, carotenoids, phenolic acids, flavonoids, coumarins, glucosinolates, and alkaloids, which naturally occur in plants and act as a protective mechanism against predators, pathogens, and competitors (Peters, 2001). Some of these compounds, like phenolic acids, flavonoids, and diterpenes are known for their potential benefits for human health. In recent years, these dietary plants have been shown to possess valuable phytochemicals of great nutritional and therapeutic value (Lai and Roy, 2004). Africa is blessed with a vast amount of vegetables, fruits, and spices which are consumed for their nutrients or for their medicinal purposes. Herbs, spices, and vegetables have been used as food and for their flavor as well as for curing diseases over many centuries in different indigenous systems of medicine as well as folk medicines. In African traditional medicine, several edible fruits and spices are thought to be effective in relieving pain.

Scientific evidence is accumulating that many of these herbs and spices do indeed have medicinal properties that alleviate inflammatory conditions. A growing body of research has demonstrated that the commonly used herbs and spices, such as cloves, ginger, rosemary, turmeric, and garlic possess antiinflammatory properties that, in some cases, can be therapeutic. Many researchers have focused attention on spices and vegetables, and a number of species have been investigated for their antiinflammatory and antinociceptive potential (Chukwujekwu et al., 2005; Agyare et al., 2013; Dzoyem et al., 2014). As a result, a number of spices have been shown to exhibit potent antiinflammatory effects in the treatment of inflammation using various models. Nonetheless, there is still a paucity of updated, comprehensive information on promising antiinflammatory dietary plants from the African continent. A comprehensive review of vegetables and spices with potential antiinflammatory efficacy is beyond the scope of this chapter. In vivo and in vitro methods used in the screening of plant extracts for antiinflammatory and antinociceptive activities are reviewed. We have also documented results from the latest research on selected spices, vegetables, and derived compounds with demonstrated antiinflammatory and antinociceptive properties, with specific reference to those used in Africa.

2 METHODS USED IN THE SCREENING OF ANTIINFLAMMATORY AND ANTINOCICEPTIVE ACTIVITY OF AFRICAN SPICES AND VEGETABLES

African spices and vegetables have been shown to exhibit potent antiinflammatory effects using various models. The testing models generally used can be classified into two categories: in vitro and in vivo methods.

2.1 In vivo methods
Different researchers have used different experimental models to evaluate the antiinflammatory and antinociceptive activities of herbs. Here, we have summarized those methods that are most commonly used in terms of animal (in vivo) models in Africa.

2.1.1 Experimental inflammatory methods
2.1.1.1 Carrageenan-induced paw edema
Carrageenan-induced paw edema is one of the most popular tests used in the screening of African spices and vegetables for antiinflammatory activity (Winter et al., 1962). It is a highly sensitive and reproducible test for nonsteroidal antiinflammatory drugs and has long been established as a valid model to study new antiinflammatory drugs (Willoughby and DiRosa, 1972). Carrageenan-induced inflammation is useful in detecting orally active antiinflammatory agents; therefore, it has significant predictive value for antiinflammatory agents acting through mediators of acute inflammation (Vinegar et al., 1969). The development of edema induced by carrageenan injection causes an acute and local inflammatory response. In the early phase (0–1 h), histamine, serotonin, and bradykinin are the first mediators involved, whereas prostaglandins and various cytokines such as IL-1β, IL-6, IL-10, and TNF-α are implicated in the second phase (Crunkhorn and Meacock, 1971).

2.1.1.2 Xylene-induced ear edema assay
Xylene, xylol, or dimethylbenzene is known to cause severe vasodilation and edematous changes of skin as signs of acute inflammation (Tang et al., 1984). The increased thickness of ear tissues is caused by these histopathological changes. After topical application of xylene, severe vasodilation and edematous changes of the skin occur, and infiltration of inflammatory cells is detected, providing signs of acute inflammation. Suppression of this response is taken as a hallmark of antiphlogistic effect (Atta and Alkofahi, 1998). In the screening of spices for antiinflammatory properties, inhibition of the xylene-induced increase in ear weight is regarded as evidence of antiinflammatory efficacy through reducing vasodilation and hence improving the edematous condition (Okoli et al., 2007; Ishola et al., 2011, 2012, 2013; Ibironke and Odewole, 2012; Sowemimo et al., 2013, 2015)

2.1.1.3 Egg albumin–induced paw edema testin rodents
This is an in vivo model of inflammation used to screen agents for antiinflammatory effect (Amos et al., 2002). The characteristic swelling of the paw is due to edema formation. Inhibition of increased vascular permeability, and hence the attendant edema, modulates the extent and magnitude of the inflammatory reaction. The paw edema induced by injection of egg albumin peaks after 30 min and then progressively declines with time. Many chemical mediators like histamine, serotonin (5-HT), kinins, and prostanoids mediate an acute inflammation response induced by phlogistic agents including egg albumin

(Marsha-Lyn et al., 2002). Inflammation occurs through three distinct phases: an early phase mediated by histamine and serotonin (up to 2 h), an intermediate phase involving the activity of bradykinin, and the later phase which occurs 3–5 h after the administration of the irritant. This later phase is induced by bradykinin protease, with prostanoid synthesis by the cyclooxygenase (COX) enzyme (Pérez-Guerrero et al., 2001). This method has been applied in the screening of antiinflammatory activity of African spices and herbs by some researchers (Otimenyin et al., 2008; Chinasa et al., 2011; Adzu et al., 2014).

2.1.1.4 Cotton pellet granuloma test

Cotton pellet–induced granuloma formation is the most suitable method for studying the efficacy of drugs against the proliferative phase of inflammation (Swingle and Shideman, 1972). The subcutaneous implantation of a cotton pellet into a rodent results in the formation of a granuloma at the site of the implant. The initial events include accumulation of fluid and proteinaceous material together with an infiltration of macrophages, neutrophils, and fibroblasts, and multiplication of small blood vessels, which are the basic sources of the highly vascularized reddish mass termed granulation tissue. This method has been widely used to assess the transudative, exudative, and proliferative phases of subacute inflammation. The fluid adsorbed by the pellet greatly influences the wet weight of the granuloma, whereas the dry weight correlates well with the amount of granulomatous tissue formed (Purnima et al., 2010). In the cotton pellet–induced chronic inflammation model, the cotton pellet, when applied in the interscapular area, induces a chronic inflammation process. In this process, monocyte migration, liquid accumulation, apoptosis, damage, and so on will occur in the surrounding tissue of the pellets, and these accumulations will produce a granulation tissue that covers the pellets (Uzkeser et al., 2012).

2.1.2 Experimental models of nociceptive pain

These methods are classical nociception models used to screen prospective antinociceptive compounds.

2.1.2.1 Acetic acid–induced writhing test

The acetic acid–induced abdominal writhing test has been used as a screening tool for assessing analgesic or antiinflammatory agents. Pain is induced by injection of irritants into the peritoneal cavity of mice (Koster et al., 1959; Singh and Majumdar, 1995). The animals react with a characteristic stretching behavior which is called writhing (Burke and Fitzgerald, 2006). Writhing is defined as a stretch, tension to one side, extension of hind legs, or contraction of the abdomen so that the abdomen of the mice touches the floor, or turning of the trunk (twist). Any writhing is considered a positive response (Mishra et al., 2011). Analgesic activity of the test compound is inferred from a decrease in the frequency of writhings. Acetic acid induces an inflammatory response in the abdominal cavity, with subsequent activation of nociceptors (Collier et al., 1968). When animals are intraperitoneally injected with acetic acid, a painful reaction and acute inflammation emerge in the peritoneal area. Constriction induced by acetic acid is considered to be a nonselective antinociceptive model, as acetic acid acts indirectly by inducing the release of endogenous mediators which stimulate the nociceptive neurons that are sensitive to nonsteroidal antiinflammatory drugs, to narcotics, and to other centrally active drugs (Bighetti et al., 1999).

2.1.2.2 Hot plate latency tests

The hot plate test involves higher brain function, and is considered to be a supraspinally organized response (Eddy and Leimbach, 1953). This test consists of introducing a rat or mouse into an open-ended cylindrical space with a floor consisting of a metallic plate that is heated by a thermode or a boiling liquid (Woolfe and MacDonald, 1944; O'Callaghan and Holzman, 1975). A plate heated to a constant temperature produces two behavioral components that can be measured in terms of their reaction times, namely paw licking and jumping. Both are considered to be supraspinally integrated responses.

2.1.2.3 Formalin-induced paw edema

The formalin test is a popular chemical assay of injury-produced inflammatory pain. It is regarded as a more satisfactory model of clinical pain and is a useful model for the screening of novel compounds, as it encompasses inflammatory, neurogenic, and central mechanisms of nociception (Hunskaar and Hole, 1987; Tjølsen et al., 1992; Tjølsen and Hole, 1997). The advantage of the formalin assay over other models of inflammatory pain is that the injection of a dilute solution of formalin into the surface of a mouse or rat's hindpaw allows modeling of both acute and tonic pain using a single chemical in a relatively limited time (approximately 1 h). As shown in Table 9.1, there have been many reports of formalin tests in the screening of African spices and herbs for antinociceptive properties (Owoyele et al., 2009; Jimoh et al., 2011; Ishola et al., 2011, 2012, 2013; Ibironke and Odewole, 2012; Sowemimo et al., 2013; Sofidiya et al., 2014).

2.1.2.4 Tail-flick test

The tail-flick test has been used in the assessment of African spices for antinociceptive properties (Bianchi and Franceschini, 1954; Ben-Bassat et al., 1959; John et al., 2009; Abdala et al., 2014). It is one of the most common tests based on a phasic stimulus of high intensity. There are two variants of the tail-flick test. The first consists of immersing the tail in water at a predetermined temperature, and the second is to apply radiant heat to a small area of the tail. Although apparently similar, the two methods differ in terms of the stimulated surface areas. Compared to other nociceptive methods, the tail-flick method is simple to perform but animals should be accustomed to being handled so that they remain calm during measurements.

Table 9.1 Other African spices and vegetables with antiinflammatory or antinociceptive potential

Spice, vegetables names (Family)	Type extract used	Anti inflammatory/antinociceptive, analgesic model	Animals/cells used	Country of plant collection	References
A. barteri Oliv. (Apocynaceae)	Ethanol extract of leaves	Carrageenan–induced paw edema, formalin–induced ear edema, xylene-induced ear edema, edema/tail immersion, and formalin	Rats, mice	Nigeria	(Sofidiya et al., 2014)
A. mannii (Oliv.) Engl. et Diels (Amonaceae)	Methanol extract of leaves	15-LOX assay	/	Cameroon	(Dzoyem et al., 2014)
B. maderaspatensis (L.) B. Heyne ex Roth (Acanthaceae)	Ethanol extract of leaves	Carrageenan-induced paw edema, formalin-induced edema, xylene-induced ear edema/tail immersion and formalin assays	Rat, mice	Nigeria	(Sowemimo et al., 2013)
C. dalzielii N.E. Br. (Apocynaceae)	Aqueous whole plant	Xylene-induced edema/writhing and tail clip tests	Rat, mice	Nigeria	(Ugwah-Oguejiofor et al., 2013)
C. alba G. Don (Polygalaceae)	Petroleum ether, dichloromethane, and 80% ethanol extracts roots	Cyclooxygenase assays	/	Nigeria	(Chukwujekwu et al., 2005)
C. colocynthis (L.) Schrad. (Cucurbitaceae)	Aqueous extract of different parts	Formalin-induced edema/subplantar formalin-induced nociception, tail-flick test	Mice	Tunisia	(Marzouk et al., 2010, 2011)
C. ferruginea Vahl ex DC. (Connaraceae)	Methanol extract of root	Carrageenan-induced paw edema, formalin-induced edema/acetic acid–induced writhing test	Rat, mice	Nigeria	(Ishola et al., 2011, 2012, 2013; Ibironke and Odewole, 2012)
C. olitorius L. (Tiliaceae)	Methanol extract of leaves	15-LOX assay	/	Cameroon	(Dzoyem et al., 2014)
C dfer Ker Gawl. (Costaceae)	70% methanol extract of leaves	Carrageenan and egg albumin-induced paw edema, formaldehyde-induced arthritis, xylene-induced ear edema, TNF-α production/abdominal constriction test, hot plate and tail-flick tests, acetic acid–induced writhing, and formalin tests	Rats	Nigeria	(Anyasor et al., 2014)

Plant (Family)	Extract	Model	Cell/Animal	Country	Reference
C. lanceolatum Forssk. (Boraginaceae)	70% (v/v) aqueous ethanol extract of root	Carrageenan, arachidonic acid and formaldehyde-induced arthritis	Rat, mice	Ethiopia	(Yu et al., 2012)
C. esculentus L. (Cyperaceae)	Methanol extract of seeds	15-LOX assay	/	Cameroon	(Dzoyem et al., 2014)
D. saxatilis Hook.f. (Leguminosae)	Methanol extract of leaves	Carrageenan-induced paw edema, dimethylbenzene-induced inflammation, fresh egg white-induced paw edema	Mice	Nigeria	(Hassan et al., 2015)
H. cheirifolia (L.) Kuntze (Compositae)	Methanol extract of leaves and stems	Carrageenan-induced paw edema/acetic acid-induced writhing and hot plate test	Rat	Tuisia	(Ammar et al., 2009)
H. sabdariffa L. (Malvaceae)	Methanol extract of flowers	15-LOX assay	/	Cameroon	(Dzoyem et al., 2014)
L. micranthus Linn (Loranthaceae)	Methanol extract of leaves	NO production inhibition and TNF-α	RAW 264.7	Nigeria	(Agbo et al., 2014)
Mixture of M. indica, C. citratus, C. sinensis, and O. gratissimum	Aqueous extract of leaves	Carrageenan-induced hind paw edema	Rat	Kenya	(Tarkang et al., 2015)
M. cecropioides R. Br. ex Tedlie (Urticaceae)	Ethanol extract of leaves	Carrageenan-induced paw edema	Mice	Nigeria	(Sowemimo et al., 2015)
O. labiatum (N.E. Br.) A.J. Paton (Lamiaceae)	Ethanol extract of leaves	Phytohemagglutinin (PHA)-induced NO production	PBMCs cells	South Africa	(Kapewangolo et al., 2015)
O. ficus-indica (L.) Mill. (Cactaceae)	Phenolics from flowers	NO production	RAW 264.7	Morocco	(Zakia et al., 2014)
P nigrescens (Afzel.) Bullock (Apocynaceae)	Aqueous extract of leaves	Carrageenan, histamine, serotonin, and xylene-induced edema tests as well as the formalin/writhing and tail clip tests.	Rat	Nigeria, Ghana	(Owoyele et al., 2009)

(Continued)

Table 9.1 Other African spices and vegetables with antiinflammatory or antinociceptive potential (cont.)

Spice, vegetables names (Family)	Type extract used	Anti inflammatory/antinociceptive, analgesic model	Animals/cells used	Country of plant collection	References
P. nitida (Stapf) T. Durand & H. Durand (Apocynaceae)	Petroleum ether, dichloromethane, and 80% ethanol extracts of roots	Cyclooxygenase assays	/	Nigeria	(Chukwujekwu et al., 2005)
R. abyssinicus Jacq. (Polygonaceae)	80% methanol extract of rhizomes	Fresh raw egg albumin/chemically using acetic acid and formalin, and mechanically using analgesy meter	Rats	Ethiopia	(Mulisa et al., 2015)
S. latifolius (Sm.) E.A. Bruce (Rubiaceae)	Aqueous extract of stem bark	Egg albumin– induced edema and acetic acid–induced writhing	Mice	Nigeria	(Otimenyin et al., 2008)
Schwenckia americana L. (Solanaceae)	Essential oil from cones	Egg albumin–induced paw (systemic) edema	Rats	Nigeria	(Jimoh et al., 2011)
S. indicum L. (Pedaliaceae)	Methanol extract of seeds	15-LOX assay	/	Cameroon	(Dzoyem et al., 2014)
S. radiatum L. (Pedaliaceae)	Methanol extract of seeds	15-LOX assay	/	Cameroon	(Dzoyem et al., 2014)
T. indica L. (Caesalpiniaceae)	Methanol extract of fruits	15-LOX assay	/	Cameroon	(Dzoyem et al., 2014)
T. articulate (Vahl) Mast. (Cupressaceae)	Essential oil from leaves	LOX inhibition assay, XO inhibition assay	/	Algeria	(Djouahri et al., 2014)
V. faba L. (Leguminosae)	Aqueous (70%) acetone and (80%) ethanol extract of cotyledons	15-LOX, COX (COX1 and COX2) inhibition	/	Algeria	(Boudjou et al., 2013)
X. parviflora Spruce (Annonaceae)	Essential oil from fruits	Nitric oxide inhibitory assay	RAW 264.7	Cameroon	(Woguem et al., 2014)
Z. capense (Thunb.) Harv. (Rutaceae)	Acetone extract of leaves	15-LOX, inhibition of NO production	/	South Africa	(Adebayo et al., 2015)
Z. album L.f. (Zygophyllaceae)	Methanol extract of aerial parts	NO production	RAW 264.7	Tunisia	(Ksouri et al., 2013)

COX, cyclooxygenas; LOX, lipoxygenase; NO, nitric oxide; TNF-α, tumor necrosis factor-α; XO, xanthine oxidase; PHA, phytohemagglutinin.

2.2 In vitro methods

In vitro studies are helpful in developing an understanding of the mechanism of antiinflammatory activity of bioactive compounds. Cell and protein-based assays are generally fast, simple, relatively cheap, and easy to perform. They involve minimal ethical considerations compared to whole animal work, and allow insight into the biochemical and physiological processes induced by the tested compound. Many pharmacological agents at different concentrations can be evaluated concurrently without the intrinsic heterogeneity associated with in vivo models (Winyard and Willoughby, 2003; Nile and Park, 2013). Therefore, a number of in vitro assays have been used for the understanding of the molecular mechanisms underlying the antiinflammatory activity of African spices and vegetables. Most of the in vitro assays used for the study of antiinflammatory activity of plant constituents target proinflammatory mediators such as cytokines, enzymes, and transcription factors. Inhibition of nitric oxide production has been used as an in vitro antiinflammatory method to screen African spices and vegetables (Ksouri et al., 2013; Kapewangolo et al., 2015). The most targeted enzymes in in vitro assays were COX (COX-1 and COX-2), lipoxygenase (LOX), and inducible nitric oxide synthase (iNOS). COX enzyme activities have been monitored as described by some authors (Noreen et al., 1998; Laufer and Luik, 2010) using a COX activity assay. LOX isozymes are involved in the metabolism of eicosanoids, such as leukotrienes and prostaglandins. Inhibition of their activity has also been assessed in the screening of plants using commercial kits or various standardized methods (Waslidge and Hayes, 1995; Anthon and Barrett, 2001; Lu et al., 2013). Detection of mRNA, or protein expression of the genes coding for the synthesis of inflammatory mediators has been of interest. The transcription factor NF-κB has been the most studied whereas IL and tumor necrosis factor (TNF) are the most common inflammatory cytokines assessed in the screening of African spices. Several biochemical techniques such as polymerase chain reaction (PCR), and immunoblotting and enzyme-linked immunosorbent assays (ELISA) have also been used.

3 SELECTED AFRICAN SPICES WITH ANTIINFLAMMATORY AND ANTINOCICEPTIVE ACTIVITY

Various spices, dietary herbs, and vegetables found in Africa have been reported to exhibit antiinflammatory and antinociceptive activity. A review of the most recent studies is presented here. However, we do not emphasize the following spices that are the subject of individual chapters in this book. They include: *Allium cepa*; *Allium sativum*; *Canarium schweinfurthii*; *Cinnamon* sp.; *Cymbopogon citratus*; *Curcuma longa*; *Lactuca sativa*; *Mangifera indica*; *Moringa oleifera*; *Myristica* sp.; *Passiflora edulis*; *Petroselinum crispum*; *Sesamum indicum*; *Thymus vulgaris*; *Syzygium aromaticum*; *Xanthosoma mafaffa*; *Zinziber officinalis*.

3.1 Aframomum melegueta

Aframomum melegueta is a Zingiberaceae family plant spice widely spread in Africa. It is an herbaceous perennial plant native to swampy habitats along the West African coast, commonly known as "grains of paradise," "melegueta pepper," "alligator pepper," "guinea grains," or "guinea pepper." In African folk medicine, *Aframomum* species are used for alleviating stomach ache and diarrhea as well as hypertension, as an aphrodisiac, and against measles and leprosy (Kokwaro, 1993) They are also taken for excessive lactation and post partem hemorrhage, and are used as a purgative, galactogogue and anthelmintic, and hemostatic agent (Kokwaro, 1993).

The seeds of *A. melegueta* possess potent antiinflammatory and antinociceptive activity. The antinociceptive activity of intraperitoneal doses of 25–100 mg/kg of *A. melegueta* has been assessed using the formalin-induced paw licking, Randall–Selitto paw pressure, and hot plate models of pain. The extract produced significant inhibition of inflammation with the formalin test, and reduced the nociceptive responses elicited by compression of the inflamed hind paw of rats (Umukoro and Ashorobi, 2001a, 2007). The antiinflammatory properties of the ethanolic extract of *Aframomum* sp. seeds have been evaluated in vitro on proinflammatory gene expression, on inflammatory enzymes such as LOX, COX-2 (Odukoya et al., 1999; Dzoyem et al., 2014; Ilic et al., 2014), and in vivo by carrageenan-induced paw edema in rats (Umukoro and Ashorobi, 2001b, 2005; Ilic et al., 2014). The effects of the aqueous seed extract on events associated with inflammatory processes such as leukocyte migration and phenylhydrazine-treated red blood cells of rats have also been reported (Umukoro and Ashorobi, 2008). Phytochemical analysis revealed the presence of alkaloids, cardiac glycosides, tannins, flavonoids, sterols, triterpenes, and oils, whereas the methanol fraction contains alkaloids, glycosides, tannins, flavonoids, sterols, and resins (Okoli et al., 2007). Gingerol, another important active compound isolated from the seed of this plant, was shown to inhibit prostaglandin and leukotriene biosynthesis (Kiuchi et al., 1992). Compounds such as 3-(S)-acetyl-1-(4′-hydroxy-3′, 5′-di methoxyphenyl)-7-(3″,4″, 5″-trihydroxyphenyl)heptane, dihydrogingerenone, paradol, and [6]-shogaol suppressed inflammatory responses by inhibition of the elevated TNF-α and interleukin-1-beta (Il-1β) (El-Halawany et al., 2014).

3.2 Adansania digitata

Adansonia digitata L. (Malvaceae), commonly known as the baobab tree, is native to Africa. Baobab is a multipurpose tree which offers protection and provides food, clothing, and medicine as well as raw material for many useful items. Every part of the baobab tree is reported to be useful. The seeds, leaves, roots, flowers, fruit pulp, and bark of the baobab are edible. The leaves of the plant are the main source of food and folk medicine for many populations in Africa, and are eaten fresh or dried (Besco et al., 2007). The fruit pulp, seeds, leaves, flowers, roots, and bark of the baobab are edible and they have been studied by scientists for their useful properties (Rahula et al., 2015). From various parts of the plant, nutritional, phytochemical constituents have been isolated, including vitamin C, steroids, flavonoids, epicatechin, campesterol, tocopherol, adansonin, and amino acids. It has many medicinal and nonmedicinal uses; it is used in the treatment of bronchial asthma, dermatitis, sickle cell anemia, as a diuretic and antidiabetic, against diarrhoea and dysentery, as a laxative, to treat hiccough in children, and as an antioxidant, antiinflammatory, antidote for poison, and antitrypanosome (Sundarambal et al., 2015). The antiinflammatory and analgesic properties of *A. digitata* are well documented. The antiinflammatory activity of the fruit aqueous extract has been tested at a dose of 400 and 800 mg/kg in vivo and it was found to inhibit formalin-induced edema. The aqueous leaf extract has been reported to exhibit significant inhibition against cytokine IL-8 (Vimalanathan and Hudson, 2009). Ramadan et al. (1994) reported the pain-relieving effect of the aqueous fruit extract of *A. digitata* in mice. Khan et al. (2006) while demonstrating the analgesic potential of the petroleum ether extract containing seed oil of *A. digitata* using the tail-flick test. Extracts from standardized commercial preparations of *A. digitata* leaves, fruit-pulp, and seeds showed inhibitory effects on cytokine secretion (IL-6 and IL-8) in human epithelial cell cultures (Selvarani and Hudson, 2009)

In vitro, the methanol extract of *A. digitata* leaf significantly inhibited iNOS activity (IC_{50} of 28.6 µg/mL), IκBα degradation, as well as NF-κB translocation from the cytosol to the nucleus in LPS-stimulated RAW 264.7 cells (Ayele et al., 2013). Phytochemical investigation revealed several classes of compounds from various parts of the baobab (fruit pulp, seed oil, leaves, and roots) including terpenoids, flavonoids, sterols, vitamins, amino acids, carbohydrates, and lipids (Shukla et al., 2001). The polysaccharide obtained from *A. digitata* inhibited COX enzymes (COX-1 and COX-2) (Abeer et al., 2014).

3.3 *Xylopia aethiopica*

Xylopia aethiopica (Dunal) A.Rich. (Annonaceae) is a valuable medicinal plant widely distributed in the West African rainforest from Senegal to Sudan in eastern Africa, and down to Angola in southern Africa (Irvine, 1961; Burkill, 1985). The plant is commonly known as "spice tree," "Africa pepper," "Ethiopian pepper," or "Guinea pepper." The fruits are reported to have high nutritive and medicinal values (Burkill, 1985). It is used for the treatment of rheumatism, headache, neuralgia, and colic pain (Igwe et al., 2003). In Nigeria, the fruits are used as a cough medicine, as well as a carminative and stimulating additive to other medicines (Oliver-Bever, 1986). The powdered root is employed as a dressing and in the local treatment of cancer (Holland, 1968). Several scientific studies have confirmed the traditional use of *X. aethiopica* against pain and inflammation. Obiri and Osafo (2013) found that the antiinflammatory actions of the ethanolic extract of the 70% aqueous ethanol extract of the fruits of *X. aethiopica* are exerted through the inhibition of histamine release from mast cells. The extract (30–300 mg/kg) was inhibited by 23–62% mouse pinnal inflammation. The analgesic and antiinflammatory properties of the methanol extract of *X. aethiopica* have been investigated by the acetic acid–induced pain (writhing) model in mice and carrageenan-induced inflammation in rats as a model of acute inflammation (Aziba and Sokan, 2009). Also, the ethanol extract of *X. aethiopica* and its major diterpene, xylopic acid, inhibited acetic acid–induced visceral nociception, formalin-induced paw pain (both neurogenic and inflammatory), and thermal pain as well as carrageenan-induced mechanical and thermal hyperalgesia in murine models (Woode et al., 2012).

The fruit of *X. aethiopica* contains kaurenoic and xylopic acid which are kauranes, a class of diterpenes with potent antiinflammatory activities. The antiinflammatory activity of the kauranes has been shown to involve the impairment of inflammation signaling through inhibition of NF-κB activity (Castrillo et al., 2001). Kaurenoic acid (*ent*-kaur-16-en-19-oic acid), an *ent*-kaurene diterpene, has been shown to exert antiinflammatory and antipyretic effects in rodents (Sosa-Sequera et al., 2010; Ameyaw et al., 2014).

3.4 *Capsicum* species

The *Capsicum* genus (Solanaceae) has a very confusing terminology with pepper, chilli, chile, chili, paprika, and capsicum all used interchangeably to describe the plants and pods of the genus *Capsicum*. A range of species of *Capsicum* has been cultivated in tropical, subtropical, and temperate regions of Asia, Africa, America, and Mediterranean countries. It is now widely accepted that the genus *Capsicum* consists of five domesticated species including *Capsicum annuum*, *Capsicum chinense*, *Capsicum pubescens*, *Capsicum frutescens*, and *Capsicum baccatum*.

Peppers have been traditionally used in many cuisines and food products due to their distinctive flavor, color, and aroma. Nowadays, peppers are consumed worldwide and their importance has increased gradually to place them among the most consumed spice crops in the world (Bown, 2001). In addition to use as spices and food vegetables, *Capsicum* species have also been used as medicines. Fruits can be eaten raw or cooked. Those used in cooking are generally varieties of the *C. annuum* and *C. frutescens* species, though a few others are also used. It is evident from the literature that chilli not only has a burning flavor, but also has a strong antiinflammatory potential (Srinivasan, 2005). Several authors investigated the antiinflammatory and antinociceptive activities of various *Capsicum* species both in vitro and in animal models. The hydroalcoholic extracts of *C. annum* at a dose level of 100 mg/kg body weight showed demonstrable antiinflammatory activity in the carrageenan-induced hind paw model in rats (Vijayalakshmi et al., 2010). The stalk extract compared to other fruit parts (pericarp and placenta) showed significant nitric oxide (NO) inhibitory effect (53.5%) in lipopolysaccharide (LPS)-stimulated RAW 264.7 cells (Chen and Kang, 2013). Green, yellow, and red varieties of *Capsicum* were studied to compare their in vitro LOX inhibitory activity and the results suggested that the green *Capsicum* extract has the most significant antiinflammatory activity (LOX inhibition of 46.12%) compared to yellow and red *Capsicum* extracts

(Khabade et al., 2012). The red pepper, *C. baccatum* var. *pendulum*, had significant anti-inflammatory activity when tested in the carrageenan-induced pleurisy model in mice (Zimmer et al., 2012). *C. baccatum* L. var. *pendulum* (Willd.) juice inhibited neutrophil migration and reduced vascular permeability in carrageenan-induced peritonitis in mice. In vitro, *C. baccatum* juice also reduced neutrophil recruitment and exudate levels of proinflammatory cytokines TNF-alpha and IL-1-beta in mouse inflammatory immune peritonitis (Spiller et al., 2008). In vitro studies demonstrated a high antiinflammatory potential of chilli pepper in a lipopolysaccharide-stimulated macrophage model (Mueller et al., 2010).

Capsicum species have a wide array of phytochemicals such as carotenoids, capsaicinoids, and phenolic compounds, particularly the flavonoids quercetin and luteolin (Asnin and Park, 2015). Carotenoids are the pigments responsible for the yellow, orange, and red colors of many types of peppers (Ornelas-Paz et al., 2013). Carotenoids extracted from dried peppers (*C. annuum* L.) exhibited significant peripheral analgesic activity at 5, 20, and 80 mg/kg and induced central analgesia at 80 mg/kg (Hernández-Ortega et al., 2012). Capsaicinoids and capsinoids have also been reported to exhibit antiinflammatory activities as well as pain-reducing properties (Luo et al., 2011). The pungency of chilli is mainly due to the presence of capsaicin. It has been found to be effective, both in vitro and in vivo, against inflammatory and pain conditions. Capsaicin has been reported to be used orally or locally for the reduction of rheumatoid arthritis pain, inflammatory heat, and noxious chemical hyperalgesia (Fraenkel et al., 2004). Jolayemi and Ojewole (2013) showed that both capsaicin and the ethyl acetate extract of *C. frutescens* produce antiinflammatory effects comparable to the nonsteroidal antiinflammatory analgesic, diclofenac, in the experimental rat model. Another animal study showed that capsaicin reduced the incidence of paw edema inflammation in rats (Joe and Lokesh, 1997). Capsaicin has also received considerable attention as a pain reliever, and the current understanding of the molecular action mechanism is the inhibition of COX activity and the translocation of NF-kB into the nucleus (Han et al., 2002; Knotkova et al., 2008; Srinivasan, 2016). Due to its tremendous analgesic and antiinflammatory activity, capsaicin, a lipophilic alkaloid, has been used in clinical practice (The Capsaicin Study Group, 1991). Capsaicin is currently used in topical ointments, in concentrations of between 0.025% and 0.075%, as a cream to provide temporary relief from minor aches and pains of muscles and joints associated with arthritis, simple backache, strains, and sprains, often in combination with other rubefacients (Derry et al., 2009).

3.5 *Nigella sativa*

Nigella sativa L. (Ranunculaceae), commonly known as black seed or black cumin, has been used for medicinal purposes for centuries. It originated from Southeastern Asia and was also used in ancient Egypt, Greece, Middle East, and Africa. It is a flowering plant which has been used for centuries as a spice and food preservative (Khan et al., 2011).

The seeds have been added as a spice to a variety of Persian foods such as bread, yoghurt, pickles, sauces, and salads (Hajhashemi et al., 2004). In Northern Africa, it has been used traditionally for thousands of years to treat headache, asthma, bronchitis, rheumatism, fever, cough, influenza, and eczema. Several therapeutic effects have been attributed to the *N. sativa* seed crude extract as well as its purified components (Al-Ghamdi, 2001; Ali and Blunden, 2003; Majdalawieh and Fayyad, 2015; Amin and Hosseinzadeh, 2016). A large number of recent scientific reports have highlighted the biological activities of *N. sativa*. Particularly, *N. sativa* seeds and its oil have been extensively studied for in vivo antinociceptive and antiinflammatory effects (Hajhashemi et al., 2004; Ghannadi et al., 2005; Pichette et al., 2012). In addition, the active ingredients dithymoquinone, thymol, thymohydroquinone, saponins, alkaloids, and vitamins as well as oligo elements contribute to the health benefits associated with black cumin seeds. In particular, thymoquinone has been extensively studied and shown to possesss antinociceptive and antiinflammatory effects (Abdel-Fattah et al., 2000; Cheh et al., 2009; Woo et al., 2012; Alemi et al., 2013). Furthermore, many studies have been conducted to unfold the molecular mechanisms underlying the antiinflammatory effect of *N. sativa* (Houghton et al., 1995; El Mezayen et al., 2006; Taka et al., 2015).

3.6 *Ocimum* species

The genus *Ocimum* is a member of the Lamiaceae family, comprising more than 150 species. It grows widely and is distributed throughout temperate regions of the world with the greatest number of species in Africa. The best known species are the strongly aromatic herb *Ocimum basilicum* (Thai basil) and *Ocimum gratissimum* (African basil) as well as the medicinal herb *Ocimum tenuiflorum,* also known as *Ocimum sanctum* (holy basil or tulsi in Hindi). Basil is the main ingredient of pesto sauce but is also used to flavor other sauces and soups. Different parts including the leaves, stems, flowers, roots, seeds, and even the whole plant are useful. The seeds are edible, and when soaked in water become mucilaginous. The leaves can be eaten as a salad. Basil is widely used in traditional medicine. It is used in Ayurveda and in traditional Chinese medicine for treating digestive system disorders, such as stomach ache and diarrhoea, kidney complaints, and infections. In African traditional medicine, basil is used for treating whooping cough and various types of fever. A leaf decoction is used for treating coughs in West Africa. Many researchers have investigated the antiinflammatory and antinociceptive potential of various *Ocimum* species. These include *O. gratissimum* (Tarkang et al., 2015; Ajayi et al., 2014; Tanko et al., 2008); *O. basilicum* (Singh, 1999; Min et al., 2009; Venâncio et al., 2011; Raina et al., 2016); *O. sanctum* (Godhwani et al., 1987; Singh et al., 1996; Manaharan et al., 2014); *O. americanum* (Yamada et al., 2013); *O. suave* (Masresha et al., 2012); *O. micranthum* (Pinho et al., 2012), and *O. lamiifolium* (Woldesellassie et al., 2011). Most of the reported studies focus on the activity of essential oils using rat and mouse models.

However, significant inhibition by the ethanolic extract of *Ocimum labiatum* of proinflammatory cytokines, namely IL-2, IL-4, IL-6, and IL-17, has been reported by Kapewangolo et al. (2015).

Several antiinflammatory compounds have been isolated from *Ocimum* species. Sieboldogenin isolated from the ethyl acetate fraction of *O. sanctum* showed in vitro and in vivo inflammatory activities by the inhibition of carrageenan-induced hind paw edema at doses of 10 and 50 mg/kg, and significant LOX inhibition (IC_{50} of 38 mM) (Mallick, 2014). Many compounds including eugenol, cirsilineol, isothymonin, isothymusin, apigenin, cirsimaritin, and rosmarinic acid obtained from the acetone extract of the leaves of *O. sanctum* also showed good COX-1 inhibitory activity. Eugenol was found to be the most potent inhibitor of COX-1, reducing the activity by 97% at 1 mM (Kelm et al., 2000).

3.7 *Piper* species

Piper plants (Piperaceae) are most commonly used as a food in almost all African regions. Two species, *Piper capensis* and *Piper guineense* (West African black pepper), are the most famous species in Africa (Noumi et al., 2011). They grow in many home gardens and crop plantations and have numerous culinary uses, including flavoring of soups, meat, fish, eggs, salads, and sauces. Another widely and frequently used *Piper* species in Africa is *Piper nigrum*, commonly known as black pepper, white pepper, green pepper, peppercorn, or Madagascar pepper (Chaveerach et al., 2006). The fruits of *Piper nigrum* are used to make black pepper. Black pepper, white pepper, and green peppercorns are all produced from *Piper nigrum* fruits, but are harvested at different times and are processed differently. Black pepper is the unripe dried fruit whereas white pepper is obtained by removing the outer coating (pericarp). It is also used as flavoring, particularly for savoury foods, meat dishes, sauces, and snack foods (Chaveerach et al., 2006).

A large number of *Piper* species have been reported to possess antiinflammatory and antinociceptive activities. Black pepper essential oil has been reported to exhibit significant antinociceptive properties in the acetic acid–induced writhing test, and it also reduced acute inflammation in the carrageenan-, dextran- and formalin-induced chronic models of inflammation (Jeena et al., 2014). Tasleem et al. (2014) reported potent analgesic and antiinflammatory activity of the hexane and ethanol extracts of *Piper nigrum* tested by the tail immersion, analgesy-meter, hot plate, and acetic acid–induced writhing tests, as well as carrageenan-induced paw inflammation. At 100 and 300 mg/kg, the essential oil from *Piper vicosanum* leaves was shown to significantly reduce edema formation and inhibited leukocyte migration in the carrageenan-induced edema and pleurisy models, respectively (Hoff et al., 2015). Similarly, the essential oil and extracts of *Piper miniatum* exhibited significant antiinflammatory activity with 85.9% inhibition in the TPA-induced mouse ear edema model, while the chloroform extract inhibited LOX activity by 94.2% (Salleh et al., 2015). Antiinflammatory activity was also reported in essential oils of the stem and leaf of *Piper flaviflorum* (Li et al., 2014). Many other studies have reported good antiinflammatory, analgesic, and antinociceptive activities of various species of the genus *Piper* (Choi and Hwang, 2003; Sireeratawong et al., 2010; Zakaria et al., 2010; Lima et al., 2012; Perazzo et al., 2013; Da Silva et al., 2016).

The phyto-constituents obtained from *Piper* species are characterized by the production of typical classes of bioactive antiinflammatory compounds such as alkaloids, amides, propenylphenols, lignans, neolignans, terpenes, steroids, kawapyrones, piperolides, chalcones, dihydrochalcones, flavones, and flavanones (Parmar et al., 1997; Gutierrez et al., 2013; Juliani et al., 2013). Natural compounds such as neolignans, piperkadsin A, piperkadsin B, futoquinol, piperlactam *S*- and *N-p*-coumaroyl tyramine isolated from the stem methanol extract of *Piper kadsura* showed potent inhibition of PMA-induced ROS production in human polymorphonuclear neutrophils with IC_{50} values of 4.3, 12.2, 13.1, 7.0, and 8.4 µM, respectively (Lin et al., 2006). Liu et al. (2010) demonstrated that capsaicinoids and alkylamides, isolated from the hot pepper, and piperine from black pepper suppressed TNF-induced NF-kappaB activation. In the same study, the extract of black pepper at 200 µg/mL and piperine at 25 µg/mL inhibited COX enzymes by 31–80%.

3.8 Other African spices and vegetables with antiinflammatory or antinociceptive potential

Many other African spices and vegetables also displayed antiinflammatory and antinociceptive effects (Table 9.1). *These include Alafia barteri* Oliv. (Apocynaceae), *Anonidium mannii* (Oliv.) Engl. & Diels (Amonaceae), *Blepharis maderaspatensis* (L.) B.Heyne ex Roth (Acanthaceae), *Caralluma dalzielii* N.E.Br. (Apocynaceae), *Carpolobia alba* G.Don (Polygalaceae), *Citrullus colocynthis* (L.) Schrad. (Cucurbitaceae), *Cnestis ferruginea* Vahl ex DC. (Connaraceae), *Corchorus olitorius* L. (Tiliaceae), *Costus afer* Ker Gawl. (Costaceae), *Cynoglossum lanceolatum* Forssk. (Boraginaceae), *Cyperus esculentus* L. (Cyperaceae), *Dalbergia saxatilis* Hook.f. (Leguminosae), *Hertia cheirifolia* (L.) Kuntze (Compositae), *Hibiscus sabdariffa* L. (Malvaceae), *Loranthus micranthus* Hook.f. (Loranthaceae), the mixture of *M. indica, C. citratus, Citrus sinensis* and *O. gratissimum, Musanga cecropioides* R.Br. ex Tedlie (Urticaceae), *O. labiatum* (N.E.Br.) A.J.Paton (Lamiaceae), *Opuntia ficus-indica* (L.) Mill. (Cactaceae), *Parquetina nigrescens* (Afzel.) Bullock (Apocynaceae), *Picralima nitida* (Stapf) T. Durand & H. Durand (Apocynaceae), *Rumex abyssinicus* Jacq. (Polygonaceae), *Sarcocephalus latifolius* (Sm.) E.A.Bruce (Rubiaceae), *Schwenckia americana* L. (Solanaceae), *S. indicum* L. (Pedaliaceae), *Sesamum radiatum* L. (Pedaliaceae), *Tamarindus indica* L. (Caesalpiniaceae), *Tetraclinis articulate* (Vahl) Mast. (Cupressaceae), *Vicia faba* L. (Leguminosae), *Xylopia parviflora* Spruce (Annonaceae), *Zanthoxylum capense* (Thunb.) Harv. (Rutaceae) and *Zygophyllum album* L.f. (Zygophyllaceae)

4 PROMINENT ANTIINFLAMMATORY ACTIVE INGREDIENTS FOUND IN AFRICAN SPICES AND VEGETABLES

Several studies provide evidence that the spices and vegetables consumed by humans possess many natural compounds with antiinflammatory activities. Numerous antiinflammmatory and antinociceptive active ingredients found in African spices are well-known natural products detailed in much literature; however in this section we summarize the most promising ones. These include thymoquinone (**1**), piperine (**2**), capsaicin (**3**), apigenin (**4**), luteolin (**5**), kaurenoic acid (**6**), xylopic acid (**7**), gingerol (**8**), and eugenol (**9**) (Fig. 9.1).

4.1 Thymoquinone

Thymoquinone (**1**) is the main active ingredient of the essential oil extracted from *N. sativa* seeds. It has been extensively reported for its antiinflammatory effect. Compound **1** was shown to completely abolish the expression of proinflammatory cytokines TNF-α and IL-1β, and to significantly reduce PDA cell synthesis of monocyte chemotactic protein-1 (MCP-1) and COX-2, as well as the intrinsic activity of the MCP-1 promoter. It also inhibited the constitutive and TNF-α-mediated activation of NF-κB in PDA cells and reduced the transport of NF-κB from the cytosol to the nucleus (Cheh et al., 2009). Another study reported that **1** was effective in reducing NO production in activated BV-2 microglial cells and the expression of iNOS (Taka et al., 2015). In

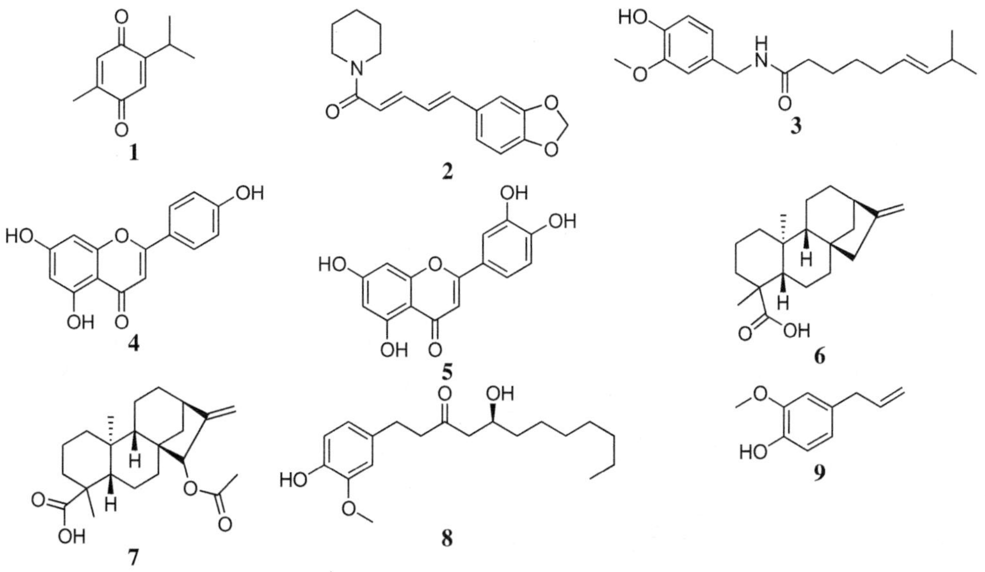

Figure 9.1 *Chemical structures of selected antiinflammatory agents from spices and vegetables.* Thymoquinone (**1**); piperine (**2**); capsaicin (**3**); apigenin (**4**); luteolin (**5**); kaurenoic acid (**6**); xylopic acid (**7**); gingerol (**8**); eugenol (**9**).

addition, compound **1** showed a significant effect in inhibiting mRNA expression and protein production of TNF-α, IL-4, IL-5, and IL-13, and also induced IFN-γ production (El Gazzar et al., 2006b; El Gazzar, 2007). Compound **1** increased the amount of the repressive NF-kappaB p50 homodimer, and simultaneously decreased the amount of transactivating NF-kappaB p65:p50 heterodimer, bound to the TNF-α promoter (El Gazzar et al., 2006a, 2007).

4.2 Piperine

Piperine (**2**) is one of the main active components isolated from the fruits of *P. nigrum*. It is responsible for the pungency of black pepper and long pepper, along with chavicine (an isomer of piperine). Compound **2** has been demonstrated in vitro and in vivo to possess potent antiinflammatory properties (Meghwal and Goswami, 2013). Bang et al. (2009) showed that **2** inhibited the expression of IL-6 and MMP13 and reduced the production of PGE2 at concentrations up to 100 μg/mL. Many researchers have demonstrated that **2** is able (1) to inhibit LPS-induced expression and production of inflammatory factors IL-1β, TNF-α, IL-6, IL-10, iNOS, MMPs mRNA expression, and NO production (Umar et al., 2013; Li et al., 2015), (2) to attenuate LPS-induced MPO activity, lung edema, and inflammatory cytokines TNF-α, IL-6, and IL-1β production, and significantly inhibit LPS-induced NF-κB activation (Hu et al., 2015; Dong et al., 2015; Lu et al., 2016), (3) to reduce edema in the submucosa, cellular infiltration, hemorrhages, and ulceration in acetic acid–induced colitis in mice (Gupta et al., 2015), (4) to inhibit the production of prostaglandin E2 (PGE2) and the gene expression of inducible NO synthase (iNOS) and COX-2 in RAW264.7 cells (Chuchawankul et al., 2012; Vaibhav et al., 2012; Kim et al., 2012; Ying et al., 2013a,b; Umar et al., 2013), (5) to alleviate synovial hyperplasia and mononuclear infiltration observed in arthritic rats (Murunikkara et al., 2012), and (6) to inhibit LPS-induced endotoxin shock, leukocyte accumulation, and the phosphorylation and nuclear translocation of IRF-3 (Bae et al., 2010).

4.3 Capsaicin

Capsaicin (**3**) is the main chemical component that makes chili peppers hot. Due to its antiinflammatory properties, compound **3** is also used to help relieve minor pain associated with rheumatoid arthritis or muscle sprains and strains. Previous studies reported that capsaicin inhibited the activity of COX-2 and the expression of the iNOS protein and completely blocked LPS-induced disappearance of IkB-α and therefore inactivated NF-kB (Kim et al., 2003). Vieira et al. (2000) showed that **3** has an irritant effect in the wiping test in guinea pig conjunctiva after local application and in the paw licking test in mice after intradermal injection. At 2.5 mg/kg i.p., it also significantly inhibited paw swelling in the rat (Jolayemi and Ojewole, 2013). Moreover, **3** inhibited NO production and inducible NO synthase protein and mRNA expression in LPS-stimulated

RAW264.7 macrophages (Kim and Lee, 2014). The antinociceptive effect of **3** has also been reported (Szolcsanyi, 1987).

4.4 Apigenin

Apigenin (**4**) is a flavonoid which is found in a wide variety of plants, especially spices and herbs. Compound **4** is also a potent antiinflammatory compound. It was shown to reduce TNF-α and IL-6 activity (Panés et al., 1996; Basios et al., 2015). It significantly inhibited the PMA-stimulated mRNA and the LPS-induced expression of iNOS, COX-2, expression of proinflammatory cytokines (IL-1β, IL-2, IL-6, IL-8, and TNF-α), and AP-1 proteins (c-Jun, c-Fos, and JunB) including nitric oxide production (Liang et al., 1999; Wang and Huang, 2013; Patil et al., 2015, 2016, Zhang et al., 2015). Compound **4** also had inhibitory effects on COX-2 gene expression and nuclear factor-kB (NF-kB) gene expression (Man et al., 2012; Wang et al., 2014)

4.5 Luteolin

Luteolin (**5**) is a common flavonoid that exists in many types of plants including spices, vegetables, and medicinal herbs. Compound **5** exerts its antiinflammatory activity by suppressing TNF-α, IL-6, IL-1β, and IL-17 (Lamy et al., 2015; Shi et al., 2015; Xu et al., 2015). As low as 0.5 µM of **5** significantly inhibited TNF-α-induced adhesion of monocytes to human EA.hy 926 endothelial cells and suppressed TNF-α-induced expression of the chemokine MCP-1 and adhesion molecules intercellular adhesion molecule-1 (ICAM-1) and vascular cell adhesion molecule-1 (VCAM-1). Furthermore, **5** inhibited TNF-α-induced nuclear factor (NF)-κB transcriptional activity, IκBα degradation, expression of IκB kinase β, and subsequent NF-κB p65 nuclear translocation in endothelial cells. Moreover, dietary intake of luteolin significantly reduced TNF-α-stimulated adhesion of monocytes to aortic endothelial cells ex vivo (Jia et al., 2015). Compound **5** was shown to be an effective HO$^-$ inducer capable of inhibiting macrophage-derived proinflammatory mechanisms (Sung and Lee, 2015) and it also attenuated adipocyte-derived inflammatory responses via suppression of the nuclear factor-κB/mitogen-activated protein kinase pathway (Nepali et al., 2015). Compound **5** also inhibited the secretion of inflammatory cytokines such as interleukin-1β and TNF-α from human mast cells (HMC-1) (Nishitani et al., 2013; Jeon et al., 2014)

4.6 Kaurenoic acid and xylopic acid

Kaurenoic acid (**6**) and xylopic acid (**7**) are kaurane diterpenes found in the fruits of *X. aethiopica*. Compound **6** is mainly present in several plants including spices and herbs. There is evidence that **6** exerts its biological effects by inhibiting the inflammatory process such as in the carrageenin-induced paw edema (Lim et al., 2009; Choi et al., 2011) and TPA-induced ear edema models (Boller et al., 2010). Furthermore, in a model of asthma in guinea pigs, it was shown that **6** inhibits ovalbumin challenge-induced airway resistance in immunized animals as well as the production of histamine and activity of phospholipase A2 (Cho et al., 2010). In vitro experiments also demonstrated that **6** inhibits LPS-induced production of nitric oxide and decreases TNF-α and IL-1 expressions (Lim et al., 2009; Choi et al., 2011; Silva et al., 2015). It also inhibits prostaglandin E2 (PGE2) as well as the expression of COX-2 and inducible iNOS in RAW 264.7 macrophages (Choi et al., 2011). This inhibition of COX-2 and iNOS expressions is probably related to the inhibition of NFκB activation (Choi et al., 2011; Mizokami et al., 2012). In addition, Lyu et al. (2011) demonstrated that **6** inhibits LPS-induced NFκB activation, nitrite production, or mRNA expression of proinflammatory cytokines such as TNF-α and IL-1β, and COX-2. It has been also proved that **6** causes marked reduction of inflammatory cell infiltration and submucosal edema formation in the colon segments of rats treated with the test compound (Paiva et al., 2002). Compound **7** has demonstrated analgesic properties in acute pain models (Ameyaw et al., 2013).

4.7 Gingerol

Gingerol (**8**) is one of the active ingredients of ginger, but is also found in *Afromomum* genus. It has various pharmacological properties, including antiinflammatory activity. The antiinflammatory properties of **8** were demonstrated in paw edema induced by carrageenan assay, and the inhibition of acetic acid–induced writhing and formalin-induced licking response (Young et al., 2005). It was also shown that **8** significantly inhibited the production of the inflammatory mediators NO and PGE (2) (Dugasani et al., 2010).

4.8 Eugenol

Eugenol (**9**) is the main chemical constituent of clove oil. Besides cloves, it can also be extracted from cinnamon and other aromatic spices. Compound **9** is reported to possess antiinflammatory and antinociceptive properties (Mallavarapu et al., 1995; Daniel et al., 2009). It was found to inhibit arachidonic acid metabolism and the synthesis of prostaglandins (Saeed et al., 1995; Bennett et al., 1988). Taher et al. (2015) studied the analgesic activity of **9** using acetic acid–induced abdominal constrictions, meanwhile the hot plate test, the carrageenan-induced paw edema, and brewer's-yeast-induced pyrexia were used to investigate the antiinflammatory activity and the antipyretic effects, respectively. Studies have also shown that **9** suppressed TNF signaling, COX-2 expression, and the production of thromboxane B2 (Rasheed et al., 1984; Lee et al., 2007; Magalhães et al., 2010; Koh et al., 2013). Compound **9** also inhibits 5-LOX activity and leukotriene-C4 in human PMNL cells (Raghavenra et al., 2006). The antinociceptive effects of **9** were described in a mono-iodoacetate-induced osteoarthritis rat model by Ferland et al. (2012). It significantly inhibited acetic acid–induced abdominal constrictions (Kurian et al., 2006). Compound **9** caused a significant response in the hot-plate test as well as in the acetic acid–induced writhing test, and also inhibited carrageenan-induced edema (Apparecido et al., 2009; Park et al., 2011).

Table 9.2 Some spice- and herb-derived compounds tested in clinical trials

Compounds	ClinicalTrials.gov Identifier	Title
Gingerol	NCT01429935	Antiinflammatory and analgesic effect of ginger powder in dental pain model (GPE)
Capsaicin	NCT02700815	Capsaicin + diclofenac gel in acute back pain or neck pain
Piperine	NCT02598726	Piperine and curcumin in reducing inflammation for ureteral stent-induced symptoms in older patients with cancer
Piperine	NCT01324089	Resveratrol with or without piperine to enhance plasma levels of resveratrol

5 CLINICAL TRIALS

Very limited clinical studies have been conducted to investigate the potential antiinflammatory and antinociceptive effects in patients of African spices and vegetables as well as active compounds isolated from them. A randomized controlled clinical trial of *O. sanctum* on gingival inflammation was performed and results showed a significant reduction in gingival bleeding over a period of 15 and 30 days compared to the control group (Gupta et al., 2014). Some spice- and herb-derived compounds tested in clinical trials are shown in Table 9.2.

6 CONCLUSIONS

In this chapter, African vegetables and spices with potent antiinflammatory and antinociceptive activities were reviewed as well as the screening methods used. Carrageenan-induced paw edema and the acetic acid–induced abdominal writhing test were found to be the most popular methods used for assessing analgesic or antiinflammatory activity of African vegetables and spices. *A. melegueta, A. digitata, N. sativa, X. aethiopica, Capsicum* species, *Piper* species, and *Ocimum* species are the most potent plant species, along with their active ingredients thymoquinone, piperine, capsaicin, apigenin, luteolin, kaurenoic acid, xylopic acid, gingerol, and eugenol. Therefore, from this review we can conclude that there is no doubt that increasing our intake of spices and vegetables maybe one of the most effective, convenient, and economical ways in which to fortify ourselves against inflammation and pain-related conditions.

REFERENCES

Abdala, S., Dévora, S., Martín-Herrera, D., Pérez-Paz, P., 2014. Antinociceptive and anti-inflammatory activity of *Sambucus palmensis* Link, an endemic Canary Island species. J. Ethnopharmacol. 155, 626–632.
Abdel-Fattah, A.M., Matsumoto, K., Watanabe, H., 2000. Antinociceptive effects of *Nigella sativa* oil and its major component, thymoquinone, in mice. Eur. J. Pharmacol. 400, 89–97.

Abeer, I.Y., Manal, M.G., Mohsen, A.M.S., 2014. Anti-inflammatory and antioxidant activities of polysaccharide from *Adansonia digitata*: an in vitro study. Int. J. Pharm. Sci. Rev. Res. 25, 174.

Adebayo, S.A., Dzoyem, J.P., Shai, L.J., Eloff, J.N., 2015. The anti-inflammatory and antioxidant activity of 25 South African plants species used traditionally to treat pain. BMC Complement. Altern. Med. 15, 159.

Adzu, B., Amizan, M.B., Okhale, S.E., 2014. Evaluation of antinociceptive and anti-inflammatory activities of standardised rootbark extract of Xeromphis nilotica. J. Ethnopharmacol 158 (Pt A), 271–275.

Agbo, M.O., Nworu, C.S., Okoye, F.B.C., Osadebe, P.O., 2014. Isolation and structure elucidation of polyphenols from *Loranthus micranthus* Linn. parasitic on *Hevea brasiliensis* with anti-inflammatory property. EXCLI J. 13, 859–868.

Agyare, C., Obiri, D.D., Boakye, Y.D., Osafo, N., 2013. Anti-Inflammatory and analgesic activities of African medicinal plants. Medicinal plants market and industry in Africa. In: Kuete, V. (Ed.), Medicinal Plants Research in Africa: Pharmacology and Chemistry. Elsevier, Amsterdam, Netherland, pp. 725–752, Chapter 19.

Ajayi, A.M., Tanayen, J.K., Ezeonwumelu, J., Dare, S., Okwanachi, A., Adzu, B., Ademowo, O.G., 2014. Anti-inflammatory, anti-nociceptive and total polyphenolic content of hydroethanolic extract of *Ocimum gratissimum* L. leaves. Afr. J. Med. Med. Sci. 43, 215–224.

Alemi, M., Sabouni, F., Sanjarian, F., Haghbeen, K., Ansari, S., 2013. Anti-inflammatory effect of seeds and callus of *Nigella sativa* L. extracts on mix glial cells with regard to their thymoquinone content. AAPS Pharm. Sci. Tech. 14, 160–167.

Al-Ghamdi, M.S., 2001. The anti-inflammatory, analgesic and antipyretic activity of *Nigella sativa*. J. Ethnopharmacol. 76, 45–48.

Ali, B.H., Blunden, G., 2003. Pharmacological and toxicological properties of *Nigella sativa*. Phytother. Res. 17, 299–305.

Ameyaw, E.O., Boampong, J.N., Kukuia, K.E., Amoateng, P., Obese, E., Osei-Sarpong, C., Woode, E., 2013. Effect of xylopic acid on paclitaxel-induced neuropathic pain in rats. J. Med. Biomed. Sci. 2, 6–12.

Ameyaw, E.O., Woode, E., Boakye-Gyasi, E., Abotsi, W.K.M., Kyekyeku, J.O., Adosraku, R.K., 2014. Anti-allodynic and Anti-hyperalgesic effects of an ethanolic extract and xylopic acid from the fruits of *Xylopia aethiopica* in murine models of neuropathic pain. Pharmacognosy Res. 6, 172–179.

Amin, B., Hosseinzadeh, H., 2016. Black cumin (*Nigella sativa*) and its active constituent, thymoquinone: an overview on the analgesic and anti-inflammatory effects. Planta Med. 82, 8–16.

Ammar, S., Edziri, H., Mahjoub, M.A., Chatter, R., Bouraoui, A., Mighri, Z., 2009. Spasmolytic and anti-inflammatory effects of constituents from *Hertia cheirifolia*. Phytomedicine 16, 1156–1161.

Amos, S., Chindo, B., Edmond, I., Akah, P., Wambebe, C., Gamaniel, K., 2002. Anti-inflammatory and anti-nociceptive effects of *Ficus platyphylla* in rats and mice. J. Herbs Spices Med. Plants 9, 47–53.

Anthon, G.E., Barrett, D.M., 2001. Colorimetric method for the determination of lipoxygenase activity. J. Agric. Food Chem. 49, 32–37.

Anyasor, G.N., Onajobi, F., Osilesi, O., Adebawo, O., Oboutor, E.M., 2014. Anti-inflammatory and antioxidant activities of *Costus afer* Ker Gawl. hexane leaf fraction in arthritic rat models. J. Ethnopharmacol. 155, 543–551.

Apparecido, N.D., Sartoretto, S.M., Schmidt, G., Caparroz-Assef, S.M., Bersani-Amado, C.A., Kenji, N.R.C., 2009. Anti-inflammatory and antinociceptive activities A of eugenol essential oil in experimental animal models. Rev. Bras. Farmacogn. 19, 212–217.

Asnin, L., Park, S.W., 2015. Isolation and analysis of bioactive compounds in *Capsicum* peppers. Crit. Rev. Food Sci. Nutr. 55, 254–289.

Atta, A.H., Alkofahi, A., 1998. Anti-nociceptive and anti-inflammatory effects of some Jordanian medicinal plant extracts. J. Ethnopharmacol. 60, 117–124.

Ayele, Y., Kim, J.A., Park, E., Kim, Y.J., Retta, N., Dessie, G., Rhee, S.K., Koh, K., Nam, K.W., Kim, H.S., 2013. A Methanol extract of *Adansonia digitata* L. leaves inhibits pro-inflammatory iNOS possibly via the inhibition of NF-κB activation. Biomol. Ther. (Seoul) 21, 146–152.

Aziba, P.I., Sokan, J.A., 2009. Anti-inflammatory and analgesic activities of methanolic extracts of the stem bark of *Alstonia boneei*, *Ficus elastica* and *Xylopia aethiopica* in rodents. Afr. J. Biomed. Res. 12, 249–251.

Bae, G.S., Kim, M.S., Jung, W.S., Seo, S.W., Yun, S.W., Kim, S.G., Park, R.K., Kim, E.C., Song, H.J., Park, S.J., 2010. Inhibition of lipopolysaccharide-induced inflammatory responses by piperine. Eur. J. Pharmacol. 642, 154–162.

Bang, J.S., Oh, D.H., Choi, H.M., Sur, B.J., Lim, S.J., Kim, J.Y., Yang, H.I., Yoo, M.C., Hahm, D.H., Kim, K.S., 2009. Anti-inflammatory and antiarthritic effects of piperine in human interleukin 1beta-stimulated fibroblast-like synoviocytes and in rat arthritis models. Arthritis Res Ther 11, R49.

Basios, N., Lampropoulos, P., Papalois, A., Lambropoulou, M., Pitiakoudis, M.K., Kotini, A., Simopoulos, C., Tsaroucha, A.K., 2015. Apigenin attenuates inflammation in experimentally induced acute pancreatitis-associated lung injury. J. Invest. Surg. 2, 1–7.

Ben-Bassat, J., Peretz, E., Sulman, F.G., 1959. Analgesimetry and ranking of analgesic drugs by the receptacle method. Arch. Int. Pharmacodyn. Thér. 122, 434–447.

Bennett, A., Stamford, I.F., Tavares, I.A., Jacobs, S., Capasso, F., Mascolo, N., Autore, G., Romano, V., Di Carlo, G., 1988. The biological activity of eugenol, a major constituent of nutmeg (*Myristica fragrans*): studies on prostaglandins, the intestine and other tissues. Phytother. Res. 2, 124–130.

Besco, E., Braccioli, E., Vertuani, S., Ziosi, P., Brazzo, F., Brun, R., Sacchetti, G., Manfredini, S., 2007. The use of photochemiluminescence for the measurement of the integral antioxidant capacity of baobab products. Food Chem. 102, 1352–1356.

Bianchi, C., Franceschini, J., 1954. Experimental observation on Haffner's method for testing analgesic drugs. Br. J. Pharmacol. 9, 280–284.

Bighetti, E.J.B., Hiruma-Lima, C.A., Gracioso, J.S., Souza, B.A.R.M., 1999. Anti-inflammatory and anti-nociceptive effects in rodents of the essential oil of *Croton cajucara* Benth. J. Pharmacol. Methods 51, 1447–1453.

Boller, S., Soldi, C., Marques, M.C., Santos, E.P., Cabrini, D.A., Pizzolatti, M.G., Zampronio, A.R., Otuki, M.F., 2010. Anti-inflammatory effect of crude extract and isolated compounds from *Baccharis illinita* DC in acute skin inflammation. J. Ethnopharmacol. 130, 262–266.

Boudjou, S., Oomah, B.D., Zaidi, F., Hosseinian, F., 2013. Phenolics content and antioxidant and anti-inflammatory activities of legume fractions. Food Chem. 138, 1543–1550.

Bown, D., 2001. Encyclopedia of Herbs and Their Uses, Kindersley Dorling, London, Herb Society of America, London, UK.

Burke, A.E.S., Fitzgerald, G.A., 2006. Analgesic-antipyretic agents: pharmacotherapy of gout. In: Brunton, L.L., Lazo, J.S., Parker, K.L. (Eds.), Goodman and Gilmans The Pharmacological Basis of Therapeutics. McGraw Hill, New York.

Burkill, H.M., 1985, second ed. Useful Plants of West Africa, 1, Royal Botanic Gardens, Kew, pp. 130–132.

Castrillo, A., De Las Heras, B., Hortelano, S., Rodriguez, B., Villar, A., Bosca, L., 2001. Inhibition of the nuclear factor kappa B (NF-kappa B) pathway by tetracyclic kaurene diterpenes in macrophages. Specific effects on NF-kappa B-inducing kinase activity and on the coordinate activation of ERK and p38 MAPK. J. Biol. Chem. 276, 15854–15860.

Chaveerach, A., Mokkamul, P., Sudmoon, R., Tanee, T., 2006. Ethnobotany of the genus *Piper* (Piperaceae) in Thailand. Ethnobot. Res. Appl. 4, 223–231.

Cheh, N., Chipitsyna, G., Gong, Q., Yeo, C.J., Arafat, H.A., 2009. Anti-inflammatory effects of the *Nigella sativa* seed extract, thymoquinone, in pancreatic cancer cells. HPB (Oxford) 11, 373–381.

Chen, L., Kang, Y.H., 2013. Anti-inflammatory and antioxidant activities of red pepper (*Capsicum annuum* L.) stalk extracts: comparison of pericarp and placenta extracts. J. Funct. Foods 5, 1724–1731.

Chinasa, E.C., Ifeoma, I.A., Obodoike, E.C., Chhukwuemeka, E.S., 2011. Evaluation of anti-inflammatory property of the leaves of Sansevieria liberica ger. and labr. (fam: Dracaenaceae). Asian Pac J Trop Med 4, 791–795.

Cho, J.H., Lee, J.Y., Sim, S.S., Whang, W.K., Kim, C.J., 2010. Inhibitory effects of diterpene acids from root of *Aralia cordata* on IgE-mediated asthma in guinea pigs. Pulm. Pharmacol. Ther. 23, 190–199.

Choi, E.M., Hwang, J.K., 2003. Investigations of anti-inflammatory and antinociceptive activities of Piper cubeba, Physalis angulata and Rosa hybrida. J. Ethnopharmacol. 1, 171–175.

Choi, R.J., Shin, E.M., Jung, H.A., Choi, J.S., Kim, Y.S., 2011. Inhibitory effects of kaurenoic acid from *Aralia continentalis* on LPS-induced inflammatory response in RAW264.7 macrophages. Phytomedicine 18, 677–682.

Chuchawankul, S., Khorana, N., Poovorawan, Y., 2012. Piperine inhibits cytokine production by human peripheral blood mononuclear cells. Genet. Mol. Res. 11, 617–627.

Chukwujekwu, J.C., van Staden, J., Smith, P., 2005. Antibacterial, anti-inflammatory and antimalarial activities of some Nigerian medicinal plants. S. Afr. J. Bot. 71, 316–325.

Collier, H.O., Dinneen, L.C., Johnson, C.A., Schneider, C., 1968. The abdominal constriction response and its suppression by analgesic drugs in the mouse. Br. J. Pharmacol. Chemother. 32, 295–310.

Crunkhorn, P., Meacock, S.C., 1971. Mediators of the inflammation induced in the rat paw by carrageenan. Br. J. Pharmacol. 42, 392–402.

Da Silva, A.J., Balen, E., Júnior, U.L., Da Silva, M.J., Iwamoto, R.D., Barison, A., Sugizaki, M.M., Leite, K.C.A., 2016. Anti-nociceptive, anti-hyperalgesic and anti-arthritic activity of amides and extract obtained from *Piper amalago* in rodents. J. Ethnopharmacol. 179, 101–109.

Daniel, A.N., Sartoretto, S.M., Schmidt, G., Caparroz-Assef, S.M., Bersani-Amado, C.A., Cuman, R.K.N., 2009. Anti-inflammatory and antinociceptive activities of eugenol essential oil in experimental animal models. Rev. Bras. Farmacogn. 19, 212–217.

Derry, S., Lloyd, R., Moore, R.A., McQuay, H.J., 2009. Topical capsaicin for chronic neuropathic pain in adults (Review). Cochrane Database Syst. Rev. 4, CD007393.

Djouahri, A., Saka, B., Boudarene, L., Benseradj, F., Aberrane, S., Aitmoussa, S., Chelghoum, C., Lamari, L., Sabaou, N., Baaliouamer, A., 2014. In vitro synergistic/antagonistic antibacterial and anti-inflammatory effect of various extracts/essential oil from cones of *Tetraclinis articulata* (Vahl) Masters with antibiotic and anti-inflammatory agents. Ind. Crop. Prod. 56, 60–66.

Dong, Y., Huihui, Z., Li, C., 2015. Piperine inhibit inflammation, alveolar bone loss and collagen fibers breakdown in a rat periodontitis model. J. Periodontal Res. 50, 758–765.

Dugasani, S., Pichika, M.R., Nadarajah, V.D., Balijepalli, M.K., Tandra, S., Korlakunta, J.N., 2010. Comparative antioxidant and anti-inflammatory effects of [6]-gingerol, [8]-gingerol, [10]-gingerol and [6]-shogaol. J. Ethnopharmacol. 127, 515–520.

Dzoyem, J.P., Kuete, V., McGaw, L.J., Eloff, J.N., 2014. The 15-lipoxygenase inhibitory, antioxidant, antimycobacterial activity and cytotoxicity of fourteen ethnomedically used African spices and culinary herbs. J. Ethnopharmacol. 156, 1–8.

Eddy, N.B., Leimbach, D., 1953. Synthetic analgesics: II. Dithenylbutenyl and dithienyl-butylamines. J. Pharmacol. Exp. Ther. 107, 385–393.

El Gazzar, M.A., 2007. Thymoquinone suppresses in vitro production of IL-5 and IL-13 by mast cells in response to lipopolysaccharide stimulation. Inflamm. Res. 56, 345–351.

El Gazzar, M., El Mezayen, R., Marecki, J.C., Nicolls, M.R., Canastar, A., Dreskin, S.C., 2006a. Anti-inflammatory effect of thymoquinone in a mouse model of allergic lung inflammation. Int. Immunopharmacol. 6, 1135–1142.

El Gazzar, M.A., El Mezayen, R., Nicolls, M.R., Dreskin, S.C., 2007. Thymoquinone attenuates proinflammatory responses in lipopolysaccharide-activated mast cells by modulating NF-kappaB nuclear transactivation. Biochim. Biophys. Acta 1770, 556–564.

El Gazzar, M., El Mezayen, R., Nicolls, M.R., Marecki, J.C., Dreskin, S.C., 2006b. Downregulation of leukotriene biosynthesis by thymoquinone attenuates airway inflammation in a mouse model of allergic asthma. Biochim. Biophys. Acta 1760, 1088–1095.

El-Halawany, A.M., El Dine, R.S., El Sayed, N.S., Hattori, M., 2014. Protective effect of *Aframomum melegueta* phenolics against CCl_4-induced rat hepatocytes damage; role of apoptosis and pro-inflammatory cytokines inhibition. Sci. Rep. 4, 5880.

El Mezayen, R., El Gazzar, M., Nicolls, M.R., Marecki, J.C., Dreskin, S.C., Nomiyama, H., 2006. Effect of thymoquinone on cyclooxygenase expression and prostaglandin production in a mouse model of allergic airway inflammation. Immunol. Lett. 106, 72–81.

Ferland, C.E., Beaudry, F., Vachon, P., 2012. Antinociceptive effects of eugenol evaluated in a monoiodoacetate-induced osteoarthritis rat model. Phytother. Res. 26, 1278–1285.

Fraenkel, L., Bogardus, S.T., Concato, J., Wittink, D.R., 2004. Treatment options in knee osteoarthritis: the patient's perspective. Arch. Int. Med. 164, 1299–1304.

Ghannadi, A., Hajhashemi, V., Jafarabadi, H., 2005. An investigation of the analgesic and anti-inflammatory effects of *Nigella sativa* seed polyphenols. J. Med. Food 8, 488–493.

Godhwani, S., Godhwani, J.L., Vyas, D.S., 1987. *Ocimum sanctum*: an experimental study evaluating its anti-inflammatory, analgesic and antipyretic activity in animals. J. Ethnopharmacol. 21, 153–163.

Gupta, D., Bhaskar, D.J., Gupta, R.K., Karim, B., Jain, A., Singh, R., Karim, W., 2014. A randomized controlled clinical trial of *Ocimum sanctum* and chlorhexidine mouthwash on dental plaque and gingival inflammation. J. Ayurveda Integr. Med. 5, 109–116.

Gupta, R.A., Motiwala, M.N., Dumore, N.G., Danao, K.R., Ganjare, A.B., 2015. Effect of piperine on inhibition of FFA induced TLR4 mediated inflammation and amelioration of acetic acid induced ulcerative colitis in mice. J. Ethnopharmacol. 164, 239–246.

Gutierrez, R.M., Gonzalez, A.M., Hoyo-Vadillo, C., 2013. Alkaloids from piper: a review of its phytochemistry and pharmacology. Mini Rev. Med. Chem. 13, 163–193.

Hajhashemi, V., Ghannadi, A., Jafarabadi, H., 2004. Black cumin seed essential oil, as a potent analgesic and antiinflammatory drug. Phytother. Res. 18, 195–199.

Han, S.S., Keum, Y.S., Chun, K.S., Surh, Y.J., 2002. Suppression of phorbol ester-induced NF-kappaB activation by capsaicin in cultured human promyelocytic leukemia cells. Arch. Pharmacal. Res. 25, 475–479.

Hassan, F.I., Zezi, A.U., Yaro, A.H., Danmalam, U.H., 2015. Analgesic, anti-inflammatory and antipyretic activities of the methanol leaf extract of *Dalbergia saxatilis* Hook.F in rats and mice. J. Ethnopharmacol. 166, 74–78.

Hernández-Ortega, M., Ortiz-Moreno, A., Hernández-Navarro, M.D., Chamorro-Cevallos, G., Dorantes-Alvarez, L., Necoechea-Mondragón, H., 2012. Antioxidant, anti-nociceptive, and anti-inflammatory effects of carotenoids extracted from dried pepper (*Capsicum annuum* L.). J. Biomed. Biotechnol. 2012, 1.

Hoff, B.D.R., Mattos Vaz, M.S., Da Silva Arrigo, J., Borges de Carvalho, L.N., Souza de Araújo, F.H., Vani, J.M., Da Silva Mota, J., Cardoso, C.A., Oliveira, R.J., Negrão, F.J., Kassuya, C.A., Arena, A.C., 2015. Toxicological analysis and anti-inflammatory effects of essential oil from *Piper vicosanum* leaves. Regul. Toxicol. Pharmacol. 73, 699–705.

Holland, J.H., 1968. Useful Plants of Nigeria, first ed. Darling and Sons Ltd, London, Part 1.

Houghton, P.J., Zarka, R., De las Heras, B., Hoult, J.R., 1995. Fixed oil of Nigella sativa and derived thymoquinone inhibit eicosanoid generation in leukocytes and membrane lipid peroxidation. Planta Med. 61, 33–36.

Hu, D., Wang, Y., Chen, Z., Ma, Z., You, Q., Zhang, X., Liang, Q., Tan, H., Xiao, C., Tang, X., Gao, Y., 2015. The protective effect of piperine on dextran sulfate sodium induced inflammatory bowel disease and its relation with pregnane X receptor activation. J. Ethnopharmacol. 169, 109–123.

Hunskaar, S., Hole, K., 1987. The formalin test in mice: dissociation between inflammatory and non-inflammatory pain. Pain 30, 103–114.

Ibironke, G.F., Odewole, G.A., 2012. Analgesic and anti-inflammatory properties of methanol extract of *Cnestis ferruginea* in rodents. Afr. J. Med. Med. Sci. 41, 205–210.

Igwe, S.A., Afonne, J.C., Ghasi, S.I., 2003. Ocular dynamics of systemic aqueous extracts of *Xylopia aethiopica* (African guinea pepper) seeds on visually active volunteers. J. Ethnopharmacol. 86, 139–142.

Ilic, N.M., Dey, M., Poulev, A.A., Logendra, S., Kuhn, P.E., Raskin, I., 2014. Anti-inflammatory activity of grains of paradise (*Aframomum melegueta* Schum) extract. J. Agric. Food Chem. 62, 10452–10457.

Irvine, F., 1961. Woody Plants of Ghana. Crown Agents for Overseas Administration, London, pp. 23–24.

Ishola, I.O., Agbaje, O.E., Narender, T., Adeyemi, O.O., Shukla, R., 2012. Bioactivity guided isolation of analgesic and anti-inflammatory constituents of *Cnestis ferruginea* Vahl ex DC (Connaraceae) root. J. Ethnopharmacol. 142, 383–339.

Ishola, I.O., Akindele, A.J., Adeyemi, O.O., 2011. Analgesic and anti-inflammatory activities of *Cnestis ferruginea* Vahl ex DC (Connaraceae) methanolic root extract. J. Ethnopharmacol. 135, 55–562.

Ishola, I.O., Chaturvedi, J.P., Rai, S., Rajasekar, N., Adeyemi, O.O., Shukla, R., Narender, T., 2013. Evaluation of amentoflavone isolated from *Cnestis ferruginea* Vahl ex DC (Connaraceae) on production of inflammatory mediators in LPS stimulated rat astrocytoma cell line (C6) and THP-1 cells. J. Ethnopharmacol. 146, 440–448.

Jeena, K., Liju, V.B., Umadevi, N.P., Kuttan, R., 2014. Antioxidant, anti-inflammatory and anti-nociceptive properties of black pepper essential oil (*Piper nigrum* Linn). J. Essent. Oil Bear. Pl. 17, 1–12.

Jeon, I.H., Kim, H.S., Kang, H.J., Lee, H.S., Jeong, S.I., Kim, S.J., Jang, S.I., 2014. Anti-inflammatory and antipruritic effects of luteolin from Perilla (*P. frutescens* L.) leaves. Molecules 19, 6941–6951.

Jia, Z., Nallasamy, P., Liu, D., Shah, H., Li, J.Z., Chitrakar, R., Si, H., McCormick, J., Zhu, H., Zhen, W., Li, Y., 2015. Luteolin protects against vascular inflammation in mice and TNF-alpha-induced monocyte adhesion to endothelial cells via suppressing IKBα/NF-κB signaling pathway. J. Nutr. Biochem. 26, 293–302.

Jimoh, A.O., Chika, A., Umar, M.T., Adebisi, I., Abdullahi, N., 2011. Analgesic effects and anti-inflammatory properties of the crude methanolic extract of *Schwenckia americana* Linn (Solanaceae). J. Ethnopharmacol. 137, 543–546.

Joe, B., Lokesh, B.R., 1997. Prophylactic and therapeutic effects of n-3 PUFA, capsaicin & curcumin on adjuvant induced arthritis in rats. J. Nutr. Biochem. 8, 397–407.

John, S., Nikhil, S., Yaswanth, J., Bhaskar, A., Amit, A., Sudha, S., 2009. Analgesic property of different extracts of *Curcuma longa* (Linn.): an experimental study in animals. J. Nat. Remedies 9, 116–120.

Jolayemi, A.T., Ojewole, J.A., 2013. Comparative anti-inflammatory properties of Capsaicin and ethyl-acetate extract of *Capsicum frutescens* Linn [Solanaceae] in rats. Afr. Health Sci. 13, 357–361.

Juliani, H.R., Koroch, A.R., Giordano, L., Amekuse, L., Koffa, S., Asante-Dartey, J., Simon, J.E., 2013. *Piper guineense* (Piperaceae): chemistry, traditional uses, and functional properties of West African black pepper. In: Juliani, H.R., Simon, J.E., Chi-Tang, H. (Eds.), African Natural Plant Products Volume II: Discoveries and Challenges in Chemistry, Health, and Nutrition. American Chemical Society, Washinton, US, pp. 33–48.

Kapewangolo, P., Omolo, J.J., Bruwer, R., Fonteh, P., Meyer, D., 2015. Antioxidant and anti-inflammatory activity of *Ocimum labiatum* extract and isolated labdane diterpenoid. J. Inflamm. 12, 4.

Kelm, M.A., Nair, M.G., Strasburg, G.M., DeWitt, D.L., 2000. Antioxidant and cyclooxygenase inhibitory phenolic compounds from *Ocimum sanctum* Linn. Phytomedicine 7, 7–13.

Khabade, V.K., Lakshmeesh, N.B., Roy, S., 2012. Comparative study on antioxidant and anti-inflammatory properties of three colored varieties of *Capsicum annuum*. Int. J. Fundam. Appl. Sci. 1, 51–54.

Khan, M.A., Chen, H.C., Tania, M., Zhang, D.Z., 2011. Anticancer activities of *Nigella sativa* (black cumin). Afr. J. Tradit. Complement. Altern. Med. 8, 226–232.

Khan, M., Shingare, M.S., Zafar, R., Ramesh, D., Siddiqui, A.R., 2006. Analgesic activity of fixed oil of *Adansonia digitata*. Indian J. Nat. Prod. 22, 20–21.

Kim, H.G., Han, E.H., Jang, W.S., Choi, J.H., Khanal, T., Park, B.H., Tran, T.P., Chung, Y.C., Jeong, H.G., 2012. Piperine inhibits PMA-induced cyclooxygenase-2 expression through downregulating NF-κB, C/EBP and AP-1 signaling pathways in murine macrophages. Food Chem. Toxicol. 50, 2342–2348.

Kim, C.S., Kawada, T., Kim, B.S., Han, I.S., Choe, S.Y., Kurata, T., Yu, R., 2003. Capsaicin exhibits anti-inflammatory property by inhibiting IkB-a degradation in LPS-stimulated peritoneal macrophages. Cell. Signal. 15, 299–306.

Kim, Y., Lee, J., 2014. Anti-inflammatory activity of capsaicin and dihydrocapsaicin through heme oxygenase-1 induction in raw264.7 macrophages. J. Food Biochem. 38, 381–387.

Kiuchi, F., Iwakami, S., Shibuya, M., Hanaoka, A., Sankawa, U., 1992. Inhibition of prostaglandin and leukotriene biosynthesis by gingerols and diarylheptaniods. Chem. Pharm. Bull. 40, 387–391.

Knotkova, H., Pappagallo, M., Szallasi, A., 2008. Capsaicin (TRPV1 Agonist) therapy for pain relief: farewell or revival? Clin. J. Pain 24, 142–154.

Koh, T., Murakami, Y., Tanaka, Y., Machino, M., Sakagami, H., 2013. Re-evaluation of anti-inflammatory potential of eugenol in IL-1β-stimulated gingival fibroblast and pulp cells. In Vivo 27, 269–273.

Kokwaro, J.O., 1993. Medicinal Plants of East Africa. East Africa Literature Bureau, Nairobi, pp. 24–67.

Koster, R., Anderson, M., De Beer, E.J., 1959. Acetic acid for analgesic screening. Fed. Proc. 18, 412.

Ksouri, W.M., Medini, F., Mkadmini, K., Legault, J., Magné, C., Abdelly, C., Ksouri, R., 2013. LC-ESI-TOF-MS identification of bioactive secondary metabolites involved in the antioxidant, anti-inflammatory and anticancer activities of the edible halophyte *Zygophyllum album* Desf. Food Chem. 139, 1073–1080.

Kurian, R., Arulmozhi, D.K., Veeranjaneyulu, A., Bodhankar, S.L., 2006. Effect of eugenol on animal models of nociception. Indian J. Pharmacol. 38, 341–345.

Lai, P.K., Roy, J., 2004. Antimicrobial and Chemopreventive properties of herbs and spices. Curr. Med. Chem. 11, 1451–1460.

Lamy, S., Moldovan, P.L., Ben Saad, A., Annabi, B., 2015. Biphasic effects of luteolin on interleukin-1(-induced cyclooxygenase-2 expression in glioblastoma cells. Biochim. Biophys. Acta. 1853, 126–135.

Lanas, A., 2009. Nonsteroidal anti-inflammatory drugs and cyclooxygenase inhibition in the gastrointestinal tract: A trip from peptic ulcer to colon cancer. Am. J. Med. Sci. 338, 96–106.

Laufer, S., Luik, S., 2010. Different methods for testing potential cyclooxygenase-1 and cyclooxygenase-2 inhibitors. Methods Mol. Biol. 644, 91–116.

Laveti, D., Kumar, M., Hemalatha, R., Sistla, R., Naidu, V.G., Talla, V., Verma, V., Kaur, N., Nagpal, R., 2013. Anti-inflammatory treatments for chronic diseases: a review. Inflamm. Allergy Drug Targets 12, 349–361.

Lee, Y.Y., Hung, S.L., Pai, S.F., Lee, Y.H., Yang, S.F., 2007. Eugenol suppressed the expression of lipopolysaccharide-induced proinflammatory mediators in human macrophages. J. Endodontics 33, 698–702.

Li, Y., Li, K., Hu, Y., Xu, B., Zhao, J., 2015. Piperine mediates LPS induced inflammatory and catabolic effects in rat intervertebral disc. Int. J. Clin. Exp. Pathol. 8, 6203–6213.

Li, R., Yang, J.J., Wang, Y.F., Sun, Q., Hu, H.B., 2014. Chemical composition, antioxidant, antimicrobial and anti-inflammatory activities of the stem and leaf essential oils from *Piper flaviflorum* from Xishuangbanna, SW China. Nat. Prod. Commun. 9, 1011–1014.

Liang, Y.C., Huang, Y.T., Tsai, S.H., Lin-Shiau, S.Y., Chen, C.F., Lin, J.K., 1999. Suppression of inducible cyclooxygenase and inducible nitric oxide synthase by apigenin and related flavonoids in mouse macrophages. Carcinogenesis 20, 1945–1952.

Lim, H., Jung, H.A., Choi, J.S., Kim, Y.S., Kang, S.S., Kim, H.P., 2009. Anti-inflammatory activity of the constituents of the roots of *Aralia continentalis*. Arch. Pharmacal. Res. 32, 1237–1243.

Lima, D.K.S., Ballico, L.J., Lapa, F.R., Gonçalves, H.P., De Souza, L.M., Iacomini, M., Werner, M.F. de P., Baggio, C.H., Pereira, I.T., Da Silva, L.M., Facundo, V.A., Santos, A.R.S., 2012. Evaluation of the antinociceptive, anti-inflammatory and gastric antiulcer activities of the essential oil from *Piper aleyreanum* C.DC in rodents. J. Ethnopharmacol. 142, 274–282.

Lin, L.C., Shen, C.C., Shen, Y.C., Tsai, T.H., 2006. Anti-inflammatory neolignans from *Piper kadsura*. J. Nat. Prod. 69, 842–844.

Liu, Y., Yadev, V.R., Aggarwal, B.B., Nair, M.G., 2010. Inhibitory effects of black pepper (*Piper nigrum*) extracts and compounds on human tumor cell proliferation, cyclooxygenase enzymes, lipid peroxidation and nuclear transcription factor-kappa-B. Nat. Prod. Commun. 5, 1253–1257.

Lu, Y., Liu, J., Li, H., Gu, L., 2016. Piperine ameliorates lipopolysaccharide-induced acute lung injury via modulating NF-κB signaling pathways. Inflammation 39, 303–308.

Lu, W., Zhao, X., Xu, Z., Dong, N., Zou, S., Shen, X., Huang, J., 2013. Development of a new colorimetric assay for lipoxygenase activity. Anal. Biochem. 441, 162–168.

Luo, X.J., Peng, J., Li, Y.J., 2011. Recent advances in the study on capsaicinoids and capsinoids. Eur. J. Pharmacol. 650, 1–7.

Lyu, J.H., Lee, G.S., Kim, K.H., Kim, H.W., Cho, S.I., Jeong, S.I., Kim, H.J., Ju, Y.S., Kim, H.K., Sadikot, R.T., Christman, J.W., Oh, S.R., Lee, H.K., Ahn, K.S., Joo, M., 2011. Ent-kaur-16-en-19-oic Acid, isolated from the roots of *Aralia continentalis*, induces activation of Nrf2. J. Ethnopharmacol. 137, 1442–1449.

Magalhães, C.B., Riva, D.R., DePaula, L.J., Brando-Lima, A., Koatz, V.L., Leal-Cardoso, J.H., Zin, W.A., Faffe, D.S., 2010. In vivo anti-inflammatory action of eugenol on lipopolysaccharide-induced lung injury. J. Appl. Physiol. 108, 845–851.

Majdalawieh, A.F., Fayyad, M.W., 2015. Immunomodulatory and anti-inflammatory action of *Nigella sativa* and thymoquinone: a comprehensive review. Int. Immunopharmacol. 28, 295–304.

Mallavarapu, G.R., Ramesh, S., Chandrasekhara, R.S., Rajeswara, R.B.R., Kaul, P.N., Bhattacharya, A.K., 1995. Investigation of the essential oil of cinnamon leaf grown at Bangalore and Hyderabad. Flavour Frag. J. 10, 239.

Mallick, P., 2014. Anti-inflammatory activities of sieboldogenin from *Ocimum sanctum*: experimental and computational studies. World J. Pharm. Pharm. Sci. 3, 1459–1465.

Man, M.Q., Hupe, M., Sun, R., Man, G., Mauro, T.M., Elias, P.M., 2012. Topical apigenin alleviates cutaneous inflammation in murine models. Evid. Based Complement. Alternat. Med. 2012, 912028.

Manaharan, T., Thirugnanasampandan, R., Jayakumar, R., Ramya, G., Ramnath, G., Kanthimathi, M.S., 2014. Antimetastatic and anti-inflammatory potentials of essential oil from edible Ocimum sanctum leaves. Sci. World J. 2014, 239508.

Marsha-Lyn, M., Mckoy, G., Everton, T., Oswald, S., 2002. Preliminary investigation of the anti-inflammatory properties of an aqueous extract from *Morinda citrifoli* (Noni). Proc. West. Pharmacol. Soc. 45, 76–78.

Marzouk, B., Marzouk, Z., Fenina, N., Bouraoui, A., Aouni, M., 2011. Anti-inflammatory and analgesic activities of Tunisian *Citrullus colocynthis* Schrad. immature fruit and seed organic extracts. Eur. Rev. Med. Pharmacol. Sci. 15, 665–672.

Marzouk, B., Marzouk, Z., Haloui, E., Fenina, N., Bouraoui, A., Aouni, M., 2010. Screening of analgesic and anti-inflammatory activities of *Citrullus colocynthis* from southern Tunisia. J. Ethnopharmacol. 128, 15–159.

Masresha, B., Makonnen, E., Debella, A., 2012. In vivo anti-inflammatory activities of *Ocimum suave* in mice. J. Ethnopharmacol. 142, 201–205.

Meghwal, M., Goswami, T.K., 2013. *Piper nigrum* and piperine: an update. Phytother. Res. 27, 1121–1130.

Min, S.S., Han, S.H., Yee, J., Kim, C., Seol, G.H., Im, J.H., Kim, H.T., Lee, K.C., Kim, H.Y., Lee, M.J., 2009. Antinociceptive effects of the essential oil of *Ocimum basilicum* in mice. Kor. J. Pain 22, 206–209.

Mishra, D., Ghosh, G., Kumar, P.S., Panda, P.K., 2011. An experimental study of analgesic activity of selective cox-2 inhibitor with conventional NSAIDs. Asian J. Pharm. Clin. Res. 4, 78–81.

Mizokami, S.S., Arakawa, N.S., Ambrosio, S.R., Zarpelon, A.C., Casagrande, R., Cunha, T.M., Ferreira, S.H., Cunha, F.Q., Verri, Jr., W.A., 2012. Kaurenoic acid from Sphagneticola trilobata inhibits inflammatory pain: effect on cytokine production and activation of the NO-cyclic GMP-protein kinase G-ATP-sensitive potassium channel signaling pathway. J. Nat. Prod. 75, 896–904.

Mueller, M., Hobiger, S., Jungbauer, A., 2010. Anti-inflammatory activity of extracts from fruits, herbs and spices. Food Chem. 122, 987–996.

Mulisa, E., Asres, K., Engidawork, E., 2015. Evaluation of wound healing and anti-inflammatory activity of the rhizomes of *Rumex abyssinicus* J. (Polygonaceae) in mice. BMC Complement. Altern. Med. 15, 341.

Murunikkara, V., Pragasam, S.J., Kodandaraman, G., Sabina, E.P., Rasool, M., 2012. Anti-inflammatory effect of piperine in adjuvant-induced arthritic rats—a biochemical approach. Inflammation 35, 1348–1356.

Nepali, S., Son, J.S., Poudel, B., Lee, J.H., Lee, Y.M., Kim, D.K., 2015. Luteolin is a bioflavonoid that attenuates adipocyte-derived inflammatory responses via suppression of nuclear factor-κB/mitogen-activated protein kinases pathway. Pharmacogn. Mag. 11, 627–635.

Nile, S.H., Park, S.W., 2013. Optimized methods for in vitro and in vivo anti-inflammatory assays and its applications in herbal and synthetic drug analysis. Mini Rev. Med. Chem. 13, 95–100.

Nishitani, Y., Yamamoto, K., Yoshida, M., Azuma, T., Kanazawa, K., Hashimoto, T., Mizuno, M., 2013. Intestinal anti-inflammatory activity of luteolin: role of the aglycone in NF-κB inactivation in macrophages co-cultured with intestinal epithelial cells. Biofactors 39, 522–533.

Noreen, Y., Ringbom, T., Perera, P., Danielson, H., Bohlin, L., 1998. Development of a radiochemical cyclo-oxygenase-1 and -2 in vitro assay for identification of natural products as inhibitors of prostaglandin biosynthesis. J. Nat. Prod. 61, 2–7.

Noumi, V.N., Zapfack, L., Sonke, B., 2011. Ecological behaviour and biogeography of endemic species of the genus *Piper* L. in Africa: a case of the Guineo-Congolean region. Afr. J. Plant Sci. 5, 248–263.

Obiri, D.D., Osafo, N., 2013. Aqueous ethanol extract of the fruit of *Xylopia aethiopica* (Annonaceae) exhibits anti-anaphylactic and anti-inflammatory actions in mice. J. Ethnopharmacol. 148, 940–945.

O'Callaghan, J.P., Holzman, S.G., 1975. Quantification of the analgesic activity of narcotic antagonists by a modified hot plate procedure. J. Pharmacol. Exp. Ther. 192, 497–505.

Odukoya, O.A., Houghton, P.J., Raman, A., 1999. Lipoxygenase inhibitors in the seeds of *Aframomum danielli* K. Schum (Zingiberaceae). Phytomedicine 6, 251–256.

Okoli, C.O., Akah, P.A., Nwafor, S.V., Ihemelandu, U.U., Amadife, C., 2007. Anti-Inflammatory activity of seed extracts of *Aframomum melegueta*. J. Herbs Spices Med. Plants 13, 11–21.

Oliver-Bever, B., 1986. Medicinal Plants in Tropical West Africa. Cambridge University Press, London.

Ornelas-Paz, J.J., Cira-Chávez, L.A., Gardea-Béjar, A.A., Guevara-Arauza, J.C., Sepúlveda, D.R., Reyes-Hernández, J., Ruiz-Cruz, S., 2013. Effect of heat treatment on the content of some bioactive compounds and free radical-scavenging activity in pungent and non-pungent peppers. Food Res. Int. 50, 519–525.

Otimenyin, S.O., Uguru, M.O., Auta, A., 2008. Anti-inflammatory and nalgesic activities of *Cassia goratensis* and *Sacrocephalus esculentus* extracts. J. Herbs Spices Med. Plants 13, 59–67.

Owoyele, B.V., Nafiu, A.B., Oyewole, I.A., Oyewole, L.A., Soladoye, A.O., 2009. Studies on the analgesic, anti-inflammatory and antipyretic effects of *Parquetina nigrescens* leaf extract. J. Ethnopharmacol. 122, 86–90.

Paiva, L.A., Gurgel, L.A., Silva, R.M., Tomé, A.R., Gramosa, N.V., Silveira, E.R., Santos, F.A., Rao, V.S., 2002. Anti-inflammatory effect of kaurenoic acid, a diterpene from *Copaifera langsdorffi* on acetic acid-induced colitis in rats. Vascul. Pharmacol. 39, 303–307.

Panés, J., Gerritsen, M.E., Anderson, D.C., Miyasaka, M., Granger, D.N., 1996. Apigenin inhibits tumor necrosis factor-induced intercellular adhesion molecule-1 upregulation in vivo. Microcirculation 3, 279–286.

Park, S.H., Sim, Y.B., Lee, J.K., Kim, S.M., Kang, Y.J., Jung, J.S., Suh, H.W., 2011. The analgesic effects and mechanisms of orally administered eugenol. Arch. Pharmacal. Res. 34, 501–507.

Parmar, V.S., Jain, S.C., Bisht, K.S., Jain, R., Taneja, P., Jha, A., Tyagi, O.D., Prasada, A.K., Wengela, J., Olsena, C.E., Bolla, P.M., 1997. Phytochemistry of the genus Piper. Phytochemistry 46, 597–673.

Patil, R.H., Babu, R.L., Naveen, K.M., Kiran, K.K.M., Hegde, S.M., Ramesh, G.T., Chidananda, S.S., 2015. Apigenin inhibits PMA-induced expression of pro-inflammatory cytokines and AP-1 factors in A549 cells. Mol. Cell. Biochem. 403, 95–106.

Patil, R.H., Babu, R.L., Naveen, K.M., Kiran, K.K.M., Hegde, S.M., Nagesh, R., Ramesh, G.T., Sharma, S.C., 2016. Anti-inflammatory effect of apigenin on LPS-induced pro-inflammatory mediators and ap-1 factors in human lung epithelial cells. Inflammation 39, 138–147.

Perazzo, F.F., Rodrigues, I.V., Maistro, E.L., Souza, S.M., Nanaykkara, N.P.D., Bastos, J.K., Carvalho, J.C.T., De Souza, G.H.B., 2013. Anti-inflammatory and analgesic evaluation of hydroalcoholic extract and fractions from seeds of *Piper cubeba* L. (Piperaceae). Pharmacogn. J. 5, 13–16.

Peters, K.V. (Ed.), 2001. Handbook of Herbs and Spices. CRC Press, New York, US.

Pichette, A., Marzouk, B., Legault, J., 2012. Antioxidant, anti-inflammatory, anticancer and antibacterial activities of extracts from *Nigella sativa* (black cumin) plant parts. J. Food Biochem. 36, 539–546.

Pérez-Guerrero, C., Herrera, M.D., Ortiz, R., Alvarez de Sotomayor, M., Fernández, M.A., 2001. A pharmacological study of Cecropia obtusifolia Bertol aqueous extract. J. Ethnopharmacol 76, 279–284.

Pinho, J.P., Silva, A.S., Pinheiro, B.G., Sombra, I., Bayma, Jde, C., Lahlou, S., Sousa, P.J., Magalhães, P.J., 2012. Antinociceptive and antispasmodic effects of the essential oil of *Ocimum micranthum*: potential anti-inflammatory properties. Planta Med. 78, 681–685.

Purnima, A., Koti, B.C., Thippeswamy, A.H.M., Tikare, V.P., Dabadi, P., Viswanathaswamy, A.H.M., 2010. Evaluation of anti-inflammatory activity of *Centratherum anthelminticum* (L) kuntze seed. Indian J. Pharm. Sci. 72, 697–703.

Raghavenra, H., Diwakr, B.T., Lokesh, B.R., Naidu, K.A., 2006. Eugenol—the active principle from cloves inhibits 5-lipoxygenase activity and leukotriene-C4 in human PMNL cells. Prostaglandins Leukot. Essent. Fatty Acids 74, 23–27.

Rahula, J., Jain, M.K., Shishu, P.S., Kamal, R.K., Anuradha, A.N., Gupta, A.K., Mrityunjay, S.K., 2015. *Adansonia digitata* L. (baobab): a review of traditional information and taxonomic description. Asian Pac. J. Trop. Biomed. 5, 79–84.

Raina, P., Deepak, M., Chandrasekaran, C.V., Agarwal, A., Wagh, N., Kaul-Ghanekar, R., 2016. Comparative analysis of anti-inflammatory activity of aqueous and methanolic extracts of *Ocimum basilicum* (basil) in RAW264.7, SW1353 and human primary chondrocytes in respect of the management of osteoarthritis. J. Herb. Med. 6, 28–36.

Ramadan, A., Harraz, F.M., El-Mougy, S.A., 1994. Anti-inflammatory, analgesic and antipyretic effects of the fruit pulp of *Adansonia digitata*. Fitoterapia LXV, 418–422.

Rasheed, A., Laekeman, G., Totté, J., Vlietinck, A.J., Herman, A.G., 1984. Eugenol and prostaglandin biosynthesis. New England J. Med. 310, 50–51.

Saeed, S.A., Simjee, R.U., Shamim, G., Gllani, A.H., 1995. Eugenol: a dual inhibitor of platelet-activating factor and arachidonic acid metabolism. Phytomedicine 2, 23–28.

Salleh, W.M., Kammil, M.F., Ahmad, F., Sirat, H.M., 2015. Antioxidant and anti-inflammatory activities of essential oil and extracts of *Piper miniatum*. Nat. Prod. Commun. 10, 2005–2008.

Selvarani, V., Hudson, J.B., 2009. Multiple inflammatory and antiviral activities in *Adansonia digitata* (Baobab) leaves, fruits and seeds. J. Med. Plants Res. 3, 576–582.

Serhan, C.N., Savill, J., 2005. Resolution of inflammation: the beginning programs the end. Nat. Immunol. 6, 1191–1197.

Shi, F., Zhou, D., Ji, Z., Xu, Z., Yang, H., 2015. Anti-arthritic activity of luteolin in Freund's complete adjuvant-induced arthritis in rats by suppressing P2X4 pathway. Chem. Biol. Interact. 226, 82–87.

Shukla, Y.N., Dubey, S., Jain, S.P., Kumar, S., 2001. Chemistry, biology and uses of *Adansonia digitata*—a review. J. Med. Aromat. Plant Sci. 23, 429–434.

Silva, J.J., Pompeu, D.G., Ximenes, N.C., Duarte, A.S., Gramosa, N.V., Carvalho Kde, M., Brito, G.A., Guimarães, S.B., 2015. Effects of Kaurenoic acid and arginine on random skin flap oxidative stress, inflammation, and cytokines in rats. Aesthetic Plast. Surg. 39, 971–917.

Singh, S., 1999. Mechanism of action of anti-inflammatory effect of fixed oil of *Ocimum basilicum* Linn. Indian J. Exp. Biol. 37, 248–252.

Singh, S., Majumdar, D.K., 1995. Analgesic activity of *Ocimum sanctum* and its possible mechanism of action. Int. J. Pharmacogn. 33, 188–192.

Singh, S., Majumdar, D.K., Rehan, H.M., 1996. Evaluation of anti-inflammatory potential of fixed oil of *Ocimum sanctum* (Holybasil) and its possible mechanism of action. J. Ethnopharmacol. 54, 19–26.

Sireeratawong, S., Vannasiri, S., Sritiwong, S., Itharat, A., Jaijoy, K., 2010. Anti-inflammatory, anti-nociceptive and antipyretic effects of the ethanol extract from root of *Piper sarmentosum* Roxb. J. Med. Assoc. Thailand 7, S1–S6.

Sofidiya, M.O., Imeh, E., Ezeani, C., Aigbe, F.R., Akindele, A.J., 2014. Antinociceptive and anti-inflammatory activities of ethanolic extract of *Alafia barteri*. Rev. Bras. Farmacogn. 24, 348–354.

Sosa-Sequera, M.C., Suarez, O., Dalo, N.L., 2010. Kaurenic acid: An in vivo experimental study of its anti-inflammatory and antipyretic effects. Indian J. Pharmacol. 42, 293–296.

Sowemimo, A., Okwuchuku, E., Muyiwa, F.S., Olowokudejo, A., 2015. MutiatI. Musanga cecropioides leaf extract exhibits anti-inflammatory and anti-nociceptive activities in animal models. Rev. Bras. Farmacogn. 25, 506–512.

Sowemimo, A., Onakoya, M., Fageyinbo, M.S., Fadoju, T., 2013. Studies on the anti-inflammatory and antinociceptive properties of *Blepharis maderaspatensis* leaves. Rev. Bras. Farmacogn. 23, 830–835.

Spiller, F., Alves, M.K., Vieira, S.M., Carvalho, T.A., Leite, C.E., Lunardelli, A., Poloni, J.A., Cunha, F.Q., De Oliveira, J.R., 2008. Anti-inflammatory effects of red pepper (*Capsicum baccatum*) on carrageenan- and antigen-induced inflammation. J. Pharm. Pharmacol. 60, 473–478.

Srinivasan, K., 2005. Role of spices beyond food flavoring: nutraceuticals with multiple health effects. Food Rev. Int. 21, 167–188.

Srinivasan, K., 2015. Biological activities of red pepper (Capsicum annuum) and its pungent principle Capsaicin: a review. Crit. Rev. Food Sci. Nutr 56, 1488–1500.

Sundarambal, M., Muthusamy, P., Radha, R., Jerad, S.A., 2015. A review on *Adansonia digitata* Linn. J. Pharmacogn. Phytochem. 4, 12–16.

Sung, J., Lee, J., 2015. Anti-inflammatory activity of butein and luteolin through suppression of NFκB activation and induction of heme oxygenase-1. J. Med. Food 18, 557–564.

Swingle, K.F., Shideman, F.E., 1972. Phases of inflammatory response to subcutaneous implantation of cotton pellet and other modification by certain anti-inflammatory agents. J. Pharmacol. Exp. Ther. 183, 226–234.

Szolcsanyi, J., 1987. Capsaicin and nociception. Acta Physiol. Hung. 69, 323–332.

Taher, Y.A., Samud, A.M., El-Taher, F.E., ben-Hussin, G., Elmezogi, J.S., Al-Mehdawi, B.F., Salem, H.A., 2015. Experimental evaluation of anti-inflammatory, antinociceptive and antipyretic activities of clove oil in mice. Libyan J. Med. 10, 28685.

Taka, E., Mazzio, E.A., Goodman, C.B., Redmon, N., Flores-Rozas, H., Reams, R., Darling-Reed, S., Soliman, K.F., 2015. Anti-inflammatory effects of thymoquinone in activated BV-2 microglial cells. J. Neuroimmunol. 286, 5–12.

Tang, X., Lin, Z., Cai, W., Chen, N., Shen, L., 1984. Anti-inflammatory effect of 3-acethylaconitine. Acta Pharmacol. Sin. 5, 85–89.

Tanko, Y., Magaji, G.M., Yerima, M., Magaji, R.A., Mohammed, A., 2008. Anti-nociceptive and anti-inflammatory activities of aqueous leaves extract of *Ocimum gratissimum* (Labiate) in rodents. Afr. J. Tradit. Complement. Altern. Med. 5, 141–146.

Tarkang, P.A., Okalebo, F.A., Siminyu, J.D., Ngugi, W.N., Mwaura, A.M., Mugweru, J., Agbor, G.A., Guantai, A.N., 2015. Pharmacological evidence for the folk use of Nefang: antipyretic, anti-inflammatory and antinociceptive activities of its constituent plants. BMC Complement. Altern. Med. 15, 174.

Tasleem, F., Azhar, I., Ali, S.N., Perveen, S., Mahmood, Z.A., 2014. Analgesic and anti-inflammatory activities of *Piper nigrum* L. Asian Pac. J. Trop. Med. 7, S461–S468.

The Capsaicin Study Group, 1991. Treatment of painful diabetic neuropathy with topical capsaicin. A multicenter, double-blind, vehicle-controlled study. Arch. Int. Med. 151, 2225–2229.

Tjølsen, A., Berge, O.G., Hunskaar, S., Rosland, J.H., Hole, K., 1992. The formalin test: an evaluation of the method. Pain 51, 5–17.

Tjølsen, A., Hole, S., 1997. In: Dickenson, A., Besson, J.M. (Eds.), Animal Models of Analgesia. Springer-Verlag, Berlin, pp. 1–20.

Ugwah-Oguejiofor, C.J., Abubakar, K., Ugwah, M.O., Njan, A.A., 2013. Evaluation of the antinociceptive and anti-inflammatory effect of *Caralluma dalzielii*. J. Ethnopharmacol. 150 (3), 967–972.

Umar, S., Golam, S.A.H., Umar, K., Ahmad, N., Sajad, M., Ahmad, S., Katiyar, C.K., Khan, H.A., 2013. Piperine ameliorates oxidative stress, inflammation and histological outcome in collagen induced arthritis. Cell. Immunol. 284, 51–59.

Umukoro, S., Ashorobi, R.B., 2001a. Evaluation of the acute toxicity and analgesic potential of *Aframomum melegueta* seed extract in mice. Afr. J. Med. Pharm. Sci. 1, 59–62.

Umukoro, S., Ashorobi, R.B., 2001b. Effect of *Aframomum melegueta* seed extract on thermal pain and carrageenin-induced oedema. Nig. Q. J. Hosp. Med. 11, 33–35.

Umukoro, S., Ashorobi, R.B., 2005. Further evaluation of the anti-inflammatory activity of *Aframomum melegueta* seed extract and its possible mechanism of action. Nig. J. Health Biomed. Sci. 4, 35–39.

Umukoro, S., Ashorobi, R.B., 2007. Further studies on the anti-nociceptive action of aqueous seed extract of *Aframomum melegueta*. J. Ethnopharmacol. 109, 501–504.

Umukoro, S., Ashorobi, B.R., 2008. Further pharmacological studies on aqueous seed extract of *Aframomum melegueta* in rats. J. Ethnopharmacol. 115, 489–493.

Uzkeser, H., Cadirci, E., Halici, Z., Odabasoglu, F., Polat, B., Yuksel, T.N., Ozaltin, S., Atalay, F., 2012. Anti-inflammatory and antinociceptive effects of salbutamol on acute and chronic models of inflammation in rats: involvement of an antioxidant mechanism. Mediators Inflamm. 2012, 438912.

Vaibhav, K., Shrivastava, P., Javed, H., Khan, A., Ahmed, M.E., Tabassum, R., Khan, M.M., Khuwaja, G., Islam, F., Siddiqui, M.S., Safhi, M.M., Islam, F., 2012. Piperine suppresses cerebral ischemia-reperfusion-induced inflammation through the repression of COX-2, NOS-2, and NF-κB in middle cerebral artery occlusion rat model. Mol. Cell. Biochem. 367, 73–84.

Venâncio, A.M., Onofre, A.S., Lira, A.F., Alves, P.B., Blank, A.F., Antoniolli, A.R., Marchioro, M., Estevam, Cdos, S., De Araujo, B.S., 2011. Chemical composition, acute toxicity, and antinociceptive activity of the essential oil of a plant breeding cultivar of basil (*Ocimum basilicum* L.). Planta Med. 77, 825–829.

Vieira, C., Evangelista, S., Cirillo, R., Terracciano, R., Lippi, A., Maggi, C.A., Manzini, S., 2000. Antinociceptive activity of ricinoleic acid, a capsaicin-like compound devoid of pungent properties. Eur. J. Pharmacol. 407, 109–116.

Vijayalakshmi, K., Shyamala, R., Thirumurugan, V., Sethuraman, M., Rajan, S., Badami, S., Mukherjee, P.K., 2010. Physico-Phytochemical investigation and anti-inflammatory screening of *Capsicum annum* L. and *Hemidesmus indicus* (Linn.) R. Br. Anc. Sci. Life 29, 35–40.

Vimalanathan, S., Hudson, J.B., 2009. Multiple inflammatory and antiviral activities in *Adansonia digitata* (Baobab) leaves, fruits and seeds. J. Med. Plants Res. 3, 576–582.

Vinegar, R., Schreiber, W., Hugo, R., 1969. Biphasic development of carrageenin edema in rats. J. Pharmacol. Exp. Ther. 66, 96–100.

Wang, Y.C., Huang, K.M., 2013. In vitro anti-inflammatory effect of apigenin in the Helicobacter pylori-infected gastric adenocarcinoma cells. Food Chem. Toxicol. 53, 376–383.

Wang, J., Liu, Y.T., Xiao, L., Zhu, L., Wang, Q., Yan, T., 2014. Anti-inflammatory effects of apigenin in lipopolysaccharide-induced inflammatory in acute lung injury by suppressing COX-2 and NF-kB pathway. Inflammation 37, 2085–2090.

Waslidge, N.B., Hayes, D.J., 1995. A colorimetric method for the determination of lipoxygenase activity suitable for use in a high throughput assay format. Anal. Biochem. 231, 354–358.

Willoughby, D.A., DiRosa, M., 1972. Studies on the mode of action of non-steroid anti-inflammatory drugs. Ann. Rheum. Dis. 31, 540.

Winter, C., Risley, E., Nuss, O., 1962. Carrageenin-induced inflammation in the hind limb of the rat. Fed. Proc. 46, 118–126.

Winyard, P.G., Willoughby, D.A. (Eds.), 2003. Methods in Molecular Biology. Humana Press Inc., Totowa, NJ.

Woguem, V., Fogang, H.P., Maggi, F., Tapondjou, L.A., Womeni, H.M., Quassinti, L., Bramucci, M., Vitali, L.A., Petrelli, D., Lupidi, G., Papa, F., Vittori, S., Barboni, L., 2014. Volatile oil from striped African pepper (*Xylopia parviflora*, Annonaceae) possesses notable chemopreventive, anti-inflammatory and antimicrobial potential. Food Chem. 149, 183–189.

Woldesellassie, M., Eyasu, M., Kelbessa, U., 2011. In vivo anti-inflammatory activities of leaf extracts of *Ocimum lamiifolium* in mice model. J. Ethnopharmacol. 134, 32–36.

Woo, C.C., Kumar, A.P., Sethi, G., Tan, K.H., 2012. Thymoquinone: potential cure for inflammatory disorders and cancer. Biochem. Pharmacol. 83, 443–451.

Woode, E., Ameyaw, E.O., Boakye-Gyasi, E., Abotsi, W.K.M., 2012. Analgesic effects of an ethanol extract of the fruits of *Xylopia aethiopica* (Dunal) A. Rich (Annonaceae) and the major constituent, xylopic acid in murine models. J. Pharm. Bioallied Sci. 4, 291–301.

Woolfe, G., MacDonald, A.L., 1944. The evaluation of the analgesic action of pethidine hydrochloride (Demerol). J. Pharmacol. Exp. Ther. 80, 300–307.

Xu, Y., Zhang, J., Liu, J., Li, S., Li, C., Wang, W., Ma, R., Liu, Y., 2015. Luteolin attenuate the D-galactose-induced renal damage by attenuation of oxidative stress and inflammation. Nat. Prod. Res. 29, 1078–1082.

Yamada, A.N., Grespan, R., Yamada, Á.T., Silva, E.L., Silva-Filho, S.E., Damião, M.J., De Oliveira, Dalalio, M.M., Bersani-Amado, C.A., Cuman, R.K., 2013. Anti-inflammatory activity of *Ocimum americanum* L. essential oil in experimental model of zymosan-induced arthritis. Am. J. Chin. Med. 41, 913–926.

Ying, X., Chen, X., Cheng, S., Shen, Y., Peng, L., Xu, H.Z., 2013a. Piperine inhibits IL-β induced expression of inflammatory mediators in human osteoarthritis chondrocyte. Int. Immunopharmacol. 17, 293–299.

Ying, X., Yu, K., Chen, X., Chen, H., Hong, J., Cheng, S., Peng, L., 2013b. Piperine inhibits LPS induced expression of inflammatory mediators in RAW 264.7 cells. Cell. Immunol. 285, 49–54.

Young, H.Y., Luo, Y.L., Cheng, H.Y., Hsieh, W.C., Liao, J.C., Peng, W.H., 2005. Analgesic and anti-inflammatory activities of [6]-gingerol. J. Ethnopharmacol. 96, 207–210.

Yu, C.H., Tang, W.Z., Peng, C., Sun, T., Liu, B., Li, M., Xie, X.F., Zhang, H., 2012. Diuretic, anti-inflammatory, and analgesic activities of the ethanol extract from *Cynoglossum lanceolatum*. J. Ethnopharmacol. 139, 149–154.

Zakaria, Z.A., Patahuddin, H., Mohamad, A.S., Israf, D.A., Sulaiman, M.R., 2010. In vivo anti-nociceptive and anti-inflammatory activities of the aqueous extract of the leaves of *Piper sarmentosum*. J. Ethnopharmacol. 128, 42–48.

Zakia, B., Martinez-Villaluenga, C., Frias, J., Gomez-Cordoves, C., Es-Safi, N.E., 2014. Phenolic composition, antioxidant and anti-inflammatory activities of extracts from Moroccan *Opuntia ficus-indica* flowers obtained by different extraction methods. Ind. Crops Prod. 62, 412–420.

Zhang, T., Yan, T., Du, J., Wang, S., Yang, H., 2015. Apigenin attenuates heart injury in lipopolysaccharide-induced endotoxemic model by suppressing sphingosine kinase 1/sphingosine 1-phosphate signaling pathway. Chem. Biol. Interact. 233, 46–55.

Zimmer, A.R., Leonardi, B., Miron, D., Schapoval, E., Oliveira, J.R., Gosmann, G., 2012. Antioxidant and anti-inflammatory properties of Capsicum baccatum: from traditional use to scientific approach. J. Ethnopharmacol. 139, 228–233.

CHAPTER 10

Anticancer Activities of African Medicinal Spices and Vegetables

V. Kuete, O. Karaosmanoğlu, H. Sivas

1 INTRODUCTION

Cancer is the leading cause of morbidity and mortality worldwide and accounts for 12.5% deaths (Desai et al., 2008; Sener and Grey, 2005). It is estimated that the number of cancer deaths will reach 11.5 million in 2030 (Lopez et al., 2006). Globally, cancer is the second leading cause of death among noncommunicable diseases, after cardiovascular diseases (Desai et al., 2008). There will be about 1.28 million new cancer cases and 970,000 cancer deaths by 2030 solely in Africa, mainly due to aging and growth of the population (Ferlay et al., 2010). The development might become even worse because of the adoption of lifestyles associated with economic development, such as smoking, unhealthy diet, and physical inactivity (World Health Organization, 2008). The most occurring cancer types in Africa are those related to infectious agents (cervix, liver, and urinary bladder carcinoma as well as Kaposi sarcoma) (http://www.cancer.org, 2012). In 2008, cervical cancer accounted for 21% of the total newly diagnosed cancers in females and liver cancer for 11% of the total cancer cases in males (http://www.cancer.org, 2012). The survival rates are considerably lower in Africa than in the developed world for most cancer types (http://www.cancer.org, 2012). For example, the 5-year survival rate for breast cancer is less than 50% in Gambia, Uganda, and Algeria, compared to nearly 90% in the United States (http://www.cancer.org, 2012). According to the World Health Organization (WHO) government survey of national capacity for cancer control programs in 2001, anticancer drugs were only available in 22% and affordable in 11% of the 39 African countries that participated in the survey (Kuete et al., 2013c). Currently, limited funding is available to tackle cancer in African countries. Awareness of this impending epidemic in Africa should be a priority today, and all possible resources should be mobilized to both prevent and efficiently treat cancers. Nonetheless, efforts are being made by African scientists to discover new alternative drugs from their most affordable resources, which are medicinal plants. Natural products are well-recognized as sources for drugs in several human ailments including cancers. Examples of natural pharmaceuticals from plants include vincristine, irinotecan, etoposide, and paclitaxel. Despite the discovery of many drugs of natural origin, the search for new anticancer agents is still relevant, in order to increase the range available and to find less toxic and more effective drugs. Several plants from the flora of Africa were found to be active against various types of cancer cells. Even if not reported in the scientific literature for their antiproliferative potential, several medicinal spices and vegetables from Africa contain known antineoplastic compounds. In this chapter, the up-to-date prominent findings on anticancer spices and vegetables, and derived products available in Africa will be discussed.

2 IN VITRO SCREENING METHODS OF PHYTOCHEMICALS FOR ANTICANCER ACTIVITIES

In the initial stages of preclinical research, it is very important to obtain accurate, reliable resultsfrom the in vitro cytotoxicity assays employed as this data may influence the success of a drug candidate to proceed into the development process (van Tonder et al., 2015).

Various assays are in use to determine the effect of a chemical on cells propagated in vitro. They include simple assays measuring cell viability after drug exposure (dye exclusion that measures membrane integrity and effect of the drug on cell growth or simply enumerating cells) as well as assays measuring cell viability, indirectly, by assessing the ability of the cell to reduce compounds, such as 3-(4,5-dimethylthiazol-2-yl)-2,5-diphenyl-2H-tetrazolium bromide (MTT), 2,3-bis (2-methoxy-4-nitro-5-sulfophenyl)-5-[(phenylamino) carbonyl] -2H-tetrazolium hydroxide (XTT), (3-(4,5-dimethylthiazol-2-yl)-5-(3-carboxymethoxyphenyl)-2-(4-sulfophenyl)-2H-tetrazolium) (MTS), sulforhodamine B assay (SRB), and resazurin assay, or to generate ATP.

2.1 MTT assays

The MTT assay is a colorimetric assay for measuring cell metabolic activity. It is based on the ability of nicotinamide adenine dinucleotide phosphate (NADPH)-dependent cellular oxidoreductase enzymes to reduce the tetrazolium dye MTT to its insoluble formazan, which has a purple color (Fig. 10.1). This assay therefore measures cell viability in terms of reductive activity as enzymatic conversion of the tetrazolium compound to water insoluble formazan crystals by dehydrogenases occurring in the mitochondria

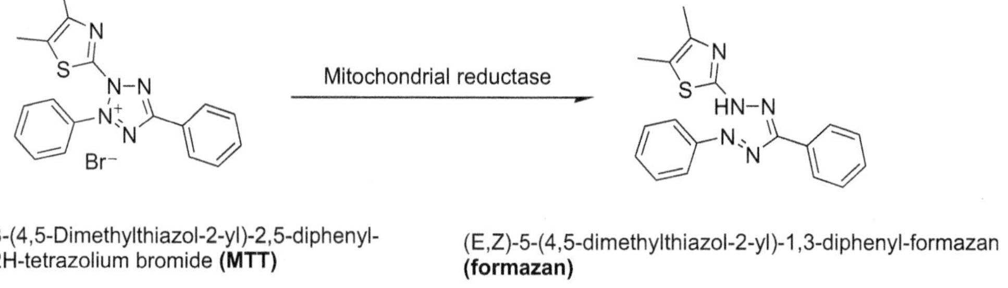

3-(4,5-Dimethylthiazol-2-yl)-2,5-diphenyl-2H-tetrazolium bromide (**MTT**)

(E,Z)-5-(4,5-dimethylthiazol-2-yl)-1,3-diphenyl-formazan (**formazan**)

Figure 10.1 *Enzymatic reduction of MTT to formazan.*

of living cells although reducing agents and enzymes located in other organelles, such as the endoplasmic reticulum are also involved (Lu et al., 2012; Stockert et al., 2012). In the MTT assay a solubilization solution (dimethyl sulfoxide or acidified ethanol solution, or a solution of the detergent sodium dodecyl sulfate in diluted hydrochloric acid) is added to dissolve the insoluble purple formazan product into a colored solution. The absorbance of this colored solution can be quantified by measuring at a certain wavelength (usually between 500 and 600 nm) by a spectrophotometer. MTT method is one of the most widely used methods to analyze cell proliferation and viability. MTT is taken up through endocytosis and is reduced by mitochondrial enzymes as well as endosomal/lysosomal compartments, and then it is transported to cell surfaces to form needle-like MTT formazans (Lu et al., 2012). It was shown that the endocytosis of MTT did not cause obvious lesion and induce cell death, but the metabolism and exocytosis of MTT could dramatically damage cells (Lu et al., 2012). MTT could activate apoptosis-related factors, such as caspase-8, caspase-3 or accelerate the leakage of cell contents after the appearance of MTT formazan crystals (Lu et al., 2012). Therefore, MTT method should be carefully chosen, otherwise the cell viability would be underestimated and incomparable (Lu et al., 2012). The main advantage of MTT assay is the gold standard for cytotoxicity testing while the disadvantage is that the conversion to formazan crystals depends on metabolic rate and number of mitochondria resulting in many known interferences (Lu et al., 2012; van Tonder et al., 2015).

2.2 XTT assays

XTT has been proposed to replace MTT, yielding higher sensitivity and a higher dynamic range. The formed formazan dye is water soluble, avoiding a final solubilization step (Fig. 10.2). XTT-microculture tetrazolium assay was designed to yield a suitably colored, aqueous-soluble, nontoxic formazan upon metabolic reduction by viable cells (Scudiero et al., 1988). The presence of two sulfonic acid groups in XTT is the key to its aqueous solubility in both the tetrazolium ion form and the formazan form (Scudiero et al., 1988). XTT has a single net negative charge at physiological pH, and bioreduction of the central positively charged tetrazolium nucleus increases the net charge to two

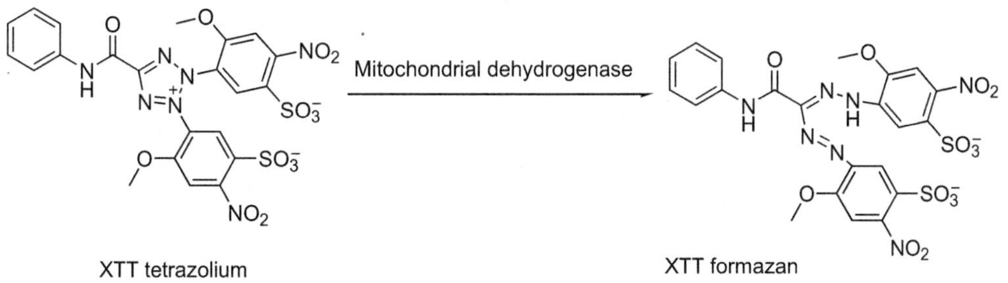

Figure 10.2 *Enzymatic reduction of XTT to formazan.*

(Scudiero et al., 1988). The corresponding reduction of MTT reduces the net positive charge to zero, and thus the MTT formazan is quite insoluble (Scudiero et al., 1988). XTT is converted to a colored formazan in the presence of metabolic activity (the primary mechanisms of XTT to formazan conversion are the mitochondrial succinoxidase and cytochrome P450 systems, as well as flavoprotein oxidases) (Altman, 1976; Kuhn et al., 2003). Since the formazan product is water soluble, it is easily measured in cellular supernatants.

2.3 MTS assays

The MTS, in the presence of phenazine methosulfate (PMS), is bioreduced by viable cells into a formazan product that is soluble in culture media (Fig. 10.3). The conversion is thought to be carried out by NADPH-dependent dehydrogenase enzymes in metabolically active cells and the formazan dye produced can be quantified by measuring the absorbance at 490–500 nm (Cory et al., 1991). The MTS assay is often described as a "one-step" MTT assay, which offers the convenience of adding the reagent straight to the cell culture without the intermittent steps required in the MTT assay. The advantage of MTS over XTT is that it is more soluble and nontoxic, allowing the cells to be returned to culture for further evaluation. The disadvantage is that like XTT it requires the presence of PMS for efficient reduction (Cory et al., 1991). However this convenience makes the MTS assay susceptible to colorimetric interference as the intermittent steps in the MTT assay remove traces of colored compounds, while these remain in the microtiter plate in the one-step MTS assay.

2.4 Sulforhodamine B assay

Sulforhodamine B (Fig. 10.4) is a fluorescent dye with uses spanning from laser-induced fluorescence to the quantification of cellular proteins of cultured cells. The SRB assay is based on binding of the dye to basic amino acids of cellular proteins, and colorimetric evaluation provides an estimate of total protein mass, which is related to cell number. SRB assay is the preferred high-throughput assay of the National Cancer Institute

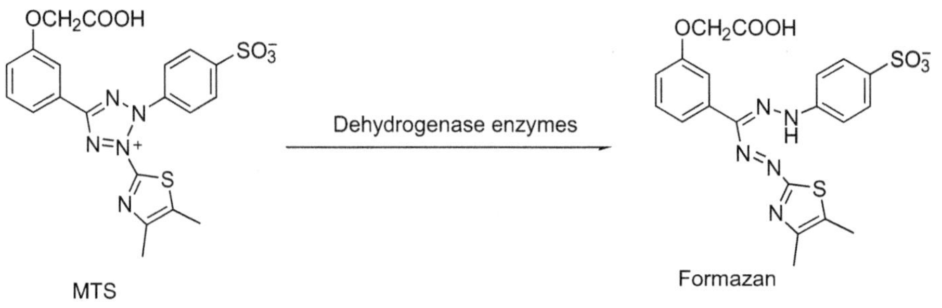

Figure 10.3 *Enzymatic reduction of MTS to formazan.*

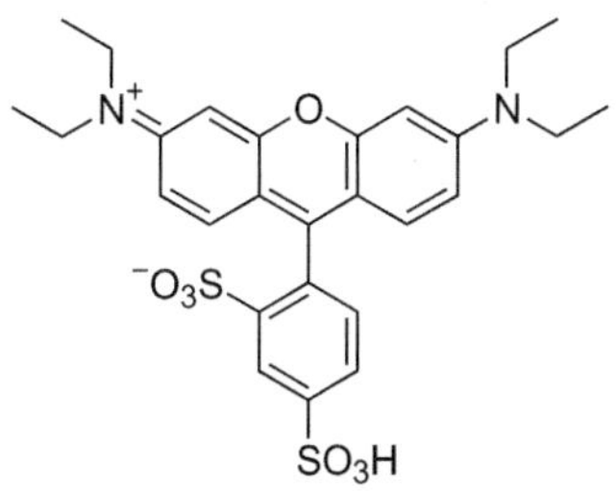

Figure 10.4 *Chemical structure of sulforhodamine B assay (SRB).*

(NCI) in the USA and is the assay used in the NCI's lead compound screening program (van Tonder et al., 2015). This assay has been widely used for the in vitro measurement of cellular protein content of both adherent and suspension cultures. The advantages of this test as compared to others include cell enumeration dependent on protein content thus no test compound interference, and it high reproducibility (van Tonder et al., 2015). The assay also has better linearity, higher sensitivity, a stable end point that does not require time-sensitive measurement, and lower cost. The disadvantages include the low sensitivity with nonadherent cells, as well as the need for the addition of trichloroacetic acid to fix the cells (van Tonder et al., 2015). This step is critical because, if not added gently, trichloroacetic acid could dislodge cells before they become fixed, generating possible artifacts that will affect the results.

2.5 Resazurin assay

The resazurin assay also known as Alamar Blue assay offers a simple, rapid, and sensitive measurement for the viability of mammalian cells and bacteria. Living cells are metabolically active and are able to reduce via mitochondrial reductase, the nonfluorescent dye resazurin to the strongly-fluorescent dye resorufin (Fig. 10.5). The fluorescence output is proportional to the number of viable cells over a wide concentration range. As Alamar Blue is not toxic, cells exposed to it can be returned to culture or used for other purposes. Proliferation measurements with Alamar Blue may be monitored using a standard spectrophotometer, a standard spectrofluorometer, or a spectrophotometric microtiter well plate reader. In resazurin assay few wash steps are involved and follow-up assays can be performed on same cells as assay is not cytotoxic (van Tonder et al., 2015). The

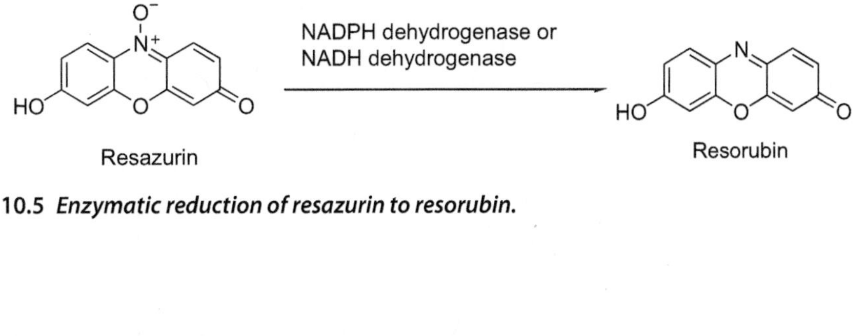

Figure 10.5 *Enzymatic reduction of resazurin to resorubin.*

major advantages of the resazurin reduction assay are that it is relatively inexpensive, uses a homogeneous format, and is more sensitive than tetrazolium assays (Riss et al., 2013). In addition, resazurin assays can be multiplexed with other methods, such as measuring caspase activity to gather more information about the mechanism leading to cytotoxicity (Riss et al., 2013). The disadvantages lie on the fact that conversion to resorufin depends on enzymatic conversion meanwhile absorbance-based method is less sensitive than fluorescence-based method (van Tonder et al., 2015).

2.6 Neutral red uptake assay

The 3-amino-7-dimethylamino-2-methylphenazine hydrochloride (Fig. 10.6) or neutral red uptake assay is a quantitative estimation of the number of viable cells in a culture and one of the most used cytotoxicity tests with many biomedical and environmental applications (Repetto et al., 2008). The assay is based on the ability of viable cells to incorporate and bind the supravital dye neutral red in the lysosomes. The uptake of neutral red depends on the cell's capacity to maintain pH gradients, through the production of ATP. At physiological pH, the dye presents a net charge close to zero, enabling it to penetrate the membranes of the cell. Inside the lysosomes, there is a proton gradient to maintain a pH lower than that of the cytoplasm. Thus, the dye becomes charged and is retained inside the lysosomes. The method is cheaper, presents less interference, and is more sensitive than other cytotoxicity tests (tetrazolium salts, enzyme leakage, or protein content) (Borenfreund et al., 1988; Repetto et al., 2008). The neutral red assay is more sensitive and requires less equipments than the estimation of cell death by enzyme leakage using lactate dehydrogenase. It also compares favorably to estimation of total cell number by assaying protein content. The neutral red uptake assay is simpler, detecting only viable cells; however, once initiated it must be completed immediately, as it is not possible to freeze the cells, as is done for the determination of total protein assay. Nevertheless, this assay is compatible with the determination of total protein content because it is possible to perform both the total protein content and the neutral red assays on the same culture, that is, neutral red estimates can be obtained and then the protein determination can be carried out (Arranz and Festing, 1990; Vichai and Kirtikara, 2006). In common to other cell culture procedures, there are certain limitations due to the character of the compounds to be tested: substances that are volatile, unstable or explosive in water, or with low solubility, present problems (Repetto et al., 2008).

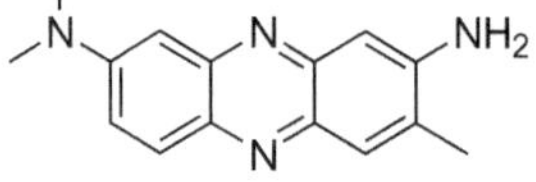

Figure 10.6 *Chemical structure of 3-amino-7-dimethylamino-2-methylphenazine hydrochloride or neutral red.*

3 ANTICANCER POTENTIAL OF PHYTOCHEMICALS

In the past decades, investigations on naturally occurring drugs have been particularly successful in the field of anticancer drug research with the discovery of antileukemic alkaloids, vinblastine and vincristine from the Madagascar periwinkle *Catharanthus roseus* L. (Apocynaceae) (Voss et al., 2006). Screenings of medicinal plants used as anticancer drugs has also provided modern medicine with established antineoplastic drugs used to date. In effect, more than 60% of the approved anticancer drugs in United States of America (from 1983 to 1994) were in one or another way from natural origin (Stevigny et al., 2005; Newman and Cragg, 2007). The well accepted criteria of potency of phytochemicals extract are the IC_{50} value of 30 µg/mL to consider cytotoxic for plant extracts to undergo purification process (Suffness and Pezzuto, 1990). Then, the IC_{50} value of 20 µg/mL is recognized for good cytotoxic extract (Suffness and Pezzuto, 1990). However, criteria are slightly different for edible plant material. In fact, the following criteria have been proposed for edible parts of plants. Significant or strong cytotoxicity: IC_{50} < 50 µg/mL; moderate cytotoxicity: 50 µg/mL < IC_{50} < 200 µg/mL; low cytotoxicity: 200 µg/mL < IC_{50} < 1000 µg/mL; no cytotoxicity: IC_{50} > 1000 µg/mL (Kuete and Efferth, 2015).

4 ANTIPROLIFERATIVE EFFECTS OF AFRICAN MEDICINAL SPICES

Considerable efforts are currently being made by scientists worldwide to evaluate the contribution of medicinal African spices and vegetables in the fight against cancers. In this section, we will report the state of the art of the most promising results documented so far. Some African medicinal spices and vegetables screened for their anticancer activities are summarized in Table 10.1. Active samples include extracts from *Aframomum* species (Zingiberaceae) namely *A. arundinaceum* (Oliv. & D. Hanb.), *Aframomum alboviolaceum* (Ridl.) K Schum., *Aframomum kayserianum* K. Schum, and *A. polyanthum* K. Schum (Kuete et al., 2014a). *A. polyanthum* and mostly *A. arundinaceum* demonstrated good cytotoxicity toward multidrug-resistant leukemia CEM/ADR5000 cells, multidrug-resistant breast adenocarcinoma MDA-MB-231/*BCRP* cells, and glioblastoma multiforme U87MG.ΔEGFR cells as well as toward their sensitive counterparts CCRF-CEM cells, MDA-MB-231 cells, and U87MG cells (Kuete et al., 2014a). Moreover, the extract from *A. arundinaceum* was less toxic to normal hepatocyte AML12 cells than to hepatocarcima HepG2 cells (Table 10.1) (Kuete et al., 2014a). The cytotoxic components of *A. arundinaceum* (Figs. 10.7 and 10.8; Table 10.1) were identified as galanals A (**1**) and B (**2**), naringenin (**6**), and kaempferol-3,7,4′-trimethylether (**7**) (Kuete et al., 2014a). Terpenoids **1** and **2** also displayed cytotoxic effects against Jurkat human T-cell leukemia cells, inducing apoptotic cell death characterized by DNA fragmentation and caspase-3 activation (Miyoshi et al., 2003). The mitochondrial damage pathway was also suggested to be involved in galanal-induced apoptosis (Miyoshi et al., 2003). The antiapoptotic Bcl-2

Table 10.1 African medicinal spices and vegetables with demonstrated cytotoxicity on cancer cell lines

Plant species (family)	Traditional use	Potential bioactive constituents	Reported cytotoxicity*
Aframomum arundinaceum (Zingiberaceae)	Laxative and as anthelmintic, toothache, fungal infections (Tane et al., 2005)	Aframodial, 8(17),12-labdadien-15,16-dial, galanolactone, 1-*p*-menthene-3,6-diol and 1,4-dimethoxybenzene, galanals A (**1**) and B (**2**), naringenin (**6**), and kaempferol-3,7,4′-trimethylether (**7**) (Kuete et al., 2014a)	Significant cytotoxicity of the crude extract on: CCRF-CEM cells (IC_{50}: 18.08 μg/mL) and CEM/ADR5000 cells (IC_{50}: 13.73 μg/mL), MDA-MB231 cells (IC_{50}: 29.98 μg/mL), MDA-MB231/*BCRP* cells (IC_{50}: 30.66 μg/mL), HCT116 (*p53*$^{+/+}$) cells (IC_{50}: 23.06 μg/mL), HCT116 (*p53*$^{-/-}$) cells (IC_{50}: 27.38 μg/mL), U87MG cells (IC_{50}: 36.70 μg/mL), U87MG.ΔEGFR cells (IC_{50}: 24.42 μg/mL), and HepG2 cells (IC_{50}: 23.15 μg/mL) (Kuete et al., 2014a). Moderate activity of: compound **1** on CCRF-CEM cells (IC_{50}: 17.32 μM) and MDA-MB231/*BCRP* cells (IC_{50}: 27.99 μM), **2** on CCRF-CEM cells (IC_{50}: 19.81 μM), **6** on CCRF-CEM cells (IC_{50}: 12.20 μM), CEM/ADR5000 cells (7.86 μM), MDA-MB231 cells (IC_{50}: 9.51 μM), MDA-MB231/*BCRP* cells (IC_{50}: 18.12 μM), HCT116(*p53*$^{+/+}$) cells (IC_{50}: 13.65 μM), HCT116 (*p53*$^{-/-}$) cells (IC_{50}: 13.86 μM), U87MG cells (IC_{50}: 29.81 μM), U87MG.ΔEGFR cells (IC_{50}: 18.02 μM) and HepG2 cells (IC_{50}: 23.46 μM) (Kuete et al., 2014a), **7** on CCRF-CEM cells (IC_{50}: 18.38 μM), CEM/ADR5000 cells (18.22 μM), MDA–MDA-MB231/*BCRP* cells (IC_{50}: 33.14 μM), and HCT116 (*p53*$^{-/-}$) cells (IC_{50}: 36.74 μM) (Kuete et al., 2014a)
Aframomum melegueta (Roscoe) K. Schum. (Zingiberaceae)	Constipation, fever, carminative (Dalziel, 1936)	Volatile oil (Oloko and Kolawole, 1988)	Significant activities with IC_{50} value above 10 μg/mL on MiaPaca-2 and CCRF-CEM cells and significant activity of the crude extract on CEM/ADR5000 cells (IC_{50}: 7.08 μg/mL) (Kuete et al., 2011)

Aframomum polyanthum (Zingiberaceae)	Bacterial infections, cancer (Kuete et al., 2014a)	Aframodial (Tane et al., 2005)	Significant activity of the crude extract on: CCRF-CEM cells (IC$_{50}$: 20.37 µg/mL) and CEM/ADR5000 cells (IC$_{50}$: 28.16 µg/mL), MDA-MB231 cells (IC$_{50}$: 33.79 µg/mL), MDA-MB231/*BCRP* cells (IC$_{50}$: 30.24 µg/mL), and U87MG.Δ*EGFR* cells (IC$_{50}$: 20.59 µg/mL) (Kuete et al., 2014a)
Anonidium mannii (oliv) Engl. et Diels. (Anonaceae)	Sore feet, spider bite, bronchitis, dysentery, sterility caused by poison, gastroenteritis (Thomas et al., 2003); syphilis, infectious diseases (Noumi and Eloumou, 2011); diarrhea, snakebite, malaria (Betti, 2004), cancer (Kuete et al., 2013a)	Alkaloids, phenols, tannins, triterpenes (Kuete et al., 2013a)	Significant activity of the leaves crude extract on: CCRF-CEM cells (IC$_{50}$: 17.32 µg/mL) and CEM/ADR5000 cells (IC$_{50}$: 16.44 µg/mL), MDA-MB231cells (IC$_{50}$: 12.65 µg/mL), MDA-MB231/*BCRP* cells (IC$_{50}$: 32.02 µg/mL), HCT116(*p53*$^{+/+}$) cells (IC$_{50}$: 13.61 µg/mL), U87MG cells (IC$_{50}$: 22.25 µg/mL), U87MG.Δ*EGFR* cells (IC$_{50}$: 9.14 µg/mL), and HepG2 cells (IC$_{50}$: 22.09 µg/mL) (Kuete et al., 2013a)
Dorstenia psilurus Welwitch (Moraceae)	Arthralgia, cardiovascular disorders, rheumatism, snakebites, headache, stomach disorders, diuretic, tonic, stimulant, analgesic (Adjanohoun et al., 1996; Dimo et al., 2001; Ngadjui et al., 1998; Ruppelt et al., 1991), spice	Psoralen, 2-sitosterol glucoside analgesic (Kuete et al., 2007a; Ngadjui et al., 1998)	Significant activity of the crude extract on: MiaPaca-2 cells (IC$_{50}$: 9.17 µg/mL), CCRF-CEM cells (IC$_{50}$: 7.18 µg/mL), and CEM/ADR5000 cells (IC$_{50}$: 7.79 µg/mL) (Kuete et al., 2011)

(Continued)

Table 10.1 African medicinal spices and vegetables with demonstrated cytotoxicity on cancer cell lines (cont.)

Plant species (family)	Traditional use	Potential bioactive constituents	Reported cytotoxicity*
Echinops giganteus var. lelyi (C. D. Adams) A. Rich. (Compositae)	Heart and gastric troubles (Tene et al., 2004), spice	Lupeol sitosteryl, β-D-glucopyranoside (Kojima et al., 1990; Kuete et al., 2007b, 2008; Tane et al., 1995), candidone (**8**), 2-(penta-1,3-diynyl)-5-(4-hydroxybut-1-ynyl)–thiophene (**12**), ursolic acid, and 4-hydroxy-2,6-di-(3′,4′-dimethoxyphenyl)–3,7dioxabicyclo-(3.3.0)octane (**9**) (Kuete et al., 2013b); echinopsolide A (**13**), 2-(penta-1,3-diynyl)-5-(3,4-dihydroxybut-1-ynyl)–thiophene (**14**), and echinopsfuroceramide (**15**) (Sandjo et al., 2016)	Significant activity of the crude extract on: MiaPaca-2 cells (IC_{50}: 9.84 μg/mL), CCRF-CEM cells (IC_{50}: 6.86 μg/mL), and CEM/ADR5000 cells (IC_{50}: 7.96 μg/mL) (Kuete et al., 2011), HL60 cells (IC_{50}: 6.38 μg/mL), HL60AR cells (IC_{50}: 9.24 μg/mL), MDA-MB231 cells (IC_{50}: 8.61 μg/mL), MDA-MB231/*BCRP* cells (IC_{50}: 6.52 μg/mL), HCT116 ($p53^{+/+}$) cells (IC_{50}: 3.58 μg/mL), HCT116 ($p53^{-/-}$) cells (IC_{50}: 3.29 μg/mL), U87MG cells (IC_{50}: 113.55 μg/mL), U87MG.Δ*EGFR* cells (IC_{50}: 14.32 μg/mL), and HepG2 cells (IC_{50}: 11.15 μg/mL) (Kuete et al., 2013b), low activity ranged from 19 to 38 μg/mL for compound **12** on all the 11 previously mentioned cell lines, low and selective activities for **8** and **9** (Kuete et al., 2013b)
Fagara leprieurii (Guill and Perr) Engl (Rutaceae)	Abdominal pain, asthma, appendicitis, toothache (Diniz et al., 2007), spice	3-Hydroxy-1-methoxy-10-methyl-9-acridone; 1-hydroxy-3-methoxy-10-methyl-9-acridone (**4**), 1-hydroxy-2,3-dimethoxy-10-methyl-9-acridone (**5**), 1,3-dihydroxy-2-methoxy-10-methyl-9-acridone (Ngoumfo et al., 2010)	Significant activities with IC_{50} value above 10 μg/mL on MiaPaca-2 cells and CCRF-CEM cells and significant activity of the crude extract on CEM/ADR5000 cells (IC_{50}: 8.13 μg/mL) (Kuete et al., 2011); compounds 3-hydroxy-1-methoxy-10-methyl-9-acridone; 1-hydroxy-3-methoxy-10-methyl-9-acridone, 1-hydroxy-2,3-dimethoxy-10-methyl-9-acridone, 1,3-dihydroxy-2-methoxy-10-methyl-9-acridone were found to be moderately active (IC_{50} ranged from 27 to 77 μM) on A549 and DLD-1 cells (Ngoumfo et al., 2010)

Imperata cylindrica Beauv. var. koenigii Durand et Schinz Gramineae (Poaceae)	Diuretic and antiinflammatory agents (Nishimoto et al., 1968), spice	Jaceidin; quercetagetin-3, 5, 6, 3′-tetramethyl ether; β-sitosterol-3-O-β-D-glucopyranosyl-6″-tetradecanoate (Mohamed et al., 2009)	Significant activity of the crude extract on: Mia-Paca-2 cells (IC_{50}: 12.11 μg/mL), CCRF-CEM cells (IC_{50}: 8.4 μg/mL), and CEM/ADR5000 cells (IC_{50}: 7.18 μg/mL) (Kuete et al., 2011), HL60 cells (IC_{50}: 11.30 μg/mL), HL60AR cells (IC_{50}: 26.64 μg/mL), MDA-MB231 cells (IC_{50}: 6.02 μg/mL), MDA-MB231/*BCRP* cells (IC_{50}: 13.08 μg/mL), HCT116 ($p53^{+/+}$) cells (IC_{50}: 3.28 μg/mL), HCT116($p53^{-/-}$) cells (IC_{50}: 4.32 μg/mL), U87MG cells (IC_{50}: 13.14 μg/mL), U87MG.$\Delta EGFR$ cells (IC_{50}: 14.79 μg/mL), and HepG2 cells (IC_{50}: 33.43 μg/mL) (Kuete et al., 2013b)
Olax subscorpioidea var. *subscorpioidea* Oliv. (Olacaceae), spice	Constipation, yellow fever, jaundice, venereal diseases, Guinea worm (Okoli et al., 2007), spice	Santalbic acid (Cantrell et al., 2003; Jones et al., 1995)	Significant activities with IC_{50} value above 10 μg/mL on MiaPaca-2 and CCRF-CEM cells and significant activity of the crude extract on CEM/ADR5000 cells (IC_{50}: 10.65 μg/mL) (Kuete et al., 2011)
Piper capense L.f. (Piperaceae)	Sleep-inducing remedy, anthelmintic (Kokowaro, 1976; Van Wyk and Gericke, 2000), spice	Kaousine; Z-antiepilepsirine (Kaou et al., 2010)	Significant activity of the crude extract on: Mia-Paca-2 cells (IC_{50}: 8.92 μg/mL), CCRF-CEM cells (IC_{50}: 7.03 μg/mL), and CEM/ADR5000 cells (IC_{50}: 6.56 μg/mL) (Kuete et al., 2011), HL60 cells (IC_{50}: 7.97 μg/mL), HL60AR cells (IC_{50}: 11.22 μg/mL), MDA-MB231 cells (IC_{50}: 4.17 μg/mL), MDA-MB231/*BCRP* cells (IC_{50}: 19.45 μg/mL), HCT116 ($p53^{+/+}$) cells (IC_{50}: 4.67 μg/mL), HCT116($p53^{-/-}$) cells (IC_{50}: 4.62 μg/mL), U87MG (IC_{50}: 13.48 μg/mL), U87MG.$\Delta EGFR$ cells (IC_{50}: 7.44 μg/mL), and HepG2 cells (IC_{50}: 16.07 μg/mL) (Kuete et al., 2013b)

(*Continued*)

Table 10.1 African medicinal spices and vegetables with demonstrated cytotoxicity on cancer cell lines (cont.)

Plant species (family)	Traditional use	Potential bioactive constituents	Reported cytotoxicity*
Piper guineense (Schum and Thonn) (Piperaceae)	Respiratory infections, female infertility, aphrodisiac (Noumi et al., 1998), spice	N-isobutyl-ll-(3,4-methylenedioxyphenyl)-2E,4E,10E-undecatrienamide; N-pyrrolidyl-12-(3,4-methylenedioxyphenyl)-2E,4E,9E,11Z-dodecatetraenamide; N-isobutyl-13-(3,4-methylenedioxyphenyl)-2E,4E,12E-tridecatrienamide; N-isobutyl-2E,4E-decadienamide; N-isobutyl-2E,4E-dodecadienamide (Gbewonyo and Candy, 1992)	Significant activities with IC_{50} value above 10 μg/mL on MiaPaca-2 and CCRF-CEM cells and significant activity of the crude extract on CEM/ADR5000 cells (IC_{50}: 8.20 μg/mL) (Kuete et al., 2011)
Xylopia aethiopica (Dunal) A. Rich. (Annonaceae)	Wounds and skin infections, fever, tapeworm, stomach ache, dysentery, stomach ulcer (Irvine, 1961; Thomas, 1965), spice	Volatile oil (N'dri et al., 2009; Tatsadjieu et al., 2003); 16α-hydroxy-ent-kauran-19-oic acid (**4**), ent-15-oxokaur-16-en-19-oic acid (**5**); 3,4′,5-trihydroxy-6″,6″-dimethylpyrano[2,3-g]flavone (**10**), *trans*-tiliroside (**11**) (Kuete et al., 2015)	Significant activity of the crude seeds extract on: MiaPaca-2 cells (IC_{50}: 6.86 μg/mL), CCRF-CEM cells (IC_{50}: 3.96 μg/mL), and CEM/ADR5000 cells (IC_{50}: 7.04 μg/mL) (Kuete et al., 2011), HL60 cells (IC_{50}: 7.94 μg/mL), HL60AR cells (IC_{50}: 30.60 μg/mL), MDA-MB231 cells (IC_{50}: 5.19 μg/mL), MDA-MB231/*BCRP* cells (IC_{50}: 10.04 μg/mL) and HCT116 (*p53*$^{+/+}$) cells (IC_{50}: 4.37 μg/mL), HCT116(*p53*$^{-/-}$) cells (IC_{50}: 4.60 μg/mL), U87MG cells (IC_{50}: 19.99 μg/mL), U87MG.ΔEGFR cells (IC_{50}: 10.68 μg/mL), and HepG2 cells (IC_{50}: 18.28 μg/mL) (Kuete et al., 2013b)
Zingiber officinale Roscoe (Zingiberaceae)	Infectious diseases, respiratory tract infections, anticancer, indigestion, diarrhea, nausea (Akoachere et al., 2002; Kato et al., 2006; Sakpakdeejaroen and Itharat, 2009)	2-(4-hydroxy-3-methoxyphenyl) ethanol and 2-(4-hydroxy-3-methoxyphenyl) ethanoic acid (Kato et al., 2006); 6-shogaol (Kim et al., 2008)	Significant activity of the crude extract on: MiaPaca-2 cells (IC_{50}: 16.33 μg/mL), CCRF-CEM cells (IC_{50}: 8.82 μg/mL), and CEM/ADR5000 cells (IC_{50}: 6.83 μg/mL) (Kuete et al., 2011); reported cytotoxicity for 6-shogaol against human A549 cells, SK-OV-3 cells, SK-MEL-2 cells, and HCT15 ells (Kim et al., 2008)

Definition of cell lines [breast adenocarcinoma cells (MDA-MB231) and the resistant subline MDA-MB-231/*BCRP*, MCF7), colon cancer cells (DlD-1 and HCT15, HCT116 (*p53*$^{+/+}$), and the resistant subline HCT116 (*p53*$^{-/-}$)), glioblastoma multiforme (U87MG and the resistant subline U87MG.ΔEGFR)], hepatocarcinoma cells (HepG2), lung carcinoma cells (A549, COR-L23), leukemia cells (CCRF-CEM and the resistant subline CEM/ADR5000, HL60 and the resistant subline HL60AR), melanoma cells (SK-MEL-2), prostate cancer cells (MiaPaca-2), ovarian cancer cells (A2780 and SK-OV-3); galanals A (**1**) and B (**2**), jaeschkeanadiol p-hydroxybenzoate (**3**), 16α-hydroxy-ent-kauran-19-oic acid (**4**); ent-15-oxokaur-16-en-19-oic acid (**5**); naringenin (**6**), kaempferol-3,7,4′-trimethylether (**7**); candidone (**8**), 4-hydroxy-2,6-di-(3′,4′-dimethoxyphenyl)-3,7dioxabicyclo-(3.3.0)octane (**9**), 2-(penta-1,3-diynyl)-5-(4-hydroxybut-1-ynyl)-thiophene (**12**), echinopsolide A (**13**), 2-(penta-1,3-diynyl)-5-(3,4-dihydroxybut-1-ynyl)-thiophene (**14**), echinopsfuroceramide (**15**); isotetrandrine (**16**).

* The criteria of classification of activities for spices as well as other edible plants are different to that of other plants as indicated in the text.

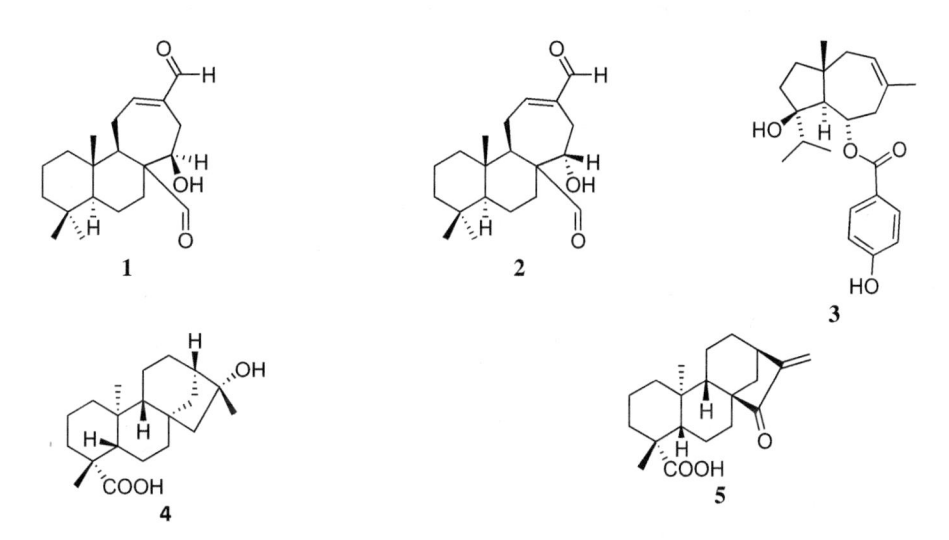

Figure 10.7 *Cytotoxic terpenoids isolated from African medicinal spices and vegetables with documented effects on cancer cells.* Galanal A (**1**); galanal B (**2**); jaeschkeanadiol *p*-hydroxybenzoate (**3**); 16α-hydroxy-ent-kauran-19-oic acid (**4**); ent-15-oxokaur-16-en-19-oic acid (**5**).

Figure 10.8 *Cytotoxic phenolics isolated from African medicinal plants with documented activity on cancer cells.* Naringenin (**6**); kaempferol-3,7,4′-trimethylether (**7**); candidone (**8**); 4-hydroxy-2,6-di-(3′,4′-dimethoxyphenyl)-3,7-dioxabicyclo-(3.3.0)octane (**9**); 3,4′,5-trihydroxy-6″,6″-dimethylpyrano[2,3-*g*]flavone (**10**); *trans*-tiliroside (**11**).

protein was downregulated by the galanal treatment together with enhancement of the Bax expression (Miyoshi et al., 2003).

The antiproliferative effects of *A. mannii* (Oliv.) Engl. (Annonaceae) harvested in Cameroon was also reported on various drug-resistant cancer cell lines (Table 10.1). This extract was cytotoxic toward CEM/ADR5000 cells and U87MGΔEGFR cells (Kuete et al., 2013a). Other African medicinal spices, such as *X. aethiopica* (Dunal) A. Rich. (Annonaceae)., *E. giganteus* A. Rich. (Asteraceae), *I. cylindrica* (L.) P. Beauv. (Poaceae), and *P. capense* L.f. (Piperaceae) also demonstrated strong antiproliferative effects toward a panel on sensitive and resistant cancer cell lines as shown in Table 10.1. *X. aethiopica, E. giganteus*, and *D. psilurus* Welw. (Moraceae) showed strong effects on both leukemia CCRF-CEM cells and its drug-resistant subline CEM/ADR5000 (Kuete et al., 2011). Besides, other spices, such as *I. cylindrica, P. capense*, and *Zingiber officinale* Roscoe (Zingiberaceae) (Table 10.1) also displayed strong activities against CCRF-CEM and CEM/ADR5000 cells (Kuete et al., 2011). BCRP-expressing MDA-MB-231 cells were reported sensitive to the extract of *P. capense* and *E. giganteus*. Also, high sensitivity was observed with extracts from *X. aethiopica, E. giganteus*, and *P. capense* toward U87MG.ΔEGFR cells and with extracts from *E. giganteus* and *P. capense* against HCT116 ($p53^{-/-}$) cells (Kuete et al., 2013b). The cytotoxic ingredients of *E. giganteus* were identified as candidone (**8**), 4-hydroxy-2,6-di-(3′,4′-dimethoxyphenyl)-3,7dioxabicyclo-(3.3.0)octane (**9**), and 2-(penta-1,3-diynyl)-5-(4-hydroxybut-1-ynyl)-thiophene (**12**) (Figs. 10.8 and 10.9; Table 10.1) (Kuete et al., 2013b), echinopsolide A (**13**), 2-(penta-1,3-diynyl)-5-(3,4-dihydroxybut-1-ynyl)-thiophene (**14**), and echinopsfuroceramide (**15**) (Sandjo et al., 2016). Compounds **13** and **14** showed moderate cytotoxicity on leukemia CCRF-CEM cells (IC_{50} of 36.78 and 46.96 µM, respectively) and CEM/ADR5000 cells (IC_{50} of 38.57 and 21.09 µM, respectively) (Sandjo et al., 2016). Besides compound **13** was toxic toward resistant glioblastoma U87MG.ΔEGFR cells (IC_{50} of 54.82 µM) (Sandjo et al., 2016). Significant cytotoxic effects were also reported with compound **15** against CCRF-CEM and CEM/ADR5000 cells (IC_{50} of 9.83 and 6.12 µM, respectively) (Sandjo et al., 2016).

It was also reported that extract prepared with 70% ethanol from *X. aethiopica* has antiproliferative activity against a panel of cancer cell lines with IC_{50} values determined as 7.5 µg/mL against U937 leukemia cells (Choumessi et al., 2012). The antiproliferative constituents of *X. aethiopica* (Figs. 10.7–10.9) were identified as 16α-hydroxy-ent-kauran-19-oic acid (**4**), 3,4′,5-trihydroxy-6″,6″-dimethylpyrano[2,3-g]flavone (**10**), *trans*-tiliroside (**11**), and isotetrandrine (**16**) (Kuete et al., 2015) and ent-15-oxokaur-16-en-19-oic acid (**5**) (Choumessi et al., 2012). Terpenoid **4** was less active, displaying IC_{50} values of 72.28 and 101.59 µM against leukemia CCRF-CEM and CEM/ADR5000 cells, respectively (Kuete et al., 2015). Flavonoid **11** was active against CCRF-CEM cells (IC_{50}: 31.95 µM), CEM/ADR5000 cells (IC_{50}: 24.47 µM), MDA-MB231 cells (IC_{50}: 41.63 µM), and its resistant subline MDA-MB231/BCRP cells (IC_{50}: 47.48 µM)

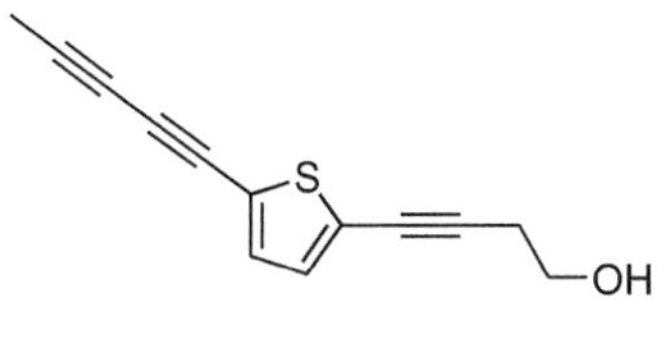

Figure 10.9 *Selected cytotoxic alkaloids and other compounds isolated from African medicinal spices with relevance to cancer cells.* 2-(Penta-1,3-diynyl)-5-(4-hydroxybut-1-ynyl)-thiophene (**12**); echinopsolide A (**13**); 2-(penta-1,3-diynyl)-5-(3,4-dihydroxybut-1-ynyl)-thiophene (**14**); echinopsfuroceramide (**15**); isotetrandrine (**16**).

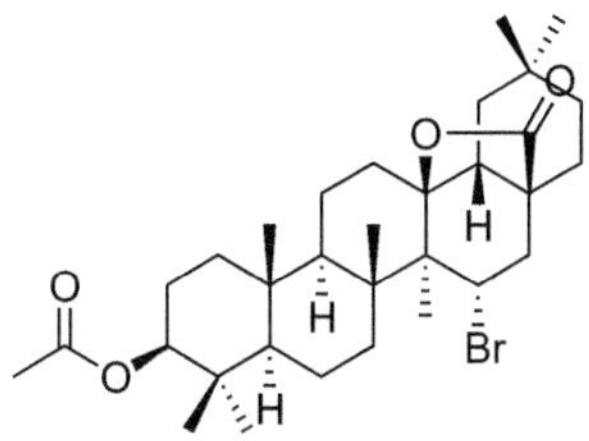

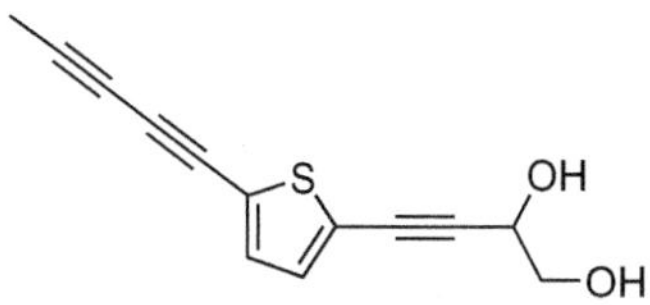

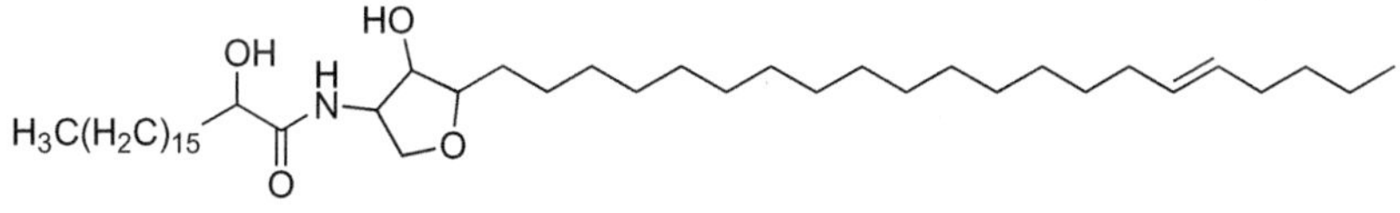

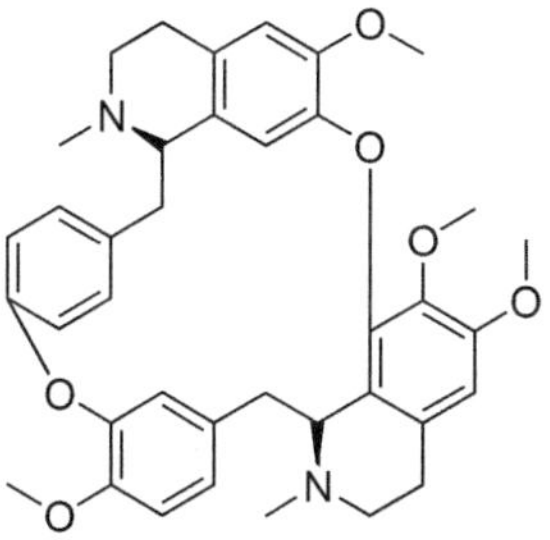

(Kuete et al., 2015). Flavonoid **10** and alkaloid **16** had detectable IC_{50} against all investigated cancer cells lines, including CCRF-CEM cells (IC_{50}: 2.61 and 1.53 µM, respectively), CEM/ADR5000 cells (IC_{50}: 9.54 and 2.36 µM, respectively), MDA-MB231 cells (IC_{50}: 14.71 and 7.28 µM, respectively) and its resistant subline MDA-MB231/BCRP cells (IC_{50}: 17.84 and 6.70 µM, respectively), colon adenocarcinoma HCT116 ($p53^{+/+}$) cells (IC_{50}: 18.01 and 2.39 µM, respectively), and its resistant counterpart HCT116 ($p53^{-/-}$) cells (IC_{50}: 17.27 and 4.55 µM, respectively), glioblastoma U87MG cells (IC_{50}: 18.01 and 3.84 µM, respectively) and its resistant subline U87MG.$\Delta EGFR$ cells (IC_{50}: 18.60 and 1.45 µM, respectively), and the hepatocarcinoma HepG2 cells (IC_{50}: 16.76 and 3.28 µM, respectively) (Kuete et al., 2015). Hence, alkaloid **16** was found to be the most active compound in *X. aethiopica*, with the significant cytotoxic (IC_{50} value below 10 µM) on all tested cancer cells (Kuete et al., 2015). It was also shown that compound **10** induced apoptosis in CCRF-CEM leukemia cells, mediated by mitochondrial membrane potential (MMP) disruption, while the isoquinoline **16** induced apoptosis mediated by reactive oxygen species (ROS) production (Kuete et al., 2015). Upon fractionation of the extract *X. aethiopica* by HPLC, the active fraction induced DNA damage, cell cycle arrest in G1 phase and apoptotic cell death (Choumessi et al., 2012).

Other African medicinal spices with good cytotoxic effects against resistant leukemia CEM/ADR5000 cells and its sensitive counterpart CCRF-CEM cells include *O. subscorpioidea* Oliv. (Olacaceae), *P. guineense* Schum. & Thonn. (Piperaceae), *F. leprieurii* (Guill. & Perr.) Engl. (Rutaceae), and *A. melegueta* K Schum. (Zingiberaceae) (Kuete et al., 2011). Essential oils of *Xylopia parviflora* from Chad and Cameroon displayed cytotoxic effects toward MCF-7 breast adenocarcinoma with IC_{50} values of 0.155 and 0.166 µL/mL, respectively (Bakarnga-Via et al., 2014). Essential oils from *Monodora myristica* from Chad and Cameroon also displayed cytotoxic effects toward MCF-7 cells (IC_{50} values of 0.295 and 0.265 µL/mL, respectively) (Bakarnga-Via et al., 2014).

5 CYTOTOXICITY OF OTHER AFRICAN MEDICINAL VEGETABLES

The popular green leafy vegetables growing in Africa include Taro (*Colocasia esculenta*), Eru (*Gnetun africanum*), Cow pea (*Vigna unguiculata*), Horseradish (*Moringa oleifera*), Groundcherry (*Physalis viscosa*), Gherkin (*Cucumis anguria*), Hell's curse (*Amaranthus cruentus*), African cabbage (*Cleome gynandra*), Chinese cabbage (*Brassica rapa*), Nightshade (*Solanum nigrum*), Jew's mallow plant (*Corchorus olitorius*), *Ferula hermonis* (Apiaceae), bitter water melon (*Citrullus lanatus*) (Randhawa et al., 2015). Among them *B. rapa*, *C. gynandra*, *C. esculenta*, *C. olitorius*, *M. oleifera*, *S. nigrum*, and *V. unguiculata* showed cytotoxic effects.

5.1 *Brassica rapa* L. (Brassicaceae)

B. rapa commonly known as Chinese cabbage or turnip is a conical, deep purple, edible root vegetable (Wu et al., 2013). In tropical Africa, *B. rapa* occurs in all countries and mostly in Cameroon and DR Congo, Eritrea, Ethiopia, Kenya, Uganda, Tanzania, Zimbabwe, and Mozambique. Compounds isolated from *B. rapa*, such as phenanthrene derivative, brassicaphenanthrene A and diarylheptanoids, 6-paradol, and *trans*-6-shogaol displayed high inhibitory activity against the growth of human cancer lines, HCT-116, MCF-7, and HeLa cervix adenocarcinoma, with IC_{50} values ranging from 15 to 35 µM (Wu et al., 2013).

5.2 *Cleome gynandra* (Cleomaceae)

C. gynandra is a green vegetable, commonly known as African cabbage, spiderwisp, and cat's whiskers, native to Africa and widespread in many tropical and subtropical parts of the world. Traditionally the plant is used in the treatment of tumor, as antiinflammatory, and in lysosomal stability actions (Bala et al., 2010). The leaves and seeds of *C. gynandra* are used indigenously in Indian traditional medicine as an anthelmentic and antimicrobial agent (Ajaiyeoba, 2000). Leaves are applied externally over the wounds to prevent the sepsis. The decoction of the root is used to treat fevers. Inhalation of the leaves also relieves headaches; leaf juice and oil are used for earache and as eye wash. The whole plant is also used in the treatment of malaria, piles, rheumatism, and tumor (Mule et al., 2008). The anticancer activity of methanol extract of *C. gynandra* was reported in Swiss albino mice against Ehrlich Ascites Carcinoma cell line (Bala et al., 2010).

5.3 *Colocasia esculenta* (Araceae)

C. esculenta commonly known as *taro* is a tropical plant that is a major dietary staple in many regions of Asia and Africa. Taro corms are very high in starch, and are a good source of dietary fiber. Raw taro contains sodium, carbohydrate, dietary fiber, sugars, protein, vitamins, and minerals (Huang et al., 2011). Some early studies suggested that poi, a pasty starch made from cooked taro might be useful for the treatment of allergies, failure to thrive in infants, and certain gastrointestinal conditions (Brown and Valiere, 2004). Fiber derived from taro could adsorb the mutagens 1,8-dinitropyrene (Ferguson et al., 1992). A soluble extract of cooked taro (poi) has been shown to inhibit proliferation of the rat YYT colon cancer cell line in vitro (Brown et al., 2005). It was also reported that a water-soluble extract of *C. esculenta* inhibits lung colonizing ability as well as spontaneous metastasis from mammary gland-implanted tumors, in a murine model of breast cancer (Kundu et al., 2012). The identified active constituent of taro was found to be highly related to three taro proteins: 12 kDa storage protein, tarin, and lectin (Kundu et al., 2012).

5.4 *Corchorus olitorius* (Malvaceae)

C. olitorius commonly known as Nalta jute or Jew's mallow, is a culinary and medicinal herb, widely used as a vegetable in several countries in Asia and Africa. The plant is native to tropical and subtropical regions throughout the world. In Africa it is reported as a wild or cultivated vegetable in many countries including Benin, Nigeria, Cameroon, Ivory Coast, Sudan, Kenya, Uganda, Zimbabwe, and Egypt. The plant is also cultivated as a leaf vegetable in the Caribbean, Brazil, India, Bangladesh, China, Japan, and the Middle East (http://uses.plantnet-project.org/en/Corchorus_olitorius_%28PROTA%29, *Corchorus olitorius* L.). Nigeria has a great potential for the production of *C. olitorius* for domestic and export market (Olawuyi et al., 2014). The leaves are demulcent, diuretic, febrifuge, and tonic; they are medicinally used in the treatment of chronic cystitis, gonorrhea, and dysuria. A cold infusion is said to restore the appetite and strength. The seeds are purgative. Injections of olitoriside, an extract from the plant, markedly improve cardiac insufficiencies and have no cumulative attributes; hence, it can serve as a substitute for strophanthin (http://www.pfaf.org/user/Plant.aspx?LatinName=Corchorus+olitorius, *Corchorus olitorius* L.). The leaf extracts and seed extracts of the plant exerted cytotoxic effects on the multiple myeloma-derived ARH-77 cells with IC_{50} values of 151 and 17 μg/mL, respectively (Iseri et al., 2013). The ethanol extract of the plant displayed a strong reduction in cell viability of the human hepatocellular carcinoma (HepG2) cells via DNA fragmentation and nuclear condensation–mediated apoptosis (Li et al., 2012). This extract triggered the activation of procaspases-3 and 9 and caused the cleavage of downstream substrate, poly-ADP-ribose polymerase, followed by downregulation of the inhibitor of caspase-activated DNase signaling (Li et al., 2012).

5.5 *Ferula hermonis* (Apiaceae)

The vegetable in the carrot family growing in Egypt, *F. hermonis*, was recently screened for its cytotoxic potential against four cell lines, namely human pancreatic cancer MiaPa-Ca-2 cells, breast cancer MCF-7 cells, leukemia CCRF-CEM cells, and their multidrug-resistant subline, CEM/ADR5000 cells (Kuete et al., 2012). The crude extract inhibited the proliferation of CCRF-CEM as well as CEM/ADR5000 cells while the fraction from *F. hermonis* and jaeschkeanadiol *p*-hydroxybenzoate (**3**) also known as ferutinin, isolated from its active fraction induced more than 50% inhibition against MiaPaCa-2 cells, MCF-7 cells, CCRF-CEM cells, and CEM/ADR5000 cells (Kuete et al., 2012).

5.6 *Moringa oleifera* (Moringaceae)

M. oleifera commonly known as moringa, drumstick tree, horseradish tree, or benzoil tree, is native to the western and sub-Himalayan tracts, India, Pakistan, Asia Minor, Africa, and Arabia and is now distributed in the Philippines, Cambodia, Central America, North and South America, and the Caribbean Islands (Anwar et al., 2007). The plant's young seed pods and leaves are used as vegetables. It grows in several African countries, such as Benin, Burkina Faso, Cameroon, Chad, Eritrea, Ethiopia, Gambia, Ghana, Guinea, Kenya, Liberia, Mali, Mauritania, Niger, Nigeria, Senegal, Sierra Leone, Sudan, Tanzania, Togo, and Uganda (Anwar et al., 2007). The moringa plant provides a rich and rare combination of zeatin, quercetin, beta-sitosterol, caffeoylquinic acid, and kaempferol. Various parts of this plant, such as the leaves, roots, seed, bark, fruit, flowers, and immature pods act as cardiac and circulatory stimulants, possess antitumor, antipyretic, antiepileptic, antiinflammatory, antiulcer, antispasmodic, diuretic, antihypertensive, cholesterol lowering, antioxidant, antidiabetic, hepatoprotective, antibacterial, and antifungal activities, and are being employed for the treatment of different ailments in the indigenous system of medicine (Anwar et al., 2007). The seed extracts have been found effective on hepatic carcinogen metabolizing enzymes and skin papillomagenesis in mice (Bharali et al., 2003). *M. oleifera* leaves and bark extracts showed cytotoxic effects against MDA-MB-231 cells and HCT-8 ileocecal adenocarcinoma cancer cell line (Al-Asmari et al., 2015).

5.7 *Solanum nigrum* (Solanaceae)

S. nigrum or black nightshade is native to Eurasia and introduced in the Americas, Australasia, and South Africa. Parts of this plant can be toxic to livestock and humans, and it is considered a weed. However, ripe berries and cooked leaves of edible strains are used as food in some locales, and plant parts are used as a traditional medicine. In European traditional medicine, the plant has been used as a strong sudorific, analgesic, and sedative with powerful narcotic properties. In traditional Indian medicines, infusions are used in case of dysentery, stomach complaints, and fever and to treat tuberculosis (Kaushik et al., 2009). The juice of the plant is used on ulcers and other skin diseases. The fruits are used as a tonic, laxative, appetite stimulant, and for treating asthma. The juice from its roots is used against asthma and whooping cough. The plant is widely used in oriental medicine where it is considered to be antitumorigenic, antioxidant, antiinflammatory, hepatoprotective, diuretic, and antipyretic (Jain et al., 2011). It was demonstrated that aqueous extract of *S. nigrum* inhibits growth of cervical carcinoma (U14) via modulating immune response of tumor bearing mice and inducing apoptosis of tumor cells (Li et al., 2008). This extract induced tumor cell cycle arrest in G0/G1 phase, as well as apoptosis with little toxicity to the animals (Li et al., 2008). It was also shown that the total alkaloid isolated from *S. nigrum* interfered with structure and function of tumor cell membrane, disturbed the synthesis of DNA and RNA, changed the cell cycle distribution in tumor cells. Hence, total alkaloids could play a role in inhibition of tumor cells, while a glycoprotein isolated from the plant exhibited anticancer abilities by blocking the antiapoptotic pathway of NF-kappaB via activation of caspase cascades reaction and increase nitric oxide production (An et al., 2006). It has also been proved that aqueous extract of *S. nigrum* can potentially be used in novel integrated chemotherapy with cisplatin or doxorubicin to treat hepatocellular carcinoma patients (Wang et al., 2015). Alpha-solanine, a naturally occurring steroidal glycoalkaloid found in *S. nigrum* was found to inhibit proliferation and induce apoptosis of tumor cells (Shen et al., 2014). It has been suggested that inhibition of PC-3 pancreatic cell invasion by α-solanine may be through blocking epithelial–mesenchymal transition and matrix metalloproteinases expression. Alpha-solanine also reduces extracellular signal-regulated kinases (ERK) and PI3K/Akt signaling pathways (signal transduction pathway that promotes survival and growth in response to extracellular signals) and regulates expression of miR-21 and miR-138 (Shen et al., 2014). These findings suggest an attractive therapeutic potential of α-solanine for suppressing invasion of prostate cancer cell (Shen et al., 2014).

5.8 *Vigna unguiculata* (Fabaceae)

V. unguiculata, commonly known as cowpeas is one of the most important food legume crops in the semiarid tropics covering Asia, Africa, Southern Europe, and Central and South America. *V. unguiculata* is grown in the African continent, particularly in Nigeria and Niger. The Sahel region also contains other major producers, such as Burkina Faso, Ghana, Senegal, and Mali. Niger is the main exporter of cowpeas and Nigeria the main importer. Outside Africa, the major production areas are Asia, Central America, and South America. Brazil is the world's second-leading producer of cowpea seed. Medicinally, the roasted seeds are used to treat neuritis, insomnia, weakness of memory, indigestion, dyspepsia, sensation of pins and needles in limbs, periodic palpitation, congestive cardiac failure, etc. It is an excellent medicine for stomatitis, corneal ulcers, coleic diseases, kwasiorkar, marasmus. Decoction of leaves is used to treat hyperacidity, nausea, and vomiting (Battu et al., 2011). A 36-kDa protein, with an N-terminal sequence highly homologous to polygalacturonase (PG) inhibiting proteins, isolated from small brown-eyed cowpea seeds, reduced [methyl-(3) H] thymidine incorporation into MBL2 lymphoma and L1210 leukemia cells with an IC_{50} of 7.4 and 5.4 µM, respectively (Tian et al., 2013). The anticarcinogenic potential of Bowman–Birk protease inhibitors identifies as black-eyed pea trypsin/chymotrypsin inhibitor on MCF-7 breast cancer cells was also reported (Joanitti et al., 2010). The free phenolic extract of whole seeds of *V. unguiculata* inhibited 65% proliferation of hormone-dependent mammary (MCF-7) cancer cells meanwhile extracts of seed coats or cotyledons also inhibited cell proliferation but to a significantly lesser extent (Gutiérrez-Uribe et al., 2011).

6 MODE OF ACTION OF AFRICAN SPICES, VEGETABLE'S EXTRACTS, AND DERIVED PRODUCTS

The mode of action of many African plant extracts and isolated compounds has been demonstrated. The documented modes of induction of apoptosis include activation of caspases, alteration of MMP, generation of ROS, and inhibition of angiogenesis.

6.1 Induction of apoptosis and cell cycle arrest

Some African medicinal spices or vegetables having strong cytotoxic effects on cancer cells were shown to induce apoptotic cell death. They include *E. giganteus*, *I. cylindrica*, *P. capense* (Kuete et al., 2013b), and *A. mannii* (Kuete et al., 2013a). A moderate to strong inductions of apoptosis were also recorded with the spice of *X. aethiopica* (Kuete et al., 2013b). It was also shown that most of the crude extracts from African medicinal plants induced cell cycle arrest mostly in G0/G1 and between G0/G1 and S phases. In fact, the cell cycle arrest in G0/G1 in leukemia CCRF-CEM cells was reported with extracts from *A. mannii* (Kuete et al., 2013a), *E. giganteus*, and *P. capense* (Kuete et al., 2013b). Arrest between G0/G1 and S phase in CCRF-CEM cells was reported with the extracts from *I. cylindrica* and *X. aethiopica* (Kuete et al., 2014b). Flavonoid **10** as well as alkaloid **16** induced a time-dependent modification of the cell cycle distribution to different extents with progressive increase of sub-G0/G1 phase cells in both cases and arrest in Go/G1 phase (Kuete et al., 2015).

6.2 Effects of African spices or vegetables and derived molecules on caspase activation

Caspases, a family of cysteine proteases, are central regulators of apoptosis (Alnemri et al., 1996). Initiator caspases (caspase-2, 8, 9, 10, 11, and 12) are closely coupled to proapoptotic signals (Alnemri et al., 1996). Upon activation, initiator caspases cleave and activate downstream effector caspases (caspase-3, 6, and 7), which in turn execute apoptosis by cleaving cellular proteins at specific aspartate residues (Alnemri et al., 1996). In general, several crude extracts having inhibitory effect on cancer cells were reported not to induce the activation of caspase enzymes (Kuete et al., 2013a,b, 2014a). However, alkaloid **16** isolated from *X. aethiopica* was found to induce apoptosis in leukemia CCRF-CEM cells through activation of caspase 3/7, caspase 8, and caspase 9 after 6-h treatment at concentrations equivalent to IC_{50} and twofold IC_{50} values (Kuete et al., 2015).

6.3 Effects of African plant extracts and derived molecules on the mitochondrial membrane potential

Apoptotic proteins target mitochondria and affect them in different ways. If cytochrome c is released from mitochondria due to formation of a channel in the outer mitochondrial membrane during the apoptosis process, it binds to apoptotic protease activating factor-1 (Apaf-1) and ATP, which then bind to procaspase-9 creating a protein complex known as apoptosome (Dejean et al., 2006). MMP disruption in cancer cells was reported as one of the likely mechanisms of induction of apoptosis by several African plant extracts and derived compounds including samples from spices and vegetables. A strong depletion of MMP in CCRF-CEM cells was reported with crude extracts from *E. giganteus*, *X. aethiopica* (Kuete et al., 2013b), and *A. mannii* (Kuete et al., 2013a). Moderate alterations of the MMP in CCRF-CEM cells were measured with extracts from *I. cylindrica* and *P. capense* (Kuete et al., 2013b). Among compounds isolated from the medicinal spice *X. aethiopica*, flavonoid 3,4′,5-trihydroxy-6″,6″-dimethylpyrano[2,3-g]flavone (**10**) as well as alkaloid **16** caused pronounced and dose-dependent depletion of MMP in leukemia CCRF-CEM cells, in a range of 42.6–48.9% and 14.0–18.2%, respectively, suggesting the alteration of mitochondrial membrane integrity is one of the mode of induction of apoptosis of the two compounds in cancer cells (Kuete et al., 2015).

6.4 Effects of African spices and vegetables and derived molecules on generation of reactive oxygen species

The appearance of malignancies resulting in gain-of-function mutations in oncogenes and loss-of-function mutations in tumor suppressor genes, leads to cell deregulation that is frequently associated with enhanced cellular stress (Raj et al., 2011). Increased ROS production in leukemia CCRF-CEM cells was reported upon treatment with extracts from some African plants. Among them were *X. aethiopica* (Kuete et al., 2013b) and *A. mannii* (Kuete et al., 2013a). The generation of ROS in CCRF-CEM cells was investigated after treatment with flavonoid **10** as well as alkaloid **16**. Compound **10** did not induce ROS production, while compound **16** increased ROS levels up to 28.9% upon treatment with a concentration equivalent to IC_{50} (Kuete et al., 2015).

6.5 Antiangiogenic effects of African plant spices and vegetable extracts

Excessive angiogenesis represents an important pathogenic factor in many industrialized western countries (Krenn and Paper, 2009). Therefore, compounds with antiangiogenic properties are of importance in the treatment and prevention of malignancies as well as other chronic diseases (Carmeliet, 2003; Paper, 1998). The extracts from *X. aethiopica*, *D. psilurus*, *E. giganteus*, and *Zingiber officinale* were reported to strongly inhibit angiogenesis in quail embryo (Kuete et al., 2011).

7 CONCLUSIONS

In this chapter, we have discussed some in vitro methods commonly used in the cytotoxicity screening of phytochemicals and we have demonstrated that Africa flora contains several cytotoxic spices and vegetables that could be used to manage cancers. The traditional use of the best plants (Table 10.1) indicated that they are not always used traditionally to treat cancers. Therefore, all plants independent of their ethnopharmacological relevance should be considered for cytotoxicity screenings in cancer cells. The most cytotoxic spices and vegetables extracts from African flora documented herein include *A. arundinaceum*, *X. aethiopica*, *E. giganteus*, *I. cylindrica*, *P. capense*, *D. psilurus*, *Z. officinale* as well as *B. rapa*, *C. gynandra*, *C. esculenta*, *C. olitorius*, *M. oleifera*, *S. nigrum*, and *V. unguiculata*. The best cytotoxic phytochemicals reported in this review include galanal A (**1**), galanal B (**2**), jaeschkeanadiol *p*-hydroxybenzoate (**3**), 16α-hydroxy-ent-kauran-19-oic acid (**4**); ent-15-oxokaur-16-en-19-oic acid (**5**), naringenin (**6**), kaempferol-3,7,4′-trimethylether (**7**), candidone (**8**), 4-hydroxy-2,6-di-(3′,4′-dimethoxyphenyl)-3,7-dioxabicyclo-(3.3.0) octane (**9**), 3,4′,5-trihydroxy-6″,6″-dimethylpyrano[2,3-g]flavone (**10**), *trans*-tiliroside (**11**), 2-(penta-1,3-diynyl)-5-(4-hydroxybut-1-ynyl)-thiophene (**12**), echinopsolide A (**13**), 2-(penta-1,3-diynyl)-5-(3,4-dihydroxybut-1-ynyl)-thiophene (**14**), echinopsfuroceramide (**15**), and isotetrandrine (**16**). It can be concluded that African spices and vegetables represent an enormous resource for the search of cytotoxic compounds. Intensified research efforts are warranted throughout the continent, for the development of novel anticancer drugs in the clinical setting.

REFERENCES

Adjanohoun, J., Aboubakar, N., Dramane, K., Ebot, M., Ekpere, J., Enow-Orock, E. (Eds.), 1996. Traditional Medicine and Pharmacopoeia: Contribution to Ethnobotanical and Floristic Studies in Cameroon. OUA/STRC, Lagos.

Ajaiyeoba, E., 2000. Phytochemical and antimicrobial studies of *Gynandropsis gynandra* and *Buchholzia coriaceae*. Afr. J. Biomed. Res. 3, 161–165.

Akoachere, J., Ndip, R., Chenwi, E., Ndip, L., Njock, T., Anong, D., 2002. Antibacterial effect of *Zingiber officinale* and *Garcinia kola* on respiratory tract pathogens. East Afr. Med. J. 79, 588–592.

Al-Asmari, A.K., Albalawi, S.M., Athar, M.T., Khan, A.Q., Al-Shahrani, H., Islam, M., 2015. *Moringa oleifera* as an anti-cancer agent against breast and colorectal cancer cell lines. PLoS One 10 (8), e0135814.

Alnemri, E.S., Livingston, D.J., Nicholson, D.W., Salvesen, G., Thornberry, N.A., Wong, W.W., 1996. Human ICE/CED-3 protease nomenclature. Cell 87 (2), 171.

Altman, F.P., 1976. Tetrazolium salts and formazans. Prog. Histochem. Cytochem. 9 (3), 1–56.

An, L., Tang, J.T., Liu, X.M., Gao, N.N., 2006. Review about mechanisms of anti-cancer of *Solanum nigrum*. Zhongguo Zhong Yao Za Zhi 31 (15), 1225–1226, 1260.

Anwar, F., Latif, S., Ashraf, M., Gilani, A.H., 2007. *Moringa oleifera*: a food plant with multiple medicinal uses. Phytother. Res. 21 (1), 17–25.

Arranz, M.J., Festing, M.F., 1990. Prior use of the neutral red assay and reduction of total protein determination in 96-well plate assays. Toxicol. In Vitro 4 (3), 211–212.

Bakarnga-Via, I., Hzounda, J.B., Fokou, P.V., Tchokouaha, L.R., Gary-Bobo, M., Gallud, A., 2014. Composition and cytotoxic activity of essential oils from *Xylopia aethiopica* (Dunal) A. Rich, *Xylopia parviflora* (A. Rich) Benth.) and *Monodora myristica* (Gaertn) growing in Chad and Cameroon. BMC Complement. Altern. Med. 14, 125.

Bala, A., Kar, B., Haldar, P.K., Mazumder, U.K., Bera, S., 2010. Evaluation of anticancer activity of *Cleome gynandra* on Ehrlich's Ascites Carcinoma treated mice. J. Ethnopharmacol. 129 (1), 131–134.

Battu, G., Anjana Male, C., Hari priya, T., Malleswari, V., Reeshma, S.K., 2011. A phytopharmacological review on *Vigna* species. Pharmanest 2 (1), 62–67.

Betti, J., 2004. An ethnobotanical study of medicinal plants among the Baka Pygmies in the Dja Biosphere Reserve, Cameroon. Afr. Study Monogr. 25 (1), 1–27.

Bharali, R., Tabassum, J., Azad, M.R., 2003. Chemomodulatory effect of *Moringa oleifera*, Lam, on hepatic carcinogen metabolising enzymes, antioxidant parameters and skin papillomagenesis in mice. Asian Pac. J. Cancer Prev. 4 (2), 131–139.

Borenfreund, E., Babich, H., Martin-Alguacil, N., 1988. Comparisons of two in vitro cytotoxicity assays—the neutral red (NR) and tetrazolium MTT tests. Toxicol. In Vitro 2 (1), 1–6.

Brown, A.C., Reitzenstein, J.E., Liu, J., Jadus, M.R., 2005. The anti-cancer effects of poi (*Colocasia esculenta*) on colonic adenocarcinoma cells In vitro. Phytother. Res. 19 (9), 767–771.

Brown, A.C., Valiere, A., 2004. The medicinal uses of poi. Nutr. Clin. Care 7 (2), 69–74.

Cantrell, C., Berhow, M., Phillips, B., Duval, S., Weisleder, D., Vaughn, S., 2003. Bioactive crude plant seed extracts from the NCAUR oilseed repository. Phytomedicine 10, 325–333.

Carmeliet, P., 2003. Angiogenesis in health and disease. Nat. Med. 9 (6), 653–660.

Choumessi, A.T., Danel, M., Chassaing, S., Truchet, I., Penlap, V.B., Pieme, A.C., 2012. Characterization of the antiproliferative activity of *Xylopia aethiopica*. Cell Div. 7 (1), 8.

Cory, A.H., Owen, T.C., Barltrop, J.A., Cory, J.G., 1991. Use of an aqueous soluble tetrazolium/formazan assay for cell growth assays in culture. Cancer Commun. 3 (7), 207–212.

Dalziel, J. (Ed.), 1936. The Useful Plants of West Tropical Africa. The Crown Agents for Overseas Governments and Administration, London.

Dejean, L.M., Martinez-Caballero, S., Kinnally, K.W., 2006. Is MAC the knife that cuts cytochrome c from mitochondria during apoptosis? Cell Death Differ. 13 (8), 1387–1395.

Desai, A.G., Qazi, G.N., Ganju, R.K., El-Tamer, M., Singh, J., Saxena, A.K., 2008. Medicinal plants and cancer chemoprevention. Curr. Drug Metab. 9 (7), 581–591.

Dimo, T., Rakotonirina, A., Tan, P., Dongo, E., Dongmo, A., Kamtchouing, P., 2001. Antihypertensive effects of *Dorstenia psilurus* extract in fructose-fed hyperinsulinemic, hypertensive rats. Phytomedicine 8, 101–106.

Diniz, M., Martins, E., Gomes, E., Silva, O., 2007. Contribution to the knowledge of medicinal plants from Guinea-Bissau. Port. Acta Biol. 19, 417–427.

Ferguson, L.R., Roberton, A.M., McKenzie, R.J., Watson, M.E., Harris, P.J., 1992. Adsorption of a hydrophobic mutagen to dietary fiber from taro (*Colocasia esculenta*), an important food plant of the South Pacific. Nutr. Cancer 17 (1), 85–95.

Ferlay, J., Shin, H., Bray, F., Forman, D., Mathers, C., Parkin, D., 2010. GLOBO-CAN 2008. Cancer Incidence and Mortality Worldwide: IARC Cancer-Base No.10. International Agency for Research on Cancer, Lyon, France. Available from: http://globocan.iarc.fr

Gbewonyo, W., Candy, D., 1992. Chromatographic isolation of insecticidal amides from *Piper guineense* root. J. Chromatogr. A 607, 105–111.

Gutiérrez-Uribe, J.A., Romo-Lopez, I., Serna-Saldívar, S.O., 2011. Phenolic composition and mammary cancer cell inhibition of extracts of whole cowpeas (*Vigna unguiculata*) and its anatomical parts. J. Funct. Foods 3 (4), 290–297.

http://www.cancer.org, 2012. Cancer in Africa. Available from: http://www.cancer.org/acs/groups/content/@epidemiologysurveilance/documents/document/acspc-031574.pdf1

Huang, A., Titchenal, C., Meilleur, B., 2011. Nutrient composition of taro corms and breadfruit. J. Food Comp. Anal. 13, 859–864.

Irvine, R., 1961. Woody Plant of Ghana. Oxford University Press, London.

Iseri, O.D., Yurtcu, E., Sahin, F.I., Haberal, M., 2013. *Corchorus olitorius* (jute) extract induced cytotoxicity and genotoxicity on human multiple myeloma cells (ARH-77). Pharm. Biol. 51 (6), 766–770.

Jain, R., Sharma, A., Gupta, S., Sarethy, I., Gabrani, R., 2011. *Solanum nigrum*: current perspectives on therapeutic properties. Altern. Med. Rev. 16 (1), 78–85.

Joanitti, G.A., Azevedo, R.B., Freitas, S.M., 2010. Apoptosis and lysosome membrane permeabilization induction on breast cancer cells by an anticarcinogenic Bowman–Birk protease inhibitor from *Vigna unguiculata* seeds. Cancer Lett. 293 (1), 73–81.

Jones, G., Rao, K., Tucker, D., Richardson, B., Barnes, A., Rivett, D., 1995. Antimicrobial activity of santalbic acid from the oil of *Santalum acuminatum* (Quandong). Int. J. Pharmacogn. 33, 120–123.

Kaou, A., Mahiou-Leddet, V., Canlet, C., Debrauwer, L., Hutter, S., Azas, N., 2010. New amide alkaloid from the aerial part of *Piper capense* L.f. (Piperaceae). Fitoterapia 81, 632–635.

Kato, A., Higuchi, Y., Goto, H., Kizu, H., Okamoto, T., Asano, N., 2006. Inhibitory effects of *Zingiber officinale* Roscoe derived components on aldose reductase activity in vitro and in vivo. J. Agric. Food Chem. 54, 6640–6644.

Kaushik, D., Jogpal, V., Kaushik, P., Lal, S., Saneja, A., Sharma, C., 2009. Evaluation of activities of *Solanum nigrum* fruit extract. Arch. Appl. Sci. Res. 1 (1), 43–50.

Kim, J., Lee, S., Park, H., Yang, J., Shin, T., Kim, Y., 2008. Cytotoxic components from the dried rhizomes of *Zingiber officinale* Roscoe. Arch. Pharm. Res. 31, 415–418.

Kojima, H., Sato, N., Hatano, A., Ogura, H., 1990. Sterol glucosides from *Prunella vulgaris*. Phytochemistry 29, 2351–2355.

Kokowaro, J., 1976. Medicinal Plants of East Africa. East African Literature Bureau, Kampala.

Krenn, L., Paper, D.H., 2009. Inhibition of angiogenesis and inflammation by an extract of red clover (*Trifolium pratense* L.). Phytomedicine 16 (12), 1083–1088.

Kuete, V., Ango, P.Y., Yeboah, S.O., Mbaveng, A.T., Mapitse, R., Kapche, G.D., 2014a. Cytotoxicity of four *Aframomum* species (*A. arundinaceum*, *A. alboviolaceum*, *A. kayserianum* and *A. polyanthum*) towards multifactorial drug resistant cancer cell lines. BMC Complement. Altern. Med. 14, 340.

Kuete, V., Efferth, T., 2015. African flora has the potential to fight multidrug resistance of cancer. Biomed. Res. Int. 2015, 914813.

Kuete, V., Eyong, K.O., Folefoc, G.N., Beng, V.P., Hussain, H., Krohn, K., 2007a. Antimicrobial activity of the methanolic extract and of the chemical constituents isolated from *Newbouldia laevis*. Pharmazie 62 (7), 552–556.

Kuete, V., Fankam, A.G., Wiench, B., Efferth, T., 2013a. Cytotoxicity and modes of action of the methanol extracts of six Cameroonian medicinal plants against multidrug-resistant tumor cells. Evid. Based Complement. Alternat. Med. 2013, 285903.

Kuete, V., Krusche, B., Youns, M., Voukeng, I., Fankam, A.G., Tankeo, S., 2011. Cytotoxicity of some Cameroonian spices and selected medicinal plant extracts. J. Ethnopharmacol. 134 (3), 803–812.

Kuete, V., Metuno, R., Ngameni, B., Tsafack, A.M., Ngandeu, F., Fotso, G.W., 2007b. Antimicrobial activity of the methanolic extracts and compounds from *Treculia obovoidea* (Moraceae). J. Ethnopharmacol. 112 (3), 531–536.

Kuete, V., Sandjo, L.P., Mbaveng, A.T., Zeino, M., Efferth, T., 2015. Cytotoxicity of compounds from *Xylopia aethiopica* towards multi-factorial drug-resistant cancer cells. Phytomedicine 22 (14), 1247–1254.

Kuete, V., Sandjo, L.P., Wiench, B., Efferth, T., 2013b. Cytotoxicity and modes of action of four Cameroonian dietary spices ethno-medically used to treat cancers: *Echinops giganteus*, *Xylopia aethiopica*, *Imperata cylindrica* and *Piper capense*. J. Ethnopharmacol. 149 (1), 245–253.

Kuete, V., Tankeo, S.B., Saeed, M.E., Wiench, B., Tane, P., Efferth, T., 2014b. Cytotoxicity and modes of action of five Cameroonian medicinal plants against multi-factorial drug resistance of tumor cells. J. Ethnopharmacol. 153 (1), 207–219.

Kuete, V., Viertel, K., Efferth, T., 2013c. Antiproliferative potential of African medicinal plants. In: Kuete, V. (Ed.), Medicinal Plant Research in Africa. Elsevier, Oxford, pp. 711–724, (Chapter 18).

Kuete, V., Wansi, J.D., Mbaveng, A.T., Kana Sop, M.M., Tadjong, A.T., Beng, V.P., 2008. Antimicrobial activity of the methanolic extract and compounds from *Teclea afzelii* (Rutaceae). S. Afr. J. Bot. 74 (4), 572–576.

Kuete, V., Wiench, B., Hegazy, M.E., Mohamed, T.A., Fankam, A.G., Shahat, A.A., 2012. Antibacterial activity and cytotoxicity of selected Egyptian medicinal plants. Planta Med. 78 (2), 193–199.

Kuhn, D.M., Balkis, M., Chandra, J., Mukherjee, P.K., Ghannoum, M.A., 2003. Uses and limitations of the XTT assay in studies of *Candida* growth and metabolism. J. Clin. Microbiol. 41 (1), 506–508.

Kundu, N., Campbell, P., Hampton, B., Lin, C.Y., Ma, X., Ambulos, N., 2012. Antimetastatic activity isolated from *Colocasia esculenta* (taro). Anticancer Drugs 23 (2), 200–211.

Li, C.J., Huang, S.Y., Wu, M.Y., Chen, Y.C., Tsang, S.F., Chyuan, J.H., 2012. Induction of apoptosis by ethanolic extract of *Corchorus olitorius* leaf in human hepatocellular carcinoma (HepG2) cells via a mitochondria-dependent pathway. Molecules 17 (8), 9348–9360.

Li, J., Li, Q., Feng, T., Li, K., 2008. Aqueous extract of *Solanum nigrum* inhibit growth of cervical carcinoma (U14) via modulating immune response of tumor bearing mice and inducing apoptosis of tumor cells. Fitoterapia 79 (7–8), 548–556.

Lopez, A.D., Mathers, C.D., Ezzati, M., Jamison, D.T., Murray, C.J., 2006. Global and regional burden of disease and risk factors, 2001: systematic analysis of population health data. Lancet 367 (9524), 1747–1757.

Lu, L., Zhang, L., Wai, M.S., Yew, D.T., Xu, J., 2012. Exocytosis of MTT formazan could exacerbate cell injury. Toxicol. In Vitro 26 (4), 636–644.

Miyoshi, N., Nakamura, Y., Ueda, Y., Abe, M., Ozawa, Y., Uchida, K., 2003. Dietary ginger constituents, galanals A and B, are potent apoptosis inducers in Human T lymphoma Jurkat cells. Cancer Lett. 199 (2), 113–119.

Mohamed, G., Abdel-Lateff, A., Fouad, M., Ibrahim, S., Elkhayat, E., Okino, T., 2009. Chemical composition and hepato-protective activity of *Imperata cylindrica* Beauv. Pharmacogn. Mag. 5, 28–36.

Mule, S., Patil, S., Naikwade, N., Magdum, C., 2008. Evaluation of antinociceptive and anti-inflammatory activity of stems of *Gynandropsis pentaphylla* Linn. Int. J. Green Pharm. 2, 87–90.

N'dri, K., Bosson, K., Mamyrbekova-Bekro, J., Jean, N., Bekro, Y., 2009. Chemical composition and antioxidant activities of essential oils of *Xylopia aethiopica* (Dunal) a. Rich. Eur. J. Sci. Res. 37, 311–318.

Newman, D.J., Cragg, G.M., 2007. Natural products as sources of new drugs over the last 25 years. J. Nat. Prod. 70 (3), 461–477.

Ngadjui, B., Dongo, E., Happi, N., Bezabih, M., Abegaz, B., 1998. Prenylated flavones and phenylpropanoid derivatives from roots of *Dorstenia psilurus*. Phytochemistry 48, 733–737.

Ngoumfo, R., Jouda, J., Mouafo, F., Komguem, J., Mbazoa, C., Shiao, T., 2010. In vitro cytotoxic activity of isolated acridones alkaloids from *Zanthoxylum leprieurii* Guill. et Perr. Bioorg. Med. Chem. Lett. 18, 3601–3605.

Nishimoto, K., Ito, M., Natori, S., Ohmoto, T., 1968. The structure of arundoin, cylindrin and fernenol triterpenoids of fernane and arborane groups of *Imperata cylindrica* var. Koenigii. Tetrahedron 24, 735–752.

Noumi, E., Amvam, Z., Lontsi, D., 1998. Aphrodisiac plants used in Cameroon. Fitoterapia 69, 125–134.

Noumi, E., Eloumou, M., 2011. Syphilis ailment: prevalence and herbal remedies in Ebolowa subdivision (South region, Cameroon). Int. J. Biomed. Pharm. Sci. 2 (1), 20–28.

Okoli, R., Aigbe, J., Ohaju-Obodo, J., Mensah, J., 2007. Medicinal herbs used for managing some common ailments among Esan People of Edo State, Nigeria. Pak. J. Nutr. 6, 490–496.

Olawuyi, P., Falusi, O., Oluwajobi, A., Azeez, R., Titus, S., JF, A., 2014. Chromosome studies in jute plant (*Corchorus olitorius*). Eur. J. Biotechnol. Biosci. 1 (1), 1–3.

Oloke, J., Kolawole, D., 1988. The antibacterial and antifungal activities of certain components of *Aframomum melegueta* fruits. Fitoterapia 59, 384–388.

Paper, D.H., 1998. Natural products as angiogenesis inhibitors. Planta Med. 64 (8), 686–695.

Raj, L., Ide, T., Gurkar, A.U., Foley, M., Schenone, M., Li, X., 2011. Selective killing of cancer cells by a small molecule targeting the stress response to ROS. Nature 475 (7355), 231–234.

Randhawa, M.A., Khan, A.A., Javed, M.S., Sajid, M.W., 2015. Green leafy vegetables: a health promoting source. In: Watson, R.R. (Ed.), Handbook of Fertility. Academic Press, San Diego, CA, pp. 205–220, (Chapter 18).

Repetto, G., del Peso, A., Zurita, J.L., 2008. Neutral red uptake assay for the estimation of cell viability/cytotoxicity. Nat. Protoc. 3 (7), 1125–1131.

Riss, T., Moravec, R., Niles, A., Benink, H., Worzella, T., Minor, L., 2013. Cell viability assays. Assay Guidance Manual. Eli Lilly & Company and the National Center for Advancing Translational Sciences, Bethesda, MD, USA.

Ruppelt, B., Pereira, E., Goncalves, L., Pereira, N., 1991. Pharmacological screening of plants recommended by folk medicine as anti-snake venom I. Analgesic and anti-inflammatory activities. Memórias do Instituto Oswaldo Cruz 86, 203–205.

Sakpakdeejaroen, I., Itharat, A., 2009. Cytotoxic compounds against breast adenocarcinoma cells (MCF-7) from Pikutbenjakul. J. Health Res. 23, 71–76.

Sandjo, L., Kuete, V., Poumale, P., Efferth, T., 2016. Unprecedented brominated oleanolide and a new tetrahydrofurano-ceramide from *Echinops giganteus*. Nat. Prod. Res. 5, 1–9.

Scudiero, D.A., Shoemaker, R.H., Paull, K.D., Monks, A., Tierney, S., Nofziger, T.H., 1988. Evaluation of a soluble tetrazolium/formazan assay for cell growth and drug sensitivity in culture using human and other tumor cell lines. Cancer Res. 48 (17), 4827–4833.

Sener, S.F., Grey, N., 2005. The global burden of cancer. J. Surg. Oncol. 92 (1), 1–3.

Shen, K.H., Liao, A.C., Hung, J.H., Lee, W.J., Hu, K.C., Lin, P.T., 2014. Alpha-solanine inhibits invasion of human prostate cancer cell by suppressing epithelial-mesenchymal transition and MMPs expression. Molecules 19 (8), 11896–11914.

Stevigny, C., Bailly, C., Quetin-Leclercq, J., 2005. Cytotoxic and antitumor potentialities of aporphinoid alkaloids. Curr. Med. Chem. Anticancer Agents 5 (2), 173–182.

Stockert, J.C., Blazquez-Castro, A., Canete, M., Horobin, R.W., Villanueva, A., 2012. MTT assay for cell viability: intracellular localization of the formazan product is in lipid droplets. Acta Histochem. 114 (8), 785–796.

Suffness, M., Pezzuto, J. (Eds.), 1990. Assays Related to Cancer Drug Discovery, vol. 6, Academic Press, London.

Tane, P., Bergquist, K., Tene, M., Ngadjui, B., Ayafor, J., Sterner, O., 1995. Cyclodione, an unsymmetrical dimeric diterpene from *Cylicodiscus gabunensis*. Tetrahedron 51, 11595–11600.

Tane, P., Tatsimo, S., Ayimele, G., Connolly, J., 2005. Bioactive metabolites from *Aframomum* species. 11th NAPRECA Symposium Book of Proceedings. Antananarivo, Madagascar, pp. 214–223.

Tatsadjieu, L., Essia Ngang, J., Ngassoum, M., Etoa, F., 2003. Antibacterial and antifungal activity of *Xylopia aethiopica*, *Monodora myristica*, *Zanthoxylum xanthoxyloides* and *Zanthoxylum leprieurii* from Cameroon. Fitoterapia 74, 469–472.

Tene, M., Tane, P., Sondengam, B.L., Connolly, J.D., 2004. Lignans from the roots of *Echinops giganteus*. Phytochemistry 65 (14), 2101–2105.

Thomas, S., 1965. Chemical basis of drug action. Drug Plants of Africa. Publications of University of Pennsylvania, Philadelphia, Pennsylvania, USA.

Thomas, J., Bahuchets, S., Epelboin, A., Furniss, S. (Eds.), 2003. Encyclopédie des Pygmées Aka: Techniques, Langage et Société des Chasseurs-Cueilleurs de la Forêt Centrafricaine (Sud-Centrafrique et Nord-Congo), vol. 11. Editions Peeters-SELAF, Paris, pp. 1981–2010.

Tian, G.T., Zhu, M.J., Wu, Y.Y., Liu, Q., Wang, H.X., Ng, T.B., 2013. Purification and characterization of a protein with antifungal, antiproliferative, and HIV-1 reverse transcriptase inhibitory activities from small brown-eyed cowpea seeds. Biotechnol. Appl. Biochem. 60 (4), 393–398.

van Tonder, A., Joubert, A.M., Cromarty, A.D., 2015. Limitations of the 3-(4,5-dimethylthiazol-2-yl)-2,5-diphenyl-2H-tetrazolium bromide (MTT) assay when compared to three commonly used cell enumeration assays. BMC Res. Notes 8, 47.

Van Wyk, B., Gericke, N., 2000. Peoples Plants: A Guide to Useful Plants in Southern Africa. Brizza Publications, Pretoria, South Africa.

Vichai, V., Kirtikara, K., 2006. Sulforhodamine B colorimetric assay for cytotoxicity screening. Nat. Protoc. 1 (3), 1112–1116.

Voss, C., Eyol, E., Berger, M.R., 2006. Identification of potent anticancer activity in *Ximenia americana* aqueous extracts used by African traditional medicine. Toxicol. Appl. Pharmacol. 211 (3), 177–187.

Wang, C.K., Lin, Y.F., Tai, C.J., Wang, C.W., Chang, Y.J., Choong, C.Y., 2015. Integrated treatment of aqueous extract of *Solanum nigrum*-potentiated cisplatin- and doxorubicin-induced cytotoxicity in human hepatocellular carcinoma cells. Evid. Based Complement. Alternat. Med. 2015, 675270.

World Health Organization., 2008. World Cancer Report 2008. International Agency for Research on Cancer, Lyon.

Wu, Q., Cho, J.G., Yoo, K.H., Jeong, T.S., Park, J.H., Kim, S.Y., 2013. A new phenanthrene derivative and two diarylheptanoids from the roots of *Brassica rapa* ssp. campestris inhibit the growth of cancer cell lines and LDL-oxidation. Arch. Pharm. Res. 36 (4), 423–429.

CHAPTER 11

Antiemetic African Medicinal Spices and Vegetables

S. Tchatchouang, V.P. Beng, V. Kuete

1 INTRODUCTION

Emesis is an unpleasant activity that results in the expulsion of stomach contents through the mouth and is clearly associated with gastrointestinal motor activity (Ahmed et al., 2014). Emesis or vomiting process can be categorized into the following three phases: nausea, retching, and vomiting (Qureshi, 2012). Vomiting and nausea may be manifestation of a wide variety of conditions, including pregnancy, obstruction, peptic ulcer, drug toxicity, myocardial infarction, renal failure, and hepatitis. It can lead to rejection of potentially curative principles during chemotherapy, dehydration, profound metabolic imbalances, and nutrient depletion (Tijani et al., 2008). Drugs used for the treatment and prevention of emesis are known as antiemetics (Hornby, 2001; Ray et al., 2009).

Despite the wide range of antiemetic drugs available, many are expensive and have a number of reported adverse effects. There is a renewed global interest in traditional medicines with medicinal plants being a good source for drug discovery. The important advantages claimed for therapeutic uses of medicinal plants in various ailments are their safety besides being economical, effective, and their easy availability (Atal and Kapur, 1989; Siddiqui, 1993). Due to these advantages, medicinal plants have been widely used in traditional medicine (Shivhare, 2011). Irrespective of race, civilization and culture, plants and their products have been used by humans in many ways, and one of the best use is as food and spices having potent pharmacological properties including antiemetic effects (Rajsekhar et al., 2012). Botanicals are economical, safe, and readily available particularly in rural communities. Phytomedicines are plant-derived medicines that contain chemicals, more usually, mixtures of chemical compounds that act individually or in combination on the human body to prevent disorders and to restore or maintain health (Coronas et al., 1975; Ben-Erik van Wyk, 2004; Miner and Sanger, 1986).

To contribute to the discovery of natural products that may be capable of preventing or inhibiting emesis, an antiemetic plant screening has been performed all over the world. The present review focuses on African spices and vegetables as source of antiemetics, even if the studies were not always performed using samples harvested in the continent.

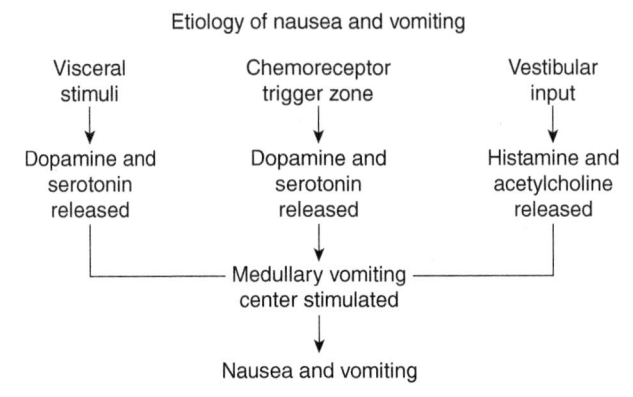

```
                    Etiology of nausea and vomiting

        Visceral            Chemoreceptor         Vestibular
        stimuli              trigger zone           input
           ↓                     ↓                    ↓
     Dopamine and          Dopamine and         Histamine and
       serotonin             serotonin          acetylcholine
       released              released              released
           └──────────────────┬─────────────────────┘
                              ↓
                      Medullary vomiting
                       center stimulated
                              ↓
                      Nausea and vomiting
```

Figure 11.1 *Physiological mechanisms that result in nausea and vomiting.*

2 MODES OF ACTION OF ANTIEMETICS

Treatment of nausea and vomiting ideally involves correcting the underlying cause. When the exact cause is not known or cannot be corrected, symptoms still can be treated. Three primary pathophysiological pathways are involved in the stimulation of the vomiting center in the medulla that directly mediates nausea and vomiting. This center can be stimulated by vestibular fibers, afferent visceral fibers, and input from the chemoreceptor trigger zone in the base of the fourth ventricle (Flake et al., 2004). The physiological mechanisms that result in nausea and vomiting are summarized in Fig. 11.1. The neurotransmitters—histamine, acetylcholine, serotonin, and dopamine—are frequently implicated in these pathways and are targets of most therapeutic modalities.

2.1 Antihistamines and anticholinergics

Histamine is a nitrogenous compound involved in local immune responses as well as regulating physiological function in the gut and acting as a neurotransmitter (a chemical released by nerve cells to send signals to other cells). It is involved in the inflammatory response and plays a central role as a mediator of pruritus (Andersen et al., 2015). Acetylcholine is a molecule that functions in the brain and body of many types of animals, including humans, as a neurotransmitter. Antihistamines inhibit the action of histamine at the histamine receptor (H1 receptor), whereas anticholinergic agents inhibit the action of acetylcholine at the muscarinic receptor. Both classes of drug limit stimulation of the vomiting center from the vestibular system (which is rich in histamine and acetylcholine) but have minimal effect on afferent visceral stimulation (Flake et al., 2004).

2.2 Dopamine antagonists

Dopamine is a molecule of the catecholamine and phenethylamine families that plays several important roles in the brain and body, including its functions as a neurotransmitter.

Dopamine antagonists minimize the effect of dopamine at the dopamine receptor D2, also known as D2R or D2 receptor in the chemoreceptor trigger zone, thereby limiting emetic input to the medullary vomiting center (Flake et al., 2004).

2.3 Serotonin antagonists

Serotonin or 5-hydroxytryptamine (5-HT) is a monoamine neurotransmitter. Selective serotonin antagonists inhibit the action of serotonin at the 5-hydroxytryptamine 3 (5-HT_3) receptor in the small bowel, vagus nerve, and chemoreceptor trigger zone. This action subsequently decreases afferent visceral and chemoreceptor trigger zone stimulation of the medullary vomiting center. Due to their diffuse blockade of serotonin, serotonin antagonists have become the primary treatment for a variety of causes of nausea. The 5-HT_3 antagonists are the newest and most expensive antiemetics (Flake et al., 2004).

3 ANTIEMETICS SCREENING METHODS

Several in vivo emesis models have been widely used even though nonanimal (in vitro) emesis models have also been reported.

3.1 In vivo emesis models

Several animal emesis models are available to evaluate the therapeutic potential of antiemetics. These emesis models include the following: ferret, mink, monkey, pig, dog (*Canis familiaris*), rat, chick, frog, house musk shrew, pigeon, and cat (Ahmed et al., 2013). Leopard and ranid frog models were the former screening methods in which antiemetics inhibited the action of orally induced emetic agent (copper sulfate pentahydrate) in leopard and ranid frogs (Kawai et al., 1994). The bioassay using frogs is useful in screening antiemetic activity in a small scale, but experimentation time is long (90 min) (Akita et al., 1998). Akita et al. have developed new assay method for screening the antiemetic properties of compounds from *Z. officinale* in young chicks instead of frogs (Akita et al., 1998; Prasad et al., 2015). This new method is easier, time economic with the observation is only consisting in counting the retching. Hence, using young chicks has advantages, such as easy handling, easy counting of retching motion, and short experimental time (30 min) (Akita et al., 1998). Other advantages include the possibility of having parallel results as well as decreasing standard errors (Prasad et al., 2015). This is because this model mimics acute emesis seen in man and serves as a useful model for evaluating the involvement of the brain in the observed antiemetic effects of botanicals (Tijani et al., 2008). The new method with young chicks has been established as a standard assay method for antiemetic effects. Young chicks also have some advantages including their easier breeding and their lower cost in comparison with other animal emesis models like pigeons (Akita et al., 1998).

3.2 In vitro emesis model (*Dictyostelium* Chemotaxis model)

Animals have been used as experimental models for centuries but ethical concerns, legislative changes, and the current economic climate encourage researchers to look for alternative nonanimal models. Animal experiments which can involve a significant level of suffering and distress due to the effects induced by emesis, for example, reduced food intake, weight loss, and dehydration have also been a major concern (Robinson, 2009). *Dictyostelium discoideum* chemotaxis model is one example of recently developed non-animal model in the analysis of the molecular effects of tastants, although it has limited utility in the identification of emetic agents in general. This in vitro chemotaxis model is simple, rapid, and inexpensive experimental design that serves as an early indicator of antiemetic compounds. The emetic response is investigated by observing the *Dictyostelium* cell behavior (shape, speed, and direction of movement) for 10 min. Usually emetic compounds blocked the motility of *Dictyostelium*. In this assay model, three bitter tasting compounds (denatonium benzoate, quinine hydrochloride, and phenylthiourea), the pungent constituent of chili peppers (capsaicin), stomach irritants (copper chloride and copper sulfate), and a phosphodiesterase IV inhibitor have been shown to strongly and rapidly block the motility of *Dictyostelium*. Such convenient in vitro bioassay models are likely to flourish in the future exploration of antiemetic agents (Robery et al., 2011; Ahmed et al., 2013).

4 IN VITRO AND IN VIVO ANTIEMETIC ACTIVITIES OF CHEMICALS

Few studies have been done in Africa on plants as source of antiemetic agents. However, some plants found in Africa with reported antiemetic activities are known and will be discussed in this section.

4.1 In vitro antiemetic activity of chemicals

To investigate the utility of employing *Dictyostelium* as a model for the study of tastants (denatonium, benzoate, phenylthiourea, quinine HCl, capsaicin, resiniferatoxin), cytotoxic agents (5-fluorouracil, actinomycin D, cisplatin cycloheximide, methotrexate, streptozotocin, vincristine), receptor agonists (5-HT, apomorphine HCl, veratridine HCl, substance P, nicotine, loperamide HCl)/antagonists (digoxin, rolipram, fluoxetine), and other emetic or aversive compounds, a standard assay is first defined (Robery et al., 2011). In this assay, *Dictyostelium* cell behavior is monitored by time lapse photography every 6 s over a 15 min period (under control conditions) within a chemotactic gradient (moving toward cAMP). Computer-generated outlines of individual cells enable the quantification of cell velocity, aspect, and angle of movement. These three measurements encapsulate the complete basic behavior of moving cells. In addition, an X, Y coordinate plot is provided illustrating the path length and direction of movement of individual cells throughout the recorded period. Under these conditions, cells exhibit stable behavior that do not significantly change over the 15 min period monitored (Robery et al., 2011). This standard assay enables the analysis of compounds with known emetic or adversive responses on *Dictyostelium* cell behavior. For each compound and concentration, at least triplicate experiments are recorded (monitoring approximately 30 cells each), establishing the behavior of cells for 5 min prior to compound addition. Following drug addition, images are then recorded for a further 10 min to monitor acute drug effects. The concentrations of compounds used in these tests are based upon concentrations used in vivo (e.g., copper sulfate), plasma concentrations (e.g., cisplatin), or concentrations shown to be active in vitro in mammalian tissues relevant to the emetic reflex [e.g., resiniferotoxin (RTX) on neurones, denatonium on intestinal epithelial cells]. A compound is then determined to have an effect on cell behavior if the average cell velocity or aspect changes significantly ($P < 0.05$) between the first 5 min period (prior to addition of the drug) and the final 5 min of the assay. Where a substance is without apparent effect at in vitro concentrations, experiments are then repeated to reduce the risk of obtaining a false-negative result (Robery et al., 2011).

4.2 In vivo antiemetic activities of chemicals

Methanol extracts are used most of the time to evaluate the antiemetic activity in young male chicks, after a period of fasting. Chemicals are emetogenic stimuli like cisplatin, copper (II) sulfate pentahydrate (copper sulfate) used as emetic drugs; metoclopramide, domperidone, and 3-(2-chloro-10*H*-phenothiazin-10-yl)-*N*,*N*-dimethyl-propan-1-amine (chlorpromazine) as standard antiemetic drugs. Dimethyl sulfoxide (DMSO), polyoxyethylene sorbitan monooleate (Tween 80) are solvents in which plant extracts are dissolved. For antiemetic activities, the chicks are usually divided into groups: the normal control group which receives saline solution; standard group which receives standard antiemetic drugs, and tested groups which receive plant extracts.

The antiemetic effect is determined by calculating the mean decrease in number of retching. Each chick is kept aside for a moment (10 min) to stabilize (at 38°C). After period of treatment (abdominary injection), emetic drugs (copper sulfate) are administered orally, and then the number of retching is recorded for 10 min. The percentage inhibition is calculated by the following formula:

$$\text{Inhibition (\%)} = [(A - B) / A] \times 100$$

where A is the frequency of retching in control group and B is the frequency of retching in test group.

For statistical analysis, antiemetic activity is expressed as mean ± standard error. The statistical significance of the difference is determined by an unpaired Student's *t*-test (Akita et al., 1998).

5 AFRICAN MEDICINAL SPICES AS SOURCES OF ANTIEMETICS

Spices are parts of the plant used for flavor, color, aroma, and preservation of food or beverages. They are different from herbs which refer to leafy and green part of the plant. Spices may be derived from many parts of the plant: bark, buds, flowers, fruits, leaves, rhizomes, roots, seeds, stigmas, and styles or the entire plant tops (Peter, 2001; Weingarten, 2011). Many spices reported for their antiemetic which can be found in Africa are listed in the subsequent sections.

5.1 *Aframomum melegueta*

A. melegueta commonly called *melegueta* pepper or grains of paradise belongs to the Zingiberaceae family and is native to West Africa where it has been found abundantly in Cameroon (Dzoyem et al., 2014). It is also an important cash crop in the Basketo district (Basketo special woreda) of southern Ethiopia. *A. melegueta* has been used in traditional herbal healing as either a stimulant or an antiemetic (bloom, 2016). However, scientific evidence of this activity has not been provided.

5.2 *Anethum graveolens*

A. graveolens L. is a popular plant widely used as a spice. It is an aromatic and annual herb of Apiaceae family used in Africa to treat emesis. *Anethum* seeds are used as spice and its fresh and dried leaves called dill weed are used as condiment and tea (Jana and Shekhawat, 2010; Ahmed et al., 2013). Phytochemical screenings of this plant showed that leaves, stems, and roots were rich in tannins, terpenoids, cardiac glycosides, and flavonoids. As the whole plant is used, there are various volatile components of *A. graveolens* seeds and herb: carvone is the predominant odorant of its seed while α-phellandrene, limonene, *A. graveolens* ether, and myristicin are the most important odorants of *A. graveolens* herb. Other compounds reported in the seeds are coumarins, flavonoids, phenolic acids, and steroids (Jana and Shekhawat, 2010). Although the plant is traditionally used as antiemetic scientific evidence is still to be provided.

5.3 *Mentha longifolia*

M. longifolia belonging to Lamiaceae family is also known as a wild mint. It is widely distributed throughout southern Africa (Petkar and Viljoen, 2008), Egypt, and Arabic countries and is frequently consumed in the form of hot beverage. *M. longifolia* is also used as spices in some food recipes (Al-Okbi et al., 2015; Bahtiti, 2015). The dried aerial parts of *M. longifolia* plant have antiemetic properties (Karousou et al., 2007; Bharat and Ajaikumar, 2009; Ahmed et al., 2012). Chemical composition of the plant revealed that the main constituent of essential oil is monoterpenes. Pulegone is the main compound of the plant responsible for most of its pharmacological effects along with menthone, isomenthone, menthol, 1,8-cineole, borneol, and piperitenone oxide (Mikaili et al., 2013).

5.4 Monodora myristica

M. myristica (Gaertn.) Dunal. commonly known as Calabash nutmeg is a tropical tree of the family Annonaceae. The seeds of *M. myristica*, chewed up are applied to the forehead for headache and for migraine in Gabon and ground up for headaches, rhino-pharygitis, or loss of voice, to apply on sores, or eaten as an antiemetic aperative and tonic in Congo. However, evidence of the antiemetic activity has not been scientifically provided. The aromatic seeds of *M. myristica* are also used as spices in Tropical Africa (Gambia to Sudan and Kenya, south to Angola and Tanzania) and also in Equatorial Africa as well as in Nigeria (Dzoyem et al., 2014).

5.5 Murraya koenigii

The Curry tree *M. koenigii* (L.) Spreng., a tropical to subtropical tree of the family Rutaceae, is a spice used in many countries in Africa including Nigeria (Kazeem et al., 2015) and Niger (Ndukwu and Ben-Nwadibia, 2016). Apart from many other properties, its leaves are used to relieve nausea and vomiting (Chowdhury et al., 2008). The major constituents responsible for the aroma and flavor have been reported as pinene, sabinene, caryophyllene, cadinol, and cadinene. In fact, the essential oils from the leaves of *M. koenigii* contained 39 compounds of which the most abundant is 3-carene followed by caryophyllene (Chowdhury et al., 2008). This plant is known to be the richest source of carbazole alkaloids (Rajendran et al., 2014).

5.6 Myristica fragrans

M. fragrans belongs to the Myristicaceae family; its seeds are used as spices in the South-Eastern part of Nigeria and are known as nutmeg (Skidmore-Roth, 2009; Chibuzor Okonkwo, 2014). It is one of the most commonly used spices in the world (Kim et al., 2013). It is widespread in Asia. Depending on the source, the essential oil of *M. fragrans* mainly has sabinene, α-pinene, and β-pinene, with myrcene, safrole, and terpinen, and also 1,8-cineole, myristicin, limonene (Chibuzor Okonkwo, 2014). The plant is used traditionally as an antiemetic (Skidmore-Roth, 2009; Chibuzor Okonkwo, 2014) but no scientific evidence of this effect has yet been published.

5.7 Ocimum species

The genus *Ocimum*, comprising more than 150 species, has been reported to grow widely throughout temperate regions of the world (Pandey et al., 2014). *Ocimum* species are widespread in Africa and India (Mbakwem-Aniebo et al., 2012). *O. basilicum* and *O. gratissimum* have been reported for their antiemetic activities (de Albuquerque et al., 2007; Ahmed et al., 2014).

O. basilicum Linn, with antiemetic flowers (de Albuquerque et al., 2007), is a spice from Lamiaceae family distributed in Africa and other continents (Abdullahi, 2011; Azhar et al., 2005; Pandey et al., 2014). The major chemical constituents of *O. basilicum* found in Africa (Cameroon, Egypt, Guinea, Mali, Nigeria, and Rwanda) are as follows: linalool, 1,8-cineole, eugenol, methyl chavicol, methyl cinnamate, (E)-α-bergamotene, thymol, methyl eugenol, and limonene (Pandey et al., 2014).

O. gratissimum L. (Lamiaceae) is an important herbal medicinal plant in Kenyan communities as well as in other sub-Saharan African countries (Matasyoh et al., 2008). In West Africa, *O. gratissimum* is commonly found around village huts and gardens (Iwu, 1993) and cultivated for medicinal and culinary purposes. The leaves have strong aromatic odor and are popularly used to flavor soup (Ijeh et al., 2005) and meat (Okoli et al., 2010). The antiemetic activity of the whole plant has been reported (Ahmed et al., 2014). The main constituents of essential oil of African (Cameroon and Rwanda) *O. gratissimum* include β-phellandrene, limonene, γ-terpinene, thymol, *p*-cymene, and eugenol (Ntezurubanza et al., 1987; Tchoumbougnang et al., 2005).

Within the *Ocimum* species, there is a clear variation in their chemical composition. The major constituents which have been isolated from different *Ocimum* oils include 1,8-cineol, linalool, pinene, eugenol, camphor, methyl chavicol, ocimene, terpinene, and limonene (Pandey et al., 2014).

5.8 *Piper nigrum*

Black pepper of commerce is the dried mature fruits of the tropical, perennial climbing plant *P. nigrum* L. which belongs to the family Piperaceae (Sruthi et al., 2013). It is used as spice and condiment around the Niger Delta area (Africa) (Ndukwu and Ben-Nwadibia, 2016) and possess antiemetic properties (Hasan et al., 2012a; Samydurai et al., 2012). The volatile oil and pungent compounds are the two main components of black pepper. The alkaloid, piperine, is the major contributor to pungency whereas essential oil constituents like α- and β-pinene, limonene, myrcene, linalool, α-phellandrene, sabinene, β-caryophyllene, germacrene D, and so on, are the major aroma and flavor compounds of pepper (Sruthi et al., 2013).

5.9 *Solanum* species

Solanum species belong to the family Solanaceae, with over 1500 species worldwide and at least 100 indigenous species spread across Africa. About 25 species of *Solanum* have been reported in Nigeria (Gbile and Adesina, 1988), some of which are wild while others are cultivated with their leaves, fruits, or both consumed as raw or cooked vegetables or used as folklore medicine (Edijala et al., 2005). *S. melongena* L., also called eggplant or aubergine, grows in a number of African countries and many other parts of the world (Kalloo, 1993). Leaves and fruits have antiemetic activities. In Cameroon, fruits are used as spices (Abdou Bouba et al., 2010; Ahmed et al., 2014) and also as vegetables (Banu Naujeer, 2009). Phytochemical analyses showed the presence of alkaloids, flavonoids, tannins, steroids, and glycosides in callus, root, or fruit extracts (Tiwari et al., 2009; Ghoson, 2015).

5.10 Syzygium aromaticum

S. aromaticum L. Merr. & Perry, a spice commonly known as clove, is a plant of Myrtaceae family also found in Africa (Abdullahi, 2011; Olalekan et al., 2003). Flowering buds showed antiemetic properties. This medicinal spice is discussed in detail in Chapter 29.

5.11 The genus *Thymus*

The genus *Thymus* belongs to the Lamiaceae family. This genus is distributed in the Old World and on the coasts of Greenland, from the Macaronesian Region, Northern Africa and the Sinai Peninsula, through the West and East Asia (Ghannadi et al., 2004). *T. transcaspicus* and *T. vulgaris* are antiemetic spices found in Africa. *T. vulgaris* is discussed in detail in Chapter 28. *T. transcaspicus* is an aromatic and medicinal plant, which has been widely distributed in the north of Africa. Thymol, carvacrol, γ-terpinene, and *p*-cymene constitute the major components of the oil of the aerial parts of this plant (Moallem et al., 2009; Arzani and Motamedi, 2013). *T. transcaspicus* extracts from aerial parts revealed antiemetic effects due to the peripheral and central mechanisms (Moallem et al., 2009).

5.12 Xylopia aethiopia

X. aethiopia or "African Pepper" is an African spice of Annonaceae family with antiemetic properties (Ijeh et al., 2005; Juliani et al., 2008). Monoterpene are the major constituents of the plant oil. It is reported that β-pinene is predominant in all parts of the plant, whereas trans-*m*-mentha-1(7),8-diene is the main compound in the essential oils of the leaves, roots, and stems barks (Karioti et al., 2004).

5.13 Zingiber officinale

Ginger, the rhizome (underground stem) of *Z. officinale*, is one of the most widely used species of the ginger family (Zingiberaceae) and is widely used around the world in foods as a spice (Mishra et al., 2012). In Africa, it is commonly used in many dishes (OParaeke et al., 2004; Abdullahi, 2011). The main constituents are sesquiterpenoids with zingiberene as the major component (Rajsekhar et al., 2012). The components in ginger that are responsible for the antiemetic effect are thought to be the gingerols, shogaols, and galanolactone, a diterpenoid of ginger (Yamahara et al., 1989; Huang et al., 1991; Bhattarai et al., 2001). Recent animal models and in vitro studies have demonstrated that ginger extract possesses antiserotoninergic 5-HT_3 receptor antagonism effects, which play an important role in the etiology of postoperative nausea and vomiting (Huang et al., 1991; Yamahara et al., 1989; Lumb, 1993). *Z. officinale* is discussed in detail in Chapter 30.

6 AFRICAN VEGETABLES AS SOURCES OF ANTIEMETICS

Vegetables refer to edible plants, commonly collected and/or cultivated for their nutritional value for humans. Most often, the botanical definition of vegetables is the "edible part of the plant" (Agudo, 2004). Some African vegetables have been reported to possess antiemetic properties.

6.1 *Abrus precatorius*

Leaves of *Abrus precatorius subsp. africanus* are consumed as a vegetable in central and east Africa. *A. precatorius subsp. africanus,* a plant of the Fabaceae family, is a common plant in Kwazulu-Natal and Limpopo Province (South Africa) and native to many tropical areas of the world (Masupa, 2009). Traditionally, leaves, roots, and seeds of *A. precatorius* L. which is in the same family with *A. precatorius subsp. Africanus* are used as antiemetic herbal medicine (Gul et al., 2013). *A. precatorius* L. seeds have been used for treating emesis in China (Ahmed et al., 2013). Several compounds like abrine, trigonelline, abruslactone A, hemiphloin, abrusoside A, abrusoside B, abrusoside C, abrusoside D, arabinose, galactose, xylose, choline, hypaphorine, precatorine, glycyrrhizin, montanyl alcohol, inositol, D monomethyl ether, and pinitol have been identified in the leaves of *A. precatorius* (Garaniya and Bapodra, 2014).

6.2 *Acalypha species*

Natives to Africa, *Acalypha* species have been reported to possess antiemetic activities (Seebaluck et al., 2015). *Acalypha* species viz. and *A. ornata* Hochst (Euphorbiaceae) are used in Tanzania as vegetables. They are widespread from Nigeria to Eritrea and in Angola, Namibia, Botswana, Zimbabwe, and Mozambique (Schmelzer and Gurib-Fakim, 2008; Pickering and Roe, 2009; Quds et al., 2012). Its leaves contain flavonoids and phenols (Ahmed and Onocha, 2013). *Acalypha* species contain flavonoids and terpenes which are reported as active principles against emesis in chick emesis model (Quds et al., 2012). Their possible antiemetic mechanism in animal models might be via $5-HT_3$, $5-HT_4$, and/or NK1 receptors antagonism (Ahmed et al., 2014).

6.3 *Adenanthera pavonina*

Endemic to Asia, *A. pavonina* is found in western and eastern Africa (Adedapo et al., 2009) and even Central Africa (DR Congo) (Opota Onya et al., 2013). Its leaves having antiemetic properties (Hasan et al., 2012b) have already been spread (Ahmed et al., 2014). *A. pavonina* (L.) is tropical deciduous tree of the Leguminosae family (Zarnowskia et al., 2004). It is known as a food tree because its seeds and young leaves are cooked and eaten (Ezeagu et al., 2004). Chemical literature survey of *A. pavonina* revealed the presence of triterpenes, flavonoids, and fatty acids (Hasan et al., 2012b). The antiemetic mode of action is not known. However, based on previous studies, the leave extracts could have a peripheral antiemetic action. However, it has been postulated that alkaloidal contents may play some role in the antiemetic effect of the plant. Nonetheless, further studies are required to determine the exact mode of action and the active compounds responsible for this effect (Hasan et al., 2012b).

6.4 *Afzelia africana*

A. africana belongs to the family Fabaceae and has been reported to possess antiemetic effect (Ahmed et al., 2013). The English name of the plant is mahogany. The tree is widely distributed in Africa and Asia, where it is used as food and for plank wood; it is also used as folklore remedies among many tribes in Africa (Aiyegoro et al., 2011). Leaves of most *Afzelia* species are eaten as vegetable (Donkpegan et al., 2014).

6.5 *Polygonum lapathifolium*

P. lapathifolium, a plant of the Polygonaceae family, is found in South Africa (Bulbul et al., 2013), North Africa (Egypt, Morocco, and Tunisia), and other parts of the world and used as a medicinal plant (Holm et al., 1979). It is eaten as a vegetable or salad by local people in Andhra Pradesh, India (Singh and Jain, 2003; Choudhary et al., 2011; Bulbul et al., 2013). Flowers extracts of *P. lapathifolium* have been reported to possess antiemetic activity. Phytochemical screening showed that flowers extract contain alkaloids, phytosterols, diterpenes and triterpenes, amino acids and proteins, flavonoids, and phenolic compounds (Bulbul et al., 2013).

6.6 *Solanum aethiopicum*

The scarlet eggplant *S. aethiopicum* L. (Solanaceae) grows mostly in Africa for its fruits and leaves. *S. aethiopicum* is an antiemetic plant (Ahmed et al., 2014) which is a delicacy in the eastern part of Nigeria. The immature fruits of *S. aethiopicum* L. are used as cooked vegetables in stews and sometimes eaten raw. The leaves and shoots are used as cooked vegetables (Ibiam and Nwigwe, 2013; Chinedu et al., 2011; Ahmed et al., 2014). There is a significant presence of alkaloids, saponins, flavonoids, tannins, and ascorbic acid in its fruits; terpenoids are found in trace amount as well as steroids (Chinedu et al., 2011).

7 CONCLUSIONS

Nausea and vomiting are uncomfortable phenomena which can be overcome using antiemetics. In spite of the rapid growth in the development and spread of modern medicine, traditional knowledge on the use of plants continues to be the most popular alternative to health problems in Africa. In the present review, we reported some antiemetics spices found in Africa, such as *A. melegueta*, *A. graveolens*, *M. longifolia*, *M. myristica*, *M. koenigii*, *M. fragrans*, *O. basilicum*, *O. gratissimum*, *P. nigrum*, *S. melongena*, *S. aromaticum*, *T. transcaspicus*, *T. vulgaris*, *X. aethiopia*, and *Z. officinale*. Antiemetic African vegetables are also reported; they include *A. precatorius*, *A. ornata*, *A. pavonina*, *A. Africana*, *P. lapathifolium*, and *S. aethiopicum*. In summary, the screening of the antiemetic activity of African medicinal plants in general needs to be intensified and the present review can boost scientists worldwide to focus more on the flora of the continent.

REFERENCES

Abdou Bouba, A., Njintang, Y.N., Scher, J., Mbofung, C.M.F., 2010. Phenolic compounds and radical scavenging potential of twenty Cameroonian spices. Agric. Biol. J. N. Am. 1 (3), 213–224.

Abdullahi, M., 2011. Biopotency role of culinary spices and herbs and their chemical constituents in health and commonly used spices in Nigerian dishes and snacks. Afr. J. Food Sci. 5 (3), 111–124.

Adedapo, A.D.A., Osude, Y.O., Adedapo, A.A., Olanrewaju Moody, J., Adeagbo, A.S., Olajide, O.A., 2009. Blood pressure lowering effect of *Adenanthera pavonina* seed extract on normotensive rats. Rec. Nat. Prod. 3 (2), 82–89.

Agudo, A., 2005. Measuring intake of fruit and vegetables: background paper for Join FAA/WHO workshop on Fruit and Vegetables for Health, September 1–3, 2004, Kobe, Japan.

Ahmed, S., Hasan, M.M., Ahmed, S.W., 2014. Natural antiemetics: an overview. Pak. J. Pharm. Sci. 27 (5 Special), 1583–1598.

Ahmed, S., Hasan, M.M., Ahmed, S.W., Mahmood, Z.A., Azhar, I., Habtemariam, S., 2013. Anti-emetic effects of bioactive natural products. Phytopharmacology 4 (2), 390–433.

Ahmed, S., Onocha, P.A., 2013. Antiemetic activity of *Tithonia diversifolia* (HEMSL.) A. Gray leaves in copper sulfate induced chick emesis model. Am. J. Phytomed. Clin. Ther. 1, 734–739.

Ahmed, S., Zahid, A., Abidi, S., Meer, S., 2012. Anti-emetic activity of four species of Genus Cassia in chicks. IOSR J. Pharm. 2 (3), I380–I384.

Aiyegoro, O., Adewusi, A., Oyedemi, S., Akinpelu, D., Okoh, A., 2011. Interactions of antibiotics and methanolic crude extracts of *Afzelia Africana* (Smith.) against drug resistance bacterial isolates. Int. J. Mol. Sci. 12 (7), 4477–4503.

Akita, Y., Yang, Y., Kawai, T., Kinoshita, K., Koyama, K., Takahashi, K., 1998. New assay method for surveying anti-emetic compounds from natural sources. Nat. Prod. Sci. 4 (2), 72–77.

Al-Okbi, S.Y., Fadel, H.H.M., Mohamed, D.A., 2015. Phytochemical constituents, antioxidant and anticancer activity of *Mentha citrata* and *Mentha longifolia*. Res. J. Pharm. Biol. Chem. Sci. 6 (1), 739–751.

Andersen, H.H., Elberling, J., Arendt-Nielsen, L., 2015. Human surrogate models of histaminergic and non-histaminergic itch. Acta Derm. Venereol. 95 (7), 771–777.

Arzani, H., Motamedi, 2013. Chemical compounds of Thyme as a medicinal herb in the mountainous areas of Iran. J. Nutr. Disorders Ther. S12, 003.

Atal, C.K., Kapur, B.M., 1989. Cultivation and Utilization of Aromatic Plants. PID, India CSIR.

Azhar, A.F., Sreeramu, B.S., Srinivasappa, K.N., 2005. Cultivation of Spice Crops. Universities Press, India.

Bahtiti, N.H.A., 2015. Biological activities of the methanol extract of cultivated Jordanian fresh and dried mint species (*Mentha Longifolia*). Int. J. Res. Stud. Biosci. 3 (1), 205–210.

Banu Naujeer, H., 2009. Morphological diversity in eggplant (*Solanum melongena* L.), their related species and wild types conserved at the National gene bank in Mauritius. (Master International Master Programme), Swedish Biodiversity Centre, Sweden.

Ben-Erik van Wyk, M.W., 2004. Medicinal Plants of the World, first ed. Briza Publications, Pretoria.

Bharat, B.A., Ajaikumar, B.K., 2009. Molecular Targets and Therapeutic Uses of Spices: Modern Uses for Ancient Medicine, first ed. World Scientific, Singapore.

Bhattarai, S., Tran, V.H., Duke, C.C., 2001. The stability of gingerol and shogaol in aqueous solutions. J. Pharm. Sci. 90 (10), 1658–1664.

bloom, c. 2016. *Aframomum melegueta*. Available from: http://crescentbloom.com/plants/specimen/AE/Aframomum%20melegueta.htm

Bulbul, L., Uddin, J.M., Sushanta, S.M., Roy, J., 2013. Phytochemical screening, anthelmintic and antiemetic activities of *Polygonum lapathifolium* flower extract. Eur. J. Med. Plants 3 (3), 333–344.

Chibuzor Okonkwo, A.O., 2014. Nutritional evaluation of some selected spices commonly used in the south-eastern part of Nigeria. J. Biol. Agric. Healthcare 15, 4.

Chinedu, S.N., Olasumbo, A.C., Eboji, O.K., Emiloju, O.C., Arinola, O.K., Dania, D.I., 2011. Proximate and phytochemical analyses of *Solanum aethiopicum* L. and *Solanum macrocarpon* L fruits. Res. J. Chem. Sci. 1 (3), 63–71.

Chowdhury, J.U., Bhuiyan, M.N.I., Yusuf, M., 2008. Chemical composition of the leaf essential oils of *Murraya koenigii* (L.) Spreng and Murraya paniculata (L.) Jack. Bangladesh J. Pharmacol. 3, 59–63.

Choudhary, R.K., Oh, S., Lee, J., 2011. An ethnomedicinal inventory of knotweeds of Indian Himalaya. J. Med. Plants Res. 5 (10), 2095–2103.

Coronas, R., Pitarch, L., Mallol, J., 1975. Blockade of reserpine emesis in pigeons by metoclopramide. Eur. J. Pharmacol. 32 (02), 380–382.

de Albuquerque, U.P., Muniz de Medeiros, P., de Almeida, A.L., Monteiro, J.M., Machado de Freitas Lins Neto, E., Gomes de Melo, J., 2007. Medicinal plants of the caatinga (semi-arid) vegetation of NE Brazil: a quantitative approach. J. Ethnopharmacol. 114 (3), 325–354.

Donkpegan, A.S.L., Hardy, O.J., Lejeune, P., Oumorou, M., Daïnou, K., Doucet, J.L., 2014. [Un complexe d'espèces d'Afzelia des forêts africaines d'intérêt économique et écologique (synthèse bibliographique)]. Biotechnol. Agron. Soc. Environ. 18 (2), 233–246.

Dzoyem, J.P., Tchuenguem, R.T., Kuiate, J.R., Teke, G.N., Kechia, F.A., Kuete, V., 2014. In vitro and in vivo antifungal activities of selected Cameroonian dietary spices. BMC Complement. Altern. Med. 14, 58.

Edijala, J.K., Asagba, S.O., Eriyamremu, G.E., Atomatofa, U., 2005. Comparative effect of garden egg fruit, oat and apple on serum lipid profile in rats fed a high cholesterol diet. Pak. J. Nutr. 4 (4), 245–249.

Ezeagu, I.E., Gopal Krishna, A.G., Khatoon, S., Gowda, L.R., 2004. Physico-chemical characterization of seed oil and nutrient assessment of *adenanthera pavonina*, l: an underutilized tropical legume. Ecol. Food Nutr. 43 (4), 295–305.

Flake, Z.A., Scalley, R.D., Bailey, A.G., 2004. Practical selection of antiemetics. Am. Fam. Physician 69 (5), 1169–1174.

Garaniya, N., Bapodra, A., 2014. Ethno botanical and phytopharmacological potential of *Abrus precatorius* L.: a review. Asian Pac. J. Trop. Biomed. 4 (Supplement 1), S27–S34.

Gbile, Z.O., Adesina, S.K., 1988. Nigerian Solanum species of economic importance. Ann. MO Bot. Gard. 75, 862–865.

Ghannadi, A., Sajjadi, S.E., Kaboucheb, A., Kabouche, Z., 2004. Thymus fontanesii Boiss. & Reut. A potential source of thymol-rich essential oil in North Africa. Z. Naturforsch. 59c, 187–189.

Ghoson, S.S., 2015. Chemical detection of some active compounds in egg plant (*Solanum melongena*) callus as compared with fruit and root contents. Int. J. Curr. Microbiol. Appl. Sci. 4 (5), 160–165.

Gul, M.Z., Ahmad, F., Kondapi, A.K., Qureshi, I.A., Ghazi, I.A., 2013. Antioxidant and antiproliferative activities of Abrus precatorius leaf extracts: an in vitro study. BMC Complement. Altern. Med. 13, 53.

Hasan, M.M., Ahmed, S., Ahmed, Z., Azhar, I., 2012a. Antiemetic activity of some aromatic plants. J. Pharm. Sci. Innov. 1 (1), 47.

Hasan, M.U.M., Azhar, I., Muzammil, S., Ahmed, S., Ahmed, S.W., 2012b. Anti-emetic activity of some leguminous plants. Pak. J. Bot. 44 (1), 389–391.

Holm, L., Pancho, J.V., Herberger, J.P., Plucknett, D.L., 1979. A Geographical Atlas of World Weeds. Wiley Interscience, J. Wiley & Sons, New York.

Hornby, P.J., 2001. Central neurocircuitry associated with emesis. Am. J. Med. 111 (Suppl 8A), 106s–112s.

Huang, Q., Iwamoto, M., Aoki, S., 1991. Anti-5-hydroxytryptamine3, effect of galanolactone, diterpenoid isolated from ginger. Chem. Pharm. Bull. 39 (2), 397–399.

Ibiam, O.F.A., Nwigwe, I., 2013. The effect of fungi associated with leaf blight of *Solanum aethiopicum* L. in the field on the nutrient and phytochemical composition of the leaves and fruits of the plant. Plant Pathol. Microbiol. 4, 7.

Ijeh, I.I., Omodamiro, O.D., Nwanna, I.J., 2005. Antimicrobial effects of aqueous and ethanolic fractions of two spices, *Ocimum gratissimum* and *Xylopia aethiopica*. Afr. J. Biotechnol. 4 (9), 953–956.

Iwu, M.M., 1993. Handbook of African Medicinal Plants, second ed. CRC Press, Boca Raton, FL.

Jana, S., Shekhawat, G.S., 2010. *Anethum graveolens*: an Indian traditional medicinal herb and spice. Pharmacogn. Rev. 4 (8), 179–184.

Juliani, R.H., Kwon, T., Koroch, A.R., Asante-Dartey, J., Acquaye, D., Simon, J.E., 2008. *Xylopia aethiopia* (Annonaceae): Chemistry, traditional uses and functional properties of an "African pepper." ACS Symposium Series.

Kalloo, G., 1993. Eggplant: *Solanum melongena* L. In: Bergh, O., Kalloo, G. (Eds.), Genetic Improvement of Vegetable Crops. Pergamon, Amsterdam, pp. 587–604.

Karioti, A., Hadjipavlou-Litina, D., Mensah, M.L.K., Fleischer, T.C., Skaltsa, H., 2004. Composition and antioxidant activity of the essential oils of *Xylopia aethiopica* (Dun) A. Rich. (Annonaceae) leaves, stem bark, root bark, and fresh and dried fruits, growing in Ghana. J. Agric. Food Chem. 52 (26), 8094–8098.

Karousou, R., Balta, M., Hanlidou, E., Kokkini, S., 2007. "Mints", smells and traditional uses in Thessaloniki (Greece) and other Mediterranean countries. J. Ethnopharmacol. 109 (2), 248–257.

Kawai, T., Kinoshita, K., Koyama, K., Takahashi, K., 1994. Anti-emetic principles of Magnolia obovata bark and *Zingiber officinale* rhizome. Planta Med. 60 (1), 17–20.

Kazeem, M.I., Ashafa, A.O.T., Nafiu, M.O., 2015. Biological activities of three Nigerian spices—*Laurus nobilis* Linn, *Murraya koenigii* (L) Spreng and *Thymus vulgaris* Linn. Trop. J. Pharm. Res. 14 (12), 2255–2261.

Kim, H., Bu, Y., Lee, B.J., Bae, J., Park, S., Kim, J., 2013. Myristica fragrans seed extract protects against dextran sulfate sodium-induced colitis in mice. J. Med. Food 16 (10), 953–956.

Lumb, A.B., 1993. Mechanism of antiemetic effect of ginger. Anaesthesia 48 (12), 1118.

Masupa, T.T., 2009. *Abrus precatorius* L. subsp. *africanus* Verdc. In *Abrus precatorius* L. subsp. *africanus* Verdc. Kew Bull. **24**, 1970, 235.

Matasyoh, L.G., Matasyoh, J.C., Wachira, F.N., Kinyua, M.G., Muigai, A.W.T., Mukiama, T.K., 2008. Antimicrobial activity of essential oils of *Ocimum gratissimum* L. from different populations of Kenya. Afr. J. Tradit. Complement. Altern. Med. 5 (2), 187–193.

Mbakwem-Aniebo, C., Onianwa, O., Okonko, I.O., 2012. Effects of *Ocimum Gratissimum* leaves on common dermatophytes and causative agent of pityriasis versicolor in Rivers State, Nigeria. J. Microbiol. Res. 2 (4), 108–113.

Mikaili, P., Mojaverrostami, S., Moloudizargari, M., Aghajanshakeri, S., 2013. Pharmacological and therapeutic effects of *Mentha Longifolia* L. and its main constituent, menthol. Anc. Sci. Life 33 (2), 131–138.

Miner, W.D., Sanger, G.J., 1986. Inhibition of cisplatin-induced vomiting by selective 5-hydroxytryptamine M-receptor antagonism. Br. J. Pharmacol. 88 (3), 497–499.

Mishra, R.K., Kumar, A., Kumar, A., 2012. Pharmacological activity of *Zingiber officinale*. Int. J. Pharm. Chem. Sci. 1 (3), 1422–1427.

Moallem, S.A., Farahmand-Darzab, A., Sahebkar, A., Iranshahi, M., 2009. Antiemetic activity of different extracts from the aerial parts of *Thymus transcaspicus* Klokov. Pharmacologyonline 1, 1284–1292.

Ndukwu, B.C., Ben-Nwadibia, N.B., 2016. Ethnomedicinal aspects of plants used as spices and condiments in the Niger delta area of Nigeria. Available from: http://www.ethnoleaflets.com/leaflets/niger.htm

Ntezurubanza, L., Scheffer, J.J., Svendsen, A.B., 1987. Composition of the essential oil of *Ocimum gratissimum* grown in Rwanda1. Planta Med. 53 (5), 421–423.

Okoli, C.O., Ezike, A.C., Agwagah, O.C., Akah, P.A., 2010. Anticonvulsant and anxiolytic evaluation of leaf extracts of *Ocimum gratissimum*, a culinary herb. Pharmacognosy Res. 2 (1), 36–40.

Olalekan, D.A., Abdulkareem, A.S., Idris, B.M., Odigure, J.O., 2003. Microbial, chemical composition evaluation and development of a technological process for the production of compound spices in Nigeria. Nat. Prod. Rad. 2 (6), 314–320.

OParaeke, A.M., Dike, M.C., Amatobi, C.I., 2004. Field evaluation of extracts of five nigerian spices for control of post-flowering insect pests of cowpea, *Vigna unguiculata* (L.) Walp. Plant Protect Sci. 41 (1), 14–20.

Opota Onya, D., Senga Kitumbe, P., Covaci, A., Cimanga Kanyanga, R., 2013. Physico-chemical study of *Adenanthera pavonina* seed oil growing in Democratic Republic of Congo. Int. J. Pharm. Tech. Res. 5 (4), 1870–1881.

Pandey, A.K., Singh, P., Tripathi, N.N., 2014. Chemistry and bioactivities of essential oils of some Ocimum species: an overview. Asian Pac. J. Trop. Biomed. 4 (9), 682–694.

Peter, K.V., 2001. Handbook of Herbs and Spices. Woodhead Publishing Limited and CRC Press LLC, Cambridge England.

Petkar, S., Viljoen, A., 2008. Indigenous South African medicinal plants. Part 8: *Mentha longifolia* (Wild mint). SA Pharm. J. 64, 1.

Pickering, H., Roe, E., 2009. Wild Flowers of the Victoria Falls Area. Helen Pickering, London.

Prasad, S.S., Desai, A., Shah, C., Vajpeyee, S.K., Bhavsar, V.H., 2015. Effect of ginger on isolated intestine of rat establishing it s action on gastro-intestinal motility and peristalsis. Int. J. Curr. Res. Biosci. Plant Biol. 2 (10), 92–97.

Quds, T., Ahmed, S., Shaiq Ali, M., Onocha, P.A., Azhar, I., 2012. Antiemetic activity of Acalypha fimbriata Schumach. & Thonn., *Acalypha ornata* Hochst., and *Acalypha wilkesiana* cv. godseffiana Muell Arg. Phytopharmacology 3 (2), 335–340.

Qureshi, Z.K., 2012. Evaluation of Some Indigenous Herbal Plant Extracts for their Antiemetic Activity. PhD thesis, Baqai Medical University, Karachi.

Rajendran, M.P., Pallaiyan, B.B., Selvaraj, N., 2014. Chemical composition, antibacterial and antioxidant profile of essential oil from *Murraya koenigii* (L.) leaves. Avicenna J. Phytomed. 4 (3), 200–214.

Rajsekhar, S., Kuldeep, B., Chandaker, A., Upmanyu, N., 2012. Spices as antimicrobial agents: a review. Int. Res. J. Pharm. 3 (2), 4–9.

Ray, A.P., Chebolu, S., Ramirez, J., Darmani, N.A., 2009. Ablation of least shrew central neurokinin NK1 receptors reduces GR73632-induced vomiting. Behav. Neurosci. 123 (3), 701–706.

Robery, S., Mukanowa, J., Percie du Sert, N., Andrews, P.L., Williams, R.S., 2011. Investigating the effect of emetic compounds on chemotaxis in Dictyostelium identifies a non-sentient model for bitter and hot tastant research. PLoS One 6 (9), e24439.

Robinson, V., 2009. Less is more: reducing the reliance on animal models for nausea and vomiting research. Br. J. Pharmacol. 157 (6), 863–864.

Samydurai, P., Jagatheshkumar, S., Aravinthan, V., Thangapandian, V., 2012. Survey of wild aromatic ethnomedicinal plants of Velliangiri Hills in the Southern Western Ghats of Tamil Nadu, India. Int. J. Med. Arom. Plants 2 (2), 229–234.

Schmelzer, G.H., Gurib-Fakim, A., 2008. Plant Resources of Tropical Africa 11(1). Medicinal plants 1. PROTA Foundation, Wageningen.

Seebaluck, R., Gurib-Fakim, A., Mahomoodally, F., 2015. Medicinal plants from the genus Acalypha (Euphorbiaceae)—a review of their ethnopharmacology and phytochemistry. J. Ethnopharmacol. 159, 137–157.

Shivhare, Y., 2011. Medicinal plants as source of antiemetic agents: a review. Asian J. Pharm. Technol. 1 (2), 25–27.

Siddiqui, H.H., 1993. Safety of herbal drugs—an overview. Drugs News Views 1 (2), 7–10.

Singh, V., Jain, A.P., 2003. Ethnobotany and Medicinal Plants of India and Nepal. vol. 2. Scientific Publishers, Jodhpur, India.

Skidmore-Roth, L., 2009. Mosby's Handbook of Herbs & Natural Supplements, fourth ed. Elsevier Health Sciences, London.

Sruthi, D., Zachariah, T.J., Leela, N.K., Jayarajan, K., 2013. Correlation between chemical profiles of black pepper (*Piper nigrum* L.) var. Panniyur-1 collected from different locations. J. Med. Plants Res. 7 (31), 2349–2357.

Tchoumbougnang, F., Zollo, P.H., Dagne, E., Mekonnen, Y., 2005. In vivo antimalarial activity of essential oils from *Cymbopogon citratus* and *Ocimum gratissimum* on mice infected with *Plasmodium berghei*. Planta Med. 71 (1), 20–23.

Tijani, A.Y., Okhale, S.E., Oga, F.E., Tags, S.Z., Salawu, O.A., Chindo, B.A., 2008. Anti-emetic activity of Grewia lasiodiscus root extract and fractions. Afr. J. Biotechnol. 7 (17), 3011–3016.

Tiwari, A., Jadon, S.R., Tiwari, P., Nayak, S., 2009. Phytochemical investigations of crown of *Solanum melongena* fruit. Int. J. Phytomed. 1, 9–11.

Weingarten, H., 2011. What's the difference between herbs and spices? Available from: http://blog.fooducate.com/2011/10/19/whats-the-difference-between-herbs-and-spices/

Yamahara, J., Rong, H.Q., Iwamoto, M., 1989. Active components of ginger exhibiting anti-serotonergic action. Phytother. Res. 3 (2), 70–71.

Zarnowskia, R., Jaromin, A., Certikc, M., Czabanyc, T., Fontained, J.L., Jakubike, T., 2004. The oil of *Adenanthera pavonina* L. seeds and its emulsions. Z. Naturforsch. 59c, 321–326.

CHAPTER 12

African Medicinal Spices and Vegetables and Their Potential in the Management of Metabolic Syndrome

V. Kuete

1 INTRODUCTION

Metabolic syndrome (MetS) can be defined as clustering of at least three of five of the following medical conditions: abdominal (central) obesity, elevated fasting plasma glucose, elevated blood pressure, high serum triglycerides, and low high-density lipoprotein (HDL) levels. MetS is associated with a twofold increased risk of cardiovascular disease and a fivefold risk of diabetes mellitus (Nikbakht-Jam et al., 2015). Hence, insulin resistance, visceral adiposity, atherogenic dyslipidemia, endothelial dysfunction, genetic susceptibility, elevated blood pressure, hypercoagulable state, and chronic stress are the several factors which constitute MetS (Kaur, 2014). Specifically, the National Cholesterol Education Program (NCEP) defines the MetS as having three or more of the following five cardiovascular risk factors: (1) central obesity (waist circumference: men >102 cm; women >88 cm); (2) elevated triglycerides ($\geq$150 mg/dL); (3) diminished HDL cholesterol (men <40 mg/dL; women <50 mg/dL); (4) systemic hypertension ($\geq$130/$\geq$85 mmHg); and (5) elevated fasting glucose ($\geq$110 mg/dL) (Mottillo et al., 2010). However, in 2004, this NCEP definition was revised (rNCEP) by lowering the threshold for fasting glucose to $\geq$100 mg/dL in concordance with American Diabetes Association criteria for impaired fasting glucose (Grundy et al., 2004; Mottillo et al., 2010). Also, thresholds for central obesity were lowered from strictly >102 cm in men and 88 cm in women to greater than or equal to these values. Finally, the rNCEP definition includes patients being treated for dyslipidemia, hyperglycemia, or systemic hypertension (Grundy et al., 2004). MetS is also associated with a twofold increase in mortality from cardiovascular disease and a 1.5-fold increase in mortality from other causes (Mottillo et al., 2010; Nikbakht-Jam et al., 2015). Worldwide prevalence of MetS ranges from <10% to as much as 84%, depending on the region, urban or rural environment, composition (sex, age, race, and ethnicity) of the population studied, and the definition of the syndrome used (Kaur, 2014). In general, the International Diabetes Federation (IDF) estimates that one-quarter of the world's adult population has the MetS (Kaur, 2014). MetS affects approximately one-quarter of North Americans and has become a leading health concern due to its link to cardiovascular disease (Mottillo et al., 2010). In a population sample aged from 20 to 25 and upward, the prevalence varied from 8% (India) to 24% (United States) in men and from 7% (France) to 46% (India) in women (Cameron et al., 2004). In Africa, the prevalence is increasing, and it tends to increase with age. This increase in the prevalence of MetS in the continent is not limited to adults but is also becoming common among the young ones (Okafor, 2012). Obesity and dyslipidemia seem to be the most common occurring components. While obesity appears more common in females, hypertension tends to be more predominant in males. Insulin resistance has remained the key underlying pathophysiology. The prevalence of MetS in African populations ranges from as low as 0% to as high as about 50% or even higher depending on the population setting (Okafor, 2012). Based on various reports, prevalence are 24.8–30.3% in Seychelles, 20.5–63.6% in Nigeria, 12.5–17.9% in Ethiopia, 18.9–27.1% in Congo, and 34% in Botswana (Okafor, 2012). The management of MetS is essentially through lifestyle modification, and drug therapy, or bariatric surgery if associated with severe obesity (Nikbakht-Jam et al., 2015). Though drugs are available to treat the different components of the syndrome, prevention is still possible by reverting to the traditional African way of life (Okafor, 2012). Clinically, the individual disorders that compose MetS are treated separately; diuretics and angiotensin-converting-enzyme (ACE) inhibitors may be used to treat hypertension. Cholesterol drugs may be used to lower low density lipoprotein (LDL) cholesterol and triglyceride levels, if they are elevated, and to raise HDL levels if they are low. Drugs that decrease insulin resistance, for example, metformin and thiazolidinediones can also be used but this is controversial; weight loss medications should be applied; weight loss and lifestyle changes in diet as well as physical activity are recommended to overcome obesity. Herbal medicines can also be combined with conventional drugs for the treatment of many conditions. In this chapter, we will discuss the role of some African medicinal spices and vegetables in the management of MetS.

2 IN VITRO SCREENING METHODS OF PHYTOCHEMICALS AGAINST METABOLIC SYNDROME

Screening methods related to diabetes, obesity as well as other components of MetS are often considered separately because of the complexity of this disease that includes several alterations of metabolic conditions (Di Lorenzo et al., 2013). MetS describes the human condition characterized by the presence of coexisting traditional risk factors for cardiovascular disease, such as hypertension, dyslipidemia, glucose intolerance, and obesity, in addition to nontraditional cardiovascular disease risk factors, such as inflammatory processes and abnormalities of the blood coagulation system. Each in vitro phytochemical screening method for these disease states can therefore be considered as far as the overall activity of the botanicals allowed it to be used against MetS. Also, MetS increases the risk of developing type 2 diabetes and hence, botanicals having antidiabetic properties together with antihypertensive, antiobesity, antiinflammatory, and other related activities are relevant in the management of disease. Reaven (1991) have proposed to study parallels between human disease and rodent model of MetS based on markers, such as insulin resistance, hyperinsulinemia, hypertriglyceridemia, and hypertension. However, some research laboratories have developed a rat model of MetS, a variant of one described by Reaven. They used rats receiving 30% sugar in their drinking water for 6 months to develop several alterations associated to the MetS, such as hypertriglyceridemia, moderate hypertension, insulin resistance, central obesity, and nephropathy (Perez-Torres et al., 2009). This model can be used to screen botanicals for their anti-MetS properties.

3 POTENTIAL OF PHYTOCHEMICALS AGAINST METABOLIC SYNDROME

Pharmacologically, several classes of drugs which include antihypertensive agents, oral glucose lowering agents, insulin sensitizers, and lipid-lowering agents are available to treat MetS. Due to the clustering of the components of the syndrome, an individual with full-blown syndrome is exposed to high pill burden and in turn increased cost. These can contribute to poor adherence or compliance (Okafor, 2012). The use of natural products from plant kingdom can efficiently reduce the treatment costs. Medicinal plants, some of which have been used for thousands of years, serve as an excellent source of bioactive compounds for the treatment of MetS because they contain a wide range of phytochemicals with diverse metabolic effects.

Obesity is known to be a risk factor for the development of metabolic disorders, type 2 diabetes, systemic hypertension, cardiovascular disease, dyslipidemia, and atherosclerosis. In an interesting study conducted by Roh and Jung, crude extracts from 400 plants purchased from a plant extract bank at Korea Research Institute of Bioscience & Biotechnology (KRIBB) were screened for their antiobesity activity using porcine pancreatic lipase assay (triacylglycerol lipase, EC 3.1.1.3) in vitro. Among the 400 plants species examined, authors found that 44 extracts from plants had high antilipase activity using 2,4-dinitrophenylbutyrate as a substrate in porcine pancreatic lipase assay. Furthermore, 44 plant extracts were investigated for their inhibition of lipid accumulation in 3T3-L1 cells. Among these 44 extracts examined, authors found that crude extracts from 4 natural plant species were active. The best activity was observed with *Salicis radicis* cortex, meanwhile *Rubi fructus*, *Corni fructus*, and *Geranium nepalense* exhibited fat inhibitory capacity higher than 30% at 100 µg/mL in 3T3-L1 adipocytes, suggesting antiobesity activity (Roh and Jung, 2012). Hence, Roh and Jung contributed to demonstrate the potential of botanicals in the management of a MetS-related symptom.

Many other plant-derived therapeutics or extracts have been reported on disease risks associated with MetS. Some of the promising extracts include Cinnamon (*Cinnamomum cassia* and *Cinnamomum verum*), Russian tarragon (*Artemisia dracunculus*), bitter melon (*Momordica charantia*), fenugreek (*Trigonella foenum-graecum*), lowbush blueberry (*Vaccinium angustifolium*), grape seed (*Vitis vinifera*), *Crataegus laevigata*, *Crataegus monogyna*, *Crataegus curvisepala* (*Crataegus oxyacantha*) and *Crataegus tanacetifolia*, and Hoodia (*Hoodia gordonii*) (Graf et al., 2010). Several compounds (Fig. 12.1) used in the treatment of MetS have been isolated from these plants. These include the polyphenol type-A polymer (cinnamon), 2′,4′-dihydroxy-4-methoxydihydrochalcone (**1**) (Russian tarragon), a cucurbitane-type triterpenoids (**2**) (bitter melon), trigonelline (**3**) (fenugreek), P57-AS3 pregnane glycoside (**4**) (Hoodia), malvidin-3-glucoside (**5**) (lowbush blueberry), hyperoside (**6**) (hawthorn), and trimeric procyanidin (**7**) (grape seed) (Graf et al., 2010). Crocin (**8**), a natural carotenoid, derivative of saffron (*Crocus sativus*), at a dose of 30 mg/day significantly reduced serum prooxidant–antioxidant balance in individuals with MetS (Nikbakht-Jam et al., 2015).

4 EFFECTS OF AFRICAN MEDICINAL SPICES ON METABOLIC SYNDROME

The beneficial effects of spices and vegetables in chronic disease or conditions including diabetes, cancer, inflammation, hypertension, and cardiovascular diseases have largely been discussed; other beneficial actions, such as antioxidant and antithrombotic effects are also well known (Panickar, 2013). Some of the aforementioned properties are associated with attenuating the MetS. Some spices and vegetables readily available commercially for MetS cure include cinnamon, black pepper, garlic, cloves, ginger, oregano, paprika, and turmeric. They have been investigated for the potential to attenuate factors associated with MetS. There is evidence to indicate that cinnamon and purified extracts from cinnamon affect insulin functioning (Anderson, 2008; Panickar, 2013). Cinnamon polyphenol extract affects immune responses by regulating anti- and proinflammatory genes and glucose transporter gene expression in mouse macrophages (Cao et al., 2008) and adipocytes (Cao et al., 2010). Cinnamic acid isolated from *C. cassia* increased glucose transport in L6 myotubes (Lakshmi et al., 2009; Panickar, 2013). Cinnamic acid increased glucose uptake in mouse liver FL83B cells made insulin-resistant by administration of tumor necrosis factor alpha (TNF-α) (Huang et al., 2009). Cinnamaldehyde appeared to have a dual action on glucose transport in L929 fibroblast cells and while it stimulated glucose uptake under basal conditions, it inhibited glucose uptake in glucose deprived condition (Plaisier et al., 2011; Panickar, 2013). In this section, an overview of spices and vegetables available in Africa and having beneficial action against MetS will be provided.

4.1 *Allium sativum*

A. sativum (Liliaceae), known as garlic is a strongly aromatic bulb crop believed to originate from Kazakhstan, Uzbekistan. and Western China. In Africa, the main producers

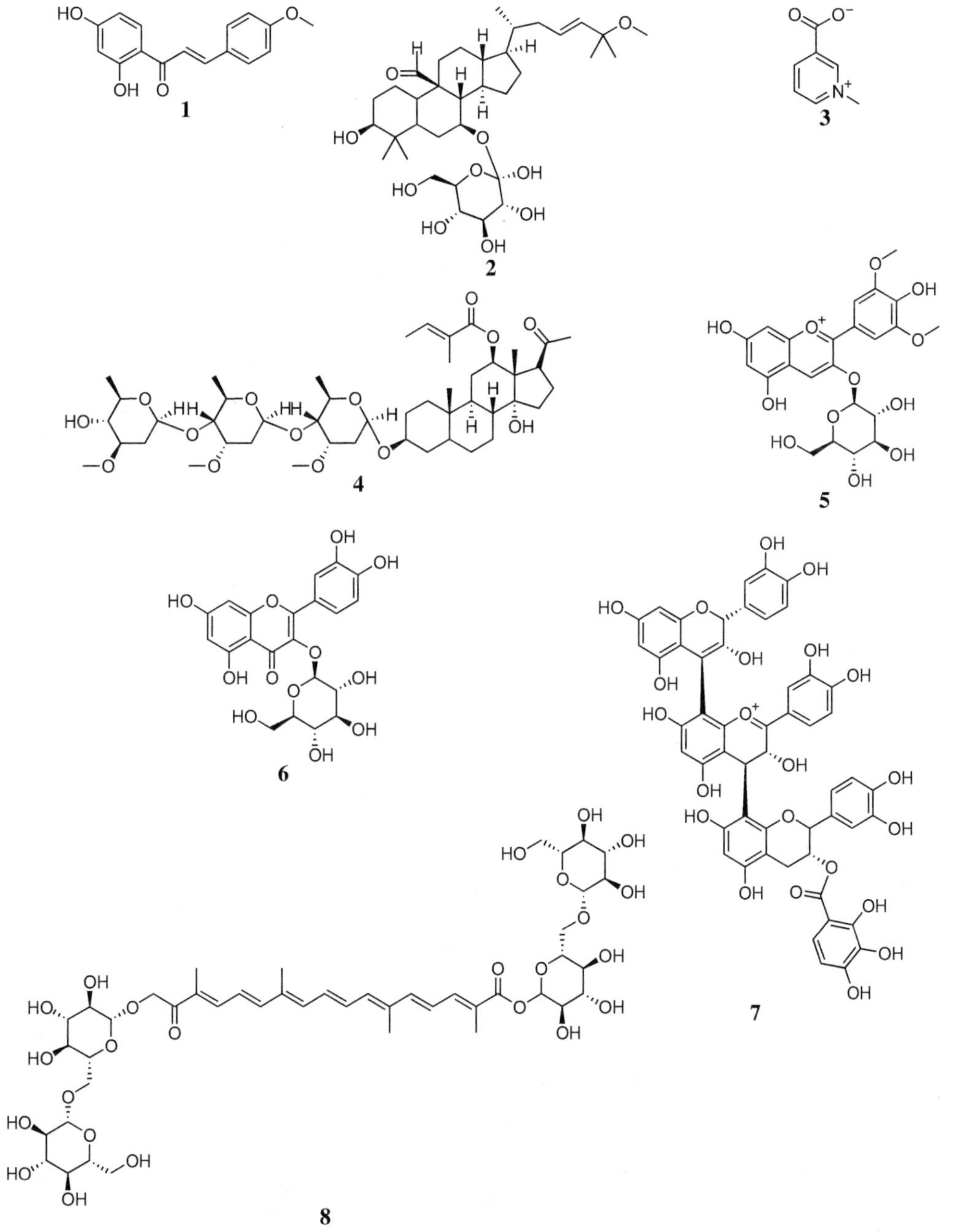

Figure 12.1 Chemical structures of phytochemicals used in the treatment of metabolic syndrome (MetS). 2′,4′-Dihydroxy-4-methoxydihydrochalcone (**1**); cucurbitane-type triterpenoids (**2**); trigonelline (**3**); P57-AS3 pregnane glycoside (**4**); malvidin-3-glucoside (**5**); hyperoside (**6**); trimeric procyanidin (**7**); crocin (**8**).

include the Mediterranean Northern region as well as tropical Africa mainly including Sudan, Niger, and Tanzania. However, most of the West African countries as well as Kenya, Tanzania, Uganda, and Zambia are also producers (http://uses.plantnet-project.org/en/Allium_sativum_%28PROTA%29). In addition to its beneficial effects against MetS, this spice has a great medicinal value (see details in Chapter 15). It was shown that raw garlic extract significantly improves insulin sensitivity while attenuating MetS and oxidative stress in fructose-fed rats for 8 weeks (Padiya et al., 2011; Panickar, 2013). In rats that were made diabetic by a single injection of streptozotocin (STZ), a reduction in hyperglycemia and dyslipidemia was observed with garlic, or with a mixture containing garlic, ginger, and turmeric (Madkor et al., 2011). Other effects of garlic or extracts include reducing blood pressure and total cholesterol and increasing HDL in spontaneously hypertensive rats (Preuss et al., 2001), diabetic rats (Eidi et al., 2006), or normal rats (Mehrzia et al., 2006), improving endothelial dysfunction in diabetic rats (Baluchnejadmojarad et al., 2003), and lowering glucose levels in fructose-fed insulin-resistant rats (Jalal et al., 2007). It was also shown that garlic extracts prevent oxidative stress and vascular remodeling in rats fed with sucrose in drinking water (Vazquez-Prieto et al., 2010). Garlic constituents with beneficial effects in reducing risk factors associated with MetS have been identified. In fact, ajoene attenuated the high fat diet-induced hepatic steatosis, occasionally an end-result of prolonged MetS (Han et al., 2011; Panickar, 2013). Thiacremonone, given to db/db mice, significantly reduced body weight and decreased blood triglyceride and glucose levels when compared with controls (Ban et al., 2012). Allicin was found to possess insulin potentiating effects (Mathew and Augusti, 1973). S-allylcysteine significantly improved glucose levels and insulin sensitivity in STZ-induced diabetic rats (Mathew and Augusti, 1973; Panickar, 2013).

4.2 *Brassica oleracea*

B. oleracea species (Brassicaceae) include many common foods as cultivars, such as cabbage, broccoli, cauliflower, kale, Brussels sprouts, collard greens, savoy, kohlrabi, and kai-lan. Various species of this vegetable are widely cultivated in Africa. El-Houri and coworkers have investigated dichloromethane and methanol extracts of broccoli in a screening platform for identification of a potential bioactivity related to insulin-dependent glucose uptake and fat accumulation. The screening platform included a series of in vitro bioassays, peroxisome proliferator-activated receptor (PPAR) γ-mediated transactivation, adipocyte differentiation of 3T3-L1 cell cultures, and glucose uptake in both 3T3-L1 adipocytes and primary porcine myotubes, as well as one in vivo bioassay, fat accumulation in the nematode *Caenorhabditis elegans* (El-Houri et al., 2014). The investigators found that methanol extract of *B. oleracea* enhanced glucose uptake in myotubes but were not able to activate PPARγ, indicating a PPARγ-independent effect on glucose uptake. However, dichloromethane extract activated PPARγ and reduced fat accumulation in *C. elegans*, suggesting a beneficial effect of this vegetable in combating MetS.

4.3 Curcuma longa

C. longa L. commonly known as turmeric is a member of the ginger family (Zingiberaceae), native to southwest India with its rhizomes being the source of a bright yellow spice and dye. In Africa it is cultivated in home gardens in many countries and is for sale in numerous markets (Jansen, 2005). Rhizomes of *C. longa* are part of numerous traditional compound medicines used as stomachic, stimulant and blood purifier, and to treat liver complaints, biliousness and jaundice (Jansen, 2005), for arthritic, muscular disorders, biliary disorders, anorexia, cough, diabetic wounds, hepatic disorders, and sinusitis. The role of the major constituent of this spice, curcumin, against MetS have been investigated in several studies. Curcumin increased the insulin-stimulated glucose uptake in 3T3-L1 cells; this compound also suppressed the transcription and secretion of TNF-α and interleukin-6 (IL-6) induced by palmitate indicating potentially beneficial antiinflammatory as well as glucoregulatory properties (Wang et al., 2009; Panickar, 2013). This compound increased glucose uptake in rat L6 myotubes via an AMP-activated protein kinase (AMPK) and (MEK)3/6-p38 mitogen-activated protein kinase (MAPK) pathway (Kim et al., 2010; Na et al., 2011). Its effects on C2C12 mouse myoblast cells in increasing glucose uptake is possibly mediated by AMPK (Kang and Kim, 2010) or phosphoinositide 3-kinase (PI3K) or MAPK signaling pathways (Deng et al., 2012). In skeletal muscle isolated from rats, curcumin induced a dose-dependent increase in glucose uptake possibly mediated by the phospholipase C or PI3K pathway (Cheng et al., 2009; Panickar, 2013). This compound significantly decreased TNF-α, IL-1β, IL-6, and cyclooxygenase-2 (COX-2) gene expression in 3T3-L1-derived adipocytes stimulated by TNF-α possibly mediated by nuclear factor κB (NFκB) (Gonzales and Orlando, 2008). Its effects on leptin signaling by reducing phosphorylation levels of leptin receptor (Ob-R) and its downstream intermediators in hepatic stellate cells (Tang et al., 2009) may also be important in its role in attenuating liver fibrogenesis activated by leptin in obesity (Panickar, 2013). It also suppressed LDL receptor gene expression in activated hepatic cells in vitro by activating PPARγ, reducing cellular cholesterol, and attenuating the stimulatory effects of LDL on cell activation (Kang and Chen, 2009; Panickar, 2013). It inhibits fatty acid synthase, an enzyme that catalyzes fatty acid synthesis, in 3T3-L1 adipocytes and also suppressed adipocyte differentiation and lipid accumulation (Panickar, 2013; Zhao et al., 2011). Other pharmacological properties of *C. longa* and curcumin are discussed in detail in Chapter 19.

4.4 Daucus carota

D. carota, commonly known as carrot is a flowering plant in the family Apiaceae, native to temperate regions of Europe and southwest Asia, and naturalized to North America and Australia. It is generally assumed that the eastern, purple-rooted carrot originated in Afghanistan in the region where the Himalayan and Hindu Kush mountains meet, and that it was domesticated in Afghanistan and adjacent regions of Russia, Iran, India,

Pakistan, and Anatolia (van der Vossen and Kahangi, 2004). The western, orange carrot probably arose in Europe or in the western Mediterranean region through gradual selection within yellow carrot populations. They have now largely replaced the eastern types because of superior taste and nutritional value, and can also be found throughout Africa (van der Vossen and Kahangi, 2004). Carrot is a worldwide important market vegetable. The roots are consumed raw or cooked, alone or in combination with other vegetables, as an ingredient of soups, dishes, sauces, juices, and in dietary compositions; large coarse roots are also used as fodder. Young leaves are sometimes eaten raw or used as fodder. In Ethiopia, fruits are used against tapeworm (van der Vossen and Kahangi, 2004). El-Houri and coworkers found that dichloromethane extract of carrot was more potent than methanol extract toward several MetS-related markers. Authors showed that dichloromethane extract was able to induce PPARγ transactivation, to stimulate glucose uptake in adipocytes and in myotubes and to reduce fat accumulation in *C. elegans* (El-Houri et al., 2014).

4.5 *Cinnamomum* species

C. cassia (syn. *Cinnamomum aromaticum*) or Cassia, *Cinnamomum burmannii* known as Korintje, *Cinnamomum loureiroi* or Saigon cinnamon, and *Cinnamomum zeylanicum* (*syn. C. verum*) known as Ceylon are four commercially used cinnamon species among up to hundred species (The Seasoning and Spice Association, 2010) (detailed information on *Cinnamomum* species are given in Chapter 17). The two widely distributed cinnamon species in Africa include *C. cassia* and *C. zeylanicum*. These spices showed positive effects on other markers of MetS in addition to blood glucose. They had antihypertensive action in both spontaneously hypertensive rats (Preuss et al., 2006) and patients with MetS (22 men and women treated with a 500-mg extract daily for 12 weeks) (Ziegenfuss et al., 2006; Graf et al., 2010). Moreover, triglyceride-lowering effects with cinnamon were observed in Wistar rats fed with a high-fructose diet (Qin et al., 2010) and in some clinical trials. For example, in the trial in Pakistani patients with type 2 diabetes, serum triglycerides, LDL-cholesterol, and total cholesterol were all significantly lowered (Khan et al., 2003), although these effects were not corroborated by results from a trial conducted in a less severely diabetic European population administered a comparable dose of cinnamon (Mang et al., 2006; Graf et al., 2010).

4.6 *Syzygium aromaticum*

S. aromaticum (L.) Merr. & L.M. Perry (syn. *Eugenia caryophyllus*) is a tree in the family Myrtaceae, native to Indonesia. The aromatic flower buds of the plant are known as cloves and are commonly used as a spice. Cloves are commercially harvested in Indonesia, India, Pakistan, Sri Lanka as well as in African countries, such as Comoro Islands, Madagascar, Seychelles, and Tanzania. Antioxidant and antiinflammatory properties of cloves have been reported (Shobana and Naidu, 2000; Lee and Shibamoto, 2001; Chaieb et al., 2007; Yoshimura et al., 2011; Panickar, 2013). The active compounds in clove may include eugenol and other polyphenols (Gulcin, 2011). Cloves also have glucoregulatory properties; in a rat epididymal fat cell assay, cloves significantly potentiated insulin activity and thus may also regulate glucose (Khan et al., 1990; Broadhurst et al., 2000). It is hypothesized that some of the glucose-regulating property of cloves might be due to its effects on PPARγ activation (Kuroda et al., 2012; Panickar, 2013). More details on other pharmacological activities of cloves are available in Chapter 29.

4.7 *Thymus vulgaris*

T. vulgaris commonly known as thyme (Lamiaceae) is a spicy herb native to Southern Europe and having worldwide distribution (Hosseinzadeh et al., 2015). The plant is indigenous to the Mediterranean and neighboring countries, Northern Africa, and parts of Asia. In Africa, the plant has been cultivated in Egypt, Morocco, Algeria, Tunisia and Libya (Stahl-Biskup and Sáez, 2002), Cameroon (Nkouaya Mbanjo et al., 2007), Nigeria (Kayode and Ogunleye, 2008), and South Africa (Schmitz, 2015). People have used thyme for many centuries as a flavoring agent, culinary herb, and herbal medicine (Stahl-Biskup and Venskutonis, 2012). The beneficial effect of thyme in the amelioration of MetS-related markers has been reported. Methanol and dichloromethane extracts of this spice induced PPARγ transactivation, stimulated glucose uptake in myotubes, and reduced fat accumulation in *C. elegans* (El-Houri et al., 2014). In addition, dichloromethane also stimulated glucose uptake in adipocytes (El-Houri et al., 2014). More details on other medicinal values of thyme are given in Chapter 28.

4.8 *Zingiber officinale*

Z. officinale Roscoe (Zingiberaceae), commonly known as ginger, is a flowering plant with root named ginger, widely used as a spice or a folk medicine. The plant is indigenous to South China and was spread eventually to the Spice Islands, other parts of Asia, and subsequently to West Africa. In Africa, other big producers of ginger include Nigeria, Cameroon, Ethiopia, Mauritius, Kenya, Uganda, Madagascar, Ghana, and Zambia (Prabhakaran Nair, 2013). Antioxidant, antiinflammatory, and glucoregulating properties of ginger have been reported (Mascolo et al., 1989). Gingerol, an active component of ginger extract, increased insulin-sensitive glucose uptake in 3T3-L1 mouse adipocytes (Sekiya et al., 2004). One mechanism of action of ginger in modulating insulin release may be through its actions on the serotonin receptor as demonstrated in N1E-115 cells (Heimes et al., 2009). Mechanistic studies indicate that ginger-enhanced-glucose transport in L6 myotubes is possibly mediated through the activation of PI3K (Noipha et al., 2010) and possibly by increasing glucose transporter type 4 (GLUT4) transport to the membrane (Kuroda et al., 2012; Rani et al., 2012). More details on other medicinal values of ginger are given in Chapter 30.

5 CONCLUSIONS

Pharmacologically, several classes of drugs are available to treat MetS. However, with regard to the low income of African population, prevention still remains the best option. Medicinal spices and vegetables currently used in the continent are also attractive strategies to combat the components of MetS. In this chapter, we have discussed the beneficial role of spices and vegetables available in Africa against MetS. Some of them include *A. sativum, B. oleracea, C. longa, D. carota, C. cassia, C. zeylanicum, S. aromaticum, T. vulgaris,* and *Z. officinale.*

REFERENCES

Allium sativum L. Available from: http://uses.plantnet-project.org/en/Allium_sativum_%28PROTA%29. *Allium sativum* L.

Anderson, R.A., 2008. Chromium and polyphenols from cinnamon improve insulin sensitivity. Proc. Nutr. Soc. 67 (1), 48–53.

Baluchnejadmojarad, T., Roghani, M., Homayounfar, H., Hosseini, M., 2003. Beneficial effect of aqueous garlic extract on the vascular reactivity of streptozotocin-diabetic rats. J. Ethnopharmacol. 85 (1), 139–144.

Ban, J.O., Lee, D.H., Kim, E.J., Kang, J.W., Kim, M.S., Cho, M.C., 2012. Antiobesity effects of a sulfur compound thiacremonone mediated via down-regulation of serum triglyceride and glucose levels and lipid accumulation in the liver of db/db mice. Phytother. Res. 26 (9), 1265–1271.

Broadhurst, C.L., Polansky, M.M., Anderson, R.A., 2000. Insulin-like biological activity of culinary and medicinal plant aqueous extracts in vitro. J. Agric. Food Chem. 48 (3), 849–852.

Cameron, A.J., Shaw, J.E., Zimmet, P.Z., 2004. The metabolic syndrome: prevalence in worldwide populations. Endocrinol. Metab. Clin. North Am. 33 (2), 351–375.

Cao, H., Graves, D.J., Anderson, R.A., 2010. Cinnamon extract regulates glucose transporter and insulin-signaling gene expression in mouse adipocytes. Phytomedicine 17 (13), 1027–1032.

Cao, H., Urban, Jr., J.F., Anderson, R.A., 2008. Cinnamon polyphenol extract affects immune responses by regulating anti- and proinflammatory and glucose transporter gene expression in mouse macrophages. J. Nutr. 138 (5), 833–840.

Chaieb, K., Hajlaoui, H., Zmantar, T., Kahla-Nakbi, A.B., Rouabhia, M., Mahdouani, K., 2007. The chemical composition and biological activity of clove essential oil, *Eugenia caryophyllata* (*Syzigium aromaticum* L. Myrtaceae): a short review. Phytother. Res. 21 (6), 501–506.

Cheng, T.C., Lin, C.S., Hsu, C.C., Chen, L.J., Cheng, K.C., Cheng, J.T., 2009. Activation of muscarinic M-1 cholinoceptors by curcumin to increase glucose uptake into skeletal muscle isolated from Wistar rats. Neurosci. Lett. 465 (3), 238–241.

Deng, Y.T., Chang, T.W., Lee, M.S., Lin, J.K., 2012. Suppression of free fatty acid-induced insulin resistance by phytopolyphenols in C2C12 mouse skeletal muscle cells. J. Agric. Food Chem. 60 (4), 1059–1066.

Di Lorenzo, C., Dell'Agli, M., Colombo, E., Sangiovanni, E., Restani, P., 2013. Metabolic syndrome and inflammation: a critical review of in vitro and clinical approaches for benefit assessment of plant food supplements. Evid. Based Complement. Alternat. Med. 2013, 10.

Eidi, A., Eidi, M., Esmaeili, E., 2006. Antidiabetic effect of garlic (*Allium sativum* L.) in normal and streptozotocin-induced diabetic rats. Phytomedicine 13 (9–10), 624–629.

El-Houri, R.B., Kotowska, D., Olsen, L.C., Bhattacharya, S., Christensen, L.P., Grevsen, K., 2014. Screening for bioactive metabolites in plant extracts modulating glucose uptake and fat accumulation. Evid. Based Complement. Alternat. Med. 2014, 156398.

Gonzales, A.M., Orlando, R.A., 2008. Curcumin and resveratrol inhibit nuclear factor-kappaB-mediated cytokine expression in adipocytes. Nutr. Metab. 5, 17.

Graf, B.L., Raskin, I., Cefalu, W.T., Ribnicky, D.M., 2010. Plant-derived therapeutics for the treatment of metabolic syndrome. Curr. Opin. Investig. Drugs 11 (10), 1107–1115.

Grundy, S.M., Brewer, H.B., Cleeman, J.I., Smith, S.C., Lenfant, C., 2004. Definition of metabolic syndrome: report of the National Heart, Lung, and Blood Institute/American Heart Association conference on scientific issues related to definition. Circulation 109 (3), 433–438.

Gulcin, I., 2011. Antioxidant activity of eugenol: a structure-activity relationship study. J. Med. Food 14 (9), 975–985.

Han, C.Y., Ki, S.H., Kim, Y.W., Noh, K., Lee da, Y., Kang, B., 2011. Ajoene, a stable garlic by-product, inhibits high fat diet-induced hepatic steatosis and oxidative injury through LKB1-dependent AMPK activation. Antioxid. Redox Signal. 14 (2), 187–202.

Heimes, K., Feistel, B., Verspohl, E.J., 2009. Impact of the 5-HT3 receptor channel system for insulin secretion and interaction of ginger extracts. Eur. J. Pharmacol. 624 (1–3), 58–65.

Hosseinzadeh, S., Kukhdan, A., Hosseini, A., Armand, R., 2015. The application of *Thymus vulgaris* in traditional and modern medicine: a review. Global J. Pharmacol. 9 (3), 260–266.

Huang, D.W., Shen, S.C., Wu, J.S., 2009. Effects of caffeic acid and cinnamic acid on glucose uptake in insulin-resistant mouse hepatocytes. J. Agric. Food Chem. 57 (17), 7687–7692.

Jalal, R., Bagheri, S.M., Moghimi, A., Rasuli, M.B., 2007. Hypoglycemic effect of aqueous shallot and garlic extracts in rats with fructose-induced insulin resistance. J. Clin. Biochem. Nutr. 41 (3), 218–223.

Jansen, P.C.M., 2005. *Curcuma longa* L. In: Jansen, P.C.M., Cardon, D. (Eds.), PROTA 3: Dyes and Tannins/Colorants et Tanins. PROTA, Wageningen, The Netherlands.

Kang, Q., Chen, A., 2009. Curcumin suppresses expression of low-density lipoprotein (LDL) receptor, leading to the inhibition of LDL-induced activation of hepatic stellate cells. Br. J. Pharmacol. 157 (8), 1354–1367.

Kang, C., Kim, E., 2010. Synergistic effect of curcumin and insulin on muscle cell glucose metabolism. Food Chem. Toxicol. 48 (8–9), 2366–2373.

Kaur, J., 2014. A comprehensive review on metabolic syndrome. Cardiol. Res. Pract. 2014, 21.

Kayode, J., Ogunleye, T., 2008. Checklist and status of plant species used as spices in Kaduna state of Nigeria. Afr. J. Gen. Agric. 4 (1), 13–18.

Khan, A., Bryden, N.A., Polansky, M.M., Anderson, R.A., 1990. Insulin potentiating factor and chromium content of selected foods and spices. Biol. Trace Elem. Res. 24 (3), 183–188.

Khan, A., Safdar, M., Ali Khan, M.M., Khattak, K.N., Anderson, R.A., 2003. Cinnamon improves glucose and lipids of people with type 2 diabetes. Diab. Care 26 (12), 3215–3218.

Kim, J.H., Park, J.M., Kim, E.K., Lee, J.O., Lee, S.K., Jung, J.H., 2010. Curcumin stimulates glucose uptake through AMPK-p38 MAPK pathways in L6 myotube cells. J. Cell. Physiol. 223 (3), 771–778.

Kuroda, M., Mimaki, Y., Ohtomo, T., Yamada, J., Nishiyama, T., Mae, T., 2012. Hypoglycemic effects of clove (*Syzygium aromaticum* flower buds) on genetically diabetic KK-Ay mice and identification of the active ingredients. J. Nat. Med. 66 (2), 394–399.

Lakshmi, B.S., Sujatha, S., Anand, S., Sangeetha, K.N., Narayanan, R.B., Katiyar, C., 2009. Cinnamic acid, from the bark of *Cinnamomum cassia*, regulates glucose transport via activation of GLUT4 on L6 myotubes in a phosphatidylinositol 3-kinase-independent manner. J. Diab. 1 (2), 99–106.

Lee, K.G., Shibamoto, T., 2001. Inhibition of malonaldehyde formation from blood plasma oxidation by aroma extracts and aroma components isolated from clove and eucalyptus. Food Chem. Toxicol. 39 (12), 1199–1204.

Madkor, H.R., Mansour, S.W., Ramadan, G., 2011. Modulatory effects of garlic, ginger, turmeric and their mixture on hyperglycaemia, dyslipidaemia and oxidative stress in streptozotocin-nicotinamide diabetic rats. Br. J. Nutr. 105 (8), 1210–1217.

Mang, B., Wolters, M., Schmitt, B., Kelb, K., Lichtinghagen, R., Stichtenoth, D.O., 2006. Effects of a cinnamon extract on plasma glucose, HbA, and serum lipids in diabetes mellitus type 2. Eur. J. Clin. Invest. 36 (5), 340–344.

Mascolo, N., Jain, R., Jain, S.C., Capasso, F., 1989. Ethnopharmacologic investigation of ginger (*Zingiber officinale*). J. Ethnopharmacol. 27 (1–2), 129–140.

Mathew, P.T., Augusti, K.T., 1973. Studies on the effect of allicin (diallyl disulphide-oxide) on alloxan diabetes. I. Hypoglycaemic action and enhancement of serum insulin effect and glycogen synthesis. Indian J. Biochem. Biophys. 10 (3), 209–212.

Mehrzia, M., Ferid, L., Mohamed, A., Ezzedine, A., 2006. Acute effects of a partially purified fraction from garlic on plasma glucose and cholesterol levels in rats: putative involvement of nitric oxide. Indian J. Biochem. Biophys. 43 (6), 386–390.

Mottillo, S., Filion, K.B., Genest, J., Joseph, L., Pilote, L., Poirier, P., 2010. The metabolic syndrome and cardiovascular risk: a systematic review and meta-analysis. J. Am. Coll. Cardiol. 56 (14), 1113–1132.

Na, L.X., Zhang, Y.L., Li, Y., Liu, L.Y., Li, R., Kong, T., 2011. Curcumin improves insulin resistance in skeletal muscle of rats. Nutr. Metab. Cardiovasc. Dis. 21 (7), 526–533.

Nikbakht-Jam, I., Khademi, M., Nosrati, M., Eslami, S., Foroutan-Tanha, M., Sahebkar, A., 2016. Effect of crocin extracted from saffron on pro-oxidant–anti-oxidant balance in subjects with metabolic syndrome: a randomized, placebo-controlled clinical trial. Eur. J. Integr. Med. 8 (3), 307–312.

Nkouaya Mbanjo, E., Tchoumbougnang, M., Jazet Dongmo, P., Sameza, M., Amvam Zollo, P., Menut, C., 2007. Mosquito larvicidal activity of essential oils of *Cymbopogon citratus* and *Thymus vulgaris* grown in Cameroon. Planta Med. 73, 329.

Noipha, K., Ratanachaiyavong, S., Ninla-Aesong, P., 2010. Enhancement of glucose transport by selected plant foods in muscle cell line L6. Diab. Res. Clin. Pract. 89 (2), e22–e26.

Okafor, C.I., 2012. The metabolic syndrome in Africa: current trends. Indian J. Endocrinol. Metab. 16 (1), 56–66.

Padiya, R., Khatua, T.N., Bagul, P.K., Kuncha, M., Banerjee, S.K., 2011. Garlic improves insulin sensitivity and associated metabolic syndromes in fructose fed rats. Nutr. Metab. 8, 53.

Panickar, K.S., 2013. Beneficial effects of herbs, spices and medicinal plants on the metabolic syndrome, brain and cognitive function. Cent. Nerv. Syst. Agents Med. Chem. 13 (1), 13–29.

Perez-Torres, I., Roque, P., El Hafidi, M., Diaz-Diaz, E., Banos, G., 2009. Association of renal damage and oxidative stress in a rat model of metabolic syndrome. Influence of gender. Free Radic. Res. 43 (8), 761–771.

Plaisier, C., Cok, A., Scott, J., Opejin, A., Bushhouse, K.T., Salie, M.J., 2011. Effects of cinnamaldehyde on the glucose transport activity of GLUT1. Biochimie 93 (2), 339–344.

Prabhakaran Nair, K., 2013. The Agronomy and Economy of Tumeric and Ginger. Elsevier, London, p. 446.

Preuss, H.G., Clouatre, D., Mohamadi, A., Jarrell, S.T., 2001. Wild garlic has a greater effect than regular garlic on blood pressure and blood chemistries of rats. Int. Urol. Nephrol. 32 (4), 525–530.

Preuss, H.G., Echard, B., Polansky, M.M., Anderson, R., 2006. Whole cinnamon and aqueous extracts ameliorate sucrose-induced blood pressure elevations in spontaneously hypertensive rats. J. Am. Coll. Nutr. 25 (2), 144–150.

Qin, B., Polansky, M.M., Anderson, R.A., 2010. Cinnamon extract regulates plasma levels of adipose-derived factors and expression of multiple genes related to carbohydrate metabolism and lipogenesis in adipose tissue of fructose-fed rats. Horm. Metab. Res. 42 (3), 187–193.

Rani, M.P., Krishna, M.S., Padmakumari, K.P., Raghu, K.G., Sundaresan, A., 2012. *Zingiber officinale* extract exhibits antidiabetic potential via modulating glucose uptake, protein glycation and inhibiting adipocyte differentiation: an in vitro study. J. Sci. Food Agric. 92 (9), 1948–1955.

Reaven, G.M., 1991. Insulin resistance, hyperinsulinemia, hypertriglyceridemia, and hypertension. Parallels between human disease and rodent models. Diab. Care 14 (3), 195–202.

Roh, C., Jung, U., 2012. Screening of crude plant extracts with anti-obesity activity. Int. J. Mol. Sci. 13 (2), 1710–1719.

Schmitz, P. *Thymus vulgaris*. Available from: http://ecoport.org/ep?Plant=2441&entityType=PLCR**&entityDisplayCategory=full

Sekiya, K., Ohtani, A., Kusano, S., 2004. Enhancement of insulin sensitivity in adipocytes by ginger. Biofactors 22 (1–4), 153–156.

Shobana, S., Naidu, K.A., 2000. Antioxidant activity of selected Indian spices. Prostaglandins Leukot. Essent. Fatty Acids 62 (2), 107–110.

Stahl-Biskup, E., Sáez, F., 2002. Thyme: The Genus Thymus. CRC Press, London, 18.

Stahl-Biskup, E., Venskutonis, R.P., 2012. Thyme. In: Peter, K.V. (Ed.), Handbook of Herbs and Spices. second ed. Woodhead Publishing, Cambridge, pp. 499–525, (Chapter 27).

Tang, Y., Zheng, S., Chen, A., 2009. Curcumin eliminates leptin's effects on hepatic stellate cell activation via interrupting leptin signaling. Endocrinology 150 (7), 3011–3020.

The Seasoning and Spice Association, 2010. Culinary Herbs and Spices. Available from: http://www.seasoningandspice.org.uk/ssa/background_culinary-herbs-spices.aspx

van der Vossen, H.A.M., Kahangi, E., 2004. *Daucus carota* L. Record from PROTA4U. In: Grubben, G.J.H, Denton, O.A. (Eds.), PROTA—Plant Resources of Tropical Africa. PROTA Foundation, Wageningen, Netherlands.

Vazquez-Prieto, M.A., Gonzalez, R.E., Renna, N.F., Galmarini, C.R., Miatello, R.M., 2010. Aqueous garlic extracts prevent oxidative stress and vascular remodeling in an experimental model of metabolic syndrome. J. Agric. Food Chem. 58 (11), 6630–6635.

Wang, S.L., Li, Y., Wen, Y., Chen, Y.F., Na, L.X., Li, S.T., 2009. Curcumin, a potential inhibitor of up-regulation of TNF-alpha and IL-6 induced by palmitate in 3T3-L1 adipocytes through NF-kappaB and JNK pathway. Biomed. Environ. Sci. 22 (1), 32–39.

Yoshimura, M., Amakura, Y., Yoshida, T., 2011. Polyphenolic compounds in clove and pimento and their antioxidative activities. Biosci. Biotechnol. Biochem. 75 (11), 2207–2212.

Zhao, J., Sun, X.B., Ye, F., Tian, W.X., 2011. Suppression of fatty acid synthase, differentiation and lipid accumulation in adipocytes by curcumin. Mol. Cell. Biochem. 351 (1–2), 19–28.

Ziegenfuss, T.N., Hofheins, J.E., Mendel, R.W., Landis, J., Anderson, R.A., 2006. Effects of a water-soluble cinnamon extract on body composition and features of the metabolic syndrome in pre-diabetic men and women. J. Int. Soc. Sports Nutr. 3, 45–53.

CHAPTER 13

Other Health Benefits of African Medicinal Spices and Vegetables

V. Kuete

1 INTRODUCTION

Spices may be defined as the dried parts of aromatic plants with the exception of the leaves; meanwhile vegetables are the fresh and edible parts of herbaceous plants (Nirmal et al., 2010; Randhawa et al., 2015). Vegetable can be more precisely defined as any plant part consumed for food that is not a fruit or seed, but includes mature fruits that are eaten as part of a main meal (Nirmal et al., 2010). Generally, the words "fruits" and "vegetables" are mutually exclusive; "fruit" is a part that is developed from the ovary of a flowering plant (e.g., peaches, plums, and oranges). However, many vegetables, such as eggplants and bell peppers, are botanically fruits. Common vegetables include cabbage (*B. oleracea*), turnip (*Brassica rapa*), radish (*Raphanus sativus*), carrot (*Daucus carota*), parsnip (*Pastinaca sativa*), beetroot (*Beta vulgaris*), lettuce (*Lactuca sativa*), beans (*Phaseolus vulgaris, Phaseolus coccineus, or Phaseolus lunatus*), peas (*Pisum sativum*), potato (*Solanum tuberosum*), aubergine/eggplant (*Solanum melongena*), tomato (*Solanum lycopersicum*), cucumber (*Cucumis sativus*), pumpkin/squash (*Cucurbita* spp.), onion (*Allium cepa*), garlic (*A. sativum*), leek (*Allium ampeloprasum*), pepper (*Capsicum annuum*), spinach (*Spinacia oleracea*), yam (*Dioscorea* spp.), sweet potato (*Ipomoea batatas*), and cassava (*Manihot esculenta*). The popular green leafy vegetables growing in Africa include Taro (*Colocasia esculenta*), Eru (*Gnetun africanum*), Cow pea (*Vigna unguiculata*), Horseradish (*Moringa oleifera*), Groundcherry (*Physalis viscose*), Gherkin (*Cucumis angora*), Hell's curse (*Amaranthus cruentus*), African cabbage (*Cleome gynandra*), Chinese cabbage (*B. rapa*), Nightshade (*Solanum nigrum*), Jew's mallow plant (*Corchorus olitorius*), and bitter water melon (*Citrullus lanatus*) (Randhawa et al., 2015).

Conventional classification of spices includes (1) hot spices [*Capsicum* sp. (chillies), cayenne pepper (*C. annuum*), black and white peppers (*P. nigrum*), ginger (*Z. officinale*)], (2) mild spices [paprika (*C. annuum*), coriander (*C. sativum*)], (3) aromatic spices [allspice (*Pimenta dioica*), cardamom (*Elettaria* sp. *and Amomum* sp.), cassia (*Cinnamomum cassia*), clove (*S. aromaticum*), cumin (*Cuminum cyminum*), dill (*Anethum graveolens*), fennel (*Foeniculum vulgare*), fenugreek (*Trigonella foenum-graecum*), nutmeg (*Myristica fragrans*)], (4) herbs [basil (*Ocimum basilicum*), bay leaves (*Laurus nobilis, Umbellularia californica, Cinnamomum tamala, Syzygium polyanthum, Litsea glaucescens*), marjoram (*Origanum majorana*), tarragon

(*Artemisia dracunculus*), thyme (*Thymus vulgaris*)], and (5) aromatic vegetables [onion, garlic, celery (*Apium graveolens*)]. Spices have been used medicinally for thousands of years in Ayurvedic and traditional Chinese medicine (Tosun and Khan, 2015) as well as in African folk medicine.

Vegetables supply dietary fiber and are important sources of essential vitamins, minerals, and trace elements. They also contain various types of secondary metabolites reported to have antioxidant, antibacterial, antifungal, antiviral, and anticarcinogenic properties (Steinmetz and Potter, 1996; Fankam et al., 2011; Djeussi et al., 2013). Besides, major constituents of food, such as carbohydrate, protein, fat, and micronutrients, many compounds in vegetables (flavonoids, carotenoids, and other polyphenols, glucosinolates, isothiocyanates, allylic sulfides, phytosterols, monoterpenes dietary fibers, and phenolic acids) have shown health-promoting effects (Kris-Etherton et al., 2002). However, vegetables also contain toxic compounds, such as α-solanine, α-chaconine, cyanide and cyanide precursors, oxalic acid, phytohaemagglutinin from some beans, and cyanogenic glycoside from cassava roots (Finotti et al., 2006; Kuete, 2014).

In this chapter, we will discuss the health effects of African medicinal spices and vegetables, such as their use in the treatment of age-related and Alzheimer's diseases, convulsive diseases, in the prevention of rheumatoid arthritis, as antioxidant and sedative, their effects on obesity, etc.

2 AFRICAN MEDICINAL SPICES AND VEGETABLES IN THE TREATMENT OF AGE-RELATED AND ALZHEIMER'S DISEASES

Aging is an intrinsic process that causes a gradual but relentless impairment of normal biological functions and the ability to resist physiological stresses, resulting in functional decline and an increased risk of morbidity and mortality (Wachtel-Galor et al., 2014). Many complex interacting mechanisms and changes are associated with aging, among which are the shortened and dysfunctional telomeres, accumulation of mutations, inflammation, disturbed hormonal pathways, mitochondrial dysfunction, cellular senescence, apoptosis, and altered gene expression (Wachtel-Galor et al., 2014). Alzheimer's disease is a progressive, neurodegenerative, and irreversible brain disorder with symptoms, such as memory loss, confusion, impaired judgments, and loss of language skills (Frydman-Marom et al., 2011; Hamidpour et al., 2015). It has been demonstrated that the accumulation of soluble oligomeric assemblies of β-amyloid polypeptides [amyloid-beta (Aβ)] plays a key role in the development of Alzheimer's disease (Frydman-Marom et al., 2011; Hamidpour et al., 2015). Vegetables and spices have been proved to exert good effect on human health, providing protection against age-related degenerative diseases, such as cardiovascular diseases, Alzheimer's disease, several forms of cancers, and cataracts (Kang et al., 2005; Hritcu et al., 2014; Randhawa et al., 2015).

2.1 *Cinnamomum* species

Cinnamomum (cinnamon) is a genus of the family Lauraceae, many of whose members are used as spices (Shan et al., 2007). There are two main varieties of cinnamon namely the Ceylon or true cinnamon (*Cinnamon zeylanicum*) and cassia (*Cinnamom aromaticum*). *C. zeylanicum* is native to Sri Lanka and tropical Asia (Hamidpour et al., 2015) and exotic to several African countries, such as Comoros, Ghana, Madagascar, Mauritius, Nigeria, Seychelles, Sierra Leone, Tanzania, and Uganda (Orwa et al., 2009). The barks are commonly used in Cameroon as spices and for the treatment of cardiovascular diseases and cancers (Kuete et al., 2011; Nkanwen et al., 2013). Cinnamaldehyde (**1**; 75%) and camphor (56%) are the major compounds in oil and stem bark, respectively (Nkanwen et al., 2013). *C. cassia* Blume is originated from China, Indonesia, and Vietnam and also African countries, such as Nigeria and Madagascar (http://jlvagro.com/products/spices/; Lockwood, 1979). Cinnamon extract inhibits the formation of toxic Aβ oligomers and prevents the toxicity of Aβ on neuronal PC12 cells (Hamidpour et al., 2015). Oral administration of cinnamon extract induced to the reduction of plaques and improvement in cognitive behavior to aggressive Alzheimer's disease transgenic mice (Frydman-Marom et al., 2011). The prevention or treatments of Alzheimer's disease are mostly based on drugs inhibition of cholinesterase function or formation of amyloid plaque. Pathologies and dementias of the nervous system, such as Alzheimer's disease and Parkinson's disease (Lei et al., 2010) can result when tau proteins (tau proteins are proteins that stabilize microtubules) become defective and no longer stabilize microtubules properly. Chemicals able to prevent the tau aggregation may be a key factor in the development of new drugs (Peterson et al., 2009). Cinnamon extract was found to inhibit the aggregation of human tau in vitro, and this was attributed to both proanthocyanidin timer and compound **1** (Peterson et al., 2009).

2.2 *Piper nigrum* L (Piperaceae)

P. nigrum L., commonly known as black pepper, is a flowering vine in the family Piperaceae, cultivated for its fruit, which is usually dried and used as a spice and seasoning. The plant is native to the southwestern coastal region of India from where it is spread to many parts of Asia, South America, and Africa, especially in Benin, Cameroon, Congo, Ethiopia, Gabon, Ivory Coast, Nigeria, Madagascar, Malawi, Sierra Leone, Zambia, and Zimbabwe (Campagnolo, 2000; Ravindran, 2000). It is used in traditional medicine of many countries as analgesic, antiinflammatory, anticonvulsant, antioxidant, antidepressant, and cognitive-enhancing agent (Hritcu et al., 2015). Methanol extract from *P. nigrum* fruits was analyzed for its possible memory-enhancing proprieties in amyloid beta (1-42) rat model of Alzheimer's disease; it was found that administration of the plant extract significantly improved memory performance and ameliorated amyloid beta (1-42)-induced spatial memory impairment by attenuation of the oxidative stress in the rat hippocampus (Hritcu et al., 2014). The methanol extract of the fruits from *P. nigrum*

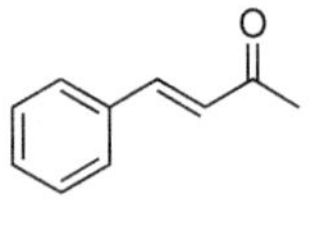

Figure 13.1 *Chemical structure of some antiaging[a] and anti-Alzheimer's[b] diseases compounds from spices and vegetable present in Africa.* Cinnamaldehyde (**1**); kaempferol (**2**); eugenol (**3**); resveratrol (**4**); quercetin (**5**); curcumin (**6**); capsaicin (**7**).

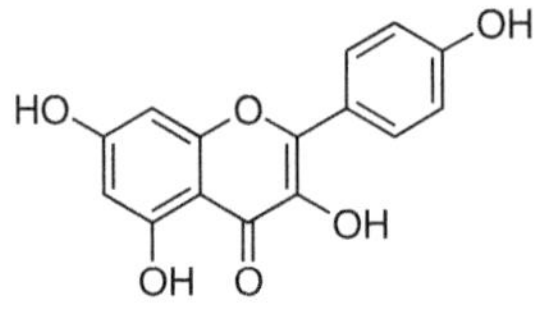

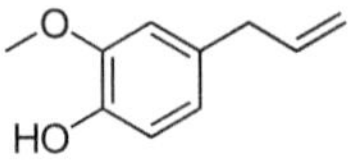

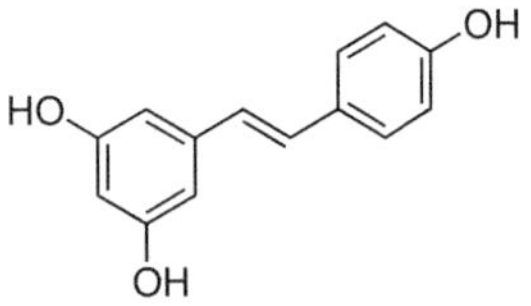

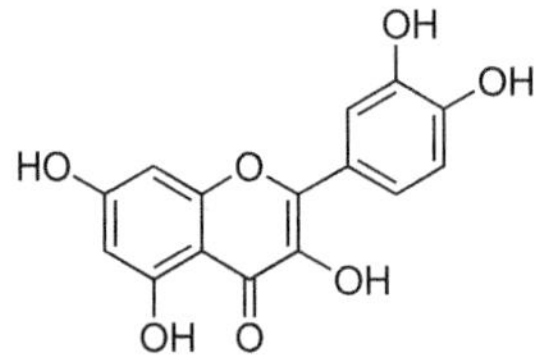

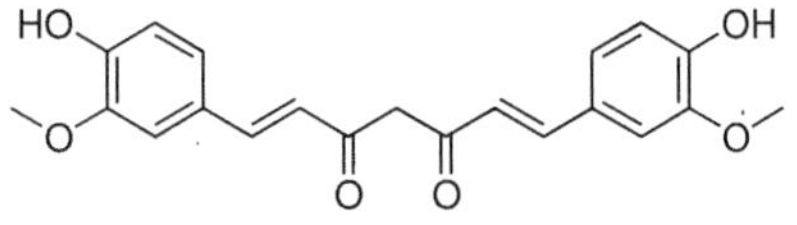

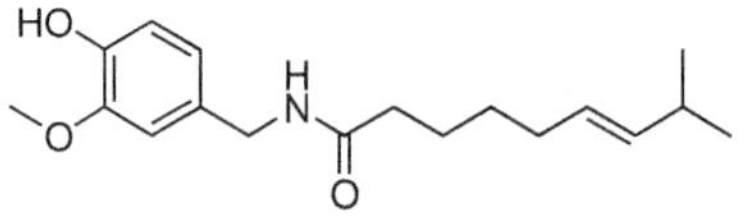

also ameliorated beta-amyloid (1-42)-induced anxiety and depression by attenuation of the oxidative stress in the rat amygdala (Hritcu et al., 2015).

2.3 *Syzygium aromaticum* (Myrtaceae)

Cloves are the aromatic flower buds of *S. aromaticum*, a plant native to the Maluku Islands in Indonesia, and commonly used as a spice. Cloves are commercially harvested primarily in Indonesia, India, Madagascar, Zanzibar, Pakistan, Sri Lanka, and Tanzania. Clove oil helps to modulate physiologic responses in aging rodents. Eugenol (**3**) (Fig. 13.1) and eugenyl acetate in cloves are antioxidative agents (Mohamed, 2014). Eugenol dose dependently binds to membranes, thus stabilizing them and protecting them against free radical attack (Kumaravelu et al., 1996). Compound **3** inhibits lipid oxidation and helps to limit structural changes to various tissues, such as the heart, kidney, and liver (Kumaravelu et al., 1996; Mohamed, 2014). Compound **3** inhibits histamine release from mast cells to reduce hypersensitivity besides having antianaphylactic and antispasmodic properties (Mohamed, 2014). Cloves inhibit oxidative tissue damage and cataract formation in the eye lens of rats (Shukri et al., 2010).

2.4 *Zingiber officinale* (Zinziberaceae)

Z. officinale Roscoe commonly known as ginger is a flowering plant whose rhizome, called ginger root or simply ginger, is widely used as a spice or a folk medicine. *Z. officinale* is indigenous to south China, and was spread eventually to other parts of Asia and subsequently to West Africa (https://unitproj.library.ucla.edu/biomed/spice/index.cfm?displayID=15). Ginger is known to be carminative, thermogenic, antiinflammatory, antioxidative, and stimulating. It is used traditionally for dyspepsia, gastroparesis, slow motility symptoms, constipation, nausea, coughs, colds or flu, vomiting, colic, pain relief, rheumatoid arthritis, osteoarthritis, and joint and muscle injury (Mohamed, 2014). Dietary ginger has antiglycating properties, inhibiting the polyol pathway, and is antihyperglycemic. It also inhibited the formation of various advanced glycation endproducts, including carboxymethyl lysine (Mohamed, 2014).

2.5 Other antiage-related extract and compounds from medicinal spices and vegetables found in Africa

A. sativum, commonly known as garlic, is native to central Asia and has long been a staple in the Mediterranean region, as well as a frequent seasoning in Asia, Africa, and Europe. Garlic was known to ancient Egyptians, and has been used for both culinary and medicinal purposes (Ensminger, 1994). Extract from *A. sativum* containing S-allycisteine, S-allymercaptocysteine, allicin, and diallosulfides was shown to increase lifespan and learning in mice (Moriguchi et al., 1997). *A. sativum* extract suppressed Aβ-induced lipid peroxidation in PC12 pheochromocytoma cell line; reduced apoptosis in cells treated with Aβ; showed antiamyloidogenic and antitangle effects in the brain of Alzheimer's disease transgenic mice (Griffin et al., 2000). Kaempferol (**2**), a flavonol present in the African vegetable, broccoli, was found to extend the lifespan of *Caenorhabditis elegans* by improving antioxidant potential and daf-16 translocation (Kampkotter et al., 2007). DAF-16 is the sole ortholog of the FOXO family of transcription factors in the nematode *C. elegans* and is notable for being the primary transcription factor required for the profound lifespan extension observed upon mutation of the insulin-like receptor daf-2.

Several other antiaging compounds were isolated in other species and vegetables naturally growing in Africa. In effect, clinical trials have been undertaken to assess the potential beneficial effects of the stilbene, resveratrol (**4**), or compound **4**-containing products on various age-related health conditions, such as obesity, insulin resistance, impaired glucose tolerance, dyslipidemia, inflammation, cognitive function, and type 2 diabetes (Si and Liu, 2007). Food sources of **4** include the skin of grapes (*Vitis vinifera*), blueberries, raspberries, and mulberries (Jasinski et al., 2013). Among these vegetables, *V. vinifera* grows in Africa, being cultivated in Egypt and South Africa.

The dietary supplementation with quercetin (**5**) has been reported to extend the lifespan of *C. elegans* by 20%. It was suggested that **5** may prolong lifespan in the worms through inhibiting, either directly or indirectly, the age-1 and daf-2-mediated pathways

(Saul et al., 2008; Pietsch et al., 2009). A recent study reported that **5** and its derivative quercetin caprylate can rejuvenate senescent fibroblasts and increase their lifespan via activating proteasome (Chondrogianni et al., 2010). Compound **5** is abundant in many spices and vegetables found in Africa, including grapes, onions, apples, and broccoli (Williamson and Manach, 2005).

Curcumin (**6**) is the active ingredient in the herbal medicine and dietary spice, turmeric (*C. longa*) (Itokawa et al., 2008; Liao et al., 2011). Turmeric, a member of the ginger root (Zingiberaceae), thrives in hot, moist climates, such as China, India, as well as in South Africa. Turmeric is under study in several human diseases, including Alzheimer's disease (Mishra and Palanivelu, 2008). It was demonstrated that compound **6** supplementation extends the lifespan of drosophila (Shen et al., 2013) and *C. elegans* (Liao et al., 2011), and dietary intake (0.2%) of tetrahydrocurcumin, a metabolite of **6**, significantly increased survival rate in mice (Kitani et al., 2007). Compound **6** and the alkaloid piperine (**16**), present in black pepper, synergistically attenuated D-galactose-induced senescence in rats (Banji et al., 2013). Besides, **6** has antiinflammatory and antioxidant properties, which may also contribute to its potential antiaging effects (Si and Liu, 2007).

3 OXIDATIVE STRESS AND ANTIOXIDANT EFFECTS OF AFRICAN MEDICINAL SPICES AND VEGETABLES

The aerobic metabolic processes, such as respiration and photosynthesis lead to the production of reactive oxygen species (ROS) in mitochondria, chloroplasts, and peroxisomes (Mekha et al., 2014). Free radicals, such as the superoxide anion (O_2^-), the hydroxyl radical (OH·), singlet oxygen ($^1O_2$), and hydrogen peroxide (H_2O_2) produced in the body are able to cause oxidative damage to deoxyribonucleic acid (DNA), proteins, and lipids. Within the mitochondria, O_2^- is continually formed while hydroxyl radicals are short-lived, but are the most damaging radicals within the body and can be formed via Haber–Weiss reaction from O_2^- and H_2O_2 (Mekha et al., 2014). Oxidative stress often results in degenerative chronic complications and premature aging affecting the eyes, heart, liver, blood vessels, nerves, and kidneys (Mohamed, 2014). Aging animals show increased lipid peroxidation activities and decreased antioxidant levels in the kidneys, skins, lenses, hearts, and aortas (Dubois et al., 2008). Prolonged oxidative stress increases protein glycation and the pathologic advanced glycation endproducts (AGEs) in organ tissues. Detoxification of hydrogen peroxide involves glutathione peroxidase through the conversion of reduced glutathione (GSH) to oxidized glutathione (GSSG), during which H_2O_2 is reduced to water (Alessio and Blasi, 1997; Mekha et al., 2014). Glutathione peroxidase is a selenoprotein, which reduces lipidic or nonlipidic hydroperoxides as well as H_2O_2 while oxidizing glutathione (Mekha et al., 2014). Circulating human erythrocytes have the ability to scavenge O_2^- and H_2O_2 generated extracellularly by activated neutrophils, superoxide dismutase, catalase, and glutathione peroxidase-dependent mechanisms.

Superoxide dismutase, the first line of defense against ROS, catalyzes the dismutation of the superoxide anion into hydrogen peroxide. Catalase can then transform hydrogen peroxide into H_2O and O_2 (Mekha et al., 2014). Spices and vegetables are of considerable interest due to their antioxidant properties and potential beneficial health effects. Some of the common antiglycaption African vegetables and spices include *A. sativum*, *Z. officinale*, *T. vulgaris*, *Petroselinum crispum*, *C. longa*, and *A. cepa* (Mohamed, 2014).

The curry tree, *M. koenigii* or *B. koenigii* (Rutaceae), is a tropical to subtropical tree which is native to India and Sri Lanka. In tropical Africa it is planted in many countries, including Nigeria, Kenya, Tanzania, and most of the Indian Ocean Islands, where Indian immigrants settled (Matu, 2011). The leaves of curry tree are used as an herb in ayurvedic medicine. They possess antidiabetic properties (Arulselvan and Subramanian, 2007). The leaves extract of *M. koenigii* displayed scavenging activity against radicals, on 2,2-diphenyl-1-picrylhydrazyl (DPPH), nitric oxide, and superoxide, in a concentration-dependent manner and inhibited $FeSO_4$-induced lipid peroxides and hydroperoxides in an erythrocyte membrane model (Mekha et al., 2014). It was shown that extract from this plant significantly decreased lipid peroxidation in plasma and erythrocytes, significantly elevated levels of nonenzymatic antioxidants (β-carotene, vitamins A, C, and E) in serum, and significantly decreased activity of catalase, elevated activities of superoxide dismutase and glutathione-S-transferase, and raised levels of reduced GSH in erythrocytes in curry leaves-treated aged subjects (Mekha et al., 2014).

C. sativum (Apiaceae), commonly known as coriander or cilantro or Chinese parsley, is an annual herb native to regions spanning from southern Europe and northern Africa to southwestern Asia. All parts of the plant are edible with fresh leaves and the dried seeds being the parts most traditionally used in cooking. Coriander is common in South Asian, Southeast Asian, Indian, Middle Eastern, Caucasian, Central Asian, Mediterranean, Tex-Mex, Latin American, Brazilian, Portuguese, Chinese, and African cuisines (Samuelsson, 2003). It was shown that supplementation of the leaves of *C. sativum* could increase levels of serum nonenzymatic antioxidants, such as β-carotene and vitamin C, and erythrocyte nonenzymatic antioxidant-glutathione, improved activity of erythrocyte antioxidant enzyme, glutathione-S-transferase subjects (Mekha et al., 2014). Lipid peroxidation was also very effectively decreased in erythrocytes and plasma in arthritis patients. Bioactive components were suggested to be **5**, apigenin (**9**), caffeic acid, chlorogenic acid (**13**), protocatechuric acid, ascorbic acid, β-carotene, rhamnetin, rutin (**17**), terpinen, *trans*-anethole, umbelliferone, borneol present in coriander leaves subjects (Mekha et al., 2014).

It was demonstrated that among vegetables, *B. oleracea* (broccoli) contained the highest total phenolic content, followed by *S. oleracea* (spinach), *A. cepa* (onion), *Capsium* sp. (red pepper), *D. carota* (carrot), *B. oleracea* (cabbage), *S. tuberosum* (potato), *L. sativa* (lettuce), *A. graveolens* (celery), and *C. sativus* (cucumber). Red pepper displayed the highest total antioxidant activity, followed by broccoli, carrot, spinach, cabbage, yellow onion, celery, potato, lettuce, and cucumber (Chu et al., 2002).

G. africanum (Gnetaceae) is one of the most popular green leafy vegetables in Cameroon, Gabon, Congo, Nigeria, and Angola (Ogbonnaya and Chinedum, 2013). It is known as "*Eru* or *Kok*" in Cameroon, "*Koko*" in the Republic of Nigeria, Central Africa, "*Ntoumou*" in Gabon, "*Afang* or *Okazi*" in Nigeria, and eru in English. The leaves are edible and used in the treatment of enlarged spleen, boils, nausea, sore throat, pain at child birth, snake poisoning, diabetes mellitus, cataracts, and worm expeller (Ogbonnaya and Chinedum, 2013). It was shown that both raw and boiled forms of *G. africanum* have good antioxidant potentials indicating that both forms of the vegetable could be useful in the management of diseases that implicate free radicals (Ogbonnaya and Chinedum, 2013).

V. unguiculata commonly known as cowpea showed antioxidant effects, being able to reduce oxidative stress in cardiac and modulate the aorta estrogen receptor-β of ovariectomized rats (Khusniyati et al., 2014). The major phenolics associated with seed coats were identified as gallic and protocatechuic acid, whereas in cotyledons, *p*-hydroxybenzoic acid was prevalent (Gutiérrez-Uribe et al., 2011). After acid hydrolysis, antioxidant flavonoids, such as myricetin, quercetin, and kaempferol were identified in seed coats (Gutiérrez-Uribe et al., 2011).

Spices and vegetable's phytochemical, such as **5** is reported as the most potent scavenger of ROS in the flavonoid family and has more than sixfold the antioxidant capacity than the reference antioxidant trolox or vitamin C (van Acker et al., 1995; Boots et al., 2008). Compounds **3**, **6**, and **7** have been reported to inhibit lipid peroxidation by quenching oxygen free radicals and by enhancing the activity of endogenous antioxidant enzymes, such as superoxide dismutase, catalase, glutathione peroxidase, and glutathione-S-transferase (Mekha et al., 2014).

Broccoli is a source of antioxidant phenolics, such as **5** and kaempferol 3-O-sophoroside, 1-sinapoyl-2-feruloylgentiobiose, 1,2-diferuloylgentiobiose, 1,2,2′-trisina poylgentiobiose, and neochlorogenic acid (Shahidi and Ambigaipalan, 2015).

Cabbage contained a mixture of phenolics, such as 3-O-sophoroside-7-O-glucosides of **2** and **5**, sinapic, ferulic, or caffeic acid, anthocyanins (red cabbage), such as acylglycosides of cyaniding (Shahidi and Ambigaipalan, 2015). Compound **2** and myricetin (**8**) (Fig. 13.2) derivatives were also found in *Brassica* vegetables. Compound **9** and luteolin (**10**) were detected in different *Brassica* vegetables, except for broccoli (Bahorun et al., 2004).

The predominant phenolic acids in carrots are compound **13**, caffeic acid, *p*-hydroxybenzoic acid, ferulic acid, and other cinnamic acid isomers (Alasalvar et al., 2001). It was suggested that purple–yellow or purple–orange carrots had high antioxidant content and capacity (Sun et al., 2009). The carrot constituent, dicaffeoylquinic acid (**11**), exerts a very strong antioxidant activity (Alasalvar et al., 2001; Shahidi and Ambigaipalan, 2015).

Onions are one of the richest sources of flavonoids in the human diet. Onions possess a high level of antioxidant activity, attributed to their flavonoid constituents, namely **2**, **5**, **8**, and catechin (**12**) (Shahidi and Ambigaipalan, 2015). Two major components,

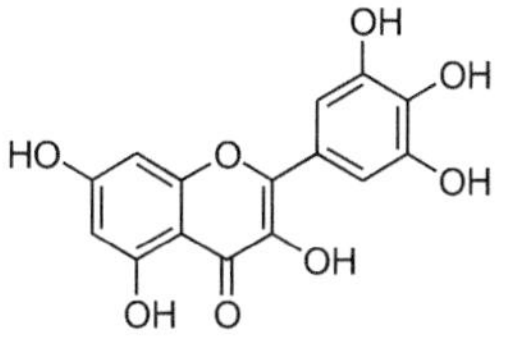

Figure 13.2 *Chemical structures of selected antioxidant compounds from spices and vegetables.* Myricetin (**8**); apigenin (**9**); luteolin (**10**); dicaffeoylquinic acid (**11**); catechin (**12**); chlorogenic acid (**13**); malvidin (**14**).

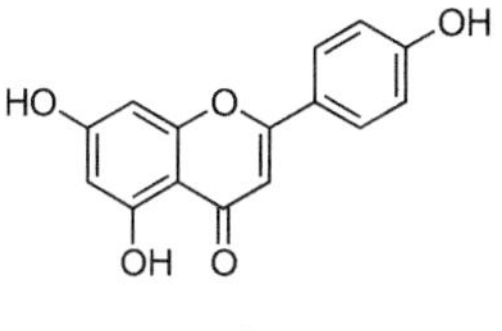

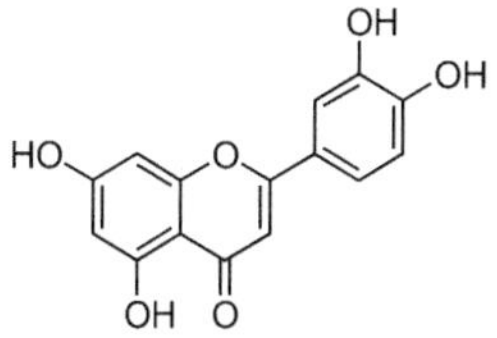

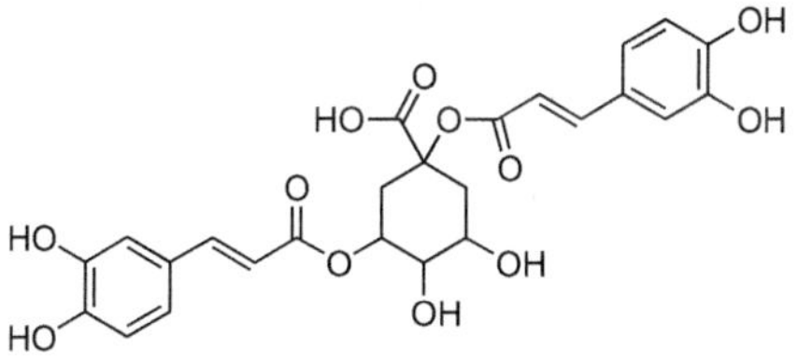

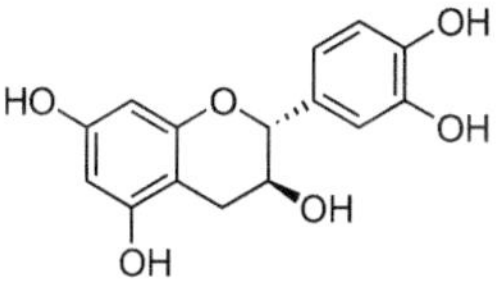

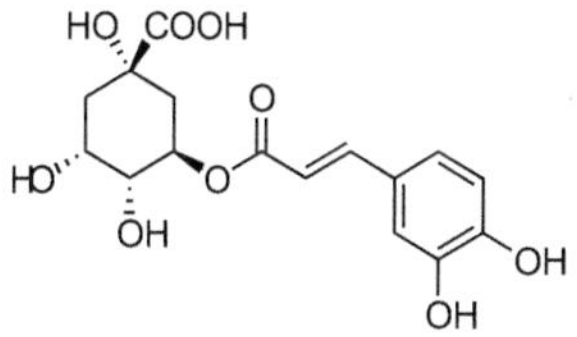

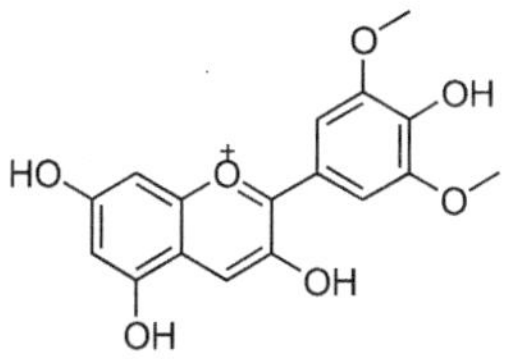

5 monoglucoside and quercetin, diglucoside account for 80% of the total flavonoids in onions (Shahidi and Ambigaipalan, 2015).

Compound **13** is the predominant phenolic in potato tuber, constituting up to 90% of its total phenolic content (Shahidi and Ambigaipalan, 2015). Major phenolics in potato peel are **13**, gallic acid, protocatechuic acid, caffeic acid, and **5** (Shahidi and Ambigaipalan, 2015). A higher anthocyanidin content and more hydroxylated anthocyanidins (malvidin, **14**) might contribute to a higher antioxidant activity of purple-fleshed potatoes (Shahidi and Ambigaipalan, 2015).

In celery 7-O-apiosylglucosides and 7-O-glucosides of **9**, **10**, and chrysoeriol as well as **13**, cinnamic acids, coumarins have identified (Shahidi and Ambigaipalan, 2015). Lettuce is a good source of flavonoids, such as **2**, **5**, **9**, **10** and chrysoeriol, quercetin conjugates [quercetin 3-O-galactoside, quercetin 3-O-glucoside, quercetin 3-O-glucuronide, quercetin 3-O-(6-O-malonyl) glucoside, and quercetin 3-O-rhamnoside] and luteolin 7-O-glucuronide, cyanidin conjugates [cyanidin 3-O-glucoside and cyanidin 3-O-[(6-O-malonyl)glucoside]] (Shahidi and Ambigaipalan, 2015).

4 MEDICINAL SPICES AND VEGETABLES FROM AFRICA USED IN THE TREATMENT OF EPILEPSY

Epilepsy is a serious and complex set of neurological conditions, which affects 65 million individuals worldwide, and more than 85% of these patients live in developing countries (Xiao et al., 2015). Epilepsy in Africa is associated with significant morbidity and mortality as well as a large treatment gap (Kariuki et al., 2015). Seizure and neurological deficits are common in people with epilepsy from Africa and may be related to perinatal complications, head injuries, and central nervous system infections (Kariuki et al., 2015). Vegetables and spices have been used in Africa since ancient times in the treatment of epilepsy or as sedatives in neurologic disorders. The scientific evidence of their effectiveness has also been reported.

4.1 *Curcuma longa* (Zingiberaceae)

C. longa or turmeric is a rhizomatous herbaceous perennial plant native to southwest India and grow in Africa, especially in Nigeria and South Africa. In traditional Indian medicine, *C. longa* has been used to treat stomach and liver ailments, as well as topically to heal sores, basically for its supposed antimicrobial property (Prasad and Aggarwal, 2011). The active compound **6** is believed to have a wide range of biological effects including antiinflammatory, antioxidant, antitumor, antibacterial, and antiviral activities, which indicate potential in clinical medicine (Prasad and Aggarwal, 2011). Curcuma oil has been recognized to possess multiple therapeutic activities for the treatment of viral infection, tumor, and inflammation, as well as substantial pharmacological properties, such as cognitive enhancement countries (Sun et al., 2008). Turmeric, obtained from the rhizomes of *C. longa*, is used in Asia as a traditional medicine for the treatment of epilepsy (Sun et al., 2008; Xiao et al., 2015). Curcumol (**15**) (Fig. 13.3) is the principal component responsible for the antiseizure activity of turmeric (Sun et al., 2008; Xiao et al., 2015). It was shown that compound **15** may suppress epileptic seizures countries by significantly

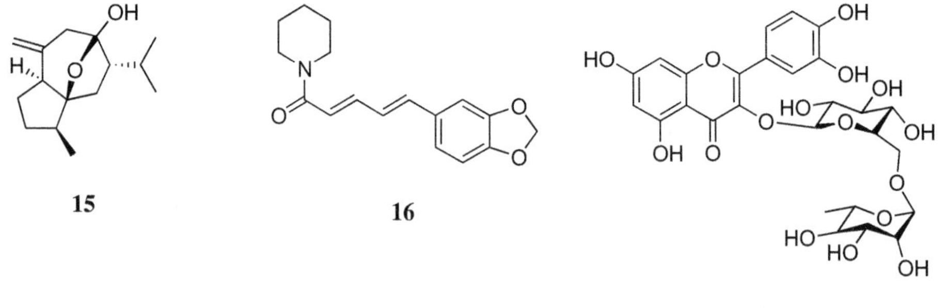

Figure 13.3 *Chemical structures of three selected compounds.* Curcumol (**15**); piperine (**16**); rutin (**17**) from spices and vegetables with potential anticonvulsant effects.

facilitating γ-aminobutyric acid (GABA$_A$) receptors in a zebrafish seizure model (Sun et al., 2008; Xiao et al., 2015). The pharmacological potential of *C. longa* is discussed in detail in Chapter 19.

4.2 *Glycyrrhiza glabra* (Fabaceae)

G. glabra, commonly known as liquorice or licorice, is an herbaceous perennial legume native to North Africa, southern Europe, India, and parts of Asia (http://www.jstor.org/stable/4114903, 1894). *G. glabra* has been used to treat sore throats, mouth ulcers, stomach ulcers, inflammatory stomach conditions, and indigestion. It is also used to combat food poisoning in modern Chinese herbalism. Liquorice rhizomes can be chewed or made into tea, which with other antispasmodic herbs is often taken for menstrual cramps. It was demonstrated that the liquorice flavonoid extract has protective effects on seizure-induced neuronal cell death and cognitive impairment through its antioxidative effects suggesting a potential use in the prevention and treatment of seizure-induced brain injury (Xiao et al., 2015; Zeng et al., 2013).

4.3 *Piper nigrum* as an antiepileptic plant

P. nigrum was used to treat epilepsy in China approximately 1400 years ago according to the traditional Chinese book *Tang Xin Xiu Ben Cao* in the Tang Dynasty (Pei, 1983; Xiao et al., 2015). Piperine (**16**) (Fig. 13.3), a piperidine alkaloid derived from *Piper* sp., exhibited effects on inhibitory amino acids and the GABAergic system, which significantly increased the latencies to first convulsion and death, as well as the percentage of survival in pilocarpine-induced convulsive mice (da Cruz et al., 2013; Xiao et al., 2015). Compound **16** and its derivatives are effective anticonvulsant drugs that antagonize convulsions induced by physical and chemical methods (Pei, 1983). Their major anticonvulsant activity as shown in animal tests lies in the modification of the maximal electroshock seizure pattern (Pei, 1983). They also have sedative-hypnotic, tranquilizing, and muscle-relaxing actions and can intensify the depressive action of other depressants, when used in combination (Pei, 1983). Antiepilepsirine, one of the derivatives of **16**, is used as an antiepileptic drug in treating different types of epilepsy (Pei, 1983).

4.4 *Zingiber officinale* as an antiepileptic plant

Glutamate has been recognized as the key excitatory transmitter in the central nervous system that underlies the expression of seizure discharges (Anwyl, 2009; Xiao et al., 2015). The expression of glutamate receptors known as metabotropic glutamate receptors is strongly implicated in the initiation of an epileptogenic process (Anwyl, 2009; Xiao et al., 2015). *Z. officinale* has neuroprotective and inhibitory effects on glutamate receptors, which may be responsible for the anticonvulsant activity and the prevention of seizure discharges. The 5-HT$_3$ receptor is an ionotropic receptor activated by serotonin which belongs to the Cys-loop receptor family, such as the GABA$_A$ receptor, acetylcholine, and glycine receptors. It has been proved that the extract of ginger and its fractions have anti-5-HT$_3$-receptor effects (Abdel-Aziz et al., 2006; Riyazi et al., 2007; Xiao et al., 2015). Also, 5-HT$_3$-receptor antagonists have been suggested to possess anticonvulsant activity (Vishwakarma et al., 2002; Xiao et al., 2015). Ginger extract, with its modulatory effect on the 5-HT$_3$-receptor, may alter pentylenetetrazole (PTZ)-induced seizures and increase the seizure threshold. A study in mouse seizure model showed that the acute administration of ginger extract had anticonvulsant effects on a timed intravenous PTZ-induced seizure and suggested that the anticonvulsant effect of ginger may have been mediated by antioxidant mechanisms, oxidative stress inhibition, and a simultaneous influence on different types of calcium channels, as well as both excitatory and inhibitory systems of neurotransmission (Hosseini and Mirazi, 2014; Xiao et al., 2015).

4.5 Other phytochemicals with antiepileptic properties

Flavonoids are a class of natural constituents that are widely distributed in spices and vegetables with many pharmacologic properties including antiseizure properties. Compound **9** is a flavone found in *C. sativum* (Mekha et al., 2014), *Brassica* vegetables (Bahorun et al., 2004), and lettuce (Shahidi and Ambigaipalan, 2015). Compound **9** was suggested to reduce the latency in the onset of PTZ-induced convulsions (Liu et al., 2012). Rutin (**17**) (Fig. 13.3) or rutoside, quercetin-3-O-rutinoside and sophorin, is the glycoside between the flavonol **5** and the disaccharide rutinose (α-L-rhamnopyranosyl-(1→6))-β-D-glucopyranose). Compound **17** is one of the constituents of *C. sativum* (Mekha et al., 2014). Compound **17** is also one of the most common flavonoid glycosides in many plants, fruits, and vegetables, and has shown anticonvulsant effects (Zhu et al., 2014). Intracerebroventricular injection of compound **17** dose-dependently diminished PTZ-induced minimal clonic seizures and tonic clonic seizures (Nassiri-Asl et al., 2010; Zhu et al., 2014). The phenolic monoterpene eugenol (**3**) was isolated from *S. aromaticum,* a spice harvested in Madagascar, Zanzibar, and Tanzania (Mohamed, 2014). Compound **3** has been found to have various uses in medicine including ameliorating epileptic seizures because of its modulating effect on neuronal excitability (Zhu et al., 2014). It was shown that the anticonvulsant effect of **3** may be associated primarily with its specific effect on ionic currents, namely increasing the degree of voltage-gated Na$^+$ currents (I_{Na}) inactivation and suppressing the noninactivating I_{Na} (Huang et al., 2012; Zhu et al., 2014).

5 PREVENTION OF RHEUMATOID ARTHRITIS WITH AFRICAN MEDICINAL SPICES AND VEGETABLES

Arthritis, a chronic, inflammatory multisystem disease, occurs in all the age groups and involves array of chronic conditions, affecting the joints and surrounding tissues. Osteoarthritis, the most common chronic medical condition in people aged 65 years and older, is a progressive rheumatic disease characterized by the deterioration of articular cartilage

(Mekha et al., 2014). Hypouricemic agents are commonly employed for the treatment of chronic gouty arthritis, including xanthine oxidase (XO) inhibitors and uricosuric agents (Gawlik-Dziki, 2012). Allopurinol is the only clinically used XO inhibitor (XOI), although it suffers from many side effects, such as hypersensitivity syndrome, Stevens Johnson syndrome, and renal toxicity (Gawlik-Dziki, 2012). The role of vegetables and spices of which many are naturally occurring in Africa has been documented.

W. somnifera (Solanaceae), commonly known as Indian ginseng, poison gooseberry, or winter cherry, is an herb distributed in Canary Islands, Africa, Middle East, India, Sri Lanka, and China (Gurib-Fakim, 2008). This plant has been used for centuries to improve overall physical and mental health and increase longevity and vitality by rejuvenating the body. It is also considered narcotic, hypnotic, aphrodisiac, liver tonic, purgative and diuretic, and is used in the treatment of tuberculosis, senile debility, nervousness, rheumatism, furuncles, sores, dropsy, cough and hiccup, as well as to induce abortion (Gurib-Fakim, 2008). In Madagascar the plant infusion is used to treat asthma. In Mauritius a poultice made from the fresh leaves and roots is applied on rheumatic limbs. The plant is also used as a tonic, aphrodisiac, and to treat skin problems. Root powder mixed with milk is drunk as an aphrodisiac. A root decoction is drunk to treat dysuria, gonorrhea, or upset stomach, whereas a maceration of the roots is given as an enema against gangrenous inflammation of the rectum. A decoction of the root bark is drunk to treat asthma. In southern Africa the hypnotic effects of an alcoholic extract are used in the treatment of alcoholism, tuberculosis, and emphysema. In Somalia a root decoction is given to children with fever and disturbed sleep, for example, from nightmares; rubbing preparations of crushed leaves has similar effects (Gurib-Fakim, 2008). It was shown that *W. somnifera* root extract was able to significantly reduce bony degenerative changes in Freund's adjuvant-induced arthritis in rats, better than the reference drug, hydrocortisone 40 (Mekha et al., 2014). The antirheumatic of *Z. officinale* has also been reported (Mekha et al., 2014).

6 AFRICAN SPICES AND VEGETABLES AND THEIR POTENTIAL EFFECTS ON OBESITY

Obesity is defined as a phenotypic manifestation of abnormal or excessive fat accumulation that alters health and increases mortality (Gonzalez-Castejon and Rodriguez-Casado, 2011). Obesity facilitates the development of metabolic disorders, such as diabetes, hypertension, and cardiovascular diseases in addition to chronic diseases, such as stroke, osteoarthritis, sleep apnea, some cancers, and inflammation-based pathologies (Gonzalez-Castejon and Rodriguez-Casado, 2011). In Western countries, people are considered obese when a measurement obtained by dividing a person's weight by the square of the person's height, known as their body mass index (BMI), exceeds 30 kg/m^2, with the range 25–30 kg/m^2 defined as overweight. The most common obesity classification is based on BMI intervals related to risk of mortality (National Institutes of Health (NIH), 1998). Hence, obesity is classified as class I for a BMI between 30 and 34.9 associated with a moderate risk, class II for a BMI between 35 and 39.9 with a high risk, and class III for a BMI 40 or above associated with a very high risk of mortality. Anatomically, the ratio of waist-to-hip circumferences (WHR) measures the degree of central (visceral, abdominal) versus peripheral (subcutaneous) adiposity (Gonzalez-Castejon and Rodriguez-Casado, 2011). Visceral fat is a major risk for metabolic disorders, whereas peripheral fat appears to be benign to metabolic complications (Gonzalez-Castejon and Rodriguez-Casado, 2011). Obesity can occur secondary to drug treatment, such as antipsychotic, antidepressant, antiepileptic, steroids, insulin (iatrogenic), or to certain diseases, such as cushing syndrome, hypothyroidism, and hypothalamic defects (Gonzalez-Castejon and Rodriguez-Casado, 2011). Dietary phytochemicals might be employed as antiobesity agents, because they may suppress growth of adipose tissue, inhibit differentiation of preadipocytes, stimulate lipolysis, and induce apoptosis of existing adipocytes, thereby reducing adipose tissue mass (Gonzalez-Castejon and Rodriguez-Casado, 2011). Compounds listed in this section were reported previously in medicinal spices or vegetables occurring in Africa.

Compound **4** was found having effects on obesity; it decreases adipogenesis and viability in maturing preadipocytes by downregulating adipocytespecific transcription factors, and by altering the expression of adipocyte specific genes like PPARγ, C/EBPα, SREBP-1c, FAS, LPL, and HSL (Baile et al., 2011; Gonzalez-Castejon and Rodriguez-Casado, 2011; Rayalam et al., 2008). Compound **4** could alter fat mass by directly affecting biochemical pathways involved in adipogenesis in maturing preadipocytes. It increases the activity of enzymes related to calorie restriction [NAD-dependent deacetylase sirtuin-3, (SIRT3), uncoupling protein 1 (UCP1), and mitofusin-2 (MFN2)] in vitro in maturing preadipocytes (Gonzalez-Castejon and Rodriguez-Casado, 2011; Rayalam et al., 2008; Zhang, 2006). In mature adipocytes, compound **4** increases lipolysis, induces apoptosis, and reduces lipogenesis and proliferation, thereby contributing to reduce lipid accumulation in vitro (Gonzalez-Castejon and Rodriguez-Casado, 2011; Picard et al., 2004; Rayalam et al., 2008). Also, **4** reduces the expression of mediators of the inflammatory response [tumor necrosis factor alpha (TNFα), interleukin-6 (IL-6), and cyclooxygenase-2 (COX-2)] in mature adipocytes, inhibits TNFα-activated nuclear factor-kappa B (NF-κB) signaling (Baile et al., 2011; Gonzalez-Castejon and Rodriguez-Casado, 2011), and reverses the TNFα-induced secretion, and mRNA expression of plasminogen activator inhibitor-1(PAI-1), IL-6, and adiponectin (Kang et al., 2010b; Olholm et al., 2010). Besides, **4** modulates adipokine expression and improves insulin sensitivity in adipocytes by modification of Ser/Thr phosphorylation of insulin receptor substrate-1, and downstream serine/threonine kinase (AKT) (Gonzalez-Castejon and Rodriguez-Casado, 2011; Kang et al., 2010a; Olholm et al., 2010). In human preadipocytes, **4** inhibits proliferation and adipogenic differentiation in a sirtuin 1 (SIRT1) that promotes fat mobilization by repressing peroxisome proliferator activated receptor-gamma (PPARγ) (Fischer-Posovszky et al., 2010). In the mature human adipocytes, **4** increases insulin-stimulated glucose uptake, but, simultaneously, inhibits lipogenesis (Fischer-Posovszky et al., 2010; Gonzalez-Castejon and Rodriguez-Casado, 2011), reverses IL-1β-stimulated secretion, and decreases gene expression of the proinflammatory adipokines IL-6, IL-8, monocyte chemotactic protein-1 (MCP-1), and PAI-1 (Gonzalez-Castejon and Rodriguez-Casado, 2011; Olholm et al., 2010).

The more abundant curcuminoids **6**, demethoxycucumin, and bisdemethoxycurcumin are nontoxic yellow pigment found in the *Curcuma* and *Zingiber* species, which are sources of the spices turmeric and ginger, respectively (Gonzalez-Castejon and Rodriguez-Casado, 2011). It was reported that **6** can regulate the expression of genes involved in energy metabolism and lipid accumulation, decreasing the level of intracellular lipids (Ejaz et al., 2009; Aggarwal, 2010; Gonzalez-Castejon and Rodriguez-Casado, 2011). In adipose tissues, compound **6** suppresses angiogenesis, which is necessary for tissue growth. Together with the effects on lipid metabolism in adipocytes, compound **6** contributes to lower body fat and body weight gain (Ejaz et al., 2009; Gonzalez-Castejon and Rodriguez-Casado, 2011). Also, **6** improves obesity-associated inflammation and associated metabolic disorders, such as insulin resistance, hyperglycemia, hyperlipidemia, and hypercholesterolemia (Gonzalez-Castejon and Rodriguez-Casado, 2011).

The flavones **9** and **10** appear to inhibit cell adhesion molecule (CAM) gene expression, increase in obesity processes, by blunting the activation of NF-κB stimulated by TNFα. These inhibitory mechanisms of the flavones appear to be independent of an antioxidant effect and may argue for transcriptional mechanisms as the major target of the antiatherogenic action of flavones. It was demonstrated that **10** inhibits low-grade chronic inflammation induced during the coculture of adipocytes and macrophages (Ando et al., 2009; Gonzalez-Castejon and Rodriguez-Casado, 2011), and prevents the phosphorylation of c-Jun N-terminal protein kinases (JNK) in macrophages activated from adipocytes (Ando et al., 2009; Gonzalez-Castejon and Rodriguez-Casado, 2011). Compound **10** exhibits antilipase activity (17.3%) and enhanced insulin sensitivity via activation of PPARγ transcriptional activity in adipocytes (Zheng et al., 2010; Gonzalez-Castejon and Rodriguez-Casado, 2011).

Capsaicin (**7**) was reported to attenuate not only obesity-induced inflammation, but also obesity-related metabolic disorders and liver diseases (Gonzalez-Castejon and Rodriguez-Casado, 2011; Kang et al., 2010b). Compound **7** suppresses obesity-induced inflammatory responses by reducing levels of TNFγ, IL-6, and MCP-1, and enhances adiponectin levels in adipose tissue and liver, which are important for insulin response (Gonzalez-Castejon and Rodriguez-Casado, 2011; Kang et al., 2010b). These beneficial effects are associated with the dual action of **7** on PPARγ/PPARα and transient receptor potential cation channel subfamily V member 1 (TrpV1) expression/activation associated with NF-κB inactivation and PPARγ activation (Gonzalez-Castejon and Rodriguez-Casado, 2011; Kang et al., 2010b). In addition, **7** was reported to suppress macrophage migration induced by the adipose tissue and macrophage activation (Gonzalez-Castejon and Rodriguez-Casado, 2011).

7 ROLE OF AFRICAN SPICES AND VEGETABLES IN HUMAN FERTILITY

Fertility is the natural capability to produce offspring and differ from fecundity, which is defined as the potential for reproduction (influenced by gamete production, fertilization, and carrying a pregnancy to term). Infertility is known as a lack of fertility while a lack of fecundity is known as sterility. Many factors, such as nutrition, sexual behavior, consanguinity, culture, instinct, endocrinology, timing, economics, way of life, and emotions influence human fertility. Women's fertility peaks in the early 20s, and drops considerably after age 35 while menopause typically occurs during a women's midlife (usually between ages 45 and 55 (http://www.ncbi.nlm.nih.gov/pubmedhealth/PMHT0025041/). During menopause (considered the end of the fertile phase of a woman's life), hormonal production by the ovaries is reduced, eventually causing a permanent cessation of the primary function of the ovaries. In men, it was demonstrated that increased age is associated with a decline in semen volume, sperm motility, and sperm morphology (Kidd et al., 2001). Sperm count declines with age, with men aged 50–80 years producing sperm at an average rate of 75% compared with men aged 20–50 years (Kidd et al., 2001). Vegetables, such as broccoli and cabbage, contain di-indolylmethane, which helps the body to get rid of excess estrogen (Randhawa et al., 2015). These types of greens should be eaten at least three times a week. Folic acid is very important for the proper development of the fetus. Eating folic acid while a woman is preparing for conception is better, because the baby will need this vital nutrient before the woman is even able to get a positive pregnancy test (Randhawa et al., 2015).

CONCLUSIONS

In this chapter, we discussed the health effects of African medicinal spices and vegetables, such as their use in the treatment of age-related and Alzheimer's diseases, convulsive diseases, in the prevention of rheumatoid arthritis, as antioxidant and sedative, their effects on obesity and fertility. It's appeared that several vegetables and spices naturally occurring in Africa could be used to improve the aforementioned health conditions. Some of these plants include *A. sativum*, *B. oleracea*, *Cinnamomum* species, *C. sativum*, *C. longa*, *G. glabra*, *P. nigrum*, *S. aromaticum*, *W. somnifera*, and *Z. officinale*. Some documented bioactive phytochemicals include cinnamaldehyde, kaempferol, eugenol, resveratrol, quercetin, curcumin, capsaicin, myricetin, apigenin, luteolin, dicaffeoylquinic acid, catechin, chlorogenic acid, malvidin, curcumol, piperine, and rutin.

REFERENCES

Abdel-Aziz, H., Windeck, T., Ploch, M., Verspohl, E.J., 2006. Mode of action of gingerols and shogaols on 5-HT3 receptors: binding studies, cation uptake by the receptor channel and contraction of isolated guinea-pig ileum. Eur. J. Pharmacol. 530 (1–2), 136–143.

Aggarwal, B.B., 2010. Targeting inflammation-induced obesity and metabolic diseases by curcumin and other nutraceuticals. Annu. Rev. Nutr. 30, 173–199.

Alasalvar, C., Grigor, J.M., Zhang, D., Quantick, P.C., Shahidi, F., 2001. Comparison of volatiles, phenolics, sugars, antioxidant vitamins, and sensory quality of different colored carrot varieties. J. Agric. Food Chem. 49 (3), 1410–1416.

Alessio, H.M., Blasi, E.R., 1997. Physical activity as a natural antioxidant booster and its effect on a healthy life span. Res. Q. Exerc. Sport 68 (4), 292–302.

Ando, C., Takahashi, N., Hirai, S., Nishimura, K., Lin, S., Uemura, T., 2009. Luteolin, a food-derived flavonoid, suppresses adipocyte-dependent activation of macrophages by inhibiting JNK activation. FEBS Lett. 583 (22), 3649–3654.

Anwyl, R., 2009. Metabotropic glutamate receptor-dependent long-term potentiation. Neuropharmacology 56 (4), 735–740.

Arulselvan, P., Subramanian, S.P., 2007. Beneficial effects of *Murraya koenigii* leaves on antioxidant defense system and ultra structural changes of pancreatic beta-cells in experimental diabetes in rats. Chem. Biol. Interact. 165 (2), 155–164.

Bahorun, T., Luximon-Ramma, A., Crozier, A., Aruoma, O., 2004. Total phenol, flavonoid, proanthocyanidin and vitamin C levels and antioxidant activities of Mauritian vegetables. J. Sci. Food Agric. 84, 1553–1561.

Baile, C.A., Yang, J.Y., Rayalam, S., Hartzell, D.L., Lai, C.Y., Andersen, C., 2011. Effect of resveratrol on fat mobilization. Ann. NY Acad. Sci. 1215, 40–47.

Banji, D., Banji, O.J., Dasaroju, S., Annamalai, A.R., 2013. Piperine and curcumin exhibit synergism in attenuating D-galactose induced senescence in rats. Eur. J. Pharmacol. 703 (1–3), 91–99.

Boots, A.W., Haenen, G.R., Bast, A., 2008. Health effects of quercetin: from antioxidant to nutraceutical. Eur. J. Pharmacol. 585 (2–3), 325–337.

Campagnolo, D., 2000. The tropical clumbing vine *Piper nigrum* (pepper). Seventh Australian Herb Conference, Townsville, July 7–9, 2000. Available from: http://rfcarchives.org.au/Next/Fruits/Pepper/GrowingPepper116-3-00.htm

Chondrogianni, N., Kapeta, S., Chinou, I., Vassilatou, K., Papassideri, I., Gonos, E.S., 2010. Anti-ageing and rejuvenating effects of quercetin. Exp. Gerontol. 45 (10), 763–771.

Chu, Y.F., Sun, J., Wu, X., Liu, R.H., 2002. Antioxidant and antiproliferative activities of common vegetables. J. Agric. Food Chem. 50 (23), 6910–6916.

da Cruz, G.M., Felipe, C.F., Scorza, F.A., da Costa, M.A., Tavares, A.F., Menezes, M.L., 2013. Piperine decreases pilocarpine-induced convulsions by GABAergic mechanisms. Pharmacol. Biochem. Behav. 104, 144–153.

Djeussi, D.E., Noumedem, J.A., Seukep, J.A., Fankam, A.G., Voukeng, I.K., Tankeo, S.B., 2013. Antibacterial activities of selected edible plants extracts against multidrug-resistant Gram-negative bacteria. BMC Complement. Altern. Med. 13 (1), 164.

Dubois, M., Canadas, D., Despres-Pernot, A.G., Coste, C., Pfohl-Leszkowicz, A., 2008. Oxygreen process applied on nongerminated and germinated wheat: role of hydroxamic acids. J. Agric. Food Chem. 56 (3), 1116–1121.

Ejaz, A., Wu, D., Kwan, P., Meydani, M., 2009. Curcumin inhibits adipogenesis in 3T3-L1 adipocytes and angiogenesis and obesity in C57/BL mice. J. Nutr. 139 (5), 919–925.

Ensminger, A., 1994. Foods and Nutrition Encyclopediavol. 1CRC Press, Boca Raton, Florida, USA, p. 750.

Fankam, A.G., Kuete, V., Voukeng, I.K., Kuiate, J.R., Pages, J.M., 2011. Antibacterial activities of selected Cameroonian spices and their synergistic effects with antibiotics against multidrug-resistant phenotypes. BMC Complement. Altern. Med. 11, 104.

Finotti, E., Bertone, A., Vivanti, V., 2006. Balance between nutrients and anti-nutrients in nine Italian potato cultivars. Food Chem. 99 (4), 698–701.

Fischer-Posovszky, P., Kukulus, V., Tews, D., Unterkircher, T., Debatin, K.M., Fulda, S., 2010. Resveratrol regulates human adipocyte number and function in a Sirt1-dependent manner. Am. J. Clin. Nutr. 92 (1), 5–15.

Frydman-Marom, A., Levin, A., Farfara, D., Benromano, T., Scherzer-Attali, R., Peled, S., 2011. Orally administrated cinnamon extract reduces beta-amyloid oligomerization and corrects cognitive impairment in Alzheimer's disease animal models. PLoS One 6 (1), e16564.

Gawlik-Dziki, U., 2012. Dietary spices as a natural effectors of lipoxygenase, xanthine oxidase, peroxidase and antioxidant agents. LWT—Food Sci. Technol. 47 (1), 138–146.

Gonzalez-Castejon, M., Rodriguez-Casado, A., 2011. Dietary phytochemicals and their potential effects on obesity: a review. Pharmacol. Res. 64 (5), 438–455.

Griffin, B., Selassie, M., Gwebu, E.T., 2000. Aged garlic extract suppresses lipid peroxidation induced by beta-amyloid in PC12 cells. Vitro Cell Dev. Biol. Anim. 36 (5), 279–280.

Gurib-Fakim, A., 2008. Withania somnifera (L.) Dunal. In: Schmelzer, G.H., Gurib-Fakim, A. (Eds.), Prota 11(1): Medicinal plants/Plantes médicinales 1. [CD-Rom]. PROTA, Wageningen.

Gutiérrez-Uribe, J.A., Romo-Lopez, I., Serna-Saldívar, S.O., 2011. Phenolic composition and mammary cancer cell inhibition of extracts of whole cowpeas (*Vigna unguiculata*) and its anatomical parts. J. Funct. Foods 3 (4), 290–297.

Hamidpour, R., Hamidpour, M., Hamidpour, S., Shahlari, M., 2015. Cinnamon from the selection of traditional applications to its novel effects on the inhibition of angiogenesis in cancer cells and prevention of Alzheimer's disease, and a series of functions such as antioxidant, anticholesterol, antidiabetes, antibacterial, antifungal, nematicidal, acaracidal, and repellent activities. J. Trad. Complement. Med. 5 (2), 66–70.

Hosseini, A., Mirazi, N., 2014. Acute administration of ginger (*Zingiber officinale* rhizomes) extract on timed intravenous pentylenetetrazol infusion seizure model in mice. Epilepsy Res. 108 (3), 411–419.

Hritcu, L., Noumedem, J.A., Cioanca, O., Hancianu, M., Kuete, V., Mihasan, M., 2014. Methanolic extract of *Piper nigrum* fruits improves memory impairment by decreasing brain oxidative stress in amyloid beta (1-42) rat model of Alzheimer's disease. Cell Mol. Neurobiol. 34 (3), 437–449.

Hritcu, L., Noumedem, J.A., Cioanca, O., Hancianu, M., Postu, P., Mihasan, M., 2015. Anxiolytic and antidepressant profile of the methanolic extract of *Piper nigrum* fruits in beta-amyloid (1-42) rat model of Alzheimer's disease. Behav. Brain Funct. 11, 13.

Huang, C.W., Chow, J.C., Tsai, J.J., Wu, S.N., 2012. Characterizing the effects of Eugenol on neuronal ionic currents and hyperexcitability. Psychopharmacology (Berl) 221 (4), 575–587.

Itokawa, H., Shi, Q., Akiyama, T., Morris-Natschke, S.L., Lee, K.H., 2008. Recent advances in the investigation of curcuminoids. Chin. Med. 3, 11.

Jasinski, M., Jasinska, L., Ogrodowczyk, M., 2013. Resveratrol in prostate diseases—a short review. Cent. Eur. J. Urol. 66 (2), 144–149.

Kampkotter, A., Gombitang Nkwonkam, C., Zurawski, R.F., Timpel, C., Chovolou, Y., Watjen, W., 2007. Effects of the flavonoids kaempferol and fisetin on thermotolerance, oxidative stress and FoxO transcription factor DAF-16 in the model organism *Caenorhabditis elegans*. Arch. Toxicol. 81 (12), 849–858.

Kang, J.H., Ascherio, A., Grodstein, F., 2005. Fruit and vegetable consumption and cognitive decline in aging women. Ann. Neurol. 57 (5), 713–720.

Kang, J.H., Goto, T., Han, I.S., Kawada, T., Kim, Y.M., Yu, R., 2010a. Dietary capsaicin reduces obesity-induced insulin resistance and hepatic steatosis in obese mice fed a high-fat diet. Obesity (Silver Spring) 18 (4), 780–787.

Kang, L., Heng, W., Yuan, A., Baolin, L., Fang, H., 2010b. Resveratrol modulates adipokine expression and improves insulin sensitivity in adipocytes: relative to inhibition of inflammatory responses. Biochimie 92 (7), 789–796.

Kariuki, S.M., White, S., Chengo, E., Wagner, R.G., Ae-Ngibise, K.A., Kakooza-Mwesige, A., 2016. Electroencephalographic features of convulsive epilepsy in Africa: A multicentre study of prevalence, pattern and associated factors. Clin. Neurophysiol 127 (2), 1099–1107.

Khusniyati, E., Sari, A.A., Yueniwati, Y., Noorhamdani, N., Nurseta, T., Keman, K., 2014. The effects of *Vigna unguiculata* on cardiac oxidative stress and aorta estrogen receptor-β expression of ovariectomized rats. Asian Pac. J. Reprod. 3 (4), 263–267.

Kidd, S.A., Eskenazi, B., Wyrobek, A.J., 2001. Effects of male age on semen quality and fertility: a review of the literature. Fertil. Steril. 75 (2), 237–248.

Kitani, K., Osawa, T., Yokozawa, T., 2007. The effects of tetrahydrocurcumin and green tea polyphenol on the survival of male C57BL/6 mice. Biogerontology 8 (5), 567–573.

Kris-Etherton, P.M., Hecker, K.D., Bonanome, A., Coval, S.M., Binkoski, A.E., Hilpert, K.F., 2002. Bioactive compounds in foods: their role in the prevention of cardiovascular disease and cancer. Am. J. Med. 113 (Suppl 9B), 71S–88S.

Kuete, V., 2014. Physical, hematological, and histopathological signs of toxicity induced by African medicinal plants. In: Kuete, V. (Ed.), Toxicological Survey of African Medicinal Plants. Elsevier, London, pp. 635–657.

Kuete, V., Krusche, B., Youns, M., Voukeng, I., Fankam, A.G., Tankeo, S., 2011. Cytotoxicity of some Cameroonian spices and selected medicinal plant extracts. J. Ethnopharmacol. 134 (3), 803–812.

Kumaravelu, P., Subramaniyam, S., Dakshinamoorthy, D., Devaraj, N., 1996. The antioxidant effect of eugenol on CCl4-induced erythrocyte damage in rats. J. Nutr. Biochem. 7 (1), 23–28.

Lei, P., Ayton, S., Finkelstein, D.I., Adlard, P.A., Masters, C.L., Bush, A.I., 2010. Tau protein: relevance to Parkinson's disease. Int. J. Biochem. Cell Biol. 42 (11), 1775–1778.

Liao, V.H., Yu, C.W., Chu, Y.J., Li, W.H., Hsieh, Y.C., Wang, T.T., 2011. Curcumin-mediated lifespan extension in *Caenorhabditis elegans*. Mech. Ageing Dev. 132 (10), 480–487.

Liu, Y.F., Gao, F., Li, X.W., Jia, R.H., Meng, X.D., Zhao, R., 2012. The anticonvulsant and neuroprotective effects of baicalin on pilocarpine-induced epileptic model in rats. Neurochem. Res. 37 (8), 1670–1680.

Lockwood, G., 1979. Major constituents of the essential oils of *Cinnamomum cassia* Blume growing in Nigeria. Planta Med. 36 (4), 380–381.

Matu, E., 2011. *Murraya koenigii* (L.) Spreng. In: Schmelzer, G.H., Gurib-Fakim, A. (Eds.), Prota 11(2): Medicinal plants/Plantes médicinales 2. [CD-Rom]. PROTA, Wageningen.

Mekha, M.S., Shobha, R.I., Rajeshwari, C.U., Andallu, B., 2014. Aging and arthritis: oxidative stress and antioxidant effects of herbs and spices. In: Preedy, V.R. (Ed.), Aging. Academic Press, San Diego, pp. 233–245, (Chapter 23).

Mishra, S., Palanivelu, K., 2008. The effect of curcumin (turmeric) on Alzheimer's disease: an overview. Ann. Indian Acad. Neurol. 11 (1), 13–19.

Mohamed, S., 2014. Herbs and spices in aging. In: Preedy, V.R. (Ed.), Aging. Academic Press, San Diego, pp. 99–107, (Chapter 10).

Moriguchi, T., Saito, H., Nishiyama, N., 1997. Anti-ageing effect of aged garlic extract in the inbred brain atrophy mouse model. Clin. Exp. Pharmacol. Physiol. 24 (3–4), 235–242.

Nassiri-Asl, M., Mortazavi, S.R., Samiee-Rad, F., Zangivand, A.A., Safdari, F., Saroukhani, S., 2010. The effects of rutin on the development of pentylenetetrazole kindling and memory retrieval in rats. Epilepsy Behav. 18 (1–2), 50–53.

National Institutes of Health (NIH), N.H., Lung, and Blood Institute (NHLBI), 1998. Clinical guidelines on the identification, evaluation, and treatment of overweight and obesity in adults—The Evidence Report. National Institutes of Health. Obes. Res. 6 (Suppl 2), 51S–209S.

Nirmal, S., Hui, Y., Özgül, E., Muhammad, S., Jasim, A., 2010. Handbook of Vegetables and Vegetable Processing. Wiley-Blackwell, New Jersey.

Nkanwen, E., Awouafack, M., Bankeu, J., Wabo, H., Mustafa, S., Ali, M., 2013. Constituents from the stem bark of *Cinnamomum zeylanicum* Welw. (Lauraceae) and their inhibitory activity toward *Plasmodium falciparum* enoyl-ACP reductase enzyme Rec. Nat. Prod. 7 (4), 296–301.

Ogbonnaya, E.C., Chinedum, E.K., 2013. Health promoting compounds and in vitro antioxidant activity of raw and decoctions of *Gnetum aficanum* Welw. Asian Pac. J. Trop. Dis. 3 (6), 472–479.

Olholm, J., Paulsen, S.K., Cullberg, K.B., Richelsen, B., Pedersen, S.B., 2010. Anti-inflammatory effect of resveratrol on adipokine expression and secretion in human adipose tissue explants. Int. J. Obes. (Lond) 34 (10), 1546–1553.

Orwa, C., Mutua, A., Kindt, R., Jamnadass, R., Anthony, S., 2009. Agroforestree database: a tree reference and selection guide. version 4.0. Available from: http://www.worldagroforestry.org/sites/treedbs/treedatabases.asp

Pei, Y.Q., 1983. A review of pharmacology and clinical use of piperine and its derivatives. Epilepsia 24 (2), 177–182.

Peterson, D.W., George, R.C., Scaramozzino, F., LaPointe, N.E., Anderson, R.A., Graves, D.J., 2009. Cinnamon extract inhibits tau aggregation associated with Alzheimer's disease in vitro. J. Alzheimers Dis. 17 (3), 585–597.

Picard, F., Kurtev, M., Chung, N., Topark-Ngarm, A., Senawong, T., Machado De Oliveira, R., 2004. Sirt1 promotes fat mobilization in white adipocytes by repressing PPAR-gamma. Nature 429 (6993), 771–776.

Pietsch, K., Saul, N., Menzel, R., Sturzenbaum, S.R., Steinberg, C.E., 2009. Quercetin mediated lifespan extension in *Caenorhabditis elegans* is modulated by age-1, daf-2, sek-1 and unc-43. Biogerontology 10 (5), 565–578.

Prasad, S., Aggarwal, B.B., 2011. Turmeric, the Golden Spice: From Traditional Medicine to Modern Medicine. CRC Press, Boca Raton, Florida, USA.

Randhawa, M.A., Khan, A.A., Javed, M.S., Sajid, M.W., 2015. Green leafy vegetables: a health promoting source. In: Watson, R.R. (Ed.), Handbook of Fertility. Academic Press, San Diego, pp. 205–220, (Chapter 18).

Ravindran, P., 2000. Black Pepper: Piper nigrum. Harwood Academic Publishers, Amsterdam, p. 12.

Rayalam, S., Della-Fera, M.A., Baile, C.A., 2008. Phytochemicals and regulation of the adipocyte life cycle. J. Nutr. Biochem. 19 (11), 717–726.

Riyazi, A., Hensel, A., Bauer, K., Geissler, N., Schaaf, S., Verspohl, E.J., 2007. The effect of the volatile oil from ginger rhizomes (*Zingiber officinale*), its fractions and isolated compounds on the 5-HT3 receptor complex and the serotoninergic system of the rat ileum. Planta Med. 73 (4), 355–362.

Samuelsson, M., 2003. Aquavit: And the New Scandinavian Cuisine. Houghton Mifflin Harcourt, Brownstown, MI, p. 12.

Saul, N., Pietsch, K., Menzel, R., Steinberg, C.E., 2008. Quercetin-mediated longevity in *Caenorhabditis elegans*: is DAF-16 involved? Mech. Ageing Dev. 129 (10), 611–613.

Shahidi, F., Ambigaipalan, P., 2015. Phenolics and polyphenolics in foods, beverages and spices: antioxidant activity and health effects—a review. J. Funct. Foods 18, 820–897.

Shan, B., Cai, Y.Z., Brooks, J.D., Corke, H., 2007. Antibacterial properties and major bioactive components of cinnamon stick (*Cinnamomum burmannii*): activity against foodborne pathogenic bacteria. J. Agric. Food Chem. 55 (14), 5484–5490.

Shen, L.R., Xiao, F., Yuan, P., Chen, Y., Gao, Q.K., Parnell, L.D., 2013. Curcumin-supplemented diets increase superoxide dismutase activity and mean lifespan in Drosophila. Age (Dordr) 35 (4), 1133–1142.

Shukri, R., Mohamed, S., Mustapha, N., 2010. Cloves protect the heart, liver and lens of diabetic rats. Food Chem. 122 (4), 1116–1121.

Si, H., Liu, D., 2007. Phytochemical genistein in the regulation of vascular function: new insights. Curr. Med. Chem. 14 (24), 2581–2589.

Steinmetz, K.A., Potter, J.D., 1996. Vegetables, fruit, and cancer prevention: a review. J. Am. Diet Assoc. 96 (10), 1027–1039.

Sun, C.Y., Hu, W., Qi, S.S., Dai, K.Y., Hu, S.W., Lou, X.F., 2008. Effect of *Rhizoma curcumae* oil on the learning and memory in rats exposed to chronic hypoxia and the possible mechanisms. Sheng Li Xue Bao 60 (2), 228–234.

Sun, T., Simon, P.W., Tanumihardjo, S.A., 2009. Antioxidant phytochemicals and antioxidant capacity of biofortified carrots (*Daucus carota* L.) of various colors. J. Agric. Food Chem. 57 (10), 4142–4147.

Tosun, A., Khan, S., 2015. Antioxidant actions of spices and their phytochemicals on age-related diseases. In: Preedy, R.R.W.R. (Ed.), Bioactive Nutraceuticals and Dietary Supplements in Neurological and Brain Disease. Academic Press, San Diego, pp. 311–318, (Chapter 32).

van Acker, S.A., Tromp, M.N., Haenen, G.R., van der Vijgh, W.J., Bast, A., 1995. Flavonoids as scavengers of nitric oxide radical. Biochem. Biophys. Res. Commun. 214 (3), 755–759.

Vishwakarma, S.L., Pal, S.C., Kasture, V.S., Kasture, S.B., 2002. Anxiolytic and antiemetic activity of *Zingiber officinale*. Phytother. Res. 16 (7), 621–626.

Wachtel-Galor, S., Siu, P.M., Benzie, I.F.F., 2014. Antioxidants, vegetarian diets and aging. In: Preedy, V.R. (Ed.), Aging. Academic Press, San Diego, pp. 81–91, (Chapter 8).

Williamson, G., Manach, C., 2005. Bioavailability and bioefficacy of polyphenols in humans. II. Review of 93 intervention studies. Am. J. Clin. Nutr. 81 (1 Suppl.), 243S–255S.

Xiao, F., Yan, B., Chen, L., Zhou, D., 2015. Review of the use of botanicals for epilepsy in complementary medical systems—Traditional Chinese Medicine. Epilepsy Behav 52 (Pt B), 281–289.

Zeng, L.H., Zhang, H.D., Xu, C.J., Bian, Y.J., Xu, X.J., Xie, Q.M., 2013. Neuroprotective effects of flavonoids extracted from licorice on kainate-induced seizure in mice through their antioxidant properties. J. Zhejiang Univ. Sci. B 14 (11), 1004–1012.

Zhang, J., 2006. Resveratrol inhibits insulin responses in a SirT1-independent pathway. Biochem. J. 397 (3), 519–527.

Zheng, C.D., Duan, Y.Q., Gao, J.M., Ruan, Z.G., 2010. Screening for anti-lipase properties of 37 traditional Chinese medicinal herbs. J. Chin. Med. Assoc. 73 (6), 319–324.

Zhu, H.L., Wan, J.B., Wang, Y.T., Li, B.C., Xiang, C., He, J., 2014. Medicinal compounds with antiepileptic/anticonvulsant activities. Epilepsia 55 (1), 3–16.

PART III

Popular African Medicinal Spices and Vegetables, and Their Health Effects

14. *Allium cepa*	353
15. *Allium sativum*	363
16. *Canarium schweinfurthii*	379
17. Cinnamon Species	385
18. *Cymbopogon citratus*	397
19. *Curcuma longa*	425
20. *Lactuca sativa*	437
21. *Mangifera indica* L. (Anacardiaceae)	451
22. *Moringa oleifera*	485
23. *Myristica fragrans*: A Review	497
24. *Passiflora edulis*	513
25. *Petroselinum crispum*: A Review	527
26. *Sesamum indicum*	549
27. African Medicinal Spices of Genus *Piper*	581
28. *Thymus vulgaris*	599
29. *Syzygium aromaticum*	611
30. *Zingiber officinale*	627

CHAPTER 14

Allium cepa

V. Kuete

1 INTRODUCTION

A. cepa L. (Liliaceae), also known as the bulb onion or common onion, is a vegetable and is the most widely cultivated species of the genus *Allium*, and probably originates from Central Asia (Messiaen and Rouamba, 2004). Onions are cultivated and used throughout the world. Bulbs from *A. cepa* are thought to have been used as a popular vegetable everywhere for millennia. They can be used raw, sliced for seasoning salads, boiled with other vegetables, or fried with other vegetables and meat. They are an essential ingredient in many African sauces and relishes (Messiaen and Rouamba, 2004). *A. cepa* is cultivated by about 170 countries for domestic use and about 8% of the global production is traded internationally. Besides its use as a condiment and spice for flavoring and enriching various cuisines, onion has been known for its high medicinal properties for thousands of years (Lawande, 2012). It acts as stimulant, diuretic and expectorant and, mixed with vinegar, it is useful in the case of sore throat. Essential oil from onion contains a heart stimulant, increasing pulse volume and frequency of systolic pressure and coronary flow. Onion consumption lowers blood sugar, lipids, and cholesterol. Fresh onion juice has antibacterial properties due to allicin, disulfide, and cysteine compounds and their interactions; an antiplatelet aggregation effect in human and animal blood has been reported due to regular consumption of onion (Mittal et al., 1974; Baghurst et al., 1977; Lawande, 2012). In this chapter, we have discussed the evidence of the pharmacological properties as well as the phytochemical composition of onion as obtained from scientific publications retrieved in Pubmed, Sciencedirect, Scopus, Web-of-knowledge, and Google Scholar.

2 CULTIVATION AND DISTRIBUTION OF *Allium cepa*

A. cepa is a biennial glabrous herb, usually growing as an annual from seed or bulbs, up to 100 cm tall. Onions (Fig. 14.1) are best cultivated in fertile soils that are well drained. They are cultivated by about 170 countries for domestic use and about 8% of the global production is traded internationally. According to the UN Food and Agriculture Organisation (FAO) records of 2012, Egypt is the first African producer with 2,208,080 tons and the fourth world producer after China (20,507,759 tons), India (13,372,100 tons), and USA (3,320,870 tons) (https://en.wikipedia.org/wiki/Onion). Most African

Figure 14.1 *Bulbs of onion. (Photo courtesy: V. Kuete).*

tropical countries import onion bulbs, either from Niger, which exports an important part of its 200,000 tons production, or from Europe or South Africa. Shallots are exported from northern Côte d'Ivoire and Mali to the neighboring countries (Messiaen and Rouamba, 2004).

3 CHEMISTRY OF *Allium cepa*

A. cepa is the source of various biologically active compounds, such as phenolic acids, thiosulfinates, and flavonoids.

3.1 Essential oil

Onion oil, with yield of extraction varying from 0.002% to 0.03%, consists of a complex mixture mostly formed by mono-, di-, tri-, and tetrasulfides with different alkyl groups (Fig. 14.2) (Hosoda et al., 2003; Vazquez-Armenta et al., 2016). The bioactive properties and characteristic flavor of onion have been attributed to sulfur-containing compounds, which are the main constituents of its essential oil (Hosoda et al., 2003; Vazquez-Armenta et al., 2016). Sulfur-containing compounds available in the scales of bulbs are responsible for the distinctive flavor and pungency of onion (Lawande, 2012). The main constituents of onion essential oil were identified as methyl 5-methylfuryl sulfide (**1**), dimethyl

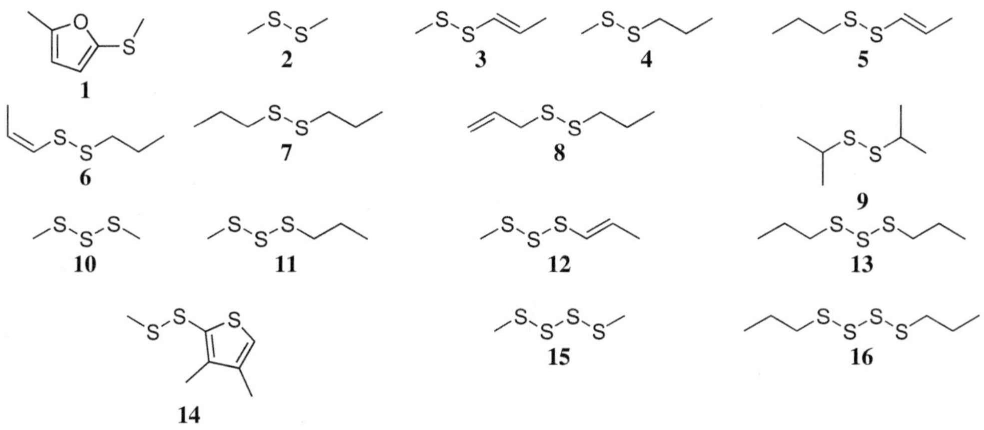

Figure 14.2 *Major volatile constituents identified in onion oil.* Methyl 5-methylfuryl sulfide (**1**); dimethyl disulfide (**2**); methyl 1-propenyl disulfide (**3**); methyl propyl disulfide (**4**); propyl *trans*-propenyl disulfide (**5**); propyl *cis*-propenyl disulfide (**6**); dipropyl disulfide (**7**); allyl propyl disulfide (**8**); isopropyl disulfide (**9**); dimethyl trisulfide (**10**); methyl propyl trisulfide (**11**); methyl 1-propenyl trisulfide (**12**); dipropyl trisulfide (**13**); methyl 3,4-dimethyl 2-thienyl disulfide (**14**); dimethyl tetrasulfide (**15**); dipropyl tetrasulfide (**16**).

disulfide (**2**), methyl 1-propenyl disulfide (**3**), methyl propyl disulfide (**4**), propyl *trans*-propenyl disulfide (**5**), propyl *cis*-propenyl disulfide (**6**), dipropyl disulfide (**7**), allyl propyl disulfide (**8**), isopropyl disulfide (**9**), dimethyl trisulfide (**10**), methyl propyl trisulfide (**11**), methyl 1-propenyl trisulfide (**12**), dipropyl trisulfide (**13**), methyl 3,4-dimethyl 2-thienyl disulfide (**14**), dimethyl tetrasulfide (**15**), and dipropyl tetrasulfide (**16**).

3.2 Phenolics

A. cepa is rich in phenolic compounds (Fig. 14.3), especially the outer layers of the bulb, which are not eaten and before thermal treatment (Bystrická et al., 2013). Some of these phenolics in onion include phenolic acids derived from benzoic acid or cinnamic acid, such as protocatechuic acid (**17**), gallic acid (**18**), *p*-hydroxybenzoic acid (**19**) as well as flavonoids, such as kaempferol (**20**), quercetin (**21**), myricetin (**22**), fisetin (**23**), and luteolin (**24**) (Bystrická et al., 2013). The red varieties of onion present the highest content of flavonols, and they also contain red anthocyans in the form of glycosides of cyanidin (**25**), peonidin (**26**), pelargonidin (**27**), etc. (Bystrická et al., 2013). Yellow onions have been reported to contain 270–1187 mg of flavonols per kilogram of fresh weight (FW), whereas red onions contain 415–1917 mg of flavonols per kilogram of FW (Slimestad et al., 2007). Flavonols are the predominant pigments of onions with at least 25 different compounds characterized, quercetin derivatives being the most important ones in all cultivars (Slimestad et al., 2007). Their glycosyl moieties are almost exclusively glucose, which is mainly attached to the 4′, 3, and/or 7-positions of the aglycones (Slimestad et al., 2007). Quercetin 4′-glucoside and quercetin 3,4′-diglucoside were also reported;

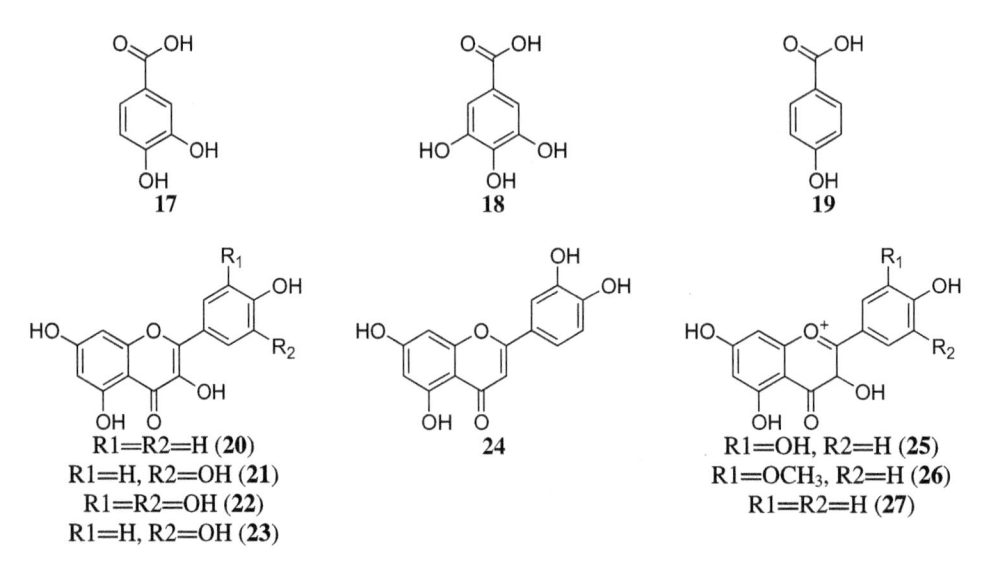

Figure 14.3 *Selected phenolic compounds identified in onion.* Protocatechuic acid (**17**); gallic acid (**18**); *p*-hydroxybenzoic acid (**19**); kaempferol (**20**); quercetin (**21**); myricetin (**22**); fisetin (**23**); luteolin (**24**); cyanidin (**25**); peonidin (**26**); pelargonidin (**27**).

analogous derivatives of kaempferol and isorhamnetin have been identified as minor pigments (Slimestad et al., 2007). Free amino acids, such as glutanic acid and arginine were reported abundant in onion (Lawande, 2012).

4 PHARMACOLOGY OF *Allium cepa*

A. cepa has a variety of pharmacological activities including anticancer, antidiabetic, cardiovascular, antioxidant, etc.

4.1 Anticancer activity

A. cepa is commonly used in daily diet and has been used as a supplementary folk remedy for cancer treatment. The cytotoxicity of crude onion extract and fractionated extracts (aqueous, methanolic, and ethyl acetate) as well as some of its constituents, **4** and **21**, were reported in Lucena multidrug resistant (MDR) human erythroleukemic cells and its K562 parental cell line (Votto et al., 2010). It was found that similar sensitivities were obtained for both tumoral cells with crude onion extract having significant effects in the cells (Votto et al., 2010). This extract provoked significant DNA damage in both cell lines, suggesting its ability to overcome cancers involving MDR cell lines (Votto et al., 2010). When investigating the effect of the onion constituent **22** on ultraviolet (UV) B-induced angiogenesis in an SKH-1 hairless mouse skin tumorigenesis model, it was found that this compound suppressed UVB-induced angiogenesis by regulating phosphoinositide 3 (PI-3) kinase activity (Jung et al., 2010). Polyphenols extract of onion-induced cytotoxicity toward leukemia U937 cells with an IC$_{50}$ value below 40 μg/mL (Han et al., 2013). The polyphenols extract induced caspase-dependent apoptosis in several human leukemic cells including U937 cells (Han et al., 2013). The apoptosis was triggered through extrinsic pathway by upregulating tumor necrosis factor (TNF)-related apoptosis-inducing ligand (TRAIL) receptor DR5 modulating as well as through intrinsic pathway by modulating B-cell lymphoma 2 (Bcl-2) and inhibitors of apoptosis proteins (IAP) family members (Han et al., 2013). It also induced caspase-dependent apoptosis through inhibition of phosphatidylinositol 3-kinase (PI3K)/Akt signaling pathway (Han et al., 2013). The role of diet on breast cancer risk was investigated in a case–control study of 345 patients diagnosed with primary breast carcinoma between 1986 and 1989 in the northeast of France (Lorraine) (Challier et al., 1998). The results suggested that higher intake of onion was correlated with lower risk of breast cancer (Challier et al., 1998).

4.2 Antidiabetic effects

The antidiabetic effect of *A. cepa* in the streptozotocin (STZ)-induced diabetic rats was evidenced by decreased blood glucose, serum lipid levels, and reduced renal oxidative stress (Bang et al., 2009). Investigations conducted by Yoshinari et al. (2012) also suggested that dietary onion extract is beneficial for improving diabetes by decreasing lipid levels. Onion peel extracts were suggested to improve glucose response and insulin resistance associated with type 2 diabetes by alleviating metabolic dysregulation of free fatty acids, suppressing oxidative stress, upregulating glucose uptake at peripheral tissues, and/or downregulating inflammatory gene expression in liver (Jung et al., 2011). The onion constituents cycloalliin, S-methyl-L-cysteine, S-propyl-L-cysteine sulfoxide, dimethyl trisulfide, especially S-methyl-L-cysteine sulfoxide were reported to be effective in inhibiting formation of oil drop in rat white preadipocyte cells, suggesting that these compounds may be involved in the antiobesity effect of the onion extract (Yoshinari et al., 2012).

4.3 Antimicrobial activities

Zohri et al. (1995) investigated the inhibitory effect of onion oil against the growth of various isolates of selected Gram-positive and Gram-negative bacteria and nine different species of dermatophytic fungi. They found that onion oil was highly active against all Gram-positive bacteria tested, namely *Bacillus anthracis, Bacillus cereus, Micrococcus luteus*, and *Staphylococcus aureus* and only one isolate of Gram-negative bacteria, *Klebsiella pneumoniae* (Zohri et al., 1995). The investigators also found that onion oil at 200 ppm completely inhibited the growth of *Microsporum canis, Microsporum gypseum*, and *Trichophyton simii* whereas the growth of both *Chrysosporium queenslandicum* and *Trichophyton mentagrophytes* was completely inhibited by 500 ppm. Onion oil at different concentrations (100, 200, and 500 ppm) tested also reduced fungal growth and aflatoxin production by *Aspergillus flavus* and *Aspergillus parasiticus* var. *globosus* strains (Zohri et al., 1995). Onions also displayed antibacterial action against *Streptococcus mutans* and *Streptococcus sobrinus*, the main causal bacteria for dental caries, and *Porphyromonas gingivalis* and *Prevotella intermedia*, considered to be the main causal bacteria of adult periodontitis (Kim, 1997). The antimicrobial assays conducted with onion against 33 clinical isolates of *Vibrio cholerae* showed the activity of extracts of two types (purple and yellow), with purple type having minimal inhibitory concentration (MIC) range of 19.2–21.6 mg/mL and yellow type having an MIC range of 66–68.4 mg/mL (Hannan et al., 2010).

4.4 Cardiovascular effects

The effects of onion peel extract in collagen-stimulated washed rat platelet aggregation were investigated and it was found that it inhibited platelet aggregation via inhibition of aggregation-inducing molecules, intracellular Ca^{2+}, and thromboxane A2 (TXA2), by blocking cyclooxygenase-1 (COX-1) and TXA2 synthase (TXAS) activities in a dose-dependent manner (Ro et al., 2015). This effective inhibition of collagen-stimulated platelet aggregation in vitro suggested that the use of onion can be a promising and safe strategy to combat cardiovascular diseases (Ro et al., 2015). Ischemic heart injury is one of the most common cardiovascular diseases; the methanolic extract of onion was shown to attenuate ischemia/hypoxia-induced apoptosis in heart-derived H9c2 cells in vitro and in rat hearts in vivo, through, at least in part, an antioxidant effect (Park et al., 2009). Evidence of the effectiveness of onion effect on acute myocardial infarction (AMI) was also provided, suggesting that a diet rich in onions may have a favorable effect on the risk of AMI (Galeone et al., 2009). Dietary onion intake as part of a typical high fat diet was also shown to improve indices of cardiovascular health in pigs (Gabler et al., 2006). S-methyl cysteine sulfoxide isolated from onion showed lipid lowering effect in high cholesterol diet in Sprague–Dawley rats (Kumari and Augusti, 2007). It was also demonstrated that the antihypertensive activity of onion disappeared during boiling, and the disappearance of the activity of raw onion after boiling might come, in part, from a decrease of the antioxidative activity, with a consequent reduction in the saving of nitric oxide (Kawamoto et al., 2004). There are evidence supporting that compound **21** should be a potential therapeutic agent against cardiovascular diseases (Gormaz et al., 2015). In fact, diets supplemented with compound **21** were reported to lower blood pressure and attenuate cardiac hypertrophy in rats with aortic constriction (Jalili et al., 2006).

4.5 Antioxidant effects

The antioxidant activity of onion and onion scales has been studied in several models including in lipid oxidation models and in radical scavenging assays (Lawande, 2012). It was found that both yellow and red onions were poor antioxidants toward oxidation on methyl linoleate (Kahkonen et al., 1999; Lawande, 2012) in contrast to their high antioxidant activities toward oxidation of low-density lipoprotein (LDL) (Vinson et al., 2001). Onion also demonstrated a poor antioxidant score in the oxygen radical absorbance capacity (ORAC) activity test (Kahkonen et al., 1999; Lawande, 2012). Content of health-promoting phenols and the deriving antioxidant activity of onion was found to considerably vary among the investigated cultivars (Lisanti et al., 2015). Onion extract showed protective effects against doxorubicin-induced hepatotoxicity due to the antioxidant properties as evidenced by the decrease of tissue malondialdehyde and glutathione levels, increase of superoxide dismutase and glutathione peroxidase in rats (Mete et al., 2013). Onion extract also antagonized the toxic effects of aluminum chloride and improved the antioxidant status of male rat via decrease in superoxide dismutase and catalase activities (Ige and Akhigbe, 2012). The antioxidant role of compound **21** derived from onion on aldehyde oxidase (OX-LDL) and hepatocytes apoptosis in streptozotocin-induced diabetic rats was reported (Bakhshaeshi et al., 2012).

4.6 Other pharmacological effects

Several other pharmacological properties were assigned to onion extracts and its constituents. The efficacy of the herbal fraction of onion against various events responsible for type I allergic reactions was evaluated and it was found that it has promising antiallergic profile that could be attributed to its potential antihistaminic, antiinflammatory, and antioxidant activities (Kaiser et al., 2009). Various onion extracts showed antiinflammatory activity with a dose-related response in the lipoxygenase inhibitor screening assay (Takahashi and Shibamoto, 2008). The aqueous extract of Welsh onion green leaves also showed antiinflammatory effects in mice (Wang et al., 2013). The antiallergic activity (type I hypersensitivity) of 11 onions along with 8 cultivars and 3 different geographical origins was comparatively determined using rat basophilic leukemia 2H3 cells. The results indicated that Hokkaido onions were the best inhibitors of type I allergy, possibly due to their high content of quercetin 4′-glucoside (Sato et al., 2015). Soo-Wang and coworkers investigated the potential beneficial effects of onion extract on brain (Hyun et al., 2013); the investigators found that onion extract prevents brain edema, blood–brain barrier (BBB) hyperpermeability, and tight junction proteins disruption, possibly through its antioxidant effects in the mouse middle cerebral artery occlusion model (Hyun et al., 2013). This study suggested that onion extract may be a beneficial nutrient for the prevention of BBB function during brain ischemia (Hyun et al., 2013). The neuroprotective potential of onion in aluminum chloride–induced neurotoxicity was also demonstrated (Bakhshaeshi et al., 2012).

5 PATENTS WITH *Allium cepa*

1. *Patent US7,893,323 B2* by Reynolds (2011), Title: Transformation of *Allium* sp. with *Agrobacterium* using embryogenic callus cultures. The invention relates to a method for transforming an *Allium* species, such as *A. cepa* or *Allium fistulosum*, with a heterologous gene (Reynolds, 2011).

2. *Patent US6,340,483 B1* by Goren et al. (2002), Title: Antiviral composition derived from *Allium cepa* and therapeutic use thereof (Goren et al., 2002). This invention relates to the use of *A. cepa* extracts in the treatment of AIDS.
3. *Patent WO2,003,059,071* A1 by Goldman et al. (2003), Title: Antiviral composition derived from *Allium cepa* and therapeutic use thereof (Goldman et al., 2003).

CONCLUSIONS

A. cepa is cultivated worldwide for culinary and therapeutic purposes. The plant is best cultivated in fertile soils that are well drained. *A. cepa* is the source of various biologically active compounds, such as phenolic acids, thiosulfinates, and flavonoids. It has a variety of pharmacological activities including anticancer, antidiabetic, antimicrobial, cardiovascular, antioxidant effects, justifying its possible use in the treatment of various human ailments.

REFERENCES

Baghurst, K.I., Raj, M.J., Truswell, A.S., 1977. Onions and platelet aggregation. Lancet 1 (8002), 101.

Bakhshaeshi, M., Khaki, A., Fathiazad, F., Khaki, A.A., Ghadamkheir, E., 2012. Anti-oxidative role of quercetin derived from *Allium cepa* on aldehyde oxidase (OX-LDL) and hepatocytes apoptosis in streptozotocin-induced diabetic rat. Asian Pac. J. Trop. Biomed. 2 (7), 528–531.

Bang, M.A., Kim, H.A., Cho, Y.J., 2009. Alterations in the blood glucose, serum lipids and renal oxidative stress in diabetic rats by supplementation of onion (*Allium cepa* Linn). Nutr. Res. Pract. 3 (3), 242–246.

Bystrická, J., Musilová, J., Vollmannová, A., Timoracká, M., Kavalcová, P., 2013. Bioactive components of onion (*Allium cepa* L): a review. Acta Aliment. 42 (1), 11–22.

Challier, B., Perarnau, J.-M., Viel, J.-F., 1998. Garlic, onion and cereal fibre as protective factors for breast cancer: a French case–control study. Eur. J. Epidemiol. 14 (8), 737–747.

Gabler, N.K., Osrowska, E., Imsic, M., Eagling, D.R., Jois, M., Tatham, B.G., 2006. Dietary onion intake as part of a typical high fat diet improves indices of cardiovascular health using the mixed sex pig model. Plant Foods Hum. Nutr. 61 (4), 179–185.

Galeone, C., Tavani, A., Pelucchi, C., Negri, E., La Vecchia, C., 2009. Allium vegetable intake and risk of acute myocardial infarction in Italy. Eur. J. Nutr. 48 (2), 120–123.

Goldman, S.P., Goldman, W.F., Goren, A., Trainin, Z., 2003. Antiviral composition derived from *Allium cepa* and therapeutic use thereof: Google Patents.

Goren, A., Goldman, W.F., Trainin, Z., Goldman, S.R., 2002. Antiviral composition derived from *Allium cepa* and therapeutic use thereof: Google Patents.

Gormaz, J.G., Quintremil, S., Rodrigo, R., 2015. Cardiovascular disease: a target for the pharmacological effects of quercetin. Curr. Top. Med. Chem. 15 (17), 1735–1742.

Han, M.H., Lee, W.S., Jung, J.H., Jeong, J.-H., Park, C., Kim, H.J., 2013. Polyphenols isolated from *Allium cepa* L. induces apoptosis by suppressing IAP-1 through inhibiting PI3K/Akt signaling pathways in human leukemic cells. Food Chem. Toxicol. 62, 382–389.

Hannan, A., Humayun, T., Hussain, M.B., Yasir, M., Sikandar, S., 2010. In vitro antibacterial activity of onion (*Allium cepa*) against clinical isolates of *Vibrio cholerae*. J. Ayub Med. Coll. Abbottabad 22 (2), 160–163.

Hosoda, H., Ohmi, K., Sakaue, K., Tanaka, K., 2003. Inhibitory effect of onion oil on browning of shredded lettuce and its active components. J. Jpn. Soc. Hortic. Sci. 72, 451–456.

Hyun, S.-W., Jang, M., Park, S.W., Kim, E.J., Jung, Y.-S., 2013. Onion (*Allium cepa*) extract attenuates brain edema. Nutrition 29 (1), 244–249.

Ige, S.F., Akhigbe, R.E., 2012. The role of *Allium cepa* on aluminum-induced reproductive dysfunction in experimental male rat models. J. Hum. Reprod. Sci. 5 (2), 200–205.

Jalili, T., Carlstrom, J., Kim, S., Freeman, D., Jin, H., Wu, T.C., 2006. Quercetin-supplemented diets lower blood pressure and attenuate cardiac hypertrophy in rats with aortic constriction. J. Cardiovasc. Pharmacol. 47 (4), 531–541.

Jung, S.K., Lee, K.W., Byun, S., Lee, E.J., Kim, J.E., Bode, A.M., 2010. Myricetin inhibits UVB-induced angiogenesis by regulating PI-3 kinase in vivo. Carcinogenesis 31 (5), 911–917.

Jung, J.Y., Lim,Y., Moon, M.S., Kim, J.Y., Kwon, O., 2011. Onion peel extracts ameliorate hyperglycemia and insulin resistance in high fat diet/streptozotocin-induced diabetic rats. Nutr. Metab. (Lond.) 8 (1), 18.

Kahkonen, M.P., Hopia, A.I., Vuorela, H.J., Rauha, J.P., Pihlaja, K., Kujala, T.S., 1999. Antioxidant activity of plant extracts containing phenolic compounds. J. Agric. Food Chem. 47 (10), 3954–3962.

Kaiser, P., Youssouf, M.S., Tasduq, S.A., Singh, S., Sharma, S.C., Singh, G.D., 2009. Anti-allergic effects of herbal product from *Allium cepa* (bulb). J. Med. Food 12 (2), 374–382.

Kawamoto, E., Sakai, Y., Okamura, Y., Yamamoto, Y., 2004. Effects of boiling on the antihypertensive and antioxidant activities of onion. J. Nutr. Sci. Vitaminol. (Tokyo) 50 (3), 171–176.

Kim, J.H., 1997. Anti-bacterial action of onion (*Allium cepa* L) extracts against oral pathogenic bacteria. J. Nihon. Univ. Sch. Dent. 39 (3), 136–141.

Kumari, K., Augusti, K.T., 2007. Lipid lowering effect of S-methyl cysteine sulfoxide from *Allium cepa* Linn in high cholesterol diet fed rats. J. Ethnopharmacol. 109 (3), 367–371.

Lawande, K.E., 2012. Onion. In: Peter, K.V. (Ed.), Handbook of Herbs and Spices. second ed. Woodhead Publishing, pp. 417–429, (Chapter 23).

Lisanti, A., Formica, V., Ianni, F., Albertini, B., Marinozzi, M., Sardella, R., 2016. Antioxidant activity of phenolic extracts from different cultivars of Italian onion (*Allium cepa*) and relative human immune cell proliferative induction. Pharm. Biol. 54 (5), 799–806.

Messiaen, C.-M., Rouamba, A., 2004. *Allium cepa* L. Available from: http://www.prota4u.org/search.asp

Mete, R., Oran, M., Topcu, B., Oznur, M., Seber, E.S., Gedikbasi, A., 2016. Protective effects of onion (*Allium cepa*) extract against doxorubicin-induced hepatotoxicity in rats. Toxicol. Ind. Health 32 (3), 551–557.

Mittal, M.M., Mittal, S., Sarin, J.C., Sharma, M.L., 1974. Effects of feeding onion on fibrinolysis, serum cholesterol, platelet aggregation and adhesion. Indian J. Med. Sci. 28 (3), 144–148.

Park, S., Kim, M.Y., Lee, D.H., Lee, S.H., Baik, E.J., Moon, C.H., 2009. Methanolic extract of onion (*Allium cepa*) attenuates ischemia/hypoxia-induced apoptosis in cardiomyocytes via antioxidant effect. Eur. J. Nutr. 48 (4), 235–242.

Reynolds, J., 2011. Transformation of *Allium* sp. with *Agrobacterium* using embryogenic callus cultures: Google Patents.

Ro, J.Y., Ryu, J.H., Park, H.J., Cho, H.J., 2015. Onion (*Allium cepa* L) peel extract has anti-platelet effects in rat platelets. Springerplus 4, 17.

Sato, A., Zhang, T., Yonekura, L., Tamura, H., 2015. Antiallergic activities of eleven onions (*Allium cepa*) were attributed to quercetin 4′-glucoside using QuEChERS method and Pearson's correlation coefficient. J. Funct. Foods 14, 581–589.

Slimestad, R., Fossen, T., Vagen, I.M., 2007. Onions: a source of unique dietary flavonoids. J. Agric. Food Chem. 55 (25), 10067–10080.

Takahashi, M., Shibamoto, T., 2008. Chemical compositions and antioxidant/anti-inflammatory activities of steam distillate from freeze-dried onion (*Allium cepa* L) sprout. J. Agric. Food Chem. 56 (22), 10462–10467.

Vazquez-Armenta, F.J., Cruz-Valenzuela, M.R., Ayala-Zavala, J.F., 2016. Onion (*Allium cepa*) essential oils. In: Preedy, V.R. (Ed.), Essential Oils in Food Preservation Flavor and Safety. Academic Press, San Diego, pp. 617–623, (Chapter 70).

Vinson, J.A., Su, X., Zubik, L., Bose, P., 2001. Phenol antioxidant quantity and quality in foods: fruits. J. Agric. Food Chem. 49 (11), 5315–5321.

Votto, A.P., Domingues, B.S., de Souza, M.M., da Silva Junior, F.M., Caldas, S.S., Filgueira, D.M., 2010. Toxicity mechanisms of onion (*Allium cepa*) extracts and compounds in multidrug resistant erythroleukemic cell line. Biol. Res. 43 (4), 429–437.

Wang, B.S., Huang, G.J., Lu, Y.H., Chang, L.W., 2013. Anti-inflammatory effects of an aqueous extract of Welsh onion green leaves in mice. Food Chem. 138 (2–3), 751–756.

Yoshinari, O., Shiojima, Y., Igarashi, K., 2012. Anti-obesity effects of onion extract in Zucker diabetic fatty rats. Nutrients 4 (10), 1518–1526.

Zohri, A.N., Abdel-Gawad, K., Saber, S., 1995. Antibacterial, antidermatophytic and antitoxigenic activities of onion (*Allium cepa* L) oil. Microbiol. Res. 150 (2), 167–172.

CHAPTER 15

Allium sativum

V. Kuete

1 INTRODUCTION

Allium sativum (Liliaceae), known as garlic, is a strongly aromatic bulb crop believed to originate from Kazakhstan, Uzbekistan, and Western China (http://uses.plantnet-project.org/en/Allium_sativum_%28PROTA%29). *A. sativum* was domesticated long ago and is mentioned in ancient Egyptian, Greek, Indian, and Chinese writings. Garlic grows in temperate and tropical regions all over the world, and many cultivars have been developed to suit different climates (http://www.kew.org/science-conservation/plants-fungi/allium-sativum-garlic). *A. sativum* is the most widely consumed bulb after onion (http://www.kew.org/science-conservation/plants-fungi/allium-sativum-garlic). Almost 10 million tons of garlic are produced each year with China, Korea, India, USA, Spain, Egypt, and Turkey being the world's largest producers (http://www.kew.org/science-conservation/plants-fungi/allium-sativum-garlic). Garlic is widely used as a medicine worldwide and is regarded as an aphrodisiac (http://www.kew.org/science-conservation/plants-fungi/allium-sativum-garlic). In India, garlic is used to relieve problems, such as coughs and fevers or applied externally to prevent graying of hair and to improve skin conditions, such as eczema and scabies as well as to treat tetanus and lungs inflammation (http://www.kew.org/science-conservation/plants-fungi/allium-sativum-garlic). In Pakistan, garlic extract is traditionally taken orally to settle the stomach, treat coughs, and reduce fever (http://www.kew.org/science-conservation/plants-fungi/allium-sativum-garlic). In Nepal, East Asia, and the Middle East, *A. sativum* is used to treat several ailments including fevers, diabetes, rheumatism, intestinal worms, colic, flatulence, dysentery, liver disorders, tuberculosis, facial paralysis, high blood pressure, and bronchitis (http://www.kew.org/science-conservation/plants-fungi/allium-sativum-garlic). In Africa, garlic is used as an antibiotic, and it has a reputation for lowering blood pressure and cholesterol, and inhibiting thrombus formation (http://uses.plantnet-project.org/en/Allium_sativum_%28PROTA%29). Leaves and bulbs are considered to have hypotensive, carminative, antiseptic, anthelmintic, diaphoretic, and expectorant properties (http://uses.plantnet-project.org/en/Allium_sativum_%28PROTA%29). Spiritual uses have been assigned to garlic; in fact, it was reported that garlic could be worn, hung in windows, or rubbed on chimneys and keyholes to ward off vampires. It was also reported that in Islam, it is generally recommended not to eat raw garlic prior to going to the mosque, since the odor could distract other Muslims during their prayer.

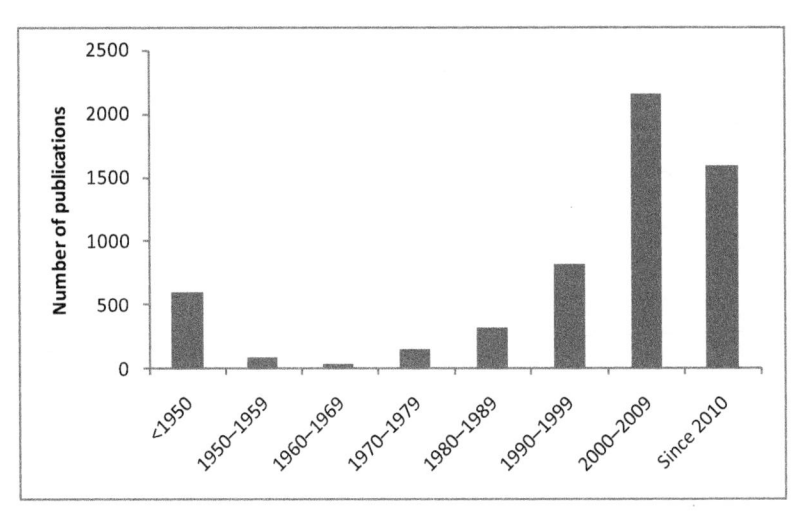

Figure 15.1 *Publication records from pubmed database related to the password* A. sativum.

A. sativum has been and continues to be subject of particular research interests for scientists of various fields with more than 5000 scientific publications recorded in Pubmed, related to the keyword *A. sativum* (Fig. 15.1).

2 CULTIVATION AND DISTRIBUTION OF *Allium sativum*

A. sativum is an herb with long, narrow, and flat leaves (Leyva et al., 2016). The edible bulbs (Fig. 15.2) are arranged in groups known as "cloves," entrapped by a white skin. Garlic is easy to grow and can be grown year-round in mild climates. Garlic plants are usually very hardy, and are not attacked by many pests or diseases. Garlic plants are said to repel rabbits and moles. Two of the major pathogens that attack garlic are nematodes and white rot disease, which remain in the soil indefinitely after the ground has become infected (http://uses.plantnet-project.org/en/Allium_sativum_%28PROTA%29). Bulbs of *A. sativum* are sold fresh or processed to produce a dry powder or oil. Garlic is popular in French, Spanish, Portuguese, and South Asian (http://www.kew.org/science-conservation/plants-fungi/allium-sativum-garlic), as well as in African cuisine. Oil from garlic is used commercially as flavoring. World production of garlic for 2001 was estimated at about 11 million tons, with China being the largest producer (7,900,000 tons), followed by India (500,000 tons), South Korea (486,000 tons), United States (336,000 tons), Egypt (215,000 tons), Russia (202,000 tons), Spain (179,000 tons), Argentina (130,000 tons), Ukraine (127,000 tons), and Turkey (103,000 tons) (Charron et al., 2016). In Africa, the main producers include the Mediterranean Northern region (excluding Egypt) with 66,000 tons, tropical Africa (29,000 tons) mainly including Sudan (17,000 tons), Niger (8,000 tons), and Tanzania (2,000 tons). However, most of the West African countries as

Figure 15.2 *Cloves of garlic. (Photo courtesy: V. Kuete).*

well as Kenya, Tanzania, Uganda, and Zambia are also producers (http://uses.plantnet-project.org/en/Allium_sativum_%28PROTA%29). Four main garlic products are available in the international market. These include garlic essential oil, garlic oil macerate, garlic powder, and aged garlic extract (AGE) (Leyva et al., 2016).

3 CHEMISTRY OF *Allium sativum*

The nutritional value of the edible parts of *A. sativum* has been reported as 410.7 kcal/100 g with 33 g of carbohydrates, 0.34 g of fat, 9.26 g of proteins, 1.2 mg of vitamin B6, 5.29 mg of iron, 36.3 mg of calcium, and 600.9 mg of phosphorus (Leustek et al., 2000). Garlic bulb contains water (65%), carbohydrates (28%), organosulfur compounds (2.3%), proteins (mostly allinase; 2%), amino acids (1.2%), and fiber (1.5%) (Nouroz et al., 2015). Fresh or crushed garlic contain the sulfur-containing compounds and enzymes, saponins, and phenolics.

3.1 Sulfur-containing compounds from *Allium sativum*

More than 20 kinds of sulfide compounds from a few sulfur-containing amino acids are produced by garlic with diverse functions. Some of them (Fig. 15.3) include alliin (**1**),

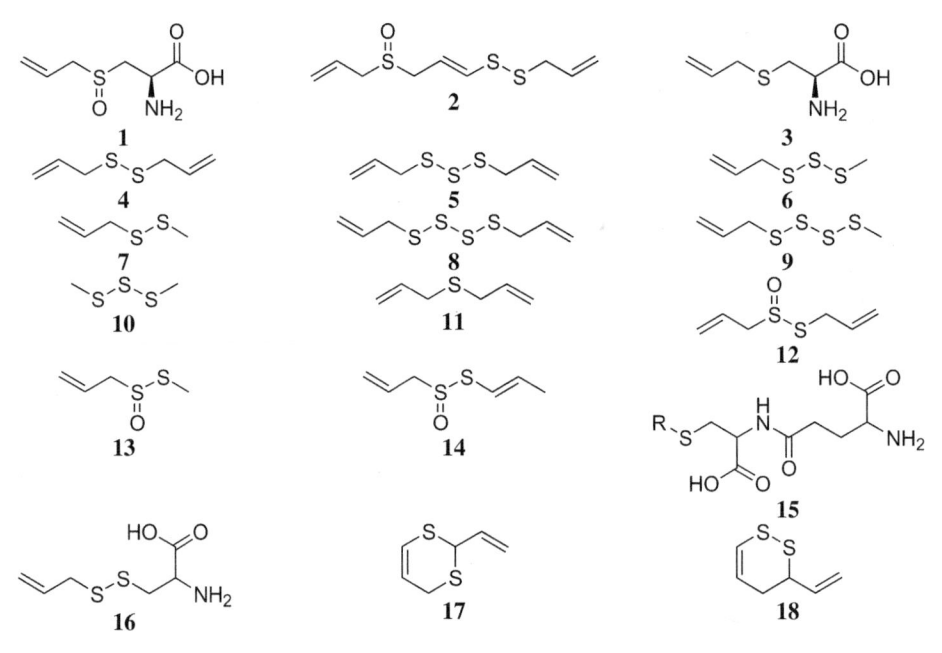

Figure 15.3 *Sulfur-containing compounds from* **A. sativum.** Alliin (**1**); ajoene (**2**); S-allyl cysteine (**3**); diallyl disulfide (**4**); diallyl trisulfide (**5**); allylmethyl trisulfide (**6**); allyl methyl disulfide (**7**); diallyl tetrasulfide (**8**); allyl methyl tetrasulfide (**9**); dimethyl trisulfide (**10**); diallyl sulfide (**11**); allicin (**12**); methyl allyl thiosulfinate (**13**); 1-propenyl allyl thiosulfinate (**14**); L-glutamyl-S-alkyl-L-cysteine (**15**); S-allyl mercaptocysteine (**16**); 2-vinyl-4-*H*-1,3-dithiin (**17**); 3-vinyl-4-*H*-1,2-dithiin (**18**).

ajoene (**2**), diallyl polysulfides, vinyldithiins (formed in the breakdown of allicin), and S-allylcysteine (**3**). The strong odor of garlic is due to the sulfur compounds, mostly diallyl disulfide (**4**), diallyl trisulfide (**5**), allylmethyl trisulfide (**6**), diallyl tetrasulfide (**8**), and diallyl sulfide (**11**) which are the main active constituents (Casella et al., 2013). The availability of organo-sulfur compounds varies in different garlic preparations. In garlic homogenate, the main constituents include allicin (**12**), methyl allyl thiosulfinate (**13**), 1-propenyl allyl thiosulfinate (**14**), and L-glutamyl-S-alkyl-L-cysteine (**15**); meanwhile compound **1** is the main constituent of powder and heat-treated garlic. AGE mostly contains compound **3** and S-allyl mercaptocysteine (**16**) whereas steam distilled garlic oil contains compounds **4, 5, 6, 7** allyl methyl disulfide (**7**), allyl methyl tetrasulfide (**9**), dimethyl trisulfide (**10**), and diallyl sulfide (**11**) (Casella et al., 2013). The major component of oil macerated or ether extracted garlic oil includes compound **2**, 2-vinyl-4-*H*-1,3-dithiin (**17**) and 3-vinyl-4-*H*-1,2-dithiin (**18**) (Casella et al., 2013).

3.2 Saponins from *Allium sativum*

Though organosulfur compounds are predominant in garlic, saponins were also isolated. In fact, steroid saponins, such as proto-eruboside B (**19**) and eruboside B (**20**) were reported in AGE (Matsuura, 2001; Peng et al., 1996). Up to 17 other steroid saponins and sapogenins were also reported in aged garlic (Matsuura, 2001). A bioassay-guided phytochemical analysis of the polar extract from the bulbs of garlic harvested in Italy led to the isolation of ten furostanol saponins namely voghieroside A1 and A2, voghieroside B1 and B2, voghieroside C1 and C2, and voghieroside D1, D2, E1 and E2 and spirostanol saponins, agigenin 3-O-trisaccharide, and gitogenin 3-O-tetrasaccharide (Lanzotti et al., 2012).

3.3 Phenolics and other compounds from *Allium sativum*

When assessing the phenolic composition of garlic cultivars harvested in China, it was found that the common flavonoids, such as myricetin, quercetin, kaempferol, and apigenin were not detected (Beato et al., 2011). Nevertheless, other phenolics, such as caffeic acid (**21**) and ferulic acid (**22**) (Fig. 15.4) were found to be the major phenolic acids (Beato et al., 2011). Vanillic (**23**), *p*-hydroxybenzoic, and *p*-coumaric (**24**) acids as well

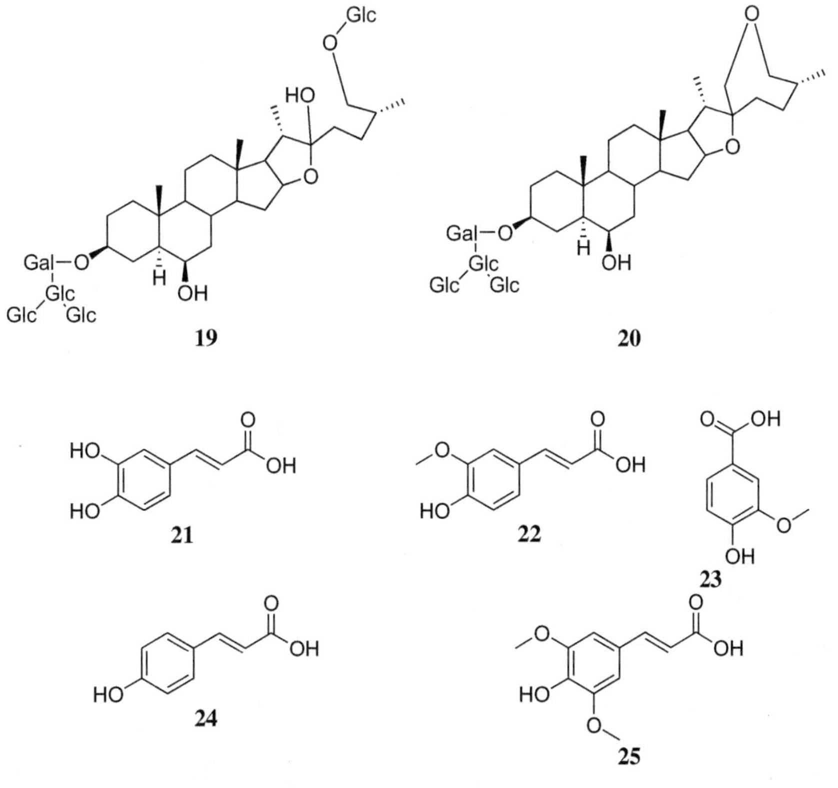

Figure 15.4 *Other phytochemicals reported in* **A. sativum**. Proto eruboside B (**19**); eruboside B (**20**); caffeic acid (**21**); ferulic acid (**22**); vanillic acid (**23**); *p*-coumaric acid (**24**); sinapic acid (**25**). Gal: beta-D-galactopyranosyl, Glc: beta-D-glucopyranosyl.

as sinapic acid (**25**) were also isolated in garlic (Beato et al., 2011). Garlic also contains enzymes, such as allinase which is activated upon injuring the garlic bulb and metabolizes compound **1** to allicin, and also peroxidases and myrosinase (Bhandari, 2012).

4 PHARMACOLOGY OF *Allium sativum*

A. sativum is an incredible plant with numerous pharmacological properties, such as antithrombotic, antimicrobial, antiarthritic, antitumor, hypoglycemic, and hypolipidemic activities (Nouroz et al., 2015).

4.1 Anticancer activity

Many in vitro and in vivo studies have suggested possible cancer-preventive effects of garlic preparations and their respective constituents. The chemopreventive potential of garlic has been attributed to the ability to modulate the activity of several metabolizing enzymes that activate (cytochrome P450) or detoxify (glutathione S-transferases) carcinogens and inhibit the formation of DNA adducts in several target tissues (Santhosha et al., 2013). It was shown that regular garlic intake is associated with low incidence of intestinal cancer with up to 35% of women showing lower risk of developing colon cancer (Santhosha et al., 2013; Yin et al., 2002). Clinical trials have demonstrated that garlic intake could reduce the risk of stomach cancer by 33% (Li et al., 2004) and colon adenocarcinoma by 50% (Tanaka et al., 2004). A meta-analysis conducted by Woo and collaborators (Woo et al., 2014) in the Korean population indicated that garlic consumption was associated with a lower risk of gastric cancer. Another clinical trial conducted on persons with basal cell carcinoma demonstrated that topical application of compound **2** for 1 month induced a reduction of 69–88% in the size of tumors (Tilli et al., 2003). This compound was found effective in inducing apoptosis in human leukemia cells through the activation of nuclear factor Kappa B (NF-κB) and stimulation of peroxide production (Dirsch et al., 1998). Compound **11** reduced breast carcinogenesis, inhibiting cell the proliferation through decrease of reactive oxygen species (ROS) production and DNA damage (Santhosha et al., 2013). Compound **11** reduced mitosis in tumor cells, decreased histone deacetylase activity, increased acetylation of histone H3 and H4, inhibited cell cycle progression, and decreased protumor markers (survivin, Bcl-2, c-Myc, mTOR, EGFR, VEGF) (Wallace et al., 2013).

4.2 Antidiabetic activity

Several reports have demonstrated that garlic can help reduce blood glucose levels in diabetic rats, mice, rabbits and increase plasma insulin secretion in diabetic rats (Santhosha et al., 2013). The active ingredient from garlic, such as S-allyl cysteine sulfoxide was reported to have hypoglycemic effect through the stimulation of insulin production or interference with dietary glucose absorption (Srinivasan, 2005). Upon 4 weeks treatment of diabetic rats with garlic extract, the final blood sugar was found to reduce significantly (Shariatzadeh et al., 2008). It was reported that long-term treatment of streptozotocin-induced diabetic rats with garlic oil was found to improve oral glucose tolerance and renal function (Liu et al., 2006). Mahmoodi and collaborators (Mahmoodi et al., 2011) demonstrated that garlic consumption helps to reduce fasting blood sugar in hyperglycemic and hypolipidemic individuals.

4.3 Antimicrobial activities

Garlic showed broad spectrum of antimicrobial activity against many species of bacteria, virus, parasites, protozoan, and fungi. Garlic extract was reported active against a panel of bacteria including *Shigella dysenteriae, Staphylococcus aureus, Pseudomonas aeruginosa, Eshcerichia coli, Streptococcus* spp., *Salmonella* spp., and *Proteus mirabilis* with the minimum inhibitory concentration (MIC) varying from 0.05 to 1.0 mg/mL (Bongiorno and Fratellone, 2008; Gull et al., 2012). Other studies demonstrated that garlic or its active ingredient, compound **12** were able to prevent the growth of *Staphylococcus, Streptococcus, Bacillus, Brucella, Vibrio, Citrobacter, Klebsiella, Mycobacterium,* and *Proteus* species (Cavallito and Bailey, 1944; Sharma et al., 1977; Elnima et al., 1983). Garlic raw juice was found to be effective against many common pathogenic intestinal bacteria, which are responsible for diarrhea in humans and animals (Sivam, 2001). Garlic oil, powder, and other constituents have been shown to exert potent antibacterial effects on *Helicobacter pylori*, explaining its epidemiological evidence for protection against gastric cancers (Sivam, 2001; You et al., 1998). Saha and collaborators (Saha et al., 2015) demonstrated that garlic extracts (with MIC value of 600 µg/mL) fulfill the criteria for therapeutic use to combat the common nosocomial infectious bacteria, *P. aeruginosa*. The antibacterial study of different concentrations of garlic extract against human dental plaque microbiota, *Streptococcus mutans, Streptococcus sanguis, Streptococcus salivarius, P. aeruginosa*, and *Lactobacillus* spp. indicated the usefulness of the 5% extract (Houshmand et al., 2013). Also, ethanol extract of garlic was shown active against multidrug resistant (MDR) clinical strains of *E. coli, P. aeruginosa, Proteus* sp., *S. aureus*, and *Bacillus* sp. (Karuppiah and Rajaram, 2012). The combination of garlic and honey was found effective against *Salmonella* strain, *S. aureus, Lyesria moncytogenes*, and *Streptococcus pneumoniar* and was suggested as an alternative natural antimicrobial drug for the treatment of pathogenic bacterial infections (Andualem, 2013).

The antifungal potency of garlic was reported on various fungal species. The aqueous extract was evaluated against *Malassezia furfur, Candida albicans*, other *Candida* sp. as well as 35 strains of various dermatophyte species and it was found that garlic might be promising in the treatment of fungal-associated diseases from important pathogenic genera including *Candida, Malassezia*, and the *dermatophytes* (Shams-Ghahfarokhi et al., 2006). AGE was reported to exhibit antifungal effects against the *Aspergillus* species involved in otomycosis (Pai and Platt, 1995). A study by Venugopal and Venugopal on the antifungal potential of garlic also demonstrated that it could be used as an effective antidermatophytic agent in the control of *Microsporum canis, M. audouinii, Trichophyton rubrum, T. mentagrophytes, T. violaceum, T. simii, T. verrucosum, T. erinacei*, and *Epidermophyton floccosum* (Venugopal and Venugopal, 1995). Garlic and its bioactive components were shown to suppress hyphae production and to affect the expression level of sirtuin 2 (SIR2) gene (Low et al., 2008). Commercial *A. sativum* extract given intravenously was found effective to patients with cryptococcal meningitis as well as patients with other types of meningitis (Davis et al., 1990). A protein from garlic, allivin, displayed antifungal activity against *Botrytis cinerea, Mycosphaerella arachidicola*, and *Physalospora piricola* (Wang and Ng, 2001). AGE applied to Czapek's broth was shown to be effective on growth of two strains of *Aspergillus flavus* (OC1 and OC10), spores, and aflatoxin B1 production (Ismaiel et al., 2012). Davis and collaborators (Davis et al., 1994) demonstrated the synergistic fungistatic activity with amphotericin B against three different isolates of *Cryptococcus neoformans*. Also, alcoholic extracts of fresh garlic elicited anticryptococcal activity against murine disseminated cryptococcosis (Khan and Katiyar, 2000). Yousuf and collaborators showed that compound **4** acts as a prooxidant to *Candida* species and hence may act as a potent antifungal in the management of candidiasis (Yousuf et al., 2010). Khodavandi et al. also demonstrated that compound **12** and fluconazole seemed to share a common anti-*Candida* mechanism through inhibition of SIR2 gene, while fluconazole appeared to also exert its fungistatic effect through another pathway that involved secreted aspartyl proteinase (SAP4) gene suppression (Khodavandi et al., 2011). The antifungal activity of compound **2** was reported on mice intratracheally infected with *Paracoccidioides brasiliensis* and the mode of action involves a direct effect on fungi and a protective proinflammatory immune response (Thomaz et al., 2008).

Garlic and its constituents have also been shown to have antiviral activity. Using direct preinfection incubation assays, the virucidal effects of fresh garlic extract, its polar fraction, as well as compounds **1, 2, 4, 5, 12, 13** and deoxyalliin were investigated against Herpes Simplex Virus type 1, herpes simplex virus type 2, parainfluenza virus type 3, vaccinia virus, vesicular stomatitis virus, and human rhinovirus type 2; The results showed that garlic polar fraction, **1, 4, 5** and deoxyalliin are devoid of antiviral activity with fresh garlic extract, in which thiosulfinates appeared to be the active components, being virucidal to each virus tested (Weber et al., 1992). The in vitro antiviral activity of garlic extract on human cytomegalovirus (HCMV) reveals a dose-dependent inhibitory effect, and the study suggested that its clinical use against HCMV infection should be persistent and the prophylactic use is preferable in immune-compromised patients (Guo et al., 1993).

4.4 Cardiovascular effects

Cardiovascular disease (CVD) is a complex and multifactorial disease which includes elevated serum lipids (cholesterol and triglyceride) increased plasma fibrinogen and coagulation factors, and increased platelet activation (Santhosha et al., 2013). CVD involves the heart or blood vessels and includes coronary artery diseases (CAD), such as angina and myocardial infarction (commonly known as a heart attack). Other CVDs are stroke, hypertensive heart disease, rheumatic heart disease, cardiomyopathy, atrial fibrillation, congenital heart disease, endocarditis, aortic aneurysms, peripheral artery disease, and venous thrombosis. It was reported that garlic could reduce lipid peroxides due to its lipid lowering effects (Sangeetha and Quine, 2006). Daily consumption of up to one clove was also associated with reduced cholesterol level to up to 9% (Tapsell et al., 2006). Garlic intake was also found to prevent hypercholesterolemia (Santhosha et al., 2013). Egg yolk enriched with garlic powder was also reported to inhibit low-density lipoprotein (LDL) oxidation, to suppress the production of peroxides and hence protects endothelial cells from injury (Yamaji et al., 2004). Daily intake of enteric-coated garlic powder was shown to decrease the level of cholesterol, LDL, high-density lipoprotein (HDL), and triglyceride (Kojuri et al., 2007). Garlic paste and oil as well as compounds **2** and **12** significantly reduced cholesterol biosynthesis in rat hepatocytes, as a result of the inhibition in the production of 3-hydroxy-3-methyl (HMG) COA-reductase and 14-alpha-demethylase (Gebhardt and Beck, 1996). Gebhardt and collaborators demonstrated that different garlic-derived organosulfur compounds interfere differently with cholesterol biosynthesis and, thus, may provoke multiple inhibition of this metabolic pathway in response to garlic consumption (Gebhardt and Beck, 1996). Yeh and Liu also proved that the cholesterol-lowering effects of garlic extract, such as AGE, come in part from inhibition of hepatic cholesterol synthesis by water-soluble sulfur compounds, especially compound **3** (Yeh and Liu, 2001). In a clinical study, it was found that treatment of person with a serum total cholesterol (TC) level greater than or equal to 220 mg/dL, with standardized garlic 900 mg/day produced a significant reduction in serum TC- and LDL cholesterol (Jain et al., 1993). The vascular protection effect of garlic as atherosclerosis prevention, by influencing the mentioned risk parameters for CVDs has been reported and it was found to induce thrombocyte aggregation inhibiting effect (Kiesewetter et al., 1991).

4.5 Antioxidant effects

Various parts of garlic have been reported to have antioxidant properties; garlic oil was shown to be effective in reducing tributyltin (TBT)-induced oxidative damage both in vivo and in vitro (Liu and Xu, 2007). TBT is the most significant pesticide which contains the $(C_4H_9)_3Sn$ group, used for various industrial purposes, such as slime control in paper mills, disinfection of circulating industrial cooling waters, antifouling agents, and the preservation of wood (Antizar-Ladislao, 2008). TBT can be transported to the human body by contaminated seafood (Liu and Xu, 2007). The possible protective mechanism may stem from the considerable ability of the garlic oil to scavenge ROS (Liu and Xu, 2007). It was also demonstrated AGE was able to inhibit the formation of glycation end products and the formation of glycation-derived free radicals (Ahmad and Ahmed, 2006). It also inhibits lipid peroxidation and oxidation of LDL (Lau, 2006). Aqueous garlic extract also showed tissue protection against nicotine-induced oxidative damage (Augusti, 1996). Garlic compounds, such as **5, 12,** S-ethyl cysteine, and *N*-acetyl cysteine also displayed antioxidant effects. In fact, compound **5** was reported to reduce liver injury caused by carbon tetra chloride (Fukao et al., 2004); compound **12** was documented as hydroxyl radical scavengers (Prasad et al., 1995) whereas S-ethyl cysteine and *N*-acetyl cysteine showed protection against lipid-related oxidation (Tsai et al., 2005).

4.6 Other pharmacological effects

Several other pharmacological potencies were assigned to garlic extracts and its constituents. Various garlic preparations showed hepatoprotective effects in many experimental models. In fact, garlic oil constituents, particularly compound **11** was found to activate constitutive androstane receptor (CAR) and nuclear factor E2-related factor 2 (Nrf2) which helps in synthesizing drug metabolizing enzymes (Fisher et al., 2007). It was shown that freshly prepared garlic homogenate has beneficial effects against isoniazid + rifampicin–induced liver injury in rats (Fisher et al., 2007). The prophylactic efficacy of garlic extract to reduce tissue lead (Pb) concentration was demonstrated experimentally in rats; in effect, concomitant use of garlic extract at the different doses was found to reduce lead concentration considerably, indicating the potential therapeutic activity of garlic against lead (Senapati et al., 2001). Garlic oil, similar to *N*-acetylcysteine, was also found to eliminate electrophilic intermediates and free radicals through conjugation and reduction reactions and therefore protected the liver from toxic doses of acetaminophen (Kalantari and Salehi, 2001).

The effect of consumption of garlic and its components on immune stimulation was reported to be due to increase in total white blood cells count and improve bone marrow cellularity (Kuttan, 2000). The organosulfur compound in garlic was also found to scavenge oxidizing agents and prevent the formation of proinflammatory messengers by interacting with sulfur-containing enzyme (Kuttan, 2000). The ability of compound **12** in preventing immune-mediated concanavalin A-induced liver damage in mice was also investigated and it was found that **12** modulates the effect of T cells and adhesion molecules and exerts an inhibitory effect on NF-κB activation; it was also found that **12** could be used therapeutically in treating chronic inflammatory diseases like inflammatory bowel disease (Kuttan, 2000); it was therefore concluded that garlic extract could help to resolve inflammation associated with inflammatory bowel disease by inhibiting T-helper-1 and inflammatory cytokines (Kuttan, 2000).

A study investigating whether prior consumption of garlic by nursing mothers modifies their infant's behaviors during breastfeeding when the mothers again consume garlic was conducted; The results demonstrated that infants who had no exposure to garlic volatiles in their mothers' milk during the experimental period spent significantly more time breastfeeding after their mothers ingested garlic capsules compared with those infants whose mothers repeatedly consumed garlic during the experimental period (Mennella and Beauchamp, 1993). Moreover, the former group of infants spent significantly more time attached to their mothers' breasts during the 4-h test session in which their mothers ingested the garlic compared with the session in which she ingested the placebo (Mennella and Beauchamp, 1993).

5 TOXICITY OF *Allium sativum*

Garlic generally poses little in terms of safety issues though some side effects were reported. Garlic may interact with warfarin, antiplatelets, saquinavir, antihypertensives, calcium channel blockers, quinolone family of antibiotics, such as ciprofloxacin and hypoglycemic drugs. Isolated cases of topical garlic burns and anaphylaxis were reported. In fact, one of garlic's adverse local effects is contact dermatitis (Friedman et al., 2006). Garlic application usually results in local inflammation, but, if applied under a pressure bandage, or if there is poor wound care or a secondary infection, it can cause a severe dermal reaction and a deep chemical burn (Friedman et al., 2006); three patients were reported in the Department of Plastic Surgery, Assaf Harofeh Medical Center, Zerifin in Israel for suspected self-inflicted lower extremity burns caused by garlic (Friedman et al., 2006). In a two case report in China, an anaphylactic reaction and a food dependent exercise-induced anaphylaxis caused by eating fresh younger garlic was reported in two patients sensitized to Artimisia pollen (Yin and Li, 2007). Even though not common, garlic allergy has been attributed to the protein alliin lyase, which was reported to induce immunoglobulin E (IgE)-mediated hypersensitivity responses from skin prick testing (Kao et al., 2004). Garlic intake has been reported to be associated with decreased platelet aggregation and bleeding events (Chagan et al., 2002), explaining why it is generally cautioned against using garlic while using anticoagulant therapy (Saw et al., 2006) as well as other medications.

6 PATENTS WITH *Allium sativum*

Some patents were published on *A. sativum* or its components; In cosmetics, the patent WO2,003,035,085 A1 (in 2001) was granted to Isabelle (Title: *Allium sativum* bulb absolutes and therapeutic or cosmetic uses); it reports the garlic bulb absolutes, compositions, and uses for therapeutic purposes, in particular in the treatment of obesity and for cosmetic purposes with emphasis on the treatment of the skin, of cellulite, and of localized dermal adipose deposits. A US patent No. 7,014,874 entitled "compositions and methods containing *Allium sativum* Linn. (garlic) naturally enriched with organic selenium compounds for nutritional supplementation" was granted to SAMI LABS LTD in 2003; it provides a safe and efficacious means of providing supplemental amounts of the essential trace mineral nutrient selenium for diverse health benefits. The EP1,968,620 A2 patent by Majeed et al. (2008) entitled "Compositions containing *Allium sativum* Linn. (garlic) naturally enriched with organic selenium compounds for nutritional supplementation" discloses selenium enriched garlic compositions that are a safe and efficacious means of providing supplemental amounts of the essential trace mineral nutrient selenium, to humans and animals.

CONCLUSIONS

Various preparations of garlic are used worldwide for culinary and therapeutic purposes. In this chapter, we have reported the beneficial effects of these spices, such as anticancer, antidiabetic, cardioprotective, antioxidant potential as well as its possible harmful effects. We have also reviewed the major constituents of the plant, such as sulfur-containing compounds, terpenoids, and phenoids. It can be concluded that garlic is a tremendous spice that should be used with caution to exploit its enormous therapeutic potential, bearing in mind that misuses could lead to potential unexpected effects.

REFERENCES

Ahmad, M.S., Ahmed, N., 2006. Antiglycation properties of aged garlic extract: possible role in prevention of diabetic complications. J. Nutr. 136 (3 Suppl), 796S–799S.

Andualem, B., 2013. Combined antibacterial activity of stingless bee (Apis mellipodae) honey and garlic (*Allium sativum*) extracts against standard and clinical pathogenic bacteria. Asian Pac. J. Trop. Biomed. 3 (9), 725–731.

Antizar-Ladislao, B., 2008. Environmental levels, toxicity and human exposure to tributyltin (TBT)-contaminated marine environment. A review. Environ. Int. 34 (2), 292–308.

Augusti, K.T., 1996. Therapeutic values of onion (*Allium cepa* L.) and garlic (*Allium sativum* L.). Indian J. Exp. Biol. 34 (7), 634–640.

Beato, V.M., Orgaz, F., Mansilla, F., Montano, A., 2011. Changes in phenolic compounds in garlic (*Allium sativum* L.) owing to the cultivar and location of growth. Plant Foods Hum. Nutr. 66 (3), 218–223.

Bhandari, P., 2012. Garlic (*Allium sativum* L.): a review of potential therapeutic applications. Int. J. Green Pharm. 6 (2), 118–129.

Bongiorno, P., Fratellone, P., & P, L. (2008). Potential Health Benefits of Garlic (*Allium Sativum*): A Narrative Review. *Journal of Complementary and Integrative Medicine, 5*(1), 1-24.

Casella, S., Leonardi, M., Melai, B., Fratini, F., Pistelli, L., 2013. The role of diallyl sulfides and dipropyl sulfides in the in vitro antimicrobial activity of the essential oil of garlic, *Allium sativum* L., and leek, *Allium porrum* L. Phytother. Res. 27 (3), 380–383.

Cavallito, C., Bailey, J., 1944. Allicin, the antibiotic principle of *Allium sativum*: isolation, physical properties and antibacterial action. J. Am. Chem. Soc. 66, 1950–1951.

Chagan, L., Ioselovich, A., Asherova, L., Cheng, J.W., 2002. Use of alternative pharmacotherapy in management of cardiovascular diseases. Am. J. Manag. Care. 8 (3), 270–285, quiz 286-278.

Charron, C.S., Milner, J.A., Novotny, J.A., 2016. Garlic. In: Caballero, B., Finglas, P.M., Toldrá, F. (Eds.), Encyclopedia of Food and Health. Academic Press, Oxford, pp. 184–190.

Davis, L.E., Shen, J.K., Cai, Y., 1990. Antifungal activity in human cerebrospinal fluid and plasma after intravenous administration of *Allium sativum*. Antimicrob. Agents Chemother. 34 (4), 651–653.

Davis, L.E., Shen, J., Royer, R.E., 1994. In vitro synergism of concentrated *Allium sativum* extract and amphotericin B against *Cryptococcus neoformans*. Planta Med. 60 (6), 546–549.

Dirsch, V.M., Gerbes, A.L., Vollmar, A.M., 1998. Ajoene, a compound of garlic, induces apoptosis in human promyeloleukemic cells, accompanied by generation of reactive oxygen species and activation of nuclear factor κB. Mol. Pharmacol. 53 (3), 402–407.

Elnima, E.I., Ahmed, S.A., Mekkawi, A.G., Mossa, J.S., 1983. The antimicrobial activity of garlic and onion extracts. Pharmazie 38 (11), 747–748.

Fisher, C.D., Augustine, L.M., Maher, J.M., Nelson, D.M., Slitt, A.L., Klaassen, C.D., 2007. Induction of drug-metabolizing enzymes by garlic and allyl sulfide compounds via activation of constitutive androstane receptor and nuclear factor E2-related factor 2. Drug Metab. Dispos. 35 (6), 995–1000.

Friedman, T., Shalom, A., Westreich, M., 2006. Self-inflicted garlic burns: our experience and literature review. Int. J. Dermatol. 45 (10), 1161–1163.

Fukao, T., Hosono, T., Misawa, S., Seki, T., Ariga, T., 2004. The effects of allyl sulfides on the induction of phase II detoxification enzymes and liver injury by carbon tetrachloride. Food Chem. Toxicol. 42 (5), 743–749.

Gebhardt, R., Beck, H., 1996. Differential inhibitory effects of garlic-derived organosulfur compounds on cholesterol biosynthesis in primary rat hepatocyte cultures. Lipids 31 (12), 1269–1276.

Gull, I., Saeed, M., Shaukat, H., Aslam, S.M., Samra, Z.Q., Athar, A.M., 2012. Inhibitory effect of *Allium sativum* and *Zingiber officinale* extracts on clinically important drug resistant pathogenic bacteria. Ann. Clin. Microbiol. Antimicrob. 11, 8.

Guo, N.L., Lu, D.P., Woods, G.L., Reed, E., Zhou, G.Z., Zhang, L.B., 1993. Demonstration of the anti-viral activity of garlic extract against human cytomegalovirus in vitro. Chin. Med. J. (Engl.) 106 (2), 93–96.

Houshmand, B., Mahjour, F., Dianat, O., 2013. Antibacterial effect of different concentrations of garlic (*Allium sativum*) extract on dental plaque bacteria. Indian J. Dent. Res. 24 (1), 71–75.

Ismaiel, A.A., Rabie, G.H., Kenawey, S.E., Abd El-Aal, M.A., 2012. Efficacy of aqueous garlic extract on growth, aflatoxin B1 production, and cyto-morphological aberrations of *Aspergillus flavus*, causing human ophthalmic infection: topical treatment of *A. flavus* keratitis. Braz. J. Microbiol. 43 (4), 1355–1364.

Jain, A.K., Vargas, R., Gotzkowsky, S., McMahon, F.G., 1993. Can garlic reduce levels of serum lipids? A controlled clinical study. Am. J. Med. 94 (6), 632–635.

Kalantari, H., Salehi, M., 2001. The protective effect of garlic oil on hepatotoxicity induced by acetaminophen in mice and comparison with N-acetylcysteine. Saudi Med. J. 22 (12), 1080–1084.

Kao, S.H., Hsu, C.H., Su, S.N., Hor, W.T., Chang, T.W., Chow, L.P., 2004. Identification and immunologic characterization of an allergen, alliin lyase, from garlic (*Allium sativum*). J. Allergy Clin. Immunol. 113 (1), 161–168.

Karuppiah, P., Rajaram, S., 2012. Antibacterial effect of *Allium sativum* cloves and *Zingiber officinale* rhizomes against multiple-drug resistant clinical pathogens. Asian Pac. J. Trop. Biomed. 2 (8), 597–601.

Khan, Z.K., Katiyar, R., 2000. Potent antifungal activity of garlic (*Allium sativum*) against experimental murine dissemenated cryptococcosis. Pharm. Biol. 38 (2), 87–100.

Khodavandi, A., Alizadeh, F., Harmal, N.S., Sidik, S.M., Othman, F., Sekawi, Z., 2011. Expression analysis of SIR2 and SAPs1-4 gene expression in *Candida albicans* treated with allicin compared to fluconazole. Trop. Biomed. 28 (3), 589–598.

Kiesewetter, H., Jung, F., Pindur, G., Jung, E.M., Mrowietz, C., Wenzel, E., 1991. Effect of garlic on thrombocyte aggregation, microcirculation, and other risk factors. Int. J. Clin. Pharmacol. Ther. Toxicol. 29 (4), 151–155.

Kojuri, J., Vosoughi, A.R., Akrami, M., 2007. Effects of *Anethum graveolens* and garlic on lipid profile in hyperlipidemic patients. Lipids Health Dis. 6, 5.

Kuttan, G., 2000. Immunomodulatory effect of some naturally occuring sulphur-containing compounds. J. Ethnopharmacol. 72 (1–2), 93–99.

Lanzotti, V., Barile, E., Antignani, V., Bonanomi, G., Scala, F., 2012. Antifungal saponins from bulbs of garlic, *Allium sativum* L. var. Voghiera. Phytochemistry 78, 126–134.

Lau, B.H., 2006. Suppression of LDL oxidation by garlic compounds is a possible mechanism of cardiovascular health benefit. J. Nutr. 136 (3 Suppl), 765S–768S.

Leustek, T., Martin, M.N., Bick, J.A., Davies, J.P., 2000. Pathways and regulation of sulfur metabolism revealed through molecular and genetic studies. Annu Rev. Plant Physiol. Plant Mol. Biol. 51, 141–165.

Leyva, J.M., Ortega-Ramirez, L.A., Ayala-Zavala, J.F., 2016. Garlic (*Allium sativum* Linn.) oils. In: Preedy, V.R. (Ed.), Essential Oils in Food Preservation, Flavor and Safety. Academic Press, San Diego, pp. 441–446, (Chapter 49).

Li, H., Li, H.Q., Wang, Y., Xu, H.X., Fan, W.T., Wang, M.L., 2004. An intervention study to prevent gastric cancer by micro-selenium and large dose of allitridum. Chin. Med. J. 117 (8), 1155–1160.

Liu, C.T., Wong, P.L., Lii, C.K., Hse, H., Sheen, L.Y., 2006. Antidiabetic effect of garlic oil but not diallyl disulfide in rats with streptozotocin-induced diabetes. Food Chem. Toxicol. 44 (8), 1377–1384.

Liu, H.G., Xu, L.H., 2007. Garlic oil prevents tributyltin-induced oxidative damage in vivo and in vitro. J. Food Prot. 70 (3), 716–721.

Low, C.F., Chong, P.P., Yong, P.V., Lim, C.S., Ahmad, Z., Othman, F., 2008. Inhibition of hyphae formation and SIR2 expression in *Candida albicans* treated with fresh *Allium sativum* (garlic) extract. J. Appl. Microbiol. 105 (6), 2169–2177.

Mahmoodi, M., Hassanshahi, S.M.H.Z.G.H., Toghroli, M.A., Khaksari, M., Hajizadeh, M.R., Mirzajani, E., 2011. The effects of consumption of raw garlic on serum lipid level, blood sugar and a number of effective hormones on lipid and sugar metabolism in hyperglycemic and/or hyperlipidemic individuals: benefit of raw garlic consumption. Adv. Biol. Chem. 1 (1), 29–33.

Majeed, M., Bammi, R.K., Badmaev, V., Prakash, S., Nagabhushanam, K. 2008. Compositions containing allium sativum linn. (garlic) naturally enriched with organic selenium compounds for nutritional supplementation. Google Patents.

Matsuura, H., 2001. Saponins in garlic as modifiers of the risk of cardiovascular disease. J. Nutr. 131 (3s), 1000S–1005S.

Mennella, J.A., Beauchamp, G.K., 1993. The effects of repeated exposure to garlic-flavored milk on the nursling's behavior. Pediatr. Res. 34 (6), 805–808.

Nouroz, F., Mehboob, M., Noreen, S., Zaidi, F., Mobin, T., 2015. A review on anticancer activities of garlic (*Allium sativum* L.). Middle-East J. Sci. Res. 23 (6), 1145–1151.

Pai, S.T., Platt, M.W., 1995. Antifungal effects of *Allium sativum* (garlic) extract against the Aspergillus species involved in otomycosis. Lett. Appl. Microbiol. 20 (1), 14–18.

Peng, J.P., Chen, H., Qiao, Y.Q., Ma, L.R., Narui, T., Suzuki, H., 1996. Two new steroidal saponins from *Allium sativum* and their inhibitory effects on blood coagulability. Yao Xue Xue Bao 31 (8), 607–612.

Prasad, K., Laxdal, V., Yu, M., Raney, B., 1995. Antioxidant activity of allicin, an active principle in garlic. Mol. Cell. Biochem. 148 (2), 183–189.

Saha, S.K., Saha, S., Hossain, M.A., Paul, S.K., 2015. In vitro assessment of antibacterial effect of garlic (*Allium sativum*) extracts on *Pseudomonas aeruginosa*. Mymensingh Med. J. 24 (2), 222–232.

Sangeetha, T., Quine, S.D., 2006. Preventive effect of S-allyl cysteine sulfoxide (alliin) on cardiac marker enzymes and lipids in isoproterenol-induced myocardial injury. J. Pharm. Pharmacol. 58 (5), 617–623.

Santhosha, S.G., Jamuna, P., Prabhavathi, S.N., 2013. Bioactive components of garlic and their physiological role in health maintenance: A review. Food Biosci. 3, 59–74.

Saw, J.T., Bahari, M.B., Ang, H.H., Lim, Y.H., 2006. Potential drug-herb interaction with antiplatelet/anticoagulant drugs. Complement Ther. Clin. Pract. 12 (4), 236–241.

Senapati, S.K., Dey, S., Dwivedi, S.K., Swarup, D., 2001. Effect of garlic (*Allium sativum* L.) extract on tissue lead level in rats. J. Ethnopharmacol. 76 (3), 229–232.

Shams-Ghahfarokhi, M., Shokoohamiri, M.R., Amirrajab, N., Moghadasi, B., Ghajari, A., Zeini, F., 2006. In vitro antifungal activities of *Allium cepa, Allium sativum* and ketoconazole against some pathogenic yeasts and dermatophytes. Fitoterapia 77 (4), 321–323.

Shariatzadeh, S., Soleimani Mehranjani, M., Mahmoodi, M., Abnosi, M., Momeni, H., Dezfulian, A., 2008. Effects of garlic (*Allium sativum*) on blood sugar and nephropathy in diabetic rats. J. Biol. Sci. 8 (8), 1316–1321.

Sharma, V.D., Sethi, M.S., Kumar, A., Rarotra, J.R., 1977. Antibacterial property of *Allium sativum* Linn.: in vivo and in vitro studies. Indian J. Exp. Biol. 15 (6), 466–468.

Sivam, G.P., 2001. Protection against *Helicobacter pylori* and other bacterial infections by garlic. J. Nutr. 131 (3s), 1106S–1108S.

Srinivasan, K., 2005. Plant foods in the management of diabetes mellitus: spices as beneficial antidiabetic food adjuncts. Int. J. Food Sci. Nutr. 56 (6), 399–414.

Tanaka, S., Haruma, K., Kunihiro, M., Nagata, S., Kitadai, Y., Manabe, N., 2004. Effects of aged garlic extract (AGE) on colorectal adenomas: a double-blinded study. Hiroshima J. Med. Sci. 53 (3–4), 39–45.

Tapsell, L.C., Hemphill, I., Cobiac, L., Patch, C.S., Sullivan, D.R., Fenech, M., 2006. Health benefits of herbs and spices: the past, the present, the future. Med. J. Aust. 185 (4 Suppl.), S4–24.

Thomaz, L., Apitz-Castro, R., Marques, A.F., Travassos, L.R., Taborda, C.P., 2008. Experimental paracoccidioidomycosis: alternative therapy with ajoene, compound from *Allium sativum*, associated with sulfamethoxazole/trimethoprim. Med. Mycol. 46 (2), 113–118.

Tilli, C.M.L.J., Stavast-Kooy, A.J.W., Vuerstaek, J.D.D., Thissen, M.R.T.M., Krekels, G.A.M., Ramaekers, F.C.S., 2003. The garlic-derived organosulfur component ajoene decreases basal cell carcinoma tumor size by inducing apoptosis. Arch. Dermatol. Res. 295 (3), 117–123.

Tsai, T.-H., Tsai, P.-J., Ho, S.-C., 2005. Antioxidant and anti-inflammatory activities of several commonly used spices. J. Food Sci. 70 (1), C93–C97.

Venugopal, P.V., Venugopal, T.V., 1995. Antidermatophytic activity of garlic (*Allium sativum*) in vitro. Int. J. Dermatol. 34 (4), 278–279.

Wallace, G.C., Haar, t., Vandergrift, C.P., 3rd, W.A., Giglio, P., Dixon-Mah, Y.N., Varma, A.K., 2013. Multitargeted DATS prevents tumor progression and promotes apoptosis in ectopic glioblastoma xenografts in SCID mice via HDAC inhibition. J. Neurooncol. 114 (1), 43–50.

Wang, H.X., Ng, T.B., 2001. Purification of allivin, a novel antifungal protein from bulbs of the round-cloved garlic. Life Sci. 70 (3), 357–365.

Weber, N.D., Andersen, D.O., North, J.A., Murray, B.K., Lawson, L.D., Hughes, B.G., 1992. In vitro virucidal effects of *Allium sativum* (garlic) extract and compounds. Planta Med. 58 (5), 417–423.

Woo, H.D., Park, S., Oh, K., Kim, H.J., Shin, H.R., Moon, H.K., 2014. Diet and cancer risk in the Korean population: a meta-analysis. Asian Pac. J. Cancer Prev. 15 (19), 8509–8519.

Yamaji, K., Sarker, K.P., Abeyama, K., Maruyama, I., 2004. Anti-atherogenic effects of an egg yolk-enriched garlic supplement. Int. J. Food Sci. Nutr. 55 (1), 61–66.

Yeh, Y.Y., Liu, L., 2001. Cholesterol-lowering effect of garlic extracts and organosulfur compounds: human and animal studies. J. Nutr. 131 (3s), 989S–993S.

Yin, M.-c., Hwang, S.-w., Chan, K.-c., 2002. Nonenzymatic antioxidant activity of four organosulfur compounds derived from garlic. J. Agric. Food Chem. 50 (21), 6143–6147.

Yin, J., Li, H., 2007. Anaphylaxis caused by younger garlic: two cases report in China. J. Allergy Clin. Immunol. 119 (1), S34.

You, W.C., Zhang, L., Gail, M.H., Ma, J.L., Chang, Y.S., Blot, W.J., 1998. *Helicobacter pylori* infection, garlic intake and precancerous lesions in a Chinese population at low risk of gastric cancer. Int. J. Epidemiol. 27 (6), 941–944.

Yousuf, S., Ahmad, A., Khan, A., Manzoor, N., Khan, L.A., 2010. Effect of diallyldisulphide on an antioxidant enzyme system in *Candida* species. Can. J. Microbiol. 56 (10), 816–821.

CHAPTER 16

Canarium schweinfurthii

V. Kuete

1 INTRODUCTION

Canarium schweinfurthii Engl. (Burseraceae), commonly known as African elemi, is a species of large tree native to tropical Africa (http://www.worldagroforestry.org/). The African elemi tree is economically useful oleoresin known as elemi and bears edible seeds with a thick, dense, hard shell, generally harvested from the wild for local use but are often sold in local markets (Ruffo et al., 2002). The tree is a major source of elemi, an oleoresin that is used in food, in medicine, and for a range of industrial applications (Ruffo et al., 2002). The plant is found throughout tropical Africa in rainforest, gallery forest, and transitional forest from Senegal to Cameroon and extending to Ethiopia, Tanzania, and Angola (Hutchinson and Dalziel, 1954; Burkill, 1985; Orwa et al., 2009; Bostoen et al., 2013). The oils can be used as vegetable oil (Shakirin et al., 2012). The oily seed kernel is eaten cooked, and in Nigeria it is sometimes prepared into a vegetable butter and eaten as a substitute for shea butter (Burkill, 1985). In West Africa, elemi is traditionally burned for fumigating dwellings and mixed with oil for body paint. *C. schweinfurthii* is used by traditional healers as a remedy for diabetes mellitus in southern Senegal (Kamtchouing et al., 2006) whereas in Congo and Central African Republic the plant is used in case of fever, as stimulant, emollient, in postpartum pain, constipation, malaria, diarrhea, sexual infections, and rheumatism (Koudou et al., 2005). A bark decoction is used against dysentery, gonorrhea, cough, chest pains, pulmonary affections, stomach complaints, food poisoning as well as a purgative and an emetic; the resin is used against roundworm infections and other intestinal parasites; it has action on skin infections and eczema. The pounded bark is used against leprosy and ulcers; root is used against adenitis whereas root scrapings are made into a poultice (http://www.worldagroforestry.org/).

2 BOTANICAL ASPECTS AND DISTRIBUTION OF *Canarium schweinfurthii*

C. schweinfurthii is a large forest tree with its crown reaching to the upper canopy of the forest, with a long clean, straight, and cylindrical bole exceeding 50 m. Diameter above the heavy root swellings can be up to 4.5 m. Bark is thick, on young tree fairly smooth, becoming increasingly scaly and fissured with age. The slash is reddish or light brown with turpentine like odor, exuding a heavy, sticky oleoresin that colors to sulfur yellow and becomes solid (http://www.worldagroforestry.org/). Leaves are pinnate, clustered at the end of the branches, and may be 15–65 cm long, with 8–12 pairs of leaflets, mostly opposite, oblong, cordate at base, 5–20 cm long, and 3-6 cm broad, with 12–24 main lateral nerves on each side of the mid-rib, prominent and pubescent beneath. The lower leaflets are bigger than the upper ones. The lower part of the petiole is winged on the upper side (http://www.worldagroforestry.org/). The creamy white unisexual flowers about 1-cm long grow in inflorescences that stand in the axils of the leaves and may be up to 28-cm long. The fruit is a small drupe, bluish-purple, glabrous, 3–4 cm long, and 1–2 cm thick. The calyx is persistent and remains attached to the fruit. The fruit contains a hard spindle-shaped, trigonous stone that eventually splits releasing three seeds (http://www.worldagroforestry.org/). The plant is native to Angola, Cameroon, Ethiopia, Ghana, Guinea-Bissau, Liberia, Mali, Nigeria, Senegal, Sierra Leone, Sudan, Tanzania, Togo, Uganda, Zambia, and exotic to Indonesia (http://www.worldagroforestry.org/). The natural habitat of *C. schweinfurthii* includes the riverine forest and forest patches, rain forest, gallery forest, and transitional forest (http://www.worldagroforestry.org/).

3 CHEMISTRY OF *Canarium schweinfurthii*

There is a great variability in the composition of the essential oils of *C. schweinfurthii*. Nagawa et al. (2015) have reported 37 compounds from the essential oil of the resin of *C. schweinfurthii* from Sango Bay Area, Southern Uganda. The main components were γ-terpinene (**1**) (32.4%), α-phellandrene (**2**) (17.9%), α-thujene (**3**) (14.0%), β-phellandrene (**4**) (12.9%), *p*-cymene (**5**) (8.5%), α-pinene (**6**) (4.9%), sabinene (**7**) (2.9%), and β-pinene (**8**) (2.3%) (Nagawa et al., 2015). *C. schweinfurthii* oil from Gabon also contained compounds **7** (19.2%), **6** (10.7%), and **5** (4.3%), while its major component was identified as limonene (**9**) (52.1%) (Edou et al., 2012). Oil from sample from the Central African Republic (Koudou et al., 2005) rather contained octyl acetate (**10**; 60%), nerolidol (**11**; 14%), and *n*-octanol (**12**; 9.5%) as major compounds, which were not present in the studied essential oil from Uganda. In another study carried out by Dongmo et al. (2010), oils from two different trees in Cameroon contained compounds **5, 9** and α-terpineol (**13**) as main constituents (Fig. 16.1). Among other compounds found in *C. schweinfurthii* oil are polyphenols, such as catechol, *p*-hydroxybenzaldehyde, dihydroxyphenylacetic acid, tyrosol, *p*-hydroxybenzoic acid, dihydroxybenzoic acid, vanillic acid, phloretic acid, pinoresinol, and secoisolariciresinol (Atawodi, 2011). Ten phenolic compounds namely catechol, *p*-hydroxybenzaldehyde, dihydroxyphenylacetic acid, tyrosol, *p*-hydroxybenzoic acid, dihydroxybenzoic acid, vanillic acid, phoretic acid, pinoresinol, and secoisolariciresinol were also reported from the fruit mesocarp oil of the plant from Plateau State in Nigeria (Atawodi, 2010).

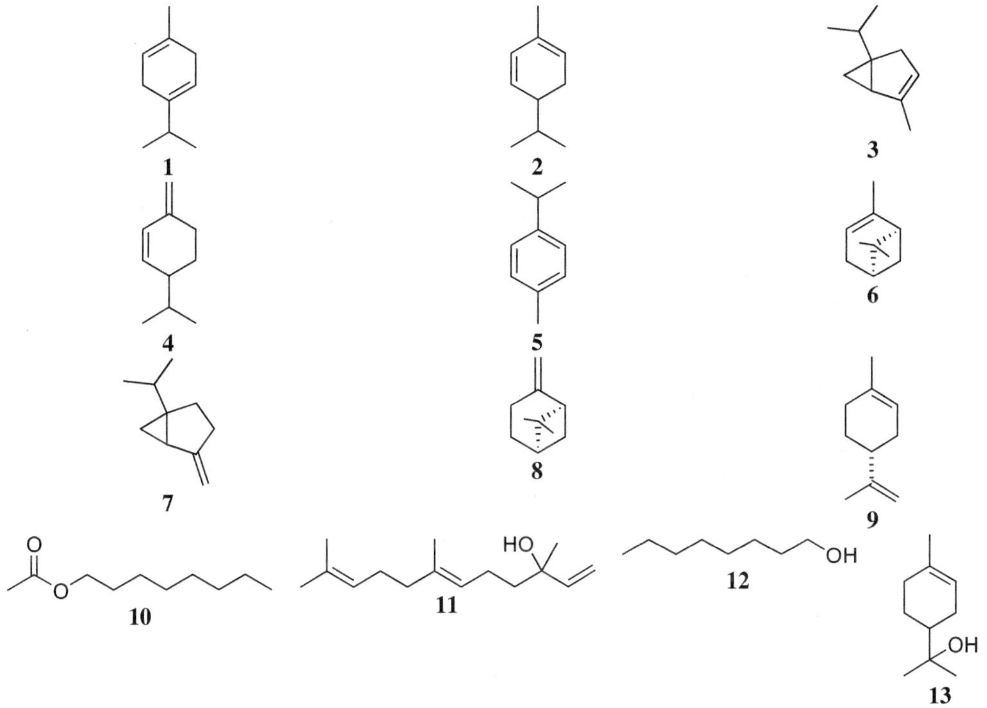

Figure 16.1 *Major constituents of the essential oil of the resin of* C. schweinfurthii. γ-Terpinene (**1**); α-phellandrene (**2**); α-thujene (**3**); β-phellandrene (**4**); p-cymene (**5**); α-pinene (**6**); sabinene (**7**); β-pinene (**8**); limonene (**9**); octyl acetate (**10**); nerolidol (**11**); *n*-octanol (**12**); α-terpineol (**13**).

4 PHARMACOLOGY OF *Canarium schweinfurthii*

4.1 Antidiabetic activity

Kamtchouing et al. have investigated the antidiabetic activity of the methanol/methylene chloride extract of stem bark of *C. schweinfurthii* on experimental male rats. The authors found that at a dose of 300 mg/kg body weight, there was 67.1% reduction in blood glucose levels after a once daily subcutaneous injection on streptozotocin-induced diabetic male rats over 14 days, versus insulin that had 76.8% reduction (Kamtchouing et al., 2006). They also observed that weight gain was only 6.6% as opposed to untreated rats that had lost 14.1% of body weight and that there was also significant reduction in food and fluid consumption by 68.5 and 79.7%. It was therefore concluded that the extract could reverse hyperglycemia, polyphagia, and polydipsia provoked by streptozotocin, thus having antidiabetic activity (Kamtchouing et al., 2006).

4.2 Antimicrobial activities

Dichloromethane extract of *C. schweinfurthii* had bactericidal activity against Gram-negative *Vibrio cholerae* with minimum inhibitory concentration (MIC) of 0.62 mg/mL whereas the ethyl acetate extract was active against Gram-positive and Gram-negative bacteria namely *Staphylococcus aureus* and *Proteus vulgaris* with MIC values of 10 and 5 mg/mL, respectively. Ethanol extract was active against *V. cholerae* and *P. vulgaris* with MIC values of 0.62 and 10 mg/mL, respectively (Moshi et al., 2009). The activity of the methanol extract of the fruits was determined on a panel of Gram-negative bacteria including sensitive and multidrug resistant (MDR) phenotypes. The inhibitory activities were obtained against various strains of *Escherichia coli* (MIC ranged from 64 to 512 μg/mL), *Enterobacter aerogenes* (MIC ranged from 128 to 1024 μg/mL), *Klebsiella pneumoniae* (MIC ranged from 256 to 1024 μg/mL), *Pseudomonas aeruginosa* PA01 (MIC value of 512 μg/mL), and *Providencia stuartii* (MIC ranged from 256 to 512 μg/mL) (Dzotam et al., 2015). The bark methanol extract of the plant also displayed good antibacterial spectrum on a panel of clinical and MDR Gram-negative bacteria; the sensitive species included various strains of *E. coli* (MIC ranged from 512 to 1024 μg/mL), *E. aerogenes* (MIC ranged from 256 to1024 μg/mL), *K. pneumoniae* (MIC ranged from 32 to 1024 μg/mL), *P. aeruginosa* PA124 (MIC value of 1024 μg/mL), *E. cloacae* (MIC value of 512 μg/mL against BM69 and ECCI strains), and *P. stuartii* (MIC ranged from 256 to 1024 μg/mL) (Seukep et al., 2015).

4.3 Antioxidant effects

The essential oil of *C. schweinfurthii* was tested for its antioxidant activities with the DPPH assay and β-carotene bleaching test; results showed it had high antioxidant activity at 150 μg/mL in both assays (Obame et al., 2007). The essential oil of the plant harvested in Cameroon also showed enzymatic inhibitory activity on lipoxygenase (Dongmo et al., 2010).

4.4 Other pharmacological effects

The resazurin reduction assay was used to evaluate the cytotoxicity of the methanol extract of bark of *C. shweinfurthii* toward leukemia CCRF-CEM cell line; it was found that the extract has low activity, with only 40% inhibition rate at 40 μg/mL (Kuete et al., 2015). The essential oil of the plant harvested in Cameroon showed enzymatic inhibitory activity on xanthine oxidase (Atawodi, 2011) as well as potent analgesic effect in the acetic acid-induced writhing and hot plate experiments (Koudou et al., 2005). A study evaluating the protective effects of aqueous and methanol extracts of stem bark on the kidney acetaminophen-induced kidney injury in rats showed improvement of renal function (Okwuosa et al., 2009). The oil was also tested for antitermitic activity against *Macrotermes bellicosus* and was found highly potent with a lethal concentration 50 (LC_{50}) (48 h) of 1.12 mg/g (Nagawa et al., 2015). *C. schweinfurthii* commercial processed mesocarp oil as well as laboratory processed cotyledon oil showed insecticidal activities against the storage pests, *Callosobruchus maculates* (Katunku et al., 2014).

5 CONCLUSIONS

In this chapter, we have reviewed the available scientific data related to the chemistry and pharmacology of *C. schweinfurthii*. The plant known as African elemi has edible fruits and has been used in traditional medicine for the treatment of various ailments. Phytochemical studies revealed that the main constituent of the oil of the plant varied with geographical area, but were mostly γ-terpinene, α-phellandrene, α-thujene, β-phellandrene, *p*-cymene, α-pinene, sabinene, β-pinene, limonene, octyl acetate, nerolidol, *n*-octanol, and α-terpineol. The reported pharmacological properties of the plant include antidiabetic, antioxidant, antibacterial, antifungal, analgesic and antiinflammatory as well as antiparasitic activities.

REFERENCES

Atawodi, S.E., 2010. Polyphenol composition and in vitro antioxydant potential of Nigerian *Canarium schweinfurthii* Engl. Oil Adv. Biol. Res. 4 (6), 314–322.

Atawodi, S.E., 2011. Nigerian foodstuffs with prostate cancer chemopreventive polyphenols. Infect. Agent Cancer 6 (Suppl 2), S9.

Bostoen, K., Grollemund, R., Koni Muluwa, J., 2013. Climate-induced vegetation dynamics and the Bantu expansion: evidence from Bantu names for pioneer trees (*Elaeis guineensis, Canarium schweinfurthii*, and *Musanga cecropioides*). Comp. Rendus Geosci. 345 (7–8), 336–349.

Burkill, H., 1985. The Useful Plants of West Tropical Africa. Royal Botanic Garden Kew, London.

Dongmo, P.M.J., Tchoumbougnang, F., Ndongson, B., Agwanande, W., Sandjon, B., Zollo, P.H.A., 2010. Chemical characterization antiradical antioxidant and anti-inflammatory potential of the essential oils of *Canarium schweinfurthii* and *Aucoumea klaineana* (Burseraceae) growing in Cameroon. Agric. Biol. J. North Am. 1, 606–611.

Dzotam, J.K., Touani, F.K., Kuete, V., 2015. Antibacterial activities of the methanol extracts of *Canarium schweinfurthii* and four other Cameroonian dietary plants against multi-drug resistant Gram-negative bacteria. Saudi J. Biol. Sci..

Edou, E.P., Abdoul-Latif, F.M., Obame, L.C., Mwenono, L., Agnaniet, H., 2012. Volatile constituents of *Canarium schwenfurthii* Engl. essential oil from Gabon. Int. J. AgriSci. 2, 200–203.

Hutchinson, J., Dalziel, J.M., 1954. Flora of West Tropical Africavol. 1Crown Agents for Oversea Governments and Administrations, London.

Kamtchouing, P., Kahpui, S.M., Dzeufiet, P.D.D., Tédong, L., Asongalem, E.A., Dimo, T., 2006. Anti-diabetic activity of methanol/methylene chloride stem bark extracts of *Terminalia superba* and *Canarium schweinfurthii* on streptozotocin-induced diabetic rats. J. Ethnopharmacol. 104 (3), 306–309.

Katunku, D., Ogunwolu, E.O., Ukwela, M.U., 2014. Contact toxicity of *Canarium schweinfurthii* Engl. tissues against *Callosobruchus maculatus* in stored bambara groundnut. Int. J. Agron. Agric. Res. 5 (5), 20–28.

Koudou, J., Abena, A.A., Ngaissona, P., Bessière, J.M., 2005. Chemical composition and pharmacological activity of essential oil of *Canarium schweinfurthii*. Fitoterapia 76 (7–8), 700–703.

Kuete, V., Sandjo, L.P., Mbaveng, A.T., Seukep, J.A., Ngadjui, B.T., Efferth, T., 2015. Cytotoxicity of selected Cameroonian medicinal plants and *Nauclea pobeguinii* towards multi-factorial drug-resistant cancer cells. BMC Complement Altern. Med. 15, 309.

Moshi, M.J., Innocent, E., Masimba, P.J., Otieno, D.F., Weisheit, A., Mbabazi, P., 2009. Antimicrobial and brine shrimp toxicity of some plants used in traditional medicine in Bukoba District, north-western Tanzania. Tanzan. J. Health Res. 11 (1), 23–28.

Nagawa, C., Böhmdorfer, S., Rosenau, T., 2015. Chemical composition and anti-termitic activity of essential oil from *Canarium schweinfurthii* Engl. Ind. Crops Prod. 71, 75–79.

Obame, L., Koudou, J., Kumulungui, B., Bassolè, I., Edou, P., Ouattara, A., 2007. Antioxidant and antimicrobial activities of *Canarium schweinfurthii* Engl. essential oil from Centrafrican Republic. Afr. J. Biotechnol. 6, 2319–2323.

Okwuosa, C., Achukwu Achukwu, P., Nwachukwu, D., Eze, A., Azubuike, N., 2009. Nephroprotective activity of stem bark extracts of *Canarium schweinfurthii* on Acetaminopheninduced renal injuries in rats. J. Coll. Med. 4 (1), 6–13.

Orwa, C., Mutua, A., Kindt, R., Jamnadass, R., Anthony, S., 2009. Agroforestree database:a tree reference and selection guide.

Ruffo, C.K., Birnie, A., Tengnas, B., 2002. Edible Wild Plants of Tanzania. Regional Land Management Unit, Nairobi.

Seukep, J.A., Ngadjui, B., Kuete, V., 2015. Antibacterial activities of *Fagara macrophylla, Canarium schweinfurthii, Myrianthus arboreus, Dischistocalyx grandifolius* and *Tragia benthamii* against multi-drug resistant Gram-negative bacteria. Springerplus 4, 567.

Shakirin, F.H., Azlan, A., Ismail, A., Amom, Z., Cheng Yuon, L., 2012. Protective effect of pulp oil extracted from canarium odontophyllum miq. fruit on blood lipids, lipid peroxidation, and antioxidant status in healthy rabbits. Oxid. Med. Cell. Longev. 2012, 840973.

CHAPTER 17

Cinnamon Species

A.T. Mbaveng, V. Kuete

1 INTRODUCTION

Cinnamon is the name for a dozen of species of trees and the commercial spice products that some of them produce, belonging to the genus *Cinnamomum* in the family Lauraceae. There are hundreds of types of Cinnamon with only four them being used for commercial purposes. These include *Cinnamomum cassia* (syn. *Cinnamomum aromaticum*) commonly known as Cassia or Chinese Cinnamon (the most common type), *Cinnamomum burmannii* known as Korintje or Indonesian Cinnamon or Padang Cassia, *Cinnamomum loureiroi* known as Saigon Cinnamon, Vietnamese cassia or Vietnamese cinnamon and *Cinnamomum zeylanicum* (syn. *Cinnamomum Verum*) known as Ceylon, true, or Mexican Cinnamon (The Seasoning and Spice Association, 2010). *C. zeylanicum* is native to Sri Lanka, India, Madagascar, Brazil, and the Caribbean and is also cultivated on a commercial scale in Seychelles and Madagascar (Mohammed, 1993). This plant is also found in West tropical Africa where it has multiple uses including its consumption as food (bark), sauces, condiments, spices, flavorings (leaf), and drink. In traditional medicine, the bark infusion is used as a remedy for pain, arthritis, rheumatism, nasopharyngeal affections, and stomach troubles whereas leaf, bark, and roots are used to heal diarrhea and dysentery (Burkill, 1985). *C. cassia* Blume is native to China and exotic to southern and eastern Asia and to Africa where it has been reported in Nigeria (Lockwood, 1979) and South Africa (Glen, 2002). Traditionally, *C. cassia* is used as an aromatic spice. Bark is used for its carminative, stomachic, antidiarrheal, and antibacterial properties (Bansode, 2012). There are several reports related to its pharmacological effects, such as antiinflammatory, antioxidant, and hepatoprotective activities (Bansode, 2012). *C. burmannii* is native to Indonesia and has been cultivated in South Africa (Glen, 2002). There is a scarcity of reports related to the cultivation of *C. burmannii* in Africa. In this chapter, we will focus on the chemistry and pharmacology of the two most popular cinnamon, *C. cassia* and *C. zeylanicum*.

2 BOTANICAL ASPECTS

The genus *Cinnamomum* consists of small, evergreen trees and shrubs of 10–15 m tall. Plants are found in south-east Asia, China, and Australia as well as in Africa. The bark is widely employed as a spice, its leaves are ovate–oblong in shape, and 7–18 cm long

(Cardoso-Ugarte et al., 2016). Flowers arranged in panicles have a greenish color and have a rather disagreeable odor. The fruit is a purple 1 cm berry containing a single seed (Cardoso-Ugarte et al., 2016; Leela, 2008). It is found in tropical rain forests, where it grows at various altitudes from highland slopes to lowland forests and occurs in both marshy places and on well-drained soils. However, in latitudes with seasonal climatic conditions, they become exceedingly rare (Jantan et al., 2008; Cardoso-Ugarte et al., 2016).

3 CHEMISTRY *OF CINNAMOMUM CASSIA* AND *CINNAMOMUM ZEYLANICUM*

Cinnamon is known for its aromatic fragrance and sweet, warm taste and is mostly used as a spice, having great medicinal value. Phytochemical studies revealed the presence of a variety of biologically active chemicals (Fig. 17.1). Leaves and bark contain cinnamaldehyde (**1**) and eugenol (**2**) as major constituents; root bark mostly contains camphor (**3**) whereas fruits contain *trans*-cinnamyl acetate (**4**) and β-caryophyllene (**5**) (Vangalapati et al., 2012). It was also shown that *C. zeylanicum* buds have terpene hydrocarbons, α-bergamotene (**6**), α-copaene (**7**), and oxygenated terpenoids as major components whereas flowers have compounds **4, 6**, and caryophyllene oxide (**8**) (Vangalapati et al., 2012). Other compounds which are present in lesser percentages are cinnamic acid, hydroxyl cinnamaldehyde, cinnamyl alcohol, coumarin (**9**), cinnamyl acetate, borneol, etc. (Vangalapati et al., 2012). However, the abundance of coumarin varies within *Cinamomum* species. Coumarin has been found to induce liver damage. Among

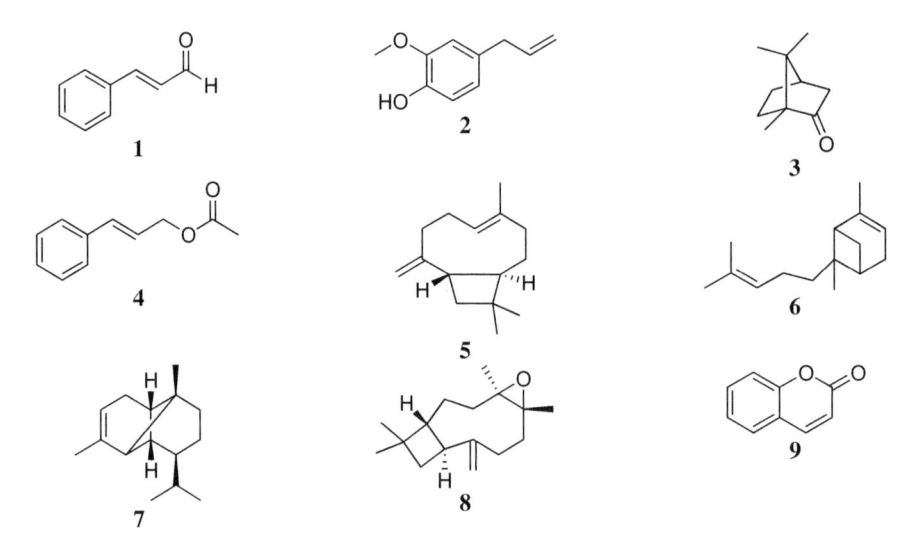

Figure 17.1 *Selected chemical structures of compounds from cinnamon species.* Cinnamaldehyde (**1**); eugenol (**2**); camphor (**3**); *trans*-cinnamyl acetate (**4**); β-caryophyllene (**5**); α-bergamotene (**6**), α-copaene (**7**); caryophyllene oxide (**8**); coumarin (**9**).

the main types of Cinnamon, *C. zeylanicum* has low levels of coumarin (17 mg/kg), whereas all other varieties, such as *C. cassia* (310 mg/kg), *C. burmannii* (2150 mg/kg), and *C. loureiroi* (6970 mg/kg) rather have high levels of coumarin (http://cinnamonvogue.com/Types_of_Cinnamon_1.html, types of cinnamon). *C. cassia* contains about 1–2% of volatile oil called cassia oil. The primary constituents of the essential oil are 65–80% compound **1** and less amount of **2** (Bansode, 2012). It also contains mucilage, starch, and tannins (Bansode, 2012).

4 PHARMACOLOGY OF *CINNAMOMUM CASSIA* AND *CINNAMOMUM ZEYLANICUM*

Several studies have focused on different pharmacological activities of *C. zeylanicum* and *C. cassia*. In this section, the synopsis of the biological effects of the two plants will be discussed.

4.1 Anticancer activity

Chulasiri and collaborators have demonstrated the cytotoxic potential of the petroleum ether and chloroform extracts from *C. zeylanicum* toward human mouth carcinoma KB cells and mouse lymphoid leukemia L1210 cells (Chulasiri et al., 1984; Ranasinghe et al., 2013). The cytotoxicity of the aqueous extract was reported in cervical cancer SiHa cells. The studied extract alters the growth of SiHa cells in a dose-dependent manner. Treated SiHa cells exhibited reduced migration potential that could be explained as due to the downregulation of matrix metalloproteinase-2 expression. Interestingly, the expression of Her-2 oncoprotein was significantly reduced in the presence of the extract. Extract also induced apoptosis in the cervical cancer cells through increase in intracellular calcium signaling as well as loss of mitochondrial membrane potential (MMP), suggesting that it could be used as a potent chemopreventive drug in cervical cancer (Koppikar et al., 2010). An investigation by Kwon et al. suggested that the antitumor effects of *C. cassia* extracts are directly linked with enhanced proapoptotic activity and inhibition of nuclear factor (NF-κB) and activator protein 1 (AP-1) activities and their target genes in vitro and in vivo in mouse melanoma model (Kwon et al., 2010). Compound **1**, a cytotoxic constituent of *C. zeylanicum* and *C. cassia*, was shown to be a potent inducer of apoptosis, transducing the apoptotic signal via reactive oxygen species (ROS) generation, thereby inducing mitochondrial permeability transition (MPT) and cytochrome c release to the cytosol (Ka et al., 2003). ROS production, mitochondrial alteration, and subsequent apoptotic cell death in cinnamaldehyde-treated cells were blocked by the antioxidant *N*-acetylcystein. Taken together, data reported by Ka and co-workers indicated that cinnamaldehyde induces the ROS-mediated MPT and resulted in cytochrome c release (Ka et al., 2003). Compound **1** also induced the death of leukemia HL-60 cells via alteration of MMP and induction of caspase-3 activity (Nishida et al., 2003).

4.2 Antidiabetic activity

The antidiabetic potential of *C. zeylanicum* and *C. cassia* were demonstrated both in vitro and in vivo. In a study conducted by Verspohl et al., experimental rats were given extracts from *C. cassia* and *C. zeylanicum* to evaluate blood glucose and plasma insulin levels under various conditions. Results showed that *C. cassia* extract was better than *C. zeylanicum* extract. A decrease in blood glucose levels was observed in a glucose tolerance test (GTT), whereas it was not obvious in rats that were not challenged by a glucose load. The elevation in plasma insulin was direct since a stimulatory in vitro effect of insulin release from INS-1 cells (an insulin secreting cell line) was also observed (Verspohl et al., 2005). *C. zeylanicum* also reduced postprandial intestinal glucose absorption by inhibiting the activity of enzymes involved in carbohydrate metabolism (pancreatic α–amylase and α-glucosidase), stimulated cellular glucose uptake by membrane translocation of glucose transporter type 4 (GLUT4), stimulated glucose metabolism and glycogen synthesis, to inhibit gluconeogenesis by effects on key regulatory enzymes and stimulated insulin release and potentiating insulin receptor activity (Ranasinghe et al., 2012, 2013). Cinnamtannin B1 was identified as the potential active compound responsible for these effects (Ranasinghe et al., 2012, 2013). In vivo, the plant attenuated weight loss associated with diabetes, reduced fasting blood glucose, low density lipoprotein (LDL), and increased high density lipoprotein (HDL)-cholesterol, reduced hemoglobin A1c (HbA1c) and increased circulating insulin levels (Ranasinghe et al., 2012, 2013). It also showed beneficial effects against diabetic neuropathy and nephropathy (Ranasinghe et al., 2012, 2013).

Cao et al. have investigated the effects of *C. cassia* on the protein and mRNA levels of insulin receptor (IR), GLUT4, and tristetraprolin (TTP/ZFP36) in mouse 3T3-L1 adipocytes. Authors found that cinnamon have the potential to increase the amount of proteins involved in insulin signaling, glucose transport, and antiinflammatory/antiangiogenesis response (Cao et al., 2007). The antidiabetic effect of bark extract in a type II diabetic animal model (C57BIKsj db/db) was also studied by Kim and collaborators. The investigators administered the crude extract at different dosages (50, 100, 150, and 200 mg/kg). They found that blood glucose concentration significantly decreased in a dose-dependent manner with the most in the 200 mg/kg group, compared with the control. In addition, serum insulin levels and HDL-cholesterol levels were significantly higher meanwhile and the concentrations of triglyceride, total cholesterol, and intestinal alpha-glucosidase activity were significantly lower after 6 weeks of administration. These data suggested that *C. cassia* extract has a regulatory role in blood glucose level and lipids and it may also exert a blood glucose-suppressing effect by improving insulin sensitivity or slowing absorption of carbohydrates in the small intestine (Kim et al., 2006). Treatment at a dose of 200 mg/kg/day for 12 weeks significantly decreased blood glucose, increased the levels of reduced glutathione and the activities of glutathione-S-transferase (GST), glutathione reductase (GR) and glutathione peroxidase (GPX), superoxide dismutase (SOD), and catalase (CAT) in the liver. Extract's treatment also significantly decreased lipid peroxidation. These results suggested that this extract may be effective for correcting hyperglycaemia and preventing diabetic complications (Kim et al., 2006).

4.3 Antiinflammatory activity

There are studies reporting the antiinflammatory activity of *C. zeylanicum* and *C. cassia*. Ethanol extract of *C. zeylanicum* demonstrated dose-dependent antinociceptive and antiinflammatory effects using two animal models: mice and rats. In effect, the extract showed an antiinflammatory effect against chronic inflammation induced by cotton pellet granuloma indicating antiproliferative effect (Atta and Alkofahi, 1998; Ranasinghe et al., 2013). *C. cassia* also inhibited nitric oxide (NO) and cyclooxygenase, thereby having antiinflammatory activity. In effect, Lee et al. investigated *C. cassia* bark extract and three of its constituents on NO production in RAW 264.7 cells. They observed potent inhibitory effects of cinnamaldehyde (**1**) against NO production at 0.1 µg/mL. Little or no activity was observed for cinnamic acid and compound **2**. Suppression effects of compound **1** on inducible NO synthase expression were revealed by Western blot analysis, suggesting that **1** could be useful for developing new types of NO inhibitors (Lee et al., 2002). The inhibitory effect of the methanol extract of *C. cassia* on cyclooxygenase (COX-2) was also reported by Hong et al. (2002).

4.4 Antimicrobial activities
4.4.1 Antibacterial activity

C. zeylanicum displayed antibacterial activity on a panel of Gram-negative and Gram positive bacteria, as well as against *Mycobacterium tuberculosis*. An interesting review by Ranasinghe and collaborators summarized the antibacterial activity against *Acinetobacter baumannii, Acinetobacter lwoffii, Bacillus cereus, Bacillus coaguiaris, Bacillus subtilis, Brucella melitensis, Clostridium difficile, Clostridium perfringens, Enterobacter aerogenes, Enterobacter cloacae, Enterococcus faecalis, Enterococcus faecium, Escherichia coli, Haemophilus influenza, Helicobacter pylori, Klebsiella pneumoniae, Listeria ivanovii, Listeria monocytogenes, Mycobacterium smegmatis, M. tuberculosis, Proteus mirabilis, Pseudomonas aeruginosa, Salmonella typhi, Salmonella typhimurium, Staphylococcus albus, Staphylococcus aureus, Streptococcus agalactiae, Streptococcus pneumoniae, Streptococcus pyogenes*, and *Yersinia enterocolitica* (Ranasinghe et al., 2013). The extract of *C. zeylanicum* also displayed good activity against MDR strain of *E. coli* AG100 with the lowest minimal inhibitory concentration (MIC) of 64 µg/mL (Voukeng et al., 2012). Sharma and coworkers also demonstrated the effect of ethanol extract of *C. cassia* against urinary tract pathogens *E. coli, K. pneumoniae, P. aeruginosa*, and *E. faecalis* (Sharma et al., 2009).

4.4.2 Antifungal activity

The antifungal activity of *C. zeylanicum* was compiled in a review by Ranasinghe et al. The sensitive strains of fungi included *Aspergillus flavus, Aspergillus fumigatus, Aspergillus nididans, Aspergillus niger, Aspergillus ochraceus, Aspergillus parasiticus, Aspergillus terreus, Candida albicans, Candida glabrata, Candida krusei, Candida parapsilosis, Candida tropicalis, Crytococcus neoformans, Epidermophyton floccosum, Hisioplasma capsulatum, Malassezia furfur, Microsporum audouini, Microsporum canis, Microsporum gypseum, Trichophyton mentagraphytes, Trichophyton rubrum,* and *Trichophyton tonsurans* (Ranasinghe et al., 2013). *C. cassia* oil as well as its major consitituent, **1** effective in inhibiting the growth of yeasts (four species of *Candida*: *C. albicans, C. tropicalis, C. glabrata,* and *C. krusei*), filamentous molds, three *Aspergillus* spp., and dermatophytes, *M. gypseum, T. rubrum,* and *T. mentagrophytes* (Ooi et al., 2006).

4.4.3 Antiviral activity

The antiviral activity of *C. cassia* bark was reported against human immodeficiency virus type 1 (HIV-1) (Premanathan et al., 2000). Fatima et al. (2016) have also synthesized silver nanoparticles using *C. cassia* and evaluated their activity against highly pathogenic avian influenza virus subtype H7N3. The silver nanoparticles derived from *Cinnamon* extract enhanced the antiviral activity and were effective in both treatments, when incubated with the virus prior to infection and introduced to cells after infection. Therefore, the biosynthesized nanoparticles may be a promising approach to provide treatment against influenza virus infections (Fatima et al., 2016). *C. cassia* also prevented airway epithelia from human respiratory syncytial virus (HRSV) infection through inhibiting viral attachment, internalization, and syncytium formation (Yeh et al., 2013). This suggests that *C. cassia* could also be a candidate to develop therapeutic modalities to manage HRSV infection. From the active fraction of *C. zeylanicum,* an isolated compound identified as cinnzeylanine inhibited the proliferation of herpes simplex virus type 1 in Vero cells (Orihara et al., 2008).

4.5 Antioxidant activity

Several authors have reported the antioxidant activity of *C. zeylanicum* and *C. cassia* using various experimental models. Jayaprakasha et al. evaluated the antioxidant activity of various extracts from *C. zeylanicum* through in vitro model systems, such as β-carotene-linoleate and 1,1-diphenyl-2-picryl hydrazyl (DPPH). Authors found that the plant had antioxidant activity and that in both model systems, the activity of the extracts was in the order of water > methanol > acetone > ethyl acetate (Jayaprakasha et al., 2007). Using β-carotene/linoleic acid system, Mancini-Filho and coworkers also showed that ether, methanol, and aqueous *C. zeylanicum* extracts inhibited the oxidative process in 68, 95.5, and 87.5%, respectively (Mancini-Filho et al., 1998). In addition to its ability to scavenge $DPPH^+$ radicals, *C. zeylanicum* extract was potent in free 2,2′-azinobis-3-ethylbenzothiazoline-6-sulfonic acid (ABTS) radical cations, hydroxyl (^+OH), and superoxide radicals (O^{2-}) and metal chelating (Mathew and Abraham, 2006). The peroxidation inhibiting activity of *C. zeylanicum* recorded using a linoleic acid emulsion system, also showed very good antioxidant activity. Essential oil obtained from the bark of *C. zeylanicum* and three of its major compounds, **1**, **2**, and linalool (representing 82.5% of the total composition), were tested in two in vitro models of peroxynitrite-induced nitration and lipid peroxidation. Essential oil and **2** showed very powerful activities, decreasing 3-nitrotyrosine formation with IC_{50} values of 18.4 μg/mL and 46.7 μM, respectively (ascorbic acid, 71.3 μg/mL and 405 μM) and also inhibiting the peroxynitrite-induced lipid peroxidation with an IC_{50} of 2 μg/mL and 13.1 μM, respectively, against 59 μg/mL (235.5 μM) for the reference compound Trolox (Chericoni et al., 2005). On the contrary, compound **1** and linalool were completely inactive. The antioxidative stress capacity of *C. zeylanicum* was studied clinically in humans when administered in a regular and controlled manner (Ranjbar et al., 2006). A total of 54 normal subjects were divided into three groups: receiving water, regular tea, or cinnamon tea for 2 weeks. Blood samples were obtained before and after treatment and analyzed for lipid peroxidation levels, total antioxidant power, and total thiol groups. The results indicated increased total antioxidant power and total thiols but a decrease in lipid peroxidation levels in individuals who received regular or cinnamon tea compared with controls. The investigator also found that the extent of increase in total antioxidant power and decrease in lipid peroxidation levels were more evident in individuals who received cinnamon tea compared with those who received regular tea. Therefore, it was concluded that *C. zeylanicum* has a marked antioxidant potential and may be beneficial in alleviating the complications of many illnesses related to oxidative stress in humans (Ranjbar et al., 2006).

C. cassia extract exhibited a higher percentage of inhibition of oxidation as tested by the lipid peroxidation assay than reference food antioxidants, butylated hydroxyanisole (BHA or E-320), butylated hydroxytoluene (BHT or E-321), and propyl gallate (E-310) (Murcia et al., 2004). This plant extract was also found to be better superoxide radical scavenger with capacity approximately equal to that of propyl gallate (Murcia et al., 2004). Wang et al. also showed that one-half milliliter of *C. cassia* leaf has free radical scavenging ability with an EC_{50} of 53 μg/mL (Wang et al., 2008). Barks, buds, and leaves of *C. cassia* extracted with ethanol and supercritical fluid extraction were tested for their DPPH scavenging activity and it was found that the bark ethanol was the best antioxidant sample (Yang et al., 2012).

4.6 Hepatoprotective activity

Eidi et al. have evaluated the protective effects of *C. zeylanicum* against carbon tetrachloride (CCl_4)-induced liver damage in male Wistar rats (Eidi et al., 2012). The authors reported that administration with *C. zeylanicum* extracts (0.01, 0.05, and 0.1 g/kg) for 28 days significantly reduced the impact of CCl_4 toxicity on the serum markers of liver damage, aspartate aminotransferase, alanine aminotransferase, and alkaline phosphatase. In addition, they found that treatment with *C. zeylanicum* extract resulted in markedly increased levels of superoxide dismutase and CAT enzymes in rats (Eidi et al., 2012).

These findings indicate that *C. zeylanicum* extract acts as a potent hepatoprotective agent against CCl4-induced hepatotoxicity in rats.

Kanuri and collaborators also assessed the protective effects of an alcoholic extract of *C. cassia* bark in a mouse model of acute alcohol-induced steatosis and in RAW 264.7 macrophages, used as a model of Kupffer cells (Kanuri et al., 2009). They found that acute alcohol ingestion caused a >20-fold increase in hepatic lipid accumulation. Pretreatment with cinnamon extract significantly reduced the hepatic lipid accumulation. This protective effect of cinnamon extract was associated with an inhibition of the induction of the myeloid differentiation primary response gene (MyD) 88, inducible NO synthase (iNOS), and plasminogen activator inhibitor 1 mRNA expression found in livers of alcohol-treated animals. They also found that in vitro prechallenge with *C. cassia* extract suppressed lipopolysaccharide (LPS)-induced MyD88, iNOS, and TNFα expressions as well as NO formation almost completely. Furthermore, LPS treatment of RAW 264.7 macrophages further resulted in degradation of inhibitor kappa B; this effect was almost completely blocked by cinnamon extract (Kanuri et al., 2009). These results suggested that an alcohol extract of *C. cassia* bark may protect the liver from acute alcohol-induced steatosis through mechanisms involving the inhibition of MyD88 expression. The effect of ethanol extract from *C. cassia* on the activation of hepatic stellate cells (HSCs) in addition to the effects of *C. cassia* powder in Sprague–Dawley rats with acute liver injury induced by dimethylnitrosamine (DMN) were investigated by Lim et al. (2010). They found that ethanol extract significantly reduced the expression of alpha-smooth muscle actin (a-SMA), connective tissue growth factor (CTGF), transforming growth factor beta 1 (TGF-β1), and tissue inhibitor of metalloproteinase-1 (TIMP-1). In vivo, the protective effect of the *C. cassia* powder was evidenced by increase in the serum total protein, albumin, total-bilirubin, direct-bilirubin, glutamic oxaloacetic transaminase, glutamic pyruvic transaminase, and alkaline phosphatase (Lim et al., 2010). From these data, authors suggested that *C. cassia* inhibits fibrogenesis, followed by HSC-T6 cell activation and increased restoration of liver function, ultimately resulting in acute liver injury.

4.7 Other pharmacological activities

Crude extract from *C. zeylanicum* demonstrated wound healing properties in rats, in a study using 32 rats where experimental excision wounds were induced and treated with topical extract containing ointments (Farahpour and Habibi, 2012). It was also shown that *C. cassia* has no adverse effects if used as myorelaxant agent, and might be an effective anxiolytic agent, regulating the serotonergic and GABAergic system (Yu et al., 2007). Bark extract of *C. cassia* at a dose of $(10^{-5}$–10^{-4} g/mL) was shown to significantly protect against glutamate-induced cell death and also inhibited glutamate-induced Ca^{2+} influx in a dose-dependent manner (Shimada et al., 2000). Crude extracts from leaf and bark of *C. zeylanicum* were tested against *Culex quinquefasciatus*, *Anopheles tessellatus*, and *Aedes aegypti*. It was shown that bark oil had good knock-down effect and mortality against

A. tessellatus and *C. quinquefasciatus* with a lethal dose 50 (LD_{50}) of 0.33 and 0.66 μg/mL, respectively; meanwhile the corresponding values for leaf oil were 1.03 and 2.1 μg/mL, respectively (Samarasekera et al., 2005). Park et al. have examined the insecticidal and fumigant activities of *C. cassia* bark-derived materials against the oak nut weevil (*Mechoris ursulus* Roelofs) and compared to those of the commercially available *Cinnamomum* bark-derived compounds (**2**, salicylaldehyde, *trans*-cinnamic acid, and cinnamyl alcohol) (Park et al., 2000). They found that test compounds had insecticidal activity, attributable to fumigant action as well as to direct contact.

5 CONCLUSIONS

In this chapter, we brought together the available data related to the phytochemical composition and to the phamacology of *C. zeylanicum* and *C. cassia*. The two *Cinnamomum* species have demonstrated various pharmacological activities, such as anticancer, antidiabetic, antiinflammatory, antibacterial, antifungal, antiviral, antioxidant activities, etc. However, there are still limited clinical studies on both species; therefore this chapter provides baseline information to motivate researchers to conduct such investigations, to optimize their use in the management of various human ailments.

REFERENCES

Atta, A.H., Alkofahi, A., 1998. Anti-nociceptive and anti-inflammatory effects of some Jordanian medicinal plant extracts. J. Ethnopharmacol. 60 (2), 117–124.

Bansode, V.J., 2012. A review on pharmacological activities of *Cinnamomum cassia* Blume. Int. J. Green Pharm. 6 (2), 102–108.

Burkill, H., 1985. The Useful Plants of West Tropical Africa. Royal Botanic Garden Kew, London.

Cao, H., Polansky, M.M., Anderson, R.A., 2007. Cinnamon extract and polyphenols affect the expression of tristetraprolin, insulin receptor, and glucose transporter 4 in mouse 3T3-L1 adipocytes. Arch. Biochem. Biophys. 459 (2), 214–222.

Cardoso-Ugarte, G.A., López-Malo, A., Sosa-Morales, M.E., 2016. Cinnamon (*Cinnamomum zeylanicum*) essential oils. In: Preedy, V.R. (Ed.), Essential Oils in Food Preservation, Flavor and Safety. Academic Press, San Diego, pp. 339–347, (Chapter 38).

Chericoni, S., Prieto, J.M., Iacopini, P., Cioni, P., Morelli, I., 2005. In vitro activity of the essential oil of *Cinnamomum zeylanicum* and eugenol in peroxynitrite-induced oxidative processes. J. Agric. Food Chem. 53 (12), 4762–4765.

Chulasiri, M.U., Picha, P., Rienkijkan, M., Preechanukool, K., 1984. The cytotoxic effect of petroleum ether and chloroform extracts from ceylon cinnamon (*Cinnamomum zeylanicum* nees) barks on tumor cells in vitro. Int. J. Crude Drug Res. 22, 177–180.

Eidi, A., Mortazavi, P., Bazargan, M., Zaringhalam, J., 2012. Hepatoprotective activity of cinnamon ethanolic extract against CCL_4-induced liver injury in rats. EXCLI J. 11, 495–507.

Farahpour, M.R., Habibi, M., 2012. Evaluation of the wound healing activity of an ethanolic extract of ceylon cinnamon in mice. Vet. Med. 57, 53–57.

Fatima, M., Zaidi, N.S., Amraiz, D., Afzal, F., 2016. In vitro antiviral activity of *Cinnamomum cassia* and its nanopartilces against H7N3 influenza A virus. J. Microbiol. Biotechnol. 26 (1), 151–159.

Glen, H.F., 2002. Cultivated Plants of Southern Africa. Jacana, Johannesburg.

Hong, C.H., Hur, S.K., Oh, O.J., Kim, S.S., Nam, K.A., Lee, S.K., 2002. Evaluation of natural products on inhibition of inducible cyclooxygenase (COX-2) and nitric oxide synthase (iNOS) in cultured mouse macrophage cells. J. Ethnopharmacol. 83 (1–2), 153–159.

Jantan, I., Karim, B., Santhanam, J., Azdina, J., 2008. Correlation between chemical composition and antifungal activity of the essential oils of eight *Cinnamomum* species. Pharm. Biol. 46 (6), 406–412.

Jayaprakasha, G.K., Negi, P.S., Jena, B.S., Jagan Mohan Rao, L., 2007. Antioxidant and antimutagenic activities of *Cinnamomum zeylanicum* fruit extracts. J. Food Compos. Anal. 20 (3–4), 330–336.

Ka, H., Park, H.J., Jung, H.J., Choi, J.W., Cho, K.S., Ha, J., 2003. Cinnamaldehyde induces apoptosis by ROS-mediated mitochondrial permeability transition in human promyelocytic leukemia HL-60 cells. Cancer Lett. 196 (2), 143–152.

Kanuri, G., Weber, S., Volynets, V., Spruss, A., Bischoff, S.C., Bergheim, I., 2009. Cinnamon extract protects against acute alcohol-induced liver steatosis in mice. J. Nutr. 139 (3), 482–487.

Kim, S.H., Hyun, S.H., Choung, S.Y., 2006. Anti-diabetic effect of cinnamon extract on blood glucose in db/db mice. J. Ethnopharmacol. 104 (1–2), 119–123.

Koppikar, S.J., Choudhari, A.S., Suryavanshi, S.A., Kumari, S., Chattopadhyay, S., Kaul-Ghanekar, R., 2010. Aqueous cinnamon extract (ACE-c) from the bark of *Cinnamomum cassia* causes apoptosis in human cervical cancer cell line (SiHa) through loss of mitochondrial membrane potential. BMC Cancer 10, 210.

Kwon, H.K., Hwang, J.S., So, J.S., Lee, C.G., Sahoo, A., Ryu, J.H., 2010. Cinnamon extract induces tumor cell death through inhibition of NFkappaB and AP1. BMC Cancer 10, 392.

Lee, H.S., Kim, B.S., Kim, M.K., 2002. Suppression effect of *Cinnamomum cassia* bark-derived component on nitric oxide synthase. J. Agric. Food Chem. 50 (26), 7700–7703.

Leela, J., 2008. Cinnamon and Cassia. In: Parthasarathy, V., Chempakam, B., Zachariah, T. (Eds.), Chemistry of Spices. CABI, USA.

Lim, C.S., Kim, E.Y., Lee, H.S., Soh, Y., Sohn, Y., Kim, S.Y., 2010. Protective effects of *Cinnamomum cassia* Blume in the fibrogenesis of activated HSC-T6 cells and dimethylnitrosamine-induced acute liver injury in SD rats. Biosci. Biotechnol. Biochem. 74 (3), 477–483.

Lockwood, G., 1979. Major constituents of the essential oils of *Cinnamomum cassia* Blume growing in Nigeria. Planta Med. 36 (4), 380–381.

Mancini-Filho, J., Van-Koiij, A., Mancini, D.A., Cozzolino, F.F., Torres, R.P., 1998. Antioxidant activity of cinnamon (*Cinnamomum zeylanicum*, Breyne) extracts. Boll. Chim. Farm. 137 (11), 443–447.

Mathew, S., Abraham, T.E., 2006. Studies on the antioxidant activities of cinnamon (*Cinnamomum verum*) bark extracts, through various in vitro models. Food Chem. 94 (4), 520–528.

Mohammed, I., 1993. International trade in non-wood forest products: an overview. FO: Misc/93/11—Working Paper. Food and Agriculture Organization of the United Nations.

Murcia, M.A., Egea, I., Romojaro, F., Parras, P., Jimenez, A.M., Martinez-Tome, M., 2004. Antioxidant evaluation in dessert spices compared with common food additives. Influence of irradiation procedure. J. Agric. Food Chem. 52 (7), 1872–1881.

Nishida, S., Kikuichi, S., Yoshioka, S., Tsubaki, M., Fujii, Y., Matsuda, H., 2003. Induction of apoptosis in HL-60 cells treated with medicinal herbs. Am. J. Chin. Med. 31 (4), 551–562.

Ooi, L.S., Li, Y., Kam, S.L., Wang, H., Wong, E.Y., Ooi, V.E., 2006. Antimicrobial activities of cinnamon oil and cinnamaldehyde from the Chinese medicinal herb *Cinnamomum cassia* Blume. Am. J. Chin. Med. 34 (3), 511–522.

Orihara, Y., Hamamoto, H., Kasuga, H., Shimada, T., Kawaguchi, Y., Sekimizu, K., 2008. A silkworm baculovirus model for assessing the therapeutic effects of antiviral compounds: characterization and application to the isolation of antivirals from traditional medicines. J. Gen. Virol. 89 (Pt 1), 188–194.

Park, I.K., Lee, H.S., Lee, S.G., Park, J.D., Ahn, Y.J., 2000. Insecticidal and fumigant activities of *Cinnamomum cassia* bark-derived materials against *Mechoris ursulus* (Coleoptera: attelabidae). J. Agric. Food Chem. 48 (6), 2528–2531.

Premanathan, M., Rajendran, S., Ramanathan, T., Kathiresan, K.N.H., Yamamoto, N., 2000. A survey of some Indian medicinal plants for anti-human immunodeficiency virus (HIV) activity. Indian J. Med. Res. 112, 73–77.

Ranasinghe, P., Jayawardana, R., Galappaththy, P., Constantine, G.R., de Vas Gunawardana, N., Katulanda, P., 2012. Efficacy and safety of 'true' cinnamon (*Cinnamomum zeylanicum*) as a pharmaceutical agent in diabetes: a systematic review and meta-analysis. Diab. Med. 29, 1480–1492.

Ranasinghe, P., Pigera, S., Premakumara, G.A., Galappaththy, P., Constantine, G.R., Katulanda, P., 2013. Medicinal properties of 'true' cinnamon (*Cinnamomum zeylanicum*): a systematic review. BMC Complement Altern. Med. 13, 275.

Ranjbar, A., Ghasmeinezhad, S., Zamani, H., Malekirad, A.A., Baiaty, A., Mohammadirad, A., 2006. Antioxidative stress potential of *Cinnamomum zeylanicum* in humans: a comparative cross-sectional clinical study. Therapy 3 (1), 113–117.

Samarasekera, R., Kalhari, K.S., Weerasinghe, I.S., 2005. Mosquitocidal activity of leaf and bark essential oils of Ceylon *Cinnamomum zeylanicum*. J. Essent. Oil Res. 17 (3), 301–303.

Sharma, A., Chandraker, S., Patel, V.K., Ramteke, P., 2009. Antibacterial activity of medicinal plants against pathogens causing complicated urinary tract infections. Indian J. Pharm. Sci. 71 (2), 136–139.

Shimada, Y., Goto, H., Kogure, T., Kohta, K., Shintani, T., Itoh, T., 2000. Extract prepared from the bark of *Cinnamomum cassia* Blume prevents glutamate-induced neuronal death in cultured cerebellar granule cells. Phytother. Res. 14 (6), 466–468.

The Seasoning and Spice Association. (2010). Culinary Herbs and Spices. http://www.seasoningandspice.org.uk/ssa/background_culinary-herbs-spices.aspx

Vangalapati, M., Sree Satya, N., Surya Prakash, D.V., Avanigadda, S., 2012. A review on pharmacological activities and clinical effects of *Cinnamon* species. Res. J. Pharm. Biol. Chem. Sci. 3 (1), 653–663.

Verspohl, E.J., Bauer, K., Neddermann, E., 2005. Anti-diabetic effect of *Cinnamomum cassia* and *Cinnamomum zeylanicum* in vivo and in vitro. Phytother. Res. 19 (3), 203–206.

Voukeng, I.K., Kuete, V., Dzoyem, J.P., Fankam, A.G., Noumedem, J.A., Kuiate, J.R., 2012. Antibacterial and antibiotic-potentiation activities of the methanol extract of some cameroonian spices against Gram-negative multi-drug resistant phenotypes. BMC Res. Notes 5, 299.

Wang, H.F., Wang, Y.K., Yih, K.H., 2008. DPPH free-radical scavenging ability, total phenolic content, and chemical composition analysis of forty-five kinds of essential oils. J. Cosmet. Sci. 59 (6), 509–522.

Yang, C.H., Li, R.X., Chuang, L.Y., 2012. Antioxidant activity of various parts of *Cinnamomum cassia* extracted with different extraction methods. Molecules 17 (6), 7294–7304.

Yeh, C.F., Chang, J.S., Wang, K.C., Shieh, D.E., Chiang, L.C., 2013. Water extract of *Cinnamomum cassia* Blume inhibited human respiratory syncytial virus by preventing viral attachment, internalization, and syncytium formation. J. Ethnopharmacol. 147 (2), 321–326.

Yu, H.S., Lee, S.Y., Jang, C.G., 2007. Involvement of 5-HT1A and GABAA receptors in the anxiolytic-like effects of *Cinnamomum cassia* in mice. Pharmacol. Biochem. Behav. 87 (1), 164–170.

CHAPTER 18

Cymbopogon citratus

O.A. Lawal, A.L. Ogundajo, N.O. Avoseh, I.A. Ogunwande

1 INTRODUCTION

The genus *Cymbopogon* belongs to one of the most important essential oil yielding and monocots grass of the Poaceae (Gramineae) family and comprises about 180 species, subspecies, varieties, and subvarieties that are widely distributed in temperate and tropical regions of the world (Bertea and Maffei, 2010). Although several species are poorly known and studied, some species of the genus have been reportedly used in traditional medicine, while some that are known to be rich in volatile oils have applications in the cosmetics, pharmaceuticals, and perfumery industries (Jeong et al., 2009; Avoseh et al., 2015).

1.1 History and distribution

Cymbopogon citratus (DC. ex Nees) Stapf. (Fig. 18.1) is one of the best known species of the genus *Cymbopogon* with the synonyms *Andropogon citratus* DC. ex Nees, *Andropogon ceriferus* Hack., *Andropogon nardus* subsp. ceriferus L. (Hack.) Hack., *Andropogon roxburghii* Nees ex Steud., and *Cymbopogon nardus* subvar. *citratus* (L.) Rendle (DC. ex Nees) Roberty. It is native to Asia (Indochina, Indonesia, and Malaysia), Africa, and the Americas, but are widely cultivated in temperate and tropical regions of the world (Jayasinha, 2001). *C. citratus* is known by numerous international common names, such as West Indian lemon grass or lemon grass (English), hierba limon or zacate de limón (Spanish), citronelle or verveine des indes (French) and xiang mao (Chinese) and locally identified with over 28 vernacular names from different countries of the world (Jayasinha, 2001; Ross, 1999). Table 18.1 summarizes the geographical distribution of *C. citratus* around the world.

C. citratus is a tall perennial fast growing grass with tuft of lemon scented leaves from the annulate and sparingly branched rhizomes. It grows to a height of 1 m and a width of 5–10 mm, and has distinct bluish-green leaves which does not produce seed. However, it has many bulbous stems that increase the clump size as the plant grows. The leaves are long, glabrous, glaucus green, linear, tapering upward and along the margins, with very short ligule and tightly clasp sheaths at the base, narrow and separating at distal end. The inflorescences are nodding with pairly racemes of spirelets and subtended by spathes of about 1 m long with peduncles of 30–60 cm long (Jayasinha, 2001; Ross, 1999).

Figure 18.1 C. citratus *(Lagos, Nigeria, February, 2016)*.

1.2 Botanical aspects and distribution

C. citratus is propagated by means of root divisions. The clump of a mature plant is divided into a number of slips and, the top and fibrous roots of each slip are trimmed off before planting. The soil must be loosened thoroughly by ploughing before planting. Holes about 20 cm deep are dug with a crowbar and the slips are planted 45–60 cm apart, in rows about 90 cm apart. One to four segments are planted in each hole depending on the fertility of soil. The planting holes are lightly filled with earth to facilitate the development of a good root system. A 100 m^2 nursery would yield sufficient root segments to plant 1 ha, with two root segments per planting hole. Thus 15,000 to 20,000 segments can be planted/ha. *C. citratus* grows in tropical climate. It grows on poor soil and is more resistant to drought. However, it flourishes on a wide variety of soils ranging from rich loam to poor laterite and a warm and humid climate with plenty of sunshine and temperatures of 24–27°C. It grows best at elevations ranging from sea level to 1200 m and prefers an annual rainfall of 200–250 cm. In regions of abundant rainfall, the plant may be harvested more regularly during the year. However, with prolonged rainfall, planting on ridges may be suggested (Jayasinha, 2001). *C. citratus* is a commercially viable plant

Table 18.1 Geographical distribution of *Cymbopogon citratus* around the world

Continent	Country	References
Asia	Bangladesh	Chowdury et al. (2015)
	China	Yang and Lei (2005)
	India	Jeong et al. (2009)
	Indonesia	Muchtaridi and Subarnas (2011)
	Japan	Abe et al. (2003)
	Malaysia	Tajidin et al. (2012)
	Nepal	Singh et al. (1980)
	North Korea	Ho-Dzun et al. (1997)
	Pakistan	Mirza et al. (2005)
	Philippines	Balboa and Lim-Sylianco (1995)
	Sri Lanka	Jayasinha (2001)
	Thailand	Detpiratmongkol et al. (2005)
	Vietnam	Nham (1993)
Africa	Algeria	Boukhatem et al. (2013)
	Angola	Soares et al. (2013a,b)
	Benin	Bossou et al. (2013)
	Burkina Faso	Bassolé et al. (2011)
	Cameroon	Ntonga et al. (2014)
	Cote d'Ivoire	Kouame et al. (2015)
	Democratic Republic of Congo	Cimanga et al. (2002)
	Egypt	Hanaa et al. (2012)
	Ethiopia	Abegaz et al. (1983)
	Ghana	Nyarko et al. (2012)
	Kenya	Matasyoh et al. (2011)
	Mali	Sidibé et al. (2001)
	Nigeria	Adeneye and Agbaje (2007)
	Tanzania	Santos et al. (2013)
	Togo	Koba et al. (2009)
	Zambia	Chisowa et al. (1998)
	Zimbabwe	Kazembe and Chauruka (2012)
Europe	Italy	Bertea et al. (2003)
	United Kingdom	Humphrey (1973)
North America	Canada	Nanon et al. (2014)
	Costa Rica	Pohl (1972)
	Cuba	Tapia et al. (2007)
	Mexico	Berlin et al. (1974, 1977)
	Panama	Covich and Nickerson (1966)
	Saint Lucia	Graveson (2012)
	Trinidad and Tobago	Clement et al. (2015)
	United States of America	Rozzi et al. (2002)
South America	Argentina	Hilgert (2001)
	Bolivia	Moore et al. (2007)
	Brazil	Costa et al. (2011)
	Colombia	Olivero-Verbel et al. (2010)
	Peru	Leclercq et al. (2000)
	Venezuela	Gonzalez et al. (2008)
Australia and Oceania	Australia	Beech (1990)
	Papua New Guinea	Wossa et al. (2004)

accounting for high foreign exchange earnings for countries, such as Brazil, Indonesia, India, and China. Although commercial activities on *C. citratus* in West Africa are low, its subsistence cultivation is numerous and scattered all across the region for foods, ornamental, and medicinal purposes.

2 CHEMICAL CONSTITUENTS OF *C. citratus*

As a consequence of several ethno-pharmacological applications of *C. citratus*, many studies were carried out aimed at enlarging the knowledge of its volatile and nonvolatile chemical components. These studies have revealed that the chemical composition of the essential oils and extracts of *C. citratus* varied according to the geographical origin. The chemical classes of isolated substances from the plant are tannins, sterols, terpenoids, phenols, ketone, flavonoids, and sugars (Chisowa et al., 1998; Matouschek and Stahl, 1991; Negrelle and Gomes, 2007).

2.1 Phenolics

Matouschek and Stahl (1991) reported the nonvolatile polyphenol compounds (Fig. 18.2) present in the *C. citratus*, namely chlorogenic (**1**), caffeic (**2**), and *p*-coumaric (**3**) acids. Elemicin (**4**), catechol (**5**), and hydroquinone (**6**) were the phenolic compounds isolated from *C. citratus* essential oils from Bangladesh (Faruq, 1994).

2.2 Steroids and fatty alcohol

Olaniyi et al. (1975) reported the isolation of a steroidal (Fig. 18.3) compound, β-sitosterol (**7**), and fatty alcohols identified as hexaconsanol (**8**) and triacontanol

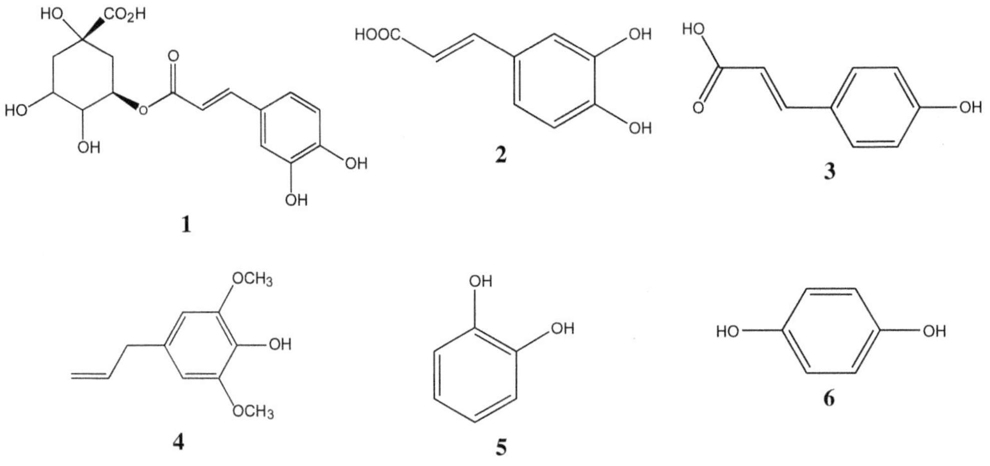

Figure 18.2 *Nonvolatile polyphenol compounds present in* **C. citratus.** chlorogenic acid (**1**); caffeic acid (**2**); *p*-coumaric acid (**3**); elemicin (**4**); catechol (**5**); hydroquinone (**6**).

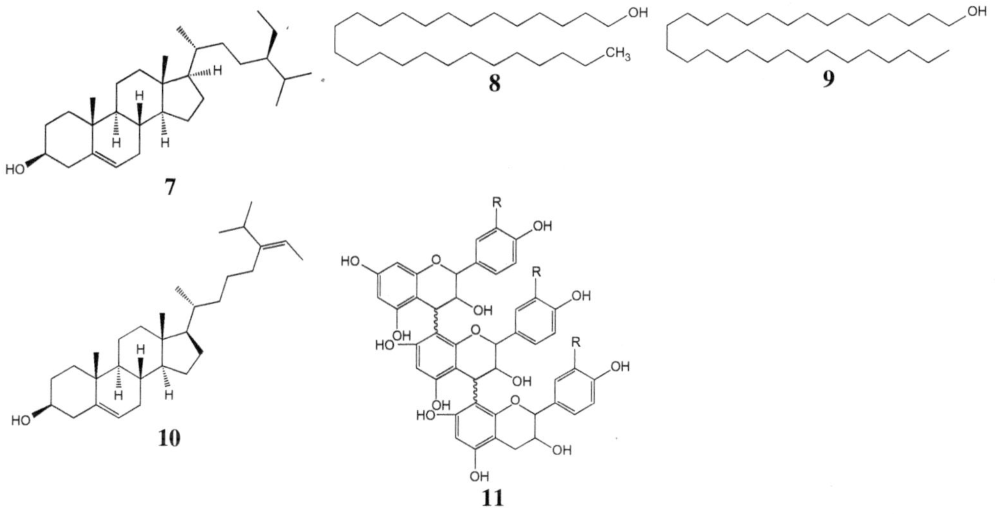

Figure 18.3 *Steroids and fatty alcohol and tannins from* **C. citratus.** β-sitosterol (**7**); hexaconsanol (**8**); triacontanol (**9**); fucosterol (**10**); a proanthocyanidin (**11**).

(**9**) from nonsaponifiable fraction of the light petroleum extract of *C. citratus*. Isolation of steroidal saponin, closely related to fucosterol (**10**), was also reported from the neutral fraction of the plant material (Olaniyi et al., 1975).

2.3 Tannins

Several literatures reported the presence of tannins in the phytochemical screening of *C. citratus* (Aftab et al. 2011; Edeoga et al., 2005). However, literature on the isolation of the tannins from *C. citratus* is scarce. Fractionated extracts of the *C. citratus* collected from Portugal were reported to contain 10 mg dry weight (DW) of hydrolysable tannins (proanthocyanidins) (**11**) (Figueirinha et al., 2008).

2.4 Flavonoids

Many literatures have reported several isolated flavonoids from *C. citratus* (Fig. 18.4). Isolation of luteolin (**12**), luteolin 6-O-glucosides (**13**), and luteolin 7-O-glucoside (cynaroside) (**14**) from *C. citratus* has been reported (Matouschek and Stahl, 1991). Leuteolin (**12**) was considered as one of the plant marker compounds (Gunasingh and Nagarajan, 1980; Matouschek and Stahl, 1991). Other isolated flavonoids reported from *C. citratus* are as follows: 2″-O-rhamnosyl isoorientin (**15**), quercetin (**16**), kaempferol (**17**), and apigenin (**18**) (Matouschek and Stahl, 1991; Miean and Mohamed, 2001). Cheel et al. (2005) also reported the isolation of isoscoparin (**19**), swertiajaponin (**20**), and orientin (**21**) in *C. citratus*.

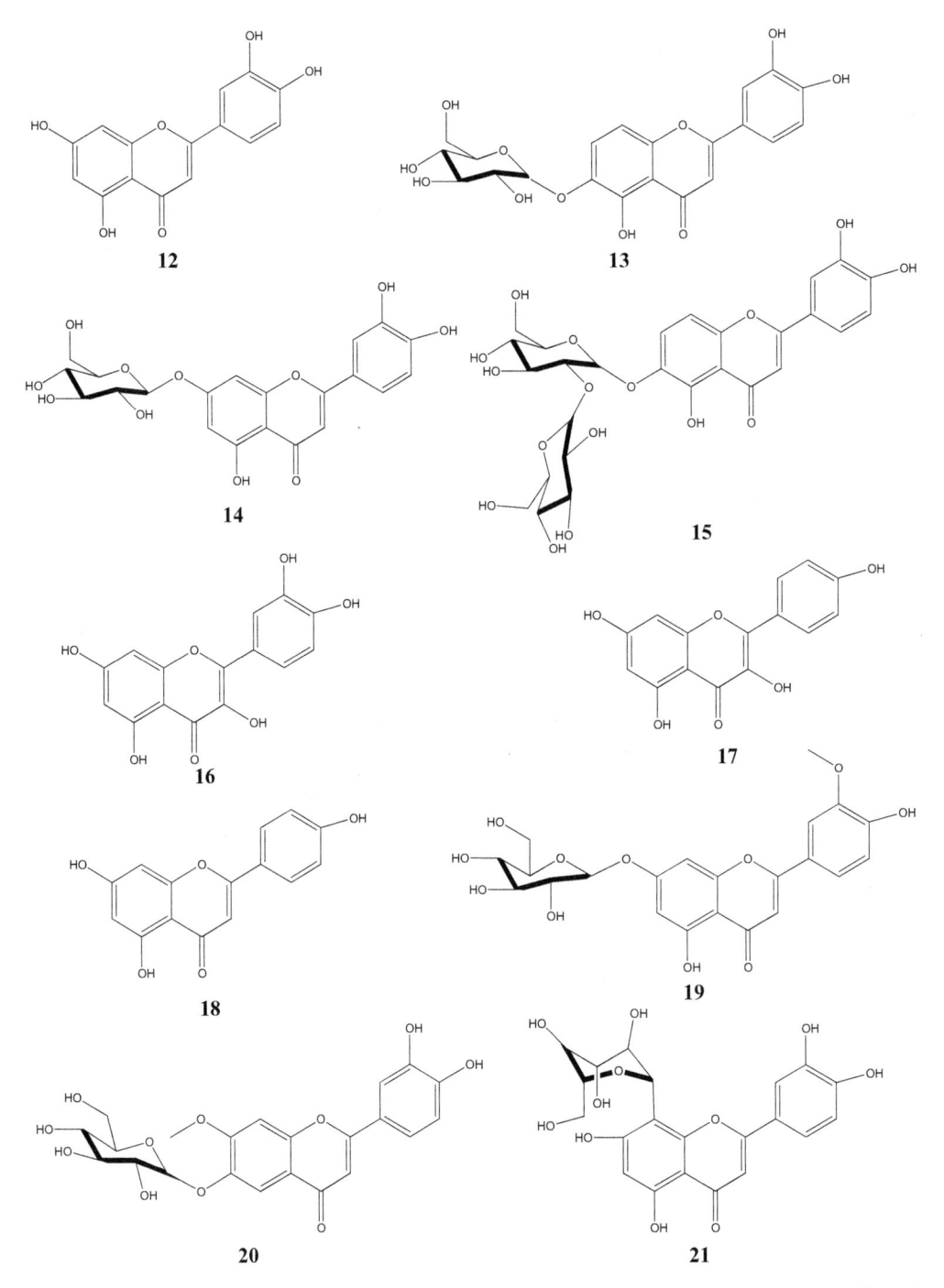

Figure 18.4 *Flavonoids from* **C. citratus.** Luteolin (**12**); luteolin 6-O-glucosides (**13**); cynaroside (**14**); 2″-O-rhamnosyl isoorientin (**15**); quercetin (**16**); kaempferol (**17**); apigenin (**18**); isoscoparin (**19**); swertiajaponin (**20**); orientin (**21**).

2.5 Terpenoids

Investigations have shown that the terpenoid constituents of *C. citratus* (Figs. 18.5–18.8) essential oils vary according to the geographical origin, genetic differences, part of the plant used, method of extraction, age/stage of maturity, and season of harvest (Ewansiha et al., 2012; Idrees et al., 2012). Despite the mentioned variation, myrcene (**22**) remains a characteristic compound of this species in a variable quantity (Adeola et al., 2013; Bossou et al., 2013; Matasyoh et al., 2011; Zannou et al., 2015).

However, *C. citratus* has been reported to contain a high percentage (about 80%) of citral (**23**), which is a mixture of terpenoids neral (**24**) and geranial (**25**), and is responsible for the lemony smell that characterizes the species (Negrelle and Gomes, 2007). Analysis of Ethiopian variety of *C. citratus* revealed high contents of geraniol (40%) (**26**),

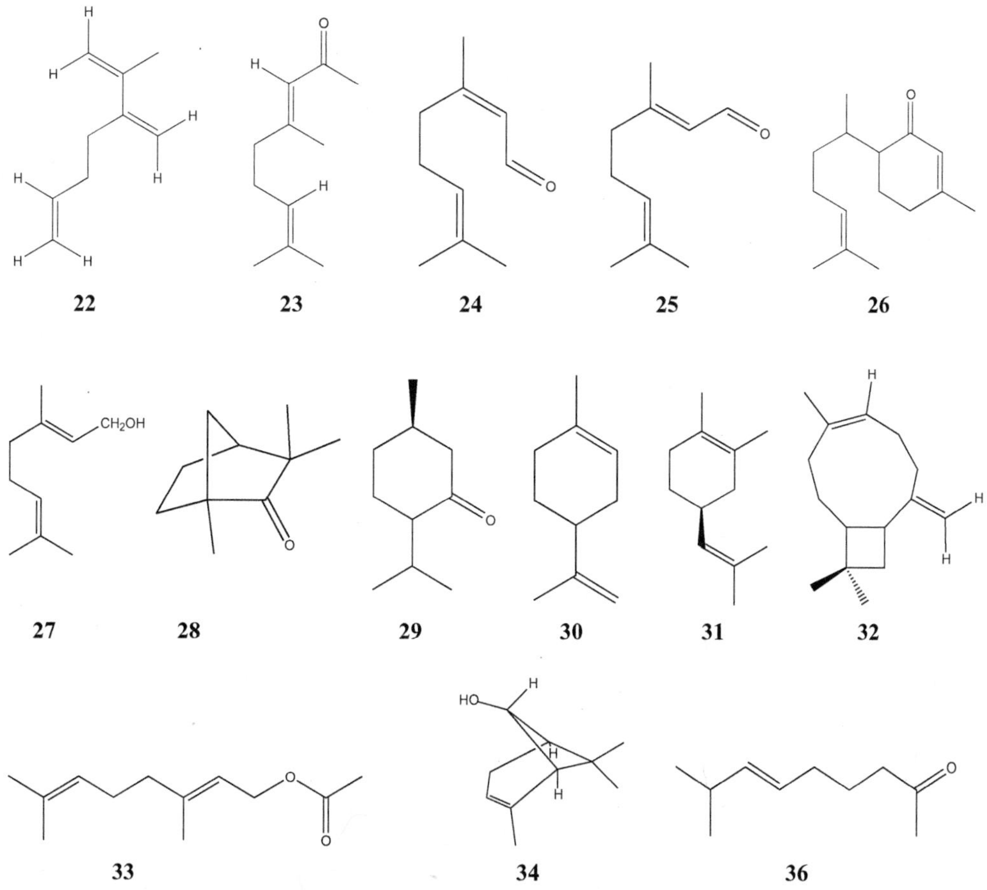

Figure 18.5 *Terpenoids from* **C. citratus.** Myrcene (**22**); citral (**23**), neral (**24**); geranial (**25**); geraniol (**26**); α-oxobisabolene (**27**); citral (**23**); fenchone (**28**); menthone (**29**); terpinolene (**30**); limonene (**31**); caryophyllene (**32**); geranyl acetate (**33**); *cis*-chrysanthenol (**34**); sulcatone (**35**).

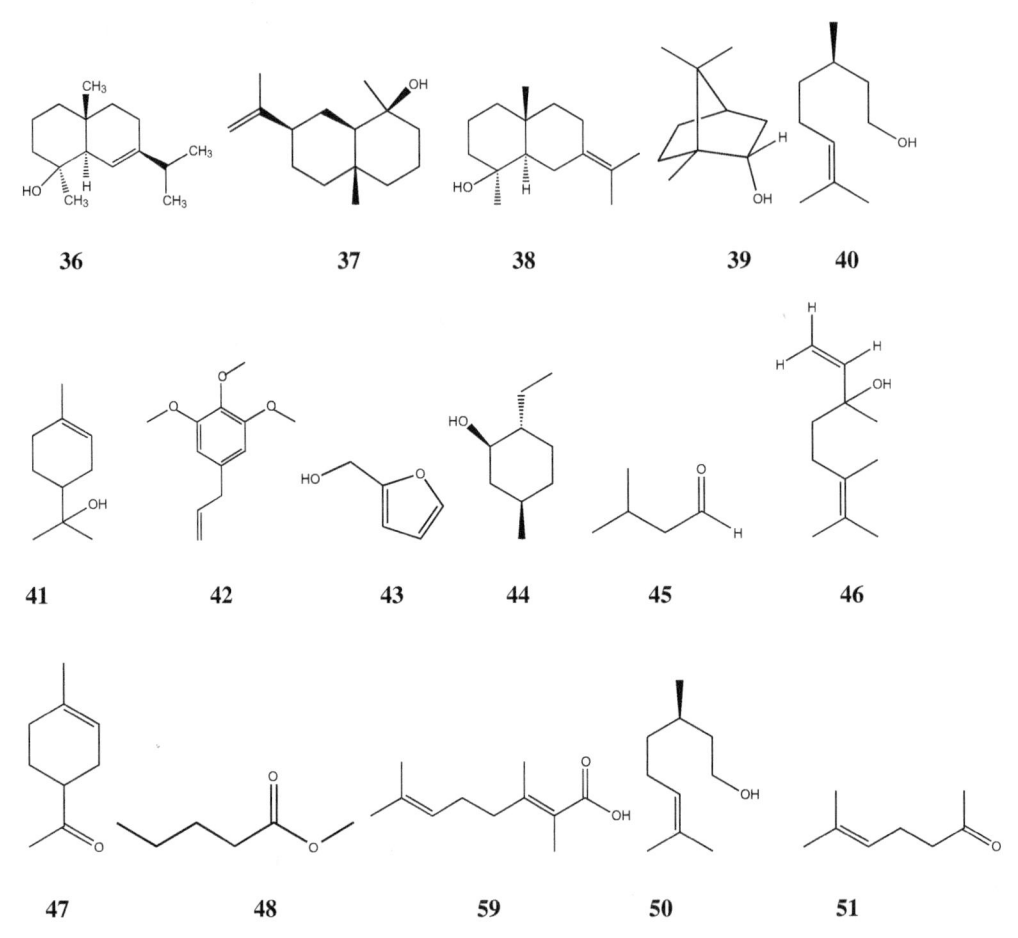

Figure 18.6 *Terpenoids from* **C. citratus.** Selina-6-en-4-ol (**36**); neointermediol (**37**); eudesma-7(11)-en-4-ol (**38**); burneol (**39**); citronellol (**40**); α-terpineol (**41**); elemicin (**42**); furfurol (**43**); isopulegol (**44**); isovaleranic aldehyde (**45**); linanool (**46**); terpineone (**47**), valeric ester (**48**); geranic acid (**49**); citronellol (**50**); methyheptenone (**51**).

α-oxobisabolene (12%) (**27**), and citral (13%) (**23**) while other minor components were also identified which include fenchone (**28**) (0.3%), menthone (0.2%) (**29**), and terpinolene (0.1%) (**30**) (Abegaz et al., 1983). The *C. citratus* oil from Kenya was dominated by monoterpene hydrocarbons which accounted for 94.25% of the total oil and characterized by a high percentage of **25** (39.53%), **24** (33.31%), and myrecene **22** (11.41%) (Matasyoh et al., 2011). Farhang et al. (2013) reported the predominant components observed in *C. citratus* essential oils from Iran which include **23** (70.11%), limonene (5.83%) (**31**), and caryophyllene (3.44%) (**32**).

Analysis of the essential oil of *C. citratus* from different parts of Benin showed four major chemotypes based on the amount of **24**, **25**, **22**, and **26** with 33, 41.3, 10.4,

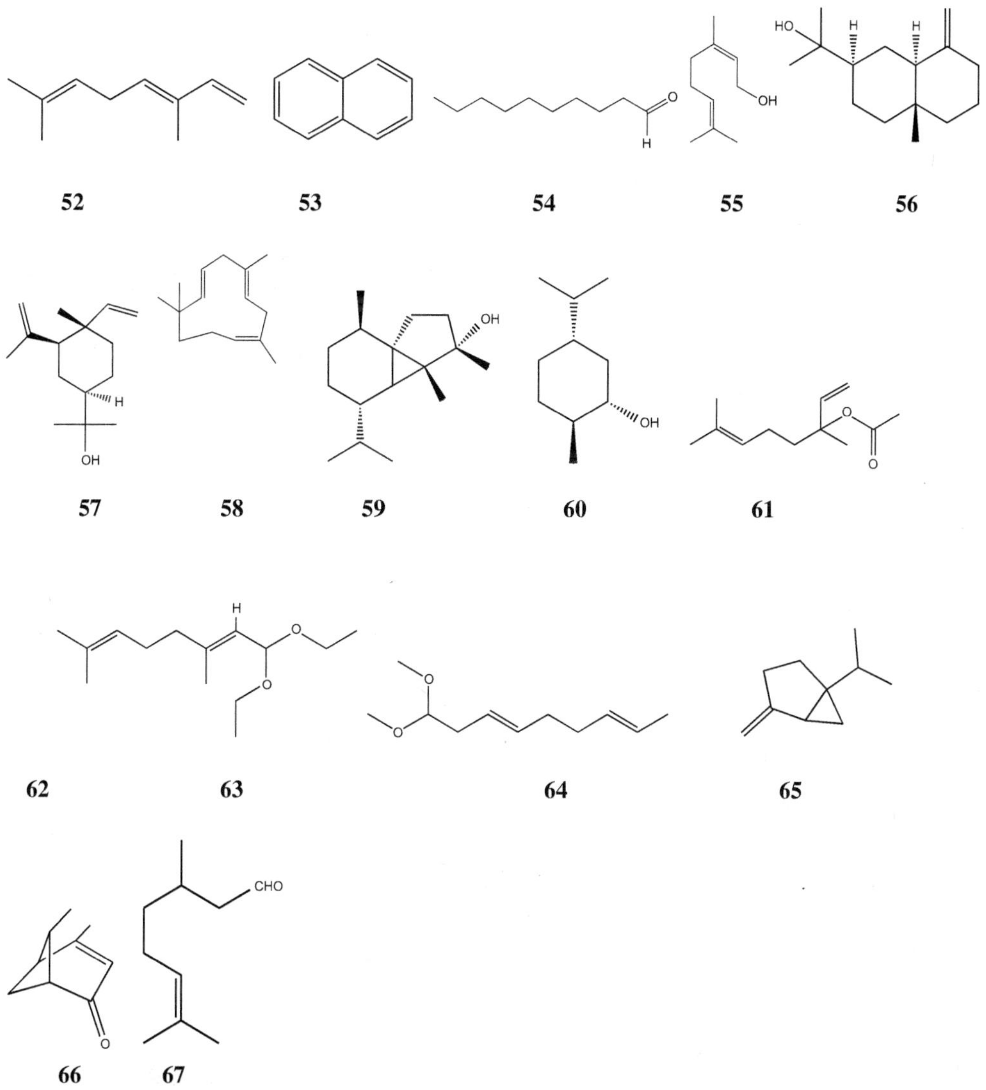

Figure 18.7 *Terpenoids from* **C. citratus.** 3,7-Dimethyl-1,3,6-octatriene (**52**); naphtalene (**53**); decanal (**54**); nerol (**55**); β-eudesmol (**56**); elemol (**57**); humulene (**58**); cubebol (**59**); neo-menthol (**60**); linalyl acetate (**61**); Z-ocimene (**62**); citral diethylacetal (**63**); citralacetate (**64**); sabinene (**65**); verbenone (**66**); citronella (**67**).

and 6.6% compositions, respectively, while geranyl acetate (2.4%) (**33**), *cis*-chrysanthenol (0.5%) (**34**), and sulcatone (0.5%) (**35**) were identified in trace quantity (Bossou et al., 2013; Zannou et al., 2015). The comparative study of the essential oil constituents of *C. citratus* from Cuba and Brazil revealed that the isomers **25** and **24** are the main components in both oils (Pinto et al., 2015). Also, the essential oils from aerial parts

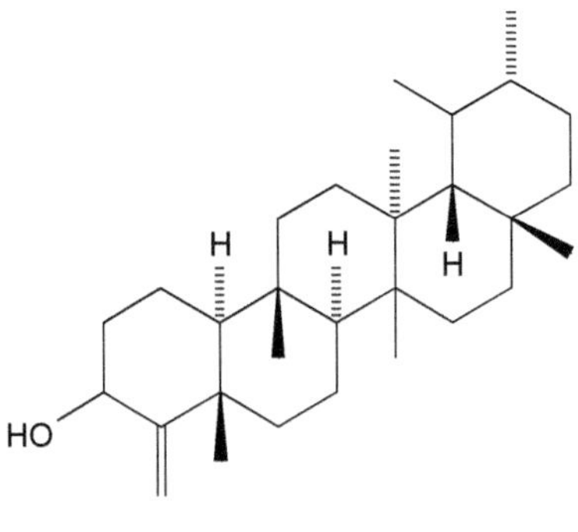

Figure 18.8 *Triterpenoids from* **C. citratus.** Cymboponol (**68**); cymbopogone (**69**).

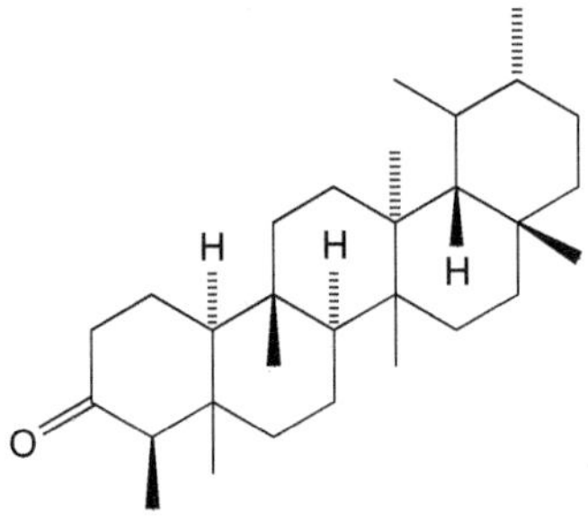

(blades and sheaths) and rizhome of *C. citratus* oil from North of Brazil were studied (Andrade et al., 2009) whereby **24** and **25** were major constituents in blades (30.1 and 39.9%, respectively) and sheats (27.8 and 50.0%, respectively). The rhizome oils displayed a different chemical profile and was characterized by a high amount of selina-6-en-4-ol (27.8%) (**36**), α-cadinol (8.2%), neointermediol (7.2%) (**37**), and eudesma-7(11)-en-4-ol (5.3%) (**38**). Similar chemotypes were reported by Costa et al. (2013) in the analysis of *C. citratus* oil from Brazilian species. Barbosa et al. (2008) reported the composition of essential oils from 12 different samples of the leaves of *C. citratus* from Brazil. The report indicated **23** as the major chemical constituent (30–93.74%). Other monoterpenes present in the essential oil include compounds **22, 24, 25, 31, 33**, borneol (**39**), citronellol (**40**), α-terpineol (**41**), and elemicin (**42**). Also, furfurol (**43**), isopulegol (**44**), isovaleranic aldehyde (**45**), linanool (**46**), terpineone (**47**), and valeric ester (**48**) have been isolated from Brazilian species of *C. citratus* oil (Akhila, 2009).

Ranitha et al. (2014) reported variation in the chemical constituents of essential oil of *C. citratus* from north-west of Malaysia, extracted by microwave-assisted hydrodistillation (MAHD) and conventional hydrodistillation (HD) methods. The study revealed that **22–25** were found in the oils obtained from both methods. However, the oil compositions revealed that higher amounts of oxygenated monoterpenes, such as linalool (**46**), geranic acid (**49**), and citronellol (**50**) were present in the essential oil isolated by MAHD and absent in HD. Similar report with little variation was obtained in another analysis of *C. citratus* oil from Malaysia (Piarua et al., 2012). The result of gas chromatography-mass spectrometry (GC-MS) analysis of *C. citratus* oil from Iran revealed that citral (30.95%) (**23**), limonene (5.83%) (**31**), and caryophyllene (3.44%) (**32**) were the main components. The minor constituents identified in the oil include methylheptenone (1.2%) (**51**), 3,7-dimethyl-1,3,6-octatriene (0.58%) (**52**), naphtalene (0.79%) (**53**), and decanal (0.25%) (**54**) (Vahid et al., 2013).

The hydrodistilled essential oil from the leaves of *C. citratus* grown in Zambia was analyzed by GC and GC-MS. Sixteen compounds representing 93.4% of the oil were identified, of which **23** (39.0%) and **22** (18.0%) were the major components. Little quantity of **26** (1.7%) and **46** (1.3%) were also detected (Esmort et al., 1998). Vera et al. (2014) reported the compositions of *C. citratus* oil from Colombia to be **25** (26%), **26** (2.9%), **28** (1.8%), **22** (20.2%), **24** (29.0%) nerol (3.0%) (**55**). Similar constituents with variation in percentage composition were also obtained from the essential oil of *C. citratus* from Burkina Faso and they are **25** (48.1%), **24** (34.6%), and **22** (11.0%) (Bassolé et al., 2011).

The phytochemical analysis of aqueous extracts of *C. citratus* from Saudi Arabia revealed the presence of β-eudesmol (45%) (**56**) and elemol (41%) (**57**) as major components while humulene (4%) (**58**) and cubebol (4.7%) (**59**) were significantly present (Halabi and Sheikh, 2014). Kasali et al. (2001) reported that the main constituents of the volatile oil of *C. citratus* from Nigeria were **25** (33.7%), **24** (26.5%), and **22** (25.3%). Small amounts of neo-menthol (3.3%) (**60**), linalyl acetate (2.3%) (**61**), and Z-ocimene (1.0%) (**62**) were also detected. Furthermore, *C. citratus* oil from Belgium had its main constituents as **23** (57.46%), citral diethylacetal (24.68%) (**63**), **31** (6.36%), citral acetate (2.05%) (**64**), **22** (1.21%), and methyl **29** (1.17%) (Katsukawa et al., 2010). Similar chemotypes with different percentage compositions compared to Belgium species were also obtained from the India species of *C. citratus* oil (Bharti et al., 2013; Kumar et al., 2013).

The comparative compositional analysis of the essential oils of *C. citratus* from Mali and the Ivory Coast was determined by GC and GC/MS. *C. citratus* oil from Mali contained a high proportion of **23** (75%), **22** (6.2–9.1%), and **26** (3.0–5.6%). The composition differed from the oil of the Ivory Coast in which the contents of **22**, **24**, and **25** each ranged between 18 and 35%. Some other trace components identified in the Malian and Ivorian species of *C. citratus* include sabinene (**65**), verbenone (**66**), citronella (**67**), and geranyl acetate (**33**) (Sidibé et al., 2001). The chemical composition of *C. citratus* essential obtained from plants grown in Angola revealed the major constituents as **23** (40.55%), **22** (10.50%), and **26** (3.37%) (Soares et al., 2013a). Furthermore, nonvolatile triterpenoids were also reported from *C. citratus* extract, namely cymbopogonol (**68**) and cymbopogone (**69**) (Ansari et al., 1996).

3 PHARMACOLOGICAL ACTIVITIES OF *C. citratus*

C. citratus is one of the most diverse and used plants in West Africa. Its use ranges from pharmaceutical application, domestic, food, shelter, and as ornamental plant. It is ethnomedically used as a folk remedy for coughs, elephantiasis, influenza, headache, arthritis, leprosy, malaria, and inflammatory disorders (Manvitha and Bidya, 2014). The medicinal application of *C. citratus* follows different treatment mediums, such as topical, infusion, concoction, and as tea preparations. Several reports had been published on the pharmacological potential of *C. citratus*, its clinical applications, and some patent products that had been produced from its extract. This section emphasizes some cutting edge application, patent, novel reports, and research of *C. citratus*.

3.1 Antidiabetic activity

Diabetics is a chronic disease that occurs when the pancreas does not produce enough insulin (a hormone that regulates blood sugar) or when the body cannot effectively use the insulin it produces. Bharti et al. (2013) investigated the antidiabetic activity of *C. citratus* in vivo and by molecular docking at a dose level of 400 and 800 mg. Their findings reveal a significant reduction in the level of insulin ($p < 0.001$), glucose ($p < 0.001$), and triglycerides ($p < 0.001$) which was also confirmed by docking of the major components. Another report showed a low or no hyperglycemic activity in wistar rats treated with *C. citratus* tea for 2 weeks (Leite et al., 1986). Furthermore with the oral administration of the *C. citratus* decoction at a dose of 125–500 mg/kg, a significant reduction in hypoglycemic rate was observed and had been alluded to the increased insulin synthesis and secretion (hyperinsulinemia) or increased peripheral glucose utilization. The discrepancies observed in both reports had been the effect of geological variation in the content and quality of the *C. citratus* components (Adewale and Oluwatoyin, 2007). The antidiabetic activity of *C. citratus* against Type II diabetes was evaluated in vitro using the α-amylase and α-glucosidase inhibitory effects. An inhibition of 99.9% at 1 mg/mL and EC_{50} of 0.31 mg/mL were reported for the α-glucosidase and α-amylase, respectively (Boaduo et al., 2014).

3.2 Antimutagenicity and carcinogenicity activity

Environmental pollution by chemicals and poisonous substances has led to serious health consequences by increasing mortality in the entire world. However, in cases where there are no direct effect of this hazardous chemicals, their carcinogenicity effects are seen several years after coming into contact. Mutagens are not only associated with carcinogenesis but also involved in the initiation of numerous chronic deteriorating diseases including hepatic disorders, cardiovascular disorders, diabetes, arthritis, chronic inflammation, and aging. Use of antimutagens in diet is thereby encouraged (Hodgson, 2004). In some cases, some materials can cause the breakage of some sections of chromosomes by deletion and rearrangement leading to carcinogenesis: these materials are called clastogens. Rabbani et al. (2005) investigated the anticlastogenic properties of citral (major component in *C. citratus*) using known mutagens, such as cyclophoshamide, mitomycin-C, and nickel metal ($NiCl_2$) in an Albino mice. At a dose of 20 mg/kg, citral prevented the nuclear damage induced by cyclophosphamide, mitomycin-C, and nickel chloride in both bone and peripheral blood micronucleus tests. In another study, the protective capacity of *C. citratus* on leukocyte DNA damage induced by *N*-methyl-*N*-nitrosurea (MNU) and carcinogenicity revealed that at a dose of 500 mg/kg, the *C. citratus* essential oils provided protective action against MNU-induced DNA damage and a potential anticarcinogenic activity against mammary carcinogenesis in female Balb/C mice (Bidinotto et al., 2011). In a related study, rats treated with lemon grass essential oil for 21 days showed no significant changes to their serum biochemical parameters, indicating there was no genotoxic effect except for the reduction of cholesterol level in the same analysis (Costa et al., 2011).

3.3 Antiinflammatory

Chronic inflammation is one of the leading causes of death in the world and had been associated with terminal diseases, such as cancer (Colotta et al., 2009). The natural products as cure to inflammation have been used since the ancient times. Traditional uses of *C. citratus* essential oil as an additive to ointments, creams and in massages to treat topical inflammation have been practiced in several countries, such as Cuba, Nigeria, Trinidad and Tobago, India, etc. Citral the main component of *C. citratus* has shown an extremely high inhibition of the inflammatory mediators. For instance, at a concentration of 0.1%, citral suppressed tumor necrosis factor (TNF)-α-induced neutrophil adherence (De Cássia da Silveira e Sá et al., 2013). Citral shows a high suppression of COX-2 (60–70% inhibition) and peroxisome proliferator-activated receptor alpha (PPAR-α), activator (4.5-fold) than *C. citratus* COX-2 suppression (20%) and PPARα (58%) (Katsukawa et al., 2010). The observed high activity has been attributed to the ability of citral (an α,β-unsaturated ketone) to covalently bind to the receptors. Extract of *C. citratus* obtained from Portugal displayed inhibition of the nuclear factor-kappaB (NF-kB) pathway by the polyphenols, such as luteolin, phenolic, and tannins present in the extract (Francisco et al., 2013). Evidenced from the inhibition of inducible nitric oxide synthase (*i*NOS), nitric oxide (NO) production and other lipopolysaccharide (LPS)-induced pathways by *C. citratus* and its phenolics provided further scientific basis for *C. citratus* as antiinflammatory agent (Francisco et al., 2013). Boukhatem et al. (2014) discovered an antiinflammatory inhibition rate of 80–90% when the essential oil of *C. citratus* was administered orally and topically.

In addition to citral isolated from *C. citratus*, certain derivative isolates, such as epoxyestragole, 6,7-epoxycitral, carvone, 8,9-epoxycarvone, and 6,7-epoxycitronellal successfully inhibited the NO and the prostagladins (PGE$_2$) associated with inflammations (Sepúlveda-Arias et al., 2013). Luteolin glycosides (Luteolin O-, C-, and O,C-glycosides) isolated *C. citratus* from Portugal successfully inhibited antiinflammatory mediators although activity was based on structure–activity relationship depending on the position of linkage between the sugar moiety and the aglycone (Francisco et al., 2014). Consequently, Nishijima et al. (2014) demonstrated that citral was highly effective in suppressing postoperative pains associated with surgery by reducing or inhibiting the expression of certain pain mediators and also enabled a gastro-protective effect against gastric ulcers associated with nonsteroidal drugs (NSAIDs).

3.4 Anti-HIV activity

Generally known as one of the deadliest diseases in the world, human immunodeficiency virus infection and acquired immune deficiency syndrome (HIV/AIDS) has claimed a high mortality rate. The sub-Saharan Africa has the highest number of people living with AIDS and in South Africa alone, 1500 individuals are infected daily with the HIV virus. Oral thrush commonly observed in HIV-AIDS patients was suppressed when extract of *C. citratus* was applied to patients. *C. citratus* simmered in water was able to cure the oral thrush within 1–5 days (Wright et al., 2009). The oral thrush was caused by the *Candida albicans* microbials.

3.5 Antioxidant property

Biochemical reactions in the body often lead to generation of reactive oxygen species (ROS), which are controlled and balanced by the antioxidant defense system of the body including some enzymes and nonenzymatic components, such as vitamins and other antioxidant micronutrients from our diet (Miguel, 2010). The resulting oxidative stress from uncontrolled ROS is often associated with various diseases, such as coronary heart disease, Parkinson's disease, rheumatoid arthritis, and cancer (Jabaut and Ckless, 2007). The antioxidant properties of *C. citratus* have been reported in several articles. Lemon grass extract inhibited the *tert*-butyl hydroperoxide (*t*BuOOH) and hemin-induced maloodiadehyde (MDA) formation by 50% in vitro in a wistar rats model (Suaeyun et al., 1997). Certain natural antioxidants, such as chlorogenic acids, caffeoylquinic acid derivative, isoorientin, and swertiajaponin isolated from lemon grass were reported to be responsible for the diminishing low-density lipoprotein (LDL) oxidation induced by Cu^{2+} (a preliminary step in endothelial dysfunction and cardiovascular disease growth) activity higher than that of ascorbic acid (Campos et al., 2014).

In Malaysia a chloroform extract of *C. citratus* displayed a low antioxidant activity in ferric reducing ability of plasma (FRAP), 1,1-diphenyl-2-picryl-hydrazyl (DPPH), and β-carotene assays (Lu et al., 2014). In a study carried out on the in vitro shoot culture of *C. citratus* from Cuba, approximately 28.6% of free radical scavenging activity was contributed by flavonoids while about 49.7% of this activity was contributed by total phenolic contents, an indication that the observed antioxidant activities in *C. citratus* are via the total phenolic contents (Tapia et al., 2007). Conversely, Cheel et al. (2005) observed that 76% of the free radical scavenging activity resulting from the flavonoids content form the aerial parts of *C. citratus* from Chile. Another study indicated that the total antioxidant capacity of *C. citratus* was about 4.41 mmol trolox/100 g of DW (Shan et al., 2005). The antioxidative action of Citrus lemon isolated in Sicily, Italy on human skin revealed a high activity (Bertuzzi et al., 2013). The essential oils of *C. citratus* collected in Lagos, Nigeria displayed a low antioxidant activity ($p < 0.05$) with an average mean value of 37.0 ± 6.9 (Adesegun et al., 2013) while the ethanolic extract was able to suppress oxidative stress in winstar rats analyzed for diabetic conditions (Ademuyiwa and Grace, 2015).

3.6 Antihypertension assay

In Brazil, *C. citratus* was used in folklore medicine to treat high blood pressure, commonly taken as tea (Shah et al., 2011). An earlier report showed an endothelium-independent vasorelaxation by citral (main component in *C. citratus*) via the blockage of Ca^{2+} influx, while the methanolic extracts of the leaves, roots, and stems displayed vasorelaxation activities via the prostacyclin (PGI_2) channel (Devi et al., 2012). This is an indication that the synergistic effect of the components in *C. citratus* can result in a different activity from that of the main components. In addition, specific activity of citral has revealed that it is a potential therapy for reduction of blood pressure as shown in the report of Shimono et al. (2010).

3.7 Insecticidal activity

The use of an environmentally benign pest and insect control has been on the increase due to environmental pollution and generation of insect resistant species. Natural product has taken the lead in this application and has been effectively used for such. Essential oils from *C. citratus* have been applied in control of pathogens and insects (Masamba et al., 2003; Sessou et al., 2012; Usha Rani, 2012). In Bogota and Satander regions of Colombia, activities of 10, 700, and 123.3 ppm, respectively, at LC_{50} were reported when the essential oil was tested against *Aedes aegypti* (Vera et al., 2014). In Zimbabwe, Kazembe and Chauruka (2012) observed a 1.5 h repellant activity against the same pathogen at a dose of 5.0 mg. A concoction containing *C. citratus* led to 90–98% mortality rate of *Phemacoccus solenopsis* insects prominent in cotton crops, vegetables, and fruits. In this report, the activity was attributed to the presence of butyl butyrate in the essential oil (Banu and Bewaraji, 2013). The plant has shown larvicidal and oviposition effects against housefly with LT_{50} values of 38.99 and 52.08 h, respectively (Soonwera, 2015). *C. citratus* essential oils from Brazil and Cuba (1 µL/larva) were applied topically to newly hatched *Musca domestica* larvae. The activity observed was dose dependent in the range of 10–80% and LC_{50} of 4.24 and 3.24 in the oil of *C. citratus* from Brazil and Cuba, respectively (Pinto et al., 2015). Mortalities from lemongrass extract at 50% concentration resulted in more than 91% mortalities of *Dermatophagoides farinae* and *Dermatophagoides pteronyssinus* (Hanifah et al., 2011).

3.8 Antimalarial potential

The *Plasmodium flasiparum* parasite still remains one of the deadliest insects on the planet earth. It is the causative agent for the deadly malaria disease (Chinsembu, 2015). Several drugs have been developed with natural products taking the central figure. *C. citratus* has been used by various communities to combat this disease. The dichloromethane extract of *C. citratus* collected from Nigeria exhibited a very high antiplasmodic activity within the range of 2–10 µg/mL against the *P. falsiparum* and also a positive activity against the schizontoids of *P. berghei* (Melariri, 2010). Tarkang et al. (2014)

analyzed ethanolic extracts of *C. citratus* against two strains of *P. falsiparium* namely the CQ-sensitive (3D7) and multidrug resistant (Dd2). The EC_{50} (3D7/Dd2) was 28.75/54.84 showing good antiplasmodial activities against both parasite strains. In a separate study, the group investigated the antioxidant of Nefang (a popular mixture in Cameroon for malaria treatment), which contains *C. citratus* and other plants. Using CCl_4-induced oxidative stress, the liver weights of the treated rats were reduced considerably via a dose-dependent effect. This finding provided a means of improving the antioxidant status of oxidative stress-associated complications during malaria (Tarkang et al., 2013).

3.9 Dermatotoxicity activity

Dermatitis describes an inflammation of the skin and it involves an itchy rash on swollen, reddened skin. The use of natural products as antidermatitis has long been established since time immemorial. A popular soap called the "Nubian Heritage's Lemongrass & Tea Tree Oil" along with Orange Peel Soap was evaluated and confirmed for dermatotoxicity potential. Phase I and Phase II clinical trials of the oil in patients with *Pityriasis versicolor* revealed a significant activity. The mycological cure was 60% ($p < 0.05$) after patients were treated for 40 days with the oil (Carmo et al., 2013). Elkhair Abou (2014) reported a high inhibition of clinically isolated *Microsporum canis* by the essential oil of *C. citratus* from Gaza Strip at the rate of 85.35% at a concentration of 1 µL/mL. A novel lemin protein from lemon grass displayed 95% activity against transgenic mouse model for dandruff (Springer et al., 1995). In another report, lemon grass from Bangkok was mixed with shampoo for the treatment of *Malassezia furfur*, a fungus associated with dandruff. A significant inhibition of about 75 µL/mL was observed (Wuthi-Udomlert et al., 2011).

3.10 Nephrotoxicity of *C. citratus*

Ullah et al. (2013) reported the prevention of renal damage linked to aminoglycosides upon the administration of 200 mg/kg of *C. citratus* for 3 weeks. In a related study, the ethanolic extracts of *C. citratus* did not influence the calculated red blood cell (RBC) indices in the kidney of rats, an indication that it cannot lead to development of anemia (Ademuyiwa et al., 2015). The prolonged effect of this extract may be harmful to the organs as observed by Tarkang et al. (2012). Prolonged use of the ethylacetate extract of *C. citratus* for 28 days resulted in liver scarring and mild tubular distortion indicating liver tissue damage. In addition to the liver scarring, there was increased level of uric acid (URIC) and triglycerides in the liver tissue. The increased level of URIC (end product of metabolism) and triglycerides in the liver after administration of 1 g/kg of body weight might be indicative of liver tissue damage. In addition, at day 28 of the same dose liver scarring and mild tubular distortion were observed indicating renal toxicity. Although further analysis will establish these findings, the prolonged use of the ethanolic extract could be discouraged.

3.11 Anticancer and antitumor activity

Cancer is an abnormal growth of cells and more than 100 types of cancer have been reported, including breast cancer, leukemia, lymphoma, skin cancer, colon cancer, prostate cancer, and lung cancer. Cancer treatments include combination of surgery, radiation therapy, chemotherapy, hormone therapy, or immunotherapy. Some plants have proved to be highly effective and therefore have been developed as drugs. Investigation into the apoptotic ability of *C. citratus* has shown some promising and interesting results. ($\rightarrow$4) linked β-D-xylofuranose-(1$\rightarrow$) polysaccharides isolated from *C. citratus* were analyzed in vitro against the Siha and LNCap reproductive cancer cells. The apoptotic observed were via the decline in mitochondrial transmembrane potential, upregulation of bax, and stimulation of caspase 3 in sampled cancer cells (Thangam et al., 2014). The essential oils of *C. citratus* from Abomey Calavi, Benin Republic were reported to induce cytotoxicity against the Chinese hamster ovary (CHO) cells (IC_{50} of 10.63 μg/mL) and the human noncancer fibroblast cell line (W138) cells (IC_{50} of 39.77 μg/mL). Furthermore, the major component of the oil also displayed high activity; citral showed 20.62 μg/mL (CHO) and 39.48 μg/mL (W138) while β-pinene was 10.10 μg/mL (CHO) and no activity for the W138 cells at > 50 μg/mL (Kpoviessi et al., 2014).

Also, Puatanachokchai et al. (2002) investigated the effect of lemon grass on the early stage of heptacarcinogenesis in rats. At a concentration of 80%, ethanolic extract of *C. citratus* effectively prevented the development of hepatocellular lesions in rats infected with diethylnitrosamine (DEN) without a significant alteration of the sizes of the lesions. The anticancer activities of *C. citratus* had been suggested to emanate from the synergistic or individualistic properties of its constituents. Citral, a major component, inhibits 12-O-tetradecanoylphorbol-13-acetate-derived promotion of mouse skin carcinogenesis (Connor, 1991), geraniol suppresses the growth of hepatoma and melanoma cells transplanted to mice and rats (Yu et al., 1995) and of colon cancer cells in vitro (Raul, 2001), while β-myrcene reduces the sister-chromatid exchange induced by aflatoxin B1 and cyclophosphamide in V79 and HTC cells in vitro (Roscheisen et al., 1991).

3.12 Antimicrobial activity

Numerous antimicrobial activities of the extracts, essential oils, and components of *C. citratus* have been reported in the literature. However, the antimicrobial activity differs from region to region, plant parts used, type of extraction, and protocols employed. Table 18.2 gives a comprehensive report of some analyses, their activities, plant parts, and the country. Few of the reports identified here showed an effective activity of *C. citratus* against major Gram-positive and Gram-negative bacteria. According to a finding Gram-negative bacteria, such as *Escherichia coli*, *Klebsiella pneumoniae*, and *Pseudomonas aeruginosa* developed an outer membrane shielding the effect of the *C. citratus* extracts (Soares et al., 2013b). However, in Gram-positive shielding is reduced.

Table 18.2 Activity, type, region, and potency of *C. citratus* extracts

Organisms	Extracts	Region	Potency	References
E. coli, *K. pneumoniae*, *P. aeruginosa*, *Proteus vulgaris*, *Bacillus subtilis* and *Staphylococcus aureus*	Alcohol and water extracts	Tamil Nadu, India	The MIC ranged between 150 and 50 mg/mL	Hindumathy (2011)
S. aureus (ATCC 25923), *Bacillus cereus* (ATCC 1247), *E. coli* (ATCC 25922) and *P. aeruginosa* (ATCC 27853)	Methanolic extract	Ahar, Iran	MBC between 25 to 200 mg/mL	Jafari et al. (2012)
E. coli CIP 105182, *Enterobacter aerogenes* CIP 104725, *Enterococcus faecalis* CIP 103907, *Listeria monocytogenes* CRBIP 13.134, *P. aeruginosa* CRBIP 19.249, *Salmonella enterica* CIP 105150, *Salmonella typhimurium* ATCC 13311, *Shigella dysenteriae* CIP 54.51 and *S. aureus* ATCC 9144.	Essential oils	Ouagadougou, Burkina faso	MIC ranging between 1 and 2.50 mg/mL	Bassolé et al. (2011)
Salmonella typhi and *E. coli*	Aqueous extract, essential oils and powder	Zaria, Nigeria	Complete growth inhibition at 24 h incubation	Akin-Osanaiye et al. (2007)
Lactobacillus rhamnosus (MTCC1408), *Lactobacillus acidophilus* (MTCC10307), *Streptococcus mutans* (MTCC 890), *Streptococcus oralis* (MTCC 2696), *Actinomyces howelii* (MTCC 3048) and *C. albicans* (MTCC4748)	Essential oils and solvent extracts	Pune, India	High inhibition rates of bacteria wth MIC at the range of 20 to 2.5 mg/mL	Ambade and Bhadbhade (2015)
C. albicans	Aqueous extract	Ghana	A high activity	Nyarko et al. (2012)
Hospital isolated multidrug resistant strains of *S. aureus*, *Staphylococcus epidermidis*, *E. coli*, *K. pneumoniae*, and their respective ATCC control strains. Pathogenic *C. albicans*, *C. parapsilosis*, and *C. tropicalis*	Essential oils	Angola	Lower activity observed for the Gram-negative while high activity in the gram-positive bacteria.	Soares et al. (2013a,b)

Strains of *Propionibacterium acnes*	Essential oils	Thailand	0.125% v/v activity against the strains	Luangnarumitchai et al. (2007)
Desulfovibrio alaskensis	Lemon grass essential oils	Sergipe, Brazil	MIC value of 0.17 mg/mL	Korenblum et al. (2013)
Crithidia deanei	Essential oils	Parana, Brazil	IC_{50} value of 120 and 60 μg/mL	Pedroso et al. (2006)
C. albicans, S. aureus, E. coli	Essential oils	Mbarara, Uganda	About 40% activity against the organisms	Armando and Rahma (2009)
Helicobacter pylori	Essential oils	Rockville, USA	At a concentration of 0.01 v/v, decreases of the number of colonies by $p < 0.001$	Ohno et al. (2003)

MBC, minimum bactericidal concentration; MIC, minimum inhibition concentration; IC_{50}, inhibitory concentration.

4 PATENTS

C. citratus was blended together with other plant materials to activate vitamin C transporter generation and its assimilation by the body because of its high content of geranyl acetate (Suhaila et al., 2010). A traditional Chinese medicine composition for treating stomach ache composed of different *C. citratus* and other plant condiments was prepared and applied to eliminate swelling, relieve pain, activate blood, invigorate the stomach, clear heat, remove toxicity, dispell the wind, and expel the cold (Wei, 2015). In addition, plant phytochemicals were used as additives in latex used for polymer coatings to prevent colonization of a treated surface by microorganisms. *C. citratus* successfully exhibits prolonged biocidal effects on treated surface (Seabrook, 2008). Lemongrass oil has also been combined with corn oil and/or other oils and organic acid, surfactants to control the germination and growth of weeds. It is benign and lacks the usual environmental hazards associated with known herbicides and fungicides (Brian et al., 2009). Another invention reveals an extractive which is extracted from a *Cymbopogon* plant and that is capable of insulin sensitizing medicine, and is thus used as an additive for insulin drugs (Wang and Li, 2015).

5 CONCLUSIONS

From the literature, we can postulate that the various parts and extracts of *C. citratus* and their components have many uses, both in pharmacology and in food. In addition, the plant possesses interesting biological activities and therapeutic potential. The observed biological potentials of the plants may be useful as natural remedies. Therefore, the economic importance of *C. citratus* extracts and the components is indisputable. It appears imperative, therefore, to preserve our natural, diverse flora and support its protection to maintain this inexhaustible source of molecules destined for multiple targets.

REFERENCES

Abe, S., Sato, Y., Inoue, S., Ishibashi, H., Maruyama, N., Takizawa, T., Yamaguchi, H., 2003. Anti-*Candida albicans* activity of essential oils including lemon grass (*Cymbopogon citratus*) oil and its component, citral. Nihon Ishinkin Gakkai Zasshi 44, 285–291.

Abegaz, B., Yohannes, P.G., Deiter, R.K., 1983. Constituents of essential oil of Ethiopian *Cymbopogon citratus* Stapf. J. Nat. Prod. 46, 424–426.

Ademuyiwa, A.J., Elliot, S.O., Olamide, O.Y., Johnson, O.O., 2015. Nephroprotective effect of aqueous extract of *Cymbopogon citratus* (lemon grass) in Wistar albino rats. Int. J. Pharm. Biomed. Res. 2015, 1–6.

Ademuyiwa, A.J., Grace, O.K., 2015. The effects of *Cymbopogon citratus* (Lemon grass) on the antioxidant profiles wistar albino rats. Merit Res. J. 3, 51–58.

Adeneye, A.A., Agbaje, E.O., 2007. Hypoglycemic and hypolipidemic effects of fresh leaf aqueous extract of *Cymbopogon citratus* Stapf in rats. J. Ethnopharmacol. 112, 440–444.

Adeola, S.A., Folorunso, O.S., Raimi, G.O., Sowunmi, A.F., 2013. Antioxidant activity of the volatile oil of *Cymbopogon citratus* and its inhibition of the partially purified and characterized extracellular protease of *Shigella sonnei*. Am. J. Res. Commun. 1, 31–45.

Adesegun, A.S., Samuel, F.O., Olawale, R.G., Funmilola, S.A., 2013. Antioxidant activity of the volatile oil of *Cymbopogon citratus* and its inhibition of the partially purified and characterized extracellular protease of *Shigella sonnei*. Am. J. Res. Commun. 1, 31–45.

Adewale, A., Oluwatoyin, E., 2007. Hypoglycemic and hypolipidemic effects of fresh leaf aqueous extract of *Cymbopogon citratus* Stapf in rats. J. Ethnopharmacol. 112, 440–444.

Aftab, K., Ali, M.D., Aijaz, P., Beena, N., Gulzar, H.J., Sheikh, K., et al., 2011. Determination of different trace and essential element in lemon grass samples by X-Ray flouresence spectroscopy technique. Int. Food Res. J. 18, 265–270.

Akhila, A., 2009. Essential Oil-Bearing Grasses: The Genus Cymbopogon. CRC Press, Boca Raton, FL.

Akin-Osanaiye, B.C., Agbaji, A.S., Dakare, M.A., 2007. Antimicrobial activity of oils and extracts of *Cymbopogon citratus* (Lemon Grass), *Eucalyptus citriodora* and *Eucalyptus camaldulensis*. J. Med. Sci. 7, 694–697.

Ambade, S.V., Bhadbhade, B.J., 2015. In-vitro comparison of antimicrobial activity of different extracts of *Cymbopogon citratus* on dental plaque isolates. Int. J. Curr. Microbiol. Appl. Sci. 4, 672–681.

Andrade, E.H.A., Zoghbi, M.G.B., Lima, M.P., 2009. Chemical composition of the essential oils of *Cymbopogon citratus* (DC.) Stapf cultivated in North of Brazil. J. Essent. Oil Bearing Plants 12, 41–45.

Ansari, S.H., Ali, M., Siddiqui, A.A., 1996. Evaluation of chemical constituents and trade potential of *Cymbopogon citratus* (Lemongrass). Hamdard Med. 39, 55–59.

Armando, C.C., Rahma, H.Y., 2009. Evaluation of the yield and the antimicrobial activity of the essential oils from: *Eucalyptus globulus*, *Cymbopogon citratus* and *Rosmarinus officinalis* in Mbarara district (Uganda). Rev. Colombiana Cienc. Anim. 1, 240–249.

Avoseh, O., Oyedeji, O., Rungqu, P., Nkeh-Chungag, B., Oyedeji, A., 2015. *Cymbopogon* species: ethnopharmacology, phytochemistry and the pharmacological importance. Molecules 20, 7438–7453.

Balboa, J.G., Lim-Sylianco, C.Y., 1995. Effect of some medicinal plants on skin tumor promotion. Philipp. J. Sci. 124, 203–207.

Banu, A.K., Bewaraji, A., 2013. A novel herbal pesticide to control adult scale insects. J. Pharmacol. Sci. Innov. 2, 22–25.

Barbosa, L.C.A., Pereira, U.A., Martinazzo, A.P., Maltha, C.R., Teiveira, R.R., De Castromelo, E., 2008. Evaluation of the chemical composition of Brazilian commercial *Cymbopogon citratus* (D.C.) stapf samples. Molecules 13, 1864–1874.

Bassolé, I.H.N., Lamien-Meda, A., Bayala, B., Obame, L.C., Ilboudo, A.J.C., Franz, J., Dicko, M.H., 2011. Chemical composition and antimicrobial activity of *Cymbopogon citratus* and *Cymbopogon giganteus* essential oils alone and in combination. Phytomedicine 18, 1070–1074.

Beech, D.F., 1990. The effect of carrier and rate of nitrogen application on the growth and oil production of lemongrass (*Cymbopogon citratus*) in the Ord Irrigation Area, Western Australia. Austr. J. Exp. Agric. 30, 243–250.

Berlin, B., Breedlove, D.E., Raven, P.H., 1974. Principles of Tzeltal Plant Classification. An Introduction to the Botanical Ethnography of a Mayan Speaking People of Highland Chiapas. Academic Press, New York.

Berlin, B., Breedlove, D.E., Raven, P.H., 1977. Principles of Tzeltal plant classification: an introduction to the botanical ethnography of a Mayan speaking people of highland Chiapas. Hum. Ecol. 5, 171–175.

Bertea, C.M., Maffei, M.E., 2010. The genus *Cymbopogon*: botany including anatomy, physiology, biochemistry, and molecular biology. Akhila, A. (Ed.), Essential Oil Bearing Grasses: The Genus Cymbopogon-CRC Press, Boca Raton, FL, pp. 1–24.

Bertea, C.M., Tesio, M., Agostino, G.D., Buffa, G., Camusso, W., Bossi, S., Maffei, M., 2003. The C4 biochemical pathway and the anatomy of lemon grass (*Cymbopogon citratus* (DC) Stapf. cultivated in temperate climates. Plant Biosyst. 137, 175–184.

Bertuzzi, G., Tirillini, B., Angelini, P., Venanzoni, R., 2013. Antioxidative action of citrus limonum essential oil on skin. Eur. J. Med. Plants 3, 1–9.

Bharti, S.K., Kumar, A., Prakash, O., Krishnan, S., Gupta, A.K., 2013. Essential oil of *Cymbopogon citratus* against diabetes: validation by in-vivo experiments and computational studies. J. Bioanal. Biomed. 5, 194–203.

Bidinotto, L.T., Costa, C.A., Salvadori, D.M., Costa, M., Rodrigues, M.A., Barbisan, L.F., 2011. Protective effects of lemongrass (*Cymbopogon citratus* STAPF) essential oil on DNA damage and carcinogenesis in female Balb/C mice. J. Appl. Toxicol. 31, 536–544.

Boaduo, N.K., Katerere, D., Eloff, J.N., Naidoo, V., 2014. Evaluation of six plant species used traditionally in the treatment and control of *diabetes mellitus* in South Africa using in vitro methods. Pharm Biol. 52, 756–761.

Bossou, A.D., Mangelinck, S., Yedomonhan, H., Boko, P.M., Akogbeto, M.C., De Kimpe, N., Sohounhloue, D.C.K., 2013. Chemical composition and insecticidal activity of plant essential oils from Benin against *Anopheles gambiae* (Giles). Parasit. Vectors 6, 1–17.

Boukhatem, M.N., Ferhat, M.A., Kameli, A., Saidi, F., Kebir, H.T., 2013. Lemon grass (*Cymbopogon citratus*) essential oil as a potent anti-inflammatory and antifungal drugs. Libyan J. Med. 8, 25431.

Boukhatem, M.N., Ferhat, M.A., Kameli, A., Saidi, F., Kebir, H.T., 2014. Lemon grass (*Cymbopogon citratus*) essential oil as a potent anti-inflammatory and antifungal drugs. Libyan J. Med. 9, 1–10.

Brian, C., Lorena, F., Huazhang, H., Marja, K., Pamela, G.M., 2009. A natural herbicide containing lemon grass essential oil. US Patent WO 2,009,049,153 A2.

Campos, J., Schmeda-Hirschmann, G., Leiva, E., Guzmán, L., Orrego, R., Fernández, P., Aguayo, C., 2014. Lemon grass (*Cymbopogon citratus* (D.C) Stapf) polyphenols protect human umbilical vein endothelial cell (HUVECs) from oxidative damage induced by high glucose, hydrogen peroxide and oxidised low-density lipoprotein. Food Chem. 151, 175–181.

Carmo, E.S., Pereira, F., de, F.de.O., Cavalcante, N.M., Gayoso, C.W., Lima, E.de.O., 2013. Treatment of *Pityriasis versicolor* with topical application of essential oil of *Cymbopogon citratus (DC)* Stapf—therapeutic pilot study. An. Bras. Dermatol. 88, 381–385.

Cheel, J., Theoduloz, C., Rodriguez, J., Schemeda-Harschmann, G., 2005. Free radical and antioxidants from Lemon grass (*Cymbopogan citratus* (DC) Staff). J. Agric. Food Chem. 53, 2511–2517.

Chinsembu, K.C., 2015. Plants as antimalarial agents in Sub-Saharan Africa. Acta Trop. 152, 32–48.

Chisowa, E.H., Hall, D.R., Farman, D.I., 1998. Volatile constituents of the essential oil of *Cymbopogon citratus* Stapf grown in Zambia. Flavour Frag. J. 13, 29–30.

Chowdury, M.I.A., Debnath, M., Ahmad, M.F., Alam, M.N., Saleh, A.M., Chowdhury, S., Kama, A.H.M., 2015. Potential phytochemical, analgesic and anticancerous activities of *Cymbopogon citratus* leaf. Am. J. Biomed. Res. 3, 66–70.

Cimanga, K.K., Kambu, T.L., Apers, S., De Bruyne, T., Hermans, N., Totte, J., Vlietinck, A.J., 2002. Correlation between chemical compositions and antibacterial activity of essential oils of some aromatic medicinal plants growing in the Democratic Republic of Congo. J. Ethnopharmacol. 79, 213–220.

Clement, Y.N., Baksh-Comeau, Y.S., Seaforth, C.E., 2015. An ethnobotanical survey of medicinal plants in Trinidad. J. Ethnobiol. Ethnomed. 11, 67.

Colotta, F., Allavena, P., Sica, A., Garlanda, C., Mantovani, A., 2009. Cancer-related inflammation, the seventh hallmark of cancer: links to genetic instability. Carcinogenesis 30, 1073–1081.

Connor, M.J., 1991. Modulation of tumor promotion in mouse skin by the food additive citral (3, 7-dimethyl-2, 6-octadienal). Cancer Lett. 56, 25–28.

Costa, C.A., Bidinotto, L.T., Takahira, R.K., Salvadori, D.M., Barbisan, L.F., Costa, M., 2011. Cholesterol reduction and lack of genotoxic or toxic effects in mice after repeated 21-day oral intake of lemongrass (*Cymbopogon citratus*) essential oil. Food Chem. Toxicol. 49, 2268–2272.

Costa, A.V., Pinheiro, P.F., Rondelli, V.M., De Queiroz, V.T., Tuler, A.C., Brito, K.B., Pratissoli, D., 2013. *Cymbopogon citratus* (Poaceae) essential oil on *Frankliniella schultzei* (Thysanoptera: Thripidae) and *Myzus persicae* (Hemiptera: Aphididae). Biosci. J., Uberlândia 29, 1840–1847.

Covich, A.P., Nickerson, N.H., 1966. Studies of cultivated plants in Choco dwelling clearings, arien, Panama. Econ. Bot. 20, 285–301.

De Cássia da Silveira e Sá, R., Andrade, L.N., de Sousa, D.P., 2013. A review on antiinflammatory activity of monoterpenes. Molecules 18, 1227–1254.

Detpiratmongkol, S., Ubolkerd, T., Yousukyingsatapron, S., 2005. Effects of irrigation frequencies and amounts on growth and yield of local lemongrass cultivar. Proceedings of 43rd Kasetsart University Annual Conference, Thailand, pp. 632–640.

Devi, R.C., Sim, S.M., Ismail, R., 2012. Effect of *Cymbopogon citratus* and citral on vascular smooth muscle of the isolated thoracic rat aorta. Evidence-Based Complement. Alternat. Med. 2012, 1–8, Available from: http://doi.org/10.1155/2012/539475.

Edeoga, H.O., Okwu, D.E., Mbaebie, B.O., 2005. Phytochemical constituents of some Nigeria medicinal plants. Afr. J. Biotech. 4, 685–688.

Elkhair Abou, K.E., 2014. Antidermatophytic activity of essential oils against locally isolated *Microsporum canis*—Gaza Strip. Nat. Sci. 6, 676–684.

Esmort, H., David, R.H., Dudley, I.F., 1998. Volatile constituents of the essential oil of *Cymbopogon citratus* Stapf. grown in Zambia. Flavour Frag. J. 13, 29–30.

Ewansiha, J.U., Garba, S.A., Mawak, J.D., Oyewole, O.A., 2012. Antimicrobial activity of *Cymbopogon citratus* (lemon grass) and its phytochemical properties. Front. Sci. 2, 214–220.

Farhang, V., Amini, J., Javadi, T., Nazemi, J., Ebadollahi, A., 2013. Chemical composition and antifungal activity of essential oil of *Cymbopogon citratus* (DC.) Stapf. against three phytophthora species. Greener J. Biol. Sci. 3, 292–298.

Faruq, M.O., 1994. TLC technique in the component characterization and quality determination of Bangladeshi lemongrass oil. (*Cymbopogon citratus* DC stapf.). Bangladesh J. Sci. Ind. Res. 29, 27–38.

Figueirinha, A., Paranhos, A., Perez-Alonso, J., Santos-buelga, C., Batista, M., 2008. *Cymbopogon citratus* leaves: characterisation of flavonoids by HPLC-PDA-ESI/MS/MS and an approach to their potential as a source of bioactive polyphenols. Food Chem. 110, 718–728.

Francisco, V., Costa, G., Figueirinha, A., Marques, C., Pereira, P., Miguel, B., Teresa, M., 2013. Anti-inflammatory activity of *Cymbopogon citratus* leaves infusion via proteasome and nuclear factor-κB pathway inhibition: contribution of chlorogenic acid. J. Ethnopharmacol. 148, 126–134.

Francisco, V., Figueirinha, A., Costa, G., Liberal, J., Lopes, M.C., García-Rodríguez, C., Batista, M.T., 2014. Chemical characterization and antiinflammatory activity of luteolin glycosides isolated from lemongrass. J. Funct. Foods 10, 436–443.

Gonzalez, A.J.C., Colmenares, N.G., Enrique, U.A., Darghan, S.L., 2008. Evaluation of agronomic traits in the cultivation of lemongrass *(Cymbopogon citratus* Stapf) for the production of essential oil. Interciencia 33, 693–699.

Graveson, R.S., 2012. Survey of invasive alien plant species on Gros Piton, Saint Lucia. Department of Forestry.

Gunasingh, C.B.G., Nagarajan, S., 1980. Flavonoids of *Cymbopogon citratus*. Ind. J. Pharm. Sci. 43, 115.

Halabi, M.F., Sheikh, B.Y., 2014. Anti-proliferative effect and phytochemical analysis of *Cymbopogon citratus* extract. BioMed. Res. Int. 2014, 8.

Hanaa, A.R.M., Sallam, Y.I., El-Leithy, A.S., Aly, S.E., 2012. Lemon grass (*Cymbopogon citratus*) essential oil as affected by drying methods. Ann. Agric. Sci. 57, 113–116.

Hanifah, A.L., Awang, S.H., Ming, H.T., Abidin, S.Z., Omar, M.H., 2011. Acaricidal activity of *Cymbopogon citratus* and *Azadirachta indica* against house dust mites. Asian Pac. J. Trop. Biomed. 1, 365–369.

Hilgert, N.I., 2001. Plants used in home medicine in the Zenta River Basin, Northwest Argentina. J. Ethnopharmacol. 76, 11–34.

Hindumathy, C.K., 2011. In-vitro study of antibacterial activity of *Cymbopogon citratus*, 5, 48–52.

Hodgson, E., 2004. A Textbook of Modern Toxicology, third ed. John Wiley & Sons Inc., Publication, NJ, pp. 248–249.

Ho-Dzun, H., Knopffer, H., Hammer, K., 1997. Additional notes to the checklist of Korean cultivated plants (5). Consolidated summary and indexes. Genet. Resour. Crop Evolut. 44, 349–391.

Humphrey, A.M., 1973. The chromatography of the spice oils. Proceedings of the Conference on Spices, London. Tropical Products Institute, pp. 123–128.

Idrees, M., Naeem, M., Khan, M., Aftab, T., Tariq, M., 2012. Alleviation of salt stress in lemon grass by salicylic acid. Protoplasma 249, 709–720.

Jabaut, J., Ckless, K., 2007. Inflammation, immunity and redox signaling. In: Dr Khatami, M. (Ed.), Cell and Molecular Biology, Immunology and Clinical Bases. InTech, Rijeka, Croatia, pp. 145–160.

Jafari, B., Ebadi, A., Aghdam, B.M., Hassanzade, Z., 2012. Antibacterial activities of lemon grass methanol extract and essence on pathogenic bacteria. Am.-Eur J. Agric. Environ. Sci. 12, 1042–1046.

Jayasinha, P., 2001. Lemon grass: a literature survey. Medicinal and Aromatic Plant Series No. 9. Industrial Technology Institute, Baudhaloka Mawatha, Colombo, Sri Lanka.

Jeong, M., Park, P.B., Kim, D., Jang, Y., Jeong, H.S., Choi, S., 2009. Essential oil prepared from *Cymbopogon citrates* exerted an antimicrobial activity against plant pathogenic and medical microorganisms. Mycobiology 37, 48–52.

Kasali, A.A., Oyedeji, A.O., Ashilokun, A.O., 2001. Volatile leaf oil constituents of *Cymbopogon citratus* (DC) Stapf. Flavour Frag. J. 16, 377–378.

Katsukawa, M., Nakata, R., Takizawa, Y., Hori, K., Takahashi, S., Inoue, H., 2010. Citral, a component of lemon grass oil, activates PPARα and γ and suppresses COX-2 expression. Biochim. Biophys. Acta 1801, 1214–1220.

Kazembe, T., Chauruka, D., 2012. Mosquito repellence of *Astrolochii hepii*, *Cymbopogon citratus* and *Ocimum gratissimum* extracts and mixtures. Bull. Environ. Pharmacol. Life' Sci. 1, 60–64.

Koba, K., Sanda, K., Guyon, C., Raynaud, C., Chaumont, J., Nicod, L., 2009. In vitro cytotoxic activity of *Cymbopogon citratus* L. and *Cymbopogon nardus* L. essential oils from Togo. Bangladesh J. Pharmacol. 4, 29–34.

Korenblum, E., Regina, F., Goulart, D.V., Rodrigues, I.D.A., Abreu, F., Lins, U., Seldin, L., 2013. Antimicrobial action and anti-corrosion effect against sulfate reducing bacteria by lemongrass (*Cymbopogon citratus*) essential oil and its major component, the citral. AMB Express 3, 1–8.

Kouame, N.M., Kamagate, M., Koffi, C., Die-Kakou, H.M., Yao, N.A.R., Kakou, A., 2015. *Cymbopogon citratus* (DC.) Stapf: Ethnopharmacology, phytochemical, pharmacological activities and toxicology. Phytothérapie 13, 1–9.

Kpoviessi, S., Bero, J., Agbani, P., Gbaguidi, F., Kpadonou-Kpoviessi, B., Sinsin, B., Leclercq, J., 2014. Chemical composition, cytotoxicity and in vitro antitrypanosomal and antiplasmodial activity of the essential oils of four *Cymbopogon* species from Benin. J. Ethnopharmacol. 151, 652–659.

Kumar, P., Mishra, S., Malik, A., Satya, S., 2013. Housefly (Musca domestia L.) control potential of *Cymbopogon citratus* Stapf. (Poales: Poaceae) essential oil and monoterpenes (citral and 1,8-cineole). Parasitol. Res. 112, 69–76.

Leclercq, P.A., Delgado, H.S., Garcia, J., Hidalgo, J.E., Cerruttti, T., Mestanza, M., Menedez, R., 2000. Aromatic plant oils of the Peruvian Amazon. Part 2. *Cymbopogon citratus* (DC) Stapf., *Renealmia* sp., *Hyptis recurvata* Poit. and *Tynanthus panurensis* (Bur.) Sandw. J. Essent. Oil Res. 12, 14–18.

Leite, J.R., Seabra, M.V., Maluf, E., Assokut, K., Sucheki, D., Tufik, S., Caclini, E.A., 1986. Pharmacology of lemon graa (*Cymbopogon citratus,* Stapf) III. Assessment of eventual toxic, hypnotic and anxiolytic effects on humans. J. Ethnopharmacol. 17, 75–81.

Lu, Y., Shipton, F.N., Khoo, T.J., Wiart, C., 2014. Antioxidant activity determination of citronellal and crude extracts of *Cymbopogon citratus* by 3 different methods. Pharmacol. Pharm. 5, 395–400.

Luangnarumitchai, S., Lamlertthon, S., Tiyaboonchai, W., 2007. Antimicrobial activity of essential oils against five strains of *Propionibacterium acnes*. Mahidol Univ. J. Pharm. Sci. 34, 60–64.

Manvitha, K., Bidya, B., 2014. Review on pharmacological activity of *Cymbopogon citratus*. Int. J. Herb. Med. 1, 5–7.

Masamba, W.R.L., Kamanula, J.F.M., Henry, E.M.T., Nyirenda, G.K.C., 2003. Extraction and analysis of lemongrass (*Cymgopogon citratus*) oil: As essential oil with potential to control the Larger Grain Borer (*Prostephanus truncatus*) in stored products in Malawi. Malawi J. Agric. Sci. 2, 56–64.

Matasyoh, J.C., Wagara, I.N., Nakavuma, J.L., Kiburai, A.M., 2011. Chemical composition of *Cymbopogon citratus* essential oil and its effect on *mycotoxigenic Aspergillus* species. Afr. J. Food Sci. 5, 138–142.

Matouschek, B.K., Stahl, B.E., 1991. Phytochemical study of non volatile substance from *Cymbopongon citratus* (D.C) Stapf (Poaceae). Pharm. Acta Helv. 66, 242–245.

Melariri, P.E., 2010. The Therapeutic Effectiveness of Some Local Nigerian Plants Used in the Treatment of Malaria. A PhD Thesis, University of Capetown, Department of Pharmacy, 147–160.

Miean, K.H., Mohamed, S., 2001. Flavonoid (myricitin, quercetin, kaempferol, luteolin, and apigenin) content of edible tropical plants. J. Agric. Food Chem. 49, 3106–3112.

Miguel, M.G., 2010. Antioxidant and anti-inflammatory activities of essential oils: a short review. Molecules 15, 9252–9287.

Mirza, M., Rizwani, G.H., Yaseen, Z., Qadri, R.B., Ahmad, M., Mehmood, S., 2005. Pharmacognostic studies of some diuretic medicinal plants of Pakistan. Hamdard Med. 48, 178–194.

Moore, S.J., Davies, C.R., Hill, N., Cameron, M.M., 2007. Are mosquitoes diverted from repellent-using individuals to non-users? Results of a field study in Bolivia. Trop. Med. Int. Health 12, 532–539.

Muchtaridi, A.D., Subarnas, A., 2011. Analysis of Indonesian spice essential oil compounds that inhibit locomotor activity in mice. Pharmaceuticals 4, 590–602.

Nanon, A., Suksombat, W., Beauchemin, K.A., Yang, W.Z., 2014. Assessment of lemon grass oil supplementation in a dairy diet on *in vitro* ruminal fermentation characteristics using the rumen simulation technique. Can. J. Anim. Sci. 94, 731–736.

Negrelle, R.R.B., Gomes, E.C., 2007. *Cymbopogon citratus* (D.C) Stapf: chemical composition and biological activities. Rev. Bras. Plantas Med. 9, 80–92.

Nham, T.N., 1993. Medicinal and aromatic plants in Vietnam. Chomchalow, N., Henle, H.V. (Eds.), Medicinal and Aromatic Plants in Asia: Breeding and ImprovementScience Publishing, Lebanon.

Nishijima, C.M., Ganev, Ellen.G., Mazzardo-Martins, L., Martins, D.F., Rocha, L.R., Santos, A.S., Hiruma-Lima, C.A., 2014. Citral: a monoterpene with prophylactic and therapeutic anti-nociceptive effects in experimental models of acute and chronic pain. Eur. J. Pharmacol. 736, 16–25.

Ntonga, P.A., Baldovini, N., Mouray, E., Mambu, L., Belong, P., Grellier, P., 2014. Activity of *Ocimum basilicum*, *Ocimum canum*, and *Cymbopogon citratus* essential oils against *Plasmodium falciparum* and mature-stage larvae of *Anopheles funestus* s.s. Parasite 21, 33.

Nyarko, H.D., Barku, V.Y.A., Batama, J., 2012. Antimicrobial examinations of *Cymbopogon citratus* and *Adiatum capillus-veneris* used in Ghanaian folkloric medicine. Int. J. Life Sci. Pharm. Res. 2, L115–L121.

Ohno, T., Kita, M., Yamaoka, Y., Imamura, S., Yamamoto, T., Mitsufuji, S., Imanishi, J., 2003. Antimicrobial activity of essential oils against *Helicobacter pylori*. Helicobacter 8, 207–215.

Olaniyi, A.A., Sofowora, E.A., Oguntimenin, B.O., 1975. Phytochemical investigation of some Nigerian plants used against fevers II *Cymbopogon citratus*. Planta Med. 28 (6), 186–189.

Olivero-Verbel, J., Nerio, L.S., Stashenko, E.E., 2010. Bioactivity against *Tribolium castaneum* Herbst (Coleoptera: Tenebrionidae) of *Cymbopogon citratus* and *Eucalyptus citriodora* essential oils grown in Colombia. Pest Manag. Sci. 66, 664–668.

Pedroso, R.B., Ueda-nakamura, T., Dias Filho, B., Cortez, A.G., Cortez, R.E.L., Morgado- díaz, J.A., Nakamura, C.V., 2006. Biological activities of essential oil obtained from *Cymbopogon citratus* on *Crithidia deanei*. Acta Protozool. 45, 231–240.

Piarua, S.P., Perumala, S., Caia, L.W., Mahmuda, R., Abdul Majida, A.M.S., Ismailb, S., Manc, C.N., 2012. Chemical composition, anti-angiogenic and cytotoxicity activities of the essential oils of *Cymbopogan citratus* (lemon grass) against colorectal and breast carcinoma cell lines. J. Essent. Oil Res. 24, 453–459.

Pinto, Z.T., Sánchez, F.F., Ramos, A., Amaral, A.C.F., Ferreira, J.L.P., Escalona-Arranz, J.C., De Carvalho Queiroz, M.M., 2015. Chemical composition and insecticidal activity of *Cymbopogon citratus* essential oil from Cuba and Brazil against housefly. Rev. Bras. Parasitol. Vet. 24 (1), 36–44.

Pohl, R.W., 1972. Keys to the genera of grasses of Costa Rica. Rev. Biolog a Trop. 20, 189–219.

Puatanachokchai, R., Kishida, H., Denda, A., Murata, N., 2002. Inhibitory effects of lemon grass (*Cymbopogon citratus*, Stapf) extract on the early phase of hepatocarcinogenesis after initiation with diethylnitrosamine in male Fischer 344 rats. Cancer Lett. 183, 9–15.

Rabbani, S.I., Devi, K., Zahra, N., 2005. Anti-clastogenic effects of citral. Iran. J. Pharmacol. Ther. 4, 28–31.

Ranitha, M., Abdurahman, H.N., Sulaiman, Z.A., Nour, A.H., Thana Raj, S.A., 2014. Comparative study of lemongrass (*Cymbopogon citratus*) essential oil extracted by microwave-assisted hydrodistillation (MAHD) and conventional hydrodistillation (HD) method. Int. J. Chem. Eng. Appl. 5 (2), 104–108.

Raul, F., 2001. Geraniol, a component of plant essential oils, inhibits growth and polyamine biosynthesis in human colon cancer cells. J. Pharmacol. Exp. Ther. 298, 197–200.

Roscheisen, C., Zamith, H., Paumgartten, F.J.R., Speit, G., 1991. Influence of β-myrcene on sister-chromatid exchanges induced by mutagens in V79 and HTC cells. Mutat. Res. 264, 43–49.

Ross, I.A., 1999. *Cymbopogon citratus* (DC.) Stapf. Medicinal Plants of the World: Volume 1 Chemical Constituents, Traditional and Modern Medicinal Uses. Humana Press, Totowa, NJ.

Rozzi, N.L., Phippen, W., Simon, J.E., Singh, R.K., 2002. Supercritical fluid extraction of essential oil components from lemon scented botanicals. Lebensm. Wiss. Technol. 35, 319–324.

Santos, G.R., Sarmento, B.R.B.C., de Castro, H.G., Gonçalves, C.G., Fidelis, R.R., 2013. Effect of essential oils of medicinal plants on leaf blotch in Tanzania grass. Rev. Ciênc. Agron. 44, 587–593.

Seabrook, Jr S.G., 2008. Polymer coatings containing phytochemical agents and methods for making and using same. US Patent EP 1,940,431 A1 9 Jul 2008.

Sepúlveda-Arias, J.C., Veloza, L.A., Escobar, L.M., Orozco, L.M., Lopera, I.A., 2013. Anti-inflammatory effects of the main constituents and epoxides derived from the essential oils obtained from *Tagetes lucida*, *Cymbopogon citratus*, *Lippia alba* and *Eucalyptus* citriodora. J. Essent. Oil Res. 25, 186–193.

Sessou, P., Farougou, S., Kaneho, S., Djenontin, S., Alitonou, G.A., Azokpota, P., Sohounhloué, D., 2012. Bioefficacy of *Cymbopogon citratus* essential oil against food borne pathogens in culture medium and in traditional cheese wagashi produced in Benin. Int. Res. J. Microbiol. 3, 406–415.

Shah, G., Shri, R., Panchal, V., Sharma, N., Singh, B., Mann, A.S., 2011. Scientific basis for the therapeutic use of *Cymbopogon citratus*, stapf (Lemon grass). J. Adv. Pharm. Technol. Res. 2, 3–8.

Shan, B., Cai, Y., Sun, M., Orke, H., 2005. Antioxidant capacity of 26 spice extracts and characterization of their phenolic constituents. J. Agric. Food Chem. 53, 7749–7759.

Shimono, K., Hiroaki, O., Masato, S., Kanae, S., Shoji, K., 2010. Aromatic antihypertensive agent, and method for lowering blood pressure in mammals. US Patent 20,100,216,891 A1.

Sidibé, L., Chalchat, J.C., Garry, R.P., Lacombe, L., Harama, M., 2001. Aromatic plants of Mali (IV): chemical composition of essential oils of *Cymbopogon citratus* (DC) Stapf and *C. giganteus* (Hochst.) Chiov. J. Essent. Oil Res. 13 (2), 110–112.

Singh, M.P., Malla, S.B., Rajbhandari, S.B., Manandhar, A., 1980. Medicinal plants of Nepal—retrospects and prospects. Econ. Bot. 33, 185–198.

Soares, M.O., Vinha, A.F., Barreira, S.V.P., Coutinho, F., 2013a. *Cymbopogon citratus* EO antimicrobial activity against multi-drug resistant Gram-positive strains and non- *Candida albicans*-species. Méndez-Vilas, A. (Ed.), Microbial Pathogens and Strategies for Combating Them: Science, Technology and Education-Formatex Research Center, pp. 1081–1086.

Soares, M.O., Vinha, A.F., Barreira, S.V.P., Coutinho, F., Aires-Gonçalves, S., Oliveira, M., Castro, A., 2013b. Evaluation of antioxidant and antimicrobial properties of the Angolan *Cymbopogon citratus* essential oil with a view to its utilization as food biopreservative. J. Agric. Sci. 5, 36–45.

Soonwera, M., 2015. Larvicidal and oviposition deterrent activities of essential oils against house fly (*Musca domestica* L.; Diptera: Muscidae). J. Agric. Technol. 11, 657–667.

Springer, M.L., Lam, V.T., Druff, D., 1995. Identification of a Novel 167 kDa lemon grass protein with anti-Dandruff properties. J. Scalp Sci. 6, 26447.

Suaeyun, R., Kinouchi, T., Arimochi, H., Vinitketkumnuen, U., 1997. Inhibitory effects of lemon grass (*Cymbopogon citratus* Stapf) on formation of azoxymethane-induced DNA adducts and *aberrant crypt foci* in the rat colon. Carcinogenesis 18, 949–955.

Suhaila, M., Farideh, N., Kheen, K.C., 2010. Agent for promoting vitamin c transporter production US Patent CN 101,912,352 A 15th December, 2010.

Tajidin, N.E., Ahmad, S.H., Rosenani, A.B., Azimah, H., Munirah, M., 2012. Chemical composition and citral content in lemongrass (*Cymbopogon citratus*) essential oil at three maturity stages. Afr. J. Biotechnol. 11, 2685–2693.

Tapia, A., Cheel, J., Theoduloz, C., Rodriguez, J., Schmeda-hirschmann, G., Gerth, A., Mendoza, E., 2007. Free radical scavengers from *Cymbopogon citratus* (DC.) Stapf. plants cultivated in bioreactors by the temporary immersion (TIS) principle. Z. Naturforsch C 62, 447–457.

Tarkang, P.A., Agbor, G.A., Tsabang, N., Tchokouaha, L.R., Tchamgoue, D.A., Kemeta, D., Weyepe, F., 2012. Effect of long-term oral administration of the aqueous and ethanol leaf extracts of *Cymbopogon citratus* (DC. ex Nees) Stapf. Ann. Biol. Res. 3, 5561–5570.

Tarkang, P.A., Franzoi, K.D., Lee, S., Lee, E., Vivarelli, D., Freitas-Junior, L., Guantai, A.N., 2014. In vitro antiplasmodial activities and synergistic combinations of differential solvent extracts of the polyherbal product, Nefang. BioMed. Res. Int. 2014, 1–10.

Tarkang, P.A., Nwachiban Atchan, A.P., Kuiate, J.R., Okalebo, F.A., Guantai, A.N., Agbor, G.A., 2013. Antioxidant potential of a polyherbal antimalarial as an indicator of its therapeutic value. Adv. Pharmacol. Sci. 2013, 9.

Thangam, R., Sathuvan, M., Poongodi, A., Suresh, V., Pazhanichamy, K., Sivasubramanian, S., Kannan, S., 2014. Activation of intrinsic apoptotic signaling pathway in cancer cells by *Cymbopogon citratus* polysaccharide fractions. Carbohydr. Polym. 107, 138–150.

Ullah, N., Khan, M.A., Ahmad, W., 2013. *Cymbopogon citratus* protects against the renal injury induced by toxic doses of aminoglycosides in rabbits. Ind. J. Pharm. Sci 75, 241–246.

Usha Rani, P., 2012. Fumigant and contact toxic potential of essential oils from plant extracts against stored product pests. J. Biopesticides 5, 120–128.

Vahid, F., Jahanshir, A., Taimoor, J., Javad, N., Asgar, E., 2013. Chemical composition and antifungal activity of essential oil of *Cymbopogon citratus* (DC.) Stapf. against three phytophthora species. Greener J. Biol. Sci. 3, 292–298.

Vera, S.S., Zambrano, D.F., Méndez-Sanchez, S.C., Rodríguez-Sanabria, F., Stashenko, E.E., Duque Luna, J.E., 2014. Essential oils with insecticidal activity against larvae of *Aedes aegypti* (Diptera: Culicidae). Parasitol. Res. 113, 2647–2654.

Wang, Y.L., Li, Y., 2015. Biological pesticide insecticide without pesticide resistance. CN 104,488,999 A, 8 April, 2015.

Wei, J.F., 2015. Traditional Chinese medicine composition for treating stomach ache. US Patent CN 105,012,765 A, 4 Nov 2015.

Wossa, S.W., Rali, T., Leach, D.N., 2004. Analysis of essential oil composition of some selected spices of Papua New Guinea. Papua New Guinea J. Agric. Forestry Fisheries 47, 17–20.

Wright, S.C., Maree, J.E., Sibanyoni, M., 2009. Treatment of oral thrush in HIV/AIDS patients with lemon juice and lemon grass (*Cymbopogon citratus*) and gentian violet. Phytomedicine 16, 118–124.

Wuthi-Udomlert, M., Chotipatoomwan, P., Panyadee, S., Gritsanapan, W., 2011. Inhibitory effect of formulated lemon grass shampoo on *Malassezia furfur*: a yeast associated with dandruff. Southeast Asian J. Trop. Med. Public Health 42, 363–369.

Yang, S.Y., Lei, Y., 2005. Antimicrobial activity of *Cymbopogon citratus* against utilized bacteria and fungus. J. Shanghai Jiaotong Univ. (Agric. Sci.) 23, 374–376.

Yu, S.G., Hildebrandt, L.A., Elson, C.E., 1995. Geraniol, an inhibitor of mevalonate biosynthesis, suppresses the growth of hepato- mas and melanomas transplanted to rats and mice. J. Nutr. 125, 2763–2767.

Zannou, A., Christian, T.R.K., Gbaguidi, M.A.N., Ahoussi-Dahouenon, E., 2015. Antimicrobial activity of extracts from *Cymbopogon citratus* L. and of *Mentha spicata* L. against fungal and bacterial strains isolated from peuhl's cheese (Waragashi) produced in Benin. Int. J. Adv. Res. 3, 1684–1695.

CHAPTER 19

Curcuma longa

L.K. Omosa, J.O. Midiwo, V. Kuete

1 INTRODUCTION

Curcuma longa L., commonly known as turmeric, is a member of the ginger family (Zingiberaceae), native to Southwest India with its rhizomes being the source of a bright yellow spice and dye. In Africa it is cultivated in home gardens in many countries and sold in numerous markets (Jansen, 2005). In West Africa its mainly used as a dye to color products, such as tanned leather, cotton cloth, thread, and palm fibers to a golden yellow. The use of the yellow color of turmeric rhizome and other plant derivatives as dyes is on the increase toward replacing synthetic additives with natural compounds (Ravindran, 2007). This yellow color is due to the presence of three main curcuminoids in the rhizome namely, curcumin, demethoxycurcumin, and *bis*-demethoxycurcumin. Rhizomes are used in Africa and Asia as a cosmetic for body and face. In Asia, turmeric is widely used as an important constituent of curry powder containing up to 25% turmeric (Jansen, 2005). In Western countries, ground turmeric rhizome is widely used in food industry, in particular as a coloring agent (E 100 in the European Union) in processed foods and sauces. Turmeric is an important medicinal and aromatic plant which is considered as one of the golden resources with immense export potential as medicine, beauty aid, cooking spice, and as a dye (Das, 2016). Rhizomes of *C. longa* are part of numerous traditional medicines used as stomachic, stimulant, and blood purifier, and to treat liver complaints, biliousness, and jaundice (Jansen, 2005), for arthritic, muscular disorders, biliary disorders, anorexia, cough, diabetic wounds, hepatic disorders, rheumatism and sinusitis (Shishodia et al., 2007). Mixed with warm milk, they are used to cure common cold, bronchitis, and asthma. Juice from fresh rhizomes is applied against many skin infections, whereas a decoction is effective against eye infections (Jansen, 2005). It has promising pharmaceutical activity against cancer, dermatitis, AIDS, inflammation, high cholesterol levels, and dyspeptic conditions (Jansen, 2005). *C. longa* also has insecticidal, fungicidal, and nematicidal properties. In Madagascar, ground rhizomes are mixed with grains to protect them from storage pests (Jansen, 2005). One of the curcuminoids in the rhizome, curcumin (**1**), has demonstrated a number of pharmacological activities including antioxidant, antineoplastic, antiviral, antiinflammatory, antibacterial, antifungal, antidiabetic, anticoagulant, antifertility, cardiovascular protective, hepatoprotective, and immunostimulant activity in animals (Bengmark et al., 2009; Singh and Sharma, 2011) and therefore could be one of the constituents responsible for bioactivities of the rhizome of this plant.

2 CULTIVATION AND DISTRIBUTION OF *Curcuma longa*

C. longa is an upright, perennial herb about 1 m tall. The rhizome is thick and ringed with the bases of old leaves; turmeric only reproduces via its rhizomes (Ravindran, 2007). The plant requires temperatures between 20 and 30°C and an annual rainfall of about 1500 mm is optimum. The rhizome matures in about 9 months and the plant is grown extensively in tropical climate (Das, 2016; Khanna, 1999; Lal, 2012). India is considered a center of domestication and the plant has been grown there since time immemorial. It reached China before the 7th century, East Africa in the 8th century, and West Africa in the 13th century. At present turmeric is widely cultivated throughout the tropics, but commercial production is concentrated in India and South-East Asia (Jansen, 2005). Trade in turmeric from Asian countries is mainly routed through Singapore. Leading importers are Iran, Sri Lanka, most of the Middle East, and North African countries (Jansen, 2005).

3 CHEMISTRY OF *Curcuma longa*

Several classes of secondary metabolites have been characterized from *C. longa*. The rhizomes were found to contain curcuminoids, such as curcumin (**1**) and derivatives, such as, demethoxycurcumin, *bis*-demthoxycurcumin, 5′-methoxycurcumin, dihydrocurcumin, and cyclocurcumin (Sambhav et al., 2014). Many sesquiterpenes have been isolated from the rhizomes, including, germacrone (**2**), turmerone, ar-(**3**)-, α-, β-turmerones, β-bisabolene (**4**), α-curcumene (**5**), zingiberene (**6**), β-sesquiphellandene, bisacurone (**7**), curcumenone (**8**), dehydrocurdinone, procurcumadiol (**9**), bisacumol (**10**), curcumenol (**11**), isoprocurcumenol, epiprocurecumenol, zedoaronediol, curlone (**12**), turmeronols A (**13**) and B (**14**). Steroids, such as stigmasterol (**15**), β-sitosterol (**16**), cholesterol as well as anthraquinone, 2-hydroxymethyl anthraquinone (**17**) were also reported in the rhizomes. Essential oil from rhizomes contains compound **6**, α-phellandrene (**18**), sabinene (**19**), cineol (**20**), borneol (**21**), and sesquiterpenes with turmerones skeleton (**22**) (Sambhav et al., 2014) (Fig. 19.1).

4 PHARMACOLOGY OF *Curcuma longa*

4.1 Anticancer activity

Antitumor activity has been reported for rhizomes of *C. longa* as well as its constituents. The cytotoxicity of the rhizomes was evaluated in vitro using tissue culture methods and in vivo in mice using Dalton's lymphoma cells grown as ascites form (Kuttan et al., 1985). It was found that crude extract inhibited the cell growth in Chinese hamster ovary (CHO) cells at a concentration of 0.4 mg/mL and was cytotoxic to lymphocytes and Dalton's lymphoma cells at the same concentration. Compound **1** obtained from rhizomes inhibited the proliferation (Wilken et al., 2011) of cell lines

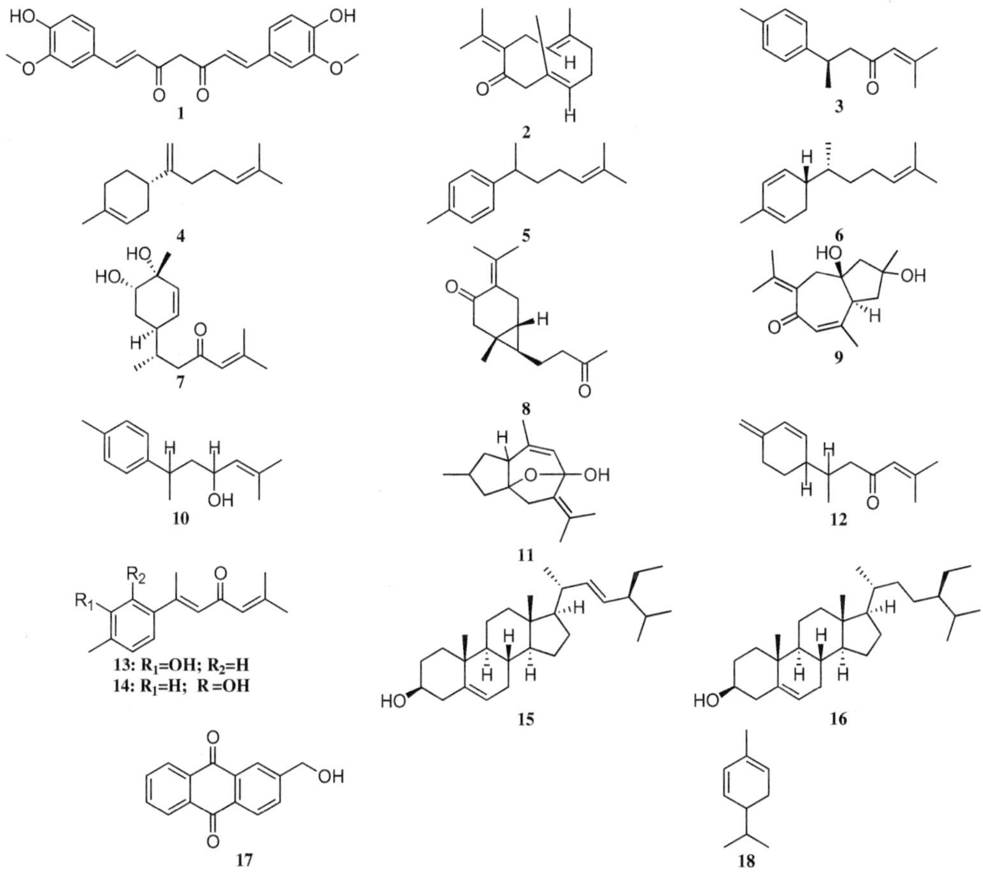

Figure 19.1 Chemical structures of selected compounds from C. longa. Curcumin (**1**); germacrone (**2**); ar-turmerone (**3**); β-bisabolene (**4**); α-curcumene (**5**); zingiberene (**6**); bisacurone (**7**); curcumenone (**8**); procurcumadiol (**9**); bisacumol (**10**); curcumenol (**11**); curlone (**12**); turmeronols A (**13**) and B (**14**); stigmasterol (**15**); β-sitosterol (**16**); 2-hydroxymethyl anthraquinone (**17**); α-phellandrene (**18**).

involved in various cancers, such as prostate carcinoma PC-3 cells (Cheng et al., 2013; Wei et al., 2013), breast adenocarcinoma MDA-MB-231 cells (Sun et al., 2012), MCF-7 cells (Liu and Chen, 2013), colon carcinoma HCT-8/VCR cells (Lu et al., 2013), HCT-15 cells (Shehzad et al., 2013), and liver cancer HepG2 cells (Fan et al., 2014). Interestingly, **1** showed protective effect in normal cell line; in effect, **1** significantly inhibited amino-1-methyl-6-phenylimidazo[4,5-b]pyridine (PhIP)-induced DNA adduct formation and DNA double stand breaks with a concomitant decrease in reactive oxygen species (ROS) production in normal breast epithelial cells (MCF-10A) (Jain et al., 2015). PhIP found in cooked meat is a known food carcinogen that causes several types of cancers, including breast cancer due to the ability of its metabolites to produce DNA adduct and DNA strand breaks (Jain et al., 2015).

4.2 Antidiabetic activity

Ahmad and coworkers demonstrated that *C. longa* can act as antidiabetic, hepatoprotective, and antioxidant in diabetes especially type 1 diabetes (Ahmad et al., 2014). They showed that administration of methanolic extract of this plant to alloxan-induced diabetic rabbits significantly improved the levels of serum glucose, serum transaminases, and antioxidant activity. Aqueous extract stimulated insulin secretion from mouse pancreatic tissues under both basal and hyperglycemic conditions, although the maximum effect was only 68% of that of the reference compound, tolbutamide (Mohankumar and McFarlane, 2011). This extract induced stepwise stimulation of glucose uptake from abdominal muscle tissues in the presence and absence of insulin; meanwhile the combination of the extract and insulin significantly potentiated the glucose uptake into abdominal muscle tissue (Mohankumar and McFarlane, 2011). In another study dealing with the effects of freeze dried rhizome powder of *C. longa* dissolved in milk on normal as well as streptozotocin-induced diabetic rats, Rai et al., 2010 found that effective dose was 200 mg/kg body weight as it increases high density lipoprotein (HDL), hemoglobin, and body weight with significant decrease in the levels of blood glucose, lipid profile, and hepatoprotective enzymes. Kuroda et al., 2005 also showed that *C. longa* ethanol extract is a promising ingredient of functional food for prevention and/or amelioration of type 2 diabetes and that compounds 1,3-demethoxycurcumin, and *bis*-demethoxycurcumin mainly contribute to the effects via peroxisome proliferator-activated receptor (PPAR)-gamma activation. Lekshmi et al., 2014 have investigated the in vitro antidiabetic and inhibitory potential of *C. longa* rhizome against cellular and low-density lipoprotein (LDL) oxidation and angiotensin converting enzyme. They found that α-glucosidase (0.4 µg/mL) and α-amylase (0.4 µg/mL) were inhibited by the ethyl acetate extract of this plant with higher effect compared to that of the reference drug acarbose (17.1 and 290.6 µg/mL, respectively). Protein glycation inhibitory potential of ethyl acetate extract was 800 times higher than that of ascorbic acid. The accumulation of advanced glycation end products (AGE's) in the body, due to the nonenzymatic glycation of proteins is associated with several pathological conditions like aging and diabetes mellitus. Hence a plant having antiglycation and antioxidation potential may serve as a therapeutic agent for diabetic complications and aging. The methanol extract of *C. longa* showed antiglycation activity with minimum inhibitory concentration 50 (MIC_{50}) of 324 µg/mL (Khan et al., 2014). The antidiabetic capacity of *C. longa* volatile oil in terms of its ability to inhibit glucosidase activities was also evaluated (Lekshmi et al., 2012). It was found that volatile oils inhibited glucosidase enzymes more effectively than the reference standard drug, acarbose. Drying of rhizomes was also found to enhance α-glucosidase (IC_{50} value of 1.32–0.38 µg/mL) and α-amylase (IC_{50} value of 64.7–34.3 µg/mL) inhibitory capacities of volatile oils. Ar-turmerone, the major volatile component in the rhizome, also showed potent α-glucosidase (IC_{50} of 0.28 µg) and α-amylase (IC_{50} of 24.5 µg) inhibition (Lekshmi et al., 2012). It was also found that curcuminoids and sesquiterpenoids from *C. longa* suppress an increase in blood glucose level in type 2 diabetic KK-Ay mice. In fact, compounds 1,3-demethoxycurcumin, and -demethoxycurcumin, exhibited hypoglycemic effects via PPAR-gamma activation as one of the mechanisms of action (Nishiyama et al., 2005).

4.3 Antiinflammatory activity

C. longa ethanol extract showed inhibitory activity on lipoxygenase from the rat lung cytosolic fraction (Bezakova et al., 2014). Illuri et al., 2015 have evaluated the antiinflammatory effects of polysaccharide fraction of *C. longa* in classical rodent models of inflammation. The investigators found that polysaccharide fraction significantly inhibited carrageenan-induced paw edema at 1 and 3 h at doses of 11.25, 22.5, and 45 mg/kg body weight in rats. They also found that this fraction at doses of 15.75, 31.5, and 63 mg/kg significantly inhibited the xylene-induced ear edema in mice. In a chronic model, they observed that 11.25, 22.5, and 45 mg/kg doses produced significant reduction of wet and dry weights of cotton pellets in rats. Overall results indicated that polysaccharide fraction of this plant significantly attenuated acute and chronic inflammation in rodent models. In an anatomico-radiological study, Taty Anna et al., 2011 have demonstrated that crude extract of *C. longa* had antiinflammatory effect on collagen-induced arthritis on male Sprague–Dawley rats. The constituent of *C. longa* **1** have demonstrated antiinflammatory action, inhibiting a number of inflammation markers, such as phospholipase, lipooxygenase, cycloxoygenase-2, leukotrienes, thromboxane, prostaglandins, nitric oxide, collagenase, elastase, hyaluronidase, interferon-inducible protein, tumor necrosis factor, and interleukin-12 (Chainani-Wu, 2003). It was also shown that a derivative of curcumin *bis*-demethylcurcuminisolated from this plant, was a potent antiinflammatory agent as indicated by suppression of tumor necrosis factor-induced nuclear factor (NF-κB) activation (Ravindran et al., 2010).

4.4 Antimicrobial activities
4.4.1 Antibacterial activity
The aqueous extract of *C. longa* rhizomes exhibited antibacterial effects against *Staphylococus epidermis* ATCC 12228, *S aureus* ATCC 25923, *Klebsiella pneumoniae* ATCC 10031, and *Escherichia coli* ATCC 25922 with the minimum inhibitory concentration (MIC) value ranging from 4–16 µg/mL (Moghadamtousi et al., 2014; Niamsa and Sittiwet, 2009). The methanol extract of this plant displayed inhibitory effects against *Bacillus subtilis* (MIC value of 16 µg/mL) and *S. aureus* (MIC value 128 µg/mL) (Ungphaiboon et al., 2005). In the study conducted by Lawhavinit and coworkers, the hexane and methanol extracts of this plant also showed antibacterial effect against an array of bacteria including, *Vibrio harveyi, Vibrio alginolyticus, Vibrio vulnificus, Vibrio parahaemolyticus, Vibrio cholerae, B. subtilis, Bacillus cereus, Aeromonas hydrophila, Streptococcus agalactiae,*

S. aureus, Staphylococcus intermedius, Staphylococcus. epidermidis, and *Edwardsiella tarda* with MIC values ranging from 125–1000 µg/mL (Lawhavinit et al., 2010). Curcumin (**1**), the active constituent of *C. longa* exhibited inhibitory activity on methicillin-resistant *S. aureus* strains (MRSA) with MIC values ranging from 125–250 µg/mL (Negi et al., 1999). This compound displayed good antibacterial activity with MIC values ranging from 5 to 50 µg/mL against 65 clinical isolates of *Helicobacter pylori* (De et al., 2009).

4.4.2 Antifungal activity

The methanol extract of *C. longa* showed antifungal activity against *Cryptococcus neoformans* (MIC value of 128 µg/mL) and *Candida albicans* (MIC of 256 µg/mL) (Ungphaiboon et al., 2005). The hexane extract at 1000 µg/mL showed antifungal effect against *Rhizoctonia solani, Phytophthora infestans,* and *Erysiphe graminis* while ethyl acetate extract prevented the growth of *R. solani, P. infestans, Puccinia recondita,* and *Botrytis cinerea* (Kim et al., 2003). The active constituent of the plant was identified as **1**, having inhibitory effects on the growth of *R. solani, P. recondita,* and *P. infestans* (Kim et al., 2003). The methanol extract of *C. longa* harvested in Thailand inhibited the growth of some clinical isolates of dermatophytes (Wuthi-udomlert et al., 2000). Compound **1** was active against a panel of *Candida* including both ATCC strains and clinical isolates with MIC values ranging from 250 to 2000 µg/mL (Neelofar et al., 2011).

4.4.3 Antiviral activity

The aqueous extract of *C. longa* rhizome showed antiviral activity against hepatitis B virus (HBV) in HepG2 cells containing HBV genomes via repression of HBsAg secretion from liver cells, without any cytotoxic effect. The HBV particles production and the rate of mRNA production of HBV on infected cells were also suppressed (Kim et al., 2009). Compound **1** was identified as the antiviral constituent of the plant against hepatitis C virus (HCV), decreasing HCV gene expression and replication through suppression of the Akt-SREBP-1 pathway (Kim et al., 2010).

4.5 Cardiovascular effects

C. longa and its major bioactive constituent, **1**, showed cardioprotective effects. Mohanty et al., 2004 have examined the protective effects of this plant on ischemia-reperfusion-induced myocardial injuries and their mechanisms in Wistar rats for 31 days. On the 31st day, rats of the control ischemia-reperfusion (I/R) and *C. longa* 100 mg/kg treated groups were subjected to 45 min of occlusion of the left anterior descending coronary artery and were thereafter reperfused for 1 h. They found that I/R resulted in significant cardiac necrosis, depression in left ventricular function, decline in antioxidant status, and elevation in lipid perodixation in the control I/R group as compared to control. They also observed that myocardial infarction produced after I/R was significantly reduced in the treated group. Treatment with the plant extract resulted in restoration of the myocardial antioxidant status and altered hemodynamic parameters as compared to control I/R. Furthermore, I/R-induced lipid peroxidation was significantly inhibited by treatment. The beneficial cardioprotective effects also translated into the functional recovery of the heart. Cardioprotective effect of treatment likely results from the suppression of oxidative stress and correlates with the improved ventricular function. In contrast Wattanapitayakul and coworkers demonstrated that *C. longa* does not have cardioprotective effect against doxorubicin-induced cardiotoxicity (Wattanapitayakul et al., 2005). However, one of the active components of turmeric, **1**, has been shown to possess cardiprotective actions, displaying antiplatelet and anticoagulant effects as well as free-radical scavenging activity, particularly against O_2^- radical and depletion of the oxidative stress (Miriyala et al., 2007). It was suggested that the antioxidant activity of compound **1** could be attributed to the phenolic and methoxy groups in conjunction with the 1,3-diketone-conjugated diene system, for scavenging of the oxygen radicals (Miriyala et al., 2007). In addition, **1** is shown to enhance the activities of detoxifying enzymes, such as glutathione-S-transferase in vivo (Miriyala et al., 2007). It was also shown that **1** induces cardioprotective effect, inhibiting oxygen free-radical generation in myocardial ischemia in rats (Miriyala et al., 2007). In vitro and in vivo studies showed a promising role of **1** as a cardioprotective agent against palmitate and high fat diet mediated cardiac dysfunction (Zeng et al., 2015).

4.6 Hepatoprotective effects

Several reports have discussed the hepatopretective role of *C. longa* and compound **1**. The rhizome ethanol extract of the plant was reported to have hepatoprotective effect via its ability to act as an antioxidant and antiinflammatory agent (Salama et al., 2013). Kim and collaborators have investigated the hepatoprotective effect of fermented *C. longa* in rats under carbon tetrachloride (CCl_4)-induced oxidative stress (Kim et al., 2014). They observed that pretreatment with plant extract at a dose of 30 or 300 mg/kg body weight orally administered for 14 days drastically prevented the elevated activities of serum aspartate aminotransferase (AST), alanine aminotransferase (ALT), alkaline phosphatase (ALP), and lactate dehydrogenase (LDH) caused by CCl_4-induced hepatotoxicity. Furthermore, extract from fermented *C. longa* enhanced antioxidant capacities with higher activities of catalase, glutathione-S-transferase, glutathione reductase, and glutathione peroxidase, and level of reduced glutathione. Hence, authors suggested that the extract could be a candidate for the prevention against various liver diseases induced by oxidative stress via elevating antioxidative potentials and decreasing lipid peroxidation. Sengupta et al., 2011 also reported the hepatoprotective and immunomodulatory properties of aqueous extract of rhizomes of this plant in CCl_4-intoxicated Swiss albino mice, indicating its ability to ameliorate hepatotoxicity. Other models of hepatoprotective action of turmeric include amelioration of effect of methanol extract on galactosamine-induced liver injury in mice (Adaramoye et al., 2010) as well as the hepatoprotective effects of ethanol soluble fraction on tacrine-induced cytotoxicity in human liver-derived HepG2 cells (Song et al., 2001). The hepatoprotective effect of compound **1** was evidenced in lipopolysaccharide/D-galactosamine model of liver injury in rats, via decrease in ALT and AST levels as well as in lipid peroxidation (Cerny et al., 2011). Compound **1** also induced recovery from hepatic injury via induction of apoptosis of activated hepatic stellate cells (Priya and Sudhakaran, 2008). This compound attenuated dimethylnitrosamine-induced liver injury in rats through NF-E2-related factor 2 (Nrf2)-mediated induction of heme oxygenase-1 (Farombi et al., 2008).

4.7 Neuroprotective effects

The neuroprotective potential of *C. longa* and its component, **1**, has been largely documented; Issuriya et al., 2014 have demonstrated the neuroprotective effects of the ethanol extract of this plant on neuronal loss induced by dexamethasone treatment in the rat hippocampus. Curcuma oil was also reported to reduce endothelial cell–mediated inflammation in postmyocardial ischemia/reperfusion in rats (Manhas et al., 2014). Curcumin **1**, was shown to mediate neuroprotective effect through attenuation of quinoprotein formation, p-p38 mitogen activated protein kinases (MAPKs) expression, and caspase-3 activation in 6-hydroxydopamine treated SH-SY5Y neuroblastoma cells (Meesarapee et al., 2014). Chronic administration of compound **1** significantly improved memory retention, attenuated oxidative damage, acetylcholinesterase activity, and aluminum concentration in aluminum treated rats indicating that this compound has neuroprotective effects against aluminum-induced cognitive dysfunction and oxidative damage (Kumar et al., 2009). Rajakrishnan et al., 1999 have also investigated the neuroprotective role of compound **1** from *C. longa* on ethanol-induced brain damage. They found that oral administration of this compound to rats caused a significant reversal in lipid peroxidation, brain lipids and produced enhancement of glutathione, a nonenzymic antioxidant in ethanol-intoxicated rats, revealing that the antioxidative and hypolipidemic action of **1** is responsible for its protective role against ethanol-induced brain injury.

5 PATENTS WITH *Curcuma Longa*

Patent US7220438 B2 granted to Almagro et al. (2007); title: Pharmacological activities of *C. longa* extracts. This invention concerns to a topical pharmaceutical composition comprising a water soluble *Curcuma* extract, and suitable excipients for topical administration; the process for obtaining said pharmaceutical compositions; the use of different *Curcuma* extracts as photosensitizing agents for the treatment of proliferative diseases; and the use of *Curcuma* extract or curcuminoids in combination with a radiation therapy for the treatment of proliferative diseases on eukaryote cells (Almagro et al., 2007).

6 CONCLUSIONS

In this chapter, we compiled data related to the medicinal potentials of *C. longa,* commonly known as turmeric, as well as that of its bioactive constituent, curcumin, **1**. Turmeric is used as a worldwide spice but only a few people are aware of its therapeutic properties. It can be regarded as a drug used for the treatment of many diseases, such as cancer, inflammations, microbial infections, and diabetes. Curcumin, **1**, is one of the bioactive principles responsible for the pharmacological activity of *C. longa*. This chapter provides a new impetus to use this plant in the management of various human ailments.

REFERENCES

Adaramoye, O.A., Odunewu, A.O., Farombi, E.O., 2010. Hepatoprotective effect of *Curcuma longa* L. in D-galactosamine induced liver injury in mice: evidence of antioxidant activity. Afr. J. Med. Med. Sci. 39 (Suppl), 27–34.

Ahmad, M., Kamran, S.H., Mobasher, A., 2014. Protective effect of crude *Curcuma longa* and its methanolic extract in alloxanized rabbits. Pak. J. Pharm. Sci. 27 (1), 121–128.

Almagro, E.Q., Bosca, A.R., Bernd, A., Zapata, J.P., Alperi, J.D., Mira, D.P., 2007. Pharmacological activities of *Curcuma longa* extracts: Google Patents.

Bengmark, S., Mesa, M.D., Gil, A., 2009. Plant-derived health-the effects of turmeric and curcuminoids. Nutr. Hosp 24 (3), 273–281.

Bezakova, L., Kostalova, D., Oblozinsky, M., Hoffman, P., Pekarova, M., Kollarova, R., 2014. Inhibition of 12/15 lipoxygenase by curcumin and an extract from *Curcuma longa* L. Ceska Slov. Farm. 63 (1), 26–31.

Cerny, D., Lekic, N., Vanova, K., Muchova, L., Horinek, A., Kmonickova, E., 2011. Hepatoprotective effect of curcumin in lipopolysaccharide/-galactosamine model of liver injury in rats: relationship to HO-1/CO antioxidant system. Fitoterapia 82 (5), 786–791.

Chainani-Wu, N., 2003. Safety and anti-inflammatory activity of curcum: a component of tumeric (*Curcuma longa*). J. Altern. Complement Med. 9 (1), 161–168.

Cheng, T.S., Chen, W.C., Lin, Y.Y., Tsai, C.H., Liao, C.I., Shyu, H.Y., 2013. Curcumin-targeting pericellular serine protease matriptase role in suppression of prostate cancer cell invasion, tumor growth, and metastasis. Cancer Prev. Res. (Phila) 6 (5), 495–505.

Das, K., 2016. Turmeric (*Curcuma longa*) oils. In: Preedy, V.R. (Ed.), Essential Oils in Food Preservation, Flavor and Safety. Academic Press, San Diego, pp. 835–841, (Chapter 95).

De, R., Kundu, P., Swarnakar, S., Ramamurthy, T., Chowdhury, A., Nair, G.B., 2009. Antimicrobial activity of curcumin against *Helicobacter pylori* isolates from India and during infections in mice. Antimicrob. Agents Chemother. 53 (4), 1592–1597.

Fan, H., Tian, W., Ma, X., 2014. Curcumin induces apoptosis of HepG2 cells via inhibiting fatty acid synthase. Target Oncol. 9 (3), 279–286.

Farombi, E.O., Shrotriya, S., Na, H.K., Kim, S.H., Surh, Y.J., 2008. Curcumin attenuates dimethylnitrosamine-induced liver injury in rats through Nrf2-mediated induction of heme oxygenase-1. Food Chem. Toxicol. 46 (4), 1279–1287.

Illuri, R., Bethapudi, B., Anandhakumar, S., Murugan, S., Joseph, J.A., Mundkinajeddu, D., 2015. Anti-inflammatory activity of polysaccharide fraction of *Curcuma longa* (NR-INF-02). Antiinflamm. Antiallergy Agents Med. Chem. 14 (1), 53–62.

Issuriya, A., Kumarnsit, E., Wattanapiromsakul, C., Vongvatcharanon, U., 2014. Histological studies of neuroprotective effects of *Curcuma longa* Linn. on neuronal loss induced by dexamethasone treatment in the rat hippocampus. Acta Histochem. 116 (8), 1443–1453.

Jain, A., Samykutty, A., Jackson, C., Browning, D., Bollag, W.B., Thangaraju, M., 2015. Curcumin inhibits PhIP induced cytotoxicity in breast epithelial cells through multiple molecular targets. Cancer Lett. 365 (1), 122–131.

Jansen, P.C.M., 2005. *Curcuma longa* L. In: Jansen, P.C.M., Cardon, D. (Eds.), PROTA 3: Dyes and Tannins/Colorants et tanins. [CD-Rom]. PROTA, Wageningen.

Khan, I., Ahmad, H., Ahmad, B., 2014. Anti-glycation and anti-oxidation properties of *Capsicum frutescens* and *Curcuma longa* fruits: possible role in prevention of diabetic complication. Pak. J. Pharm. Sci. 27 (5), 1359–1362.

Khanna, M.M., 1999. Turmeric—nature's precious gift. Curr. Sci. 76 (10), 1351–1356.

Kim, M.K., Choi, G.J., Lee, H.S., 2003. Fungicidal property of *Curcuma longa* L. rhizome-derived curcumin against phytopathogenic fungi in a greenhouse. J. Agric. Food Chem. 51 (6), 1578–1581.

Kim, K., Kim, K.H., Kim, H.Y., Cho, H.K., Sakamoto, N., Cheong, J., 2010. Curcumin inhibits hepatitis C virus replication via suppressing the Akt-SREBP-1 pathway. FEBS Lett. 584 (4), 707–712.

Kim, H.J., Yoo, H.S., Kim, J.C., Park, C.S., Choi, M.S., Kim, M., 2009. Antiviral effect of *Curcuma longa* Linn extract against hepatitis B virus replication. J. Ethnopharmacol. 124 (2), 189–196.

Kim, Y., You, Y., Yoon, H.G., Lee, Y.H., Kim, K., Lee, J., 2014. Hepatoprotective effects of fermented *Curcuma longa* L. on carbon tetrachloride-induced oxidative stress in rats. Food Chem. 151, 148–153.

Kumar, A., Dogra, S., Prakash, A., 2009. Protective effect of curcumin (*Curcuma longa*), against aluminium toxicity: Possible behavioral and biochemical alterations in rats. Behav. Brain Res. 205 (2), 384–390.

Kuroda, M., Mimaki, Y., Nishiyama, T., Mae, T., Kishida, H., Tsukagawa, M., 2005. Hypoglycemic effects of turmeric (*Curcuma longa* L. rhizomes) on genetically diabetic KK-Ay mice. Biol. Pharm. Bull. 28 (5), 937–939.

Kuttan, R., Bhanumathy, P., Nirmala, K., George, M.C., 1985. Potential anticancer activity of turmeric (*Curcuma longa*). Cancer Lett. 29 (2), 197–202.

Lal, J., 2012. Turmeric, curcumin and our life: a review. Bull. Environ. Pharmacol. Life Sci. 1 (7), 11–17.

Lawhavinit, O.A., Kongkathip, N., Kongkathip, B., 2010. Antimicrobial activity of curcuminoids from *Curcuma longa* L. on pathogenic bacteria of shrimp and chicken. Kasetsart J. Nat. Sci. 44 (3), 364–371.

Lekshmi, P.C., Arimboor, R., Indulekha, P.S., Menon, A.N., 2012. Turmeric (*Curcuma longa* L.) volatile oil inhibits key enzymes linked to type 2 diabetes. Int. J. Food Sci. Nutr. 63 (7), 832–834.

Lekshmi, P.C., Arimboor, R., Nisha, V.M., Menon, A.N., Raghu, K.G., 2014. In vitro anti-diabetic and inhibitory potential of turmeric (*Curcuma longa* L) rhizome against cellular and LDL oxidation and angiotensin converting enzyme. J. Food Sci. Technol. 51 (12), 3910–3917.

Liu, D., Chen, Z., 2013. The effect of curcumin on breast cancer cells. J. Breast Cancer 16 (2), 133–137.

Lu, W.D., Qin, Y., Yang, C., Li, L., Fu, Z.X., 2013. Effect of curcumin on human colon cancer multidrug resistance in vitro and in vivo. Clinics (Sao Paulo) 68 (5), 694–701.

Manhas, A., Khanna, V., Prakash, P., Goyal, D., Malasoni, R., Naqvi, A., 2014. Curcuma oil reduces endothelial cell-mediated inflammation in postmyocardial ischemia/reperfusion in rats. J. Cardiovasc. Pharmacol. 64 (3), 228–236.

Meesarapee, B., Thampithak, A., Jaisin, Y., Sanvarinda, P., Suksamrarn, A., Tuchinda, P., 2014. Curcumin I mediates neuroprotective effect through attenuation of quinoprotein formation, p-p38 MAPK expression, and caspase-3 activation in 6-hydroxydopamine treated SH-SY5Y cells. Phytother. Res. 28 (4), 611–616.

Miriyala, S., Panchatcharam, M., Rengarajulu, P., 2007. Cardioprotective effects of curcumin. Adv. Exp. Med. Biol. 595, 359–377.

Moghadamtousi, S.Z., Kadir, H.A., Hassandarvish, P., Tajik, H., Abubakar, S., Zandi, K., 2014. A review on antibacterial, antiviral, and antifungal activity of curcumin. Biomed. Res. Int. 2014, 186864.

Mohankumar, S., McFarlane, J.R., 2011. An aqueous extract of *Curcuma longa* (turmeric) rhizomes stimulates insulin release and mimics insulin action on tissues involved in glucose homeostasis in vitro. Phytother. Res. 25 (3), 396–401.

Mohanty, I., Singh Arya, D., Dinda, A., Joshi, S., Talwar, K.K., Gupta, S.K., 2004. Protective effects of *Curcuma longa* on ischemia-reperfusion induced myocardial injuries and their mechanisms. Life Sci. 75 (14), 1701–1711.

Neelofar, K., Shreaz, S., Rimple, B., Muralidhar, S., Nikhat, M., Khan, L.A., 2011. Curcumin as a promising anticandidal of clinical interest. Can. J. Microbiol. 57 (3), 204–210.

Negi, P.S., Jayaprakasha, G.K., Jagan Mohan Rao, L., Sakariah, K.K., 1999. Antibacterial activity of turmeric oil: a byproduct from curcumin manufacture. J. Agric. Food Chem. 47 (10), 4297–4300.

Niamsa, N., Sittiwet, C., 2009. Antimicrobial activity of *Curcuma longa* aqueous extract. J. Pharmacol. Toxicol. 4 (4), 173–177.

Nishiyama, T., Mae, T., Kishida, H., Tsukagawa, M., Mimaki, Y., Kuroda, M., 2005. Curcuminoids and sesquiterpenoids in turmeric (*Curcuma longa* L.) suppress an increase in blood glucose level in type 2 diabetic KK-Ay mice. J. Agric. Food Chem. 53 (4), 959–963.

Priya, S., Sudhakaran, P.R., 2008. Curcumin-induced recovery from hepatic injury involves induction of apoptosis of activated hepatic stellate cells. Indian J. Biochem. Biophys. 45 (5), 317–325.

Rai, P.K., Jaiswal, D., Mehta, S., Rai, D.K., Sharma, B., Watal, G., 2010. Effect of *Curcuma longa* freeze dried rhizome powder with milk in STZ induced diabetic rats. Indian J. Clin. Biochem. 25 (2), 175–181.

Rajakrishnan, V., Viswanathan, P., Rajasekharan, K.N., Menon, V.P., 1999. Neuroprotective role of curcumin from *Curcuma longa* on ethanol-induced brain damage. Phytother. Res. 13 (7), 571–574.

Ravindran, P.N., 2007. Turmeric: the Genus Curcuma. CRC Press, London.

Ravindran, J., Subbaraju, G.V., Ramani, M.V., Sung, B., Aggarwal, B.B., 2010. Bisdemethylcurcumin and structurally related hispolon analogues of curcumin exhibit enhanced prooxidant, anti-proliferative and anti-inflammatory activities in vitro. Biochem. Pharmacol. 79 (11), 1658–1666.

Salama, S.M., Abdulla, M.A., AlRashdi, A.S., Ismail, S., Alkiyumi, S.S., Golbabapour, S., 2013. Hepatoprotective effect of ethanolic extract of *Curcuma longa* on thioacetamide induced liver cirrhosis in rats. BMC Complement Altern. Med. 13, 56.

Sambhav, J., Rohit, R., Ankit Raj, U., Garima, M., 2014. *Curcuma longa* in the management of inflammatory diseases: a review. Int. Ayur. Med. J. 2 (2), 33–40.

Sengupta, M., Sharma, G.D., Chakraborty, B., 2011. Hepatoprotective and immunomodulatory properties of aqueous extract of *Curcuma longa* in carbon tetra chloride intoxicated Swiss albino mice. Asian Pac. J. Trop. Biomed. 1 (3), 193–199.

Shehzad, A., Lee, J., Huh, T.L., Lee, Y.S., 2013. Curcumin induces apoptosis in human colorectal carcinoma (HCT-15) cells by regulating expression of Prp4 and p53. Mol. Cells 35 (6), 526–532.

Shishodia, S., Chaturvedi, M.M., Aggarwal, B.B., 2007. Role of curcumin in cancer therapy. Curr. Prob. Cancer 31 (4), 243–305.

Singh, R., Sharma, P., 2011. Hepatoprotective effect of curcumin on lindane-induced oxidative stress in male wistar rats. Toxicol. Int. 18 (2), 124.

Song, E.K., Cho, H., Kim, J.S., Kim, N.Y., An, N.H., Kim, J.A., 2001. Diarylheptanoids with free radical scavenging and hepatoprotective activity in vitro from *Curcuma longa*. Planta Med. 67 (9), 876–877.

Sun, X.D., Liu, X.E., Huang, D.S., 2012. Curcumin induces apoptosis of triple-negative breast cancer cells by inhibition of EGFR expression. Mol. Med. Rep. 6 (6), 1267–1270.

Taty Anna, K., Elvy Suhana, M.R., Das, S., Faizah, O., Hamzaini, A.H., 2011. Anti-inflammatory effect of *Curcuma longa* (turmeric) on collagen-induced arthritis: an anatomico-radiological study. Clin. Ter. 162 (3), 201–207.

Ungphaiboon, S., Supavita, T., Singchangchai, P., Sungkarak, S., Rattanasuwan, P., Itharat, A., 2005. Study on antioxidant and antimicrobial activities of turmeric clear liquid soap for wound treatment of HIV patients. Songklanakarin J. Sci. Technol. 27 (2), 269–578.

Wattanapitayakul, S.K., Chularojmontri, L., Herunsalee, A., Charuchongkolwongse, S., Niumsakul, S., Bauer, J.A., 2005. Screening of antioxidants from medicinal plants for cardioprotective effect against doxorubicin toxicity. Basic Clin. Pharmacol. Toxicol. 96 (1), 80–87.

Wei, X., Zhou, D., Wang, H., Ding, N., Cui, X.X., Verano, M., 2013. Effects of pyridine analogs of curcumin on growth, apoptosis and NF-kappaB activity in prostate cancer PC-3 cells. Anticancer Res. 33 (4), 1343–1350.

Wilken, R., Veena, M.S., Wang, M.B., Srivatsan, E.S., 2011. Curcumin: a review of anti-cancer properties and therapeutic activity in head and neck squamous cell carcinoma. Mol. Cancer 10, 12.

Wuthi-udomlert, M., Grisanapan, W., Luanratana, O., Caichompoo, W., 2000. Antifungal activity of *Curcuma longa* grown in Thailand. Southeast Asian J. Trop. Med. Public Health 31 (Suppl 1), 178–182.

Zeng, C., Zhong, P., Zhao, Y., Kanchana, K., Zhang, Y., Khan, Z.A., 2015. Curcumin protects hearts from FFA-induced injury by activating Nrf2 and inactivating NF-kappaB both in vitro and in vivo. J. Mol. Cell Cardiol. 79, 1–12.

CHAPTER 20

Lactuca sativa

J.A.K. Noumedem, D.E. Djeussi, L. Hritcu, M. Mihasan, V. Kuete

1 INTRODUCTION

Lactuca sativa (Lettuce) is the most important crop in the group of leafy vegetables. It belongs to the family of Asteraceae which comprises between 23,000 and 30,000 species and is thought to be the largest family of plants (Bayer and Starr, 1998). Cultivated lettuce is believed to have been domesticated in the Mediterranean region from the wild species *Lactuca serriola* L. with Southwest Asia apparently being the center of origin. The market demand of lettuce was first fulfiled by the Europe and North America. Nowadays the consumption of lettuce has spread tremendously throughout the world due to its high nutritive value and medicinal importance. The species *L. sativa* is characterized by a high genetic diversity resulting from its polyphyletic origin and a complex domestication process (Kesseli et al., 1991). A survey of lettuce cultivars and classification of types was provided by Rodenburg (1960). More comprehensive overviews of taxonomic and phenotypic analyses of lettuce cultivars were later presented (DeVries and Van Raamsdonk, 1994; DeVries, 1997; Mou, 2008). The crop comprises seven main groups of cultivars usually described as morphotypes. They are classified as butterhead lettuce (var. capitata *L. nidus* tenerrima Helm) (Kopfsalat, Laitue pommé), crisphead lettuce (var. capitata L. nidus jäggeri Helm) (Iceberg type, Eissalat, Batavia), cos lettuce (var. longifolia Lam., var. romana Hort. in Bailey) (Römischer Salat, Laitue romaine), cutting lettuce (var. acephala Alef., syn. var. secalina Alef., syn. var. crispa L.) (Gathering lettuce, Loose-leaf, Picking lettuce, Schnittsalat, Laitue à couper), stalk (Asparagus) lettuce (var. angustana Irish ex Bremer, syn. var. asparagina Bailey, syn. L. angustana Hort. in Vilm.) (Stem lettuce, Stengelsalat, Laitue-tige), Latin lettuce (without scientific name) (Lebeda et al., 2007). Many studies have focused on domesticated lettuce (Wei et al., 2015). Lettuce is a source of the quasi-totality of vitamins and contains many minerals including calcium, phosphorus, iodine, iron, copper, and arsenic. Due to its high content of vitamin C, lettuce is known for its resistance to infections and its ability to fight anemia. The phytochemicals of the plant *L. sativa* belong mainly to secondary metabolites that are synthesized during the normal growth of plants or in response to a number of environmental conditions. The plants have been used in traditional medicine since many decades for many ailments including inflammation, pain, stomach problems including indigestion and lack of appetite, bronchitis, and urinary tract infections (Ismail and Mirza, 2015). Different reports documented the scientific evidence of its biological activities including antimicrobial, antioxidant, and neuroprotective (Nazri et al., 2011; Noumedem et al., 2013; Moharib et al., 2014; Ismail and Mirza, 2015).

2 HISTORY

Ancient Egypt is known as the place where *L. sativa* was first cultivated with evidence of its cultivation appearing as early as 2680 BC. The Egyptians cultivated the plant for the production of seed oil. It is thought that the Egyptians selectively bred it into a plant grown for its edible leaves. The lettuce was seen as a sacred plant of the reproduction God Min, and it was carried during his festivals and placed near his images. The plant was thought to help the God perform the sexual act untiringly (Hart, 2005). Its use in religious ceremonies resulted in the creation of many images in tombs and wall paintings. The cultivated variety appears to have been about 30 inches (76 cm) tall and resembled a large version of the modern romaine lettuce. These lettuces developed by the Egyptians were further introduced in the Greece and then in Roma. The plant was further introduced in Americas by Christopher Columbus in the late 15th century (Subbarao and Koike, 2007). Many medieval authors described *L. sativa* and its uses especially as a medicinal herb. These included Hildegard of Bingen (1098 and 1179) and Joachim Camerarius (1586) who described three basic modern lettuces including head lettuce, loose-leaf lettuce, and romaine or cos lettuce (Bennett and Hollister, 2001). Between the late 16th century and the early 18th century, many varieties were developed in Europe.

3 WORLD AND AFRICAN DISTRIBUTION

China is the first world producer of lettuce with 13,500,000 tons in 2014. It produces almost four times the lettuce produced by USA which is classified as the second most important producer in the world. They are followed in this classification by India. Spain, ranked as the fourth world main producer, appeared as the European country which produces the highest amount of lettuce with 904,300 metric tons in 2014. Italy, Iran, Japan, Turkey, Mexico, and Germany complete the list of the top ten world lettuce producing countries in 2013 as shown in Fig. 20.1A (FAO, 2013). Fourteen African countries are found among the top 100 world lettuce producing countries (Fig. 20.1B) (FAO, 2013). The top 10 world (Fig. 20.2A) and African (Fig. 20.2B) lettuce cultivating countries and area harvested in 2013 are also summarized in FAO (2013).

4 CHEMISTRY

Many studies have reported the chemical composition of *L. sativa*. One hundred grams of the leaves of the plants are composed of 96% water, 1.4 g protein, 2.2 g carbohydrate, 1.1 g dietary fiber, 0.2 g fat, and 1.2 g ash. Lettuce is a rich source of almost all vitamins and is particularly rich in vitamins A, C, K, niacin, and folate. It also contains many

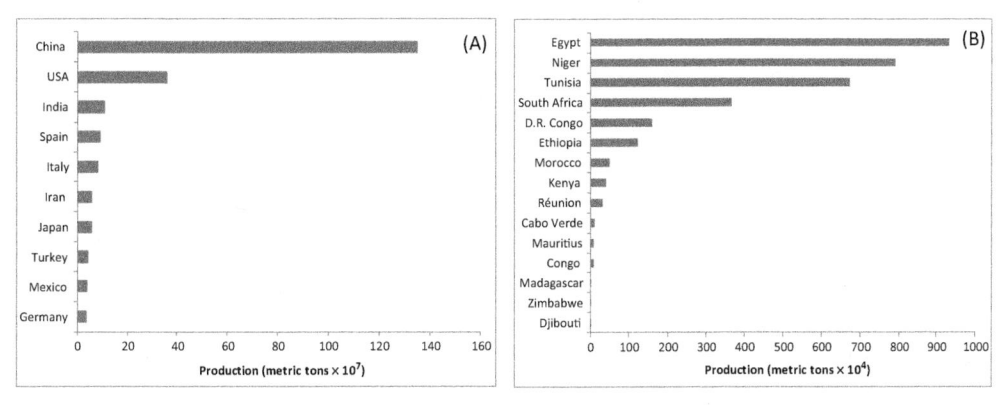

Figure 20.1 Top world (A) and African (B) lettuce producing countries in 2013 (FAO, 2013).

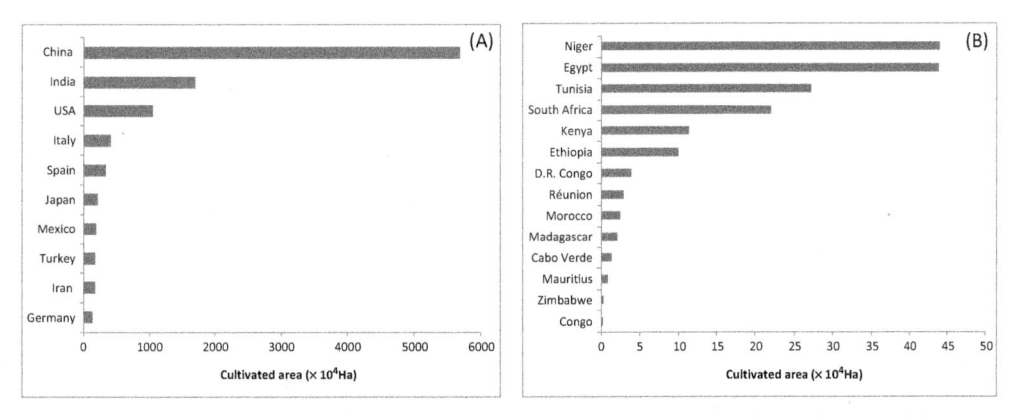

Figure 20.2 Top world (A) and African (B) lettuce cultivating countries and area harvested for 2013 (FAO, 2013).

minerals, such as calcium, iron, phosphorous, sodium, and potassium (USDA Nutrient Database, 2016). Several studies focused on the secondary metabolites content of *L. sativa*, leading to the identification of many compounds (Fig. 20.3) which justify the variety of the biological activities attributed to the plant.

A comparative high performance liquid chromatography (HPLC) study of 11 *Lactuca* species, based on distributional data of 8 sesquiterpene lactones, revealed that 5 species of the subsection *Lactuca*, that is, *Lactuca dregeana*, *L. sativa*, *L. serriola* f. serriola, *Lactuca saligna* L., and *Lactuca virosa* L. var. virosa, contained the germacranolide lactuside A (**1**) and the guaianolides lactucin (**2**), and lactucopicrin (**3**) as major constituents (Michalska et al. 2009). In *L. sativa* extracts the major components include 15-oxalyl and 8-sulfate conjugates of guaianolides: compounds **2, 3**, and deoxylactucin (Harsha and Anilakumar, 2013a). Quantitative studies showed that in *L. sativa* roots, the amount of 8-deoxylactucin and crepidiaside B (**4**) was less than 0.025 mg/g of the dry plant material, whereas the amounts of **1, 2, 3**, and dihydrolactucin were between 0.025 and 0.250 mg/g. In leaves,

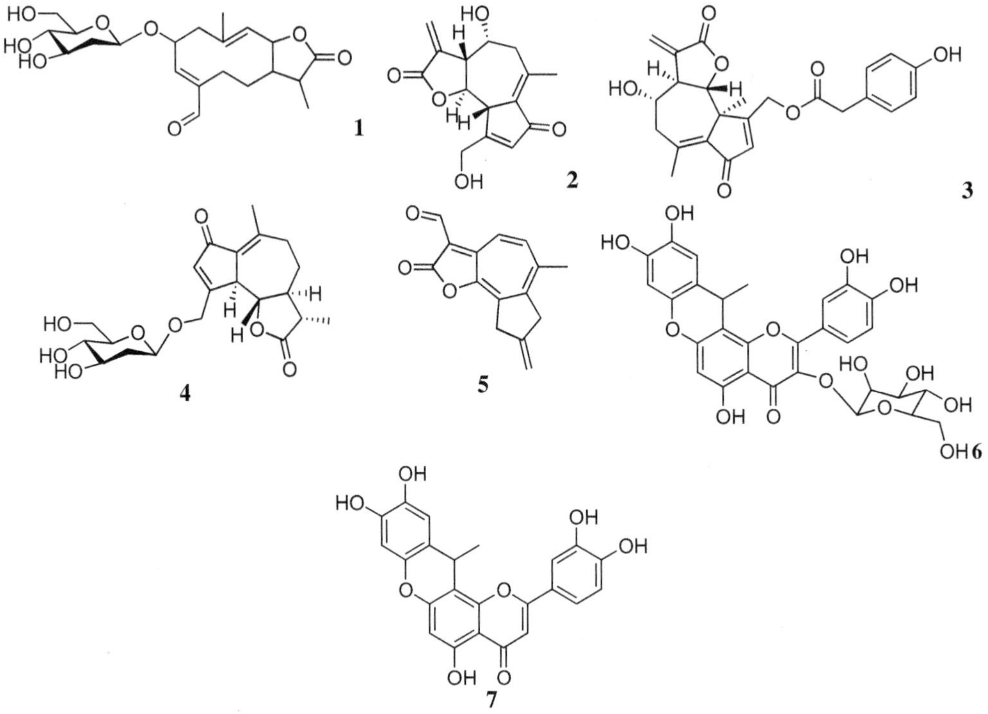

Figure 20.3 *Chemical structures of selected compound isolated in* **L. sativa. 1:** lactuside A; **2:** lactucin; **3:** lactucopicrin; **4:** crepidiaside B; **5:** lettucenin A; **6:** lactucasativoside A; **7:** japonicin A.

4, 8-deoxylactucin, and dihydrolactucin concentrations are less than 0.025 mg/g of the dry plant material, while **2** was in concentrations ranging from 0.025 to 0.250 mg/g. Compounds **1** and **3** were not detected in leaves (Michalska et al., 2009). It was shown that when leaves of *L. sativa* are innoculated with the pathogenic bacterium *Pseudomonas cichorii*, the plant produces lettucenin A (**5**) and another phytoalexin with antifungal properties known as costunolide (Takasugi et al., 1985). *L. sativa* also contains eudesmane sesquiterpene lactones. The first isolation of two of them was carried out for the first time from the *n*-butanol fraction of the methanol extract of stalk of *L. sativa* var. anagustata collected in Zangzhou city (Zhejiang province of China). These compounds were identified as 1b-O-β-D-glucopyranosyl-4a-hydroxyl-5a,6b,11bH-eudesma-12,6a-olide and 1b-hydroxyl-15-O-(*p*-methoxyphenylacetyl)-5a,6b, 11bH-eudesma-3-en-12,6a-olide (Han et al., 2009).

The chemical constituents of the seeds of *L. sativa* (collected at the Miquan Vegetable Planting Base in Xinjiang, China in 2008) were investigated using 95 and 50% ethanol extracts. This led to the identification of four compounds including a new flavonol glycoside with a rare structure type, lactucasativoside A (**6**) and three known compounds, japonicin A (**7**), isoquercitrin, and caffeic acid (Xu et al., 2012).

Chemical investigation of the methanol extract of *L. sativa* L var. anagustata led to the isolation of 11 compounds belonging to sesquiterpenes among which three were new sesquiterpenes. The eight known compounds were identified as 9b-hydroxyl-4b,11b,13,15-tetrahydrozaluzanin C, 10b,14-dihydroxyl-11bH-guaiane4(15)-ene-12,6a-olide, 1b-hydroxyl-5a,6bH-eudesman-3-ene-12,6a-olide, macrocliniside A, 11b,13-dihydrolactucin, cichorioside B, 11b,13-dihydrolactucopicrin and 10b,14-dihydroxy-10(14),11b(13)-tetrahydro-8,9-didehydro-3-deoxyzaluzanin C-10-O- β-glucopyranoside. The three new compounds were identified as 1b-O-β-D-glucopyranosyl-4ahydroxyl-5a,6b,11bH-eudesma-12,6a-olide; 1b-hydroxyl-15-O-(p-methoxyphenylacetyl)-5 a,6b,11bH-eudesma-3-en-12,6a-olide and 4a-O-β-D--glucopyranosyl15-hydroxyl-5 a,6bH-guaiane-10(14), 11(13)-dien-12,6a-olide.

Another study reported different chemicals in *L. sativa* from the MeOH/AcOH/H_2O (25:4:21) extract. These compounds included phenolic acids: caffeoyltartaric acid, chlorogenic acid, dicaffeoyltartaric acid, dicaffeoylquinic acid; flavonoids: quercetin 3-glucuronide, quercetin 3-glucoside, quercetin 3-(6-malonylglucoside) and quercetin 3-(6-malonylglucoside) 7-glucoside (identified as a new compound); and one anthocyanin namely cyanidin 3-malonylglucoside (Ferreres et al., 1997). Phenolic compounds were also isolated from *L. sativa* by Mai and Glomb (Mai and Glomb, 2013). They included syringin which was identified as a new compound and ten known compounds: caffeic acid, chlorogenic acid, phaseolic acid, chicoric acid, isochlorogenic acid, luteolin-7-O-glucuronide, quercetin-3-O-glucuronide, quercetin-3-O-galactoside, quercetin-3-O-glucoside, and quercetin-3-O-(6″-malonyl)-glucoside (Mai and Glomb, 2013).

The analysis of the essential oils and volatile compounds from *L. sativa* showed that it contains different compounds whose proportions vary according to the state of the plant (fresh or dry). Compounds identified in these essential oils included α-pinene, γ-cymene, thymol, durenol, α-terpinene, thymol acetate, caryophyllene, spathulenol, camphene, limonene representing the constituents. The minor constituents were β-pinene, α-terpinolene, linalool, 4-terpineol, α-terpineol, O-methylthymol, *L*-alloaromadendrene, and viridiflorene. These findings suggested that preparations from lettuce could be used as candidates in the search of antioxidant, antifungal, or antimicrobial agents (Nomaani et al., 2013). A diterpenoid, phytol was also isolated from *L. sativa* (Bang et al., 2002). Polysaccharides with cytotoxic activities were also isolated from *L. sativa* (Moharib et al., 2014), as well as glutathione S-transferase (GST) having acting on 1-chloro-2,4-dinitrobenzene (a general GST substrate), cumene hydroperoxide, and ethacrynic acid (Park et al., 2005).

5 PHARMACOLOGICAL ACTIVITIES

5.1 Antioxidant effects

Evidence of antioxydant activity of *L. sativa* has been provided in vitro and in vivo using different experimental models, such as the measurement of total phenolic compounds, radical scavenging activity, as well as the effect on the level of endogenous antioxydant enzymes. We demonstrated that methanol extract of *L. sativa* possesses total reducing power capacity of 6.37 mg equivalent of ascorbic acid per gram of extract, a total flavonoid content of 5.59 µg equivalent of gallic acid per gram of dry extract, and a total phenol content 6.82 equivalent of rutin acid per gram of dry extract (personal unpublished data). A phenol content (mg/100 g dry) of 168.56 mg equivalent of /100 g gallic acid per dry of weight was also reported (Souri et al., 2008). The radical scavenging activity of *L. sativa* extract was studied against 2,2-diphenyl-1-picrylhydrazyl (DPPH) radical. The first study showed that ethanol and dichloromethane extracts of *L. sativa* possess moderate scavenging activity against DPPH free radical with percentage of inhibition of 40–55% at a concentration of 1200 ppm (Nazri et al., 2011), while the second study used different organic extracts including methanol, ethyl acetate, hexane, chloroform, and butanol. It showed that all the extracts exert concentration-depending free-radical scavenging activity and methanol extract which emerged as the most active scavenge up to 81.90% of DPPH free radical at 200 µg/mL (Nomaani et al., 2013). Other effects observed in vitro include the antioxidant activity of methanol extract of the plant against linoleic acid peroxidation with an IC_{50} value of 14.28 µg/mL (Souri et al., 2008) and the inhibition of oxidative damage induced by UV radiations in *Salmonella typhi* (Garg et al., 2004).

The antioxidant activities of *L. sativa* were also demonstrated in vivo in rat models. Methanol extract of leaves increases in vivo the level of two antioxidant enzymes, catalase and superoxide dismutase (SOD) in blood and brain of male albino wistar rats at 400 mg/kg body weight (b.w). It also reduces malondialdehyde (MDA) formation in blood and brain of rats (Garg et al., 2004). Ethanol extract had protective effect against D-galactose induced oxidative stress in testes and epididymis by preventing excessive phospholipid peroxydation in the testes and epididymis of D-galactose stressed rats at 40 mg/kg b.w. (Patil et al., 2009).

Weekly administration of lettuce ethanol extract during 10 weeks exerted hepatoprotective effects against carbon tetrachloride (CCl_4)-induced oxidative damage by decreasing the thiobarbituric acid reactive substances and nitrite, and increasing the contents of glutathione (GSH), catalase, peroxyde dismutase, superoxyde dismutase, and gluthatione peroxydase. It also increased the synthesis of testosterone, luteinizing hormone, follicle stimulating hormone, estradiol, prolactin, histology, body weight, and relative testis weight (Hefnawy and Ramadan, 2013). The antioxidant properties of lettuce and similar plants have been attributed to their polyphenol contents as reported for five genotypes of lettuce belonging to *L. sativa*, *Cicorium intybus*, *Plantago coronopus*, *Eruca sativa*, and *Diplotaxis tenuifolia*. Quercetin, kaempferol, luteolin, apigenin, and crysoeriol have been identified in the investigated cultivars. Another antioxidant compound detected in lettuce is ascorbic acid which also contributes to the total antioxidant capacity of leafy vegetables (Altunkaya et al., 2009). Moreover, tannins, amino acids, saponins, terpenoids, and steroids have been detected, but not alkaloids (Harsha and Anilakumar, 2013b; Harsha et al., 2013).

All these antioxidant parameters may vary according to different parameters, as it was shown that lettuce cells undergoing an hypersensitive reaction experience a prolonged oxidative stress, primarily through an increase in prooxidant activities initially occurring in the absence of enhanced antioxidant activities (Bestwick et al., 2001). After characterization of enzymes regulating aspects of reactive oxygen metabolism in expanded lettuce leaf tissue in different conditions, catalase activity was optimal between pH 7.0 and 8.0, predominantly located within peroxisomes. Of three principal SOD isoenzymes detected, CuZn SOD represented 51.7%, Mn SOD: 14.6%, and a putative Fe SOD: 33.7% of total activity. Lipoxygenase activity was optimal at pH 5. Inoculation of *L. sativa* with the wild-type *Pseudomonas syringae* pv. *phaseolicola* and a related *hrpD* mutant as well as water infiltration induced quite different effects with hypersensitive reaction and increase in lipid peroxidation, lipoxygenase, and catalase activities (Bestwick et al., 2001). It was also demonstrated that among 25 cultivars of lettuce tested, leaf lettuce had the highest total phenolic content (TPC) and highest DPPH scavenging ability, followed by romaine, butterhead, and Batavia (crisphead subtype); it was also shown that within a lettuce type, red pigmented lettuce cultivars had higher TPC [up to 85.7 gallic acid equivalent (GAE) mg/g] and antioxidant capacity than green cultivars grown under similar conditions. In addition, lettuce harvested in July possessed higher TPC and antioxidant capacity than that harvested in September, suggesting that environmental conditions could influence its phenolic content and antioxidant activity (Liu et al., 2007).

5.2 Antimicrobial activities

Many studies have demonstrated the antimicrobial activities of *L. sativa*. A decoction of leaves was used in the green synthesis of nanoparticles which avoid toxic, flammable, and unstable properties of chemical synthesis protocol of nanoparticles by exploiting the reduction capabilities of phytochemicals present in the plant, such as silver salts. The resulting silver nanoparticles were evaluated for their antimicrobial activity against *Bacillus subtilis*, *Staphylococcus aureus* ATCC 6538P, *Pseudomonas aeruginosa* ATCC 9027, and clinical isolate of *Klebsiella pneumoniae*. The antibacterial activities of these nanoparticles synthesized using *L. sativa* extract were observed on all these bacteria and the diameters of inhibition zones (IZ) obtained varied between 10 and 12 mm. The antibacterial activity of methanol extract of *L. sativa* has also been shown against clinical isolates of *S. aureus*, *Streptococcus pyogenes*, *B. subtillis*, *Escherichia coli*, and *P. aeruginosa* (IZ: 11.50–14 mm); aqueous extract was active against *B. subtillis* and *E. coli* while the juice extract was active against *S. aureus*, *S. pyogenes*, and *E. coli*. Dichloromethane extract of the plant exerted weak inhibition effects against *P. aeruginosa* and *S. aureus* while ethanol extract had weak inhibitory effect on *S. pyogenes* and no effect on *E. coli* and *Candida albicans* (Nazri et al., 2011). The antibacterial activity of methanol and aqueous extracts of *L. sativa* were evaluated in another study against Gram-negative and Gram-positive bacteria using microdilution assay. The Gram-negative bacteria included one strain of *E. coli* ATCC25922 and clinical isolates of *E. coli*, *K. pneumoniae*, *Enterobacter cloacae*, *Serratia marcescens*, and *Acinetobacter baumannii* while Gram-positive bacteria included *B. subtilis* ATCC 6633, *Enterococcus faecalis* ATCC 29212, and clinical isolates of *S. aureus*, *Enterococcus faecium*, and *Corynebacterium* spp. This study demonstrated that methanol extract had better antibacterial activities with a minimal inhibitory concentration (MIC) of 2.5 mg/mL against all the tested bacteria, while aqueous had MIC values of 5 mg/mL against all the Gram-negative bacteria and 2.5 mg/mL against Gram-positive bacteria (Edziri et al., 2011). The antibacterial activities of the methanol extract of *L. sativa* were also investigated on a panel of 29 Gram-negative bacteria using microdilution. The bacteria included both reference (from the American Type Culture Collection) and clinical multidrug resistant strains of *Providencia stuartii*, *P. aeruginosa*, *K. pneumoniae*, *E. coli*, *E. aerogenes*, and *E. cloacae*. This study revealed a better antibacterial activity of *L. sativa* with MIC of 256–1024 µg/mL against 28 of the 29 bacteria. These results demonstrated that *L. sativa* could be used as a candidate in search of new solutions against antibiotics resistance of Gram-negative bacteria (Noumedem et al., 2013).

The antifungal effect of latex saps from *L. sativa* was shown against *Candida albicans*. The study of its mechanism of action through scanning and transmission electron microscopy observations showed that active compounds of the latex saps deform and empty the cytoplasmic content of yeasts, and act on their cell wall (Moulin-Traffort et al., 1990). It has also been demonstrated that after inoculation with *P. cichorii*, *L. sativa* produces compound **5** and a phytoalexin called costunolide. Compound **5** and costunolide inhibited the germination of the spore of *Ceratocystis fimbriata* at the concentrations of 2 and 32 pg/mL, respectively (Takasugi et al., 1985).

The antiviral activities of the methanol and aqueous extracts of *L. sativa* were also evaluated against human cytomegalovirus (HCMV) strain AD-169 (ATCC Ref.VR 538) and coxsackie virus type B3 by analyzing the cytopathic effect (CPE) on the virus-infected confluent monolayer of MRC-5 cells: extracts were not toxic against MRC-5 cells ($IC_{50} > 500$ g/mL) while they had antiviral activities against coxsackie B3 and HCMV with IC_{50} of 200 g/mL (Edziri et al., 2011).

5.3 Effect on neurodegenerative disorders

Several animal models of anxiety (elevated plus maze test, open field test, Marble burying test) were used to investigate *L. sativa* extracts (hydro-alcoholic), and the results indicated that various classes of substances, and mostly polyphenols with concerted action and synergisms, play a significant role in the anxiolytic effects; in the elevated plus maze test, administration of the extract at 100, 200, and 400 mg/kg b.w., orally, for 15 or 30 days period, but especially 400 mg/kg, significantly improved time spent in the open arms and the number to entries to opens arms in mice. The dose of 400 mg/kg b.w. exhibited a near to equivalent anxiolytic-like behavior closely to diazepam. This effect has been attributed to the presence of polyphenols that may possess anxiolytic activity (Harsha et al., 2013). Additionally, it has been reported that the same doses of extract administrated in mice for 7 days increased locomotor activity in the open field test and the time spent in the open arms within elevated plus-maze test, and decrease neophobia within marble burying test, suggesting strong anxiolytic profile. These results are related to decreased levels of MDA and nitrites, along with increased catalase and glutathione levels in mice treated with *L. sativa* extracts. The obtained data suggested that the extract can afford significant protection against anxiolytic activity (Harsha et al., 2013).

In our laboratory (at the Department of Biology, Alexandru Ioan Cuza University), we administered the methanol extract from *L. sativa* leaves in doses of 100 and 200 mg/kg b.w. for 7 days in amyloid beta (1-42)-induced Alzheimer's disease rat model. The amyloid-treated rats given extract significantly exhibited an improvement of spatial memory within Y-maze and radial arm-maze tests along with an anxiolytic- and antidepressant-like behaviors within elevated plus maze and forced swimming tests (personal data). The data suggested that our high containing polyphenols extract could effectively reverse memory impairment and improve neurological deficits in an amyloid beta (1-42)-induced Alzheimer's disease rat model.

In vitro studies suggested that polyphenol-rich extracts from *L. sativa* (250–2.5 µg/mL) were able to reduce both the inflammatory and oxidative stresses in lipopolysaccharides (LPS)-stimulated J774A.1 murine monocyte macrophage cells, by lowering the release of nitric oxide (NO) and reactive oxygen species (ROS), and promoting nuclear translocation of nuclear factor (erythroid-derived 2)-like 2; (Nrf2) and nuclear factor-κB (NF-κB). These data demonstrated the enhanced antiinflammatory and antioxidant activities of the *L. sativa* extract with high amounts of polyphenols. It was also shown in vitro that ethyl acetate fraction of *L. sativa* exerts neuroprotection effect and has the potential to be used as a new therapeutic strategy for common neurodegenerative disorders as it prevents toxic effect of glucose/serum deprivation on rat pheochromocytoma-derived cell line PC12 and mouse neuroblastoma-like cell line N2a (two neuronlike cells) (Sadeghnia et al., 2012).

5.4 Anticancer activity

It has been shown that aqueous extract of *L. sativa* inhibits the growth of HL-60 leukemia cells and MCF-7 breast cancer cells by activating checkpoint kinase 2, inducting the tumor suppressor p21, and downregulating the proto-oncogene cyclin D1; and the ethyl acetate extract inhibited HL-60 cell death by acetylating alpha-tubulin (Gridling et al., 2010). Polysacharides were isolated from the plant and paper chromatographic analysis showed that it is composed of glucose (34.18%), galactose (26.53%), mannose (24.4%), arabinose (10.52%), xylose (1.16%), and rhamnose (2.84%) (Moharib et al., 2014). In vitro evaluation of the cytotoxicity using sulforhodamine B of this polysaccharide was performed toward colon (HCT 116), liver (HEPG2), cervical (Hela), and breast (MCF7) carcinoma cell lines. The results showed that the polysacharide at

5–50 μg/mL inhibits in a concentration-dependant manner the growth of these cancer cells with more pronounced effect on HCT-116 (IC_{50} of almost 4 μg/mL) (Moharib et al., 2014).

5.5 Hypnoptic effects

It has been shown that the hydro-alcoholic extract of *L. sativa* prolonged the pentobarbital-induced sleep duration at 400 mg/kg b.w. The *n*-butanol fraction of this extract increased also the sleep duration and decreased sleep latency at levels comparable to diazepam. It was shown that this extract and its fraction are devoid of neurotoxic effect suggesting that lettuce potentiates pentobarbital hypnosis without major toxic effect, and that compounds responsible for this effect are most likely to be nonpolar agents found in *n*-butanol fraction (Ghorbani et al., 2013). Phytol, a diterpenoid isolated from the ethanolic fraction of lettuce exerted sedative effect on the body by raising the levels of gamma-amino butyric acid (GABA) in the central nervous system (CNS), then inhibiting the action of one of the enzymes responsible for degradation of this neurotransmitter, succinic semialdehyde dehydrogenase (SSADH) (Bang et al., 2002). Extracts of leaves and seeds exhibited moderate antidepressant activities (Ismail and Mirza, 2015).

6 OTHER PHARMACOLOGICAL ACTIVITIES

Scientific evidence of the analgesic and antiinflammatory activities of *L. sativa* was shown. These activities were observed with methanol/petroleum ether extract of seeds of the plant using experimental rat models (Sayyah et al., 2004). Also, it was shown that leaves possess higher activity than seeds and that the activity is higher with aqueous than methanol/chloroform extract (Ismail and Mirza, 2015). These activities may be due to the presence of sesquiterpenes, as analgesic and antiinflammatory activities were found in an independent study with compound **2** as well as its derivatives (Wesołowska et al., 2006). The methanol and chloroform extracts exerted strong anticoagulant activity comparable to that of aspirin (Ismail and Mirza, 2015). The hydro-alcoholic and aqueous extracts of the seeds of the plant were reported for their antispermatogenic effects (Ahangarpour et al., 2014), but it has been shown that weekly administration of *L. sativa* ethanol extract during 10 weeks caused and increased the synthesis of sexual hormones including testosterone, luteinizing hormone, follicle stimulating hormone, estradiol, prolactin, as well as body weight and relative testis weight (Hefnawy and Ramadan, 2013).

Elmarzugi et al. showed that lettuce leave granules possess good flow properties including particle size distribution, bulk density, tapped density, Hausners index, Carr's index, friability, flowability, and water absorption (Elmarzugi et al., 2013). These results suggested that the plant can be used as a candidate for the formulation of solid dosage forms particularly tablets.

7 CLINICAL TRIALS

It was investigated whether storage under modified-atmosphere packaging may affect the antioxidant properties of fresh *L. sativa*. The study included 11 healthy volunteers with 6 men and 5 women aged 29–45 years and body weight of 48·0 –86 0 kg. Each participant consumed 250 g fresh lettuce, and blood was collected at different periods : 0 h (before consumption), 2, 3, and 6 h after consumption. Three days later, the same protocol was repeated with the same lettuce stored at 58°C under modified-atmosphere packaging (O_2–N (5:95 v/v)). This study demonstrated that after the first treatment (ingestion of fresh leaves of *L. sativa*), plasma total radical-trapping antioxidant potential (TRAP) increases significantly after 6 h compared to the value obtained with modified-atmosphere packaging-stored lettuce. Plasma TRAP, quercetin, and *p*-coumaric acid significantly increased from baseline values 2 and 3 h after fresh lettuce ingestion. A significant increase was also obtained in many blood parameters including caffeic acid at 3 h, plasma beta-carotene after 6 h, vitamin C concentration after consumption of fresh leaves of *L. sativa* while no changes were observed after ingestion of the same quantity of the same leaves stored under modified-atmosphere packaging for all the measured markers (Serafini et al., 2002). The clinical data also showed that the *L. sativa* seed oil could be useful pharmacological tools for the patients with sleep disorders and may be a hazard-free line of treatment, especially in geriatric patients suffering from mild-to-moderate forms of anxiety and sleeping difficulties (Yakoot et al., 2011). This was measured after a 1 week treatment where patients suffering from insomnia with or without anxiety were randomized to receive capsules containing 1000 mg of *L. sativa* seed oil.

8 CONCLUSIONS

L. sativa or lettuce, the most important crop in the group of leafy vegetables belongs to the family of Asteraceae and is cultivated worldwide. China and USA are the world main producers whereas Egypt and Niger are the best African producers. The plant is used in traditional medicine for the treatment of many health troubles like anemia and infections. Lettuce is a source of vitamins and many minerals. Many studies already evidenced the plant as antimicrobial, antioxidant, neuroprotective, and antiinflammatory agent.

REFERENCES

Ahangarpour, A., Oroojan, A.A., Radan, M., 2014. Effect of aqueous and hydro-alcoholic extracts of lettuce (*Lactuca sativa*) seed on testosterone level and spermatogenesis in NMRI mice Iran. J. Reprod. Med. 12 (1), 65–72.

Altunkaya, A., Becker, E.M., Gökmen, V., Skibsted, L.H., 2009. Antioxidant activity of lettuce extract (*Lactuca sativa*) and synergism with added phenolic antioxidants. Food Chem. 115 (1), 163–168.

Bang, M.H., Choi Sy Fau-Jang, T.-O., Jang To Fau-Kim, S.K., Kim Sk Fau-Kwon, O.-S., Kwon Os Fau-Kang, T.-C., Kang Tc Fau-Won, M.H., 2002. Phytol, SSADH inhibitory diterpenoid of *Lactuca sativa*. Arch. Pharm. Res. 25 (5), 643–646.

Bayer, R., Starr, I., 1998. Tribal phylogeny of the Asteraceae based on two noncoding chloroplast sequences, the tnrl intron and tnrl/trnF intergenie spacer. Ann. Missouri Bot. Gard. 85, 242–256.

Bennett, J.M., Hollister, W.C., 2001. Medieval Europe: A Short History. McGraw-Hill, New York.

Bestwick, C.S., Adam, A.L., Puri, N., Mansfield, J.W., 2001. Characterisation of and changes to pro and antioxidant enzyme activities during the hypersensitive reaction in lettuce (*Lactuca sativa* L.). Plant Sci. 161 (3), 497–506.

DeVries, I.M., 1997. Origin and domestication of *Lactuca sativa* L. Genet. Resour. Crop Evol. 44, 165–174.

DeVries, I.M., Van Raamsdonk, L.W., 1994. Numerical morphological analysis of Lettuce cultivars and species (Lactuca sect. Lactuca, Asteraceae). Plant Syst. Evol. 193 (1–4), 125–141.

Edziri, H.L., Smach, M.A., Ammar, S., Mahjoub, M.A., Mighri, Z., Aouni, M., 2011. Antioxidant, antibacterial, and antiviral effects of *Lactuca sativa* extracts. Ind. Crops Prod. 34 (1), 1182–1185.

Elmarzugi, N.A., Keleb, E.I., Mohamed, A.T., Salama, M., Hamid, M.A., Sulaiman, W.R.W., 2013. Preformulation studies on *Lactuca sativa* nutraceuticals granules. J. Teknol. 64 (2), 35–43.

FAO, 2013. Food And Agriculture Organization Of The United Nations Statistics Division. 22 March 2016. Available from: http://faostat3.fao.org/download/Q/QC/E

Ferreres, F., Gil, M.I., Castañer, M., Tomás-Barberán, F.A., 1997. Phenolic metabolites in red pigmented lettuce (*Lactuca sativa*). Changes with minimal processing and cold storage. J. Agric. Food Chem. 45 (11), 4249–4254.

Garg, M., Garg, C., Mukherjee, P.K., Suresh, B., 2004. Antioxidant potential of *Lactuca sativa*. Anc. Sci. Life 24 (1), 1–4.

Ghorbani, A., Rakhshandeh, H., Fau-Sadeghnia, H.R., Sadeghnia, H.R., 2013. Potentiating effects of *Lactuca sativa* on pentobarbital-induced sleep. Iran. J. Pharm. Res. 12 (2), 401–406.

Gridling, M., Popescu, R., Kopp, B., Wagner, K., Krenn, L., Krupitza, G., 2010. Anti-leukaemic effects of two extract types of *Lactuca sativa* correlate with the activation of Chk2, induction of p21, downregulation of cyclin D1 and acetylation of alpha-tubulin. Oncol. Rep. 23 (4), 1145–1151.

Han, Y.F., Cao, G.X., Xia, M., 2009. Two new eudesmane sesquiterpenes from *Lactuca sativa* var. anagustata L. Chin. Chem. Lett. 20 (10), 1211–1214.

Harsha, S.N., Anilakumar, K.R., 2013a. Anxiolytic property of hydro-alcohol extract of *Lactuca sativa* and its effect on behavioral activities of mice. J. Biomed. Res. 27 (1), 37–42.

Harsha, S.N., Anilakumar, K.R., 2013b. Anxiolytic property of *Lactuca sativa*, effect on anxiety behaviour induced by novel food and height. Asian Pac. J. Trop. Med. 6 (7), 532–536.

Harsha, S.N., Anilakumar, K.R., Mithila, M.V., 2013. Antioxidant properties of *Lactuca sativa* leaf extract involved in the protection of biomolecules. Biomed. Prev. Nutr. 3 (4), 367–373.

Hart, G., 2005. The Routledge Dictionary of Egyptian Gods and Goddesses, second ed. Routledge, London, UK.

Hefnawy, H.T.M., Ramadan, M.F., 2013. Protective effects of *Lactuca sativa* ethanolic extract on carbon tetrachloride induced oxidative damage in rats. Asian Pac. J. Trop. Dis. 3 (4), 277–285.

Ismail, H., Mirza, B., 2015. Evaluation of analgesic, anti-inflammatory, anti-depressant and anti-coagulant properties of *Lactuca sativa* (CV. Grand Rapids) plant tissues and cell suspension in rats. BMC Complement. Altern. Med. 15, 199.

Kesseli, R., Ochoa, O., Michelmore, R., 1991. Variation at RFLP loci in *Lactuca* spp. and origin of cultivated lettuce (*L. sativa*). Genome 34, 430–436.

Lebeda, A., Ryder, E.J., Dolezalovk, G.I., Kristova, E., 2007. Lettuce (Asteraceae; Lactuca spp.). In: Singh R. (Ed), Genetic Resources, Chromosome Engineering, and Crop Improvement. vol. 3, pp. 377-472. Vegetable Crops. CRC Press, Taylor and Francis Group, Boca Raton, FL.

Liu, X., Ardo, S., Bunning, M., Parry, J., Zhou, K., Stushnoff, C., 2007. Total phenolic content and DPPH radical scavenging activity of lettuce (*Lactuca sativa* L.) grown in Colorado. LWT-Food Sci. Technol. 40 (3), 552–557.

Mai, F., Glomb, M.A., 2013. Isolation of phenolic compounds from iceberg lettuce and impact on enzymatic browning. J. Agric. Food Chem. 61 (11), 2868–2874.

Michalska, K., Stojakowska, A., Malarz, J., Doležalová, I., Lebeda, A., Kisiel, W., 2009. Systematic implications of sesquiterpene lactones in *Lactuca* species. Biochem. Syst. Ecol. 37 (3), 174–179.

Moharib, S.A., Maksoud, N.A.E., Ragab, H.M., Mahmoud, Shehata, M., 2014. Anticancer activities of mushroom polysaccharides on chemically-induced colorectal cancer in rats. J. Appl. Pharm. Sci. 4 (7), 054–063.

Mou, B., 2008. Lettuce. In: Prohens, J., Nuez, F. (Eds.), Handbook of Plant Breeding. Vegetables I. Asteraceae, Brassicaceae, Chenopodiaceae, and Cucurbitaceae. Springer Science, New York, pp. 75–116.

Moulin-Traffort, J., Giordani, R., Fau-Regli, P., Regli, P., 1990. Antifungal action of latex saps from *Lactuca sativa* L. and Asclepias curassavica L. Mycoses 33 (7–8), 383–392.

Nazri, N.A.A.M., Ahmat, N., Adnan, A., Mohamad, S.A.S., Ruzaina, S.A.S., 2011. In vitro antibacterial and radical scavenging activities of Malaysian table salad. Afr. J. Biotechnol. 10 (30), 5728–5735.

Nomaani, R.S.S.A., Hossain, M.A., Weli, A.M., Al-Riyami, Q., Al-Sabahi, J.N., 2013. Chemical composition of essential oils and in vitro antioxidant activity of fresh and dry leaves crude extracts of medicinal plant of *Lactuca sativa* L. native to Sultanate of Oman. Asian Pac. J. Trop. Biomed. 3 (5), 353–357.

Noumedem, J., Mihasan, M., Lacmata, S., Stefan, M., Kuiate, J., Kuete, V., 2013. Antibacterial activities of the methanol extracts of ten Cameroonian vegetables against Gram-negative multidrug-resistant bacteria. BMC Complement. Altern. Med. 13 (1), 26.

Park, H.-J., Cho, H.-Y., Kong, K.-H., 2005. Purification and biochemical properties of glutathione S-transferase from *Lactuca sativa*. J. Biochem. Mol. Biol. 38 (2), 232–237.

Patil, R.B., Vora, S.R., Pillai, M.M., 2009. Antioxidant effect of plant extracts on phospholipids levels in oxidatively stressed male reproductive organs in mice. Iran. J. Reprod. Med. 7 (1), 35–39.

Rodenburg, C.M., 1960. Varieties of Lettuce. An International Monograph. W.E.J. Tjeenk Willink, Zwolle, The Netherland.

Sadeghnia, H.R., Farahmand, S.K., Asadpour, E., Rakhshandeh, H., Ghorbani, A., 2012. Neuroprotective effect of *Lactuca sativa* on glucose/serum deprivation-induced cell death. Afr. J. Pharm. Pharmacol. 6 (33), 2464–2471.

Sayyah, M., Hadidi, N., Kamalinejad, M., 2004. Analgesic and anti-inflammatory activity of *Lactuca sativa* seed extract in rats. J. Ethnopharmacol. 92, 325–329.

Serafini, M., Rossana, B., Monica, S., Elena, A., Anna, R., Maiani, G., 2002. Effect of acute ingestion of fresh and stored lettuce (*Lactuca sativa*) on plasma total antioxidant capacity and antioxidant levels in human subjects. Br. J. Nutr. 88, 615–623.

Souri, E., Amin, G., Farsam, H., Barazandeh Tehrani, M., 2008. Screening of antioxidant activity and phenolic content of 24 medicinal plant extracts. DARU J. Pharm. Sci. 16 (2), 83–87.

Subbarao, K.V., Koike, S.T., 2007. Lettuce diseases: ecology and control. Pimentel, D. (Ed.), Encyclopedia of Pest Management, vol. 2, CRC Press, Boca Raton, FL.

Takasugi, M., Okinaka, S., Katsui, N., Masamune, T., Shirata, A., Ohuchi, M., 1985. Isolation and structure of lettucenin A, a novel guaianolide phytoalexin from *Lactuca sativa* var. capitata(Compositae). J. Chem. Soc., Chem. Commun. (10), 621–622.

USDA Nutrient Database. National Nutrient Database for Standard Reference Release 28. Basic Report: 11251, Lettuce, cos or romaine, raw. Available from: http://ndb.nal.usda.gov/ndb/foods/show/3001?fgcd=&manu=&format=&offset=&sort

Wei, Z., Zhu, S.-X., Bakker, F.T., Berg, R.G.V.d., Schranz, M.E., 2015. Phylogenetic relationships within Lactuca L. (Asteraceae), including African species, based on chloroplast DNA sequence comparisons. Genet. Resour. Crop Evol..

Wesołowska, A., Nikiforuk, A., Michalska, K., Kisiel, W., Chojnacka-Wójcik, E., 2006. Analgesic and sedative activities of lactucin and some lactucin-like guaianolides in mice. J. Ethnopharmacol. 107 (2), 254–258.

Xu, F., Zou, G.A., Liu, Y.Q., Aisa, H.A., 2012. Chemical constituents from seeds of *Lactuca sativa*. Chem. Nat. Compds. 48 (4), 574–576.

Yakoot, M., Helmy, S., Fawal, K., 2011. Pilot study of the efficacy and safety of lettuce seed oil in patients with sleep disorders. Int. J. Gen. Med. 4, 451–456.

CHAPTER 21

Mangifera indica L. (Anacardiaceae)

S. Derese, E.M. Guantai, Y. Souaibou, V. Kuete

1 INTRODUCTION

Medicinal plants have been sources of a number of important compounds which have been discovered during the last century (Sajad and Saeid, 2016). Due to their established therapeutic efficacy, the pharmaceutical industries are using nowadays crude extracts of medicinal plants for manufacturing drugs (Chidozie et al., 2014). Natural product research has been conducted for many purposes; the major ones are to bring up new therapeutic agents and to provide useful lead molecules for studies directed toward the synthesis of drugs. Many modern pharmaceutical industries still rely on bioactive principles obtained from plants. The anticancer agent, taxol, isolated from the pacific yew, *Taxus brevifolia*, and the antimalarial agent artemisinin obtained from the Chinese herb *Artemisia annua* constitute few examples (Chidozie et al., 2014).

Mangifera indica belong to the genus *Mangifera* of the family Anacardiaceae, commonly known as Mango family. The genus *Mangifera* contains several species that bear edible fruit. Most of the fruit trees that are commonly known as mangos belong to the species *M. indica* (Anacardiaceae). There are other edible *Mangifera* species with lower quality fruit and are referred to as wild mangos. The genus *Mangifera* consists of about 30 species of tropical fruiting trees in the flowering plant family Anacardiaceae (Bally, 2006). *M. indica* has a long history of use in the home as an herbal remedy since ancient times. The plant is common in different parts of the world and is commonly known locally as Mango. The plant and its fruits are also referred to by vernacular names including manga, mangueira, skin mango in Brazil; mangguo in China; Aamin in Fiji; embe, mwembe in Kenya and Tanzania, and bowen mango in United States (Kritikar and Basu, 1993; Ross, 1999).

M. indica is an important medicinal plant in the Ayurvedic and indigenous medical systems for over 4000 years. Phytochemical study of *M. indica* showed the presence of highly effective bioactive compounds (Anjaneyulu and Radhika, 2000; Takashi et al., 1984; Pulak and Paridhavi, 2013). The plant chemicals are reported to have antibacterial, antifungal, antioxidant, antidiabetic, and antiviral activity (Jayanti, 2015; Rajan et al., 2011). *M. indica* grows in tropical and subtropical regions and its parts are commonly used in folk medicine as remedies to a variety of disease conditions (Kalita, 2014). Various parts of *M. indica* have been used traditionally for treatment of various complaints, including gastrointestinal problems (dysentery, piles, stomach upset, biliousness, habitual constipation), respiratory ailments (bronchitis, asthma, hiccough, throat problems), genitourinary problems (urinary discharges, leucorrhoea, vaginal problems), and ophthalmic complaints. It is also used as aphrodisiac, tonic, appetizer, beautifier of complexion, laxative, diuretic, antisyphilitic, and for tanning purposes in various parts of the world (Kalita, 2014; Alkizima et al., 2012; Ahmed et al., 2005; Zheng and Lu, 1990; Gupta and Gupta, 2011).

The widely available leaves of *M. indica* are traditionally known to be useful for the treatment of wide panel of diseases like malaria, throat infection, burns, and scalds, and to possess antidiabetic, antioxidant, antimicrobial, antiviral, and antibacterial activities (Pulak and Paridhavi, 2013; Saotoing et al., 2011; Shashi et al., 2009). Mango stem bark has been traditionally used for the treatment of menorrhagia, scabies, diarrhea, syphilis, diabetes, cutaneous infections, and anemia (Sellés et al., 2002).

Many studies provided evidence that the plant *M. indica* possesses antidiabetic, antioxidant, antiviral, cardiotonic, hypotensive, and antiinflammatory properties. Various effects like antibacterial (Awad El-Gied et al., 2012; Islam et al., 2010), antifungal (Pranay and Gulhina, 2011), anthelmintic, antiparasitic, antitumor (Fenglei et al., 2012), anti-HIV, antibone resorption, antispasmodic, antipyretic, antidiarrhoeal, antiallergic, immunomodulation (Fenglei et al., 2012; Shah et al., 2010), antihyperlipidemic (Khyati et al., 2010), antimicrobial (Islam et al., 2010), hepatoprotective (Malami et al., 2014; Mutiat et al., 2015), analgesic and antiinflammatory (Islam et al., 2010), and gastroprotective (Shah et al., 2010) have also been studied.

Although several review articles on this *M. indica* have already been reported (Shah et al., 2010; Kalita, 2014; Abdelnaser and Shinkichi, 2010), this chapter is presented to compile all the updated information on the plant widely known by scientists. This chapter focused in detail on the botanical description, the pharmacological activity, and phytochemistry of *M. indica*.

2 BOTANICAL DESCRIPTION

M. indica is a large evergreen tree in the Anacardiaceae family that grows up to 45 m, dome shaped with dense foliage, typically heavy branched from a stout trunk (Shah et al., 2010). In brief, the leaves are simple with alternate arrangement. The fruit is a drupe, and hangs from the tree on long stems when it matures. It has various shapes and sizes. The ripe fruit is variably colored yellow, orange, and red; green color usually indicates that the fruit is not yet ripe. This also depends on the varieties (Igoli et al., 2005).

M. indica trees typically branch 0.6–2 m above the ground and develop an evergreen, dome-shaped canopy. Canopy shape and openness are variable occurring among varieties and with competition from other trees (Shah et al., 2010).

The plant *M. indica* has a long taproot that often branches just below ground level, forming between two and four major anchoring taproots that can reach 5.5 m down to the water table. Fibrous finer roots are found from the surface down to approximately 1 m in most of the cases and usually extend just beyond the canopy diameter (Bally, 2006).

Flowers of *M. indica* are borne on terminal inflorescences that are broadly conical and can be up to 65 cm long on some varieties. Inflorescences have primary, secondary, and tertiary pubescent, cymose branches that are pale green to pink or red and bear hundreds of flowers (Bally, 2006). *M. indica* has two flower forms, hermaphrodite and male, with both forms occurring on the same inflorescence. The ratio of hermaphrodite to male flowers on an inflorescence varies with variety and season and is influenced by the temperature during inflorescence development. Hermaphrodite flowers are small (7 mm) with four to five ovate, pubescent sepals and four to five oblong, lanceolate, thinly pubescent petals. Only one or two of the four to five stamens that arise from the inner margin of the disk are fertile (Prakart et al., 2009). The single ovary is born centrally on the disk with the style arising from one side. The disk is divided into a receptacle of four or five fleshy lobes that form the nectaries. The male flowers are similar to the hermaphrodite flowers but are without the pistil, which has been aborted (Bally 2006).

Leaves of the plant are simple, without stipules, and alternate, with petioles up to 12 cm long and are variable in shape and size but usually are oblong with tips varying from rounded to acuminate. A range of leaf sizes can be seen on a single tree and leaf form differs among varieties but is more consistent within a variety. Leaves are dark green when they are mature with a shiny upper surface and glabrous lighter green lower surface. They emerge green, turning tan-brown to purple during leaf expansion and then gradually changing to dark green as the leaves mature. Young leaf color varies with variety and can be from light tan to deep purple; this can be used as a distinguishing character among varieties (Prakart et al., 2009; Bally, 2006).

Fruits of *M. indica* from different varieties can be highly variable in shape, color, taste, and flesh texture. Mango fruit is classed as a drupe and their shapes vary from round to ovate to oblong and long with variable lateral compression. Fruits can weigh from 50 g or less to over 2 kg. Usually, the fruit has a dark green background color when developing on the tree that turns lighter green to yellow as it ripens. In some varieties a red background color is observed at fruit set that remains until the fruits ripen. Many varieties also have an orange, red, or burgundy blush background color that develops later in the fruit development, when the rind is exposed to direct sunlight. The mesocarp is the fleshy, edible part of the fruit that usually has a sweet and slightly turpentine flavor. When ripe, its color varies from yellow to orange and its texture from smooth to fibrous (Jayanti, 2015; Bally, 2006).

Seed of mango can be classified as having either monoembryonic or polyembryonic seed embryos. The seed contains only one embryo in monoembryonic varieties that is a true sexual (zygotic) embryo. Monoembryonic seeds are a cross between the maternal and paternal (pollen) parents. Fruit from monoembryonic seedlings will often vary from the parent trees, so propagation by grafting is used to produce true-to-type monoembryonic trees. In contrast, molyembryonic seeds contain many embryos, most of which are asexual (nucellar) in origin and genetically identical to the maternal parent.

Polyembryonic seeds also contain a zygotic embryo that is the result of crosspollination. The monoembryonic seedling usually has less vigor than a nucellar seedling for use as a rootstock (Prakart et al., 2009; Bally, 2006).

3 ORIGIN AND DISTRIBUTION

M. indica is native from Southern Asia, especially Eastern India, Burma, and the Andaman Islands. The plant has been cultivated, praised, and even revered in its homeland since ancient times. Buddhist monks are believed to have taken the plant on voyages to Malaya and Eastern Asia in the fourth and fifth centuries BC. In the same light, Persians are said to have taken mangos to East Africa around the 10th century AD (Wauthoz et al., 2007). The fruit was grown in the East Indies before the earliest visits of the Portuguese who apparently introduced it to West Africa and Brazil in the early 16th century. *M. indica* was then carried to the West Indies, being first planted in Barbados in about 1742 and later in the Dominican Republic; it reached Jamaica in about 1782 and, early in the 19th century, reached Mexico from the Philippines and the West Indies (Morton, 1987). In this day and age, *M. indica* resides in most tropical biotopes in India, Southeast Asia, Malaysia, Himalayan regions, Sri Lanka, Africa, America, and Australia (Jayanti, 2015; Wauthoz et al., 2007).

Mango is the most important commercial fruit in some continents like Asia and is popular worldwide. There are studies on *Mangifera* elucidating its current genetic diversity and geographical distribution (Prakart et al., 2009). The genus Mangifera originates in tropical Asia, with the greatest number of species found in Borneo, Java, Sumatra, and the Malay Peninsula. The most-cultivated *Mangifera* species, *M. indica* (mango), has its origins in India (Bally, 2006; Shah et al., 2010). Mango is now cultivated throughout the tropical and subtropical world for commercial fruit production, as a garden tree, and as a shade tree for stock. The genus *Mangifera* has about 69 species restricted in their native range to tropical Asia, extending as far north as 27°N latitude and as far east as the Caroline Islands. Exceptional distribution ranges of some species can extend to north of the Tropic of Cancer, such as to Sikkim in India and southern China. Wild mango species occur in India, Sri Lanka, Bangladesh, Myanmar, Thailand, Cambodia, Vietnam, Laos, southern China, Malaysia, Singapore, Indonesia, Brunei, the Philippines, Papua New Guinea, and the Solomon Islands and Micronesia (Jayanti, 2015; Prakart et al., 2009). The greatest diversity, with approximately 28 species, occurs in western Malaysia and Sumatra, a region considered to be the center of diversity of the genus. Fifteen species were described for the flora of Malaya and about 16 species occur in Thailand (Prakart et al., 2009). *M. indica* is widely distributed in the whole African continent and it can be found in Western, Southern, and Eastern Africa including Senegal, Mali, Nigeria, South Africa, Kenya, South Africa, Zambia, Uganda, Tanzania, Ethiopia, and Rwanda (Kokwaro, 1986).

4 CLASSIFICATION AND CULTIVARS OF *Mangifera indica*

In general, mango trees can be classified into two major categories: local and commercial varieties. Local varieties, in most of the cases, are tasty and possess the disadvantage of having a very short shelf life. In addition to the known local varieties, some new varieties have emerged on a continuous basis because some varieties are compatible with each other producing fruits of undesirable nature (Gillett, 2015). *M. indica* plants are usually grafted so as to provide the growers with a plant that have defined properties and qualities. The second class of Mango includes commercial varieties normally grafted and this class consists of Haden, Kent, Keitt, and Tommy Atkins. This type of varieties normally have taste, good size, appearance, and a long shelf life which makes it the preferred fruit for the market (Gillett, 2015).

M. indica plants are capable of tolerating various climatic conditions. They can survive even in swampy or very hot and very humid climatic conditions. More than 1000 cultivars of mango are cultivated at different regions throughout India (Ritu et al., 2013). Many types of mangos are cultivated in Africa (Vayssieres et al., 2008). Mangos are grown over a wide area in Asia, America, Europe, and Africa. However, production decreases at higher altitudes. Different cultivars are cultivated all over the world including Tommy Atkins, medium to large sized thick skinned mangos, greenish yellow sweet juicy variety with fiberless flesh; Keitt, the rounded oval ones yellow in color with red blush and sensation, oval, oblique, and less beaked type (Ritu et al., 2013; Githure et al., 1998). Productivity depends on many factors, including quantity of previous crop, soil conditions, weather, altitude, control of pests and diseases, fertilization, and cultivar (Jurgen, 2003).

Production is based across approximately 100 countries in tropical and subtropical regions of the world. Although about 100 countries grow mangos, about 80% of production comes from the top nine countries in order of production, India, China, Indonesia, Mexico, Thailand, Pakistan, Brazil, Philippines, and Nigeria (Bally, 2011). In South Africa, approximately 84% is planted under micro, drip, sprinkler, or flood irrigation. Dry land production is no longer favored unless the annual rainfall supplements the irrigation programme during critical periods. In the South African mango industry approximately 20% of the producers produce approximately 80% of the total annual production. Data on the total hectares planted to specific cultivars are currently not readily available from the industry but the cultivars discussed previously are believed to be the main cultivars planted in South Africa (Oosthuyse, 1994).

5 PHYTOCHEMISTRY OF *Mangifera indica*

This subsection focuses on phytochemicals isolated from *M. indica*. The chemical composition of *M. indica* varies with the location, variety, and stage of maturity (Fenglei et al., 2012). The chemical constituents of the *M. indica* include aliphatic compounds, terpenoids, flavonoids, alkaloids, coumarins, terpenoidal saponins, polyphenolics, tanins,

Figure 21.1 M. indica *tree.*

and essential oils. The crude extract of stem bark of *M. indica* was also found to contain alkaloids, phenols, tannins, saponins, and glycosides (Chidozie et al., 2014) (Fig. 21.1).

5.1 Triterpenoids, steroids, and saponins

A number of terpenoids (Fig. 21.2) and steroids have been isolated from *M. indica* (Kalita, 2014; Shahid et al., 2014; Sutomo et al., 2013; Juceni and Jorge, 2006). The structures of isolated triterpenes from the plant are diversified and Fig. 21.2 illustrates some triterpenoids and steroids isolated from *M. indica*. Most of these triterpenoids are tetracyclic triterpenoids like 2-hydroxy mangiferonic acid (Chidozie et al., 2014; Shahid et al., 2014; Kaur et al., 2010; Anjaneyulu and Radhika, 2000). From the roots, the major triterpenoids include friedelin (**1**); mangiferolic acid (**2**); hydroxymangiferonic acid (**3**); mangiferonic acid (**4**); isomangiferolic acid (**5**); hydroxymangiferolic acid (**6**); cycloartenol (**7**); 24-methyl-3β-hydroxycycloart-24-en-26-al (**8**); 3β-hydroxycycloart-24-en-26al (**9**); cycloart-24-en-3β-26-diol (**10**); 3α, 27-dihydroxy cycloart-24E-en-26-oic acid (**11**); cycloartenone (**12**); taraxerol (**13**); taraxerone (**14**); lupeol (**15**); friedelan-3β-ol (**16**); α-amyrin (**17**); β-amyrin (**18**); β-sitosterol (**19**); 3-oxo-lanost-cis-1,20(22)-diene-27-oic acid (**20**); 3-oxo-lanostcis-1,24-diene-27-oic acid (**21**);

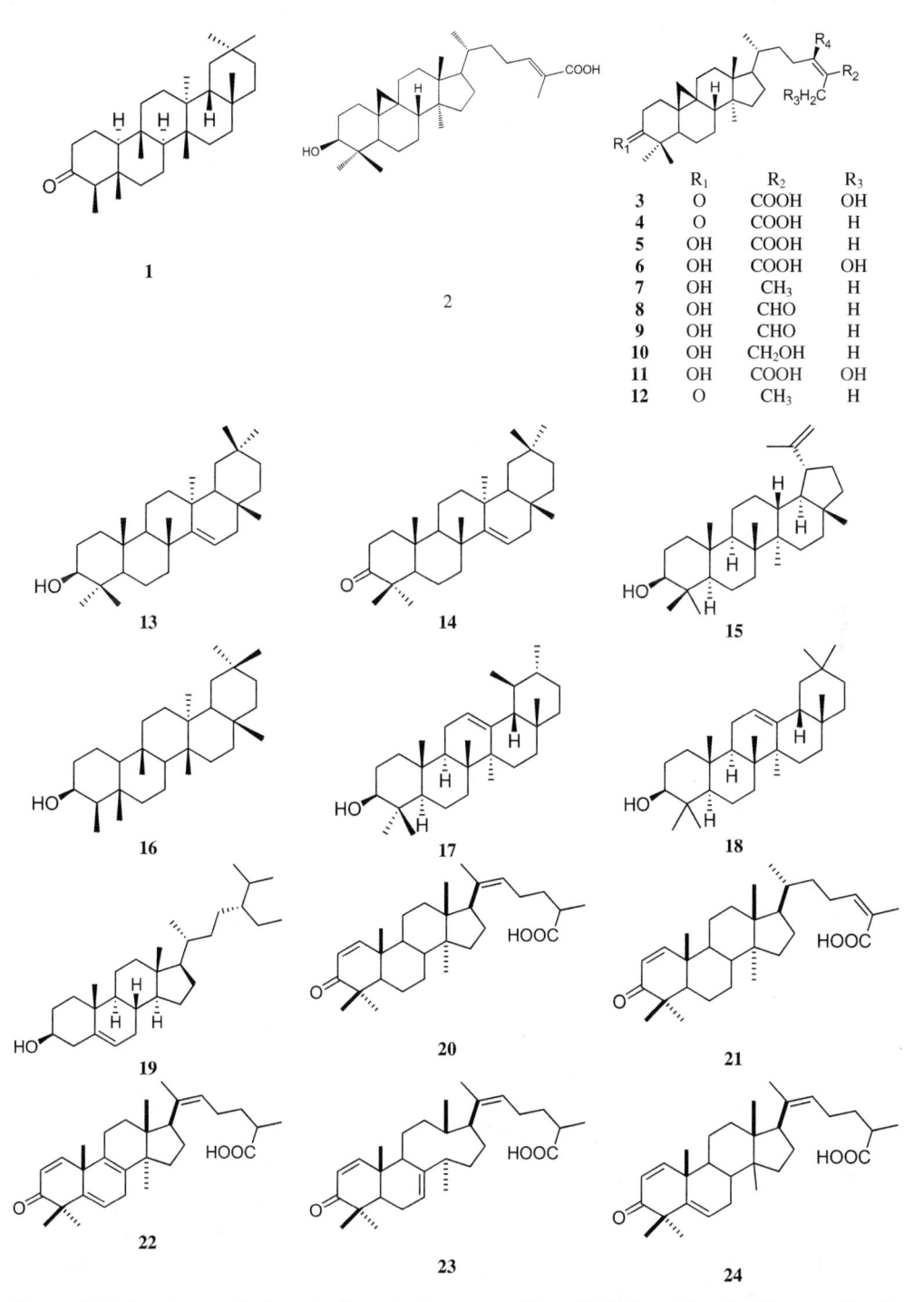

	R_1	R_2	R_3
3	O	COOH	OH
4	O	COOH	H
5	OH	COOH	H
6	OH	COOH	OH
7	OH	CH_3	H
8	OH	CHO	H
9	OH	CHO	H
10	OH	CH_2OH	H
11	OH	COOH	OH
12	O	CH_3	H

Figure 21.2 *Structures of selected triterpenoids reported from M. indica.* Friedelin (**1**); mangiferolic acid (**2**); hydroxymangiferonic acid (**3**); mangiferonic acid (**4**); isomangiferolic acid (**5**); hydroxymangiferolic acid (**6**); cycloartenol (**7**); 24-methyl-3β-hydroxycycloart-24-en-26-al (**8**); 3β-hydroxycycloart-24-en-26al (**9**); cycloart-24-en-3β-26-diol (**10**); 3α, 27-dihydroxy cycloart-24*E*-en-26-oic acid (**11**);

Figure 21.2 (*cont.*)

Figure 21.2 (cont.) cycloartenone (**12**); taraxerol (**13**); taraxerone (**14**); lupeol (**15**); friedelan-3β-ol (**16**); α-amyrin (**17**); β-amyrin (**18**); β-sitosterol (19); 3-oxo-lanost-cis-1,20(22)-diene-27-oic acid (**20**); 3-oxo-lanostcis-1,24-diene-27-oic acid (**21**); 3-oxo-lanost-cis-1,5,8,20(22)-tetra-ene-27-oic acid (**22**); 3-oxo-lanost-1,7,20(22)-triene-13,14-seco-27-oic acid (**23**); 3-oxo-lanosta-cis-1,5,24-trien-27-oic acid (**24**); campesterol (**25**); manghopanal (**26**); mangoleanone (**27**); 5-dehydro-avenasterol (**28**); α-amyrenone (**29**); α-tocopherol (**30**); β-amyrenone (**31**); cholesterol (**32**); cycloartenone (**33**); cyclobranol (**34**); dammaradienol (**35**); dammarendiol ii (**36**); daucosterol (**37**); friedelinol (**38**); germanicol (**39**); glochidonol (**40**); gramisterol (**41**); hopane-1β -3β-22-triol (**42**); isomangiferolic acid (**43**); lophenol (**44**); lupenone (**45**); mangiferonic acid methyl ester (**46**); ocotillol (**47**); octilloll (**48**); stigmast-7-en-3-β-ol (**49**); stigmasterol (**50**); ursolic acid (**51**); indicoside a (**52**); indicoside B (**53**).

3-oxo-lanost-cis-1,5,8,20(22)-tetra-ene-27-oic acid (**22**); 3-oxo-lanost-1,7,20(22)-triene-13,14-seco-27-oic acid (**23**); 3-oxo-lanosta-cis-1,5,24-trien-27-oic acid (**24**); campesterol (**25**); manghopanal (**26**); mangoleanone (**27**); 5-dehydro-avenasterol (**28**); α-amyrenone (**29**); α-tocopherol (**30**); β-amyrenone (**31**); cholesterol (**32**); cycloartenone (**33**); cyclobranol (**34**); dammaradienol (**35**); dammarendiol ii (**36**); daucosterol (**37**); friedelinol (**38**); germanicol (**39**); glochidonol (**40**); gramisterol (**41**); hopane-1-beta-3-beta-22-triol (**42**); isomangiferolic acid (**43**); lophenol (**44**); lupenone (**45**); mangiferonic acid methyl ester (**46**); ocotillol (**47**); octilloll (**48**); stigmast-7-en-3-beta-ol (**49**); stigmasterol (**50**); ursolic acid (**51**); cycloart-24-en-3,2β-26-diol, 3-keto-dammar-24(E)-en-20S,26-diol, C-24 epimers of cycloart-25-en-3β,24,27-triol, cycloartan-3β,24,27-triol, 29-hydroxymangiferonic acid, 3α-22ξ-dihydroxycycloart-24E-en-26-oic acid, sitosterol arachidate, epi-ψ-taraxastane-3β,20-diol, 6β-hydroxystigmast-4-en-3-one, 6β-hydroxycampest-4-en-3-one, 6β-hydroxystigmasta-4,22-dien-3-one, 5α-stigrnastane-3β, 6α-diol, cycloartane-3β,24,25-triol, 25 (R)-3-oxo-methylene cycloartan-26-ol, ψ-taraxastanonol, cycloart-24-ene-3β,26-diol, C-24 epimers of cycloart-25-ene-3β,24,27-triol, the C-24-epimers of cycloartane-3β,24,25-trio1, 3-ketodammar-24(E)-ene-20S,26-diol, hopane-1β,3β,22-triol, mangsterol, and manglupenone (Shahid et al., 2014; Anjaneyulu et al., 1985; Jayanti 2015; Sellés et al., 2002; Madan et al., 2014; Chidozie et al., 2014; Alkizima et al., 2012; Pulak and Paridhavi, 2013; Sutomo et al., 2013). Apart from these aforementioned compounds, it also contained saponins like indicoside A (**52**) and indicoside B (**53**) (Kalita, 2014; Pulak and Paridhavi, 2013).

5.2 Polyphenolic and phenolic compounds

5.2.1 Flavonoids

Many flavonoids (Fig. 21.3) have been isolated from Mango (Anjaneyulu and Radhika, 2000; Takashi et al., 1984). Some isolated flavonoids are illustrated in Fig. 21.3. The major bioactive constituent of *M. indica* was found to be mangiferin, a xanthone glycoside (Pulak and Paridhavi, 2013). The plant was also reported to contain mangiferin (**54**); isomangiferin (**55**); catechin (**56**); epicatechin (**57**); 5-hydroxy mangiferin (**58**); 6-O-galloyl-5'-hydroxymangiferin (**59**); quercetin-3-sulfate (**60**); (-)-epicatechin-3-O-β-glucopyranoside (**61**); 5-hydroxy-3-(4-hydroxylphenyl) pyrano [3,2-g]chromene-4(8H)-one (**62**); 6-(phydroxybenzyl) taxifolin-7-O-β-D-glucoside (tricuspid) (**63**); quercetin-3-O-α-glucopyranosyl-(1→2)-β-D-glucopyranoside (**64**); 5,7-dihydroxy-2-(2- hydroxy-5- methoxyphenyl)-4H-chromen-4-one (**65**); 5-hydroxy-2-(2,4- dihydroxy- methoxyphenyl)-7-methoxy 4H- chromen-4-one (**66**); 1-3-5-6-7-pentamethoxyxanthone (**67**); 1-3-6-7-tetramethoxyxanthone (**68**); amentoflavone (**69**); euxanthone (**70**); homomangiferin (**71**); isoquercitrin (**72**); kaempferol methyl ether (**73**); kaempferol (**74**); rutin (**75**); mangiferene (**76**); myricetin methyl ester (**77**); peonidin-3-galactoside (**78**); quercetin (**79**); quercetin methyl ether (**80**); quercitrin methyl ether (**81**).

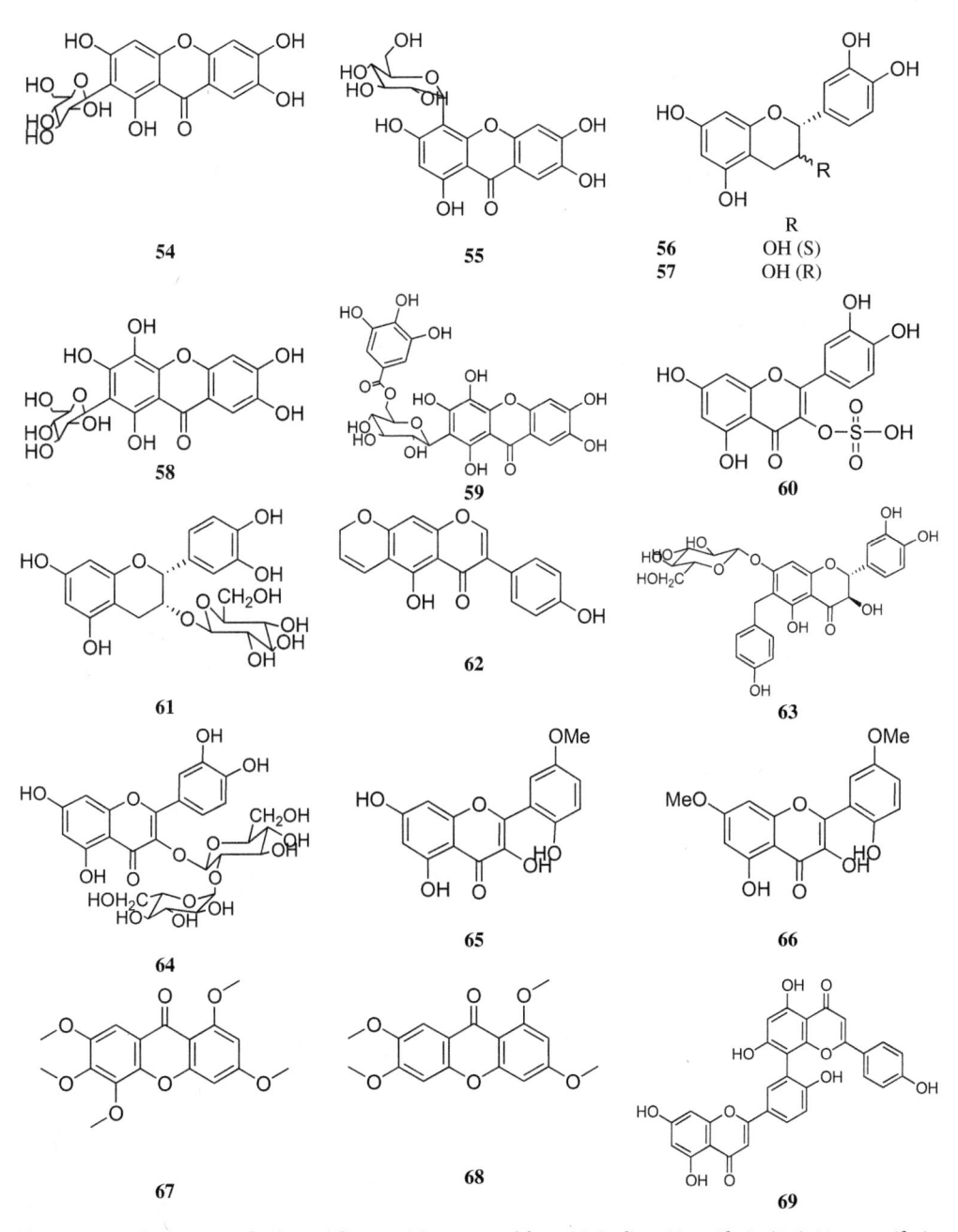

Figure 21.3 *Structures of selected flavonoids reported from* **M. indica.** Mangiferin (**54**); isomangiferin (**55**); catechin (**56**); epicatechin (**57**); 5-hydroxy mangiferin (**58**); 6-O-galloyl-5′-hydroxymangiferin (**59**); quercetin-3-sulfate (**60**); (-)-epicatechin-3-O-β-glucopyranoside (**61**); 5-hydroxy-3-(4-hydroxylphenyl) pyrano [3,2-g]chromene-4(8*H*)-one (**62**); 6-(phydroxybenzyl) taxifolin-7-O-β-D-glucoside

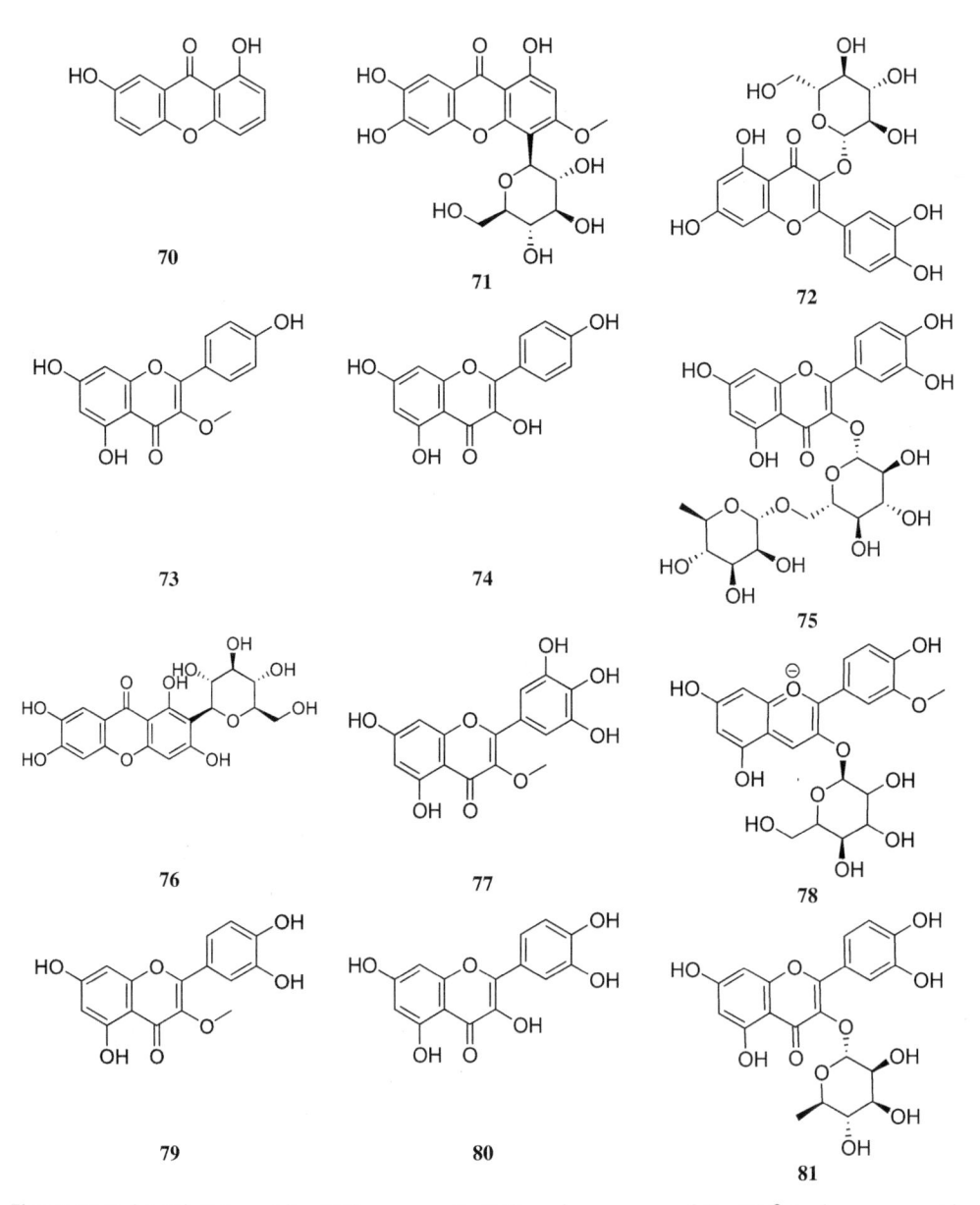

Figure 21.3 (cont.) (tricuspid) (**63**); quercetin-3-O-α-glucopyranosyl-(1→2)-β-D-glucopyranoside (**64**); 5,7-dihydroxy-2-(2- hydroxy-5- methoxyphenyl)-4H-chromen-4-one (**65**); 5-Hydroxy-2-(2,4-dihydroxy- methoxyphenyl)-7-methoxy 4H- chromen-4-one (**66**); 1-3-5-6-7-pentamethoxyxanthone (**67**); 1-3-6-7-tetramethoxyxanthone (**68**); amentoflavone (**69**); euxanthone (**70**); homomangiferin (**71**); isoquercitrin (**72**); kaempferol methyl ether (**73**); kaempferol (**74**); rutin (**75**); mangiferene (**76**); myricetin methyl ester (**77**); peonidin-3-galactoside (**78**); quercetin (**79**); quercetin methyl ether (**80**); quercitrin methyl ether (**81**).

There are also other flavonoids isolated from *M. indica* and these include mangiferin gallate, isomangiferin gallate, quercetin 3-O-gallate, quercetin 3-O-glycoside, quercetin 3-O-xylose, quercetin 3-O-arabinopyranoside, quercetin 3-O-arabinofuranoside, quercetin 3-O-rhamnoside, kaempferol 3-O-glycoside, quercetin pentoside, quercetin pentoside, quercetin-3-O-rhamnoside (quercitrin), quercetin-3-O-rutinoside, and quercetin pentoside to enumerate some (Amian et al., 2014; Bhuvaneswari, 2013; Nicolai et al., 2005).

5.2.2 Tannins

Tannins are a large diverse group of complex polyphenolic compounds of medium to large molecular weight that are widely distributed among medicinal plants in all plant parts, where they are ascribed a protective function. The principal chemical property of tannins is the ability to form strong complexes with proteins, starches, and other macromolecules. The tannins are applied widely in medicine especially in natural healing. Tannins from plant extracts are used as astringents against diarrhea, antiviral, antibacterial, as antiinflammatory agents, antiseptic, and hemostatic pharmaceuticals (Karthy and Ranjitha, 2011),

Recent reports (Nicolai et al., 2004) have demonstrated that the tannins obtained from mango leaves are a mixture of penta- to undecagalloylglucoses with a 1,2,3,4,6-penta-O-galloyl-β-D-glucose core (Fig. 21.4). Moreover, the presence of several galloyl and *p*-hydroxybenzoyl esters of benzophenone C-glycosides has been highlighted; some of them include penta-O-galloylglucose (**82**); epigallocatechin-3-O-gallate-(4β→8)-epigallocatechin-(4β→8)-catechin (**83**) and maclurin-3-C′-(2″,3″, 6″-O-galloyl)-β-D-glucoside (**84**).

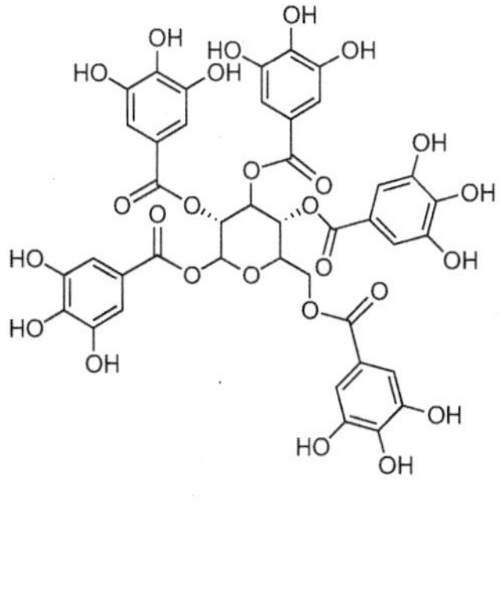

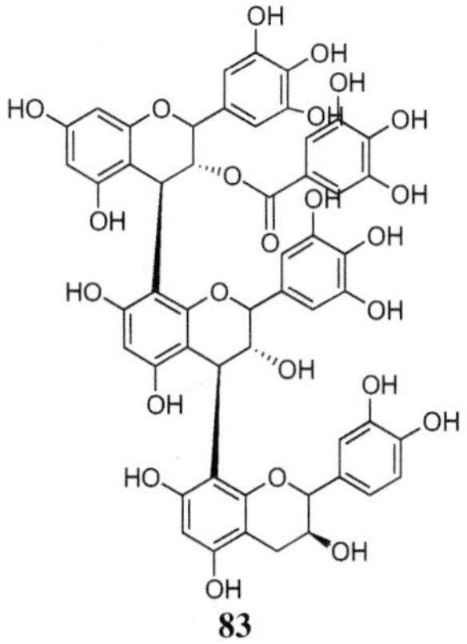

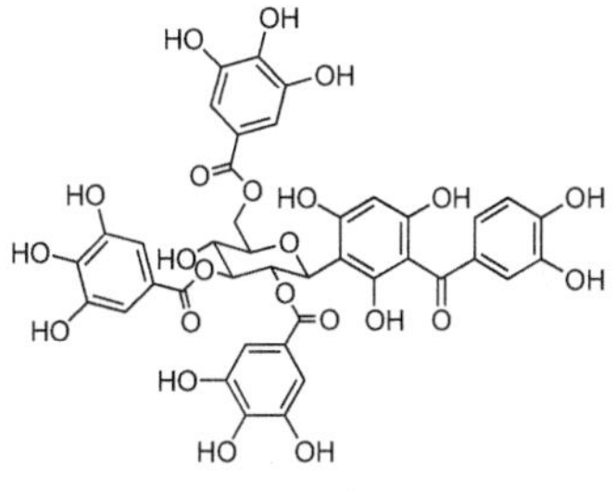

Figure 21.4 Structures of selected tannins isolated from M. indica. Penta-O-galloylglucose (**82**); epigallocatechin-3-O-gallate-(4β→8)-epigallocatechin-(4β→8)-catechin (**83**) and maclurin-3-C′-(2″,3″, 6″-o-galloyl)-β-D-glucoside (**84**).

Maclurin 3-C-β-D-glucoside, maclurin 3-C-(6″-O-p-hydroxybenzoyl)-β-D-glucoside, maclurin 3-C-(2″-O-galloyl-6″-O-p-hydroxybenzoyl)-β- D-glucoside, maclurin 3-C-(2″-O-p-hydroxybenzoyl-6″-O-galloyl)-β- D-glucoside, maclurin 3-C-(2″,3″,6″-tri-O-galloyl)-β-D-glucoside, iriflophenone 3-C-(2″,6″-di-O-galloyl)-β- D-glucoside and Iriflophenone 3-C-(2″,3″,6″-tri-O-galloyl)-β- D-glucoside have also been isolated and characterized by NMR spectroscopy. The new trimeric proanthocyanidin, epigallocatechin-3-O-gallate-(4β→8)-epigallocatechin-(4β→8)-catechin (**83**), was isolated from the air-dried leaves of *M. indica*. Some of phenolic compounds are illustrated in Fig. 21.4. Karthy and Ranjitha (2011) concluded in their work that antimethicillin resistance against *Staphylococcus aureus* activity of the seed of *M. indica* could be due to the tannins.

5.2.3 Other phenolic derivatives

Nowadays, there is now much interest in polyphenolic compounds as they have considerable antioxidant activity in vitro and are ubiquitous in our diet (Massimo et al., 2007). These compounds are usually found in plants including tea, vegetables, and fruits (Mango). There is little convincing epidemiological evidence that intakes of polyphenols are inversely related to the incidence of cancer. However, numerous cell culture and animal models indicate potent anticarcinogenic activity by certain polyphenols mediated through a range of mechanisms including antioxidant activity, enzyme modulation, gene expression, apoptosis, upregulation of gap junction communication and P-glycoprotein activation (Sellés et al., 2002). Polyphenolic composition of different parts of the plant *M. indica* was established in the literature (Shah et al., 2010; Masibo and He, 2008). The compounds include benzoic acid (**85**); gallic acid (**86**); 3,4-dihydroxy benzoic acid (**87**); propyl benzoate (**88**); propyl gallate (**89**); methyl gallate (**90**); syringic acid (**91**); ferulic acid (**92**); ethyl gallate (**93**); pyrogallol (**94**); *p*-hydroxy benzoic acid (**95**); vanillic acid (**96**); theogallin (**97**), ethyl 2,4-dihydroxy-3-(3,4,5-tridydroxybenzoyl) oxybenzoate (**98**); valoneic acid dilactone (**99**); 1-heptyl-(1-phenyl)-3-xylose benzoate (**100**); 3-C-β-D-glucopyranosyl-4′,2,4,6-tetrahydroxy benzophenone (**101**); iriflophenone-3-β-C-glucoside (**102**); 4-phenyl-*n*-butyl gallate (**103**); 5-(12-cis-heptadecenyl) resorcinol (**104**); 5-heptadec-*cis*-2-enyl resorcinol (**105**); 5-pentadecyl resorcinol (106), 6-phenyl-n-hexyl gallate (**107**); ellagic acid (**108**); gentisic acid (**109**); *n*-octyl gallate (**110**); *n*-pentyl gallate (**111**); acetophenone (**112**); benzaldehyde (**113**); bis-2-ethyl hexanyl-phthalate (**114**); elemicin (**115**); estragole (**116**); eugenol methyl ester (**117**), (Abdelnaser and Shinkichi, 2010; Gururaja et al., 2015; Madan et al., 2014; Nicolai et al., 2005; Masibo and He 2008; Anjaneyulu and Radhika, 2000). Fig. 21.5 illustrates some phenolic compounds isolated from *M. indica*.

Phenolic antioxidants and polyols were also reported from the stem bark of *M. indica* (Anjaneyulu et al., 1989). The flower yielded alkyl gallates, such as *n*-pentyl gallate, *n*-octyl gallate, 4-phenyl gallate, 6-phenyl-*n*-hexyl gallate, and dihydrogallic

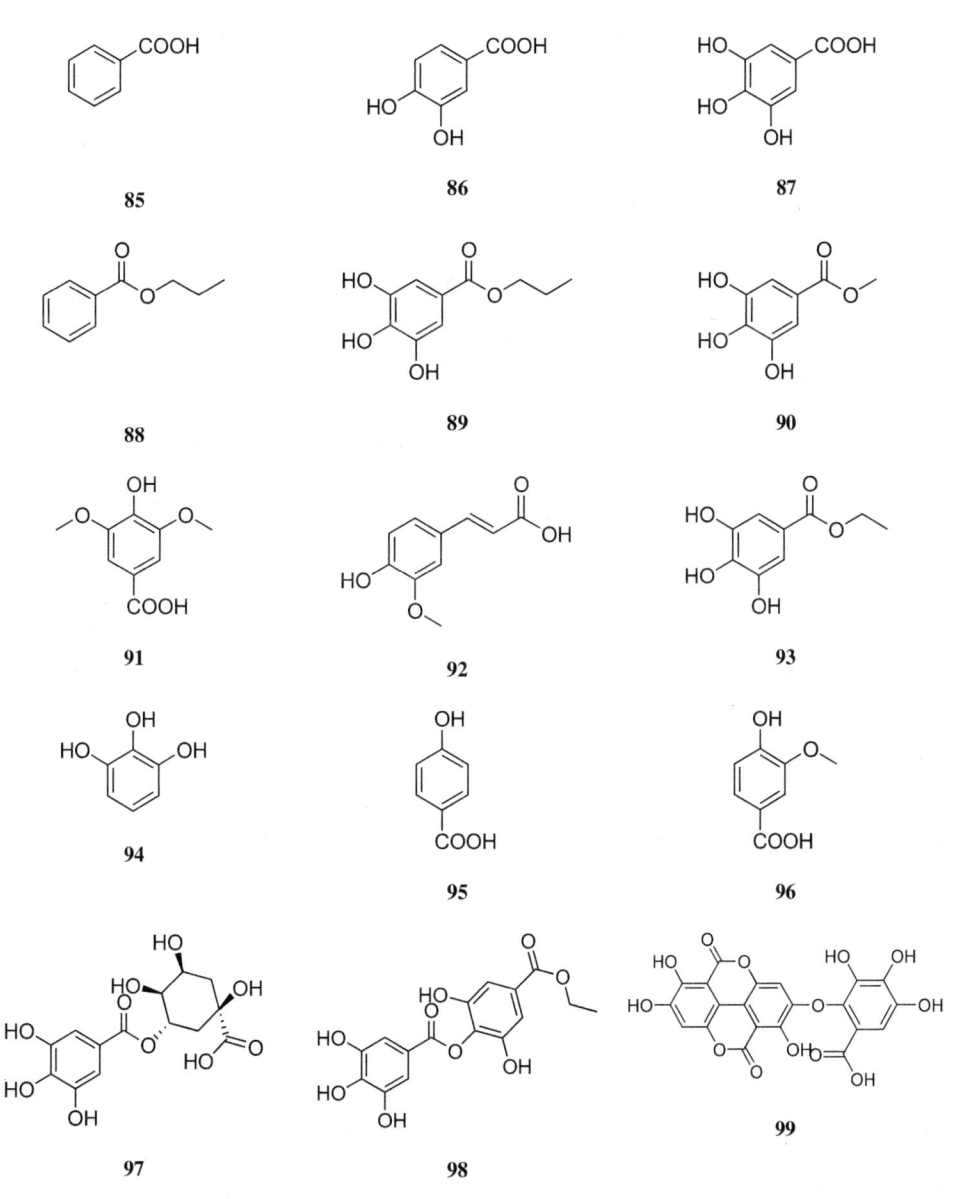

Figure 21.5 *Structures of other phenolic derivatives reported from* **M. indica.** benzoic acid (**85**); gallic acid (**86**); 3,4-dihydroxy benzoic acid (**87**); propyl benzoate (**88**); propyl gallate (**89**); methyl gallate (**90**); syringic acid (**91**); ferulic acid (**92**); ethyl gallate (**93**); pyrogallol (**94**); *p*-hydroxy benzoic acid (**95**); vanillic acid (**96**); theogallin (**97**), ethyl 2,4-dihydroxy-3-(3,4,5-tridydroxybenzoyl) oxybenzoate (**98**); valoneic acid dilactone (**99**); 1-heptyl-(1-phenyl)-3-xylose benzoate (**100**); 3-*C*-β-D-glucopyranosyl-4′,2,4,6-tetrahydroxy benzophenone (**101**); iriflophenone-3-β-*C*-glucoside (**102**); 4-phenyl-*n*-butyl gallate (**103**); 5-(12-*cis*-heptadecenyl) resorcinol (**104**); 5-heptadec-*cis*-2-enyl resorcinol (**105**); 5-pentadecyl resorcinol (106), 6-phenyl-*N*-hexyl gallate (**107**); ellagic acid (**108**); gentisic acid (**109**); *n*-octyl gallate (**110**); *n*-pentyl gallate (**111**); acetophenone (**112**); benzaldehyde (**113**); bis-2-ethyl hexanyl-phthalate (**114**); elemicin (**115**); estragole (**116**); eugenol methyl ester (**117**).

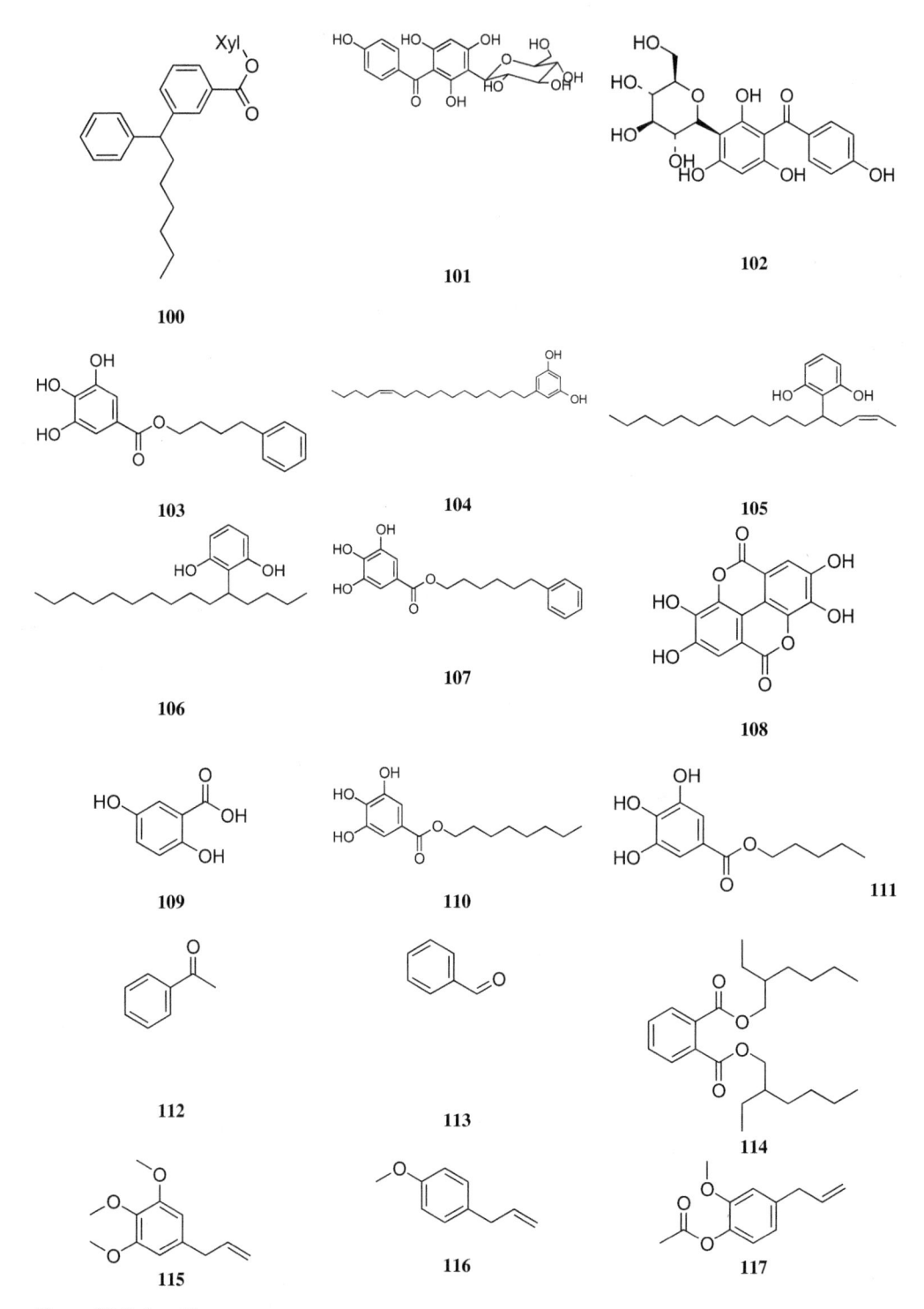

Figure 21.5 (*cont.*)

acid (Bandyopadhyay et al., 1985). Dorta et al. (2014) identified a number of phenolic compounds from *M. indica* by using HPLC-ESI-QTOF-MS method. These include galloyl glucose, di-galloyl-glucose, maclurin-3-*C*-(2-O-galloyl)-β-D-glucoside, iriflophenone-3-*C*-(2-O-galloyl)-β-D-glucoside, and maclurin-3-*C*-(2,3-di-O-galloyl)-β-D-glucoside.

5.3 Volatile components

More than 270 volatile compounds have been identified from mango. Structures of some of them are represented in Fig. 21.6. Of these, monoterpenes and sesquiterpenes were the most abundant (Sâmya et al., 2011) while esters and lactones were found to play an important role in a unique flavor of certain cultivars (Bello-Perez et al., 2007). The main constituents of the oil were α-gurjunene (24.0%), β-selinene (24.0%), β-caryophyllene (11.2%), α-humulene (7.2%), caryophyllene oxide (5.5%), and humulene epoxide (2.4%) (Bello-Perez et al., 2007). Sâmya et al. (2011) evaluated the chemical composition of essential oils from fruits (immature and mature) and leaf (immature and mature) of *M. indica* extracted by using solid phase microextraction (HS-SPME) and hydrodistillation (HD) and the summary is combined in Table 21.1. Structures of some volatile compounds identified from *M. indica* are presented in Fig. 21.6.

5.4 Carbohydrates and other class of compounds

In an unripe stage, the main carbohydrate of *M. indica* is starch that during the ripening process is hydrolysed to glucose. This monosaccharide participates in the biosynthesis of sucrose and fructose in the ripe stage of the fruit. Thus most of the varieties of mango contain high level of both sucrose and fructose and low level of glucose (Bello-Perez et al., 2007). Starch is a carbohydrate consisting of large number of glucose units which are linked by glycosidic bond. It is a food reserve substance in plants and is widely used as a diluent and binder in the pharmaceutical industry. It is also used as a food additive usually as thickeners and stabilizers (Manisha and Sikdar 2015). Related isolated compounds include polygalacturonase and fructose-1-6-diphosphatase. Some isolated compounds from bark consist of, alanine, glycine, γ-amino butyric acid, kinic acid, shikimic acid. The fruit pulp contains vitamins A and C and β-carotene. Some compounds including 5-alkyl and 5-alkenylresorcinols, as well as their hydroxylated derivatives were identified from peels (Pulak and Paridhavi, 2013). Dietary fiber, pectin, and reducing sugars were also identified from the plant (Bello-Perez et al., 2007) Mangocoumarin, mangostin, and an unusual fatty acid, *cis*-9,*cis*-15-octadecadienoic acid were also isolated (Pulak and Paridhavi, 2013). Fig. 21.7 illustrates few of other types of compounds identified from *M. indica*. These include *cis*-zeatin riboside (**151**); *cis*-zeatin (**152**); laccase (**153**); *trans*-zeatin ribose (**154**), and *trans*-zeatin (**155**).

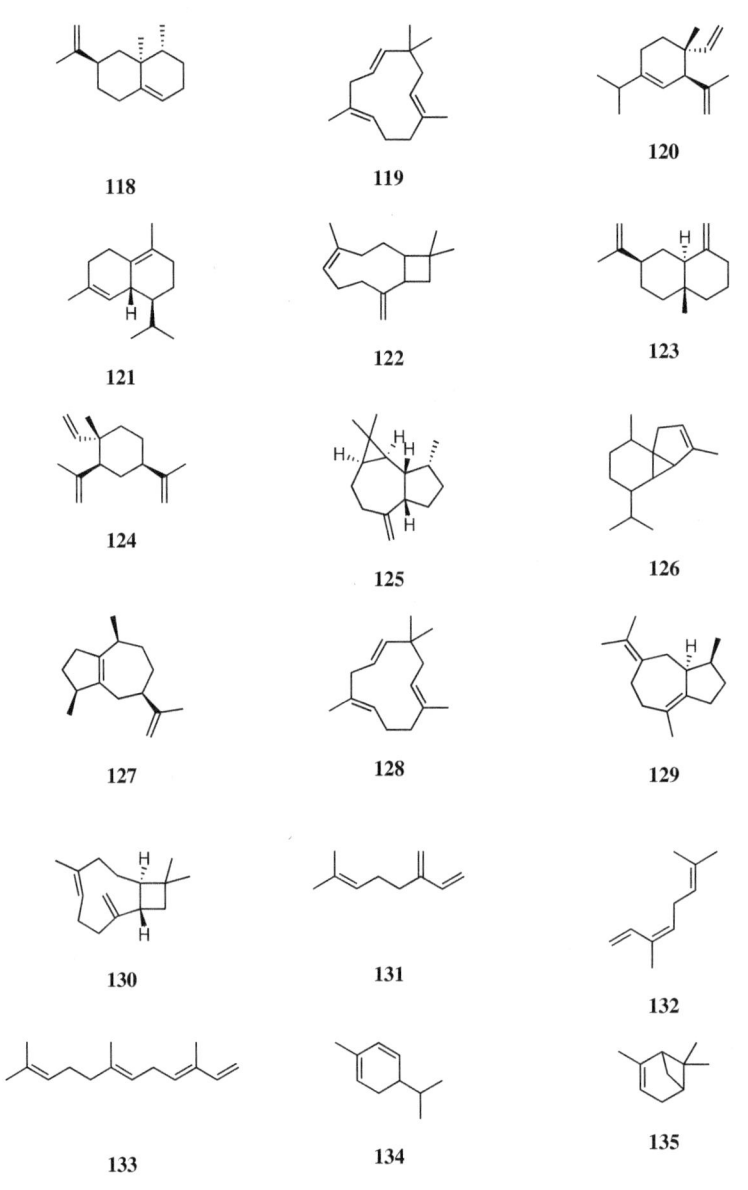

Figure 21.6 *Structures of some volatile compounds identified from* **M. indica.** Valencene (**118**); humulene (**119**); δ-elemene (**120**); δ-cadinene (**121**); caryophyllene (**122**); β-selinene (**123**); β-elemene (**124**); allo-aromadendrene (**125**); α-cubebene (**126**); α-guaiene (**127**); α-humulene (**128**); β-bulnesene (**129**); β-caryophyllene (**130**); β-myrcene (**131**); β-ocimene (**132**); α-farnesene (**133**); α-phellandrene (**134**); α-pinene (**135**); 2-ethylhexanol (**136**); 5-methylfurfur-2-al (**137**); ascorbic acid (**138**); furfural (**139**); meso inositol (**140**); linalool (**141**); linoleic acid (**142**); linolenic acid (**143**); octacosan-1-ol (**144**); octadeca-6-9-dien-1-oic acid (**145**); octadeca-cis-9-cis-15-dienoic acid (**146**); octdeca-3-6-9-trien-1-oic acid (**147**); oleic acid (**148**); palmitic acid (**149**); *n*-octane (**150**).

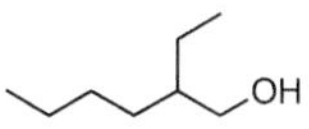

136

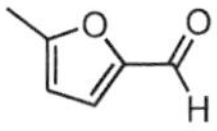

137

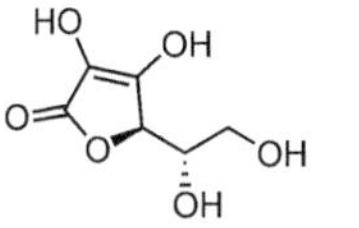

138

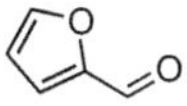

139

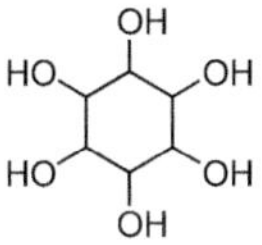

140

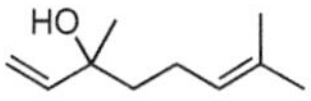

141

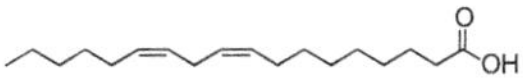

142

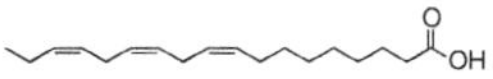

143

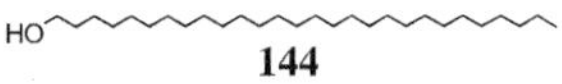

144

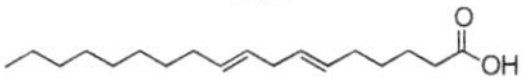

145

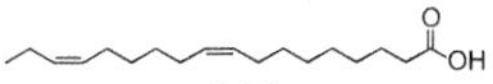

146

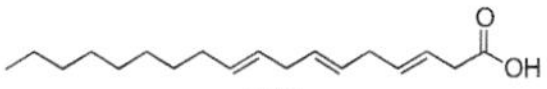

147

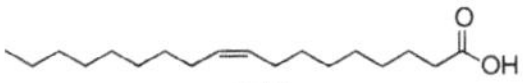

148

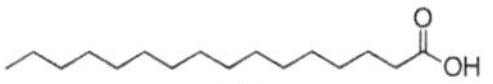

149

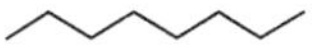

150

Figure 21.6 (*cont.*)

Table 21.1 Relative composition (%) of essential oil from *M. indica*.

	Relative composition							
	Fruit				Leaf			
	Immature		Mature		Immature		Mature	
Substance	**SPME**	**HD**	**SPME**	**HD**	**SPME**	**HD**	**SPME**	**HD**
Total	99.1	99.7	98.5	97.7	99.2	97.3	98.0	95.7
Monoterpenes	78.2	17.6	66.2	–	28.2	7.2	10.9	1.3
Oxygenated monoterpenes	0.2	23.9	–	–	0.1	3.6	–	1.1
Sesquiterpenes	20.2	28.5	29.2	63.2	60.3	76.8	74.2	80.2
Oxygenated sesquiterpenes	–	11.1	—	5.4	0.1	8.5	0.2	11.7

SPME, solid phase microextraction; HD, hydrodistillation. $t < 0.1\%$; (–), not identified.

6 PHARMACOLOGICAL ACTIVITY OF *Mangifera indica*

The chemical composition of *M. indica* has been studied extensively over the past few years and isolated chemical constituents have been shown to have numerous therapeutic uses (Shah et al., 2010). Although a lot of pharmacological investigations have been carried out on the isolated bioactive compounds, a lot more of the chemical components of *M. indica* remain unknown or inadequately investigated, and as such the opportunity remains for them to be explored and potentially utilized (Kalita, 2014). A summary of the findings of the pharmacological studies on *M. indica* is presented in the subsequent sections.

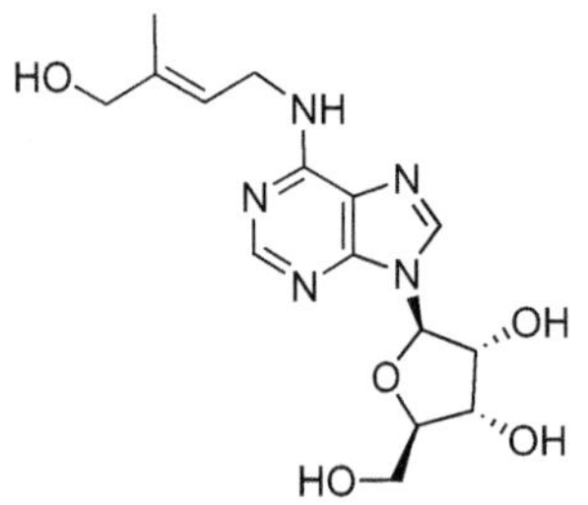

Figure 21.7 *Structures of some alkaloids reported from* **M. indica.** *Cis*-zeatin riboside (**151**); *cis*-zeatin (**152**); laccase (**153**); *trans*-zeatin ribose (**154**); *trans*-zeatin (**155**).

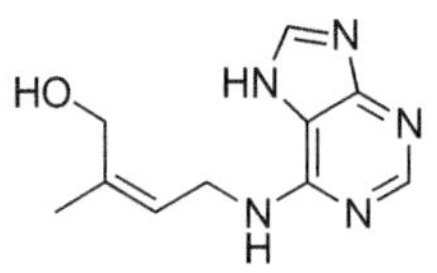

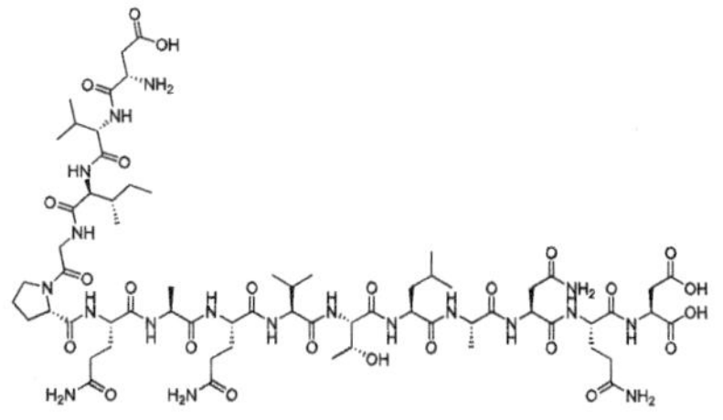

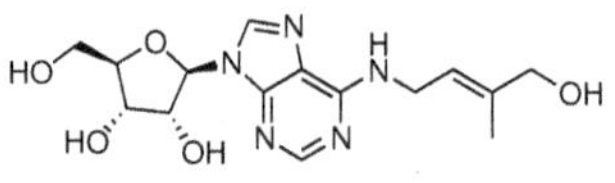

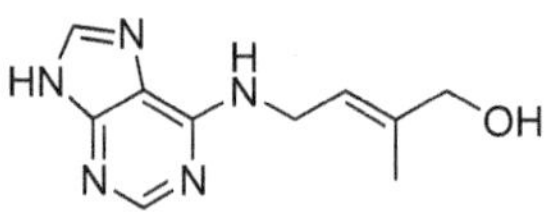

6.1 Antioxidant activity

Recent studies have demonstrated that intracellular reactive oxygen species (ROS), such as superoxide anion radicals, hydroxyl radical (OH), and lipid peroxides play an important role in the pathogenesis of acute gastric damage induced by ethanol as well as many chronic diseases including diabetes mellitus and cancers. The extreme reactivity of ROS and the resultant oxidative stress provokes severe damage at the cellular level leading to cell death (Juliana et al., 2009).

Antioxidants have been demonstrated to play important roles in ROS scavenging via several mechanisms, thereby reducing the adverse outcomes of ROS-induced injury. It is therefore not surprising that epidemiological studies have demonstrated that increased consumption of fruits and vegetables is associated with reduced risks of chronic diseases like cancers, likely due to their antioxidant-rich contents including phenolic compounds (Al-Shwyeh et al., 2015).

The mango plant has been the focus of attention of many researchers searching for potent antioxidants. This is informed by the finding that mango parts, such as stem bark, leaves, and pulp are known for possessing antioxidant and free radical scavenging and anticancer activities (Jayanti, 2015; Kassi et al., 2014). A recent study showed that *M. indica* waterlily kernel exhibited a strong antioxidant potency as determined by different antioxidant assays [1,1-diphenyl-2-picrylhydrazyl (DPPH) radical scavenging assay and ferric-reducing antioxidant power (FRAP) assay]. This was largely due to the presence of considerable amounts of phenols (up to 86 mg/g *M. indica* waterlily kernel) and associated polyphenols. The major phenolic compounds identified and quantified included epicatechin (**57**) and rutin (**75**) (Al-Shwyeh et al., 2015).

The mango is therefore a potential source of polyphenolic compounds with high antioxidative activity that help protect the body against damage linked to oxidative stress. The quantities and characteristics of the different phenolics expectedly differ in the different plant parts besides being affected by the geographic locations of the plants. Mangiferin (Fig. 21.3, compound **54**), which is mainly concentrated in the bark and leaves of the mango tree, is a unique polyphenol to the mango with notable pharmacological activity including antioxidant activity, a potential which has been exploited in medicine and food supplements (Jayanti, 2015). Whole mango extracts are more potent than pure isolated mangiferin highlighting the possible synergism between mangiferin and other mango polyphenols for enhanced activity (Masibo and He, 2008) (Table 21.2).

Table 21.2 Phytochemicals distribution in *Mangifera indica* plant parts

Compounds	Plant part	Reference
19, 54, 69	Bk	Khan et al. (1992), Pharm and Pharm (1991), Yang and Pengo (1981), Lu et al. (1982), Anjaneyulu et al. (1982)
13, 14, 15, 19, 54, 55, 70, 71, 74, 75, 86, 109	Lf	Griffiths (1959), Lu et al. (1982), Proctor and Creasy (1969), Shaft and Ikram (1982), Tanaka et al. (1984), Pharm and Pharm (1991), Yang and Pengo (1981), Anjaneyulu et al. (1982)
76, 118-135	Lf EO	Nigam (1962), Macleod and De (1982), Craveiro et al. (1980)
1, 4, 16, 17, 18, 19	Rt Bk	Pharm and Pharm (1991), Yang and Pengo (1981), Lu et al. (1982), Anjaneyulu et al. (1982), Cojocaru et al. (1986), Anjaneyulu et al. (1993), Anjaneyulu et al. (1985)
19, 54	St	Pharm and Pharm (1991), Yang and Pengo (1981), Lu et al. (1982), Anjaneyulu et al. (1982)
19, 138, 153	Fr	Hermano and Sepulveda (1934), Joel et al. (1978)
19, 78	Fr Pe	Proctor and Creasy (1969)
19, 86, 112, 113	Fr PU	Macleod and De (1982)
19, 103, 107	Fl	Khan and Khan (1989)
79	Pl	Proctor and Creasy (1969)
19, 67, 68, 86, 99, 114	Pn	Maheshwari and Mukerjee (1975), Ghosal et al. (1978)
15, 40, 45, 51, 72, 73, 77, 80, 81	Sh	Ghosal et al. (1978), Kolhe et al. (1982)
151, 152, 154, 155	Sd	Chen (1983)
15, 25, 28, 32, 34, 35, 38, 39, 41, 44, 49, 50, 142, 143, 145, 147, 149	Ker	Gaydou and Bouchet (1984)
4, 5, 7, 17, 18, 19, 36, 37, 42, 46, 47, 52, 53	St Bk	Anjaneyulu et al. (1993), Anjaneyulu et al. (1989), Anjaneyulu et al. (1985)
12, 29, 31, 45	Sd oil	Kolhe et al. (1982), Ghosal et al. (1978)

Bk, bark; EO, essential oils; Fr, fruit; Fl, flower; Ke, kernel; Lf, leaf; Pe, peel; Pn, plant; Pn, panicle; Pu, pulp; Rt, root; Sd, seed; St, stem; Sh, shoot.

In recent years there has been increased interest in the use of natural antioxidants in preference to synthetic antioxidants. This is because continuous use of synthetic antioxidants may cause health hazards, such as teratogenic and carcinogenic effects. Studies have suggested the involvement of high doses of synthetic antioxidants in cancer tissue initiation and propagation, chromosomal aberrations, and tissue damage (Sarma et al., 2010). Food legislators in many countries are hesitant to allow the addition of synthetic antioxidants to foods and wish to protect consumers from the risk of such chemicals (Emerton and Choi, 2008; Masibo and He, 2008). Mango seed kernel can be a potential source of these much sought after natural antioxidants. Being a very widespread plant, especially within the tropics, and owing to the variety of plant parts (pulp, peel, seed, bark, leaves, and flowers) being utilized domestically or industrially, the mango thus could be a cheap and readily available supplier of dietary polyphenols with great antioxidative potential that will help reduce degenerative diseases, such as cancer, atherosclerosis, diabetes, and obesity (Masibo and He, 2008).

6.2 Antimicrobial activity

This subsection focuses on antibacterial, antifungal, antiviral, and antiparasitic activities of *M. indica*.

6.2.1 Antibacterial activity

Ahmed et al. (2005) investigated the antimicrobial activity of 200 mg/mL solutions of chloroform, methanol, and aqueous extracts of *M. indica* seed kernels against two Gram-positive bacteria (*Bacillus subtilis, S. aureus*) and three Gram-negative bacteria (*Escherichia coli, Proteus vulgaris, Pseudomonas aeruginosa*). The chloroform and methanol extracts showed activity against all tested bacteria; the methanol extract was particularly active. The aqueous extract exhibited high activity against *B. subtilis, S. aureus,* and *P. vulgaris,* but showed low activity against *E. coli* and was inactive against *P. aeruginosa*. In a similar study, 100 mg/mL solutions of the methanol and ethyl acetate extracts of *M. indica* seeds showed high antibacterial activity against *B. subtilis, S. aureus, E. coli,* and *P. aeruginosa* (Intisar et al., 2010). Similar antibacterial activity of the plant has been well documented elsewhere (Jayanti, 2015; De and Arna Pal, 2014; Mushore and Matuvhunye, 2013; Doughari and Manzara, 2008; Prakash et al., 2013).

It is instructive to note that the *M. indica* extracts showed activity against a range of both Gram+ and Gram− bacteria, suggesting a broad spectrum antibacterial action, raising the possibility of the isolation of broad spectrum antibacterial compounds from this plant. In 2005, Stoilova et al. reported the antibacterial activity of the major compound of *M. indica*, mangiferin, against a range of Gram+ and Gram− bacteria using the in vitro agar diffusion technique. Mangiferin showed activity against seven bacterial species: *Bacillus pumilus, Bacillus cereus, S. aureus, E. coli, Salmonella agona, Klebsiella pneumoniae.* Separately, good activity of the same compound was also reported against *S. aureus* and *Salmonella typhi* (Taniya et al., 2015).

Kanwal et al. (2009) evaluated the antibacterial activity of five flavonoids isolated from *M. indica* leaves, which were (-)-epicatechin-3-O-β-glucopyranoside, 5-hydroxy-3-(4-hydroxyphenyl) pyrano [3,2-g]chromene-4(8*H*)-one, 6-(phydroxybenzyl) taxifolin-7-O-β-D-glucoside (tricuspid), quercetin-3-O-α-glucopyranosyl-(1→2)-β-D-glucopyranoside and (-)-epicatechin(2-(3,4-dihydroxyphenyl)-3,4-dihydro-2*H*-chromene-3,5,7-triol. The five compounds were tested against four bacterial species, namely *Lactobacillus* sp., *E. coli*, *Azospirillium lipoferum*, and *Bacillus* sp. and it was found that all the five isolated flavonoids significantly suppressed the growth of all the four tested bacterial species. The flavanoid (−)-epicatechin(2-(3,4-dihydroxyphenyl)-3,4-dihydro-2*H*-chromene-3,5,7-triol exhibited the most potent antibacterial activity, with *A. lipoferum* and *Bacillus* sp. showing the highest susceptibility to this compound. Kassi et al. (2014) reported in his study a significant antibacterial activity of methyl gallate (**90**) and penta-O-galloylglucose (**82**) against *S. aureus* and *P. aeruginosa*. These two were among several compounds isolated from the *M. indica* kernel seed cake.

The effects of the aqueous extract of young *M. indica* leaves against different bacteria that cause gastro-intestinal disorders—*E. coli*, *S. typhi*, *V. cholera*, and *Shigella sonnei*—were evaluated. The growth of all the tested organisms was inhibited in a dose-dependent manner, lending credence to the claims of antidiarrhoeal properties of *M. indica* (De and Arna Pal, 2014; Prakash et al., 2013).

6.2.2 Antifungal activity

Ahmed et al. (2005) also investigated the antifungal activity of the 200 mg/mL solutions of chloroform, methanol, and aqueous extracts of *M. indica* seed kernels against two fungal species (*Aspergillus niger* and *Candida albicans*). The chloroform and methanol extracts showed high activity against *C. albicans*, but were inactive against *A. niger*. The aqueous extract did not exhibit any inhibitory activity against either of the fungal species. Separately, the methanol extract of *M. indica* seeds was shown to exhibit high antifungal activity against *C. albicans* (Intisar et al., 2010).

Kanwal et al. (2010) also tested the five flavonoids isolated from *M. indica* leaves against the five fungal species *Alternaria alternata* (Fr.) Keissler, *Aspergillus fumigatus* Fresenius, *A. niger* van Tieghem, *Macrophomina phaseolina* (Tassi) Goid., and *Penicillium citrii*. All the compounds showed good antifungal activity which was found to be dose-dependent, increasing with increasing concentrations of the compounds. Mangiferin has also been shown to possess antifungal activity. Its strongest antifungal activity was observed against *Thermoascus aurantiacus*. Weak antifungal activity was observed against *Saccharomyces cerevisiae*, *Trichoderma reesei*, *Aspergillus flavus*, and *A. fumigatus*. No inhibitory activity was observed against *C. albicans* and *A. niger* (Stoilova et al., 2005).

6.2.3 Antiprotozoal and antiparasitic activity

In a neonatal mouse model, mangiferin at 100 mg/kg had a similar inhibitory activity against *C. parvum* as the same dose (100 mg/kg) of a known active drug, paromomycin.

Despite these high activities, both mangiferin and paromomycin only reduced intestinal colonization of *C. parvum* but were not able to completely inhibit it (Gehad and Samir 2013; Perrucci et al., 2006). These findings were very promising as mangiferin could be considered as a hit compound for optimization in the development of new anticryptosporidial agents. Further studies are also required on the primary compound to evaluate the effectiveness of the compound against *Cryptosporidium* infections using other animal models.

Antiplasmodial activity of the stem bark extract of *M. indica* against *Plasmodium yoelii nigeriensis* was evaluated (Awe et al., 1998). A marked schizontocidal effect during early infection was observed, along with notable repository activity (Awe et al., 1998).

Tarkang et al. (2014) also evaluated the antiplasmodial activity of constituent solvent extracts of Nefang, a polyherbal preparation that contains *M. indica* (bark and leaf) among other herbs. Of the 16 extracts tested, 9 showed significant antiplasmodial activity at concentrations less than 50 µg/mL. Interestingly, they found six extract pairs with apparent synergistic interactions with antiplasmodial activities >5-fold that of Nefang or the respective extract activities when tested alone. These promising synergistic interactions observed in the above study are for exploitation toward a rational phytotherapeutic antimalarial drug discovery.

The aqueous extracts of immature fruits of *M. indica* have showed high inhibition of larval development when evaluated against *Strongyloides stercoralis*, a human pathogenic parasitic roundworm, supporting its use as an antihelmintic (El-Sherbini and Osman, 2013).

García et al. (2003) report that treatment with 50 mg per kg body weight per day of mangiferin has modest and stage-dependent anthelmintic effects in the mouse model of *Trichinella spiralis*, a nematode parasite that infects humans; however, the compound was not effective against adult worms in the gut.

6.2.4 Antiviral activity

In vitro study of the effect of mangiferin and isomangiferin against *Herpes simplex* virus type 2 showed notable antiviral activity of these compounds, with isomangiferin exhibiting HSV2 inhibitory activity comparable to the control dugs acyclovir and idoxuridine (Zheng and Lu, 1990). The anti-HIV-1 activity of mangiferin was evaluated in 2014 by Wang and coworkers. They observed mangiferin to exhibit good activities against a variety of laboratory-derived, clinically isolated and resistant HIV-1 strains. Their studies suggested that mangiferin might inhibit HIV-1 protease (Wang et al., 2011).

6.3 Antitumor activity

Al-Shwyeh et al. (2014) evaluated the cytotoxicity of ethanolic kernel extract of *M. indica* L on MCF-7 and MDA-MB- 231 cell lines and concluded that the extract is significantly cytotoxic to these cell lines in a dose-dependent manner, and considerably less so toward normal breast cells MCF-10A. These findings highlighted the potential of *M. indica* L. extract in the treatment of breast cancer. Mangiferin has been investigated as a potential biological response modifier with antitumor, and has been reported to have in vivo growth-inhibitory activity against ascitic fibrosarcoma in Swiss mice. Mangiferin was found to induce antitumor effect irrespective of the size of tumor inoculum. It was also observed to have cytotoxic effect on tumor cells (Guha et al., 1996; Wauthoz et al., 2007).

Reports from the literature also suggested that mangiferin can be used to prevent carcinogenesis or reverse tumor promotion, a very promising approach to cancer control (Shashi et al., 2009). The results obtained by Joona et al. (2013) also revealed that 90% methanolic leaf extracts of *M. indica* exhibit significant anticancer activity due to the increased flavonoids and terpenoids level. The essential oils of *M. indica* presented a low toxicity to mammalian cells and the anticancer activity in vitro of these oils was considered promising (Ramos et al., 2014).

6.4 Antidiabetic activity

The leaves and stem barks of *M. indica* plant have been claimed to possess antidiabetic properties by traditional healers (Bhowmik et al., 2009). The hypoglycemic effects of *M. indica* have also been demonstrated experimentally. For example, both ethanol and water extracts of leaves and stem barks of *M. indica* showed significant hypoglycemic effect in type 2 model rats when fed simultaneously with glucose load (Bhowmik et al., 2009).

Gupta and Gupta (2011) evaluated the antidiabetic as well as antioxidant efficacy of *M. indica* ethanolic extract in streptozotocin-induced diabetic rats. Administration of ethanol extract of *M. indica* significantly lowered the elevated blood glucose levels and restored the levels of glycated hemoglobin back to normal range following 14 and 21 days treatment with dose of 300 mg/kg b.w./day of the extract.

Rajesh and Rajasekhar (2014) screened the extracts of *M. indica* seed kernel for its antidiabetic activity; they discovered that the long-term (21 days) administration of methanolic and aqueous extract of *M. indica* was effective in decreasing the blood glucose level and normalizing the other biochemical parameters in diabetic rats. Incidentally, a single dose study of the extract did not have hypoglycemic effect on normal rats.

In the study carried out by Kemasari et al. (2011) ethanol leaf extracts of *M. indica* were also found to significantly reduce the elevated levels of glucose as well as urea, uric acid, creatinine, and circulate liver enzyme levels in alloxan-induced diabetic rats. This supports the claim that *M. indica* may be beneficial for the treatment of diabetes mellitus.

Irondi et al. (2016) also observed in their study that *M. indica* kernel flour-supplemented diets improved the levels of fasting blood glucose, hepatic glycogen, glycosylated hemoglobin, lipid profile, plasma electrolytes, malonaldehyde, and the liver function biomarkers of the diabetic rats. *M. indica* kernel flour could therefore be a promising nutraceutical therapy for the management of type 2 diabetes and its associated complications.

6.5 Antiinflammatory and antiallergic activity

The analgesic and antiinflammatory effects of Vimang, a *M. indica* aqueous extract formulation which majorly contains mangiferin, were reported for the first time by Gabino et al. (2001) and later by Garrido-Suárez et al. (2010). The *M. indica* extract exhibited a potent and dose-dependent antinociceptive effect using acetic acid-induced abdominal constriction mouse model for assessment of analgesia. It also significantly inhibited both carrageenan- and formalin-induced edema in rat, guinea pigs, and mice. The inhibitions were similar to those produced by indomethacin and sodium naproxen. Since then, several reports of the analgesic and antiinflammatory activity of various *M. indica* extracts have been published (Bhadrapura and Sudharshan, 2016; Hassan et al., 2013; Oluwafemi, 2015; Olorunfemi et al., 2012).

6.6 Immunomodulatory effect

Savant et al. (2014) reported potent immunostimulant effects of a petroleum ether extract of *M. indica* and rationalized this activity by the presence of catechin, epicatechin, and oxyresveratrol in the extract based on the GC-MS analysis report, the study therefore supported the claims of the potent immunomodulatory effect of *M. indica*.

Investigations of the alcoholic extract of *M. indica* suggested that the immunostimulatory activity of mangiferin may result from an increase in humoral antibody titers and delayed type hypersensitivity in mice (Makare et al., 2001). Other studies have proposed that mangiferin may exert its immunomodulatory effects through a variety of mechanisms, including suppression of the proliferative response of splenocytes and thymocytes (Chattopadhyay et al., 1987), suppression of antibody production (Leiro et al., 2003; Rajendran et al., 2013), inhibition of T-cell activation (Garrido et al., 2005), and inhibition of macrophage function (Leiro et al., 2002).

6.7 Other pharmacological properties

A recent report of Neelapu et al. (2012) indicated that the leaves of *M. indica* possess antiulcer activity in animal models and concluded that the antiulcer activity is probably due to the presence of bioactive compounds like flavonoids, saponin, and tannins.

Nwinuka et al. (2008) conclusively noticed that extract of *M. indica* stem bark possesses hematopoietic effects manifested by increase in hematocrit and a rise in erythrocyte, leukocyte, platelet, and lymphocytes counts, though hemoglobin and neutrophil levels were found to decrease. This study seemed to discount the possibility of *M. indica* inducing any adverse complications on hematological system.

7 CLINICAL TRIALS

Strong efforts have been made to provide standard aqueous extract of *M. indica* in Cuba under the brand name of Vimang which is available as antidiabetic, antioxidant, antiinflammatory, and analgesic. Although Vimang is probably used in Cuba only; this is a good step toward new, standardized formulation of *M. indica*. Moreover, despite the fact that there is no other information available to show if any other compound or product derived from *M. indica* is on clinical trials, there are certainly many research groups focusing on the therapeutic effects of this plant. Hence, there is need to further compile all available data and information about *M. indica* and to investigate possible ways by which formulations, such as Vimang could be developed.

The information currently available on *M. indica* would go a long way in justifying why further therapeutic and toxicological studies should be done on formulations of *M. indica*, such as Vimang to validate their clinical usage. Other studies also demonstrated that *M. indica* leaves (powdered part, aqueous, or alcoholic extract) can be combined with oral hypoglycaemic agents to control blood glucose levels inpatients whose diabetes is not optimally controlled with oral hypoglycaemic agents alone, or in those patients in whom these agents produce intolerable adverse effects (Akbar and Ahmad, 2006).

Further emphases should also be made toward identifying the bioactive constituents responsible for the observed pharmacological effects. This would facilitate their incorporation into modern drug discovery approaches that would likely yield lead compounds and drug candidates that could be accepted, either alone or as combination therapies, for the treatment of human ailments in hospitals and other clinical settings (Oluwafemi, 2015).

8 TOXICITY STATUS OF *Mangifera indica*

Ashalatha et al. (2015) reported a study of the acute toxicity of seed kernel of *M. indica* Linn in which they observed *M. indica* to be minimally toxic. Similar reports of the lack of toxicity, and even cytoprotective effects of *M. indica* have been published (Ahmed et al., 2015; Zhang et al., 2014; Alkizima et al., 2012; Izunya et al., 2010).

9 CONCLUSIONS

Being a very widespread plant, especially within the tropics, and owing to the variety of plant parts (pulp, peel, seed, bark, leaves, and flowers) being utilized domestically or industrially, the mango thus could be a cheap and readily available supplier of dietary polyphenols with great antioxidative potential that will help reduce degenerative diseases, such as cancer, atherosclerosis, diabetes, and obesity.

An attempt to compile the reported information about medicinal values of *M. indica* has been made in this chapter. Although all information have not been included in this chapter, it can be initiative for further phytochemical and pharmacological investigations about the medicinal use of the plant and probable herbal formulations. The review seems to be a step ahead toward the new drug development. This may be useful to the health professionals, scientists, and scholars working in the field of natural product chemistry, pharmacology, and therapeutics to develop evidence-based alternative medicine to cure different kinds of diseases.

REFERENCES

Abdelnaser, A.E., Shinkichi, T., 2010. Preliminary phytochemical investigation on Mango (*Mangifera indica* L.) leaves. World J. Agric. J. 6, 735–739.

Ahmed, S.O.M., Omer, M.A.A., Almagboul, A.Z., Abdrabo, A.N., 2005. In vitro antimicrobial activity of *Mangifera indica* L. Gezira J. Health Sci. 1, 82–91.

Ahmed, I.A., Sohair, R.F., Fathi, M.A., Elaskalany, S.M., 2015. Renoprotective effect of *Mangifera indica* polysaccharides and silymarin against cyclophosphamide toxicity in rats. J. Basic Appl. Zool. 72, 154–162.

Akbar, W.M.G.A., Ahmad, S.I., 2006. Clinical investigation of hypoglycemic effect of leaves of *Mangifera indica* in type-2 (niddm) diabetes mell1tus. Pak. J. Pharmacol. 23, 13–18.

Alkizima, F.O., Matheka, D.M., Muriithi, A.W., 2012. Dose-dependent myocardial toxicity of *Mangifera indica* during diarrhoea treatment. Afr. J. Pharmacol. Ther. 1, 67–70.

Al-Shwyeh, H.A., Abdulkarim, S.M., Rasedee, A., Mohamed, E., Saeed, M., Al-Qubaisi, Mothanna, 2014. Cytotoxic effects of *Mangifera indica* L. kernel extract on human breast cancer (MCF-7 and MDA-MB-231 cell lines) and bioactive constituents in the crude extract. BMC Complement. Altern. Med. 14, 199.

Al-Shwyeh, H.A., Abdulkarim, S.M., Rasedee, A., 2015. Identification and quantification of phenolic compounds in *Mangifera indica* waterlily kernel and their free radical scavenging activity. J. Adv. Agric. Technol. 2, 1–7.

Amian, B.B.K., Yaya, S., Bamba, F., Koffi, J.G., Siaka, S., Amadou, S.T., Coustard, J.M., 2014. Isolation and identification of bioactive compounds from kernel seed cake ofthe Mango (*Mangifera indica* Lam). Int. J. Biol. Chem. Sci. 8, 1885–1895.

Anjaneyulu, V., Babu, J.S., Krishna, M.M., Connolly, J.D., 1993. 3-Oxo- 20S, 24R, epoxy -dammarane- 2 5 zeta, 26-diol from *Mangifera indica*. Phytochemistry 32, 469–471.

Anjaneyulu, V.K., Harischandra, P., Sambasiva, R.G., 1982. Triterpenoids of the leaves of *Mangifera indica*. Ind. J. Pharm. Sci. 44, 58–59.

Anjaneyulu, V., Prasad, K.H., Ravi, K., Connolly, J.D., 1985. Triterpenoids from *Mangifera indica*. Phytochemistry 24, 2359–2367.

Anjaneyulu, V., Radhika, P., 2000. The triterpenoids and steroids from *Mangifera indica* Linn. Ind. J. Chem. 39, 883–893.

Anjaneyulu, V., Ravi, K.K., Harischanrda, P., Connolly, J.D., 1989. Triterpenoids from *Mangifera indica*. Phytochemistry 28, 1471–1477.

Ashalatha, M., Shivakumar, Suresh, J., Shyla, R.J.B.Y., 2015. Acute toxicity study of seed kernel of *mangifera indica linn* (skmi). Intern. Ayur. Med. J. 3, 2825–2827.

Awad El-Gied, A.A., Joseph, M.R.P., Mahmoud, I.M., Abdelkareem, A.M., Al Hakami, A.M., Hamid, M.E., 2012. Antimicrobial activities of seed extracts of mango (*Mangifera indica* L.). Adv. Microbiol. 2, 571–576.

Awe, S.O., Olajide, O.A., Oladiran, O.O., Makinde, J.M., 1998. Antiplasmodial and antipyretic screening of *Mangifera indica* extract. Phytother. Res. 12, 437–438.

Bally, I.S.E., 2006. *Mangifera indica* (mango), ver. 3.1. In: Elevitch, C.R. (Ed.), Species Profiles for Pacific Island Agroforestry. Permanent Agriculture Resources (PAR), Hōlualoa, Hawai, p. 25.

Bally, I.S.E., 2011. Advances in research and development of mango industry. Rev. Bras. Frutic., Jaboticabal E. 33, 57–63.

Bandyopadhyay, C., Gholap, A.S., Mamdapur, V.R., 1985. Characterization of alkenylresorcinol in mango (*Mangifera indica* L) latex. J. Agric. Food Chem. 33, 377–379.

Bello-Perez, L.A., Garcia-Suarez, F.J.L., Agama-Acevedo, E., 2007. Mango carbohydrates. Food 1, 36–40.

Bhadrapura, L.D., Sudharshan, S., 2016. The anti-inflammatory activity of standard aqueous stem bark extract of *Mangifera indica* L. as evident in inhibition of Group IA sPLA2. An. Acad. Bras. Ciênc. 88, 197–209.

Bhowmik, A., Khan, L.A., Akhter, M., Rokeya, B., 2009. Studies on the antidiabetic effects of *Mangifera indica* stem-barks and leaves on nondiabetic, type 1 and type 2 diabetic model rats. Bangladesh J. Pharmacol. 4, 110–114.

Bhuvaneswari, K., 2013. Isolation of mangiferin from leaves of *Mangifera indica* l. var alphonso. Asian J. Pharm. Clin. Res. 6, 173–174.

Chattopadhyay, U., Das, S., Guha, S., Ghosal, S., 1987. Activation of lymphocytes of normal and tumor bearing mice by mangiferin, a naturally occurring glucosylxanthone. Cancer Lett. 37, 293–299.

Chen, W.S., 1983. Cytokinins of the developing mango fruit. Isolation, identification and changes in levels during maturation. Plant Physiol. 71 (2), 356–361.

Chidozie, V.N., Adoga, G.I., Chukwu, O.C., Chukwu, I.D., Adekeye, A.M., 2014. Antibacterial and toxicological effects of the aqueous extract of *Mangifera indica* stem bark on albino rats. Global J. Biol., Agric. Health Sci. 3, 237–245.

Cojocaru, M., Droby, S., Glotter, E., Goldman, A., Gottlieb, H.E., Jacoby, B., Prusky, D., 1986. 5-(12-Heptadecenyl)resorcinol, the major component of the antifungal activity in the peel of mango fruit. Phytochemistry 25, 1093–1095.

Craveiro, A.A., Andrade, C.H., Matos, F.J., Alencar, J.W., Machado, M.1., 1980. Volatile constituents of *Mangifera indica* linn. Rev. Latinoamer. Quim. 11, 129.

De, P.K., Arna Pal, A., 2014. Effects of aqueous young leaves extract of *Mangifera indica* on gm (-) microorganisms causing gastro-intestinal disorders. Asian J. Plant Sci. Res. 4, 23–27.

Dorta, E., González, M., Lobo, M.Gloria, Sánchez-Moreno, Concepción, de Ancos, Begoña, 2014. Screening of phenolic compounds in by-product extracts from mangoes (*Mangifera indica* L.) by HPLC-ESI-QTOF-MS and multivariate analysis for use as a food ingredient. Food Res. Int. 57, 51–60.

Doughari, J.H., Manzara, S., 2008. In vitro antibacterial activity of crude leaf extracts of *Mangifera indica* Linn. Afr. J. Microbiol. Res. 2, 067–072.

El-Sherbini, G.T., Osman, S.M., 2013. Anthelmintic activity of unripe *Mangifera indica* L. (mango) against *Strongyloides stercoralis*. Int. J. Curr. Microbiol. Appl. Sci. 2, 401–409.

Emerton, V., Choi, E., 2008. Essential Guide to Food Additives, third ed. Leatherhead Food International Ltd, Surrey, UK, 336p.

Fenglei, L., Qiang, L., Yuqin, Z., Guibing, H., Guodi, H., Jiukai, Z., Chongde, S., Xian, L., Kunsong, C., 2012. Quantification and purification of mangiferin from Chinese mango (*Mangifera indica* L.) cultivars and its protective effect on human umbilical vein endothelial cells under H_2O_2-induced stress. Int. J. Mol. Sci. 13, 11260–11274.

Gabino, G., Deyarina, G., Delporte, C., Backhouse, N., Gypsy, Q., Núñez-Sellés, A.J., Miguel, A.M., 2001. Analgesic and anti-inflammatory effects of *Mangifera indica* L. extract (Vimang). Phytother. Res. 15, 18–21.

García, D., Escalante, M., Delgado, R., Ubeira, F.M., Leiro, J., 2003. Anthelminthic and antiallergic activities of *Mangifera indica* L. stem bark components Vimang and mangiferin. Phytother. Res. 17, 1203–1208.

Garrido-Suárez, B.B., Garrido, G., Delgado, R., Bosch, F., del, C.R.M., 2010. A *Mangifera indica* L. extract could be used to treat neuropathic pain and implication of mangiferin. Molecules 15, 9035–9045.

Garrido, G., Molina, M.B., Sancho, R., Macho, A., Delgado, R., Munoz, E., 2005. An aqueous stem bark extract of *Mangifera indica* (Vimang®) inhibits T cell proliferation and TNF-induced activation of nuclear transcriptionfactor NF-κB. Phytother. Res. 19, 211–215.

Gaydou, E.M., Bouchet, P., 1984. Sterols, methyl sterols, triterpene alcohols and fatty acids of the kernel fat of different Malagasy mango *(Mangifera indica)* varieties. J. Amer. Oil Chem. Soc. 61 (10), 1589–1593.

Gehad, T.E., Samir, M.O., 2013. Anthelmintic activity of unripe *Mangifera indica* L. (mango) against *Strongyloides stercoralis*. Int. J. Curr. Microbiol. Appl. Sci. 2 (5), 401–409, (2013).

Ghosal, S., Biswas, K., Chattopadhyay, B.K., 1978. Differences in the chemical constituents of *Mangifera indica* infected with *Aspergillus niger* and *Fusarium moniliformae*. Phytochemistry 17, 689–694.

Gillett, N., 2015. Botanical Descriptions of Mango Varieties Found in the Belize District, Belize. Thesis, University of Belize, Belize, 34p.

Githure, C.W., Schoeman, A.S., McGeoch, M.A., 1998. Differential susceptibility of mango cultivars in South Africa to galling by the mango gall fly, *Procontarinia matteiana* Kieffer & Cecconi (Diptera: Cecidomyiidae). Afr. Entomol. 6, 33–40.

Griffiths, L.A., 1959. On the distribution of gentisic acid in green plants. J. Exp. Biol. 9 (10), 437.

Guha, S., Ghosal, S., Chattopadhyay, U., 1996. Antitumor, immunomodulatory and anti-HIV effect of mangiferin, a naturally occurring glucosylxanthone. Chemotherapy 42, 443–451.

Gupta, R., Gupta, R.S., 2011. Antidiabetic efficacy of *Mangifera indica* seed kernels in rats: a comparative study with glibenclamide evaluate the antidiabetic as well as antioxidant efficacy of *Mangifera indica* ethanolic extract in streptozotocin induced diabetic rats. Diabetol. Croatica 40, 107–112.

Gururaja, G.M., Deepak, M., Shekhar, M.D., Gopala, K.S., Abhilash, K., Amit, A., 2014. Cholesterol esterase inhibitory activity of bioactives from leaves of *Mangifera indica* L. Pharmacogn. Res. 7, 355.

Hassan, M.M., Khan, S.A., Shaikat, A.H., Hossain, M.E., Hoque, M.A., Ullah, M.H., Islam, S., 2013. Analgesic and anti-inflammatory effects of ethanol extracted leaves of selected medicinal plants in animal model. Vet. World 6 (2), 68–71.

Hermano, A.J., Sepulveda, G.J.R., 1934. The vitamin content of Philippine foods. II. Vitamin C in various fruits and vegetables. Philipp. J. Sci. 19 (53), 379.

Igoli, J.O., Ogaji, O.G., Tor-Anyiin, T.A., Igoli, N.P., 2005. Traditional medicine practice amongst the igede people of Nigeria. part II. Afr. J. Trad. CAM 2, 134–152.

Intisar, S.A., Mahgoub Shareif El, T., Aisha, Z.A., Verpoorte, R., 2010. Characterization of anti-microbial compounds isolated from *Mangifera indica* L seed kernel. Univ. Afr. J. Sci. 2, 77–91.

Irondi, E.A., Oboh, G., Akindahunsi, A.A., 2016. Antidiabetic effects of *Mangifera indica* kernel flour-supplemented diet in streptozotocin-induced type 2 diabetes in rats. Food Sci. Nutr. 1, 1–12.

Islam, M.R., Mannan, M.A., Kabir1, M.H.B., Islam, A., Olival, K.J., 2010. Analgesic, anti-inflammatory and antimicrobial effects of ethanol extracts of mango leaves. J. Bangladesh Agric. Univ. 8, 239–244.

Izunya, A.M., Nwaopara, A.O., Aigbiremolen, A., Odike, M.A.C., Oaikhena, G.A., Bankole, J.K., Ogarah, P.A., 2010. *Mangifera indica* L. (mango) stem bark on the liver in Wistar rats. Res. J. Appl. Sci., Eng. Technol. 2, 460–465.

Jayanti, G., 2015. Ethanomedical, chemical, pharmacological, toxicological properties of *Mangifera indica*: a review. Int. J. Pharm. Res. Rev. 4, 51–64.

Joel, D.M., Harbach, I., Mayer, A.M., 1978. Laccase in Anacardiaceae. Phytochemistry 17, 796–797.

Joona, K., Sowmia, C., Dhanya, K.P., Divya, M.J., 2013. Preliminary phytochemical investigation of *Mangifera indica* leaves and screening of antioxidant and anticancer activity. Res. J. Pharm., Biol. Chem. Sci. 4, 1112–1118.

Juceni, P.D., Jorge, M.D., 2006. Metabólitossecundários de espéciesde anacardiaceae. Quim. Nova 29, 1287–1300.

Juliana, A.S., Zeila, P.L., Hélio, K., Alba, R.M.S.B., Lourdes, C., dos Santos, Wagner, V., Clélia, A.H.L., 2009. Polyphenols with antiulcerogenic action from aqueous decoction of mango leaves (*Mangifera indica* L.). Molecules 14, 1098–1110.

Jurgen, Griesbach, 2003. Mango Growing in Kenya. World Agroforestry Centre, Kul Graphics, Nairobi, 118p.

Kalita, P., 2014. An overview on *Mangifera indica*: importance and its various pharmacological action. Pharma. Tutor 2, 72–76.

Kanwal, Q., Hussain, I., Siddiqui, H.L., Javaid, A., 2009. Flavonoids from mango leaves with antibacterial activity. J. Serb. Chem. Soc. 74, 1389–1399.

Kanwal, Q., Hussain, I., Siddiqui, H.L., Javaid, A., 2010. Antifungal activity of flavonoids isolated from mango (*Mangifera indica* L.) leaves. Nat. Prod. Res. 24, 1907–1914.

Karthy, E.S., Ranjitha, P., 2011. Screening of antibacterial tannin compound from mango (*Mangifera indica*) seed kernel extract against methicillin resistant *Staphylococcus aureus* (MRSA). Elixir Pharm. 40, 5251–5255.

Kassi, A.B.B., Soro, Y., Fante, B., Golly, K.J., Sohro, S., Toure, A.S., Coustard, J.-M., 2014. Isolation and identification of bioactive compounds from kernel seed cake of the mango (*Mangifera indica* Lam). Int. J. Biol. Chem. Sci. 8, 1885–1895.

Kaur, J., Rathinam, X., Kasi, M., Miew, L.K., Ayyalu, R., Kathiresan, S., Subramaniam, S., 2010. Preliminary investigation on the antibacterial activity of mango (*Mangifera indica* L: Anacardiaceae) seed kernel. Asian Pac. J. Trop. Med. 3, 707–710.

Kemasari, P., Sangeetha, S., Venkatalakshmi, P., 2011. Antihyperglycemic activity of *Mangifera indica Linn*. alloxan induced diabetic rats. J. Chem. Pharm. Res. 3 (5), 653–659.

Khan, M.A., Khan, M.N.I., 1989. Alkyl gallates of flowers of *Mangifera indica*. Fitoterapia 60, 284.

Khan, M.A., Nizami, S.S., Khan, M.N.I., Azeem, S.W., 1992. Biflavone from *Mangifera indica*. Pak. J. Pharm. Sci. 5, 155–159.

Khyati, A.S., Mandev, B.P., Shreya, S.S., Kajal, N.C., Parul, K.P., Natavarlal, M.P., 2010. Antihyperlipidemic activity of *Mangifera indica* l. leaf extract on rats fed with high cholesterol diet. Der. Pharm. Sin. 1, 156–161.

Kokwaro, J.O., 1986. Anacardiaceae. In: Polhill, R.M. (Ed.), Flora of Tropical East Africa. A. A. Balkema, Rotterdam, 59p.

Kolhe, J.N., Bhaskar, A., Brongi, N.V., 1982. Occurrence of 3-oxo triterpenes in the unsaponifiable matter of some vegetable fats. Lipids 17, 166–168.

Kritikar, K.R., Basu, B.D., 1993. Indian Medicinal Plantsvol. ILalit Mohan Basu, Allahabad, India.

Leiro, J., García, D., Delgado, R., Sanmartin, M.L., Ubeira, F.M., 2003. *Mangifera indica* L. extract (Vimang) and mangiferin modulate mouse humoral immune responses. Phytother. Res. 17, 1182–1187.

Leiro, J., García, D., Delgado, R., Ubeira, F.M., 2002. Modulation of rat macrophage function by the *Mangifera indica* L. extracts Vimang and mangiferin. Int. Immunopharmacol. 2, 797–806.

Lu, Z.Y., Mao, H.D., He, M.R., Lu, S.Y., 1982. Studies on the chemical constituents of mangguo *(Mangifera indica)* leaf. Chung Ts'ao Yao 13, 3–6.

Macleod, A.J., De, T.R.N.G., 1982. Volatile flavour components of mango fruit. Phytochemistry 21, 2523–2526.

Madan, K., Shukla, D.S., Tripathi, R., Tripathi, A., Singh, R., Dwivedi, H.D., 2014. Isolation of three chemical constituents of *Mangifera indica* wood extract and their characterization by some spectroscopic techniques. Int. J. Emerg. Technol. Comput. Appl. Sci. 8, 217–218.

Maheshwari, M.L., Mukerjee, S.K., 1975. Lipids and phenolics of healthy and malformed panicles of *Mangifera indica*. Phytochemistry 14, 2083–2084.

Makare, N., Bodhankar, S., Rangari, V., 2001. Immunomodulatory activity of alcoholic extract of *Mangifera indica* L. in mice. J. Ethnopharmacol. 78, 133–137.

Malami, I., Muhammad, S.M., Alhasan, M.A., Muhammad, K.D., Kabiru, A., 2014. Hepatoprotective activity of stem bark extract of *Mangifera indica* L. on carbon tetrachloride-induced hepatic injury in wistar albino rats. IJPSR 5, 1240–1245.

Manisha, S., Sikdar, D.C., 2015. Production of starch from mango (*Mangifera indica*.L) seed kernel and its characterization. Int. J. Tech. Res. Appl. 3, 346–349.

Masibo, M., He, Q., 2008. Major mango polyphenols and their potential significance to human health. Comprehen. Rev. Food Sci. Food Saf. 7, 309–319.

Massimo, D'Archivio, Carmela, F., Roberta Di, B., Raffaella, G., Claudio, G., Roberta, M., 2007. Polyphenols, dietary sources and bioavailability. Ann. Ist Super Sanità 43, 348–361.

Morton, J., 1987. *Mango Mangifera indica* L. Fruits of Warm Climates. J. F. Morton, Miami.

Mushore, Joshua, Matuvhunye, Takudzwa, 2013. Antibacterial properties of *Mangifera indica* on Staphylococcus aureus. Afr. J. Clin. Exp. Microbiol. 14, 62–74.

Mutiat, A.O., Oyinlade, C.O., Adeteju, Oluwafunmilayo, L., 2015. Hepatoprotective effect of *Mangifera indica* stem bark extracts on paracetamol-induced oxidative stress in albino rats. Eur. Sci. J. 11, 299–309.

Neelapu, N., Muvvala, S., Mrityunjaya, B.P., Lakshmi, B.V.S., 2012. Anti-ulcer activity and HPTLC analysis of *Mangifera indica* L. leaves. Int. J. Pharm. Phytopharmacol. Res. 1, 146–155.

Nicolai, B., Ramona, F., Jurgen, C., Uwe Beifuss, Reinhold, C., Andreas, S., 2005. Screening of mango (*Mangifera indica* L.) cultivars for their contents of flavonol O- and xanthone C-glycosides, anthocyanins, and pectin. J. Agric. Food Chem. 53, 1563–1570.

Nicolai, B., Reinhold, C., Andreas, S., 2004. Characterization of gallotannins and benzophenone derivatives from mango (*Mangifera indica* L. cv. 'Tommy Atkins') peels, pulp and kernels by high-performance liquid chromatography/electrospray ionization mass spectrometry. Rapid Commun. Mass Spectrom. 18, 2208–2216.

Nigam, I.C., 1962. Studies of some Indian essential oils. Agra Univ. J. Res. Sci. 11, 147–152.

Nwinuka, Nwibani, M., Monanu, Michael, O., Barine, I., 2008. Effects of aqueous extract of *Mangifera indica* L. (mango) stem bark on haematological parameters of normal Albino rats. Pak. J. Nutr. 7, 663–666.

Olorunfemi, O.J., Nworah, D.C., Egwurugwu, J.N., Hart, V.O., 2012. Evaluation of anti-inflammatory, analgesic and antipyretic effect of *Mangifera indica* leaf extract on fever-induced Albino rats (Wistar). Br. J. Pharmacol. Toxicol. 3, 54–57.

Oluwafemi, G.O., 2015. Bioactive compounds in *Magnifera indica* demonstrates dose-dependent anti-inflammatory effects. Inflamm. Cell Signal. 2, 1–7.

Oosthuyse, S.A., 1994. Recent mango-cultivar introductions to South Africa. S.A. Mango Grower's Assoc. Yearbook 13, 22–23.

Perrucci, S., Fichi, G., Buggiani, C., Rossi, G., Flamini, G., 2006. Efficacy of mangiferin against *Cryptosporidium parvum* in a neonatal mouse model. Parasitol. Res. 99, 184–188.

Pharm, X.S., Pharm, G.K., 1991. The extraction and determination of the flavanoid mangiferin in the bark and leaves of *Mangifera indica*. Tap Chi Duoc Hoc 5, 8–19.

Prakart, S., Paul, J.G., David, L.D., 2009. Tertiary leaf fossils of *Mangifera* (Anacardiaceae) from Li basin, Thailand as examples of the utility of leaf marginal venation characters. Am. J. Bot. 96, 2048–2061.

Prakash, A., Keerthana, V., Jha, C., Kumar, k., Agrawal Dinesh Chand, R., 2013. Antibacterial property of two different varieties of Indian Mango *(Mangifera indica)* kernel extracts at various concentrations against some human pathogenic bacterial strains. Int. Res. J. Biol. Sci. 2, 28–32.

Pranay, J., Gulhina, N., 2011. Antifungal activity of crude aqueous and methanolic amchur (*Mangifera indica*) extracts against *candida* species. Int. J. Pharm. Sci. Rev. Res. 9, 85–87.

Proctor, T.A., Creasy, L.L., 1969. The anthocyanin of the mango fruit. Phytochemistry 8, 2108.

Pulak, M., Paridhavi, M., 2013. An ethno-phyto-pharmacological overview of two novel indian medicinal herbs used in polyherbal formulations. Asian J. Pharm. Life Sci. 3, 146–156.

Rajan, S., Thirunalasundari, T., Jeeva, S., 2011. Anti-enteric bacterial activity and phytochemical analysis of the seed kernel extract of *Mangifera indica* Linnaeus against *Shigella dysenteriae* (Shiga, corrig.) Castellani and Chalmers. Asian Pac. J. Trop. Med. 4, 294–300.

Rajendran, P., Jayakumar, T., Nishigaki, I., Ganapathy, E., Yutaka, N., Jayabal, V.D.S., 2013. Immunomodulatory effect of mangiferin in experimental animals with benzo(a)pyrene induced lung carcinogenesis. Int. J. Biomed. Sci. 9, 68–74.

Rajesh, M.S., Rajasekhar, J., 2014. Assessment of antidiabetic activity of *Mangifera indica* seed kernel extracts in streptozotocin induced diabetic rats. J. Nat. Remed. 14, 33–40.

Ramos, E.H.S., Moraes, M.M., Laís L. de, A.N., Nascimento, S.C., Militão, G.C.G., de Figueiredo, R.C.B.Q., da Câmara, C.A.G., Teresinha, G.S., 2014. Chemical composition, Leishmanicidal and cytotoxic activities of the essential oils from *Mangifera indica* L. var. Rosa and Espada. BioMed. Res. Int. 2014, 9p.

Ritu, J., Manoj, K., Singh, c.p., 2013. Morphological characters: efficient tool for identification on different mango cultivars. Environ. Ecol. 31, 385–388.

Ross, I.A., 1999. Medicinal Plants of the World. Human Press Inc, New Jersey, pp. 199-202.

Sajad, K., Saeid, M.M.G., 2016. Application of electronic nose systems for assessing quality of medicinal and aromatic plant products: a review. J. Appl. Res. Med. Aromat. Plants 3, 1–9.

Sâmya, S.G., Wellyta, D.O.F., Nilva Ré-Poppi, Euclésio, S., Eduardo, C., 2011. Volatile compounds of leaves and fruits of *Mangifera indica* var. coquinho (Anacardiaceae) obtained using solid phase microextraction and hydrodistillation. Food Chem. 127, 689–693.

Saotoing, P., Vroumsia, T., Tchobsala, F.-N., Tchuenguem, F., Njan Nloga, A.-M., Messi, J., 2011. Medicinal plants used in traditional treatment of malaria in Cameroon. J. Ecol. Nat. Environ. 3, 104–117.

Sarma, A.D., Anisur, R.M., Ghosh, A.K., 2010. Free radicals and their role in different clinical conditions: an overview. Int. J. Pharm. Sci. Res. 1, 185–192.

Savant, C., Kulkarni, A.R., Mannasaheb, B.A., Gajare, R., 2014. Immunomostimulant phytoconstituents from *Mangifera indica* L. bark oil. J. Phytopharmacol. 3, 139–148.

Sellés, A.J.N., Velázquez-Castro, H., Agüero-Agüero, J., González-González, J., Naddeo, F., De Simone, F., Rastrelli, L., 2002. Isolation and quantitative analysis of phenolic antioxidants, free sugars, fatty acids and polyols from mango (*Mangifera indica* L.) stem bark aqueous decoction used in Cuba as nutritional supplement. Agric. Food Chem. 50, 762–766.

Shaft, N., Ikram, M., 1982. Quantitative survey of rutin-containing plants. Part 1. Int. J. Crude Drug Res. 20 (4), 183–186.

Shah, K.A., Patel, R.J., Parmar, P.K., 2010. *Mangifera indica* (mango). Pharmacogn. Rev. 4, 42–48.

Shahid, H.A., Mohd., A., Kamran, J.N., 2014. New manglanostenoic acids from the stem bark of *Mangifera indica* var. "Fazli". J. Saudi Chem. Soc. 18, 561–565.

Shashi, K.S., Vijay, K.S., Yatendra, K., Shanmugam, S.K., Saurabh, K.S., 2009. Phytochemical and pharmacological investigations on mangiferin. Herba polica 5, 126–139.

Stoilova, I., Gargova, S., Stoyanova, A., Ho, L., 2005. Antimicrobial and antioxidant activity of the polyphenol mangiferin. Herba Polon. 51, 37–44.

Sutomo, Wahyuono, S., Rianto, S., Setyowati, E.P., 2013. Isolation and identification of active compound of *n*-hexane fraction from Kasturi (*Mangifera casturi* Konsterm.) against antioxidant and immunomodulatory activity. J. Biol. Sci. 13, 596–604.

Takashi, T., Tokiko, S., Gen-Ichiro, N., 1984. Tanins and related compounds. XXI. Isolation and characterization of galloyl and o-hydroxybenzoyl esters of benzophenone and xanthone C-glucosides from *Mangifera indica* L. Chem. Pharm. Bull. 32, 2676–2686.

Tanaka, T., Sueyasu, T., Nonaka, G.I., Nishioka, I., 1984. Tannins and related compounds. XXI. Isolation and characterization of galloyl and *p*-hydroxybenzoyl esters of benzophenone and xanthone C-glucosides from *Mangifera indica* L. Chem. Pharm. Bull. 32 (7), 2676–2686.

Taniya, B., Argha, S., Rini, R., Sushomasri, M., Himangshu, S.S.M., 2015. Isolation of mangiferin from flowering buds of *Mangifera indica* L and its evaluation of in vitro antibacterial activity. J. Pharm. Anal. 3, 49–56.

Tarkang, P.A., Kathrin, D.F., Sukjun, L., Eunyoung, L., Diego, V., Freitas-Junior, F., Liuzzi, M., Tsabang, N., Ayong, L.S., Agbor, G.A., Okalebo, F.A., Guantai, A.N., 2014. In vitro antiplasmodial activities and synergistic combinations of differential solvent extracts of the polyherbal product, Nefang. BioMed. Res. Int. 2014, 1–10.

Vayssieres, J.F., Korie, S., Colibaly, O., Temple, L., Boueyi, S.P., 2008. The mango tree in central and northern Benin: cultivar inventory, yield assessment, infested stages and loss due to fruit flies (Diptera Tephritidae). Fruits 63, 335–348.

Wang, R.-R., Yue-Dong, G., Chun-Hui, M., Xing-Jie, Z., Cheng-Gang, H., Jing-Fei, H., Yong-Tang, Z., 2011. Mangiferin, an anti-HIV-1 agent targeting protease and effective against resistant strains. Molecules 16, 4264–4277.

Wauthoz, N., Aliou, B., Elhadj, Saïdou Balde, Marc, V.D., Duez, P., 2007. Ethnopharmacology of *Mangifera indica* L. Bark and pharmacological studies of its main C-glucosylxanthone, Mangiferin. Int. J. Biomed. Pharm. Sci. 1, 112–119.

Yang, T.H., Pengo, A., 1981. Studies on the constituents of the peels of *Mangifera indica* L. Tai-wan K'o Hsueh 35 (3), 69–73.

Zhang, Y., Jian Li, ZhizhenW., Erwei, L., Pingping, S., Lifeng, H., Lingling, G., Xiumei, G., Tao, W., 2014. Acute and long-term toxicity of mango leaves extract in mice and rats. Hindawi Publishing Corporation evidence-based. Complement. Altern. Med. 2014, 8p.

Zheng, M.S., Lu, Z.Y., 1990. Antiviral effect of mangiferin and isomangiferin on *Herpex simplex* virus. Chin. Med. J. 103, 160–165.

CHAPTER 22

Moringa oleifera

V. Kuete

1 INTRODUCTION

Moringa oleifera commonly known as Moringa, drumstick tree, horseradish tree, ben oil tree, or benzoil tree is the only genus in the family Moringaceae. The plant is native to northwestern India, and widely cultivated in tropical and subtropical areas (Flora and Pachauri, 2011). It is the most widely cultivated species of the genus Moringa, and its young seed pods and leaves are used as vegetables. All parts of the Moringa tree are edible and have long been consumed by humans (Prabhu et al., 2011). *Moringa* is used worldwide in the traditional medicine, for various health conditions, such as skin infections, anemia, anxiety, asthma, blackheads, blood impurities, bronchitis, catarrh, chest congestion, cholera, infections, fever, glandular, swelling, headaches, abnormal blood pressure, hysteria, pain in joints, pimples, psoriasis, respiratory disorders, scurvy, semen deficiency, sore throat, sprain, tuberculosis, for intestinal worms, lactation, diabetes, and pregnancy (Fuglie, 2001; Mahmood et al., 2010; Sairam, 1999). In many regions of Africa, it is widely consumed for self-medication by patients affected by diabetes, hypertension, or HIV/AIDS (Mbikay, 2012). Moringa oil has tremendous cosmetic value and is used in body and hair care as a moisturizer and skin conditioner. It has been shown that aqueous, hydroalcohol, or alcohol extracts of *M. oleifera* leaves possess a wide range of additional biological activities including antioxidant, tissue protective (liver, kidneys, heart, testes, and lungs), analgesic, antiulcer, antihypertensive, radioprotective, and immunomodulatory actions (Stohs and Hartman, 2015). Phytochemical analyses have shown that *M. oleifera* is a rich source of potassium, calcium, phosphorous, iron, vitamins A and D, essential amino acids, as well as known antioxidants, such as β-carotene, vitamin C, and flavonoids (Bennett et al., 2003; Mbikay, 2012). A wide variety of polyphenols and phenolic acids as well as flavonoids, glucosinolates, and possibly alkaloids are believed to be responsible for the effects of the plant (Stohs and Hartman, 2015). This chapter is intended to compile the pharmacological effects of *M. oleifera* as well as its phytochemical constituents (Mahmood et al., 2010).

2 CULTIVATION AND DISTRIBUTION OF *MORINGA OLEIFERA*

M. oleifera is a plant that grows well at lower elevations. In East Africa it is found up to 1350 m altitude, but its adaptability is shown by a naturalized stand at over 2000 m in Zimbabwe. It is highly drought tolerant and is cultivated in semiarid and arid regions of

India, Pakistan, Afghanistan, Saudi Arabia, and Eastern Africa as well as in locations with as little as 500 mm annual rainfall. It can be grown in a wide range of soils but fertile, well-drained soils are most suitable (Bosch, 2004). Moringa is native to South Asia, where it grows in the southern foothills of the Himalayas in northwestern India. The plant has been naturalized by Pakistan, West Asia, the Arabian Peninsula, East and West Africa, Southern Florida and from Mexico to Peru, Paraguay, and Brazil (Bosch, 2004). There is considerable international trade, mostly from India, in canned and fresh fruits, oil, seeds and leaf powder, but statistics on the volumes and value are not available. In Africa local trade is mainly restricted to the leaves. In Kenya, some 2000 mostly small-scale farmers produce *M. oleifera* green fruits for the Asian community. In Tanzania an enterprise has started with the aim of producing oil and a flocculating agent (Bosch, 2004).

3 CHEMISTRY OF *MORINGA OLEIFERA*

A large number of potentially bioactive compounds are present in *M. oleifera* (Stohs and Hartman, 2015).

3.1 Essential oil

Chuang et al. (2007) have determined the chemical composition of the volatile constituents of the ethanol extract of *M. oleifera* leaves using gas chromatography–mass spectroscopy. The investigators identified a total of 44 compounds with pentacosane (17.4%), hexacosane (11.2%), *(E)*-phytol (**1**; 7.7%), and 1-[2,3,6-trimethylphenyl]-2-butanone (3.4%) as major constituents. Saleem previously identified most of these ingredients in the pods of *M. oleifera* (Saleem, 1995).

3.2 Nonvolatile constituents

The methanol extract of the leaves of *M. oleifera* was reported to contain chlorogenic acid (**2**), rutin (**3**), quercetin glucoside or isoquercetin (**4**), and kaempferol rhamnoglucoside, whereas in the root and stem barks, several procyanidins were isolated (Atawodi et al., 2010). Vongsak and coauthors also quantified crypto-cholorgenic acid (**5**), **4**, and astragalin (**6**) contains in the dried extract of the leaves and the obtained amounts were 0.081, 0.120, and 0.153%, respectively (Vongsak et al., 2014). Various derivatives of salicylic acid, gallic acid, coumarin acid, and caffeic acid were found in *M. oleifera*. Also, indole alkaloid N,α-L-rhamnopyranosyl vincosamide (VR) has been isolated from *M. oleifera* leaves. Waterman et al. isolated and characterized four isothiocyanates from *M. oleifera* leaves (Waterman et al., 2014). An aqueous extract contained 1.66% isothiocyanates and 3.82% total polyphenols (Waterman et al., 2014; Stohs and Hartman, 2015). Other prominent ingredients found in the seeds, flowers, pods, and stems include myricetin (**7**), benzylamine (moringinine; **8**), and the glycosides niaziminin and niazinin (Mbikay, 2012). Various other constituents like thiocarbamate, such as moringin

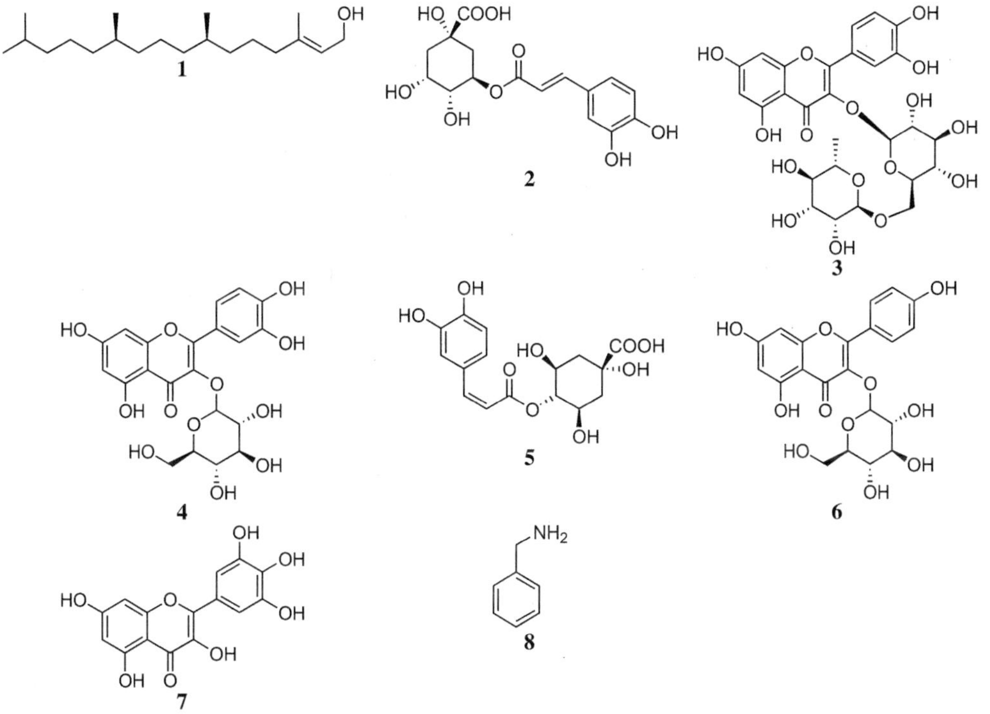

Figure 22.1 *Chemical structures of selected compounds from M. oleifera.* Phytol (**1**); chlorogenic acid (**2**); rutin (**3**); isoquercetin (**4**); crypto-cholorgenic acid (**5**); astragalin (**6**); myricetin (**7**); moringinine (**8**).

[4-(α-L-rhamnopyranosyloxy)benzyl isothiocyanate], niazinin A, niazinin B, niazimicin, niaziminin A, niaziminin B, carbamates like niazimin A, niazimin B, niazicin A, niazicin B, 4-(β-D-glucopyranosyl-1→4-α-L-rhamnopyranosyloxy)-benzyl isothiocyanate, lutein, and β-sitosterol were reported in this plant (Purwal et al., 2010; Rajan et al., 2016). The chemical constituents of selected compounds of *M. oleifera* are illustrated in Fig. 22.1.

4 PHARMACOLOGY OF *MORINGA OLEIFERA*

4.1 Anticancer activity

The cytotoxicity of *M. oleifera* and some of its constituents on various human malignancies has been reported. Tiloke and coworkers have examined the antiproliferative effect of *M. oleifera* crude aqueous leaf extract on human esophageal cancer cells (SNO) (Tiloke et al., 2016). They found that *M. oleifera* extract significantly increased lipid peroxidation and DNA fragmentation in SNO cells. They also found that the induction of apoptosis was mediated by the increase in phosphatidylserine externalization, caspase-9 and caspase-3/7 activities, and decreased ATP levels. In addition, the extract significantly increased both the expression of Smac/DIABLO protein and cleavage of poly [ADP-ribose] polymerase 1 (PARP-1), resulting in an increase in the 24 kDa fragment. These data confirmed that *M. oleifera* extract possesses antiproliferative effects on SNO esophageal cancer cells by increasing lipid peroxidation, DNA fragmentation, and induction of apoptosis. When tested against malondialdehyde (MDA)-MB-231 breast adenocarcinoma and intestinal adenocarcinoma cancer cell lines, extracts of leaves and bark showed remarkable anticancer properties (Al-Asmari et al., 2015). In vitro evaluation of cytotoxic activities of essential oil from *M. oleifera* seeds on HeLa cervix cancer cells, HepG2 hepatocarcinoma cells, MCF-7 breast adenocarcinoma, Caco-2 colon cancer cells, and L929 mouse fibroblasts was performed by Elsayed et al. (2015). All treated cell lines showed a significant reduction in cell viability in response to the increasing oil concentration. Moreover, the reduction depended on the cell line as well as the oil concentration applied. Additionally, HeLa cells were the most affected cells followed by HepG2, MCF-7, L929, and Caco-2, where the percentages of cell toxicity recorded were 76.1, 65.1, 59.5, 57.0, and 49.7%, respectively. Furthermore, the IC_{50} values obtained for MCF-7, HeLa, and HepG2 cells were 226.1, 422.8, and 751.9 µg/mL, respectively. Callus and leaf extracts of *M. oleifera* significantly decreased the viability of Hela cervix cancer cells in a concentration-dependent manner; however, leaf extract was more potent than callus extract (Jafarain et al., 2014). A crude aqueous leaf extract was prepared and A549 lung cancer cells were treated with 166.7 µg/mL extract (IC_{50}) for 24 h and assayed for oxidative stress (2-thiobarbituric acid reactive substances (TBARS) and glutathione assays), DNA fragmentation (comet assay), and caspase (3/7 and 9) activity. In addition, the expressions of Nrf2, p53, Smac/DIABLO, and PARP-1 were determined by Western blotting whereas mRNA expressions of Nrf2 and p53 were assessed using qPCR (Tiloke et al., 2013). Results indicated that extract exerts antiproliferative effects in A549 lung cells by increasing oxidative stress, DNA fragmentation, and inducing apoptosis. Leaf extract also inhibited the growth of human pancreatic cancer cells Panc-1, p34, and Colo 357, the cells NF-κB signaling pathway, and increased the efficacy of chemotherapy in human pancreatic cancer cells (Berkovich et al., 2013).

Akami and coauthors have attempted to determine the chemopreventive and antileukemic activities of ethanol extracts of *M. oleifera* leaves on benzene-induced leukemia bearing rats (Akanni et al., 2014). HeFor instance, leukemia was induced by intravenous injection of 0.2 mL benzene solution every 48 h for 4 weeks in appropriate rat groups. Ethanol extract of leaves was administered at 0.2 mL of 100 mg/mL to respective treatment rat groups. A standard antileukemic drug (cyclophosphamide) was also used to treat appropriate rat groups. Clinical examination of liver and spleen with hematological parameters was employed to assess the leukemia burden following analysis of the rat blood samples on Sysmex KX-21N automated instrument. Leukemia induction reflected in severe anemia and a marked leukocytosis over the control/baseline group. Liver and spleen enlargements were also observed in group exposed to benzene carcinogen. The in vivo antioxidative potential of extract was evaluated using MDA and reduced glutathione (GSH) levels. The liver MDA and GSH levels obtained in benzene-induced leukemic rats treated with extract compared favorably with those obtained in similar treatments with the standard drug. The extract demonstrated chemopreventive and antileukemic activities as much as the standard antileukemic drug by ameliorating the induced leukemic condition in the affected rat groups. This study revealed that *M. oleifera* extract might be an active, natural, and nontoxic anticancer drug lead.

Moringin showed to be effective in inducing apoptosis through p53 and Bax activation and Bcl-2 inhibition. In addition, oxidative stress related Nrf2 transcription factor and its upstream regulator CK2 alpha expressions were modulated at higher doses, which indicated the involvement of oxidative stress-mediated apoptosis induced by moringin. Moreover, significant reduction in 5S rRNA was noticed with moringin treatment (Rajan et al., 2016). By the analysis of apoptotic signals, including the induction of caspase or poly(ADP-ribose) polymerase cleavage, and the Annexin V and terminal deoxynucleotidyl transferase-mediated dUTP nick end labeling assays, it was demonstrated that *M. oleifera* leaf extracts induce the apoptosis of HepG2 cells (Jung et al., 2015). In the hollow fiber assay, oral administration of the leaf extracts significantly reduced (44–52%) the proliferation of the HepG2 cells and A549 non-small cell lung cancer cells (Jung et al., 2015). These results support the potential of soluble extracts of *M. oleifera* leaf as orally administered therapeutics for the treatment of human liver and lung cancers. Quercetin-3-O-glucoside and 4-(β-D-glucopyranosyl-1→4-α-L-rhamnopyranosyloxy)-benzyl isothiocyanate showed significant cytotoxicity against the Caco-2 cell line with an IC_{50} of 79 μg/mL and moderate cytotoxicity against the HepG2 cell line (IC_{50} of 150 μg/mL), while 4-(β-D-glucopyranosyl-1→4-α-L-rhamnopyranosyloxy)-benzyl isothiocyanate showed significant cytotoxicity against the Caco-2 and HepG2 hepatocarcinoma cell lines (IC_{50} of 45 and 60 μg/mL, respectively). Comparatively both compounds showed much lower cytotoxicity against the human embryonic kidney 293 (HEK293) cell line with IC_{50} values of 186 and 224 μg/mL, respectively (Maiyo et al., 2016).

4.2 Antidiabetic activity

Glucose-lowering effects of *M. oleifera* extracts have been reported. Olayaki and coauthors have investigated the effect of oral administration of methanol extracts of this plant on glucose tolerance, glycogen synthesis, and lipid metabolism in rats with alloxan-induced diabetes (Olayaki et al., 2015). Diabetes was induced by intraperitoneal injection of 120 mg/kg b.w. alloxan. Normal and diabetic control rats received saline, whereas rats in other groups received 300 or 600 mg/kg body weight of extract or metformin (100 mg/kg body weight of metformin) for 6 weeks. Intraperitoneal glucose tolerance was assessed and serum glucose, insulin, and lipids were measured at the end of the experiment. Liver and muscle glycogen synthase activities, glycogen content, and glucose uptake were determined. Authors found that administration of extract significantly improved glucose tolerance, and increased serum insulin levels. Extract treatment also significantly reduced serum concentrations of triglyceride, total cholesterol, and low-density lipoprotein (LDL)-cholesterol and enhanced serum level of high-density lipoprotein (HDL). Glycogen synthase activities and glycogen contents were higher in extract-treated rats compared with rats receiving metformin or saline and the extract improved glucose uptake. These results showed that hypoglycemic effects of *M. oleifera* extract might be mediated through the stimulation of insulin release leading to enhanced glucose uptake and glycogen synthesis.

The antidiabetic effect of low doses of *M. oleifera* seeds was reported on streptozotocin-induced diabetes and diabetic nephropathy male rats (Al-Malki and El Rabey, 2015). Waterman et al. have demonstrated that isothiocyanate-rich fraction of this plant reduces insulin resistance and hepatic gluconeogenesis in mice (Waterman et al., 2015). Abd El Latif et al. have investigated how an aqueous extract from the leaves of *M. oleifera* exerts hypoglycemia in diabetic rats (Abd El Latif et al., 2014). They found that leaf extract counteracted the alloxan-induced diabetic effects in rats as it normalized the elevated serum levels of glucose, triglycerides, cholesterol, and MDA, and normalized mRNA expression of the gluconeogenic enzyme pyruvate carboxylase in hepatic tissues. It also normalized the reduced mRNA expression of fatty acid synthase in the liver of diabetic rats. Moreover, it restored the normal histological structure of the liver and pancreas damaged by alloxan in diabetic rats. Their study revealed that the aqueous extract of *M. oleifera* leaves possesses potent hypoglycemic effects through the normalization of elevated hepatic pyruvate carboxylase enzyme and regeneration of damaged hepatocytes and pancreatic β-cells via its antioxidant properties. Yassa and Tohamy also demonstrated that leaf extract ameliorates streptozotocin-induced diabetes mellitus in adult rats (Gupta et al., 2012; Yassa and Tohamy, 2014). Adisakwattana and Chanathong also showed that *M. oleifera* leaf extract exerts alpha-glucosidase inhibitory activity, and may be used for the control of blood glucose and lipid concentration and prevention of hyperglycemia and hyperlipidemia (Adisakwattana and Chanathong, 2011). It was also shown that chronic administration of benzylamine, a constituent of *M. oleifera*, in the drinking water improved glucose tolerance and circulating cholesterol in high-fat diet-fed mice (Iffiu-Soltesz et al., 2010).

4.3 Antiinflammatory and analgesic activities

M. oleifera is a widely grown plant in most tropical countries and it has been recognized traditionally for several medicinal benefits. Fard et al. have investigated the antiinflammatory effect of *M. oleifera* hydroethanolic bioactive leaf extracts by assessing the inhibition of nitric oxide (NO) production during Griess reaction and the expression of proinflammatory mediators in macrophages (Fard et al., 2015). They found that extract significantly inhibited the secretion of NO production and other inflammatory markers, such as prostaglandin E2, tumor necrosis factor-α (TNF-α), interleukin (IL)-6, and IL-1β. Meanwhile, the bioactive extract induced the production of IL-10 in a dose-dependent manner. In addition, *M. oleifera* hydroethanolic bioactive leaf extract effectively suppressed the protein expression of inflammatory markers inducible NO synthase (iNOS), cyclooxygenase-2, and nuclear factor kappa (NF-κ)-light-chain-enhancer of activated B-cells p65 in lipopolysaccharide (LPS)-induced RAW264.7 macrophages in a dose-dependent manner. These findings supported the traditional use of the plant as an effective treatment for inflammation associated diseases. Alhakmani and collaborators have also investigated the antiinflammatory activity of ethanol extract of flowers of *M. oleifera* grown in Oman using protein denaturation method. They found that the antiinflammatory activity of plant extract was significant and comparable with the standard drug diclofenac sodium (Alhakmani et al., 2013). Flower extract was shown to suppress the activation of inflammatory mediators in LPS-stimulated RAW 264.7 macrophages via NF-κB pathway (Tan et al., 2015). The antiinflammatory effect of *M. oleifera* seeds was also reported on acetic acid-induced acute colitis in rats (Minaiyan et al., 2014). The isothiocyanates from the plant were shown to exhibit antiinflammatory activity in an in vitro macrophage cell system (Waterman et al., 2014). The antiinflammatory activity of glucomoringin (4(α-L-rhamnosyloxy)-benzyl glucosinolate), a constituent of *M. oleifera* was reported in a mouse model of experimental autoimmune encephalomyelitis (Galuppo et al., 2014). In fact, treatment of mice with this compound was able to counteract the inflammatory cascade that underlies the processes leading to severe multiple sclerosis. In particular, the compound was effective against proinflammatory cytokine TNF-α. Oxidative species generation including the influence of iNOS, nitrotyrosine tissue expression, and cell apoptotic death pathway was also evaluated resulting in a lower Bax/Bcl-2 unbalance (Galuppo et al., 2014). Extracts from root, leaf, and fruit of moringa were shown to reduce LPS-induced NO release in a dose-dependent manner. The moringa fruit extract most effectively inhibited LPS-induced NO production and levels of iNOS; the fruit extract was also shown to suppress the production of inflammatory cytokines including IL-1β, TNF-α, and IL-6; furthermore, fruit extract inhibited the cytoplasmic degradation of inhibitor of kappa B (IκB)-α and the nuclear translocation of p65 proteins, resulting in lower levels of NF-κB transactivation (Lee et al., 2013). Collectively, the results of this study demonstrate that Moringa fruit extract reduces the levels of proinflammatory mediators including NO, IL-1β, TNF-α, and IL-6 via the inhibition of NF-κB activation in RAW264.7 cells. Protective effect of ethanol extract of seeds of *M. oleifera* was also demonstrated against inflammation associated with development of arthritis in rats (Mahajan et al., 2007). The methanol extracts of the root or leaf of *M. oleifera* were found effective in the reduction of pain induced by complete Freund's adjuvant (CFA) in rats; a comparison of single and combination therapies of root and leaf extracts also showed a synergistic effect on pain reduction (Manaheji et al., 2011). Cheenpracha et al. have identified 4-[(2′-O-acetyl-alpha-l-rhamnosyloxy) benzyl]isothiocyanate, 4-[(3′-O-acetyl-alpha-l-rhamnosyloxy)benzyl]isothiocyanate, and phenolic glycosides as NO inhibitors of *M. oleifera* fruits (Cheenpracha et al., 2010).

4.4 Antimicrobial activities

Dzotam et al. have evaluated the antibacterial and antibiotic-modifying activities of *M. oleifera* against multidrug-resistant (MDR) Gram-negative bacteria (Dzotam et al., 2016). They found that methanol extract had selective activities, its inhibitory activity being recorded on 13/19 (68.4%) tested Gram-negative bacteria. The lowest MIC value of 128 μg/mL on *Escherichia coli* AG100 strain was recorded. Authors also showed that when combined with the extract, phenylalanine-arginine-ß-naphthylamide (PAßN), an efflux pumps inhibitor (EPI) PAβN improves the activity of the extract against *E. coli* AG100. They also found that synergistic effects were acheived with the associations of each of *M. oleifera* extract and chloramphenicol against *Enterobacter aerogenes* EA27 strain. The methanol, chloroform, ethyl acetate, and aqueous bark extracts of *M. oleifera* were evaluated for their antibacterial activity against four bacteria, namely *Staphylococcus aureus, Citrobacter freundii, Bacillus megaterium*, and *Pseudomonas fluorescens* (Zaffer et al., 2014). Results revealed that all the bark extracts irrespective of their types, in different concentrations inhibited growth of the test pathogens to varying degrees. Ethyl acetate extract showed maximum activity against all the bacterial strains followed in descending order by chloroform, methanol, and aqueous extracts. *S. aureus* was found to be the most sensitive test organism to different extracts. Ndhlala has evaluated the antimicrobial properties of 13 *M. oleifera* cultivars obtained from different locations across the globe (Ndhlala et al., 2014). They found that acetone extracts of all cultivars exhibited good antibacterial activity against *Klebsiella pneumoniae* (MIC values of 0.78 mg/mL). Flavonoids extracted from *M. oleifera* seed coat also displayed antibiofilm potential against *S. aureus, Pseudomonas aeruginosa*, and *Candida albicans* (Onsare and Arora, 2015). The antibiotic activity of glucomoringin on two strains of pathogens affecting the health of patients in hospital, namely *S. aureus* and *Enterococcus casseliflavus,* and on the yeast *C. albicans* was also reported (Galuppo et al., 2013). Results showed that the sensibility of *S. aureus* BAA-977 strain and *E. casseliflavus* to glucomoringin treatment was efficient. 4-(alpha-L-rhamnosyloxy) benzyl isothiocyanate and 4-(4′-O-acetyl-alpha-L-rhamnosyloxy)-benzyl isothiocyanate isolated from *M. oleifera* seeds were screened for their antibacterial activities against *S. aureus, Staphylococcus epidermidis, Bacillus subtilis, E. coli, E. aerogenes, K. pneumoniae,* and *P. aeruginosa*, and for their antifungal activities against *C. albicans, Trichophyton rubrum*, and *Epidermophyton floccosum* using the disk diffusion method (Padla et al., 2012). The two isothiocyanates were found active at the lowest inhibitory concentration of 1 mg/mL against all Gram-positive bacteria tested (*S. aureus, S. epidermidis, B. subtilis*) and against the dermatophytic fungi *E. floccosum* and *T. rubrum*. Ethanol extracts of the plant displayed antifungal activities in vitro against dermatophytes, such as *T. rubrum, Trichophyton mentagrophytes, E. floccosum*, and *Microsporum canis* (Chuang et al., 2007).

When assessing the inhibitory effects of crude extracts from some edible Thai plants against replication of hepatitis B virus and human liver cancer cells, Waiyaput and collaborators found that *M. oleifera* had anti-HBV activity, combined with a mild cytotoxicity effect on the HepG2 cells (Waiyaput et al., 2012). The anti-herpes simplex virus type 1 (HSV-1) effect of *M. oleifera* extract was also reported. In fact, the extract of *M. oleifera* at a dose of 750 mg/kg per day significantly delayed the development of skin lesions, prolonged the mean survival times, and reduced the mortality of HSV-1 infected mice as compared with 2% DMSO in distilled water (Lipipun et al., 2003).

4.5 Other pharmacological effects

Panda et al. have evaluated the protective effect of N,α-L-VR, isolated from *M. oleifera* leaves in isoproterenol (ISO)-induced cardiac toxicity in rats. They found that in treated rats, a reduction in myocardial necrosis was observed as evidenced by the triphenyl tetrazolium chloride (TTC) stain (Panda et al., 2013). These findings highlighted the cardio-protective potential of the isolated alkaloid. Aqueous and ethanol extracts of the whole pod and its parts (seeds) revealed a pronounced blood-pressure lowering effect; isolation of thiocarbamate, isothiocyanate glycosides, beta-sitosterol, and methyl *p*-hydroxybenzoate from extract identified them as the principal hypotensive constituents (Faizi et al., 1998).

Using xanthine oxidase model system, various extracts of *M. oleifera* exhibited strong in vitro antioxidant activity, with 50% inhibitory concentration (IC_{50}) values of 16, 30, and 38 μL for the roots, leaves, and stem bark, respectively (Atawodi et al., 2010). Similarly, potent radical scavenging capacity was observed when extracts were evaluated with the 2-deoxyguanosine assay model system, with IC_{50} values of 40, 58, and 72 μL for methanol extracts of the leaves, stem, and root barks, respectively (Atawodi et al., 2010). Also, the flowers of *M. oleifera* have been reported to contain various types of antioxidant compounds, such as ascorbic acid and carotenoids, as well as tannins, flavonoids, alkaloids, and cardiac glycosides (Alhakmani et al., 2013).

5 CONCLUSIONS

In this chapter, we reviewed the pharmacological properties as well as the phytochemical composition of *M. oleifera*. The plant is used in many regions of Africa by patients affected by diabetes, hypertension, or HIV/AIDS. Its reported pharmacological activities mainly include anticancer, antidiabetic, antiinflammatory, and antimicrobial effects. The plant contains several bioactive constituents among which are isothiocyanates.

REFERENCES

Abd El Latif, A., El Bialy Bel, S., Mahboub, H.D., Abd Eldaim, M.A., 2014. *Moringa oleifera* leaf extract ameliorates alloxan-induced diabetes in rats by regeneration of beta cells and reduction of pyruvate carboxylase expression. Biochem. Cell Biol. 92 (5), 413–419.

Adisakwattana, S., Chanathong, B., 2011. Alpha-glucosidase inhibitory activity and lipid-lowering mechanisms of *Moringa oleifera* leaf extract. Eur. Rev. Med. Pharmacol. Sci. 15 (7), 803–808.

Akanni, E.O., Adedeji, A.L., Adedosu, O.T., Olaniran, O.I., Oloke, J.K., 2014. Chemopreventive and anti-leukemic effects of ethanol extracts of *Moringa oleifera* leaves on wistar rats bearing benzene induced leukemia. Curr. Pharm. Biotechnol. 15 (6), 563–568.

Al-Asmari, A.K., Albalawi, S.M., Athar, M.T., Khan, A.Q., Al-Shahrani, H., Islam, M., 2015. *Moringa oleifera* as an anti-cancer agent against breast and colorectal cancer cell lines. PLoS One 10 (8), e0135814.

Al-Malki, A.L., El Rabey, H.A., 2015. The antidiabetic effect of low doses of *Moringa oleifera* Lam. seeds on streptozotocin induced diabetes and diabetic nephropathy in male rats. Biomed. Res. Int. 2015, 381040.

Alhakmani, F., Kumar, S., Khan, S.A., 2013. Estimation of total phenolic content, in-vitro antioxidant and anti-inflammatory activity of flowers of *Moringa oleifera*. Asian Pac. J. Trop. Biomed. 3 (8), 623–627.

Atawodi, S.E., Atawodi, J.C., Idakwo, G.A., Pfundstein, B., Haubner, R., Wurtele, G., 2010. Evaluation of the polyphenol content and antioxidant properties of methanol extracts of the leaves, stem, and root barks of *Moringa oleifera* Lam. J. Med. Food 13 (3), 710–716.

Bennett, R.N., Mellon, F.A., Foidl, N., Pratt, J.H., Dupont, M.S., Perkins, L., 2003. Profiling glucosinolates and phenolics in vegetative and reproductive tissues of the multi-purpose trees *Moringa oleifera* L. (horseradish tree) and *Moringa stenopetala* L. J. Agric. Food Chem. 51 (12), 3546–3553.

Berkovich, L., Earon, G., Ron, I., Rimmon, A., Vexler, A., Lev-Ari, S., 2013. *Moringa oleifera* aqueous leaf extract down-regulates nuclear factor-kappaB and increases cytotoxic effect of chemotherapy in pancreatic cancer cells. BMC Complement. Altern. Med. 13, 212.

Bosch, C.H., 2004. *Moringa oleifera* Lam. [Internet] Record from PROTA4U. PROTA (Plant Resources of Tropical Africa/Ressources végétales de l'Afrique tropicale).

Cheenpracha, S., Park, E.J., Yoshida, W.Y., Barit, C., Wall, M., Pezzuto, J.M., 2010. Potential anti-inflammatory phenolic glycosides from the medicinal plant *Moringa oleifera* fruits. Bioorg. Med. Chem. 18 (17), 6598–6602.

Chuang, P.H., Lee, C.W., Chou, J.Y., Murugan, M., Shieh, B.J., Chen, H.M., 2007. Anti-fungal activity of crude extracts and essential oil of *Moringa oleifera* Lam. Bioresour. Technol. 98 (1), 232–236.

Dzotam, J.K., Touani, F.K., Kuete, V., 2016. Antibacterial and antibiotic-modifying activities of three food plants (*Xanthosoma mafaffa* Lam., *Moringa oleifera* (L.) Schott and *Passiflora edulis* Sims) against multidrug-resistant (MDR) Gram-negative bacteria. BMC Complement. Altern. Med. 16, 9.

Elsayed, E.A., Sharaf-Eldin, M.A., Wadaan, M., 2015. In vitro evaluation of cytotoxic activities of essential oil from *Moringa oleifera* seeds on HeLa, HepG2, MCF-7 CACO-2 and L929 cell lines. Asian Pac. J. Cancer Prev. 16 (11), 4671–4675.

Faizi, S., Siddiqui, B.S., Saleem, R., Aftab, K., Shaheen, F., Gilani, A.H., 1998. Hypotensive constituents from the pods of *Moringa oleifera*. Planta Med. 64 (3), 225–228.

Fard, M.T., Arulselvan, P., Karthivashan, G., Adam, S.K., Fakurazi, S., 2015. Bioactive extract from *Moringa oleifera* inhibits the pro-inflammatory mediators in lipopolysaccharide stimulated macrophages. Pharmacogn. Mag. 11 (Suppl 4), S556–S563.

Flora, S.J.S., Pachauri, V., 2011. Moringa (*Moringa oleifera*) seed extract and the prevention of oxidative stress. In: Preedy, V.R., Watson, R.R., Patel, V.B. (Eds.), Nuts and Seeds in Health and Disease Prevention. Academic Press, San Diego, pp. 775–785, Chapter 92.

Fuglie, L.J., 2001. The Miracle Tree: *Moringa oleifera*—Natural Nutrition for the Tropics. Church World Service, Dakar, Senegal.

Galuppo, M., Giacoppo, S., De Nicola, G.R., Iori, R., Navarra, M., Lombardo, G.E., 2014. Antiinflammatory activity of glucomoringin isothiocyanate in a mouse model of experimental autoimmune encephalomyelitis. Fitoterapia 95, 160–174.

Galuppo, M., Nicola, G.R., Iori, R., Dell'utri, P., Bramanti, P., Mazzon, E., 2013. Antibacterial activity of glucomoringin bioactivated with myrosinase against two important pathogens affecting the health of long-term patients in hospitals. Molecules 18 (11), 14340–14348.

Gupta, R., Mathur, M., Bajaj, V.K., Katariya, P., Yadav, S., Kamal, R., 2012. Evaluation of antidiabetic and antioxidant activity of *Moringa oleifera* in experimental diabetes. J. Diabetes 4 (2), 164–171.

Iffiu-Soltesz, Z., Wanecq, E., Lomba, A., Portillo, M.P., Pellati, F., Szoko, E., 2010. Chronic benzylamine administration in the drinking water improves glucose tolerance, reduces body weight gain and circulating cholesterol in high-fat diet-fed mice. Pharmacol. Res. 61 (4), 355–363.

Jafarain, A., Asghari, G., Ghassami, E., 2014. Evaluation of cytotoxicity of *Moringa oleifera* Lam. callus and leaf extracts on Hela cells. Adv. Biomed. Res. 3, 194.

Jung, I.L., Lee, J.H., Kang, S.C., 2015. A potential oral anticancer drug candidate, leaf extract, induces the apoptosis of human hepatocellular carcinoma cells. Oncol. Lett. 10 (3), 1597–1604.

Lee, H.J., Jeong, Y.J., Lee, T.S., Park, Y.Y., Chae, W.G., Chung, I.K., 2013. Moringa fruit inhibits LPS-induced NO/iNOS expression through suppressing the NF-kappa B activation in RAW264.7 cells. Am. J. Chin. Med. 41 (5), 1109–1123.

Lipipun, V., Kurokawa, M., Suttisri, R., Taweechotipatr, P., Pramyothin, P., Hattori, M., 2003. Efficacy of Thai medicinal plant extracts against herpes simplex virus type 1 infection in vitro and in vivo. Antiviral Res. 60 (3), 175–180.

Mahajan, S.G., Mali, R.G., Mehta, A.A., 2007. Protective effect of ethanolic extract of seeds of *Moringa oleifera* Lam. against inflammation associated with development of arthritis in rats. J. Immunotoxicol. 4 (1), 39–47.

Mahmood, K.T., Mugal, T., Ul Haq, I., 2010. *Moringa oleifera*: a natural gift—a review. J. Pharm. Sci. Res. 2 (11), 775–781.

Maiyo, F.C., Moodley, R., Singh, M., 2016. Cytotoxicity, antioxidant and apoptosis studies of quercetin-3-O-glucoside and 4-(beta-D-glucopyranosyl-1-->4-alpha-L-rhamnopyranosyloxy)-benzyl isothiocyanate from *Moringa oleifera*. Anticancer Agents Med. Chem. 16 (5), 648–656.

Manaheji, H., Jafari, S., Zaringhalam, J., Rezazadeh, S., Taghizadfarid, R., 2011. Analgesic effects of methanolic extracts of the leaf or root of *Moringa oleifera* on complete Freund's adjuvant-induced arthritis in rats. Zhong Xi Yi Jie He Xue Bao 9 (2), 216–222.

Mbikay, M., 2012. Therapeutic potential of *Moringa oleifera* leaves in chronic hyperglycemia and dyslipidemia: a review. Front. Pharmacol. 3, 24.

Minaiyan, M., Asghari, G., Taheri, D., Saeidi, M., Nasr-Esfahani, S., 2014. Anti-inflammatory effect of *Moringa oleifera* Lam. seeds on acetic acid-induced acute colitis in rats. Avicenna J. Phytomed. 4 (2), 127–136.

Ndhlala, A.R., Mulaudzi, R., Ncube, B., Abdelgadir, H.A., du Plooy, C.P., Van Staden, J., 2014. Antioxidant, antimicrobial and phytochemical variations in thirteen *Moringa oleifera* Lam. cultivars. Molecules 19 (7), 10480–10494.

Olayaki, L.A., Irekpita, J.E., Yakubu, M.T., Ojo, O.O., 2015. Methanolic extract of *Moringa oleifera* leaves improves glucose tolerance, glycogen synthesis and lipid metabolism in alloxan-induced diabetic rats. J. Basic Clin. Physiol. Pharmacol. 26 (6), 585–593.

Onsare, J.G., Arora, D.S., 2015. Antibiofilm potential of flavonoids extracted from *Moringa oleifera* seed coat against *Staphylococcus aureus Pseudomonas aeruginosa* and *Candida albicans*. J. Appl. Microbiol. 118 (2), 313–325.

Padla, E.P., Solis, L.T., Levida, R.M., Shen, C.C., Ragasa, C.Y., 2012. Antimicrobial isothiocyanates from the seeds of *Moringa oleifera* Lam. Z. Naturforsch. C 67 (11–12), 557–564.

Panda, S., Kar, A., Sharma, P., Sharma, A., 2013. Cardioprotective potential of N,alpha-L-rhamnopyranosyl vincosamide, an indole alkaloid, isolated from the leaves of *Moringa oleifera* in isoproterenol induced cardiotoxic rats: in vivo and in vitro studies. Bioorg. Med. Chem. Lett. 23 (4), 959–962.

Prabhu, K., Murugan, K., Nareshkumar, A., Ramasubramanian, N., Bragadeeswaran, S., 2011. Larvicidal and repellent potential of *Moringa oleifera* against malarial vector, *Anopheles stephensi* Liston (Insecta: Diptera: Culicidae). Asian Pac. J. Trop. Biomed. 1 (2), 124–129.

Purwal, L., Pathak, A.K., Jain, U.K., 2010. In vivo anticancer activity of the leaves and fruits of *Moringa oleifera* on mouse melanoma. Pharmacologyonline 1, 655–665.

Rajan, T.S., De Nicola, G.R., Iori, R., Rollin, P., Bramanti, P., Mazzon, E., 2016. Anticancer activity of glucomoringin isothiocyanate in human malignant astrocytoma cells. Fitoterapia 110, 1–7.

Sairam, T.V., 1999. Home Remedies, vol II: A Handbook of Herbal Cures for Commons Ailments. Penguin, New Delhi, India.

Saleem, R., 1995. Studies in the Chemical Constituents of *Moringa oleifera* Lam and Preparation of the Potential Biologically Significant Derivatives of 8-Hydroxyquinoline. A thesis. 357 pp.

Stohs, S.J., Hartman, M.J., 2015. Review of the safety and efficacy of *Moringa oleifera*. Phytother. Res. 29 (6), 796–804.

Tan, W.S., Arulselvan, P., Karthivashan, G., Fakurazi, S., 2015. *Moringa oleifera* flower extract suppresses the activation of inflammatory mediators in lipopolysaccharide-stimulated RAW 264.7 macrophages via NF-kappaB pathway. Mediators Inflamm. 2015, 720171.

Tiloke, C., Phulukdaree, A., Chuturgoon, A.A., 2013. The antiproliferative effect of *Moringa oleifera* crude aqueous leaf extract on cancerous human alveolar epithelial cells. BMC Complement. Altern. Med. 13, 226.

Tiloke, C., Phulukdaree, A., Chuturgoon, A.A., 2016. The antiproliferative effect of *Moringa oleifera* crude aqueous leaf extract on human esophageal cancer cells. J. Med. Food 19 (4), 398–403.

Vongsak, B., Sithisarn, P., Gritsanapan, W., 2014. Simultaneous HPLC quantitative analysis of active compounds in leaves of *Moringa oleifera* Lam. J. Chromatogr. Sci. 52 (7), 641–645.

Waiyaput, W., Payungporn, S., Issara-Amphorn, J., Panjaworayan, N.T., 2012. Inhibitory effects of crude extracts from some edible Thai plants against replication of hepatitis B virus and human liver cancer cells. BMC Complement. Altern. Med. 12, 246.

Waterman, C., Cheng, D.M., Rojas-Silva, P., Poulev, A., Dreifus, J., Lila, M.A., 2014. Stable, water extractable isothiocyanates from *Moringa oleifera* leaves attenuate inflammation in vitro. Phytochemistry 103, 114–122.

Waterman, C., Rojas-Silva, P., Tumer, T.B., Kuhn, P., Richard, A.J., Wicks, S., 2015. Isothiocyanate-rich *Moringa oleifera* extract reduces weight gain, insulin resistance, and hepatic gluconeogenesis in mice. Mol. Nutr. Food Res. 59 (6), 1013–1024.

Yassa, H.D., Tohamy, A.F., 2014. Extract of *Moringa oleifera* leaves ameliorates streptozotocin-induced diabetes mellitus in adult rats. Acta Histochem. 116 (5), 844–854.

Zaffer, M., Ahmad, S., Sharma, R., Mahajan, S., Gupta, A., Agnihotri, R.K., 2014. Antibacterial activity of bark extracts of *Moringa oleifera* Lam. against some selected bacteria. Pak. J. Pharm. Sci. 27 (6), 1857–1862.

CHAPTER 23

Myristica fragrans: A Review

V. Kuete

1 INTRODUCTION

The genus *Myristica* comprises 72 tropical species occurring from Asia to Australia (Mabberly, 1987). *Myristica fragrans* (Houtt.) is an evergreen tree indigenous to the Moluccas, or Spice Islands, and exotic to Grenada, India, Mauritius, Singapore, South Africa, Sri Lanka, and United States of America (Gomathi et al., 2016; Orwa et al., 2009). Indonesia and Grenada dominate world production and exports of both nutmegs and mace with a world market share of 75 and 20%, respectively. Other producers include India, Malaysia, Papua New Guinea, Sri Lanka, and St.Vincent.Though African countries are minor traders of both products, the plant is widely cultivated throughout the continent (Iwu, 1993; Ogunwande et al., 2003;Yinyang et al., 2014). It is the main source of nutmeg and mace and the most important commercial species of the genus *Myristica*. Many countries use nutmeg and mace as spices. Nutmeg has a distinctive, pungent fragrance and a warm, slightly sweet taste (Gomathi et al., 2016); it is medicinally used as a stomachic, stimulant, carminative as well as for intestinal catarrh and colic, headaches, diarrhea, vomiting, nausea, fever, bad breath, to stimulate appetites and to control flatulence. It is also used for its aphrodisiac and antiinflammatory properties. Nutmeg is also alleged to have hallucinogenic properties (Orwa et al., 2009). Nutmeg oil is helpful to dissolve kidney stones and alleviate infections of the kidney, as well as for the treatment of diarrhea (Al-Jumaily and Al-Amiry, 2012; Sanghai-Vaijwade et al., 2011), rheumatism, and cholera (Gomathi et al., 2016). Mace, a common spice, which is the aril surrounding the shell enclosing the seed, is used in Indonesian folk medicine as aromatic stomachics, analgesics, and a medicine for rheumatism (Ozaki et al., 1989). Several pharmacological properties of *M. fragrans*, such as anticancer, antiinflammatory, antimicrobial, antioxidant, antidepressant and so on. have been reported. In this chapter, a synopsis of the pharmacological and phytochemical studies of this spicy plant will be provided, with emphasis on nutmegs and mace.

2 BOTANICAL ASPECTS AND DISTRIBUTION OF *Myristica fragrans*

M. fragrans is a spreading, medium to large sized, aromatic evergreen tree usually growing to around 5–13 m high, occasionally 20 m. The leaves are alternate, pointed, dark green 5–15 cm × 2–7 cm arranged along the branches and are borne on leaf stems about

1 cm long, shiny on the upper surface. The fruit is oval or pyriform, drooping, yellow, smooth, 6–9 cm long with a longitudinal ridge and a fleshy husk. When ripe, husk splits into two halves revealing a purplish-brown, shiny seed surrounded by a leathery red or crimson network of tissue. The shiny, brown seed inside, and the kernel of the seed is the nutmeg. The brown seed has a red cover that makes another spice called mace (Orwa et al., 2009). The plant grows wild on rich volcanic soils in lowland tropical rain forests. Nutmeg needs a warm and humid tropical climate. *M. fragrans* is indigenous native to Indonesia, and is exotic to Grenada, India, Mauritius, Singapore, Sri Lanka, United States of America, South Africa, as well as the majority of African countries. Nutmeg is a mild, delicious baking spice commonly added to sausages, meats, fish, soups, fruit pies, eggnog, puddings, vegetables and cakes, biscuits, custards, buns, etc. (Orwa et al., 2009). World production of nutmegs is estimated to average between 10,000 and 12,000 tons per year with annual world demand estimated at 9000 tons; production of mace is estimated from 1500 to 2000 tons. Indonesia and Grenada are the first two producers and exporters of both products with a world market share of 75 and 20%, respectively (http://www.fao.org/docrep/v4084e/v4084e0b.htm, Nutmeg and mace-world overview).

3 CHEMISTRY OF *Myristica fragrans*

Nutmeg contains a volatile oil, a fixed oil, proteins, fats, starch, and mucilage. Nutmeg yields 5–15% of volatile oil, containing sabinene, camphene, myristin (**8**), elemicin (**9**), isoelemicin, eugenol, isoeugenol, methoxyeugenol, safrole, diametric phenylpropanoids, lignans, neolignans, etc. (Gomathi et al., 2016). The essential oil isolated from the seeds of *M. fragrans* harvested in Nigeria was found to contain sabinene (**1**; 49.09%), α-pinene (**2**; 13.19%), α-phellandrene (**4**; 6.72%), and γ-terpinol (6.43%) as major constituents (Ogunwande et al., 2003). Power and Solway have identified compound **2** and β-pinene (**3**) as the main constituents of the volatile oil; meanwhile Bejnarowicz and Kirch have identified the monoterpenes, terpinen-4-ol (14.2–24.7%), **3** (12.3–19.1%), and limonene (**6**; 2.3–11.9%) as the dominant compounds common to all the species studied (Bejnarowicz and Kirch, 1963; Power and Solway, 1908). γ-terpinol (31.3%), γ-terpinene (**7**; 7.8%), and **8** (7.1%) were identified as major constituents in the essential oil of *M. fragrans* from Pakistan (Atta-ur-Rahman et al., 2000). The intoxicating effects of the seeds have been assigned to the presence of compounds **8** and **9** (Satyavathy et al., 1987). However, compound **8** generally constitutes 4–6% of nutmeg and mace essential oil and is responsible for most of its pharmacological effects (Latha et al., 2005). It was shown that the seeds also contain 25–30% fixed oils (myristic, stearic, palmitic, oleic, linoleic, and lauric acids) (Gopalakrishnan, 1992), saponins, polyphenols, tannins, epicatechin, triterpenic sapogenins, and fats (Varshney et al., 1968; Satyavathy et al., 1987), licarin B (Kim and Park, 1991), malabaricone B (Orabi et al., 1991), and malabaricone C (**10**) (Shinohara et al., 1999). The red pigment of mace was identified as lycopene

(11) (Gopalakrishnan et al., 1979). Antifungal lignans erythro-austrobailignan-6 (12), *meso*-dihydroguaiaretic acid (13), and nectandrin B (14) were isolated from the methanol extract of seeds (Cho et al., 2007). Four lignans 13, 14, macelignan (15), and fragransin A$_2$ (16) were also isolated from the seeds of Vietnamese nutmeg (Thuong et al., 2013). The neolignans, fragnasol C and D, and myristicanol A and D have been isolated from mace (Miyasawa et al., 1996; Rastogi and Mehrotra, 1995). Another neolignan, dihydro-di-isoeugenol was isolated from the arils (Purushothaman and Sarada, 1980). Compound 13 has been isolated from the mace of nutmeg harvested in other parts of the world (Park et al., 1998). Seven 2,5-bis-aryl-3,4-dimethyltetrahydrofuran lignans, compound 14, tetrahydrofuroguaiacin B, saucernetindiol, verrucosin, nectandrin A, fragransin C, and galbacin (Nguyen et al., 2010) as well as several other lignans, such as 13–15, machilin A, machilin F, myristargenol A, licarin A, and licarin B (Lee et al., 2009) were isolated from the ethanol extract of nutmeg harvested in Korea. Six dihydrobenzofuran type neolignans were also isolated from the dried ripe seeds of *M. fragrans* purchased in Anguo market for Chinese Medicinal Materials in Anguo city of Hebei Province of China and identified as licarin B, dehydro-di-isoeugenol, 3′-methoxylicarin B, myrisfrageal A, isodihydrocainatidin, and myrisfrageal B (Cao et al., 2013). The phytochemical studies of *M. fragrans* purchased from Indonesia led to the isolation of five new 8-O-4′ type neolignans, named myrifralignan A–E, together with (7S,8R)-2-(4-allyl-2,6-dimethoxyhenoxy)-1-(3,4,5-trimethoxyphenyl)-propan-1-ol, myrislignan, (7R,8S)-2-(4-propenyl-2-methoxyphenoxy)-1-(3,4,5-trimethoxyphenyl)-propan-1-ol, (7S,8R)-2-(4-allyl-2,6-dimethoxyphenoxy)-1-(4-hydroxy-3,5-dimethoxyphenyl)-propan-1-ol, and machilin D (Cao et al., 2015). The chemical structures of some common constituents of the plant are given in Fig. 23.1.

4 PHARMACOLOGY OF *Myristica fragrans*
4.1 Anticancer activity
Moteki and collaborators have demonstrated that the 80% ethanol extract of nutmeg significantly suppresses the growth of Molt 4B human lymphoid leukemia cells, at 50 µg/mL (Moteki et al., 2002). The cytotoxicity of the essential oil from fresh fruits of *M. fragrans* was reported on colon adenocarcinoma HCT116 cells and breast adenocarcinoma MCF-7 cells with IC$_{50}$ values of 78.61 and 66.45 µg/mL, respectively (Piaru et al., 2012a). The antiproliferative activity of lignans 13–16 was evaluated on a panel of human cancer cells including H1299 human nonsmall cell lung carcinoma, H358 human bronchiolar lung cancer cells, H460 large cell lung cancer, Hela human cervical cancer cells, HepG2 hepatocarcinoma cells, KPL4 and MCF-7 breast adenocarcinoma cells, RD rhabdomyosarcoma cells, and MDCK multicystic dysplastic kidney cells (Thuong et al., 2013). Compound 13 displayed the best cytotoxic effects in most of the cell lines with IC$_{50}$ values of 10.1 µM against H338 cells, 15.1 µM against HepG2 cells, 16.7

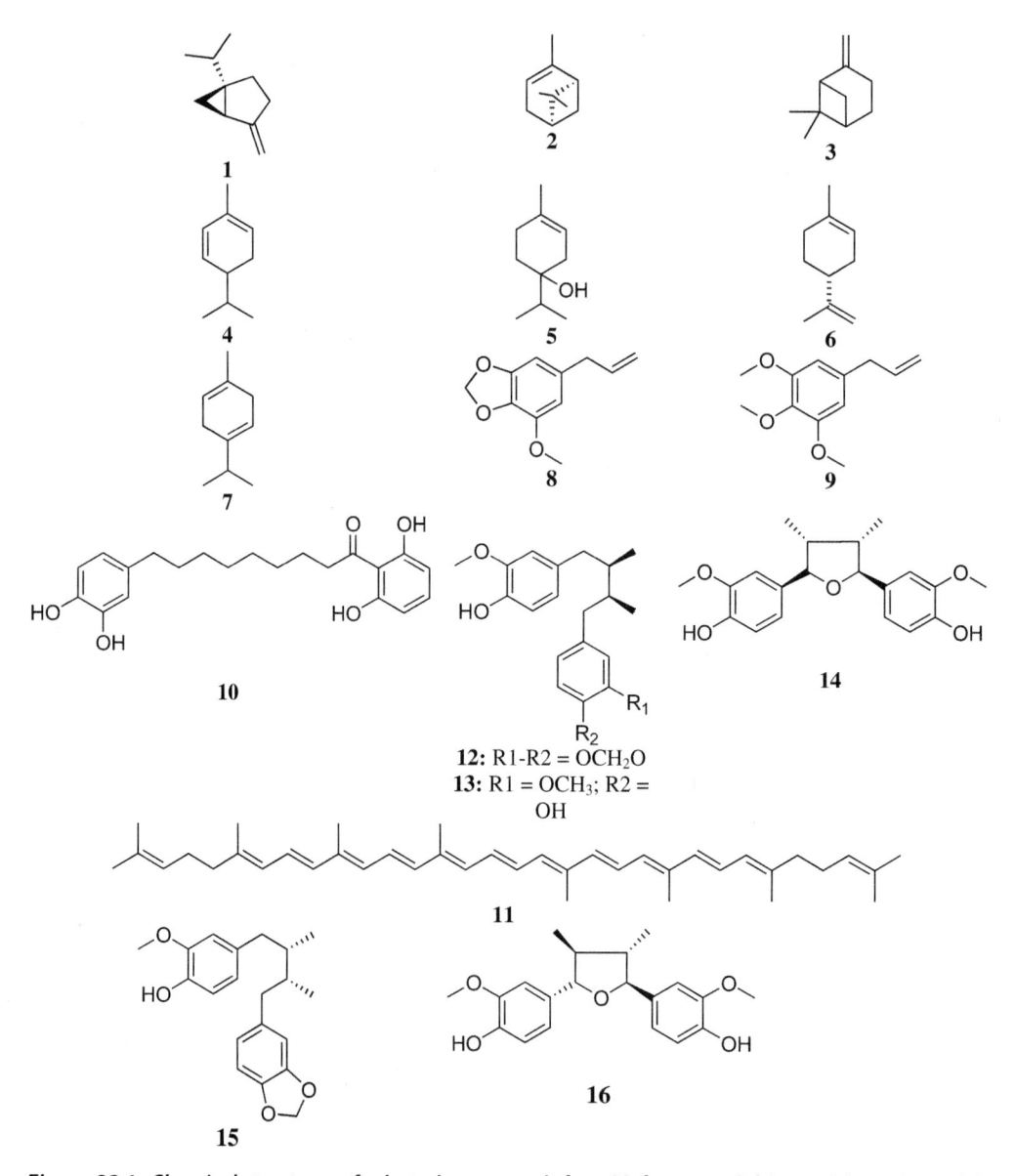

Figure 23.1 *Chemical structures of selected compounds from M. fragrans.* Sabinene (**1**); α-pinene (**2**); β-pinene (**3**); α-phellandrene (**4**); γ-terpinol (**5**); limonene (**6**); γ-terpinene (**7**); myristicin (**8**); elemicin (**9**); malabaricone C (**10**); lycopene (**11**); erythro-austrobailignan-6 (**12**); *meso*-dihydroguaiaretic acid (**13**); nectandrin B (**14**); macelignan (**15**); and fragransin A$_2$ (**16**).

and 16.9 µM against RD cells and MCF-7 cells, respectively, 22.1 µM against KPL4 cells, and 27.7 µM against H1299 cells and H460 cells; IC_{50} values were obtained with lignans **15** on H358 cells (10.2 µM) and HeLa cells (25.1 µM) and **14** against HepG2 cells (27.7 µM) while no value at up to 30 µM was obtained with **16**. Compound **8** found in the volatile oil was also reported as a potential cancer chemopreventive agent (Zheng et al. 1992). Park and collaborators have also investigated the effect of compound **13**, isolated from the aryls of *M. fragrans*, on the transcription factor (fos-jun dimer) action via in vitro assay (Park et al., 1998). Compound **13** showed an inhibitory effect against the complex formation of the fos-jun dimer and the DNA consensus sequence with an IC_{50} value of 0.21 µM. This compound suppressed leukemia, lung cancer, and colon cancer in an in vitro bioassay.

4.2 Antidepressant effect

Depression is an affection characterized by change in mood, lack of interest in the surroundings, psychomotor retardation, and melancholia. Dhingra and Sharma have investigated the effect of an *n*-hexane extract of *M. fragrans* seeds on depression in mice by using the forced swim test (FST) and the tail suspension test (TST) (Dhingra and Sharma, 2006). They administered orally different dose of extract for 3 successive days to different groups of Swiss male young albino mice. Authors found that extract significantly decreased immobility periods of mice in both the FST and the TST. The 10 mg/kg dose was found to be most potent, as indicated by the greatest decrease in the immobility period compared with the control. Furthermore, they observed that the dose of 10 mg/kg had comparable potency to reference drug, imipramine (15 mg/kg i.p.) and fluoxetine (20 mg/kg i.p.). Authors concluded that extract of *M. fragrans* elicited a significant antidepressant-like effect in mice and that the observed effect could be mediated by interaction with the adrenergic, dopaminergic, and serotonergic systems. Moinuddin et al. also evidenced the antidepressant effect of nutmeg extract (Moinuddin et al., 2012). For instance, male Wistar rats were subjected to imipramine and extract for their antidepressant activity using FST, reserpine reversal test (RRT), haloperidol-induced catalepsy (HIC), and pentobarbitone sleeping time (PST). They observed that administration of extract and imipramine induced significant reduction in immobility time in FST, RRT, and protection against HIC, compared to the control group. Authors therefore confirmed the antidepressant potential of nutmeg.

4.3 Antidiabetic and antiobesity effects

M. fragrans extract was assessed for antidiabetic activity in streptozotocin-induced diabetic rats (Patil et al., 2011). In the in vitro insulin secretion studies on isolated islets of Langerhans, extract showed dose-dependent insulin secretion. At 1 mg/mL, it also demonstrated significant in vitro alpha-glucosidase inhibitory activity with IC_{50} value of 0.85 mg/mL. These data suggested that regular use of *M. fragrans* may prevent postprandial rise in glucose levels through inhibition of intestinal alpha-glucosidase and may maintain blood glucose level through insulin secretagogue action. Inhibition of protein tyrosine phosphatase 1B (PTP1B) has been proposed as one of the drug targets for treating type-2 diabetes and obesity. Yang et al. have shown that compound **13** and otobaphenol isolated from the methanol extract of the semen of *M. fragrans* have inhibitory effect on PTP1B with IC_{50} values of 19.6 and 48.9 µM, respectively (Yang et al., 2006). Treatment with **13** on 32D murine IL-3-dependent myeloid cells overexpressing the insulin receptor (IR) resulted in a dose-dependent increase in the tyrosine phosphorylation of IR, suggesting that **13** can act as an enhancing agent in intracellular insulin signaling, possibly through the inhibition of PTP1B activity (Yang et al., 2006). Compound **15** was found to enhance insulin sensitivity and to improve lipid metabolic disorders by activating peroxisome proliferator-activated receptor (PPAR) alpha/gamma and attenuating endoplasmic reticulum (ER) stress, suggesting that it has potential as an antidiabetic agent for the treatment of type-2 diabetes (Han et al., 2008). AMP-activated protein kinase (AMPK) is a potential therapeutic target for the treatment of metabolic syndrome including obesity and type-2 diabetes. Nutmeg extract activated the AMPK enzyme in differentiated C2C12 mouse myoblast (Nguyen et al., 2010). Its active constituents, identified as compound **14**, tetrahydrofuroguaiacin B, and nectandrin A at 5 µM produced strong AMPK stimulation in differentiated C2C12 cells. In addition, the preventive effect of a tetrahydrofuran mixture (THF) of nutmeg on weight gain in a diet-induced animal model was further reported on high-fat diet (HFD)-induced mice (Nguyen et al., 2010). These results suggest that nutmeg and its active constituents can be used not only for the development of agents to treat obesity and possibly type-2 diabetes but may also be beneficial for other metabolic disorders.

4.4 Antiinflammatory and analgesic effects

Ozaki et al. have investigated the antiinflammatory effect of methanol extract, fractions, and compounds obtained from mace of *M. fragrans* on carrageenin-induced edema in rats and acetic acid-induced vascular permeability in mice (Ozaki et al., 1989). They found that methanol extract at 1.5 g/kg, ether fraction at 0.9 g/kg, *n*-hexane fraction at 0.5 g/kg, as well as the isolated compound **8** at 0.17 g/kg had a lasting antiinflammatory activity, and the potencies of these fractions were approximately the same as that of indomethacin at 10 mg/kg. The chloroform extract of nutmeg showed antiinflammatory, analgesic, and antithrombotic activities in mice (Olajide et al., 1999). In fact, this extract inhibited the carrageenan-induced rat paw edema, produced a reduction in writhings induced by acetic acid in mice, and offered protection against thrombosis induced by ADP/adrenaline mixture in mice (Olajide et al., 1999). The analgesic effect of alkaloids extract obtained from the seeds of *M. fragrans* was evaluated in a mouse model of acetic acid-induced. Alkaloid extract at a dose of 1 g/kg significantly reduced the number of writhing responses in female, but not male mice; the medium lethal dose (LD_{50}) was 5.1 g/kg. Signs of abnormal behavior were seen in animals given a dose of 4 g/kg or higher; abnormal behavior lasted for several hours after administration of the alkaloid extract. Hence, *M. fragrans* seed alkaloids have analgesic activity and are slightly toxic (Hayfaa et al., 2013). It was also reported that macelignan (**15**) isolated from *M. fragrans* has antiinflammatory properties in hippocampal neuronal and primary microglial cells (Jin et al., 2005; Ma et al., 2009). In effect, studies using neuronal cells and primary microglial cells have demonstrated that **15** suppresses nitric oxide (NO) production by inhibiting inducible nitric oxide (iNOS) expression at the transcriptional level; it also significantly suppressed the production of proinflammatory cytokine tumor necrosis factor (TNF)-alpha and interleukin-6 (IL-6) (Jin et al., 2005). The antiinflammatory effects of **15** were also evaluated using animal model with a chronic infusion of lipopolysaccharide (LPS), one of the well-characterized animal models incorporating important neuropathological features seen in Alzheimer's disease. Oral administration of lignan **15** reduced hippocampal microglial activation and the impairments of spatial memory that are induced by chronic infusions of LPS into the fourth ventricle in rat brains (Cui et al., 2008). These results indicate that **15** possesses therapeutic potential against neurodegenerative diseases that involve neuroinflammation (Jin et al., 2005; Cui et al., 2008; Ma et al., 2009). Ma et al. also showed that **15** suppresses both the phosphorylations of mitogen-activated protein kinase (MAPK) and the degradation of inhibitory-kappa B (IκB-α) and increases nuclear factor-kappa B (NF-κB) in LPS-stimulated BV-2 microglial cells (Ma et al., 2009). These results confirm that **15** has antiinflammatory effects on the affected brain through regulation of the inflammation through the MAPK signal pathway.

The antiinflammatory effect of myristicin (**8**) on double-stranded RNA (dsRNA)-stimulated macrophages was evaluated by Lee and Park (2011). They found that **8** significantly inhibited the production of calcium, NO, IL-6, IL-10, interferon inducible protein-10, monocyte chemotactic protein (MCP)-1, MCP-3, granulocyte-macrophage colony-stimulating factor, macrophage inflammatory protein (MIP)-1α, MIP-1β, and leukemia inhibitory factor in dsRNA[polyinosinic-polycytidylic acid]-induced RAW 264.7 cells. These results suggested that compound **8** has antiinflammatory properties related to its inhibition of NO, cytokines, chemokines, and growth factors in dsRNA-stimulated macrophages via the calcium pathway (Lee and Park, 2011). Myrislignan also isolated from *M. fragrans* inhibited LPS-induced production of NO, inhibited mRNA expression, and release of IL-6 and TNF-α (Jin et al., 2012). This compound significantly inhibited mRNA and protein expressions of inducible iNOS and cyclooxygenase-2 (COX-2) dose-dependently in LPS-stimulated macrophage cells. It was also shown that myrislignan decreased the cytoplasmic loss of IκB-α protein and the translocation of NF-κB from cytoplasm to the nucleus (Jin et al., 2012). These results suggest that myrislignan may exert its antiinflammatory effects in LPS-stimulated macrophages cells by inhibiting the NF-κB signaling pathway activation.

Another lignan found in the fruit of *M. fragrans*, dehydro-di-isoeugenol also showed antiinflammatory activity (Li and Yang, 2012). Licarin B, 3′-methoxylicarin B, myrisfrageal A, isodihydrocainatidin, dehydro-di-isoeugenol, and myrisfrageal B showed inhibition of NO production in LPS-activated murine monocyte-macrophage RAW264.7 with IC_{50} values of 53.6, 48.7, 76.0, 36.0, 33.6, and 45.0 µM, respectively (Cao et al., 2013). Also, myrisfrageal A, dehydro-di-isoeugenol, and myrisfrageal B suppressed LPS-induced iNOS mRNA expression in a dose-dependent manner in RAW 264.7 cells (Cao et al., 2013). Hence, these three compounds may inhibit NO overproduction via inhibition of iNOS mRNA expression. Myrislignan and machilin D also exhibited potent inhibitory activity against the NO production in the RAW264.7 cell line stimulated by lipopolysaccharide with IC_{50} values of 21.2 and 18.5 µM, respectively (Cao et al., 2015).

4.5 Antimicrobial activities

A qualitative study of the antibacterial activity of *M. fragrans* using agar diffusion indicated that ethanol and acetone extracts of the seed pulps and crust were active against Gram-positive bacteria, *Staphylococcus aureus*, and *Bacillus subtilis* contrary to aqueous extracts (Kadhim et al., 2013). However, no effect of the aforementioned extracts was observed in Gram-negative bacteria, *Escherichia coli* and *Pseudomonas aeruginosa* (Kadhim et al., 2013). The extracts from the seeds at 160 µg/mL inhibited human rotavirus by 90% indicating that it can be useful in the treatment of human diarrhea if the etiologic agent is a rotavirus (Goncalves et al., 2005). The methanol extract of fruit inhibited the growth of multidrug resistant *Salmonella typhi* B330 and M531 strains with identical minimal inhibitory concentration (MIC) value of 64 µg/mL (Rani and Khullar, 2004). Narasimhan and Dhake have evaluated the antibacterial activity of trimyristin, myristic acid, and myristicin isolated from nutmeg against Gram-positive and Gram-negative microorganisms (Narasimhan and Dhake, 2006). Results showed that trimyristin was highly active against *E. coli* and *Micrococcus luteus* (MIC: 1.250 µg/mL), *S. aureus* (MIC: 1 µg/mL), *B. subtilis*, and *P. aeruginosa* (MIC: 0.6 µg/mL); myristic acid was also highly active against *E. coli* and *B. subtilis* (MIC: 1.250 µg/mL), *S. aureus* and *M. luteus* (MIC: 0.75 µg/mL) and *P. aeruginosa* (MIC: 0.650 µg/mL); myristicin was active against *E. coli* (MIC: 1.250 µg/mL), *B. subtilis* (MIC: 1.00 µg/mL), *S. aureus* (MIC: 0.75 µg/mL), *M. luteus* (MIC: 0.625 µg/mL), and *P. aeruginosa* (MIC: 0.6 µg/mL). The constituent of nutmeg, **10** displayed antimicrobial effects against *Porphyromonas gingivalis* ATCC33277 with a MIC of 0.39 µg/mL, *P. gingivalis* 381 (MIC: 0.098 µg/mL), *Fusobacterium nucleatum* ATCC25586, *Streptococcus mutans* IFO13955 (MIC: 50 µg/mL), *Actinomyces viscosus* ATCC15987 (MIC: 13 µg/mL), various strains of *Helicobacter pylori* (MIC: 13–25 µg/mL), and *S. aureus* (MIC: 13 µg/mL) (Shinohara et al., 1999). Malabaricone C irreversibly inhibited Arg-gingipain and selectively suppressed the growth of *P. gingivalis* growth (Shinohara et al., 1999). Malabaricone B and malabaricone C exhibited strong antifungal and antibacterial activities against *S. aureus* and *Streptococcus durans* (MICs of 1 and 4 μg/mL, respectively), *B. subtilis* (MICs of 1 and 2 μg/mL, respectively), and against various strains of *C. albicans* (MIC of 8–32 μg/mL) (Orabi et al., 1991). Lignans isolated from seeds, **12–14** displayed moderate antimicrobial activity against *Alternaria alternata, Colletotrichum coccodes, Colletotrichum gloeosporioides, Magnaporthe grisea, Agrobacterium tumefaciens, Acidovorax konjaci,* and *Burkholderia glumae* (Cho et al., 2007).

The efficacy of compound **15** (macelignan), isolated from nutmeg, was evaluated; results showed that at 24 h of biofilm growth, *S. mutans, A. viscosus,* and *Streptococcus sanguis* biofilms were reduced by up to 30, 30, and 38%, respectively, after treatment with 10 μg/mL of **15** for 5 min; increasing the treatment time to 30 min resulted in a reduction of more than 50% of each of the single primary biofilms (Rukayadi et al., 2008). These results indicate that **15** is a potent natural antibiofilm agent against oral primary colonizers.

4.6 Antioxidant effects

Murcia and collaborators have evaluated the antioxidant properties of some spices including nutmeg and compared with some food antioxidants like propyl gallate, butylated hydroxyanisole (BHA), and butylated hydroxytoluene (BHT). They found that nutmeg has good activity in the deoxyribose assay, increasing the stability of some fixed oils like olive, sunflower and corn oil and fats like margarine and butter and prevented oxidation at 110°C. The antioxidant activity of nutmeg was found to be higher than BHT in the Trolox equivalent antioxidant capacity (TEAC) assay (Murcia et al., 2004). Assa et al. also showed the antioxidant activity of methanol extract of nutmeg on 1,1-diphenyl-2-picrylhydrazyl (DPPH) radical and ferric reducing antioxidant power (FRAP) tests (Assa et al., 2014). Essential oil of nutmeg was also shown to inhibit the oxidation of linoleic acid by 88.68% (Piaru et al., 2012b).

4.7 Hepatoprotective effects

The hepatoprotective activity nutmeg was evaluated on rats with liver damage caused by LPS plus D-galactosamine (D-GalN) by Morita and collaborators (Morita et al., 2003). As assessed by plasma aminotranferase activities, authors found that nutmeg had potent hepatoprotective activity. Compound **8** isolated from the extract, also showed hepatoprotective activity. In effect, compound **8** markedly suppressed LPS/D-GalN-induced enhancement of serum TNF-alpha concentrations and hepatic DNA fragmentation in mice (Morita et al., 2003). Authors therefore suggested that the hepatoprotective activity of **8** might be, at least in part, due to the inhibition of TNF-alpha release from macrophages.

4.8 Memory enhancing effects

Parle et al. have investigated the effect of the seeds on learning and memory in mice. For instance, authors administered *n*-hexane extract orally in three doses (5, 10, and

20 mg/kg p.o.) for 3 successive days to different groups of young and aged mice (Parle et al., 2004). The learning and memory parameters were then assessed using elevated plus-maze and passive-avoidance apparatus. The effect of extract on scopolamine (0.4 mg/kg i.p.)- and diazepam (1 mg/kg i.p.)-induced impairment in learning and memory was also studied. Results showed that extract at the lowest dose of 5 mg/kg p.o. significantly improved learning and memory of young and aged mice. It also reversed scopolamine- and diazepam-induced impairment in learning and memory of young mice. It enhanced learning and retention capacities of both young and aged mice. Authors finally suggested that the memory-enhancing effect may be attributed to a variety of properties of the plant, such as antioxidant, antiinflammatory, or perhaps procholinergic activity (Parle et al., 2004). The central cholinergic pathways play a prominent role in the learning and memory processes. Acetylcholinesterase is an enzyme that inactivates acetylcholine. Dhingra et al. have evaluated the acetylcholinesterase-inhibiting activity of extracts of *M. fragrans* seeds. They found that *n*-hexane extract at 5 mg/kg p.o. for 3 successive days, administered to young male Swiss albino mice significantly decreased acetylcholinesterase activity (Dhingra et al., 2006).

4.9 Other pharmacological effects

Lee et al. have investigated the stimulatory effects of machilin A and other related lignans isolated from *M. fragrans* on osteoblast differentiation (Lee et al., 2009). Authors found that machilin A stimulated osteoblast differentiation via activation of p38 MAP kinase. Lignans **13–15**, machilin F, licarin A, licarin B, and myristargenol stimulated osteoblast differentiation in mouse osteoblastic (MC3T3-E1) cells; these data suggested that lignans isolated from *M. fragrans* have anabolic activity in bone metabolism.

López et al. demonstrated that nutmeg exerts antihelmintic effects on *Anisakis simplex*, compound **8** being one of the active constituents; in fact, the extract induced a high rate of dead *A. simplex* at concentrations between 0.5 and 0.7 mg/mL without being considered cytotoxic; however, inhibition of acetylcholinesterase was discarded as the molecular mechanism involved in the activity (Lopez et al., 2015). The anxiolytic effects of **8** and its potential interaction with the gamma-aminobutyric acid (GABA)A receptor in male Sprague–Dawley rats were investigated by Leiter and coworkers (Leiter et al., 2011). Compound **8** does not decrease anxiety by modulation of the GABAA receptor but may promote anxiogenesis. When **8** was combined with midazolam, an antagonist-like effect similar to the flumazenil and **8** combination was exhibited by a decrease in anxiolysis. Hence, **8** may antagonize the anxiolytic effects of midazolam, increase anxiety, and affect motor movements (Leiter et al., 2011).

To evaluate the aphrodisiac activity of nutmeg, Tajuddin et al. administrated 50% ethanolic extract at 500 mg/kg; p.o to different groups of male Swiss mice (Tajuddin et al., 2003). Mounting behavior, mating performance, and general short-term toxicity of the test drugs were determined and compared with the standard drug Penegra

(Sildenafil citrate). They noticed that extracts of the nutmeg stimulate the mounting behavior of male mice, and also significantly increase their mating performance; the drugs were devoid of any conspicuous general short-term toxicity.

Nutmeg extract showed platelet antiaggregatory ability in mice and also induced significant decrease in total cholesterol level in heart and liver (Ram et al., 1996).

Wahab et al. assessed the anticonvulsant activities of the volatile oil of nutmeg, and its potential for acute toxicity and acute neurotoxicity (Wahab et al., 2009). Thus, volatile oil was tested for its effects in maximal electroshock, subcutaneous pentylenetetrazole, strychnine, and bicuculline seizure tests. Nutmeg oil showed a rapid onset of action and short duration of anticonvulsant effect. It was found to possess significant anticonvulsant activity against electroshock-induced hind limb tonic extension. It exhibited dose-dependent anticonvulsant activity against pentylenetetrazole-induced tonic seizures. It delayed the onset of hind limb tonic extensor jerks induced by strychnine. It was anticonvulsant at lower doses, whereas weak proconvulsant at a higher dose against pentylenetetrazole and bicuculline induced clonic seizures. Nutmeg oil was found to possess wide therapeutic margin, as it did not induce motor impairment when tested up to 600 µL/kg in the inverted screen acute neurotoxicity test. Furthermore, the LD_{50} (2150 µL/kg) value was much higher than its anticonvulsant doses (50–300 µL/kg). These results indicate that nutmeg oil may be effective against grand mal and partial seizures, as it prevents seizure spread in a set of established animal models. However, slight potentiation of clonic seizure activity limits its use for the treatment of myoclonic and absence seizures (Wahab et al., 2009).

Gotke et al. (1990) have reported the nematicidal activity of *M. fragrans* seed against *Meloidogyne incognita*. The essential oil and its constituent **9** showed repellent effect against cigarette beetle *Lasioderma serricorne* (Du et al., 2014). The insecticidal constituents of hexane-soluble fraction from a methanolic extract of the seeds against adult females of *Blattella germanica* (L.) (Dictyoptera: Blattellidae) were identified as compound **3**, camphor, dipentene, γ-pinene, and α-terpineol (Jung et al., 2007).

5 TOXICITY OF *Myristica fragrans*

The toxic effects of nutmeg have been purported to be due mainly to myristicin oil (Ehrenpreis et al., 2014). In a case report, it was highlighted that nutmeg can cause serious toxic effects like status epilepticus. In effect, a developmentally normal infant had repeated episodes of afebrile status epilepticus following nutmeg ingestion (Sivathanu et al., 2014). He had developed two episodes of afebrile status epilepticus and had received different treatments earlier, but the details of treatment were not available. On admission to the hospital, he redeveloped convulsions and loading doses of phenytoin, phenobarbitone, and midazolam were administered. However, seizures persisted and extrapyramidal movements, nystagmus, and visual dysfunction were noted. Iatrogenic phenytoin toxicity was considered and confirmed by drug levels. His symptoms completely disappeared after discontinuation of phenytoin therapy. The initial seizures were attributed to myristicin, because of the temporal association. However, the subsequent seizures were due to phenytoin toxicity caused by administration of multiple loading doses.

An excellent review of Illinois Poison Center (IPC) data regarding nutmeg exposures from January 2001 to December 2011 was published by Ehrenpreis (2014). They reported 32 cases of poisoning following nutmeg ingestion. Of the 17 (53.1%) unintentional exposures, 10 subjects (58.8%) were under the age of 13. Four of the exposures in children under the age of 13 were ocular exposures. Fifteen exposures (46.9%) were intentional exposures. Of these intentional exposures, five (33.3%) were recorded to have combined drug intoxication. All of these were between the ages of 15 and 20. One patient with polypharmaceutical exposure required ventilatory support in the hospital. Mixing of nutmeg with other drugs was seen and required more intervention in adolescents. Authors therefore advised more education about these two factors, that is, nutmeg exposures as intentional polypharmacy in adolescents and unintentional exposures in young children.

6 PATENTS WITH *Myristica fragrans*

1. Patent WO2,008,096,998 A1 by Choi et al. (2008); Title: Use of a seeds extract of *Myristica fragrans* or active compounds isolated therefrom for preventing or treating osteoporosis. The invention relates to a use of a compound selected from the group consisting of machilin A, macelignan, machilin F, nectandrin B, safrole, licarin A, licarin B, myristagenol A, meso-dihydroguaiaretic acid and a mixture thereof, or a seeds extract of *M. fragrans* comprising at least one of the compounds for preventing or treating osteoporosis (Choi et al., 2008).
2. Patent EP2,689,806 A1 by Harsh and Gittins (2014); Title: Oral compositions containing extracts of *Myristica fragrans* and related methods. The patent describes compositions comprising a combination of extracts including *M. fragrans* useful as a dentifrice (Trivedi and Gittins, 2014).

7 CONCLUSIONS

In this chapter, we have provided pharmacological and phytochemical information on *M. fragrans* which can justify its use as a traditional medicinal plant. Several compounds were identified in nutmeg and mace of the plant with terpinen-4-ol, β-pinene, and limonene being the dominant compounds common to volatile oil in all species. Although myristicin was found to be responsible for most of the pharmacological effects of nutmeg and mace, it was also reported, together with elemicin, to cause toxic effects.

Some reported pharmacological properties of *M. fragrans* include anticancer, antidepressant, antidiabetic, antiobesity, antiinflammatory, analgesic, antimicrobial, antioxidant, hepatoprotective, and memory enhancing. However, the clinical efficacy of this plant on various ailments is still to be investigated.

REFERENCES

Al-Jumaily, E.F., Al-Amiry, M.H.A., 2012. Extraction and purification of terpenes from nutmeg (*Myristica fragrans*). JNUS 15 (3), 151–160.

Assa, J.R., Widjanarko, S.B., Kusnadi, J., Berhimpon, S., 2014. Antioxidant potential of fresh seed and mace of nutmeg (*Myristica fragrans* Houtt). Int. J. ChemTech Res. 6, 2460–2468.

Atta-ur-Rahman, Choudhary, M.L., Farooq, A., Ahmed, A., Zafar, M.I., Demirci, B., 2000. Antifungal activities and essential oil constituents of some spices from Pakistan. Pak. J. Chem. Soc. 22 (1), 60–65.

Bejnarowicz, E.A., Kirch, E.R., 1963. Gas chromatographic analysis of oil of Nutmeg. J. Pharm. Sci. 52, 988.

Cao, G.Y., Xu, W., Yang, X.W., Gonzalez, F.J., Li, F., 2015. New neolignans from the seeds of *Myristica fragrans* that inhibit nitric oxide production. Food Chem. 173, 231–237.

Cao, G.Y., Yang, X.W., Xu, W., Li, F., 2013. New inhibitors of nitric oxide production from the seeds of *Myristica fragrans*. Food Chem. Toxicol. 62, 167–171.

Cho, J.Y., Choi, G.J., Son, S.W., Jang, K.S., Lim, H.K., Lee, S.O., 2007. Isolation and antifungal activity of lignans from *Myristica fragrans* against various plant pathogenic fungi. Pest Manag. Sci. 63 (9), 935–940.

Choi, J.S., Choi, Y.H., Hong, K.S., Kim, S.H., Kim, Y.S., Lee, B.H., 2008. Use of a seeds extract of *Myristica fragrans* or active compounds isolated therefrom for preventing or treating osteoporosis: Google Patents.

Cui, C.A., Jin, D.Q., Hwang, Y.K., Lee, I.S., Hwang, J.K., Ha, I., 2008. Macelignan attenuates LPS-induced inflammation and reduces LPS-induced spatial learning impairments in rats. Neurosci. Lett. 448 (1), 110–114.

Dhingra, D., Parle, M., Kulkarni, S.K., 2006. Comparative brain cholinesterase-inhibiting activity of *Glycyrrhiza glabra*, *Myristica fragrans*, ascorbic acid, and metrifonate in mice. J. Med. Food 9 (2), 281–283.

Dhingra, D., Sharma, A., 2006. Antidepressant-like activity of n-hexane extract of nutmeg (*Myristica fragrans*) seeds in mice. J. Med. Food 9 (1), 84–89.

Du, S.S., Yang, K., Wang, C.F., You, C.X., Geng, Z.F., Guo, S.S., 2014. Chemical constituents and activities of the essential oil from *Myristica fragrans* against cigarette beetle *Lasioderma serricorne*. Chem. Biodivers. 11 (9), 1449–1456.

Ehrenpreis, J.E., DesLauriers, C., Lank, P., Armstrong, P.K., Leikin, J.B., 2014. Nutmeg poisonings: a retrospective review of 10 years experience from the Illinois Poison Center, 2001–2011. J. Med. Toxicol. 10 (2), 148–151.

Gomathi, P., Aman, K., Mebrahtom, G., Gereziher, G., Anwar-ul-Hassan, G., 2016. Nutmeg (*Myristica fragrans* Houtt.) oils. In: Preedy, V.R. (Ed.), Essential Oils in Food Preservation, Flavor and Safety. Academic Press, San Diego, pp. 607–616, (Chapter 69).

Goncalves, J.L., Lopes, R.C., Oliveira, D.B., Costa, S.S., Miranda, M.M., Romanos, M.T., 2005. In vitro anti-rotavirus activity of some medicinal plants used in Brazil against diarrhea. J. Ethnopharmacol. 99 (3), 403–407.

Gopalakrishnan, M., 1992. Chemical composition of nutmeg in the Spice Islands. J. Spices Aromat. Crops 1, 49–54.

Gopalakrishnan, M., Rajaraman, K., Mathew, A.G., 1979. Identification of the mace pigments. J. Food Sci. Technol. 16, 261–262.

Gotke, N., Maheswari, M.L., Mathur, V.K., 1990. Nematicidal activity of *M. fragrans* against *Meloidogyne incognita*. Indian Perfumer 34, 105–107.

Han, K.L., Choi, J.S., Lee, J.Y., Song, J., Joe, M.K., Jung, M.H., 2008. Therapeutic potential of peroxisome proliferators—activated receptor-alpha/gamma dual agonist with alleviation of endoplasmic reticulum stress for the treatment of diabetes. Diabetes 57 (3), 737–745.

Hayfaa, A.A., Sahar, A.M., Awatif, M.A., 2013. Evaluation of analgesic activity and toxicity of alkaloids in *Myristica fragrans* seeds in mice. J. Pain Res. 6, 611–615.

Iwu, M., 1993. Handbook of African Medicinal Plants. CRC Press, Boca Raton, Florida, USA.
Jin, D.Q., Lim, C.S., Hwang, J.K., Ha, I., Han, J.S., 2005. Anti-oxidant and anti-inflammatory activities of macelignan in murine hippocampal cell line and primary culture of rat microglial cells. Biochem. Biophys. Res. Commun. 331 (4), 1264–1269.
Jin, H., Zhu, Z.G., Yu, P.J., Wang, G.F., Zhang, J.Y., Li, J.R., 2012. Myrislignan attenuates lipopolysaccharide-induced inflammation reaction in murine macrophage cells through inhibition of NF-kappaB signalling pathway activation. Phytother. Res. 26 (9), 1320–1326.
Jung, W.C., Jang, Y.S., Hieu, T.T., Lee, C.K., Ahn, Y.J., 2007. Toxicity of *Myristica fragrans* seed compounds against *Blattella germanica* (Dictyoptera: Blattellidae). J. Med. Entomol. 44 (3), 524–529.
Kadhim, M.I., Naem, R.K., Abd-Sahib, A.S., 2013. Antibacterial activity of nutmeg (*Myristica fragrans*) seed extracts against some pathogenic bacteria. J. Al-Nahrain Univ. 16 (2), 188–192.
Kim, Y.B., Park, J.Y., 1991. The crystal structure of Licarin B, a component of seeds of *M. fragrans*. Arch. Pharmacol. Res. 14, 1–6.
Latha, P.G., Sindhu, P.G., Suja, S.R., Geetha, B.S., Pushpangadan, P., Rajasekharan, S., 2005. Pharmacology and chemistry of *Myristica fragrans* Houtt.—a review. J. Spices Aromat. Crops 14 (2), 94–101.
Lee, J.Y., Park, W., 2011. Anti-inflammatory effect of myristicin on RAW 264.7 macrophages stimulated with polyinosinic-polycytidylic acid. Molecules 16 (8), 7132–7142.
Lee, S.U., Shim, K.S., Ryu, S.Y., Min, Y.K., Kim, S.H., 2009. Machilin A isolated from *Myristica fragrans* stimulates osteoblast differentiation. Planta Med. 75 (2), 152–157.
Leiter, E., Hitchcock, G., Godwin, S., Johnson, M., Sedgwick, W., Jones, W., 2011. Evaluation of the anxiolytic properties of myristicin, a component of nutmeg, in the male Sprague–Dawley rat. AANA J. 79 (2), 109–114.
Li, F., Yang, X.W., 2012. Analysis of anti-inflammatory dehydrodiisoeugenol and metabolites excreted in rat feces and urine using HPLC-UV. Biomed. Chromatogr. 26 (6), 703–707.
Lopez, V., Gerique, J., Langa, E., Berzosa, C., Valero, M.S., Gomez-Rincon, C., 2015. Antihelmintic effects of nutmeg (*Myristica fragans*) on *Anisakis simplex* L3 larvae obtained from *Micromesistius potassou*. Res. Vet. Sci. 100, 148–152.
Ma, J., Hwang, Y.K., Cho, W.H., Han, S.H., Hwang, J.K., Han, J.S., 2009. Macelignan attenuates activations of mitogen-activated protein kinases and nuclear factor kappa B induced by lipopolysaccharide in microglial cells. Biol. Pharm. Bull. 32 (6), 1085–1090.
Mabberley, D.J., 1987. The plant-book. Cambridge University Press, New York.
Miyasawa, M., Kasaga, H., Kameoka, H., 1996. Antifungal activity of neolignans from *Myristica fragrans*. Nat. Prod. Lett. 8, 271–273.
Moinuddin, G., Devi, K., Kumar Khajuria, D., 2012. Evaluation of the anti-depressant activity of *Myristica fragrans* (Nutmeg) in male rats. Avicenna J. Phytomed. 2 (2), 72–78.
Morita, T., Jinno, K., Kawagishi, H., Arimoto, Y., Suganuma, H., Inakuma, T., 2003. Hepatoprotective effect of myristicin from nutmeg (*Myristica fragrans*) on lipopolysaccharide/d-galactosamine-induced liver injury. J. Agric. Food Chem. 51 (6), 1560–1565.
Moteki, H., Usami, M., Katsuzaki, H., Imai, K., Hibasami, H., Komiya, T., 2002. Inhibitory effects of spice extracts on the growth of human lymphoid leukaemia, Molt 4B cells. J. Jpn. Soc. Food Sci. Tech. 49, 688–691.
Murcia, M.A., Egea, I., Romojaro, F., Parras, P., Jimenez, A.M., Martinez-Tome, M., 2004. Antioxidant evaluation in dessert spices compared with common food additives. Influence of irradiation procedure. J. Agric. Food Chem. 52 (7), 1872–1881.
Narasimhan, B., Dhake, A.S., 2006. Antibacterial principles from *Myristica fragrans* seeds. J. Med. Food 9 (3), 395–399.
Nguyen, P.H., Le, T.V., Kang, H.W., Chae, J., Kim, S.K., Kwon, K.I., 2010. AMP-activated protein kinase (AMPK) activators from *Myristica fragrans* (nutmeg) and their anti-obesity effect. Bioorg. Med. Chem. Lett. 20 (14), 4128–4131.
Ogunwande, I.A., Olawore, N.O., Adeleke, K.A., Ekundayo, O., 2003. Chemical composition of essential oil of *Myristica fragrans* Houtt (Nutmeg) from Nigeria. Jeobp 6 (1), 21–26.
Olajide, O.A., Ajayi, F.F., Ekhelar, A.I., Awe, S.O., Makinde, J.M., Alada, A.R., 1999. Biological effects of *Myristica fragrans* (nutmeg) extract. Phytother. Res. 13 (4), 344–345.
Orabi, K.Y., Mossa, J.S., el-Feraly, F.S., 1991. Isolation and characterization of two antimicrobial agents from mace (*Myristica fragrans*). J. Nat. Prod. 54 (3), 856–859.

Orwa, C., Mutua, A., Kindt, R., Jamnadass, R., Anthony, S., 2009. *Myristica fragrans* Agroforestree Database: a tree reference and selection guide version 4.0. Available from: http://www.worldagroforestry.org/sites/treedbs/treedatabases.asp

Ozaki, Y., Soedigdo, S., Wattimena, Y.R., Suganda, A.G., 1989. Antiinflammatory effect of mace, aril of *Myristica fragrans* Houtt., and its active principles. Jpn. J. Pharmacol. 49 (2), 155–163.

Park, S., Lee, D.K., Yang, C.H., 1998. Inhibition of fos-jun-DNA complex formation by dihydroguaiaretic acid and in vitro cytotoxic effects on cancer cells. Cancer Lett. 127 (1–2), 23–28.

Parle, M., Dhingra, D., Kulkarni, S.K., 2004. Improvement of mouse memory by *Myristica fragrans* seeds. J. Med. Food 7 (2), 157–161.

Patil, S.B., Ghadyale, V.A., Taklikar, S.S., Kulkarni, C.R., Arvindekar, A.U., 2011. Insulin secretagogue, alpha-glucosidase and antioxidant activity of some selected spices in streptozotocin-induced diabetic rats. Plant Foods Hum. Nutr. 66 (1), 85–90.

Piaru, S.P., Mahmud, R., Abdul Majid, A.M., Ismail, S., Man, C.N., 2012a. Chemical composition, antioxidant and cytotoxicity activities of the essential oils of *Myristica fragrans* and *Morinda citrifolia*. J. Sci. Food Agric. 92 (3), 593–597.

Piaru, S.P., Mahmud, R., Abdul Majid, A.M., Mahmoud Nassar, Z.D., 2012b. Antioxidant and antiangiogenic activities of the essential oils of *Myristica fragrans* and *Morinda citrifolia*. Asian Pac. J. Trop. Med. 5 (4), 294–298.

Power, F.B., Solway, A.H., 1908. The constituents of the essential oil of nutmeg. Proc. Chem. Soc. 23, 285.

Purushothaman, K.K., Sarada, A., 1980. Chemical examination of the aril of *Myristica fragrans* (Jathipathri). Ind. J. Chem. 19 B, 236–238.

Ram, A., Lauria, P., Gupta, R., Sharma, V.N., 1996. Hypolipidaemic effect of *Myristica fragrans* fruit extract in rabbits. J. Ethnopharmacol. 55 (1), 49–53.

Rani, P., Khullar, N., 2004. Antimicrobial evaluation of some medicinal plants for their anti-enteric potential against multi-drug resistant *Salmonella typhi*. Phytother. Res. 18 (8), 670–673.

Rastogi, R.P., Mehrotra, B.N., 1995. Compendium of Indian medicinal plants. New Delhi: Central Drug Research Institute, Lucknow and Publication Information Directorate, Council of Scientific and Industrial Research.

Rukayadi, Y., Kim, K.H., Hwang, J.K., 2008. In vitro anti-biofilm activity of macelignan isolated from *Myristica fragrans* Houtt. against oral primary colonizer bacteria. Phytother. Res. 22 (3), 308–312.

Sanghai-Vaijwade, D.N., Kulkarni, S.R., Sanghai, N.N., 2011. Nutmeg: a promising antibacterial agent for stability of sweets. IJRPC 1 (3), 403–407.

Satyavathy, G.V., Gupta, A.K., Tandon, N., 1987. *Myristica Boehmer* (Myristicaceae). Medicinal Plants of Indiavol. 2Indian Council of Medical Research, New Delhi.

Shinohara, C., Mori, S., Ando, T., Tsuji, T., 1999. Arg-gingipain inhibition and anti-bacterial activity selective for *Porphyromonas gingivalis* by malabaricone C. Biosci. Biotechnol. Biochem. 63 (8), 1475–1477.

Sivathanu, S., Sampath, S., David, H.S., Rajavelu, K.K., 2014. Myristicin and phenytoin toxicity in an infant. BMJ Case Rep. 2014.

Tajuddin, Ahmad, S., Latif, A., Qasmi, I.A., 2003. Aphrodisiac activity of 50% ethanolic extracts of *Myristica fragrans* Houtt. (nutmeg) and *Syzygium aromaticum* (L) Merr. & Perry. (clove) in male mice: a comparative study. BMC Complement. Altern. Med. 3, 6.

Thuong, P.T., Hung, T.M., Khoi, N.M., Nhung, H.T.M., Chinh, N.T., Quy, N.T., 2013. Cytotoxic and antitumor activities of lignans from the seeds of Vietnamese nutmeg *Myristica fragrans*. Arch. Pharm. Res. 37 (3), 399–403.

Trivedi, H.M., Gittins, E.K., 2014. Oral compositions containing extracts of *Myristica fragrans* and related methods: Google Patents.

Varshney, I.P., Sharma, S.C., Houtt, 1968. Saponins and sapogenins. Part XXII. Chemical investigation of seeds of *Myristica fragrans*. Ind. J. Chem. 6, 474–476.

Wahab, A., Ul Haq, R., Ahmed, A., Khan, R.A., Raza, M., 2009. Anticonvulsant activities of nutmeg oil of *Myristica fragrans*. Phytother. Res. 23 (2), 153–158.

Yang, S., Na, M.K., Jang, J.P., Kim, K.A., Kim, B.Y., Sung, N.J., 2006. Inhibition of protein tyrosine phosphatase 1B by lignans from *Myristica fragrans*. Phytother. Res. 20 (8), 680–682.

Yinyang, J., Mpondo Mpondo, E., Tchatat, M., Ndjib, R.C., Mvogo Ottou, P.B., Dibong, S.D., 2014. [Les plantes à alcaloïdes utilisées par les populations de la ville de Douala (Cameroun)]. J. Appl. Biosci. 78, 6600–6619.

Zheng, G.Q., Kenney, P.M., Zhang, J., Lam, L.K., 1992. Inhibition of benzo[a]pyrene-induced tumorigenesis by myristicin, a volatile aroma constituent of parsley leaf oil. Carcinogenesis 13, 1921–1923.

CHAPTER 24

Passiflora edulis

G.S. Taïwe, V. Kuete

1 INTRODUCTION

The passion fruit is so called because it is one of the many species of passionflower, leading to the English translation of the Latin genus name, *Passiflora*. Several species have a long history of use as traditional herbal medicines. The name was given by Spanish missionaries to South America as an expository aid while trying to convert the indigenous inhabitants to Christianity. The genus *Passiflora*, comprising about 520 species, is the largest in the family Passifloraceae (Arbonnier, 2000; Wohlmuth et al., 2010). The species of this genus are distributed in the warm temperate and tropical regions of the New World; they are much rarer in Asia, Australia, and tropical Africa (Sacco, 1980). In Brazil there are a number of native plants of the genus *Passiflora*, known as maracujas. Two types of *Passiflora edulis* Sims (Passifloraceae) are grown commercially, the purple form (*P. edulis* Sims) and a yellow form (*P. edulis* var. *flavicarpa* Degenerer) (Spencer and Seigler, 1983). In Brazil, *Passiflora alata* is the official *Passiflora* species in the Brazilian Pharmacopoeia and *P. edulis* is the species most employed as a flavoring and as a juice in the food industries.

This plant exhibits various pharmacological properties and possesses a complex phytochemistry. Most of the pharmacological investigations of *P. edulis* have been addressed to its central nervous system (CNS) activities, such as anxiolytic, anticonvulsant, and sedative actions. In several preclinical experiments, *P. edulis* extracts have exhibited potential effects for the treatment of inflammation, pain, and insomnia as well as for attention-deficit hyperactivity disorder, hypertension, and cancer (Spencer and Seigler, 1983). Several mechanisms, including the inhibition of proinflammatory cytokines, enzyme (myeloperoxidase) and mediators (bradykinin, histamine, substance P, nitric oxide) release and/or action, appear to account for *P. edulis*'s actions. The leaves and stems of *P. edulis* have shown antinociceptive, antitumor, antimicrobial, and antioxidant activities (Patel, 2009).

This plant-based, traditional medicine system still acquires an important place in the health care system. The pulp of the fruit acts as a stimulant and tonic. The putative clinical efficacy of *P. edulis* has been evaluated for the treatment of a variety of diseases, but the current most common use in clinical practice is in the treatment of anxiety and sleep disorders. *P. edulis* has been used as an ethnic remedy for the cure of numerous infectious disorders of bacterial, fungal, viral, mycobacterium, and protozoal origin. Although a variety of other preparations are available, dried extracts are the most important product derived from *P. edulis*. Many practitioners actually use *P. edulis* extracts alone or in combination with other herbal medicines to treat depression and insomnia in a wide range of patients (Newall et al., 1996; Zhou et al., 2008).

This chapter aims to evaluate and comment on the scientific evidence regarding the therapeutic use and basis for future research on *P. edulis*, and its real potential for the development of the market for herbal medicinal products; to summarize the chemical constituents of therapeutic preparations; to analyze the pharmacological aspects of the plant by examining both preclinical and clinical research, and to assess the toxicity and safety profile.

2 BOTANICAL DESCRIPTION

P. edulis is also known as passion fruit, grenadelle, grenadine, passionflower, purple granadilla, or purple passion fruit. *P. edulis* is a vigorous, herbaceous, long-lived (perennial) climber, widely cultivated for its edible fruit. Stems up to 15 m long, striate, with axillary simple tendrils up to 10 cm long. Leaves alternate, up to 13 × 15 cm, more or less deeply 3-lobed, slightly leathery, glossy green or yellow–green above, paler and duller green below, with two glands at the apex of the petiole; margin finely toothed; linear stipules present about 1 cm long. Flowers solitary, up to 7 cm in diameter. Petals white, corona with filaments up to 2.5 cm long in 4–5 rows, white, purple at base. Fruit ovoid to spherical, 4–5 cm in diameter, yellow, greenish–yellow, or purplish (Souza et al., 2004). The *P. edulis* is known for its sweet purple fruit. It grows to 20–30 ft. (6-9 m) in length. It is a tangled vine. That means that while it grows it will entangle all in itself. The flower will be more noticeable by examining its fruit. The fruit will appear green when unripened, and it will be a dull purple upon ripening. The flesh of the fruit is light orange, and it is a delicatessen to people in South America. The passion fruit is a pepo, a type of berry, round to oval, either yellow or dark purple at maturity, with a soft to firm, and has a juicy interior filled with numerous seeds. The fruit is both eaten and juiced; passion fruit juice is often added to other fruit juices to enhance aroma. *P. edulis* Sims has often been taken to be synonymous with *Passiflora incarnata* L. because the plants possess identical morphological and microscopic characteristics. This plant reproduces by seed that are animal dispersed. Although one article attempted to eliminate potential confusion between these two similar plants (Dhawan et al., 2001), this confusion remains, potentially leading to the selection of the wrong plant, thus accounting for the inconclusive and contradictory pharmacological reports on these two plants. Dhawan et al. established key identification parameters to differentiate between the two plants: various leaf constants, the vein-islet number, the vein termination number, the stomatal number, and the stomatal index; as well as physicochemical parameters such as h values of the extracts of *P. incarnata* and *P. edulis* (Dhawan et al., 2001).

3 PROPAGATION

The discovery of several thousand years old seeds of *Passiflora* from the archaeological sites at Virginia and North America provides strong evidences of the prehistoric use of the fruits by the ancient "Red Indian" people (Gremillion, 1989). Native to southern Brazil, Paraguay to northern Argentina, *P. edulis* is a medicinal plant distributed in warm temperatures and tropical regions. The plant is found exclusively in tropical regions. It is cultivated commercially in tropical and subtropical areas for its sweet, seedy fruit and is widely grown in several countries of South America, Central America, the Caribbean, Africa, Southern Asia, Vietnam, Israel, Australia, South Korea, Hawaii, and the mainland United States (Zibadi and Watson, 2004). Locations within which *P. edulis* is naturalized include eastern and southern Australia, southern Africa, New Zealand, southeastern USA, and some oceanic islands with warm climates. Wild plants can smother trees and shrubs and can naturalize in disturbed forests, along river banks, fencerows, abandoned farms, and urban open spaces. Young plants are eaten by livestock, so *P. edulis* is almost never found in areas that are moderately to heavily grazed. *P. edulis* is invasive in parts of Kenya and naturalized in parts of Tanzania, South Africa, and Uganda (Henderson, 2001; Ntuli et al., 2012). The species is widely grown in Uganda for its fruits and has escaped and naturalized in most of the forests (both natural and plantation) where it continues to be dispersed by humans and primates through eating its fruits and passing out the seed which passes through the digestive system unharmed, though it does not seem to be a serious threat as a weed (Neville et al., 2003; Schmelzer and Gurib-Fakim, 2008).

4 TRADITIONAL OR ETHNOMEDICINAL USES

Several species of *Passiflora* have been employed widely as a folk medicine because of sedative and tranquillizer activities (Barbosa et al., 2008). The early European travelers in North America noted that Algonkian Indians in Virginia and Creek people in Florida ate fruits of *P. edulis* from cultivated as well as wild sources (Beverley, 1947). The then European settlers also consumed the fruit and praised its flavor, thereby, suggesting the prehistoric consumption of *Passiflora* as a fruit crop (Brickell, 1968). The use of *Passiflora* as a medicine was lauded for the first time by a Spanish researcher Monardus in Peru in 1569 (Taylor, 1996).

Various species of *Passiflora* have been used extensively in the traditional system of therapeutics in many countries. In South America, leaf extracts of *P. edulis* have been popularly used for the treatment of symptoms of alcoholism, anxiety, migraine, nervousness, and insomnia. A drink from the flower was considered to treat asthma, bronchitis, and whooping cough. In traditional medicine, the plant is used as heart tonic, mild diuretic, digestive stimulant, and treatment for urinary infections. Passion fruit seed oil has been used as a stimulating lubricant and massage oil (Zibadi and Watson, 2004). In Brazil, the said species, known as "maracuja" has been put to use as an anxiolytic, sedative, diuretic, and an analgesic (Oga et al., 1984). *P. edulis* has been used as a sedative, diuretic, anthelmintic, antidiarrheal, stimulant, tonic, and also in the treatment of hypertension, menopausal symptoms, colic of infants in South America (Chopra et al., 1956; Kirtikar and Basu, 1975). In Madeire, the fruit of *P. edulis* is regarded as a digestive stimulant and is used as a remedy for gastric carcinoma (Watt and Breyer-Brandwijk, 1962). In Nagaland (India), fresh leaves of *P. edulis* are boiled in little amount of water and the extract is drunk for the treatment of dysentery and hypertension (Jamir et al., 1999). Fruits are eaten to get relief from constipation. *P. edulis* leaf infusion has been used to treat hysteria and insomnia in Nigeria (Nwosu, 1999). The plant is widely cultivated in India (Kirtikar and Basu, 1975). The leaves are applied on the head for giddiness and headache; a decoction is given in biliousness and asthma. The fruit is used as an emetic. The plant has been used as an analgesic, antispasmodic, antiasthmatic, wormicidal, and sedative in Brazil; as sedative and narcotic in Iraq; in diseased conditions like dysmenorrhea, epilepsy, insomnia, neurosis, and neuralgia in Turkey; to cure hysteria and neurasthenia in Poland; in diarrhea, dysmenorrhea, neuralgia, burns, hemorrhoids, and insomnia in America (Taylor, 1996). This plant is widely used by the South African traditional healers. These traditional uses include alcohol withdrawal, antibacterial, antiseizure, antispasm, aphrodisiac, asthma, attention-deficit hyperactivity disorder, burns (skin), cancer, chronic pain, cough, drug addiction, Epstein–Barr virus, fungal infections, gastrointestinal discomfort (nervous stomach), *Helicobacter pylori* infection, hemorrhoids, high blood pressure, menopausal symptoms (hot flashes), nerve pain, pain (general), skin inflammation, tension, and wrinkle prevention (Barbosa et al., 2008; Ingale and Hivrale, 2010).

5 PHYTOCHEMISTRY

P. edulis is phytochemically characterized by the presence of a pattern of several primary constituents. The constituents of different extracts include flavonoids, alkaloids, phenols, cyanogenic compounds, glycosides, vitamins, minerals, and terpenoid compounds. Some compounds isolated from this plant are illustrated in Fig. 24.1.

5.1 Flavonoids

The passion fruit pulp contained 16.23 mg/L of isoorientin (**1**) and 158.04 mg/L of total flavonoid, suggesting that *P. edulis* fruits may be comparable with other flavonoid food sources such as orange juice or sugarcane juice (Zeraik and Yariwake, 2010). Previous studies have described the presence of flavonoids as the major constituents of *P. edulis*, mainly C-glycosylflavones. The flavonoids compound **1**, schaftoside (**2**), isoschaftoside (**3**), orientin (**4**), isovitexin (**5**), luteolin-6-C-chinovoside, and luteolin-6-C-fucoside have been identified in the fruit (Dhawan et al., 2004; Pereira and Vilegas, 2000; Mareck et al., 1990).

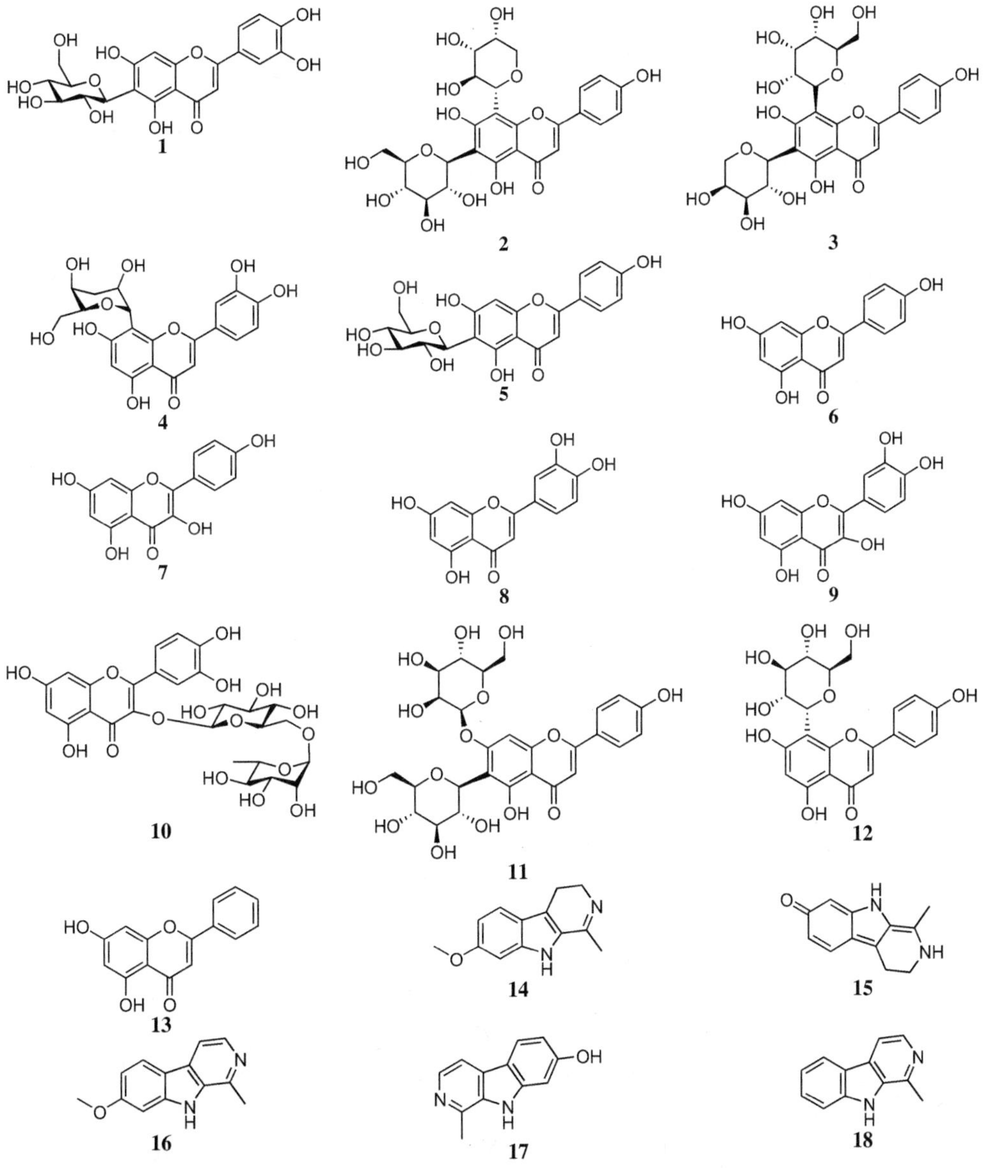

Figure 24.1 *Chemical structures of selected compounds isolated from P. edulis.* Isoorientin (**1**); schaftoside (**2**); isoschaftoside (**3**); orientin (**4**); isovitexin (**5**); apigenin (**6**); kaempferol (**7**); luteolin (**8**); quercetin (**9**); rutin (**10**); saponarin (**11**); vitexin (**12**); chrysin (**13**) ; harmaline (**14**); harmalol (**15**); harmine (**16**); harmol (**17**); harman (**18**).

Many flavonoids and their glycosides have been found in *P. edulis*, including compound **1, 2, 3, 5,** apigenin (**6**), benzoflavone, homoorientin, kaempferol (**7**), lucenin, luteolin (**8**), passiflorine (named after the genus), quercetin (**9**), rutin (**10**), saponarin (**11**), and vicenin and vitexin (**12**). Chrysin (**13**) is a naturally occurring flavone chemically extracted from *P. edulis* flower. Also documented to occur at least in some *Passiflora* in quantity are the hydrocarbon nonacosane and the anthocyanidin pelargonidin-3-diglycoside (Dhawan et al., 2004; Pereira and Vilegas, 2000; Mareck et al., 1990).

5.2 Alkaloids

The most common of these alkaloids is harmane (**18**), but harmaline (**14**), harmalol (**15**), harmine (**16**), and harmol (**17**) are also present. Many species have been found to contain beta-carboline harmala alkaloids. The most common of these alkaloids is **18**, but **15–17** are also present (Dhawan et al., 2004; Pereira and Vilegas, 2000).

5.3 Terpenoids

4-Hydroxy-β-ionol, 4-oxo-β-ionol, 4-hydroxy-7,8-dihydro-β-ionol, 4-oxo-7,8-dihydro-β-ionol, 3-oxo-α-ionol, isomeric 3-oxo retro-α-ionols, 3-oxo-7,8-dihydro-α-ionol, 3-hydroxy-1,1,6- trimethyl-1,2,3,4-tetrahydronaphthalene vomifoliol and dehydrovomifoliol, terpene alcohols linalool, and α-terpeneol, terpene diols (*E*) and (*Z*)-2,6-dimethyl-octa-2,7-diene-1,6-diol, 2,6-dimethyl-octa-3,7-dien-2,6-diol, 2,6-dimethyl-1,8-octanediol, 2,6-dimethyl-octa-1,7-diene-3,6-diol, ionol derivatives oxygenated in position 3, and 2,5-dimethyl-4-hydroxy-3-(2*H*)-furanone (furaneol) have been identified (Chassagne et al., 1996). Two new ionones I and II were isolated (Näf et al., 1977) for the first time from *P. edulis*.

5.4 Glycosides

From the methanol extract of air dried leaves, a cyclopropane triterpene glycoside, named passiflorine was isolated. *P. edulis* has been reported to be rich in glycosides which include flavonoid glycosides, namely, luteolin-6-C-chinovoside, luteolin-6-C-fucoside; cyclopentenoid cyanohydrin glycosides passicapsin and passibiflorin; cyanogenic glycosides passicoriacin, epipassicoriacin and epitetraphyllin B cyanogenic-β-rutinoside {(*R*)–mandelonitrile-α-L-rhamnopyranosyl-β-D-glucopyranoside} amygdalin, prunasin, mandelonitrile rhamnopyranosyl-β-D-glucopyranoside, sambunigrin; 6-O-α-L-arabinopyranosyl-β-D-glucopyranosides of linalool, benzyl alcohol, and 3 methyl-but-2en-1-ol; β-D-glucopyranoside and 6-O-α-L-rhamnopyranosyl-β-D-glucopyranoside of methyl salicylate and β-D-glucopyranoside of eugenol (Dhawan et al., 2004; Zucolotto et al., 2012).

5.5 Miscellaneous and phytoconstituents

The genus is rich in organic acids including formic, butyric, linoleic, linolenic, malic, myristic, oleic, and palmitic acids as well as phenolic compounds, and the amino acid α-alanine. Esters like ethyl butyrate, ethyl caproate, *n*-hexyl butyrate, and *n*-hexyl caproate give the fruits their flavor and appetizing smell. Sugars, contained mainly in the fruit, are most significantly D-fructose, D-glucose, and raffinose. Among enzymes, *Passiflora* was found to be rich in catalase, pectin methylesterase, and phenolase (Duke, 2008; Ingale and Hivrale, 2010). Other compounds found in passionflowers are coumarins (e.g., scopoletin and umbelliferone), maltol, phytosterols (e.g., lutenin), and cyanogenic glycosides (Duke, 2008). There are mainly cycloartane triterpenoids and their saponins isolated from this plant, including two new cycloartane triterpenoid saponins named cyclopassifloside XII (1) and XIII (2), together with six known cycloartane triterpenoids, cyclopassifloic acids B and E, cyclopassiflosides II, VI, IX, and XI (Wang et al., 2013).

6 IN VITRO AND IN VIVO PHARMACOLOGICAL STUDIES

6.1 Antioxidant activity

The leaves of *P. edulis*, traditionally used in American countries to treat both inflammation and nociception by folk medicine, are rich in polyphenols, which have been reported as natural antioxidant. The antioxidant activity of *P. edulis* hydro-alcoholic leaf extracts was verified in in vitro and ex vivo assays. The antioxidant activity of *P. edulis* leaf extracts was significantly correlated with polyphenol contents. In addition, *P. edulis* attenuated ex vivo iron-induced cell death, quantified by lactate dehydrogenase leakage, and effectively protected against protein damage induced by iron and glucose. The rind extracts of *P. edulis* possessed higher and dose-dependent inhibitory effects on lucigenin-enhanced chemiluminescence response and on the peroxidase activity of the neutrophil and the neutrophil granule enzyme myeloperoxidase (Zeraik et al., 2011). 1,1-Diphenyl-2-picrylhydrazyl (DPPH) offers a convenient and accurate method for titrating the oxidizable groups of natural or synthetic antioxidants. The antioxidative capacity of *P. edulis* leaves was also checked against DPPH radical and several reactive oxygen species (superoxide radical, hydroxyl radical, and hypochlorous acid), revealing it to be concentration-dependent, although a prooxidant effect was noticed for hydroxyl radical. These findings demonstrated that the *P. edulis* leaf extract has potent in vitro and ex vivo antioxidant properties and might be considered as possible new sources of natural antioxidants. Further studies are needed to examine the potential use of *P. edulis* extract in the prevention of pathologies, such as diabetes mellitus and neurodegenerative diseases, where oxidative stress damage to protein seems to play a major role (Ferreres et al., 2007; Rudnicki et al., 2007; Patel, 2009). Previous results obtained by Kandandapani et al. revealed that subacute administration of *P. edulis* extracts significantly controlled the blood glucose level in the diabetic rats. In addition, *P. edulis* extracts protected the end organs by restoring the antioxidants enzyme, significantly increasing super oxide dismutase level and decreasing catalase and thiobarbituric acid reactive substances level in visceral organs.

In conclusion, *P. edulis* extracts showed antidiabetic and antioxidant potential against streptozotocin-induced diabetes (Kandandapani et al., 2015).

6.2 Antifungal activity

These peptides are commonly characterized by having low molecular masses and cationic charges. Pelegrini et al. worked on the purification and characterization of a novel plant peptide of 5.0 kDa, Pe-AFP-1 (antifungal peptide), purified from the seeds of passion fruit (*P. edulis*). In vitro assays indicated that Pe-AFP-1 was able to inhibit the development of the filamentous fungi *Trichoderma harzianum*, *Fusarium oxysporum*, and *Aspergillus fumigatus* with the respective IC_{50} values of 32, 34, and 40 μg/mL (Pelegrini et al., 2006). The discovery of Pe-AFP1 could contribute, in a near future, to the development of biotechnological products as antifungal drugs and transgenic plants with enhanced resistance to pathogenic fungi.

6.3 Antitumor activity

Fruit's decoction of *P. edulis* has been evaluated for the inhibition of activity of gelatinase matrix metalloproteinases (MMP-2 and MMP-9), two metalloproteases involved in the tumor invasion, metastasis, and angiogenesis. Water extract of *P. edulis*, at different concentrations, inhibited the enzymes (Puricelli et al., 2003).

6.4 Cytotoxic activity

Brine shrimp lethality bioassay is widely used in bioassay for bioactive compounds (Meyer et al., 1982; Zhao et al., 1992). Simple zoological organism (*Artemia salina*) was used as a convenient monitor for the screening. The eggs of the brine shrimp were collected and hatched in artificial seawater (3.8% NaCl solution) for 48 h to mature shrimp called nauplii. The cytotoxicity assay was performed on brine shrimp nauplii using Meyer method (Meyer et al., 1982). The lethality of the crude petroleum ether and chloroform extracts of *P. edulis* leaf and stem to brine shrimp was determined on *A. salina* after 24 hr of exposure of the samples with the positive control, vincristine sulfate. This technique was applied for the determination of general toxic property of the plant extractive. The chloroform extract of stem showed the lowest LC_{50} value and the petroleum ether extract of leaf showed highest value which was 6.63 and 11.17 g/mL, respectively (Meyer et al., 1982; Zhao et al., 1992).

6.5 Antiinflammatory activity

The aqueous leaves extract of *Passiflora* species exhibited potent antiinflammatory action in the experimental model in vivo (Benincá et al., 2007). The aqueous leaf extracts of *P. edulis* possess a significant antiinflammatory activity on mice. The systemic administration of *P. edulis* exhibited pronounced antiinflammatory actions, characterized by inhibition of leukocyte influx to the pleural cavity and associated with marked blockade of myeloperoxidase, nitric oxide, tumor necrosis factors, and interleukin-1 levels in the acute model of inflammation caused by intrapleural injection of mice (Montanher et al., 2007). In one experiment, *P. edulis* was more effective in suppressing the tumor necrosis factors and interleukin-1 levels than dexamethasone (Capasso and Sorrentino, 2005). *P. edulis*, therefore, may be a source of new therapeutic candidates with a spectrum of activity similar to the current antiinflammatory steroids such as dexamethasone.

6.6 Antianxiety activity

Anxiety is a very common mental health problem in the general population. *P. edulis* is a folk remedy used for anxiety. Several species of *Passiflora* have been employed widely as a folk medicine because of sedative and tranquillizer activities (Barbosa et al., 2008). Antianxiety activity of *P. edulis* has been evaluated on the performance of mice in the elevated plus maze, open field, and horizontal-wire tests (Coleta et al., 2001). The aqueous extract presented an anxiolytic-like activity without any significant effect upon the motor activity. During a comparison between diazepam (6 mg/kg) and chrysin (**13**) for their myorelaxant effects, compound **13** did not exhibit myorelaxant effect in the horizontal-wire test even at the dose range of 0.6–30 mg/kg, suggesting that **13** was an anxiolytic devoid of sedative or muscle relaxant counter effects, unlike that of diazepam which exhibited a myorelaxant effect. Compound **13** was found to be a ligand for central as well as peripheral benzodiazepine receptors (Medina et al., 1990).

6.7 Antihypertensive activity

Despite improved pharmacotherapies and mechanical treatments, cardiovascular disease remains a principal cause of morbidity and mortality worldwide. *P. edulis*, which is an allied species of *Passiflora*, has already been reported to possess antihypertensive effects and it is used in folklore medicine for treating hypertension. Ichimura et al. (2006) reported that the orally administered methanol extract of this plant (10–50 mg/kg) or compound **8** (50 mg/kg), which is one of the consistent polyphenols of the extract, significantly lower systolic and diastolylic blood pressure in spontaneously hypertensive rats (SHRs). Quantitative analysis by liquid chromatography tandem mass spectrometry (LC–MS/MS) showed that the extract contained 20 g/g dry weight of **8** and 41 g/g dry weight of luteolin-6-C-glucoside. It also contained gamma amino butyric acid (GABA, 2.4 mg/g dry weight by LC–MS/MS), which has been reported to be an antihypertensive material. As the extract contained a relatively high concentration of GABA, the antihypertensive effects of the extract in SHRs might be due mostly to the GABA-induced antihypertensive effect and partially to the vasodilatory effect of polyphenols including luteolin (Ichimura et al., 2006).

7 CLINICAL TRIALS

Irrespective of the presence of a large variety of phytoconstituents in the genus *Passiflora*, only a few reports regarding the pharmacological investigations on the plants of this genus are available. Most of the pharmacological works have been carried out on the CNS-depressant effects of various species. Aqueous extract of *P. edulis* has been reported to exhibit nonspecific CNS-depressant effects in mice, rats, and healthy human volunteers, whereas, it was also noted that some samples of *P. edulis* had a *"nonspecific"* CNS-depressant effect (Maluf et al., 1991). In another report on CNS-depressant effects of *P. edulis*, it was noted that the aqueous extract of the plant prolonged barbiturate-induced as well as morphine-induced sleep time in mice and also *"partially"* blocked the amphetamine-induced stimulant effects (Do et al., 1983).

8 SAFETY PROFILE AND PHARMACOVIGILANCE DATA

Passion flower is generally considered to be a safe herb with few reported serious side effects. In cases of side effects, the products being used have rarely been tested for contamination, which may have been the cause. Cyanide poisoning has been associated with *Passiflora* fruit, but this has not been proven in human studies. Rapid heart rhythm, nausea, and vomiting have been reported. Side effects may also include drowsiness/sedation and mental slowness. Patients should be cautious when driving or operating heavy machinery. Passion flower may theoretically increase the risk of bleeding and affect blood tests that measure blood clotting (Kapadia et al., 2002). There is a reported case of liver failure and death of a patient taking a preparation of passionflower with kava. Caution should be applied in taking any kava-containing products, as kava has been associated with liver damage. It has been suggested that the cause of the liver damage is less likely related to the presence of passionflower.

An extract containing passionflower and hawthorn has been studied as a possible treatment for shortness of breath and difficult use of exercise in patients with congestive heart failure. Although the results are promising, the effects of passionflower alone are unclear. The high quality human research of passionflower alone compared to prescription drugs used for this condition is needed before a strong recommendation can be made (Capasso and Sorrentino, 2005). Flavonoids exhibit significant hormone activity (Zand et al., 2000); apigenin and luteolin (another flavonoid) were found to be more effective at preventing pregnancy than ethinyl estradiol (Hiremath et al., 2000). There is not enough scientific evidence to recommend the safe use of passionflower in any dose during pregnancy or breastfeeding. During the 1930s, animal studies found uterine stimulant action in components of *Passiflora*. Many tinctures contain high levels of alcohol and should be avoided during pregnancy. Most herbs and supplements have not been thoroughly tested for interactions with other herbs, supplements, drugs, or foods. The interactions listed in the subsequent sections are based on reports in scientific publications, laboratory experiments, or traditional use. One should always read product labels. If one has a medical condition, or is taking other drugs, herbs, or supplements, he/she should speak with a qualified health care provider before starting a new therapy.

In Brazil, the fruits are commonly known as "maracuja" and the fruit pulp yields a delicious juice, which is exported to the several countries (Machado et al., 2008; Dhawan et al., 2004). *Passiflora* is available on the market in a range of different preparations, mainly in tablet form (500 mg) of the dried herb for oral use or by infusion, as liquid extract or as tincture (Ingale and Hivrale, 2010). In addition to variation in preparation, several different manufacturers produce formulations of *Passiflora*, making it even more difficult to compare the efficacy of the distinct preparations. A number of species of *Passiflora* are cultivated outside their natural range because of their beautiful flowers. *P. incarnatea* L. commonly used in many herbal remedies is well known for its sedative properties, while several other species are cultivated for the production of fruit juice *Maracujá* (*P. edulis*) and a few other species are used in Central and South America for similar purposes (Dhawan et al., 2004). There is not enough scientific data to recommend *P. edulis* to be used for children at any dose (Dhawan et al., 2004).

9 CONCLUSIONS

P. edulis is commonly found throughout world. Studies have revealed its use in antiinflammatory, antimicrobial, lipid lowering, antioxidant, anxiolytic, and antitumor. Various types of preparations, extracts, and individual compounds derived from this species have been found to possess a broad spectrum of pharmacological effects on several organs such as the brain, blood, cardiovascular, and nervous systems as well as on different biochemical processes and physiological functions including proteosynthesis, work capacity, motor coordination, and exploration. Further studies are needed to examine the potential use of *P. edulis* extract in the prevention of pathologies, such as diabetes mellitus, cardiac ischemia, renal ischemia, and neurodegenerative diseases, where oxidative stress damage to protein seems to play a major role. Therefore further studies may be carried out to prove the potential of this plant. *P. edulis* is becoming an endangered species now so more work needs to be done on agricultural and climatic condition to grow this plant.

REFERENCES

Arbonnier, M. (Eds.), 2000. *[Arbres, arbustes et lianes des zones sèches d'Afrique de l'Ouest]*. Paris: Centre de Coopération Internationale en Recherche Agronomique pour le développement/ Muséum national d'histoire naturelle/Union mondiale pour la nature (CIRAD/MNHN/ UICN).

Barbosa, P.R., Valvassori, S.S., Bordignon, Jr., C.L., Kappel, V.D., Martins, M.R., Gavioli, E.C., Reginatto, F.H., 2008. The aqueous extracts of *Passiflora alata* and *Passiflora edulis* reduce anxiety-related behaviors without affecting memory process in rats. J. Med. Food 11 (2), 282–288.

Benincá, J.P., Montanher, A.B., Zucolotto, S.M., Schenkel, E.P., Fröde, T.S., 2007. Evaluation of the antiinflammatory efficacy of *Passiflora edulis*. Food Chem. 104 (3), 1097–1105.

Beverley, L. (Ed.), 1947. The History and the Present State of Virginia. University of North Carolina press, Chapel Hill.

Brickell, J. (Ed.), 1968. The Natural History of North-Carolina. Johnson Publishing Company, Murfreesboro, NC.

Capasso, A., Sorrentino, L., 2005. Pharmacological studies on the sedative and hypnotic effect of *Kava kava* and *Passiflora* extracts combination. Phytomedicine 12 (1), 39–45.

Chassagne, D., Crouzet, J.C., Bayonove, C.L., Baumes, R.L., 1996. Identification and quantification of passion fruit cyanogenic glycosides. J. Agric. Food Chem. 44 (12), 3817–3820.

Chopra, R.N., Nayar, S.L., Chopra, I.C., 1956. Glossary of Indian Medicinal Plants. CSIR, New Delhi, India, 186-187.

Coleta, M., Campos, M.G., Cotrim, M.D., Proença, D.C.A., 2001. Comparative evaluation of *Melissa officinalis* L., *Tilia europaea* L., *Passiflora edulis* Sims. and *Hypericum perforatum* L. in the elevated plus maze anxiety test. Pharmacopsychiatry 34, S20–S21.

Dhawan, K., Dhawan, S., Sharma, A., 2004. *Passiflora*: a review update. J. Ethnopharmacol. 94 (1), 1–23.

Dhawan, K., Kumar, S., Sharma, A., 2001. Anxiolytic activity of aerial and underground parts of *Passiflora incarnata*. Fitoterapia 72, 922–926.

Do, V., Nitton, B., Leite, J.R., 1983. Psychopharmacological effects of preparations of *Passiflora edulis* (Passion flower). Cienc. Cult. 35, 11–24.

Duke, J.A., 2008. Duke's Handbook of Medicinal Plants of Latin America. CRC Press, Boca Raton, FL, USA.

Ferreres, F., Sousa, C., Valentão, P., Andrade, P.B., Seabra, R.M., Gil-Izquierdo, Á., 2007. New C-deoxyhexosyl flavones and antioxidant properties of *Passiflora edulis* leaf extract. J. Agric. Food Chem. 55 (25), 10187–10193.

Gremillion, K.J., 1989. The development of a mutualistic relationship between humans and maypops (*Passiflora incarnata* L.) in the southeastern United States. J. Ethnobiol. 9, 135–155.

Henderson, L., 2001. Alien Weeds and Invasive Plants. A Complete Guide to Declared Weeds and Invaders in South Africa. Plant Protection Research Institute, South Africa.

Hiremath, S.P., Badami, S., Hunasagatta, S.K., Patil, S.B., 2000. Antifertility and hormonal properties of flavones of *Striga orobanchioides*. Eur. J. Pharmacol. 391 (1), 193–197.

Ichimura, T., Yamanaka, A., Ichiba, T., Toyokawa, T., Kamada, Y., Tamamura, T., Maruyama, S., 2006. Antihypertensive effect of an extract of *Passiflora edulis* rind in spontaneously hypertensive rats. Biosci. Biotechnol. Biochem. 70 (3), 718–721.

Ingale, A.G., Hivrale, A.U., 2010. Pharmacological studies of *Passiflora* sp. and their bioactive compounds. Afr. J. Plant Sci. 4 (10), 417–426.

Jamir, T.T., Sharma, H.K., Dolui, A.K., 1999. Folklore medicinal plants of Nagaland, India. Fitoterapia 70 (4), 395–401.

Kandandapani, S., Balaraman, A.K., Ahamed, H.N., 2015. Extracts of passion fruit peel and seed of *Passiflora edulis* (Passifloraceae) attenuate oxidative stress in diabetic rats. Chin. J. Nat. Med. 13 (9), 680–686.

Kapadia, G.J., Azuine, M.A., Tokuda, H., Hang, E., Mukainaka, T., Nishino, H., Sridhar, R., 2002. Inhibitory effect of herbal remedies on 12-O-tetradecanoylphorbol-13-acetate-promoted Epstein–Barr virus early antigen activation. Pharmacol. Res. 45 (3), 213–220.

Kirtikar, K.R., Basu, B.D., 1975. Indian Medicinal Plants. Bishen Singh Mahendra Pal Singh, New Delhi, India, pp. 1103.

Machado, L.L., Monte, F.J.Q., Maria da Conceição, F., de Mattos, M.C., Lemos, T.L., Gotor-Fernández, V., Gotor, V., 2008. Bioreduction of aromatic aldehydes and ketones by fruits' barks of *Passiflora edulis*. J. Mol. Catal. B 54 (3), 130–133.

Maluf, E., Barros, H.M.T., Frochtengarten, M.L., Benti, R., Leite, J.R., 1991. Assessment of the hypnotic/sedative effects and toxicity of *Passiflora edulis* aqueous extract in rodents and humans. Phytother. Res. 5 (6), 262–266.

Mareck, U., Galensa, R., Herrmann, K., 1990. Identifizierung von Passionsfruchtsaft in Fruchtprodukten mittels HPLC. Z. Lebensm. Unters. Forsch. 191 (4–5), 269–274.

Medina, J.H., Paladini, A.C., Wolfman, C., de Stein, M.L., Calvo, D., Diaz, L.E., Peña, C., 1990. Chrysin (5,7-di-OH-flavone), a naturally-occurring ligand for benzodiazepine receptors, with anticonvulsant properties. Biochem. Pharmacol. 40 (10), 2227–2231.

Meyer, B.N., Ferrigni, N.R., Putnam, J.E., Jacobsen, L.B., Nichols, D.E., McLaughlin, J.L., 1982. Brine shrimp: a convenient general bioassay for active plant constituents. Planta Med. 45 (5), 31–34.

Montanher, A.B., Zucolotto, S.M., Schenkel, E.P., Fröde, T.S., 2007. Evidence of anti-inflammatory effects of *Passiflora edulis* in an inflammation model. J. Ethnopharmacol. 109 (2), 281–288.

Näf, F., Decorzant, R., Willhalm, B., Velluz, A., Winter, M., 1977. Structure and synthesis of two novel ionones identified in the purple passionfruit (*Passiflora edulis* Sims). Tetrahedron Lett. 18 (16), 1413–1416.

Macdonald, I.A.W., Reaser, J.K., Bright, C., Neville, L.E., Howard, G.W., Murphy, S.J., Preston G. (Eds.), 2003. Invasive Alien Species in Southern Africa: National Reports & Directory of Resources. Global Invasive Species Programme, Cape Town, South Africa.

Newall, C.A., Anderson, L.A., Phillipson, J.D. (Eds.), 1996. Herbal Medicines: A Guide for Health-Care Professionals. The Pharmaceutical Press, London, UK.

Ntuli, N.R., Zobolo, A.M., Siebert, S.J., Madakadze, R.M., 2012. Traditional vegetables of northern KwaZulu-Natal, South Africa: has indigenous knowledge expanded the menu? Afr. J. Agric. Res. 7, 6027–6034.

Nwosu, M.O., 1999. Herbs for mental disorders. Fitoterapia 70 (1), 58–63.

Oga, S., de Freitas, P.C.D., da Silva, A.C.G., Hanada, S., 1984. Pharmacological trials of crude extract of *Passiflora alata*. Planta Med. 50 (4), 303–306.

Patel, S.S., 2009. Morphology and pharmacology of *Passiflora edulis*: a review. J. Herb. Med. Toxicol. 3, 1–6.

Pelegrini, P.B., Noronha, E.F., Muniz, M.A.R., Vasconcelos, I.M., Chiarello, M.D., Oliveira, J.T.A., Franco, O.L., 2006. An antifungal peptide from passion fruit (*Passiflora edulis*) seeds with similarities to 2S albumin proteins. Biochim. Biophys. Acta 1764 (6), 1141–1146.

Pereira, C.A.M., Vilegas, J.H.Y., 2000. Chemical and pharmacological constituents of the genus *Passiflora*, with emphasis on *P. alata* Dryander, *P. edulis* Sims and *P. incarnata* L. Rev. Bras. Plantas Med. 3 (1), 1–12.

Puricelli, L., Dell'Aica, I., Sartor, L., Garbisa, S., Caniato, R., 2003. Preliminary evaluation of inhibition of matrix-metalloprotease MMP-2 and MMP-9 by *Passiflora edulis* and *P. foetida* aqueous extracts. Fitoterapia 74 (3), 302–304.

Rudnicki, M., de Oliveira, M.R., da Veiga Pereira, T., Reginatto, F.H., Dal-Pizzol, F., Moreira, J.C.F., 2007. Antioxidant and antiglycation properties of *Passiflora alata* and *Passiflora edulis* extracts. Food Chem. 100 (2), 719–724.

Sacco, J.C., 1980. Passifloraceas. In: Reitz, R. (Ed.), Flora Ilustrada Catarinense. Herbario Barbosa Rodrigues, Itajaı, Brazil, p. 132.

Schmelzer, G.H., Gurib-Fakim, A. (Eds.), 2008. Medicinal Plants, vol. 1, Backhuys Publishers CTA PROTA, Leiden, Netherlands.

Souza, M.M., Pereira, T.N.S., Viana, A.P., Pereira, M.G., do Amaral Júnior, A.T., Madureira, H.C., 2004. Flower receptivity and fruit characteristics associated to time of pollination in the yellow passion fruit *Passiflora edulis* Sims f. flavicarpa Degener (Passifloraceae). Sci. Hortic. 101 (4), 373–385.

Spencer, K.C., Seigler, D.S., 1983. Cyanogenesis of *Passiflora edulis*. J. Agric. Food Chem. 31 (4), 794–796.

Taylor, L. (Ed.), 1996. Maracuja, Herbal Secrets of the Rainforest. Prime Publishing Inc., Austin, TX.

Wang, C., Xu, F.Q., Shang, J.H., Xiao, H., Fan, W.W., Dong, F.W., Zhou, J., 2013. Cycloartane triterpenoid saponins from water soluble of *Passiflora edulis* Sims and their antidepressant-like effects. J. Ethnopharmacol. 148 (3), 812–817.

Watt, J.M., Breyer-Brandwijk, M.G. (Eds.), 1962. The Medicinal and Poisonous Plants of Southern and Eastern Africa. Livingston, Edinburgh, pp. 826–830.

Wohlmuth, H., Penman, K.G., Pearson, T., Lehmann, R.P., 2010. Pharmacognosy and chemotypes of passionflower (*Passiflora incarnata* L.). Biol. Pharm. Bull. 33 (6), 1015–1018.

Zand, R.S.R., Jenkins, D.J., Diamandis, E.P., 2000. Steroid hormone activity of flavonoids and related compounds. Breast Cancer Res. Treat. 62 (1), 35–49.

Zeraik, M.L., Serteyn, D., Deby-Dupont, G., Wauters, J.N., Tits, M., Yariwake, J.H., Franck, T., 2011. Evaluation of the antioxidant activity of passion fruit (*Passiflora edulis* and *Passiflora alata*) extracts on stimulated neutrophils and myeloperoxidase activity assays. Food Chem. 128 (2), 259–265.

Zeraik, M.L., Yariwake, J.H., 2010. Quantification of isoorientin and total flavonoids in *Passiflora edulis* fruit pulp by HPLC-UV/DAD. Microchem. J. 96 (1), 86–91.

Zhao, G., Hui, Y., Rupprecht, J.K., McLaughlin, J.L., Wood, K.V., 1992. Additional bioactive compounds and trilobacin, a novel highly cytotoxic acetogenin, from the bark of *Asimina triloba*. J. Nat. Prod. 55 (3), 347–356.

Zhou, Y., Tan, F., Deng, J., 2008. Update review of *Passiflora*. China J. Chin. Mater. Med. 33, 1789.

Zibadi, S., Watson, R.R., 2004. Passion fruit (*Passiflora edulis*). Evid. Base. Integr. Med. 1 (3), 183–187.

Zucolotto, S.M., Fagundes, C., Reginatto, F.H., Ramos, F.A., Castellanos, L., Duque, C., Schenkel, E.P., 2012. Analysis of C-glycosyl Flavonoids from South American *Passiflora* Species by HPLC-DAD and HPLC-MS. Phytochem. Anal. 23 (3), 232–239.

CHAPTER 25

Petroselinum crispum: a Review

C. Agyare, T. Appiah, Y.D. Boakye, J.A. Apenteng

1 INTRODUCTION

Petroselinum crispum (Mill.) Nym. ex A.W. Hill belongs to the family Apiaceae or Umbelliferae and the genus *Petroselinum*. It is commonly called parsley/garden parsley in English, "patraseli," "patrasoli," or "potrasoli" in Indonesia, "phakchi-farang" in Thailand, "vannsuy baraing" in Cambodia, "paseri" in Japan, "pietersielie" in Africa, "persil" in France and "bagdouness or "maadnous" in Arab (Ipor and Oyen, 1999; Quattrocchi, 2012). Synonyms of *P. crispum* are *Apium crispum* Mill., *Apium petroselinum* L., *Petroselinum hortense* Hoffm., and *Petroselinum sativum* Hoffm.

P. crispum is believed to be originally grown in Sardinia (Mediterranean area) and was cultivated from c. 3rd century BC. Linnaeus stated its wild habitat to be Sardinia, whence it was brought to England and apparently first cultivated in Britain in 1548; Bentham considered it a native of the Eastern Mediterranean regions; De Candolle of Turkey, Algeria and the Lebanon. Since its introduction into these islands in the 16th century it has been completely naturalized in various parts of England and Scotland, on old walls and rocks.

In ancient times, parsley was not only used for culinary and medical purposes, it was subjected to wide variety of superstitious beliefs by the Greeks and ancient Romans. The ancient Greeks mainly used parsley as a form of decoration for funeral wreaths and crowns of parsleys to honor the winners of Nenena and Isthmain sport games (Tucker and DeBaggio, 2009). The ancient Roman used parsley for deodorizing the corpse and cover up the alcohol on their breath. Parsley was used in Hebrew celebration of Passover as the symbol of spring and rebirth. It was rumored that Catherine de'Medici (Queen consort of France) was responsible for popularizing parsley in the 16th century, when she brought it back to France from Italy. Later, Christianity carried on this tradition by associating parsley with the Apostle Peter because of his designation as warder of the gates of heaven. The ancient Greeks and Romans did not commonly eat parsley. However, they did grow it in their gardens as a border, and it was thought to be wonderful fodder for chariot horses.

Parsley was appreciated for its medicinal properties long before it became accepted as a food or spice. It was probably first commonly eaten in Europe in the Middle Ages. The flat leafed variety was not initially eaten because it was easily confused with false-parsley (a noxious weed). However, the curly leafed variety soon found its way to plates and dishes since it has the ability to cleanse the breath and the palate (Yanardağ et al., 2003). It was soon commonly used as a garnish (Grieve, 2014). Today parsley is found in a wide variety of dishes. It is commercially sold in both fresh and dried forms (Rayment, 2016).

2 PLANT DESCRIPTION

P. crispum is a bright green, annual herb in subtropical and tropical areas. In temperate climates, it grows as a biennial, where in the first year, it forms a rosette of tripinnate leaves with numerous leaflets and a taproot used as a food store over the winter. In the second year, it grows as a flowering plant with sparser leaves and flat-topped diameter umbels with numerous yellow to yellowish-green flowers (Simon, 1990).

It is an erect copiously branched, herb that can grow up to 30–100 cm tall, aromatic in all parts and smooth. The stem is cylindrical, grooved, and hollow. The leaves are arranged alternately, 1–3-pinnately compound, dark green, glossy, flat, or curled, and with sheath at the base. The petiole is longest in the lower leaves. The pinnae are long-stalked, with obovate–cuneate to finely linear leaflets, which are divided into acute segments. The higher leaves are gradually less divided while the topmost leaf consists of a few acute segments only.

The inflorescence is a terminal or axillary compound umbel. The 1–3 foliolate bracts are rather short. There are 3–15 secondary rays (pedicels) which are 2–5 mm long. The flowers are small, yellow-green, and bisexual. The sepal is obscure. The petal consists of five petals which are suborbicular to obovate, measuring up to 1 mm $\times$ 0.5 mm and submarginate with an inflexed apical lobe. There are five stamens. The pistil is with an inferior and two-carpelled ovary where each carpel is with a thickened stylopodium, a style, and a spherical stigma. The fruit is a schizocarp, measuring 2–3 mm long, ovoid and it splits into two mericarps when ripen with each having five narrow ribs.

The root system is slender, fibrous with taproot measures up to 1 m long, sometimes thickened, and with a radical rosette of leaves when young (Ipor and Oyen, 1999). The seeds are ovoid, 2–3 mm long, with prominent style remnants at the apex. The plant normally dies after seed maturation (Huxley, 1992).

Though several cultivated varieties exist, the three main varieties of *P. crispum* are: *P. crispum* var. neapolitanum, *P. crispum* var. tuberosum, and *P. crispum* var. crispum. *P. crispum* var. crispum is curled-leaf parsley and *P. crispum* var. neapolitanum is flat-leafed (Italian) parsley. Flat-leafed parsley is generally harder than the curled-leaf (Herbst, 2001). *P. crispum* var. tuberosum is grown as a root vegetable. It is commonly known as "Hamburg parsley" or "turnip-root parsley." This type of parsley produces much thicker roots than types cultivated for their leaves, with a root as much as 6 times the size as that of garden parsley (Hanrahan and Frey, 2005). Many cultivars exist for both the curled-leaf and flat-leaf types. The curly leaf and plain leaf types are cultivated for their foliage, whereas root parsley is grown as a root vegetable (Tucker and DeBaggio, 2009).

3 GEOGRAPHICAL DISTRIBUTION

P. crispum probably originated in the Western Mediterranean. It occurs naturally in most Mediterranean and many temperate countries. It is an old crop, which was already well-known in classical Greece and Rome. It has now widely grown in many tropical areas including East and West Africa. *P. crispum* is widely grown for its leaves in most Mediterranean countries, Europe, and North America. In the tropics, including Southeast Asia, it is cultivated on a small scale. Varieties with thickened, edible taproot are of recent origin and probably developed around 1500 AD in Northern Germany. Their cultivation is concentrated in Northwestern and Eastern Europe and among North Americans. Parsley is widely distributed in Turkey, and grown in gardens and fields (Yanardağ et al., 2003).

In Africa, *P. crispum* is occasionally found as an escape or relic of cultivation. It is cultivated in Eritrea, Ethiopia, Mozambique, South Africa, Morocco, and Tunisia as a medicinal herb used in treating cardiovascular diseases, such as arterial hypertension (Gadi et al., 2009). In most African countries, *P. crispum* is usually grown on small plots for market gardening, though no statistical information on areas under production or market volumes are available. The plant prefers a sunny to half-shady environment on fresh to moist soil. The substrate used is usually sandy loamy soil with a pH between 6.5 and 7.5. *P. crispum* tolerates temperatures down to $-29°C$.

4 ETHNOMEDICINAL USES

P. crispum has been used as a medicinal plant for ailments and complaints of the gastrointestinal tract, as well as the kidney and lower urinary tract, and for stimulating digestion (Blumenthal et al., 2000). The root of *P. crispum* is used as a powerful diuretic (Pharmacopoeia Jugoslavica, 1951). Furthermore, it is used for the treatment of dyspepsia, cystitis, dysmenorrhea, functional amenorrhea, and myalgia (Wichtl and Bisset, 1994). *P. crispum* is used for the management of menstrual disorders, and as emmenagogue, galactagogue, and stomachic. It is also applied externally against head lice (Wichtl and Bisset, 1994).

Apart from its wide usage as a green vegetable and garnish, *P. crispum* is used for different medicinal purposes in traditional and folklore medicine of different countries. Seeds have been used as antimicrobial, antiseptic, antispasmodic, and sedative agents and in the treatment of gastrointestinal disorders, inflammation, halitosis, kidney stones, and amenorrhea, and also as carminative, astringent, and gastrotonic in Iran (Behtash et al., 2008; Moazedi et al., 2007; Aghili et al., 2009; Tonkaboni et al., 2007; Avicenna, 1983). It is used as diuretic in Turkey and carminative, as well as treatment for gastritis in Peru (Rehecho et al., 2011).

Leaves of *P. crispum* have been employed as food flavor, antitussive, and diuretic and also in the treatment of kidney stones, hemorrhoids, gastrointestinal disorder, blurred vision, and dermatitis (Aghili et al., 2009; Tonkaboni et al., 2007; Avicenna, 1983). The leaves are also used to manage bleeding, hypertension, hyperlipidemia, hepatic disorders, and diabetes in Turkey. Leaves are employed as food flavor (Wong and Kitts, 2006) and for treatment for skin diseases (Aljanaby, 2013) in China and Iraq, respectively. In Moroccan traditional healing system, the leaves are used in arterial hypertension, diabetes, cardiac disease, renal disease, lumbago, eczema, and nose bleed (Ziyyat et al., 1997; Eddouks et al., 2002; Jouad et al., 2001; Merzouki et al., 1997). The leaves are also used for the treatment of amenorrhoea, dysmenorrhea, kidney stones, prostatitis, diabetes, halitosis, anaemia, hypertension, hyperuricaemia, constipation, odontalgy, pain, baldness, and induction of abortion in Spain (Benítez et al., 2010) and urinary tract diseases and management of fluid retention in Serbia (Savikin et al., 2013). Its aerial parts are used as an abortifacient in Italy (Montesano et al., 2012).

5 PYTOCHEMICAL CONSTITUENTS

The healing properties and medical use of *P. crispum* are mostly related to a wide range of active biomolecules present in the plant. Phytochemical constituents and compounds have been isolated from seeds, roots, leaves, or petioles through bioassay-guided separation, essential oils obtained by methods, such as simultaneous distillation–extraction (SDE) and analyzed by techniques, such as multilayer coil countercurrent chromatography (MCCC), gas chromatography (GC), nuclear magnetic resonance (NMR) analysis, gas chromatography–mass spectrometry (GC–MS), ultraviolet–visible spectroscopy (UV–VIS), or high performance liquid chromatography (HPLC). These phytochemical constituents can be grouped into the flavonoids, carbohydrates, coumarins, essential oils, and other miscellaneous compounds.

5.1 Essential oil components

Seeds of *P. crispum* produce high amount of essential oils. Root and leaf also contain essential oils (Bruneton, 1999). Myristicin (phenylpropene) (**1**) and apiol (phenylpropanoid) (**2**) are the two main components of *P. crispum* essential oil which are responsible for its antioxidant activity (Zhang et al., 2006). α-pinene (sesquiterpene hydrocarbon) (**3**), monoterpene hydrocarbons [sabinene (**4**), β-pinene (**5**), ρ-cymene (**6**), limonene (**7**), β-phellandrene (**8**), and γ-terpinene (**9**)], phenylpropenes [1-allyl-2,3,4,5-tetramethoxy-benzene (**10**), eugenol (**11**), and elemicin (**12**)] have also been isolated from seeds of *P. crispum* (Zhang et al., 2006; Wagner and Bladt, 1996). Zhang et al. (2006) and Wagner and Bladt (1996) also reported the presence of carotol (alcohol sesquiterpene) (**13**), myristicin and apiol in *P. crispum* seeds. Roots of *P. crispum* have been found to contain two C_{17} polyacetylenic alcohols [heptadeca-1,9(Z)-diene-4,6-diyn-3-ol (**14**) and heptadeca-1,9(Z)-diene-4,6-diyn-3,8-diol (**15**)] (Nitz et al., 1990; Christensena and Brandtb, 2006).

Essential oil obtained from the leaves of *P. crispum* have been revealed to contain sesquiterpene hydrocarbons [β-caryophyllene (**16**), γ-elemene (**17**) and β-elemene (**18**)], aldehydes [phenylacetaldehyde (**19**), benzaldehyde (**20**) and hexanal (**21**)], monoterpene hydrocarbons [β-pinene, sabinene, 3-carene (**22**), camphene (**23**), α-thujene (**24**), myrcene (**25**), α-phellandrene (**26**), β-phellandrene, α-terpinene (**27**), *cis*-β-ocimene (**28**), *trans*-β-ocimene (**29**), α-terpinolene (**30**) and ρ-1,3,8-menthatriene (**31**)], monoterpene alcohols [α-terpineol (**32**) and 2-(ρ-Tolyl) propan-2-ol (**33**)], aromatic compounds [toluene (**34**) and m-xylene (**35**) and/or ρ-xylene (**36**)], sesquiterpene hydrocarbons [α-cubebene (**37**), α-copaene (**38**), β-bisabolene (**39**) and α-elemene (**40**)], 2-pentylfuran (ether) (**41**), *cis*-Hex-3-en-1-ol (alcohol) (**42**), cryptone (ketone) (**43**), δ-cadinol (sesquiterpene alcohol) (**44**), elemicin, α-pinene, limonene, γ-terpinene and ρ-cymene (MacLeod et al., 1985).

Analysis of volatile oil from *P. crispum* plant, cell culture, and callus showed that monoterpenes were the main constituent. ρ-1,3,8-menthatriene was the most abundant compound among the monoterpenes followed by β-phellandrene and apiol. Aldehydes [nonanal (**45**) and decanal (**46**)] and also fatty acids (Free and bound) were found in the volatile oil (López et al., 1999). The triacylglycerol, tripetroselinin (**47**) has been isolated from the seeds of *P. crispum* (Destaillats et al., 2009). Guieta et al. (2003) showed that seeds of *P. crispum* contain fatty acid and petroselinic acid (**48**). The chemical structures of compounds from essential oils are shown in Fig. 25.1.

5.2 Other constituents

The most dominant compounds of *P. crispum* are the flavonoids (Pápay et al., 2012). Flavonoids (Fig. 25.2): isorhamnetin (**49**), apigenin (**50**), quercetin (**51**), luteolin (**52**), and chrysoeriol (**53**) were identified in cell suspension cultures of *P. crispum* (Kreuzaler and Hahlbrock, 1973; Hempel et al., 1999). Gadi et al. (2012) isolated kaempferol (**54**) and apigenin in *P. crispum* leaf extract. Flavonoids apigenin and flavonoid glycosides [apiin (**55**) and cosmosiin (**56**)] were obtained from aqueous leaf extract of *P. crispum* (Chaves et al., 2011). A flavone glycoside, 6-acetylapiin (**57**) and petroside (**58**), its monoterpene glucoside, the furanocoumarin cnidilin (**59**) and the flavone glycosides [diosmetin 7-O-β-D-glucopyranoside (**60**), and kaempferol 3-O-β-D-glucopyranoside (**61**)] have been isolated from the methanol aerial part of *P. crispum* (Yoshikawa et al., 2000). However, the main reported flavonoids in *P. crispum* are apiin and luteolin (Fejes et al., 1998; Nielsen et al., 1999; Fejes et al., 2000).

Apiose (**62**) is a sugar detected in the stem, seed, and leaf of *P. crispum* (Hudson, 1949). Apiose and D-glucose (**63**) (Fig. 25.2) have also been identified in cell suspension cultures of *P. crispum* (Kreuzaler and Hahlbrock, 1973). These sugars mostly contribute to the structure of flavonoid glycosides (Farzaei et al., 2013).

Furocoumarins (Fig. 25.2) including oxypeucedanin hydrate (**64**) oxypeucedanin (**65**), psoralen (**66**), isopimpinellin (**67**), 8-methoxypsoralen (**68**), 5-methoxypsoralen (**69**), and imperatorin (**70**) have been isolated from the leaves and roots of *P. crispum*.

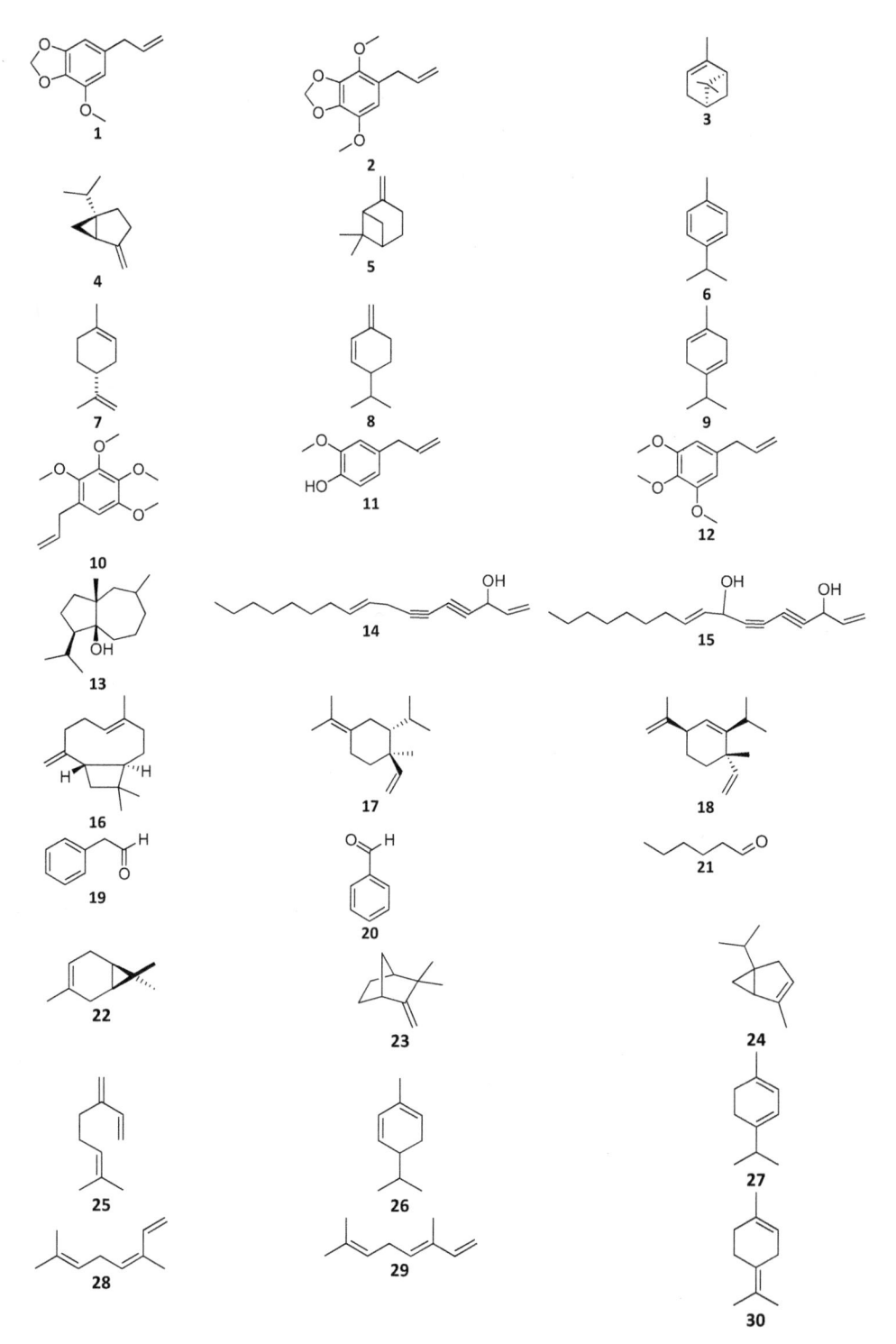

Figure 25.1 *Chemical structures of compounds from essential of* **P. crispum.** Myristicin (**1**); apiol (**2**); α-pinene (**3**); sabinene (**4**); β-pinene (**5**); ρ-cymene (**6**); limonene (**7**); β-phellandrene (**8**); γ-terpinene (**9**); 1-allyl-2,3,4,5-tetra-methoxy-benzene (**10**); eugenol (**11**); elemicin (**12**); carotol (**13**); heptadeca-1,9(Z)-diene-4,6-diyn-3-ol (**14**); heptadeca-1,9(Z)-diene-4,6-diyn-3,8-diol (**15**); β-caryophyllene (**16**); γ-elemene (**17**); β-elemene (**18**); phenylacetaldehyde (**19**); benzaldehyde (**20**); hexanal (**21**); 3-carene

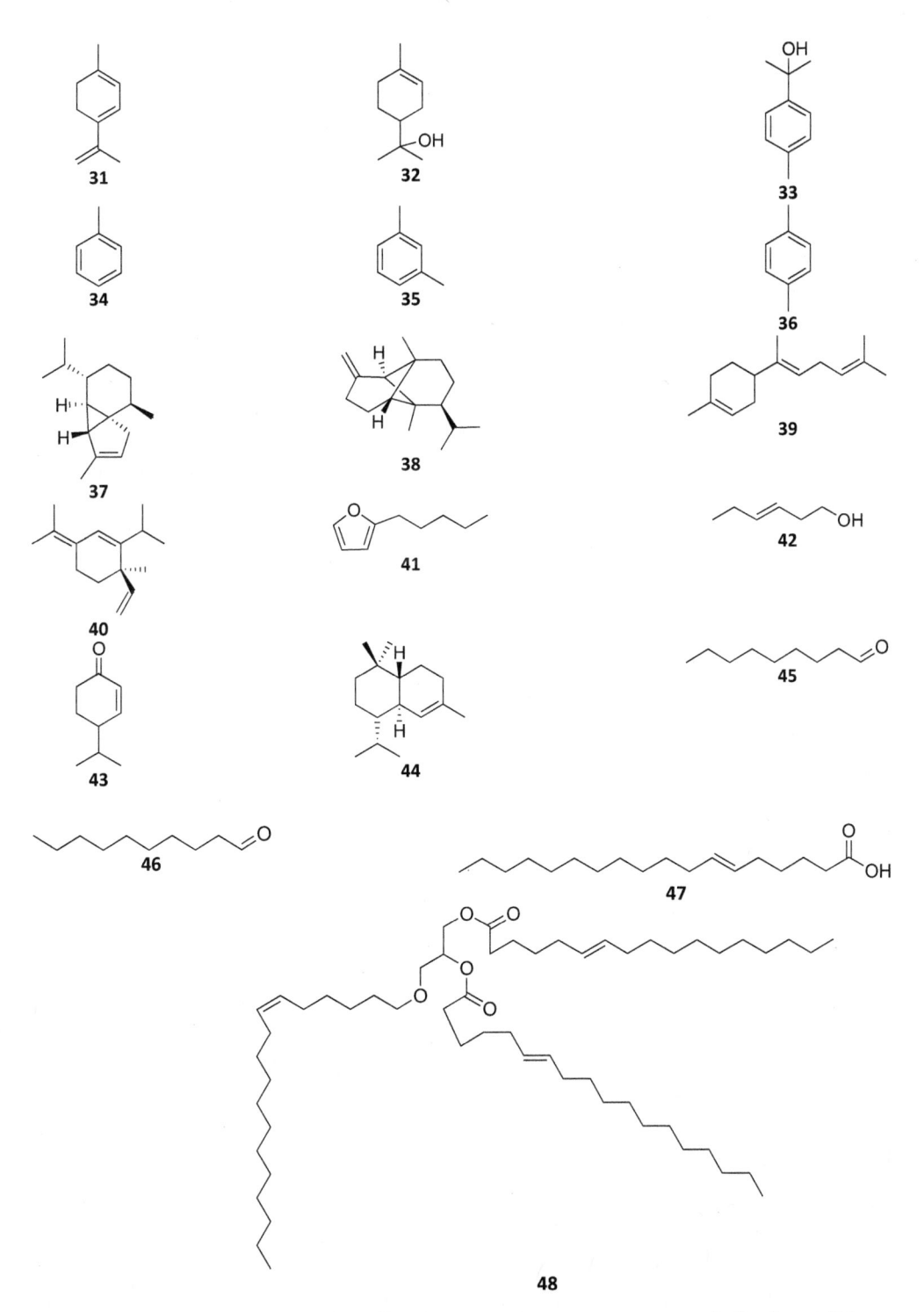

Figure 25.1 *(cont.)* (**22**); camphene (**23**); α-thujene (**24**); myrcene (**25**); α-phellandrene (**26**); α-terpinene (**27**); *cis*-β-ocimene (**28**); *trans*-β-ocimene (**29**); α-terpinolene (**30**); ρ-1,3,8-menthatriene (**31**); α-terpineol (**32**); 2-(ρ-Tolyl) propan-2-ol (**33**); toluene (**34**); *m*-xylene (**35**); ρ-xylene (**36**); α-cubebene (**37**); α-copaene (**38**); β-bisabolene (**39**); α-elemene (**40**); 2-pentylfuran (**41**); *cis*-Hex-3-en-1-ol (**42**); cryptone (**43**); δ-cadinol (**44**); nonanal (**45**); decanal (**46**); tripetroselinin (**47**); petroselinic acid (**48**).

Oxypeucedanin is the major furocoumarin of *P. crispum* and is reported to be mainly responsible for contact photodermatitis induced by this plant (Chaudhary et al., 1986).

Davey et al. (1996) isolated ascorbic acid (**71**) (Fig. 25.2) from the whole aerial parts of *P. crispum*. Leung and Foster (1996) reported high levels of vitamins A, C, some vitamins of the B complex, calcium, and iron in *P. crispum*. It is a well-known herb used to give fragrance to different food products (Fig. 25.3). The use of *P. crispum* as a natural deodorant is related to the presence of a high amount of chlorophyll (Leung and Foster, 1996). The sesquiterpenes: crispane (**72**) and crispanone (**73**) have been isolated from the ethanol seed extract of *P. crispum* (Spraul et al., 1992). Carotenoids, such as neoxanthin (**74**), β-carotene (**75**), lutein (**76**), and violaxanthin (**77**) were isolated from the leaf and stem acetone extracts of *P. crispum* (Francis and Isaksen, 1989). The oxygenated derivative of monoterpens, 1-methyl-4-(methylethenyl)-2,3-dioxabicyclo [2.2.2] oct-5-ene (**78**) and 4-methyl-7-(methylethenyl)-3,8- dioxatricyclo [5.1.0] octane (**79**) (Behtash et al., 2008; Moazedi et al. 2007; Aghili et al., 2009) were isolated from ethanol leaf extract of *P. crispum* (Nitz et al., 1989).

6 PHARMACOLOGICAL PROPERTIES

P. crispum has been found to possess various pharmacological activities, such as antibacterial, antifungal, antioxidant, antidiabetic, hypotensive, hepatoprotective, neuroprotective, analgesic, spasmolytic, immunosuppressant, anticoagulant, antiulcer, and estrogenic properties (Table 25.1).

6.1 Antimicrobial activity

The antibacterial and antifungal activities of some isolated compounds and extracts from *P. crispum* have been reported (Manderfield et al., 1997; Wong and Kitts, 2006; Aljanaby, 2013; Kim et al., 1998; Holton and Basset, 2005). Seyyednejad et al. (2008) reported that 0.1 and 0.2 mg/mL of ethanol seed extract of *P. crispum* exhibited antibacterial activity against *Brucella melitensis*. Hot-water extract of *P. crispum* leaves (250 mg/mL) has been shown to possess antimicrobial activity against *P. Aeruginosa* (Aljanaby, 2013). Furocoumarins **65–69** isolated from aqueous extracts of *P. crispum* leaves (0.12–8.0%) have been found to exhibit inhibitory activity against *Escherichia coli*, *Listeria monocytogenes*, *Erwinia carotovora*, and *Listeria innocua* (Manderfield et al., 1997) using a media-modified and photobiological assay (Table 25.1).

6.2 Antioxidant activity

Using the 2,2,1-diphenyl-1-picrylhydrazyl (DPPH) radical scavenging and potassium ferricyanide–ferric chloride assay, Marín et al. (2016) reported that essential oils extracted from parsley flowers by hydrodistillation exhibited antioxidant activity at 500, 1000, 2000, and 5000 mg/mL with the highest concentration exhibiting inhibition of

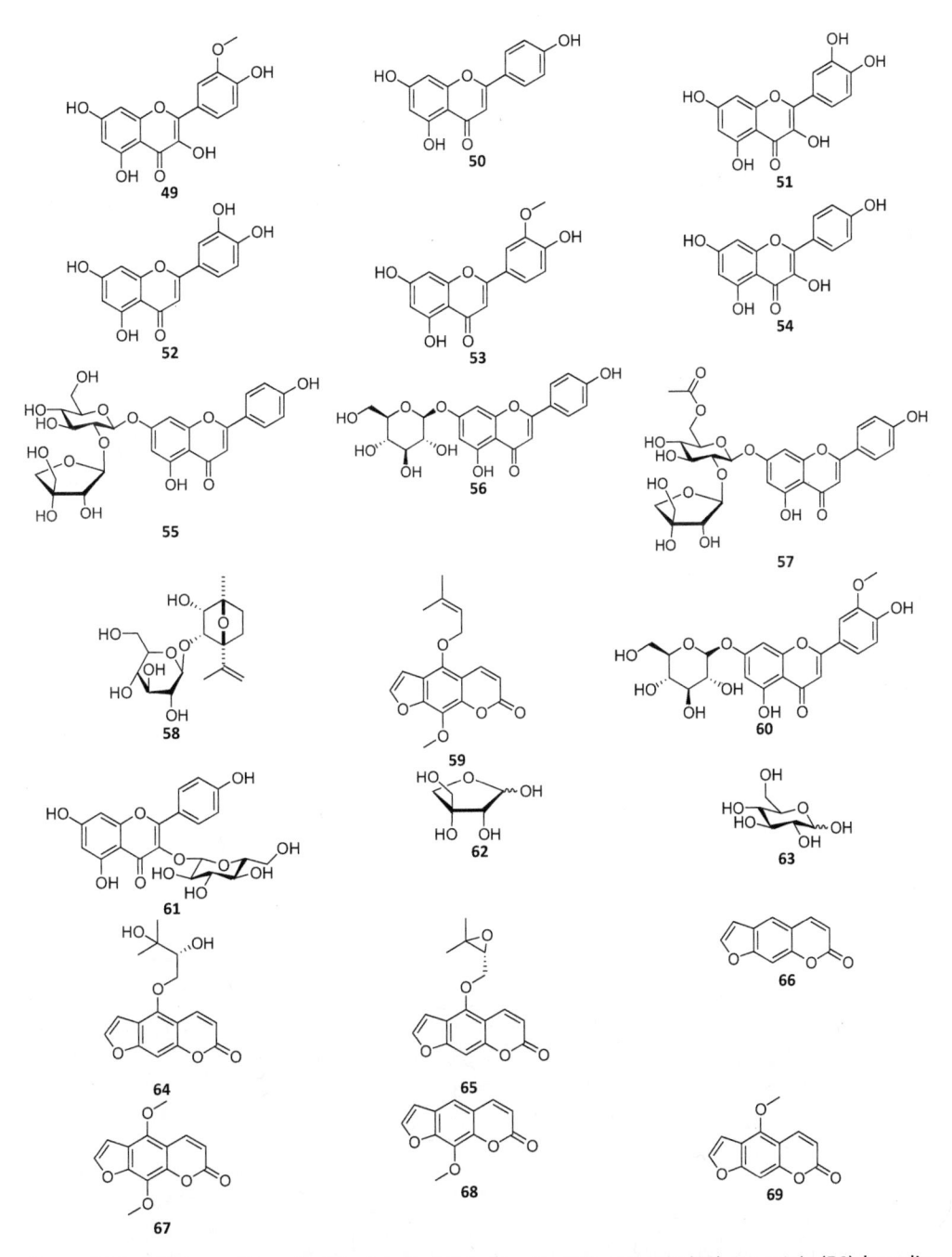

Figure 25.2 *Other constituents of* P. crispum. Isorhamnetin (**49**); apigenin (**50**); quercetin (**51**); luteolin (**52**); chrysoeriol (**53**); kaempferol (**54**); apiin (**55**); cosmosiin (**56**); 6-acetylapiin (**57**); petroside (**58**); cnidilin (**59**); diosmetin 7-*O*-β-D-glucopyranoside (**60**); kaempferol 3-O-β-D-glucopyranoside (**61**); apiose (**62**); D-glucose (**63**); oxypeucedanin hydrate (**64**); oxypeucedanin (**65**); psoralen (**66**); isopimpinellin

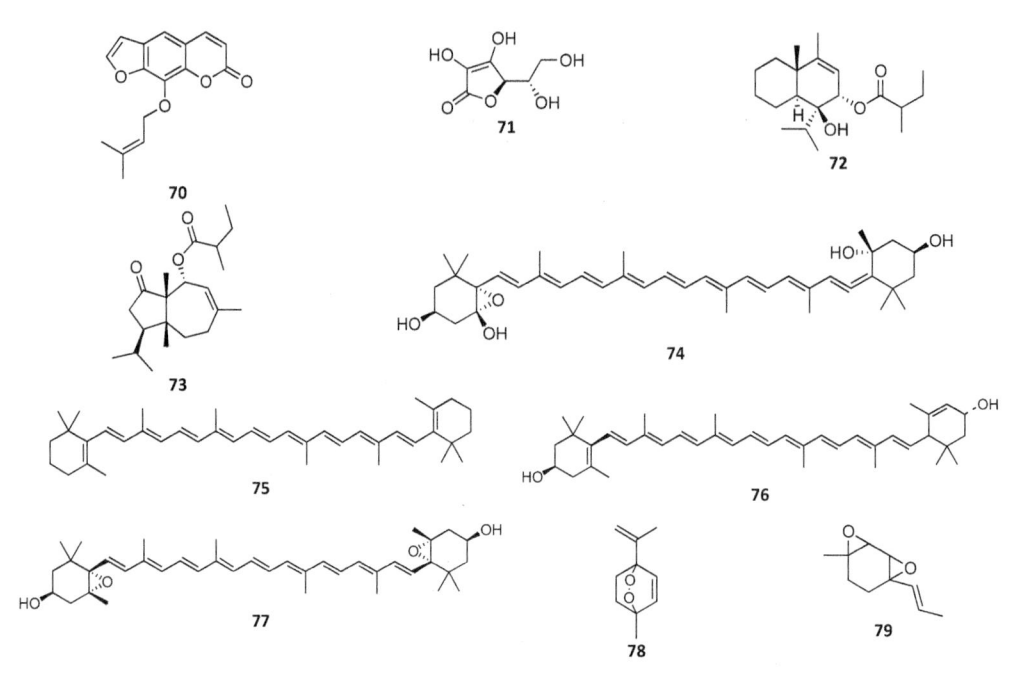

Figure 25.2 *(cont.)* (**67**); 8-methoxypsoralen (**68**); 5-methoxypsoralen (**69**); imperatorin (**70**); ascorbic acid (**71**); crispane (**72**); crispanone (**73**); neoxanthin (**74**); β-carotene (**75**); lutein (**76**); violaxanthin (**77**); 1-methyl-4-(methylethenyl)-2,3-dioxabicyclo [2.2.2] oct-5-ene (**78**); 4-methyl-7-(methylethenyl)-3,8-dioxatricyclo [5.1.0] octane (**79**).

Figure 25.3 *Leaves and aerial parts of P.* **crispum** *(www.calflora.net/losangelesarboretum/whatsbloomingmay07F.html).*

Table 25.1 Pharmacological properties of *P. crispum*

Plant part	Extractive solvent	Pharmacological activity
Leaf and stem	Hot and cold water	Antimicrobial (Aljanaby, 2013)
	Water	Antimicrobial (Manderfield et al., 1997)
	Ethanol	Antimicrobial (Kim et al., 1998)
	Methanol	Antimicrobial (Ojala et al., 2000)
	Methanol and water	Antioxidant (Fejes et al., 1998)
	Water	Hyperuricemia and antioxidant (in vivo) (Haidari et al., 2011)
	Ethanol	Brain protective (in vivo) (Vora et al., 2009)
	Water	Antidiabetic and skin damage (in vivo) (Tunali et al., 1999)
	Water	Antidiabetic (in vivo) (Yanardağ et al., 2003)
	Water	Antidiabetic and heart damage (in vivo) (Sener et al., 2003)
	Water	Antidiabetic and hepato-protective (in vivo) (Bolkent et al., 2004)
	Water	Antiplatelet (in vitro, ex vivo and in vivo) (Gadi et al., 2009)
	Water	Antiplatelet (in vitro) (Gadi et al., 2012)
	Water	Antiplatelet (in vitro) (Chaves et al., 2011)
	Ethanol	Peptic ulcer protection (in vivo) (Al-Howiriny et al., 2003)
	Methanol and water	Antioxidant (in vitro) (Wong and Kitts, 2006)
	Diethyl ether extract	Antioxidant (in vitro) (Al-juhaimi and Ghafoor, 2011)
Leaf and root	Methanol	Antioxidant (in vitro) (Popović et al., 2007)
Seeds	Essential oil	Antioxidant (in vitro) (Zhang et al., 2006)
	Ethanol	Antimicrobial (Seyyednejad et al., 2008)
	Essential oil	Antioxidant and Hepato-protection (in vivo) (Ozsoy-Sacan et al., 2006)
	Essential oil	Immunosuppressant (in vitro) (Yousofi et al., 2012)
	Ethanol	Spasmolytic (in vitro) (Moazedi et al., 2007)
	Hydroalcoholic extract	Analgesic (in vivo) (Behtash et al., 2008)
	Water	Laxative (in vitro and in vivo) (Kreydiyyeh et al., 2001)
	Water	Diuretic (in vitro and in vivo) (Kreydiyyeh et al., 2001)
	Alcohol and oil	Anticancer (Farshori et al., 2013)
Aerial part	Methanol	Estrogenic function (in vitro) (Yoshikawa et al., 2000)
	Water and ethanol	Spasmolytic (in vitro) (Branković et al., 2010)
	Hot water	Cytotoxic (in vitro) (Lantto et al., 2009)
Flowers	Essential oil	Antimicrobial (Marín et al., 2016)
	Essential oil	Antioxidant (in vitro) (Marín et al., 2016)

DPPH radical at 64.28% and ferric reducing power of 0.93 mmol/L Trolox. Haidari et al. (2011) by means of the ferric reducing ability of plasma (FRAP), lipid peroxidation, and spectrophotometry [HPLC and bicinchoninic acid kit] assay reported that aqueous extracts of *P. crispum* leaves and its isolated flavonoids (quercetin and kaempferol) at a concentration of 5 mg/g significantly ($p < 0.001$) increased the total antioxidant capacity and decreased malondialdehyde concentration in hyperuricemic rats.

Leaf and stem aqueous and methanol extracts of *P. crispum* have been identified to possess antioxidant activity in vitro via the DPPH radical-scavenging, ion-chelating, and hydroxyl radical assays (Wong and Kitts, 2006). Methanol-derived leaf extracts exhibited significantly ($p < 0.05$) greater radical-scavenging activity toward both lipid- and water-soluble radicals, which was attributed to the total phenolic content. Ferrous ion-chelating activity was significantly ($p < 0.05$) greater in the stem methanol extracts.

Sęczyk et al. (2015) using the Folin–Ciocalteu assay reported that wheat pasta fortified with powdered *P. crispum* leaves [1–4% (w/w)] exhibited antioxidant activity in vitro. Essential oil from seeds of *P. crispum* exhibited antioxidant activity using β-carotene bleaching, DPPH free radical scavenging and Fe^{2+} metal–chelating assays. The EC_{50} values of the β-carotene bleaching assay and DPPH free radical scavenging assay of the crude *P. crispum* oil dissolved in methanol were 5.12 and 80.21 mg/mL, respectively (Zhang et al., 2006) (Table 25.1).

6.3 Antidiabetic activity

Tunali et al. (1999) reported that aqueous extract of *P. crispum* leaves (2 g/kg) prevented an increase in blood glucose level in rats by using the *o*-toluidine and 2-thiobarbituric acid assays. Aqueous extract of *P. crispum* leaves (2 g/kg) was identified to increase lipid peroxidation and decrease glutathione levels in hyperglycemia-induced heart and aorta oxidative damage in rats via its antioxidant activity in the heart and aorta tissue (Sener et al., 2003). Using the *o*-toluidine and two-point assay, Bolkent et al. (2004) reported that aqueous extract of *P. crispum* leaves (2 g/kg) demonstrated significant hepatoprotective effect in diabetic rats. Yanardağ et al. (2003) showed that experimental rats administered with 2 g/kg of *P. crispum* extract by intragastric intubation containing water for 28 days, significantly ($p < 0.0001$) reduced blood glucose in streptozotocin-induced diabetic rats using the *o*-toluidine assay (Table 25.1).

6.4 Cardiovascular activity

Crude aqueous extract of *P. crispum* has been identified to exhibit antiplatelet activity in experimental animals on platelet aggregation in vitro and ex vivo, and on bleeding time in vivo. The crude extract which contained aglycone flavonoids **50, 54,** and **56** as the active compounds significantly ($p < 0.001$) inhibited platelet aggregation at 3 g/kg body weight ex vivo and prolonged bleeding time ($p < 0.001$) without changes in the amount of platelet (Gadi et al., 2009, 2012).

Chaves et al. (2011) reported that flavonoids including **50** and **56** isolated from aqueous extracts of *P. crispum* leaves in the platelet aggregation model exhibited strong in vitro antiplatelet aggregation activity (IC_{50} of 0.036 mg/mL for **50** and IC_{50} of 0.18 mg/mL for **56**). Though the aqueous *P. crispum* extract showed no inhibition on clotting activity when compared with the control, it exhibited strong antiplatelet aggregation activity (IC_{50} of 1.81 mg/mL).

6.5 Immunomodulating activity

Essential oil from seeds of *P. crispum* at concentrations of 0.01—100 μg/mL blocked humoral and cellular immune response by inhibiting splenocytes and macrophages function in the 3-(4,5-dimethylthiazol-2-yl)-2,5-diphenyltetrazolium bromide (MTT) assay (Yousofi et al., 2012).

6.6 Gastrointestinal activity

Aqueous seed extract of *P. crispum* showed laxative activity in rat by reducing the absorption of sodium and water on net fluid absorption from rat colon using a perfusion technique. The extract also enhanced $NaKCl_2$ transporter activity in the rat colon (Kreydiyyeh et al., 2001).

Ethanol leaf extract of *P. crispum* at doses of 1 and 2 g/kg body weight has been reported to exhibit beneficial effects on different peptic ulcer models in rats via its antisecretory and cytoprotective activities using the cold-restraint ulcer (CRU) technique (Al-Howiriny et al., 2003).

6.7 Genitourinary activity

Methanol aerial parts extract of *P. crispum* (1.0 and 10 mg/mL) showed proliferative activity in estrogen-sensitive MCF-7 breast cancer cell line using the MTT assay. This estrogenic activity was related to these isolated compounds including aglycones of compound **60**, compounds **50** and **54**. The EC_{50} values of these aglycones were as follows, **50** (1.0 mM), aglycone of **60** (2.9 mM) and **54** (0.56 mM). The methanol extract and compound **50** restored the uterus weight in ovariectomized mice when orally administered for consecutive 7 days (Yoshikawa et al., 2000).

P. crispum oil (0.6 mL/kg body weight.) showed protective activity against zearalenone-induced reproductive toxicity and improved testosterone levels in matured male mice (Abdel-Wahhab et al., 2006). Ethanol seed extract of *P. crispum* (5 mg/kg) reduced the dysfunction in rat kidney caused by prostadin-induced abortion via immunohistochemical and immunofluorescent staining and biochemical analysis (Rezazad and Farokhi, 2014).

6.8 Analgesic activity

Ethanol seed extract of *P. crispum* showed significant ($p < 0.001$) analgesic activity by reducing KCl and $CaCl_2$-induced contractions on rat isolated ileum (Moazedi et al., 2007) via the pressure transducer test. The ethanol leaf extract of *P. crispum* at doses of 100, 150, and 200 mg/kg body weight has been found to exhibit analgesic effects on mice by formalin and acetic acid tests (Eidi et al., 2009).

6.9 Spasmolytic activity

Ethanol seed extract of *P. crispum* has been found to exhibit relaxation effect on isolated ilea from adult male Wistar rat in a concentration-dependent manner ($p < 0.01$) by measuring contractions of the isolated ilea, induced by 60 mM potassium chloride (KCl) in the presence of two antagonists of α- and β-adrenoceptors (Damabi et al., 2010). Branković et al. (2010) reported that aqueous and ethanol leaf extracts of *P. crispum* in dose dependent manner decreased the tonus of spontaneous contractions of isolated rat ileum by 62.22 and 79.16% respectively, thereby exhibiting antispasmodic activity on rat ileum in the pressure transducer test.

6.10 Anticancer activity

The ethanol seed extract and oil of *P. crispum* in the MTT and neutral red uptake (NRU) assays showed that seed extract and oil of *P. crispum* significantly reduced cell viability, and altered the cellular morphology of MCF-7 cells in a concentration dependent manner. Cell viability at 50, 100, 250, 500, and 1000 μg/mL of seed extract was recorded as 81, 57, 33, 8, and 5%, respectively, whereas at 100, 250, 500, and 1000 μg/mL of seed oil values were 90, 78, 62, and 8%, respectively. Concentrations of 50 μg/mL and above of *P. crispum* seed extract, and above 100 μg/mL of *P. crispum* seed oil were found to be cytotoxic in MCF-7 cells (Farshori et al., 2013).

Compound **1**, an essential oil constituent isolated from *P. crispum* through glutathione S-transferase (GST) assay-guided fractionation inhibited (65% inhibition of the tumor multiplicity in the lung) benzo[*a*]pyrene (B[*a*]P)-induced tumor formation in female mice (Zheng et al., 1992).

6.11 Nutraceuticals

P. crispum has been used as a nutraceutical intervention in inflammatory bowel disease (IBD) via multiomics evaluation using dextran sodium sulphate (DSS)-induced colitis. Seven-week-old male C57BL/6J mice fed either 2% *P. crispum* leaves or basal diet and drank normal-drinking-water for 1 week after which colitis was induced by administering 1.5% (w/v) DSS-drinking-water for 9 days. *P. crispum* supplementation improved colon shortening and increased disease activity index (Huijuan et al., 2014).

Al-Daraji et al. (2012) reported that supplementing the ration of geese with different levels of fresh parsley (*P. crispum*) leaves (control diet + 80, control diet + 160 and control diet + 240 g/d parsley) resulted in significant ($p < 0.05$) improvement in most of the blood plasma traits [concentrations of glucose, total protein, albumen, globulin, uric acid, total cholesterol, triglycerides, high density lipoprotein (HDL), low density lipoprotein

(LDL), very low density lipoprotein (VLDL), calcium, phosphorus, and creatinine, and blood plasma activities of aspartate aminotransferase (AST) and alanine aminotransferase] of Iraqi geese using various biochemical assays.

6.12 Neuroprotective effect

P. crispum leaf juice (10 g/kg body weight per day) has been found to exhibit significant effects in neutralizing and reducing the deleterious changes due to cadium exposure during pregnancy on the behavioral activities, neurotransmitters, oxidative stress, and brain neurons morphology of newborn mice using the inductively coupled plasma mass spectrometer, grip-strength meter, rota-rod, acetylcholine determination, lipid peroxidation, glutathione, and peroxidase assays (Allam et al., 2016).

7 TOXICITY

The toxicity of *P. crispum* and its essential oil has not been thoroughly investigated. In ethnomedicine, it has been claimed that *P. crispum* is abortifacient. Photodermatitis due to furocoumarins particularly **55** are responsible for its contact photodermatitis activity in pigs exposed to *P. crispum* (Chaudhary et al., 1986). Eighteen sows of mixed age from an outdoor herd of 400 sows and boars were put in a field of parsley for 4–5 days and after this period, vesicles were noted on the snouts with erythema and skin fissures. In an adjoining paddock of parsley, 14 out of 18 gilts were affected with lesions, principally on their ears. In other paddocks, up to 16 out of 18 sows showed similar lesions; suckling sows and those about to furrow were most severely affected. History, clinical signs, and pathology were consistent with phytophotodermatitis (Griffiths and Douglas, 2000). Awe and Banjoko (2013) reported that ethanol leaf extract of *P. crispum* exhibited hepatotoxic and nephrotoxic activities determined by colorimetric method using bromocresol green and urease cleavage (Berthelot's reaction) at continued oral doses equal to or more than 1000 mg/kg, but no obvious toxicity when used at lower doses (Awe and Banjoko, 2013).

8 CLINICAL TRIALS

Randomized crossover clinical trial involving seven men and seven women was carried out to study the effect of intake of parsley (*P. crispum*), containing high levels of the flavone apigenin, on the urinary excretion of flavones and biomarkers for oxidative stress (Nielsen et al., 1999).

The subjects received a strictly controlled diet low in flavones and other naturally occurring antioxidants during the 2 weeks of intervention. This basic diet was supplemented with parsley providing 3.73–4.49 mg apigenin/MJ in one of the intervention weeks. Urinary excretion of apigenin (**50**) was 1.59–409.09 µg/MJ per 24 h during intervention with parsley and 0 to 112.27 µg/MJ per 24 h on the basic diet ($p < 0.05$). The fraction of apigenin intake excreted in the urine was 0.58% during parsley intervention. Erythrocyte glutathione reductase (GR) and superoxide dismutase activities increased during intervention with parsley ($p < 0.005$) as compared with the levels on the basic diet, whereas erythrocyte catalase and glutathione peroxidase activities did not change. No significant changes were observed in plasma protein 2-adipic semialdehyde residues, a biomarker of plasma protein oxidation.

Nielsen et al. (1999) also observed an overall decreasing trend in the activity of antioxidant enzymes during the 2-week study. The decreased activity of SOD was strongly correlated at the individual level with an increased oxidative damage to plasma proteins. However, the intervention with parsley seemed, partly, to overcome this decrease and resulted in increased levels of GR and SOD.

9 PATENTS

The following are some patents secured on *P. crispum*. The patented products contain either *P. crispum* alone or in combination with other pharmacologically active agents. The products are: composition for the treatment of halitosis; good living tea; breath scent camouflage spray; formulation for alleviation of kidney stone and gallstone symptoms; nutraceutical for the prevention and treatment of cancers and diseases affecting the liver; skin care product; parsley variety 'Fidelio'; Caffeoyl-CoA 3-O-methyltransferase genes from parsley (*P. crispum*) (Table 25.2).

Table 25.2 Patents secured on *P. crispum*

Patent application number	Publication number	Product name	Composition
WO99/39686	US 6350435 B1	Composition for the treatment of halitosis	Mixture of olive oil (*Oleaeuropea* L.) and parsley oil (*P. crispum*) (Hernandez, 2002)
US 10/726,146	US 20050118324 A1	Good living tea	Dried bitter melon leaves, ground fenugreek, ground cinnamon, dried parsley (*P. crispum*) flakes, and pathimukham (Anna and Mathew, 2005)
US 10/771,063	US 20050169854 A1	Breath scent camouflage spray	Chlorophyll, parsley (*P. crispum*) and dandelion extracts (Carlos, 2005)

(*Continued*)

Table 25.2 Patents secured on *P. crispum* (cont.)

Patent application number	Publication number	Product name	Composition
US 13/605,602	US 20130064912 A1	Formulation for alleviation of kidney stone and gallstone symptoms	Chancapiedra (*Phyllanthusniruri*), gravel root, hydrangea root, marshmallow root, juniper berry, corn silk uvaursi, parsley (*P. crispum*) root, agrimony dandelion leaf, horsetail, orange peel, peppermint and goldenrod extract (Barron, 2013)
US 10/560,558	US 8012510 B2	Nutraceutical for the prevention and treatment of cancers and diseases affecting the liver	*Brassica oleracea*, *Daucuscarota*, *P. crispum*, *Spinaciaoleracea*L, Beta vulgaris, aloe vera, and honey (Can, 2011)
US 13/723,906	US 8790720 B2	Skin care product	Camellia and feverfew serum fractions and/or kelp and parsley (*P. crispum*) serum fractions
US 61/974,900	US 20150282449	Parsley variety "Fidelio" (Schieder and Ladenburg, 2015)	
US 08/988,054	US 6160205 A	Caffeoyl-CoA 3-O-methyltransferase genes from parsley (*P. crispum*)	

10 CONCLUSIONS

P. crispum has several traditional uses including antiinflammatory, treatment of gastrointestinal disorder, hypertension, cardiac disease, urinary disease, diabetes, and various dermal disease in traditional and folklore medicines. Phytochemical constituents; flavonoids, and phenolic compounds especially apiin, apigenin, and 6-acetylapiin; essential oil including myristicin and apiol, as well as coumarins have been isolated from *P. crispum*. It has several pharmacological activities, such as antibacterial and antifungal, antioxidant, hepatoprotective, antidiabetic, analgesic, spasmolytic, immunosuppressant, antiplatelet, gastroprotective, and estrogenic effects in in vitro, in vivo, and ex vivo models. Several patents have been secured on *P. crispum*, which is either used alone or in combination with other pharmacologically active agents. It can be concluded that *P. crispum* is a useful and important medicinal plant with wide range of proven medicinal activity.

REFERENCES

Abdel-Wahhab, M.A., Abbes, S., Salah-Abbes, J., Hassan, A., Oueslati, R., 2006. Parsley oil protects against Zearalenone-induced alteration in reproductive function in male mice. Toxicol. Lett. 164, S266.

Aghili, M.H., Makhzan-al-Advia, R.R., Shams, A.M.R., 2009. In: Farjadmand, F. et al., (Ed.), Makhzan-al-Advia. Tehran University of Medical Sciences, Tehran, pp. 329–330.

Al-Howiriny, T., Al-Sohaibani, M., El-Tahir, K., Rafatullah, S., 2003. Prevention of experimentally-induced gastric ulcers in rats by an ethanolic extract of "Parsley" *Petroselinum crispum*. Am. J. Chinese Med. 31 (5), 699–711.

Al-Daraji, H.J., Al-Mashadani, H.A., Mirza, H.A., Al-Hassani, A.S., Al-Hayani, W.K., 2012. The effect of utilization of parsley (*Petroselinum crispum*) in local Iraqi geese diets on blood biochemistry. J. Am. Sci. 8 (8), 427–432.

Aljanaby, A.A.J.J., 2013. Antibacterial activity of an aqueous extract of *Petroselinum crispum* leaves against pathogenic bacteria isolated from patients with burns infections in Al-najaf Governorate, Iraq. Res. Chem. Intermed. 39 (8), 3709–3714.

Al-juhaimi, F., Ghafoor, K., 2011. Total phenols and antioxidant activities of leaf and stem extracts from coriander, mint and parsley grown in Saudi Arabia. Pakistan J. Bot. 43 (4), 2235–2237.

Allam, A.A., Maodaa, S.N., Abo-eleneen, R., Ajarem, J., 2016. Protective effect of parsley juice (*Petroselinum crispum*, Apiaceae) against cadmium deleterious changes in the developed albino mice newborns (*Mus musculus*) brain. Oxid. Med. Cell. Longev. 2016, 2646840.

Anna, M., Mathew, T., 2005. Good living tea—a diabetic dietary supplement drink; Google patents, pp. 2–4.

Avicenna, 1983. The cannon of medicine, translated from Arabic to Persian by Abdulrahman Sharaf-kandi. Tehran, So-rush Publication, p. 141.

Awe, E.O., Banjoko, S.O., 2013. Biochemical and haematological assessment of toxic effects of the leaf ethanol extract of *Petroselinum crispum* (Mill) Nyman ex A.W. Hill, BMC Complem. Alter. Med., ISCMR, 1–10.

Barron, J., 2013. Formulation for alleviation of kidney stone and gallstone symptoms; Google Patent, pp. 1–10.

Behtash, N., Kargarzadeh, F., Shafaroudi, H., 2008. Analgesic effects of seed extract from *Petroselinum crispum* (*Tagetes minuta*) in animal models. Toxicol. Lett. 180 (Suppl. 5), S127–S128.

Benítez, G., González-Tejero, M.R., Molero-Mesa, J., 2010. Pharmaceutical ethnobotany in the western part of Granada province (southern Spain): ethnopharmacological synthesis. J. Ethnopharmacol. 129 (1), 87–105.

Blumenthal, M., Goldberg, A., Brinckman, J., 2000. Expanded Commission E Monographs (Newton, Mass. Integr. Med. Commun.). Herbal Med. 10, 218–220.

Bolkent, S., Yanardag, R., O. Ozsoy-Sacan, O., Karabulut, B., 2004. Effects of parsley (*Petroselinum crispum*) on the liver of diabetic rats: a morphological and biochemical study. Phytother. Res. 18 (12), 996–999.

Branković, S., Kitic, D., Radenkovic, M., Ivetic, V., Veljkovic, S., Nesic, M., 2010. Relaxant activity of aqueous and ethanol extracts of parsley (*Petroselinum crispum* (Mill.) Nym. ex A.W. Hill, Apiaceae) on isolated ileum of rat. Med. Pregl. 63, 475–478.

Bruneton, J., 1999. Pharmacognosy, Phytochemistry Medicinal Plants, second edition Intercept Ltd, London, pp. 519–520.

Can, V.B., 2011. Nutraceutical for the prevention and treatment of cancers and diseases affecting the liver; Google Patent.

Carlos, C., 2005. Breath scent camouflage spray; Google Patent, pp. 9–11.

Chaudhary, S.K., Ceska, O., Têtu, C., Warrington, P.J., Ashwood-Smith, M.J., Poulton, G.A., 1986. Oxypeucedanin, a major furocoumarin in Parsley, *P. crispum*. Planta Med. 52 (6), 462–464.

Chaves, D.S., Frattani, F.S., Assafim, M., de Almeida, A.P., de Zingali, R.B., Costa, S.S., 2011. Phenolic chemical composition of *P. crispum* extract and its effect on haemostasis. Nat. Prod. Commun. 6 (7), 961–964.

Christensena, L.P., Brandtb, K., 2006. Bioactive polyacetylenes in food plants of the Apiaceae family: occurrence, bioactivity and analysis. J. Pharm. Biomed. Anal. 41 (3), 683–693.

Damabi, N.M., Moazedi, A., Seyyednejad, S., 2010. The role of α- and β-adrenergic receptors in the spasmolytic effects on rat ileum of *Petroselinum crispum*. Asian Pac. J. Trop. Med. 3 (11), 866–870.

Davey, M.W., Bauw, G., Montagu, M.V., 1996. Analysis of ascorbate in plant tissue by high performance capillary zone electrophoresis. Anal. Biochem. 239 (1), 8–19.

Destaillats, F., Keskitalo, M., Arul, J., Angers, P., 2009. Triacylglycerols of Apiaceae seed oils: Composition and regiodistribution of fatty acids. Eur. J. Lipid Sci. Technol. 111, 164–169.

Eddouks, M., Maghrani, M., Lemhadri, A., Ouahidi, M.L., Jouad, H.A., 2002. Ethnophar-macological survey of medicinal plants used for the treatment of diabetes mellitus, hypertension and cardiac diseases in the south-east region of Morocco (Tafilalet). J. Ethnopharmacol. 82 (2–3), 97–103.

Eidi, A., Eidi, M., Badiei, L., 2009. Antinociceptive effects of ethanolic extract of parsley (*Petroselinum crispum* L.) leaves in mice. Med. Sci. J. Islamic Azad Univ. 19 (3), Pe181–Pe186.

Farzaei, M.H., Abbasabadi, Z., Reza, M., Ardekani, S., Rahimi, R., 2013. Review Parsley: a review of ethnopharmacology, phytochemistry and biological activities. J. Tradit. Chinese Med. 33 (6), 815–826.

Fejes, S.Z., Blázovics, A., Lemberkovics, E., Petri, G., Szoke, E., Kery, A., 2000. Free radical scavenging and membrane protective effects of methanol extracts from *Anthriscus cerefolium* L. (Hoffm.) and *P. crispum* (Mill.) nym. ex A.W. Hill. Phytother. Res. 14 (5), 362–365.

Fejes, S., Kéry, A., Blázovics, A., Lugasi, A., Lemberkovics, E., Petri, G., Szöke, E., 1998. Investigation of the in vitro antioxidant effect of *Petroselinum crispum* (Mill.) Nym. ex A.W. Hill. Acta Pharm. Hung. 68 (3), 150–156.

Farshori, N.N., Al-Sheddi, E.S., Al-Oqail, M.M., Musarra, J., Al-Khedhairy, A.A., Siddiqui, M.A., 2013. Anticancer activity of *Petroselinum sativum* seed extracts on MCF-7 human breast cancer cells. Asian Pac. J. Cancer Prevent 14 (10), 5719–5723.

Francis, G.W., Isaksen, M., 1989. Droplet counter current chromatography of the carotenoids of parsley *P. crispum*. Chromatographia 27 (11–12), 549–551.

Gadi, D., Bnouham, M., Aziz, M., Ziyyat, A., Legssyer, A., Legran, C., Lafeve, F.F., Mekhfi, H., 2009. Parsley extract inhibits in vitro and ex vivo platelet aggregation and prolongs bleeding time in rats. J. Ethnopharmacol. 125 (1), 170–174.

Gadi, D., Bnouham, M., Aziz, M., Ziyyat, A., Legssyer, A., Bruel, A., Berrabah, M., Legrand, C., Fauvel-Lafeve, F., Mekhfi, H., 2012. Flavonoids purified from parsley inhibit human blood platelet aggregation and adhesion to collagen under flow. J. Complement. Integr. Med. 9, 19.

Grieve, M., 2014. https://www.botanical.com/botanical/mgmh/p/parsle09.html

Griffiths, I.B., Douglas, R.G.A., 2000. Phytophotodermatitis in pigs exposed to parsley (*Petroselinum crispum*). Vet. Rec. 146 (3), 73–74.

Guieta, S., Robinsa, R.J., Leesb, M., Billaulta, I., 2003. Phytochemistry quantitative 2 H NMR analysis of deuterium distribution in petroselinic acid isolated from parsley seed. Phytochemistry 64 (1), 227–233.

Haidari, F., Ali, S., Mohammad, M., Mahboob, S., 2011. Effects of parsley (*Petroselinum crispum*) and its flavonol constituents, kaempferol and quercetin, on serum uric acid levels, biomarkers of oxidative stress and liver xanthine oxidoreductase activity in oxonate-induced hyperuricemic rats. Iran. J. Pharm. Res. 10, 811–819.

Hanrahan, C., Frey, R.J., 2005. Parsley. In: Longe, J.L. (Ed.), The Gale Encyclopedia of Alternative Medicine. Thomson/Gale, Farmington Hills, MI.

Hempel, J., Pforte, H., Raab, B., Engst, W., Bo, H., Jacobasch, G., 1999. Flavonols and flavones of parsley cell suspension culture change the antioxidative capacity of plasma in rats. Nahrung 43 (3), 201–204.

Herbst, S.T., 2001. The New Food Lover's Companion: Comprehensive Definitions of Nearly 6,000 Food, Drink, and Culinary Terms. Barron's Cooking Guide. Hauppauge, NY: Barron's Educational Series.

Hernandez, M.A., 2002. Composition for the treatment of halitosis; Google Patents, pp. 1–6.

Holton, J., Basset, C., 2005. Bactericidal and anti-adhesive properties of culinary and medicinal plants against *Helicobacter pylori*. World J. Gastroenterol. 11 (47), 7499–7507.

Hudson, C.S., 1949. Apiose and the glycosides of the parsley plant. Adv. Carbohydr. Chem. 4, 57–74.

Huijuan, J., Awa, W., Hanated, M., Takahashid, S., Saitoa, K., Tanakab, H., Tomitac, M., Kato, H., 2014. Multi-faceted integrated omics analysis revealed parsley (*Petroselinum crispum*) as a novel dietary intervention in dextran sodium sulphate induced colitic mice. J. Funct. Foods 11, 438–448.

Huxley, A., 1992. New RHS Dictionary of Gardening, third ed. Macmillan, London; p. 532.

Ipor, I.B., Oyen, L.P.A., 1999. *Petroselinum crispum* (Miller) Nyman ex A.W. Hill. In: de Guzman, C.C., Siemonsma, J.S. (Eds.). Plant Resources of South-East Asia, No. 13: Spices. Leiden: Backhuys Publishers, pp. 172–176.

Jouad, H., Haloui, M., Rhiouani, H., El Hilalyb, J., Eddouk, M., 2001. Ethnobotanical survey of medicinal plants used for the treatment of diabetes, cardiac and renal diseases in the North centre region of Morocco (Fez-Boulemane). J. Ethnopharmacol. 77 (2–3), 175–182.

Kim, O.M., Kim, M.K., Lee, S.O., Lee, K.R., Kim, S.D., 1998. Antimicrobial effect of ethanol extracts from spices against *Lactobacillus plantarum* and *Leuconostoc mesenteroides* isolated from kimchi. J. Korean Soc. Food Sci. Nutr. 27 (3), 455–460.

Kreuzaler, F., Hahlbrock, K., 1973. Flavonoid glycosides from illuminated cell suspension cultures of *Petroselinum hortense*. Phytochemistry 12 (5), 1149–1152.

Kreydiyyeh, S.I., Usta, J., Kaouk, I., Al-Sadi, R., 2001. The mechanism underlying the laxative properties of Parsley extract. Phytomedicine 8 (5), 382–388.

Lantto, T.A., Colucci, M., Závadová, V.H.R., Raasmaja, A., 2009. Cytotoxicity of curcumin, resveratrol and plant extracts from basil, juniper, laurel and parsley in SH-SY5Y and CV1-P cells. Food Chem. 117 (3), 405–411.

Leung, A., Foster, S., 1996. Encyclopedia of Common Natural Ingredients Used in Foods, Drugs and Cosmetics. John Wiley & Sons, New York.

López, M.G., Sánchez-Mendoza, I.R., Ochoa-Alejo, N., 1999. Compartive study of volatile components and fatty acids of plants and in vitro cultures of parsley (*Petroselinum crispum* (Mill) nym ex hill). J. Agric. Food Chem. 47 (8), 3292–3296.

MacLeod, A.J., Snyder, C.H., Subramanian, G., 1985. Volatile aroma constituents of parsley leaves. Phytochemistry 24 (11), 2623–2627.

Manderfield, M.M., Schafer, H.W., Davidson, P.M., Zottola, E.A., 1997. Isolation and identification of antimicrobial furocoumarins from parsley. J. Food Prot. 60, 72–77.

Marín, I., Sayas-Barberá, E., Viuda-Martos, M., Navarro, C., Sendra, E., 2016. Chemical composition, antioxidant and antimicrobial activity of essential oils from organic fennel, parsley, and lavender from Spain. Foods 5 (18), 1–10.

Merzouki, A., Ed-Derfoufi, El-Aallau, A., Molero-mesa, F., 1997. Wild medicinal plants used by local Bouhmed population (Morocco). Fitoterapia 68 (5), 444–460.

Moazedi, A.A., Mirzaie, D.N., Seyyednejad, S.M., Zadkarami, M.R., Amirzargar, A., 2007. Spasmolytic effect of *Petroselinum crispum* (Parsley) on rat's ileum at different calcium chloride concentrations. Pakistan J. Biol. Sci. 10 (22), 4036–4042.

Montesano, V., Negro, D., Sarli, G., De Lisi, A., Laghetti, G., Hammer, K., 2012. Notes about the uses of plants by one of the last healers in the Basilicata region (South Italy). J. Ethnobiol. Ethnomed. 8, 15.

Nielsen, S.E., Young, J.F., Daneshvar, B., Lauridsen, S.T., Knuthsen, P., Sandström, B., Dragsted, L.O., 1999. Effect of parsley (*P. crispum*) intake on urinary apigenin excretion, blood antioxidant enzymes and biomarkers for oxidative stress in human subjects. Br. J. Nutr. 81 (6), 447–455.

Nitz, S., Kollmannsberger, H., Spraul, M.H., Drawert, F., 1989. Oxygenated derivatives of menthatriene in parsley leaves. Phytochemistry 28 (11), 3051–3054.

Nitz, S., Spraul, M.H., Drawert, F., 1990. C17 polyacetylenic alcohols as the major constituents in roots of *Petroselinum crispum* Mill. ssp. *tuberosum*. J. Agric. Food Chem. 38 (7), 1445–1447.

Ojala, T., Remes, S., Haansuu, P., Vuorela, H., Hiltunen, R., Haahtela, K., Vuorela, P., 2000. Antimicrobial activity of some coumarins containing herbal plants growing in Finland. J. Ethnopharmacol. 73 (1), 299–305.

Ozsoy-Sacan, O., Yanardag, R., Orak, H., Ozgey, Y., Yarat, A., Tunali, T., 2006. Effects of parsley (*Petroselinum crispum*) extract versus glibornuride on the liver of streptozotocin-induced diabetic rats. J. Ethnopharmacol. 104 (1–2), 175–181.

Pápay, Z.E., Kósa, A., Boldizsár, I., Ruszkai, A., Balogh, E., Klebovich, I., 2012. Pharmaceutical and formulation aspects of *P. crispum* extract. Acta Pharm. Hung. 82 (1), 3–14.

Pharmacopoeia Jugoslavica II, 1951: Farmakopeja FNRJ, Pharmacopoea jugoslavica, editio secunda, Medicinska knjiga, Beograd.

Popović, M., Kaurinovic, B., Jakovljevic, V., Mimica-Dukic, N., Bursac, M., 2007. Effect of parsley (*Petroselinum crispum* (Mill.) Nym. ex A.W. Hill, Apiaceae) extracts on some biochemical parameters of oxidative stress in mice treated with CCl_4. Phytother. Res. 21 (8), 717–723.

Quattrocchi, U., 2012. CRC World Dictionary of Medicinal and Poisonous Plants: Common Names, Eponyms, Synonyms and Etymology. Volume IV M-Q. CRC Press, Boca Raton, Florida, p. 504.

Rayment, W.J., 2016. http://www.indepthinfo.com/parsley/history.shtml

Rehecho, S., Uriate-Pueyo, I., Calvo, J., Calvo, M.I., 2011. Ethnopharmacological survey of medicinal plants in Nor-Yauyos, a part of the Landscape Reserve Nor-Yauyos-Cochas. Peru. J. Ethnopharmacol. 133 (1), 75–85.

Rezazad, M., Farokhi, F., 2014. Protective effect of *Petroselinum crispum* extract in abortion using prostadin—induced renal dysfunction in female rats. Avicenna J. Phytomed. 4 (5), 312–319.

Savikin, K., Zdunic, G., Menkovic, N., Zivkovic, J., Cujic, N., Terescenko, M., Bigovic, D., 2013. Ethnobotanical study on traditional use of medicinal plants in South-Western Serbia, Zlatibor district. J. Ethnopharmacol. 146 (3), 803–810.

Sęczyk, Ł., Świeca, M., Gawlik-Dziki, U., 2015. Changes of antioxidant potential of pasta fortified with parsley (*Petroselinum crispum* Mill.) leaves in the light of protein-phenolics interactions. Acta. Sci. Pol. Technol. Aliment. 14 (1), 29–36.

Sener, G.K., Sacan, O., Yanardag, R., Ayanoglu-Du, G.L., 2003. Effects of parsley (*Petroselinum crispum*) on the aorta and heart of stz-induced diabetic rats. Plant Foods Hum. Nutr. 58 (3), 1–7.

Seyyednejad, S.M., Maleki, S., Mirzaei Damab, N., Motamedi, H., 2008. Antibacterial activity of prunus mahaleb and parsley (*Petroselinum crispum*) against some pathogen. Asian J. Biol. Sci. 1 (1), 51–55.

Simon, J.E., 1990. Essential oils and culinary herbs. In: Janick, J., Simon, J.E. (Eds.), Advances in New Crops. Timber Press, Portland, pp. 472–483.

Spraul, M.H., Nitz, S., Drawert, F., Duddeck, H., Hiegemann, M., 1992. Crispane and crispa—none, two compounds from *P. crispum* with a new carbon skeleton. Phytochemistry 31 (9), 3109–3111.

Tonkaboni, M.M., Tohfeh-al-Momenin, R.R.S., Ardekani, M.R., 2007. In: Farjadmand, F. et al., (Ed.), Tohfeh al-Momenin. Shahid Be-heshti University of Medical Sciences, Tehran, p. 129.

Tucker, A.O., DeBaggio, O., 2009. Encyclopedia of Herbs. Timber Press, Portland, Oregon, Pp393-394.

Tunali, T., Yarat, A., Yanardag, R., elik, F.O., Zsoy, O., Ergenekon, G., Emekli, N.O., 1999. Effect of parsley (*Petroselinum crispum*) on the skin of STZ induced diabetic rats. Phytother. Res. 13 (2), 138–141.

Vora, S.R., Patil, R.B., Pillai, M.M., 2009. Protective effects of *Petroselinum crispum* (Mill) Nyman ex A.W. Hill leaf extract on D-galactose-induced oxidative stress in mouse brain. Indian J. Exp. Biol. 47 (5), 338–342.

Wagner, H., Bladt, S., 1996. Plant drug analysis. Springer-Verlag, Berlin-Heidelberg, pp. 154–175.

Wichtl, M., Bisset, N.G. (Eds.), 1994. Herbal Drugs and PhytopharmaceuticalsMedpharm Scientific Publishers, Stuttgart.

Wong, P.Y.Y., Kitts, D.D., 2006. Studies on the dual antioxidant and antibacterial properties of parsley (*P. crispum*) and cilantro (*Coriandrum sativum*) extracts. Food Chem. 97 (3), 505–515.

Yanardağ, R., Bolkent, S., Tabakoglu-Oguz, A., Ozsoy-Sacan, O., 2003. Effects of *Petroselinum crispum* extract on pancreatic B cells and blood glucose of streptozotocin-induced diabetic rats. Biol. Pharm. Bull. 26 (8), 1206–1210.

Yoshikawa, M., Uemura, T., Shimoda, H., Kishi, A., Kawahara, Y., Matsuda, H., 2000. Medicinal foodstuffs. XVIII. Phytoestrogens from the aerial part of *P. crispum* MIll. (Parsley) and structures of 6-acetylapiin and a new monoterpene glycoside, petroside. Chem. Pharm. Bull. 48 (7), 1039–1044.

Yousofi, A., Daneshmandi, S., Soleimani, N., Bagheri, K., Karimi, M.H., 2012. Immunomodulatory effect of Parsley (*Petroselinum crispum*) essential oil on immune cells: mitogen-activated splenocytes and peritoneal macrophages. Immunopharmacol. Immunotoxicol. 34 (2), 303–308.

Zhang, H., Chen, F., Wang, X., Yao, H.Y., 2006. Evaluation of antioxidant activity of parsley (*P. crispum*). Essential oil and identification of its antioxidant constituents. Food Res. Int. 39 (8), 833–839.

Zheng, G.Q., Kenney, P.M., Zhang, J., Lam, L.K., 1992. Inhibition of benzo [a] pyrene-induced tumorigenesis by myristicin, a volatile aroma constituent of parsley leaf oil. Carcinogenesis 13 (10), 1921–1923.

Ziyyat, A., Legssyer, A., Mekhfi, H., Dassouli, A., Serhrouchni, M., Benjelloun, W., 1997. Phytotherapy of hypertension and diabetes in oriental Morocco. J. Ethnopharmacol. 58 (1), 45–54.

CHAPTER 26

Sesamum indicum

S.O. Amoo, A.O.M. Okorogbona, C.P. Du Plooy, S.L. Venter

1 INTRODUCTION

Apart from the nutritional role played by plants in the provision of food to human and animals, plants also play significant roles in human health maintenance. Prior to the contemporary world of medicine—together with its pharmaceutical world of man-made medications including pills, tablets, and drugs—plants were selected from the wild in the primeval days by man, who used vegetation advantageously for the treatment of common ailments and diseases considered to be threatening human life. Thus, medicinal plants form an important part of the human natural wealth, as the plants serve as food while providing relevant therapeutic agents and raw materials useful in traditional and contemporary medicine (Motaleb et al., 2011). Among the many multipurpose plant species consumed as food for nutritional and medicinal purposes and identified for industrial use, *Sesamum indicum* L. is an important plant. It is commonly known internationally as sesame, and as benniseed or simsim in Africa.

S. indicum (family: Pedaliaceae) is an erect, herbaceous, annual plant that is considered to be one of the oldest oilseed crops, which has been under cultivation for centuries particularly in Africa and Asia (Hegde, 2012). It produces teardrop-shaped, small, and flat seeds. As mentioned by Were et al. (2006), the International Plant Genetic Resources Institute (IPGRI) listed it among neglected and underutilized crop species and as "a crop with high potential." The plant, especially its seed (Fig. 26.1), is traditionally used for various purposes, including culinary (Table 26.1) and medicinal purposes, and as a dietary supplement in different parts of the world. The seeds are used for decorating bread and cookies, and in making desserts, while paste made from sesame seeds are added to certain dishes (Elleuch et al., 2011). Oil extracted from sesame seeds serves as a solvent, as an oleaginous catalyst for medications, skin softeners, and as an ingredient in the production of soaps and margarine (Anilakumar et al., 2010).

In terms of its ethnomedicinal value in South Africa, hot-water extract of the aerial parts of the plant is used for sexual stimulation, while the leaves are used for treating malaria (Watt and Breyer-Brandwijk, 1962). Oil and paste from the seeds are topically applied for treating wounds and burns (Kiran and Asad, 2008). Other ethnomedicinal value of the seeds in different parts of the world, include their use for treating cholera, scorpion poison, respiratory infections, tinnitus, diarrhea, dysentery, ulcers, amenorrhea, dysmenorrhea, constipation, anemia, bleeding piles, dizziness, for memory enhancement, etc.

Figure 26.1 *Sesame seeds.*

Table 26.1 Gastronomic uses of ingredients obtained from sesame crop and countries of consumption

Food type	Country of consumption
Sesame cakes, wine, and brandy	Biblical Babylon (Iraq)
Bread stick, cracker, salad, and cooking oil	Africa, Europe, Asia, Australia, and North and South America
Raw, powdered, and roasted seed	India
Substitute for olive oil	Europe
Bread	Italy
Cakes	Greece
Soup, spice, and seed oil	Africa
Salad and fish oil	Japan
Confectionery	China
Sesame seed buns and chips	USA

Source: Anilakumar, K.R., Pal, A., Khanum, F. and Bawa, A.S., 2010. Nutritional, medicinal and industrial uses of sesame (*Sesamum indicum* L.) seeds—an overview. Agric. Conspect. Sci. 75(4), 159–168.

(Hegde, 2012; Kapoor, 2001; Khan et al., 2014; Sharififar et al., 2012). Sesame fruits are traditionally used as a laxative, poultice, and for treating cough (Dzoyem et al., 2014). The leaves are used for treating inflamed membranes of the mouth, diarrhea, dysentery, catarrh, acute cystitis, dandruff, rabies, diabetes, and bladder ailments (Anis and Iqbal, 1994; Hegde, 2012; Reddy et al., 1989; Singh and Ali, 1989). The roots are used to promote hair growth, prevent premature gray hair, and treat cough and asthma (Hegde, 2012). Sesame oil is used for treating tuberculosis, eye diseases, backache, migraines, ulcers, snake bites, constipation, hair loss, cough, burns, boils, to promote menstruation, induce lactation, as a purgative, galactogogue, demulcent, and as an antitussive (Hegde 2012; Ross, 2005). Thus, it is clear that all parts of the sesame plant traditionally have varied medicinal uses, embedded in the culture of different ethnic groups worldwide. In the light of sesame's innumerable traditional uses, it is pertinent to ask: how much of its claimed potency in traditional medicine has been evaluated and verified? In this chapter, information gathered on sesame plants, including its origin, cultivation, nutritional composition, pharmacological evaluation of different plant parts, and isolated chemical constituents are coherently synthesized to highlight its health benefits and therapeutic potential against different diseases, as well as the need for further exploration at clinical levels and development of identified chemotherapeutic candidates.

2 ORIGIN OF THE CROP

Various research endeavors on sesame plant have generated closely related opinions regarding the origin of the crop. Basically, suggestions from various research works related to the origin of the crop have been mostly inclined with the use of plants collected from the wild in the African and Asian continents, which triggered its cultivation. There have been controversies over the exact origin of sesame crop. As Nayar (1995) pointed out, the clarity as to whether *S. indicum* originates from Africa or Asia has not been well documented. De Candolle (1886) postulated that India received sesame from the Malayan and Indonesian region in the pre-Aryan period, probably through the South-Indian navigators. The origin of the crop in the Central Asian region was suggested by Esquinas-Alcazar (2004). Earlier studies carried out on sesame suggested that the domestication of the crop commenced on the African shore (Hiltebrandt, 1932). This ideology was later supported by other scholars, such as Nayar and Mehra (1970), Mehra (2000), and Bedigion (2010). The review work carried out by Fuller (2003) indicated that there have been bodies of scientific evidence to substantiate claims that the crop is of South-Asian origin. According to Fuller (2003), studies of wild population of *Sesamum* spp. and analysis of seed protein profiles have supported a South-Asian origin.

3 CHEMICAL PROPERTIES OF *SESAMUM INDICUM*

All crops, including medicinal, food, and those used as animal feeds exhibit different chemical compositions, which vary with the forms (such as raw, dry, or roasted form) in which they are used. For example, the crude protein and crude fiber contents of the raw pulverized form of *S. indicum* were higher than those of the roasted-milled form (roasted for 20 min at 80°C) of the crop (Yusuf et al., 2008). On the other hand, according to the authors, the protein concentrate extract of roasted flour had higher carbohydrate concentration than that of the raw form.

Table 26.2 Mineral composition of *S. indicum* L. seeds

Mineral element	Composition of element in mg 100 g^{-1} of *S. indicum* seed
Sodium	122.50
Calcium	415.38
Magnesium	579.53
Phosphorus	647.25
Potassium	851.35

Source: Nzikou, J.M., Matos, L., Bouanga-Kalou, G., Ndangui, C.B., Pambou-Tobi, N.P.G., Kimbonguila, A., Silou, T., Linder, M. and Desobry, S., 2009. Chemical composition on the seeds and oil of sesame (*Sesamum indicum* L.) grown in Congo-Brazzaville. Adv. J. Food Sci. Technol. 1(1), 6–11.

The seed of the crop, which is the most-utilized part, has been found to be a high source of minerals. Nzikou et al. (2009), for example, reported a high level of potassium among the mineral elements available in the seed, which is followed in descending order by phosphorus, magnesium, calcium, and sodium (Table 26.2).

Sesame seeds contain oil, ranging from 37% to 63%, depending on the variety/cultivar and growing season (Hegde, 2012). The differences in the oil content of the various varieties of the crop is associated with the differential influence by ecological factors, which affect the composition of the seed, particularly the differences in precipitation or rainfall and radiation of the light or sunshine (Asghar and Majeed, 2013). The size and color of the seeds can also affect seed-oil content. For instance, smaller, light-colored seeds tend to have a high oil content than dark, relatively large seeds (Seegeler, 1983). In general, unsaturated fatty acids form a large part (about 80%) of sesame oil (Hegde, 2012). Factors, such as differences in genetics and environmental conditions, as well as the seed developmental stage can alter the fatty acid composition of sesame oil (Li et al., 2008). In an experiment investigating the fatty acid composition of four different varieties of sesame crop, Asghar and Majeed (2013) observed differences in the percentage content of caprilic, linoleic, palmitic, and stearic acids in the seed oil; ranging from 16.9% to 28.8%, 4.7% to 12.5%, 3.2% to 19.3%, and 4.7% to 21.5%, respectively. Table 26.3 indicates variation in other fatty acid composition as influenced by varietal differences.

4 PHYTOCHEMICAL STUDIES

A number of chemical compounds have been isolated mainly from sesame seed and leaves and tested in different biological assays both in vitro and in vivo (Table 26.4). In particular, the recorded pharmacological activities of sesamin, sesamol, and sesamolin (Fig. 26.2) isolated from sesame seeds revealed their therapeutic potential. Sesamin is arguably the most abundant lignan occurring in sesame seed oil (Jeng et al., 2005). Although the reported in vitro antidiabetic, antioxidant, and antibacterial activities were not very interesting (Table 26.4), other studies have shown the protective effect of sesamin against

Table 26.3 Fatty acid composition in different varieties of sesame crop

Type of fatty acid	Varieties of Sesame crop			
	Til-90	P-37	Til-93	S-17
Caprilic acid (C:8:0)	—	28.8	16.9	—
Capric acid (C:10:0)	11.3	—	—	—
Lauric acid (C:12:0)	2.8	0.2	—	—
Myristic acid (C:14:0)	2.6	33.4	0.5	—
Myristoleic acid (C:14:1)	9.9	—	—	6.2
Eicosanoic acid (C:20:0)	—	0.6	5.0	17.0
Eladic acid (C:18:1)	—	0.4	—	—
Palmitic acid (C:16:0)	19.3	3.2	6.0	—
Palmitoleic acid (C:16:1)	1.7	—	—	—
Pentadecanoic acid (C:15:0)	—	28.3	—	—
Heptadecanoic acid (C:17:0)	2.2	—	—	—
Stearic acid (C:18:0)	13.9	4.7	—	21.5
Oleic acid (C:18:1)	10.2	—	—	—
Linoleic acid (C:18:2)	12.5	—	4.7	—
Linolenic acid (C:18:3)	11.0	—	3.2	—
Erucic acid (C:22:1)	—	—	15.6	—
Behenic acid (C:20:1)	—	—	14.6	—

Source: Asghar, A. and Majeed, M.N., 2013. Chemical characterization and fatty acid profile of different sesame verities in Pakistan. Am. J. Sci. Ind. Res. 4, 540–545.

oxidative stress and liver damage, as well as its suppressive effect against chemically induced mammary carcinogenesis (Akimoto et al., 1993; Hirose et al., 1992). The chondroprotective and antiinflammatory effects of sesamin were established using different in vitro– and in vivo mechanism–based assays (Chavali et al., 1998; Phitak et al., 2012). Sesamin was also shown to stimulate osteoblast differentiation in adipose stem cells (Wanachewin et al., 2012). Thus, Phitak et al. (2012) opined that "sesamin may be a natural drug of choice for treatment of arthritic diseases." In addition, sesamin's anticholesterolemic and antihypertensive properties have been highlighted (Kang et al., 1999; Nakai et al., 2003; Penalvo et al., 2006; Visavadiya and Narasimhacharya, 2008). Clinical trials revealed that sesamin does not increase cardiovascular disease risk markers in overweight men and women (Wu et al., 2009a,b). Sesamin was also reported to have antibacterial and insecticidal properties, resulting in its utilization as a synergist for pyrethrum insecticide (Anilakumar et al., 2010; Morris, 2002).

Sesamol, another lignan in sesame seed oil, has been demonstrated to exhibit potent antioxidant activity by many researchers (Kanimozhi and Prasad, 2009; Mahendra Kumar et al., 2011; Mahendra Kumar and Singh, 2015; Parihar et al., 2006). Other reported pharmacological activities of sesamol observed through in vitro and in vivo assays, include its antiaging, anticlastogenic, antimutagenic, chemopreventive, wound-healing, and antiinflammatory properties (Hsu et al., 2008b; Kapadia et al., 2002; Kaur

Table 26.4 Isolated compounds from different parts of S. indicum and their biological activities

Isolated compound	Part from which compound was isolated	Test system used	Assay/model used	Positive control	Noteworthy activity	References
3-Epibartogenic acid	Leaves	In vitro	α-Amylase inhibition	Acarbose (IC$_{50}$ value of 124 µM)	IC$_{50}$ value of 146.7 µM	Dat et al. (2016)
Epigallocatechin	Leaves	In vitro	α-Amylase inhibition	Acarbose (IC$_{50}$ value of 124 µM)	IC$_{50}$ value of 303.9 µM	Dat et al. (2016)
Kaempferol 3-O-[2-O-(trans-p-coumaroyl)-3-O-α-L-rhamnopyranosyl]-β-D-glucopyranoside	Leaves	In vitro	α-Amylase inhibition	Acarbose (IC$_{50}$ value of 124 µM)	Inactive	Dat et al. (2016)
Pinoresinol	Seed	In vitro	α-Glucosidase for antidiabetic activity	Acarbose	IC$_{50}$ value of 492 µM	Wikul et al. (2012)
Sesamin	Seed	In vitro	α-Glucosidase for antidiabetic activity	Acarbose	IC$_{50}$ value of 450 µM	Wikul et al. (2012)
Sesamin	Seed	In vitro	DPPH and β-carotene–linoleate model system	BHT (IC$_{50}$ value of 5.81 µg mL^{-1} and 97% antioxidant activity)	30% radical scavenging activity at 250 µg mL^{-1}, 68% antioxidant activity at 200 µg mL^{-1}	Mahendra Kumar and Singh (2015)

Sesamin	Seed	In vitro	Agar plate	—	69, 69, and 59 % growth inhibition at 2 mg mL^{-1} concentration against *Bacillus cereus*, *Staphylococcus aureus*, and *Pseudomonas aeruginosa*, respectively	Mahendra Kumar and Singh (2015)
Sesamol	Seed	In vitro	Monophenolase activity of tyrosinase	Kojic acid (IC$_{50}$ value of 59.72 μM)	IC$_{50}$ value of 3.2 μM (monophenolase inhibition)	Mahendra Kumar et al. (2011)
Sesamol	Seed	In vitro	DPPH and β-carotene–linoleate model system	BHT (IC$_{50}$ value of 5.81 μg mL^{-1} and 97% antioxidant activity)	IC$_{50}$ value of 5.44 μg mL^{-1} in DPPH and 78.5% antioxidant activity at 200 μg mL^{-1}	Mahendra Kumar et al. (2011); Mahendra Kumar and Singh (2015)
Sesamol	Seed	In vitro	Agar plate	—	MIC of 2 mg mL^{-1} against *B. cereus* and *Staphylococcus aureus*; 80% growth inhibition of *P. aeruginosa* at 2 mg mL^{-1} concentration	Mahendra Kumar and Singh (2015)

(Continued)

Table 26.4 Isolated compounds from different parts of S. indicum and their biological activities (cont.)

Isolated compound	Part from which compound was isolated	Test system used	Assay/model used	Positive control	Noteworthy activity	References
Sesamol	Seed	In vivo	Incision, excision, and dead space model	—	Significantly increased tensile strength when compared to control in normal and dexamethasone suppressed healing; significant increase in percentage normal wound contraction on day 7 when compared to control; faster wound closure when compared to control for the normal healing	Shenoy et al. (2011)

Sesamol	Seed	In vitro	Hydroxyl radical, superoxide anion, nitric oxide, and DPPH free radical and ABTS cation radical scavenging	Ascorbic acid (IC_{50} of 30.3, 36.85, 32.27, 5.85, and 6.63 µg mL^{-1}; hydroxyl radical, superoxide anion, nitric oxide, DPPH free radical, and ABTS cation radical scavenging activities, respectively)	IC_{50} of 31.29, 40.72, 36.36, 3.23, and 3.65 µg mL^{-1} (hydroxyl radical, superoxide anion, nitric oxide, DPPH free radical, and ABTS cation radical scavenging activities, respectively)	Kanimozhi and Prasad (2009)
Sesamol	Seed	In vivo	γ Radiation–induced DNA damage with comet assay	None	Pretreatment with 100 mg kg^{-1} body weight significantly decreased the percentage of tail DNA, tail length, tail moment, and Olive tail moment in the peripheral blood of whole-body irradiated mice	Kanimozhi and Prasad (2009)

(Continued)

Table 26.4 Isolated compounds from different parts of S. indicum and their biological activities (cont.)

Isolated compound	Part from which compound was isolated	Test system used	Assay/model used	Positive control	Noteworthy activity	References
Sesamol	Seed	In vivo	Dinitrochlorobenzene-induced model	Sulfasalazine	Significant decrease in tissue nitrite concentration, colon weight, myeloperoxidase concentration, lipid concentration, and a significant increase in α-tumor necrosis factor	Kondamudi et al. (2013)
Sesamolin	Seed	In vitro	α-Glucosidase for antidiabetic activity	Acarbose	IC_{50} value of 200 μM	Wikul et al. (2012)
Sesamolin	Seed	In vitro	DPPH and β-carotene–linoleate model system	BHT (IC_{50} value of 5.81 μg mL^{-1} in DPPH and 97% antioxidant activity)	32% radical scavenging activity at 250 μg mL^{-1}; 62.5% antioxidant activity at 200 μg mL^{-1}	Mahendra Kumar and Singh (2015)
Sesamolin	Seed	In vitro	Agar plate	—	61, 62, and 53 % growth inhibition at 2 mg mL^{-1} concentration against B. cereus, S. aureus, and P. aeruginosa, respectively	Mahendra Kumar and Singh (2015)

ABTS, 2,2′-Azino-bis-(3-ethylbenzthiazoline-6-sulfonic acid); BHT, butylated hydroxytoluene; DPPH, 1,1-diphenyl-2-picryl-hydrazyl; MIC, minimum inhibitory concentration.

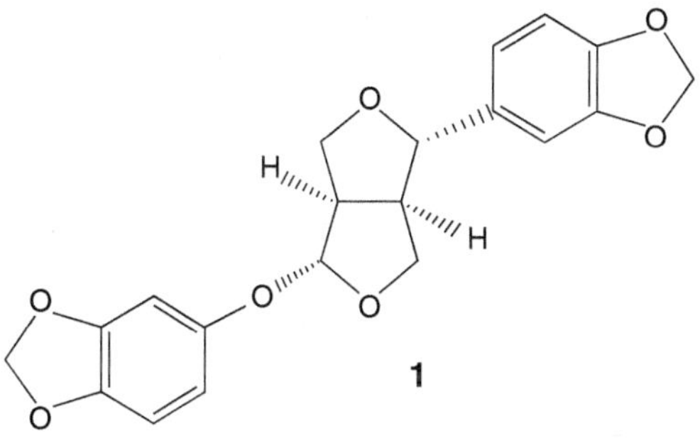

Figure 26.2 *Chemical structures of common bioactive compounds isolated from* **S. indicum.** Sesamolin (**1**); sesamol (**2**); sesamin (**3**); pinoresinol (**4**).

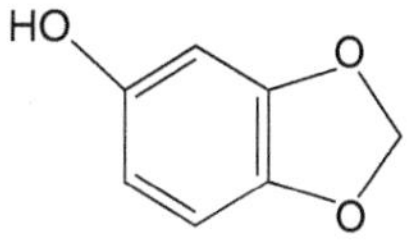

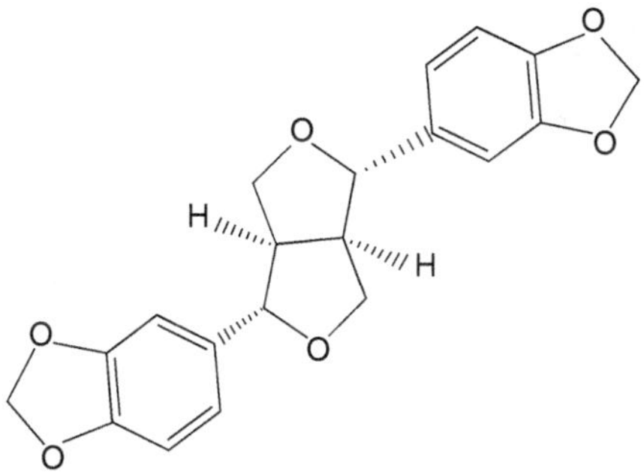

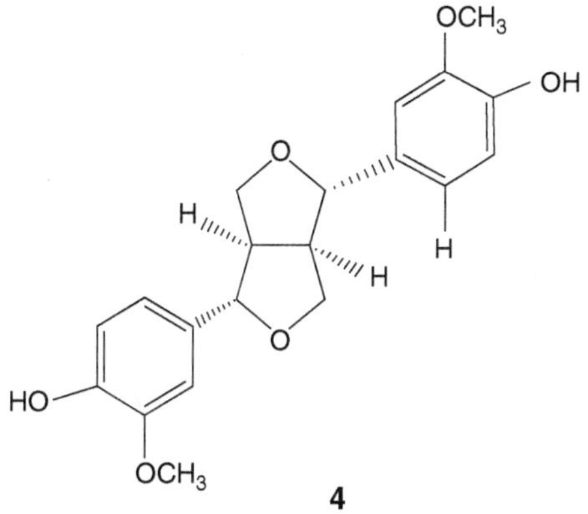

and Saini, 2000; Kondamudi et al., 2013; Parihar et al., 2006; Sharma and Kaur, 2006; Shenoy et al., 2011). Another known sesame lignan, sesamolin, showed antidiabetic and lipid peroxidation properties, as well as anticancer activity against human lymphoid leukemia cells (Kang et al., 1998; Miyahara et al., 2001; Wikul et al., 2012). Taken together, the pharmacological activities of the isolated compounds, mainly from sesame seed, provide a logical explanation for some of the activities reported with the seed crude extracts. The isolation and characterization of biologically active compounds from other parts of sesame plants and further clinical trials of potent characterized compounds are areas needing scientific exploration.

5 PHARMACOLOGICAL EVALUATION OF PLANT EXTRACTS

In accordance with the various ethnobotanical uses of different parts of sesame plant in traditional medicine all over the world, many researchers have evaluated the pharmacological potential or activity of different solvent extracts obtained from different plant parts using a battery of biological assays (Table 26.5). It is noteworthy that many of the studies were mechanism-based and advanced to the in vivo level, validating some of the claimed uses in traditional medicine.

The overproduction of free radicals and their reactions often result in oxidative stress, and this has been implicated in the pathology of many disease states, such as cancer, diabetes, and inflammation (Houghton et al., 2007). Thus, the use of antioxidants has been recommended to reduce incidences of certain diseases (Howes and Houghton, 2003). As shown in Table 26.5, sesame seed and fruit polar extracts were demonstrated to exhibit in vitro antioxidant activity using different assays with differing antioxidant mechanisms. Particularly noteworthy is the antioxidant activity of the fruit acetone extract, which was 6 times stronger than the positive standard (ascorbic acid) in ferric reducing antioxidant power (FRAP) assay (Dzoyem et al., 2014). Nonetheless, the demonstrated antioxidant activity needs to be established in vivo with due attention to the aspect of absorption and bioavailability of antioxidant compounds present in the extract. Although sesame leaves and roots are used in the treatment of different diseases involving oxidative stress (Hegde, 2012), the evaluation of their antioxidant activity still requires an in-depth scientific investigation.

Owing to the documented uses of all parts of sesame in treating infection-related ailments, a number of studies evaluated the antimicrobial activity of different extracts (largely polar extracts) obtained from the seed, fruit, leaves, and root of sesame (Table 26.5). The studies demonstrated the antimicrobial activity of the extracts against different strains of pathogenic fungi (Uniyal et al., 2012), as well as Gram-positive and Gram-negative bacteria, including drug-resistant strains (Abdelgawad et al., 2015). Antimicrobial activity of the fruit was demonstrated against three fast growing *Mycobacterium* species: *M. aurum*, *M. fortuitum*, and *M. smegmatis* (Dzoyem et al., 2014). As the authors mentioned, the activity demonstrated against *M. aurum* [minimum inhibitory concentration (MIC) of 0.62 mg mL^{-1}] is noteworthy in relation to its predictive activity against *Mycobacterium tuberculosis*, as it was reported that the two species have similar drug sensitivity profiles (Chung et al., 1995). However, in some of the documented studies on sesame, the antimicrobial activities were quantified using inhibition zone diameters only. This makes it difficult to benchmark the reported activity against other potent extracts in the literature. Studies reporting inhibition zone diameters only were often conducted using agar diffusion technique, which has been highlighted by many researchers to have some pitfalls. For example, the diffusion characteristics of nonpolar extracts (which do not diffuse easily into agar) may influence the antimicrobial potency of medicinal plant extracts (Cos et al., 2006).

The process of inflammation has been implicated in the pathogenesis of many diseases, such as asthma, rheumatoid arthritis, multiple sclerosis, and ulcerative colitis (Howes and Houghton, 2003; Polya, 2003). Inflammation is an immunological response often elicited by tissue injury from microbial infection, wounding, and other sources of

Table 26.5 Biological activities of solvent extracts from different parts of *S. indicum*

Plant part	Extracting solvent	Test system used	Assay used	Positive control and activity (where stated)	Noteworthy activity	References
Antioxidant activity						
Fruit	ACE	In vitro	DPPH, FRAP, and ABTS	Ascorbic acid (IC_{50} of 4.37 and 1.21 µg mL^{-1} in DPPH and ABTS assays, respectively; 2.62 TEAC in FRAP assay)	IC_{50} values of 13.43 and 14.10 µg mL^{-1} in DPPH radical and ABTS radical scavenging assays, respectively; 0.41 TEAC in FRAP assay	Dzoyem et al. (2014)
Seed	EtOH	In vitro	DPPH, ABTS, FRAP, nitric oxide, superoxide anion, and hydroxyl radicals scavenging, and lipid peroxidation	α-Tocopherol (IC_{50} of 43.2 and 29.8 µg mL^{-1} in superoxide anion and DPPH radical scavenging assays); trolox (IC_{50} of 59 µg mL^{-1} in ABTS radical scavenging assay)	IC_{50} values of 98.8, 45.6, 87.0, and 100.8 µg mL^{-1} in nitric oxide, superoxide anion, and DPPH radical and ABTS radical scavenging assays, respectively	Visavadiya et al. (2009)

(Continued)

Medicinal Spices and Vegetables from Africa

Table 26.5 Biological activities of solvent extracts from different parts of *S. indicum* (cont.)

Plant part	Extracting solvent	Test system used	Assay used	Positive control and activity (where stated)	Noteworthy activity	References
Seed	Water	In vitro	DPPH, ABTS, FRAP, nitric oxide, superoxide anion, and hydroxyl radicals scavenging, and lipid peroxidation	—	IC_{50} values of 238.0, 135.4, 203.3, and 249.5 µg mL^{-1} in nitric oxide, superoxide anion, and DPPH radical and ABTS radical scavenging assays, respectively	Visavadiya et al. (2009)
Seed coat	EtOH	In vitro	LDL oxidation and nitric oxide production	Trolox	74% inhibition (at 0.5 mg mL^{-1}) of LDL oxidation; 89% inhibitory effect (at 0.08 mg mL^{-1}) on nitric oxide production	Wang et al. (2007)
Acetylcholinesterase inhibitory activity						
Seed	EtOH	In vitro	—	Galantamine (IC_{50} value of 0.002 mg mL^{-1})	IC_{50} value of 10.75 mg mL^{-1}	Shariffar et al. (2012)

Gastroprotective activity

Leaves	Hex	In vivo	Necrotizing agent–induced alteration	Misoprostol (99.9% ulcer protection at 200 μg kg^{-1} body weight)	47.1 and 99.8% ulcer protection at 200 and 400 mg kg^{-1} body weight, respectively	Okwuosa et al. (2011)

Larvicidal activity

Root	DCM	In vitro	*Aedes aegypti* larvae mortality	Diazinon (LC$_{100}$ of 0.1 μg mL^{-1})	LC$_{100}$ of 125 μg mL^{-1} after 24 h	Cepleanu et al. (1994)

Cytotoxicity/Toxicity

Fruit	ACE	In vitro	MTT assay using Vero monkey kidney cells	Doxorubicin (IC$_{50}$ value of 4.51 μg mL^{-1})	IC$_{50}$ value of 140 μg mL^{-1}	Dzoyem et al. (2014)
Leaves	DCM	In vitro	MTT colorimetric	Navelbine (ED$_{50}$ of 0.9 μg mL^{-1} against human colon carcinoma cell line SW480)	ED$_{50}$ of 6.1 μg mL^{-1} against human colon carcinoma cell line SW480	Cepleanu et al. (1994)
Root	DCM	In vitro	MTT colorimetric	Navelbine (ED$_{50}$ of 0.9 μg mL^{-1} against human colon carcinoma cell line SW480)	ED$_{50}$ of 3.6 μg mL^{-1} against human colon carcinoma cell line SW480	Cepleanu et al. (1994)
Leaves	Hex	In vivo	Acute toxicity	Not stated	LD$_{50}$ > 5000 mg kg^{-1} in rats	Okwuosa et al. (2011)

(Continued)

Table 26.5 Biological activities of solvent extracts from different parts of S. indicum (cont.)

Plant part	Extracting solvent	Test system used	Assay used	Positive control and activity (where stated)	Noteworthy activity	References
Seed	EtOH	In vitro	Brine shrimp	Potassium dichromate (IC$_{50}$ value of 2.5 µg mL^{-1})	IC$_{50}$ value of 4430.9 µg mL^{-1}	Sharififar et al. (2012)
Seed	PE	In vitro	Brine shrimp	Potassium dichromate	IC$_{50}$ of 855.9 µg mL^{-1}	Abushama et al. (2014)
Seed	Hex	In vitro	Brine shrimp	Not stated	11% mortality at 500 µg mL^{-1}	Amara et al. (2008)
Seed	EtOH	In vitro	Brine shrimp	Not stated	80% mortality at 500 µg mL^{-1}	Amara et al. (2008)
Seed coat	EtOH	In vitro	Protective effect on hydrogen peroxide–induced cytotoxicity in 3T3 cells	Tocopherol	60% protective effect at 0.1 mg mL^{-1}	Wang et al. (2007)

Antimicrobial activity

Plant part	Extracting solvent	Test system used	Assay used	Positive control and activity (where stated)	Noteworthy activity	References
Aerial part	Oil	In vitro	Agar well diffusion	Amphotericin B 100 µg disc^{-1} (inhibition zone of 10 mm against both Aspergillus fumigatus and Aspergillus niger)	Inhibition zone of 30 and 20 mm against A. fumigatus and A. niger, respectively	Uniyal et al. (2012)

Beans	MeOH	In vitro	Microdilution	Chloramphenicol (MIC values range of 4–512 µg mL^{-1} for different strains of *Providencia stuartii*, *Enterobacter cloacae*, *Escherichia coli*, *Enterobacter aerogenes*, and *Klebsiella pneumoniae*)	MIC range of 512–1024 µg mL^{-1} for different strains of *P. stuartii*, *E. cloacae* and *E. coli*; MIC range of 256–1024 µg mL^{-1} for different strains of *E. aerogenes* and *K. pneumoniae*	Seukep et al. (2013)
Fruit	ACE	In vitro	Microdilution	Rifampicin (MIC of 50, 3.12, and 12.5 µg mL^{-1} against *M. smegmatis*, *M. fortutium*, and *M. aurum*, respectively)	MIC of 0.62 mg mL^{-1} against *M. smegmatis*, *M. fortutium*, and *M. aurum*.	Dzoyem et al. (2014)
Fruit	Oil	In vitro	Disc diffusion	Not stated	26-mm zone of inhibition against *Ganoderma lucidum*	Hanif et al. (2010)
Leaves	EtOH	In vitro	Not stated	Ciprofloxacin (IC$_{50}$ of 0.091 µg mL^{-1} against methicillin-resistant *S. aureus*)	IC$_{50}$ of 19.32 µg mL^{-1} against methicillin-resistant *S. aureus*	Abdelgawad et al. (2015)

(*Continued*)

Table 26.5 Biological activities of solvent extracts from different parts of S. indicum (cont.)

Plant part	Extracting solvent	Test system used	Assay used	Positive control and activity (where stated)	Noteworthy activity	References
Leaves	MeOH	In vitro	Agar well diffusion	Ampicillin at 1 mg mL^{-1} (inhibition zone of 2.4, 2.6, 3.1, 3.0, and 4.0 cm against E. coli, B. cereus, P. aeruginosa, Xanthomonas compestris, and S. aureus, respectively)	Inhibition zone of 2.2 cm against P. aeruginosa at 100 µg mL^{-1}	Sharma et al. (2014)
Root	MeOH	In vitro	Agar well diffusion	Ampicillin at 1 mg mL^{-1} (inhibition zone of 2.4, 2.6, 3.1, 3.0, and 4.0 cm against E. coli, B. cereus, P. aeruginosa, X. compestris, and S. aureus, respectively)	Inhibition zone of 0.8 and 1.4 cm against P. aeruginosa at 25 and 100 µg mL^{-1}, respectively	Sharma et al. (2014)
Root	DCM	In vitro	Bioautography on thin layer chromatograms	Propiconazole against Cladosporium cucumerinum (minimum inhibitory dose = 0.01 µg)	Active at 100 µg against C. cucumerinum	Cepleanu et al. (1994)

Seed	MeOH	In vitro	Agar well diffusion	Ampicillin at 1 mg mL^{-1} (inhibition zone of 2.4, 2.6, 3.1, 3.0, and 4.0 cm against E. coli, B. cereus, P. aeruginosa, X. compestris, and S. aureus, respectively)	Inhibition zone of 1.2 cm against P. aeruginosa at 25 μg mL^{-1}, 1.8, 2.6, and 3.2 cm at 100 μg mL^{-1} against P. aeruginosa, X. compestris, and S. aureus, respectively	Sharma et al. (2014)

Antihyperlipidemic activity

Seed	Oil	In vivo	High-fat fed rabbit model	None	Significantly lower circulating concentrations of total cholesterol, HDL cholesterol, LDL cholesterol, and liver enzymes (comprising serum glutamate oxaloacetate transaminase and serum glutamate pyruvate transaminase) compared to the hypercholesterolemic diet group	Asgary et al. (2013)

(Continued)

Table 26.5 Biological activities of solvent extracts from different parts of S. indicum (cont.)

Plant part	Extracting solvent	Test system used	Assay used	Positive control and activity (where stated)	Noteworthy activity	References
Seed	—	In vivo	Hypercholesteremic rats	—	Significant decline in plasma, hepatic total lipid and cholesterol levels, and plasma LDL cholesterol levels with an increase in plasma HDL cholesterol levels; increased fecal excretion of cholesterol, neutral sterol, and bile acid.	Visavadiya and Narasimhacharya (2008)

Hepatoprotective activity

Plant part	Extracting solvent	Test system used	Assay used	Positive control and activity (where stated)	Noteworthy activity	References
Seed	EtOH	In vivo	Carbon tetrachloride–induced hepatotoxicity in rats	—	Significant restoration toward normalization of highly elevated enzymatic activities of serum glutamate oxaloacetate transaminase and serum glutamate pyruvate transaminase	Kumar et al. (2011)

Antiinflammation and wound healing

Fruit	ACE	In vitro	LOX	Quercetin (>80% 15-LOX inhibition at 100 μg mL^{-1})	<50% 15-LOX inhibition at 100 μg mL^{-1}	Dzoyem et al. (2014)
Root	MeOH	In vivo	Excision wound model	Povidone–iodine	Faster contraction of wound diameter compared to the control	Dhumal and Kulkarni (2007)
Seed	Oil	In vivo	Incision wound model	Gentamycin	Significant reduction in wound length, faster closure rate, and shorter epithelialization period compared to the control	Sharif et al. (2013)
Seed	EtOH	In vivo	Acetic acid–induced writhing	Ibuprofen (71.8% writhing inhibition at 25 mg kg^{-1})	48.2 and 75.5% writhing inhibition at 25 and 500 mg kg^{-1}, respectively	Nahar and Rokonuzzaman (2009)

(Continued)

Table 26.5 Biological activities of solvent extracts from different parts of S. indicum (cont.)

Plant part	Extracting solvent	Test system used	Assay used	Positive control and activity (where stated)	Noteworthy activity	References
Seed coat	EtOH	In vitro	Cell-based assay	—	33% inhibition (at 0.08 mg mL^{-1}) of prostaglandin E2 production; 53% inhibitory effect (at 0.08 mg mL^{-1}) on COX-2 protein expression	Wang et al. (2007)

Anthelmintic activity

Plant part	Extracting solvent	Test system used	Assay used	Positive control and activity (where stated)	Noteworthy activity	References
Seed	MeOH	In vitro	—	Levamisole (at 1 mg mL^{-1}, time taken for paralysis and death were 3 and 6 min, respectively)	At 10 mg mL^{-1}, time taken for paralysis and death were 12 and 22 min, respectively	Kamal et al. (2015)

Antileishmanial activity

Plant part	Extracting solvent	Test system used	Assay used	Positive control and activity (where stated)	Noteworthy activity	References
Leaves	EtOH	In vitro	Alamar Blue assay	Amphotericin B (98.7%)	75.9% inhibition at 80 µg mL^{-1}	Abdelgawad et al. (2015)

Vasorelaxant activity

Root	MeOH (PE-soluble fraction of the extract)	In vivo	—	IC_{50} of 77.3 and 121.4 on phenylephrine or potassium chloride induced contraction, respectively	Kumar et al. (2008)

Anticonvulsant activity

Plant	Oil	In vivo	Locomotion and rearing test, inverted screen test, wire hanging test, catalepsy test, and maximum electric shock test	Chlorpromzine and diazepam	Significant reduction in the duration of tonic hind limb extension, delayed clonic seizures, delayed onset of seizures, and reduced severity of seizures	Advani et al. (2011)

ABTS, 2,2′-Azino-bis-(3-ethylbenzthiazoline-6-sulfonic acid); ACE, acetone; COX, cyclooxygenase; DCM, dichloromethane; DPPH, 1,1-diphenyl-2-picryl-hydrazyl; EtOH, ethanol; FRAP, ferric reducing antioxidant power; HDL, high-density lipoprotein; Hex, hexane; LDL, low-density lipoprotein; LOX, lipoxygenase; MeOH, methanol; MIC, minimum inhibitory concentration; MTT, 3-(4,5-dimethylthiazol-2-yl)-2,5-diphenyltetrazolium bromide); TEAC, trolox equivalent antioxidant capacity; PE, petroleum ether.

damage (Byeon et al., 2008; Polya, 2003). The antiinflammatory activity of fruit and seed coat extracts, as well as the wound-healing activity of root and seed extracts of sesame plant were demonstrated using in vitro and in vivo approaches, respectively (Table 26.5). Although the reported in vitro antiinflammatory activity was relatively low, the wound-healing potential of the extracts was established. The antiinflammatory activity of the leaves, which are used ethnobotanically for treating inflammation, as well as the use of different solvent extracts still require scientific exploration. In conducting further antiinflammatory assays, it may be pertinent to use mechanism-based assays targeting different key molecules at the cellular level in the inflammatory cascade, such as nuclear factor-κB, cytokines, and cyclooxygenase (COX) (Talhouk et al., 2007).

Studies evaluating the safety and toxicology of sesame fruit, leaf, and seed extracts have been documented (Table 26.5). Majority of the studies conducted in vitro suggested that the extracts were not toxic, especially at low concentrations. In vivo acute toxicity study of the hexane extract of sesame leaves indicated the lethal dose LD_{50} to be greater than 5000 mg kg^{-1} in rats (Okwuosa et al., 2011). According to Hsu et al. (2008a), long-term ingestion of a large dose of sesame oil significantly decreased the antioxidant effect, produced no cumulative antioxidant effect, and resulted in a lower body weight increase compared to controls in animal studies. Further toxicity studies, including chronic toxicity, conducted at an in vivo level would be necessary to confirm the safety threshold level of different extracts from different parts of sesame plant as used in traditional medicine.

In addition to the aforementioned pharmacological activities, other biological activities, such as the acetylcholinesterase inhibitory, antihyperlipidemic, hepatoprotective, and anthelmintic activities of sesame seed extracts; antileishmanial and gastroprotective activities of leaf extracts; as well as larvicidal and vasorelaxant activities of root extracts have been demonstrated in different assays (Table 26.5). All of the documented studies, taken together, clearly indicate the therapeutic potential of different parts of sesame plant against various metabolic diseases. Follow-up clinical studies on the noteworthy in vivo biological activity reported in the literature become necessary to establish the therapeutic activity of different parts of the sesame plant.

6 PRODUCTION AND CULTIVATION OF *SESAMUM INDICUM*

S. indicum is known to be used as a raw material for the production of industrial products. According to Anilakumar et al. (2010), extracts from the sesame plant flower are used in making perfumes and cologne, and the myristic acid available in the oil extracted from the seed of this crop forms part of the ingredients used in making cosmetics. The cultivation of sesame thus becomes crucial for its continuous availability as an industrial raw material, and for its nutritional and medicinal uses, among others. Several empirical and review works carried out on *S. indicum* have indicated that the crop is cultivated

Table 26.6 Top twenty producers of sesame seeds for the year 2013

Country	Production (tonnes)
India	636000
China (mainland)	623492
Nigeria	584980
Myanmar	539800
United Republic of Tanzania	420000
Ethiopia	220216
Burkina Faso	137347
Uganda	124200
Somalia	90000
Guatemala	52885
Cameroon	49130
Bangladesh	45000
Mexico	41522
Niger	41332
Mozambique	39400
Chad	39000
Mali	35000
Vietnam	33223
Egypt	32847
Pakistan	32423

Source: FAO, 2013a. Food and Agriculture Organization of the United Nations, Statistics Division. Available from: http://faostat3.fao.org/browse/rankings/countries_by_commodity/E

worldwide, as some researchers have pointed out that the products obtained from the crop have uses in various countries across Africa, Europe, America, Middle East, together with the Australasia regions. The global use of the crop has thus encouraged its cultivation in almost all parts of the world.

A most recent report by FAO (2013a) using 2013 statistics, listed India, China, Nigeria, Myanmar (Burma), and the United Republic of Tanzania as the top five producers of sesame seeds worldwide. Following these countries and among the top 10 producers are 4 African countries, which are Ethiopia, Burkina Faso, Uganda, and Somalia (Table 26.6). Average production of world sesame seeds between 2008 and 2013 showed that Africa and Asia produced 40 and 56.2%, respectively (FAO, 2013b). India, Ethiopia, Sudan, Nigeria, and Burkina Faso were listed as the top five exporters of sesame seeds for 2013; while China, Japan, European Union countries, Turkey, and the Republic of Korea were the top five importers for 2013 (FAO, 2013c). When compared to data from previous years (Hegde, 2012), Africa and Asia remain the top producers and exporters of sesame seeds.

Although sesame production has been recorded in areas up to 40°N (in China, Russia, and USA) and 30°S (in Australia), its main distribution is between 25°S and 25°N, being a tropical/subtropical crop (Hegde, 2012). Sesame is a drought-resistant crop that thrives well in warmer climate with annual rainfall up to 1000 mm, although there are reports of its successful cultivation in areas with annual rainfall above 1000 mm without any significant decline in yield when compared to those cultivated in the traditional growing areas (Agboola, 1979; Ogunremi, 1985; Ogunremi and Ogunbodede, 1986). It grows well in well-drained, moderately fertile soils of medium texture, with a soil pH ranging from 5.5 to 8.0. In an experiment carried out to evaluate the performance of the sesame crop in three different sites having clay, sand, and silt together with different soil depths, Ofosuhene Sintim and Yeboah-Badu (2010) reported that sites with sandy soil within the 10 cm upper soil depth and with a good nutrient status between the 10–30 cm soil depth gave the highest yield (1176 kg ha^{-1}), as this facilitated seedling emergence, high plant growth, and optimized plant population density (250,000 plants ha^{-1}).

Like every other agronomic crop, *S. indicum* has been observed to respond positively to the addition of nutrients to low fertility soil. For example, Shehu (2014) while evaluating the response of the crop to addition of nitrogen, phosphorus, and potassium fertilizers to a nutrient deficient soil, observed that seed and dry matter yields of the crop were obtained at certain optimum rates of fertilizers used in the soil-amendment process. The author noted that optimum seed yield of 678.89 kg ha^{-1} was obtained at 75 kg N ha^{-1} fertilization while the application of 45 kg P_2O_5 ha^{-1} produced an optimum seed yield of 654 kg ha^{-1}. Soil amendment with potassium fertilizer did not significantly influence the growth or yield of the crop (Shehu, 2014).

Spacing at the time of planting is known to influence the growth and yield of the sesame crop. As an example, while examining the effect of spacing on sesame growth and yield, Jakusko et al. (2013) planted the seeds using three different spacing in the form of 60 × 15 cm^2, 60 × 10 cm^2, and 75 × 10 cm^2. The authors observed that spacing had a significant effect on the number of seed per capsule, capsule per plant, and length of capsule, as well as seed weight and yield per hectare. Based on their findings, they suggested using 75 × 10 cm^2 spacing for sowing sesame seeds. On the other hand, Öztürk and Şaman (2012) recommended 30 cm interrow and 5 cm intrarow spacing following their observation of decreased seed and oil yield with the use of wider spacing (70 × 30 cm^2). Clearly, other factors, such as the growth habit of the variety, season, location, and growing conditions can influence the optimum spacing determination (Hegde, 2012).

The growth and yield of sesame can also be influenced by planting date or period of the year in which the planting is carried out and the weeding regime (Abdel Rahman et al., 2007; Ahmed et al., 2009; Mulkey et al., 1987). In Africa, sesame is normally sown at the start of the rains or timed in such a way that harvesting occurs in the dry season to reduce possible huge seed loss associated with increased disease susceptibility due to rainfall (Hegde, 2012). As the author further mentioned, waterlogging must be avoided at all times during sesame growth so as not to increase the plant susceptibility to fungal diseases despite variation in the degree of resistance to waterlogging among different varieties and/or cultivars.

In addition to influencing growth and yield, different agronomic factors can also influence the seed oil quality. Alpaslan et al. (2001), for example, observed significant variation in total oil content arising from the use of different row spacing. The authors further reported significant variation in protein, oleic acid, and linoleic acid contents of sesame seed cultivated with different irrigation and row-spacing treatments. Similarly, Öztürk and Şaman (2012) reported a decrease in oil yield and protein content of sesame seeds as a result of using wider interrow and intrarow spacing (70×30 cm^2). The application of nitrogen fertilizer significantly increased seed yield but reduced oil content (Shakeri et al., 2016). The authors further noted that the application of nitrogen fertilizer and plant growth-promoting rhizobacteria significantly decreased saturated fatty acids (palmitic and stearic acids) but caused a significant increase in unsaturated fatty acid content (oleic and linoleic acid). It becomes imperative, therefore, that cultivation efforts should focus not only on improving yield quantity, but also the quality.

7 CONCLUSIONS

As one of the oldest oilseeds known to man, sesame is undoubtedly a multipurpose crop. Its nutritional, industrial, and medicinal values are well established, cutting across different ethnic groups worldwide. Through both in vitro and in vivo assays, sesame extracts and isolated compounds demonstrated acetylcholinesterase inhibitory, antihyperlipidemic, hepatoprotective, anthelmintic, antileishmanial, gastroprotective, larvicidal, vasorelaxant, antidiabetic, antioxidant, antibacterial, wound-healing, antiinflammatory, antimutagenic, and chemopreventive properties. The various pharmacological activities of different extracts and isolated compounds from different parts of sesame plant demonstrate its therapeutic potential in treating metabolic, inflammatory, and infectious diseases. However, follow-up clinical studies on the noteworthy biological activities reported in the literature are necessary to establish the therapeutic activity of different parts of this plant. With recent drought conditions experienced in many parts of Africa, increasing the cultivation of drought-resistant crop, such as sesame, with a growing market demand is certainly a step in the right direction toward ensuring food security with the potential for increased income generation and job creation especially in the developing countries.

ACKNOWLEDGMENTS

The authors are most grateful to all the researchers and scholars whose research findings were cited in this chapter. We acknowledge that the findings presented in this chapter are indicative, and not necessarily a total compilation of the findings on this plant species. We apologize to authors whose work was not cited to keep the manuscript concise. Opinions expressed and conclusions arrived at, are those of the authors and are not necessarily attributable to the Agricultural Research Council, South Africa.

REFERENCES

Abdel Rahman, A.E., Saif Eldin, M.E., Faisl, G.A., 2007. Effect of sowing date on the performance of sesame (*Sesamum indicum* L.) genotypes under irrigation conditions in northern Sudan. ACSS Conference Proceedings 8, 1943–1946.

Abdelgawad, S.M., Hetta, M.H., Ross, S.A., Badria, F.A., 2015. Antiprotozoal and antimicrobial activity of selected medicinal plants growing in upper Egypt, Beni-Suef region. World J. Pharm. Pharm. Sci. 4, 1720–1740.

Abushama, M.F., Hilmi, Y.I., AbdAlgadir, H.M., Fadul, E., Khalid, H.E., 2014. Lethality and antioxidant activity of some Sudanese medicinal plants' fixed oils. Eur. J. Med. Plants 4, 563–570.

Advani, U., Ansari, A., Menghani, E., 2011. Anticonvulsant potentials of *Sesamum indicum* and *Allium sativum* oil alone and in combination in animal models. Int. J. Pharm. Pharm. Sci. 3, 154–158.

Agboola, S.A., 1979. The Agricultural Atlas of Nigeria. Oxford University Press, Oxford.

Ahmed, H.G., Aliyu, U., Haruna, A.B., Isa, Y.S., Muhammad, A.S., 2009. Effects of planting date and weeding regimes on growth and yield of sesame (*Sesamum indicum* L.) in Sokoto, North-Western Nigeria. Nig. J. Basic Appl. Sci. 17, 202–206.

Akimoto, K., Kitagawa, Y., Akamatsu, T., Hirose, N., Sugano, M., Shimizu, S., Yamada, H., 1993. Protective effects of sesamin against liver damage caused by alcohol or carbon tetrachloride in rodents. Ann. Nutr. Metab. 37, 218–224.

Alpaslan, M., Boydak, E., Hayta, M., Gerçek, S., Simsek, M., 2001. Effect of row spacing and irrigation on seed composition of Turkish sesame (*Sesamum indicum* L.). J. Am. Oil Chem. Soc. 78, 933–935.

Amara, A.A., El-Masry, M.H., Bogdady, H.H., 2008. Plant crude extracts could be the solution: extracts showing in vivo antitumorigenic activity. Pak. J. Pharm. Sci. 21, 159–171.

Anilakumar, K.R., Pal, A., Khanum, F., Bawa, A.S., 2010. Nutritional, medicinal and industrial uses of sesame (*Sesame indicum* L.) seeds—an overview. Agric. Conspec. Sci. 75, 159–168.

Anis, M., Iqbal, M., 1994. Medicinal plantlore of Aligard, India. Int. J. Pharmacognosy 32, 59–64.

Asgary, S., Rafieian-Kopaei, M., Najafi, S., Heidarian, E., Sahebkar, A., 2013. Antihyperlipidemic effects of *Sesamum indicum* L. in rabbits fed a high-fat diet. ScientificWorldJournal 2013, 365892.

Asghar, A., Majeed, M.N., 2013. Chemical characterization and fatty acid profile of different sesame varieties in Pakistan. Am. J. Sci. Ind. Res. 4, 540–545.

Bedigion, D., 2010. Sesame: The Genus *Sesamum*. CRC Press, Boca Raton, FL.

Byeon, S.E., Chung, J.Y., Lee, Y.G., Kim, B.H., Kim, K.H., Cho, J.Y., 2008. In vitro and in vivo anti-inflammatory effects of taheebo, a water extract from the inner bark of *Tabebuia avellanedae*. J. Ethnopharmacol. 119, 145–152.

Cepleanu, F., Hamburger, M.O., Sordat, B., Msonthi, J.D., Gupta, M.P., Saadou, M., Hostettmann, K., 1994. Screening of tropical medicinal plants for molluscicidal, larvicidal, fungicidal and cytotoxic activities and brine shrimp toxicity. Int. J. Pharmacognosy 32, 294–307.

Chavali, S.R., Zhong, W.W., Forse, R.A., 1998. Dietary alpha-linolenic acid increases TNF-alpha, and decreases IL-6, IL-10 in response to LPS: effects of sesamin on the delta-5 desaturation of omega6 and omega3 fatty acids in mice. Prostaglandins Leukot. Essent. Fatty Acids 58, 185–191.

Chung, G.A., Aktar, Z., Jackson, S., Duncan, K., 1995. High-throughput screen for detecting antimycobacterial agents. Antimicrob. Agents Chemother. 39, 2235–2238.

Cos, P., Vlietinck, A.J., Berghe, D.V., Maes, L., 2006. Anti-infective potential of natural products: how to develop a stronger in vitro "proof-of-concept". J. Ethnopharmacol. 106, 290–302.

Dat, N.T., Dang, N.H., Thanh, L.N., 2016. New flavonoid and pentacyclic triterpene from *Sesamum indicum* leaves. Nat. Prod. Res. 30, 311–315.

De Candolle, A., 1886. Origin of Cultivated Plants. Hafner Publishing Co., New York.

Dhumal, S.V., Kulkarni, S.R., 2007. Antibacterial and wound healing activity of roots of *Sesame indicum*. Indian Drugs 44, 937–944.

Dzoyem, J.P., Kuete, V., McGaw, L.J., Eloff, J.N., 2014. The 15-lipoxygenase inhibitory, antioxidant, antimycobacterial activity and cytotoxicity of fourteen ethnomedicinally used African spices and culinary herbs. J. Ethnopharmacol. 156, 1–8.

Elleuch, M., Bedigian, D., Zitoun, A., 2011. Sesame (*Sesame indicum* L.) seeds in food, nutrition, and health. In: Preedy, V.R., Watson, R.R., Patel, V.B. (Eds.), Nuts and Seeds in Health and Disease Prevention. Academic Press, London, UK, pp. 1029–1036.

Esquinas-Alcazar, J., 2004. International Treaty on Plant Genetic Resources for food and agriculture. Plant Genet. Resources Food Agric. 134, 1–6.

FAO, 2013a. Food and Agriculture Organization of the United Nations, Statistics Division. Available from: http://faostat3.fao.org/browse/rankings/countries_by_commodity/E

FAO, 2013b. Food and Agriculture Organization of the United Nations, Statistics Division. Available from: http://faostat3.fao.org/browse/Q/QC/E

FAO, 2013c. Food and Agriculture Organization of the United Nations, Statistics Division. Available from: http://faostat3.fao.org/browse/T/TP/E

Fuller, D.Q., 2003. Further evidence on the history of sesame. Asian Agri-History 7, 127–137.

Hanif, M.A., Bhatti, H.N., Jamil, M.S., Anjum, R.S., Jamil, A., Khan, M.M., 2010. Antibacterial and antifungal activities of essential oils extracted from medicinal plants using CO_2 supercritical fluid extraction technology. Asian J. Chem. 22, 7787–7798.

Hegde, D.M., 2012. Sesame. In: Peter, K. (Ed.), Handbook of Herbs and Spices. Woodhead Publishing Limited, Abington, Cambridge, UK, pp. 449–486.

Hiltebrandt, V.M., 1932. Sesame (*Sesamum indicum* L.) (In Russian.). Bull. Appl. Botany Genet. Plant Breeding 9, 1–114.

Hirose, N., Doi, F., Ueki, T., Akazawa, K., Chijiiwa, K., Sugano, M., Akimoto, K., Shimizu, S., Yamada, H., 1992. Suppressive effect of sesamin against 7,12-dimethylbenz[a]-anthracene induced rat mammary carcinogenesis. Anticancer Res. 12, 1259–1265.

Houghton, P.J., Howes, M-J., Lee, C.C., Steventon, G., 2007. Uses and abuses of in vitro tests in ethnopharmacology: visualizing an elephant. J. Ethnopharmacol. 110, 391–400.

Howes, M-J.R., Houghton, P.J., 2003. Plants used in Chinese and Indian traditional medicine for improvement of memory and cognitive function. Pharmacol. Biochem. Behav. 75, 513–527.

Hsu, D.Z., Chien, S.P., Li, Y.H., Liu, M.Y., 2008a. Sesame oil does not show accumulatively enhanced protection against oxidative-stress-associated hepatic injury in septic rats. J. Parenter. Enteral Nutr. 32, 276–280.

Hsu, D.Z., Chu, P.Y., Li, Y.H., Liu, M.Y., 2008b. Sesamol attenuates diclofenac-induced acute gastric mucosal injury via its cyclooxygenase-independent anti-oxidative effect in rats. Shock 30, 456–462.

Jakusko, B.B., Usman, B.D., Mustapha, A.B., 2013. Effect of row spacing on growth and yield of Sesame (*Sesamum indicum* l.) in Yola, Adamawa State Nigeria. J. Agric. Vet. Sci. 2, 36–39.

Jeng, K.C., Hou, R.C., Wang, J.C., Ping, L.I., 2005. Sesamin inhibits lipopolysaccharide induced cytokine production by suppression of p38 mitogen-activated protein kinase and nuclear factor-kappaB. Immunol. Lett. 97, 101–106.

Kamal, A.T.M., Chowdhury, K.A.A., Chy, M.M.H., Shill, L.K., Chowdhury, S., Chy, M.A.H., Habib, M.Z., 2015. Evaluation of anthelmintic activity of seeds of *Sesamum indicum* L. and fruits of *Capsicum frutescens* L. J. Pharmacogn. Phytochem. 3, 256–259.

Kang, M.H., Naito, M., Tsujihara, N., Osawa, T., 1998. Sesamolin inhibits lipid peroxidation in rat liver and kidney. J. Nutr. 128, 1018–1022.

Kang, M.H., Naito, M., Sakai, K., Uchida, K., Osawa, T., 1999. Mode of action of sesame lignans in protecting low-density lipoprotein against oxidative damage in vitro. Life Sci. 66, 161–171.

Kanimozhi, P., Prasad, N.J., 2009. Antioxidant potential of sesamol and its role on radiation-induced DNA damage in whole-body irradiated Swiss albino mice. Env. Toxicol. Pharmacol. 28, 192–197.

Kapadia, G.J., Azuine, M.A., Tokuda, H., Takasaki, M., Mukainaka, T., Konoshima, T., Nishino, H., 2002. Chemopreventive effect of resveratrol, sesamol, sesame oil and sunflower oil in the Epstein-Barr virus early antigen activation assay and the mouse skin two-stage carcinogenesis. Pharmacol. Res. 45, 499–505.

Kapoor, L.D., 2001. Handbook of Ayurvedic Medicinal Plants. CRC Press, New York.

Kaur, I.P., Saini, A., 2000. Sesamol exhibits antimutagenic activity against oxygen species mediated mutagenicity. Mutat. Res. 470, 71–76.

Khan, A.V., Ahmed, Q.U., Khan, M.W., Khan, A.A., 2014. Herbal cure for poisons and poisonous bites from Western Uttar Pradesh, India. Asian Pac. J. Trop. Dis. 4, S116–S120.

Kiran, K., Asad, M., 2008. Wound healing activity of *Sesamum indicum* L. seed and oil in rats. Indian J. Exp. Biol. 46, 777–782.

Kondamudi, P.K., Kovelamudi, H., Mathew, G., Nayak, P.G., Rao, C.M., Shenoy, R.R., 2013. Modulatory effects of sesamol in dinitrochlorobenzene-induced inflammatory bowel disorder in albino rats. Pharmacol. Rep. 65, 658–665.

Kumar, P.S., Patel, J.S., Saraf, M.N., 2008. Mechanism of vasorelaxant activity of a fraction of root extract of *Sesamum indicum* Linn. Indian J. Exp. Biol. 46, 457–464.

Kumar, M., Kamboj, A., Sisodia, S.S., 2011. Hepatoprotective activity of *Sesamum indicum* Linn. against CCl_4-induced hepatic damage in rats. Int. J. Pharm. Biol. Arch. 2, 710–715.

Li, X., Xiao, L.G., Wu, G., Wu, Y., Zhang, X., Lu, C., 2008. Accumulation pattern of fatty acids during the seed development of sesame (*Sesamum indicum* L.). Chin. J. Oil Crop Sci. 30, 84–89.

Mahendra Kumar, C., Singh, S.A., 2015. Bioactive lignans from sesame (*Sesamum indicum* L.): evaluation of their antioxidant and antibacterial effects for food applications. J. Food Sci. Technol. 52, 2934–2941.

Mahendra Kumar, C., Sathisha, U.V., Dharmesh, S., Appu Rao, A.G., Singh, S.A., 2011. Interaction of sesamol (3,4-methylenedioxyphenol) with tyrosinase and its effect on melanin synthesis. Biochimie 93, 562–569.

Mehra, K.L., 2000. History of sesame in India and its cultural significance. Asian Agri-History 4, 5–19.

Miyahara, Y., Hibasami, H., Katsuzaki, H., Imai, K., Komiya, T., 2001. Sesamolin from sesame seed inhibits proliferation by inducing apoptosis in human lymphoid leukemia Molt 4B cells. Int. J. Mol. Med. 7, 369–371.

Morris, J.B., 2002. Food, industrial nutraceutical uses of sesame genetic resources. In: Janick, A., Whipkey (Eds.), Tends in New Crops and New Uses. ASDHS Press, Alexandria, VA, pp. 153–156.

Motaleb, M.A., Hossain, M.K., Sobhan, I., Alam, M.K., Khan, N.A., & Firoz, R., 2011. Selected Medicinal Plants of Chittagong Hill Tracts. International Union for Conservation of Nature. Dhaka, Bangladesh, pp. 12–116.

Mulkey, J.R., Drawe, H.J., Elledge, R.E., 1987. Planting date effects on plant growth and development in sesame. Agron. J. 79, 701–703.

Nahar, L., Rokonuzzaman, 2009. Investigation of the analgesic and antioxidant activity from an ethanolic extract of seeds of *Sesamum indicum*. Pak. J. Biol. Sci. 12 (7), 595–598.

Nakai, M., Harada, M., Nakahara, K., Akimoto, K., Shibata, H., Miki, W., Kiso, Y., 2003. Novel antioxidative metabolites in rat liver with ingested sesamin. J. Agric. Food Chem. 51, 1666–1670.

Nayar, M.N., 1995. Sesame *Sesamum indicum* L. (Pedaliaceae). In: Smart, J., Simmonds, N.W. (Eds.), Evolution of Crops Plants. second ed. Longman Scientific and Technical, Essex, UK, pp. 404–407.

Nayar, M.N., Mehra, K.L., 1970. Sesame: its uses, botany, cytogenetics and origins. Econ. Bot. 24, 20–31.

Nzikou, J.M., Matos, L., Bouanga-Kalou, G., Ndangui, C.B., Pambou-Tobi, N.P.G., Kimbonguila, A., Silou, T., Linder, M., Desobry, S., 2009. Chemical composition on the seeds and oil of sesame (*Sesamum indicum* L) grown in Congo-Brazzaville. Adv. J. Food Sci. Technol. 1, 6–11.

Ofosuhene Sintim, H., Yeboah-Badu, V.I., 2010. Evaluation of Sesame (*Sesamum indicum*) production in Ghana. J. Anim. Plant Sci. 6, 653–662.

Ogunremi, E.A., 1985. Cultivation of early season sesame (*Sesamum indicum* L.) in south western Nigeria. Period of sowing. E. Afr. Agr. Forestry J. 51, 82–88.

Ogunremi, E.A., Ogunbodede, B.A., 1986. Path coefficient analysis of seed yield in early season sesame (*Sesamum indicum* L.). Life J. Sci. 1, 27–32.

Okwuosa, C.N., Okoi-Ewa, R., Achukwu, P.U., Onuba, A.C., Azubuike, N.C., 2011. Gastro-protective effect of crude hexane leaf extract of *Sesamum indicum* in rabbits. Niger. J. Physiol. Sci. 26, 49–54.

Öztürk, Ö., Şaman, O., 2012. Effects of different plant densities on the yield and quality of second crop sesame. Int. J. Biol. Biomol. Agric. Food Biotechnol. Eng. 6, 644–649.

Parihar, V.K., Prabhakar, K.R., Veerapur, V.P., Kumar, M.S., Reddy, Y.R., Joshi, R., Unnikrishnan, M.K., Rao, C.M., 2006. Effect of sesamol on radiation-induced cytotoxicity in Swiss albino mice. Mutat. Res. 611, 9–16.

Penalvo, J.L., Hopia, A., Adlercreutz, H., 2006. Effect of sesamin on serum cholesterol and triglycerides level in LDL-receptor deficient mice. Eur. J. Nutr. 45, 439–444.

Phitak, T., Pothacharoen, P., Settakorn, J., Poompimol, W., Caterson, B., Kongtawelert, P., 2012. Chondroprotective and anti-inflammatory effects of sesamin. Phytochemistry 80, 77–88.

Polya, G., 2003. Biochemical Targets of Plant Bioactive Compounds. A Pharmacological Reference Guide to Sites of Action and Biological Effects. CRC Press, Florida.

Reddy, M.B., Reddy, K.R., Reddy, M.N., 1989. A survey of plant crude drugs of Anantapur District, Andhra Pradesh. India. Int. J. Crude Drug Res. 27, 145–155.

Ross, I.A., 2005. *Sesamum indicum* L. In: Ross, I.A. (Ed.), Medicinal Plants of the World, vol. 3: Chemical Constituents, Traditional and Modern Medicinal Uses. Humana Press, New Jersey, pp. 487–505.

Seegeler, C.J.P., 1983. Oil Plants in Ethiopia: Their Taxonomy and Agricultural Significance. Centre for Agricultural Publishing and Documentation, Wageningen.

Seukep, J.A., Fankam, A.G., Djeussi, D.E., Voukeng, I.K., Tankeo, S.B., Noumdem, J.A.K., Kuete, A.H.L.N., Kuete, V., 2013. Antibacterial activities of the methanol extracts of seven Cameroonian dietary plants against bacteria expressing MDR phenotypes. SpringerPlus 2, 363.

Shakeri, E., Modarres-Sanavy, S.A.M., Dehaghi, M.A., Tabatabaei, S.A., Moradi-Ghahderijani, M., 2016. Improvement of yield, yield components and oil quality in sesame (*Sesamum indicum* L.) by N-fixing bacteria fertilizers and urea. Arch. Agron. Soil Sci. 62, 547–560.

Sharif, M.Z., Alizargar, J., Sharif, A., 2013. Evaluation of the wound healing activity of sesame oil extract in rats. World J. Med. Sci. 9, 74–78.

Sharififar, F., Moshafi, M.H., Shafazand, E., Koohpayeh, A., 2012. Acetyl cholinesterase inhibitory, antioxidant and cytotoxic activity of three dietary medicinal plants. Food Chem. 130, 20–23.

Sharma, S., Kaur, I.P., 2006. Development and evaluation of sesamol as an antiaging agent. Int. J. Dermatol. 45, 200–208.

Sharma, S., Gupta, P., Kumar, A., Ray, J., Aggarwal, B.K., Goyal, P., Sharma, A., 2014. In vitro evaluation of roots, seeds and leaves of *Sesamum indicum* L. for their potential antibacterial and antioxidant properties. Afr. J. Biotechnol. 13, 3692–3701.

Shehu, H.E., 2014. Uptake and agronomic efficiencies of nitrogen, phosphorus and potassium in sesame (*Sesamum indicum* L.). Am. J. Plant Nutr. Fertil. Technol. 4, 41–56.

Shenoy, R.R., Sudheendra, A.T., Nayak, P.G., Paul, P., Kutty, N.G., Rao, C.M., 2011. Normal and delayed wound healing is improved by sesamol, an active constituent of *Sesamum indicum* (L.) in albino rats. J. Ethnopharmacol. 133, 608–612.

Singh, V.K., Ali, Z.A., 1989. Folk medicines of Aligarh (Uttar Pradesh), India. Fitoterapia 60, 483–490.

Talhouk, R.S., Karam, C., Fostok, S., El-Jouni, W., Barbour, E.K., 2007. Anti-inflammatory bioactivities in plant extracts. J. Med. Food 10, 1–10.

Uniyal, V., Bhatt, R.P., Saxena, S., Talwar, A., 2012. Antifungal activity of essential oils and their volatile constituents against respiratory tract pathogens causing Aspergilloma and Aspergillosis by gaseous contact. J. Appl. Nat. Sci. 4, 65–70.

Visavadiya, N.P., Narasimhacharya, A.V.R.L., 2008. Sesame as a hypocholesteraemic and antioxidant dietary component. Food Chem. Toxicol. 46, 1889–1895.

Visavadiya, N.P., Soni, B., Dalwadi, N., 2009. Free radical scavenging and antiatherogenic activities of *Sesamum indicum* seed extracts in chemical and biological model systems. Food Chem. Toxicol. 47, 2507–2515.

Wanachewin, O., Boonmaleera, K., Pothacharoen, P., Reutrakul, V., Kongtawelert, P., 2012. Sesamin stimulates osteoblast differentiation through p38 and ERK1/2MAPK signaling pathways. BMC Complement. Altern. Med. 12, 71.

Wang, B.-S., Chang, L.-W., Yen, W.-J., Duh, P.-D., 2007. Antioxidative effect of sesame coat on LDL oxidation and oxidative stress in macrophages. Food Chem. 102, 351–360.

Watt, J., Breyer-Brandwijk, M., 1962. The Medicinal and Poisonous Plants of Southern and Eastern Africa: Being an Account of Their Medicinal and Other Uses, Chemical Composition, Pharmacological Effects and Toxicology in Man and Animal. Livingstone, Ltd, London, UK.

Were, A., Onkware, B., Gudu, S., Welander, M., Carlsson, A.S., 2006. Seed oil content and fatty acid composition in East African sesame (*Sesamum indicum* L.) accessions evaluated over 3 years. Field Crops Res. 97, 254–260.

Wikul, A., Damsud, T., Kataoka, K., Phuwapraisirisan, P., 2012. (+)-Pinoresinol is a putative hypoglycemic agent in defatted sesame (*Sesamum indicum*) seeds though inhibiting α-glucosidase. Bioorg. Med. Chem. Lett. 22, 5215–5217.

Wu, J.H., Hodgson, J.M., Clarke, M.W., Indrawan, A.P., Barden, A.E., Puddey, I.B., Croft, K.D., 2009a. Inhibition of 20-hydroxyeicosatetraenoic acid synthesis using specific plant lignans: in vitro and human studies. Hypertension 54, 1151–1158.

Wu, J.H., Hodgson, J.M., Puddey, I.B., Belski, R., Burke, V., Croft, K.D., 2009b. Sesame supplementation does not improve cardiovascular disease risk markers in overweight men and women. Nutr. Metab. Cardiovasc. Dis. 19, 774–780.

Yusuf, A.A., Ayedun, H., Sanni, L.O., 2008. Chemical composition and functional properties of raw and roasted Nigerian benniseed (*Sesamum indicum*) and bambara groundnut (*Vigna subterranean*). Food Chem. 111, 277–282.

CHAPTER 27

African Medicinal Spices of Genus *Piper*

I.A. Oyemitan

1 INTRODUCTION

Spices are vegetative morphological parts of plants that have been used from time immemorial to date in various forms as food, food additives, carminatives, culinary, article of trade, and medicinal recipes. Some unique features of spices are; they tend to possess strong aroma, may contain volatile constituents, astringent, variously classified as "hot" or "pepperish," frequently stimulatory or depressant after ingesting large quantity. Spices have the tendency to become addictive after prolonged use suggesting central nervous system activity, for example, the narcotic pepper. The narcotic pepper was actually a form of drinks obtained from the root of *Piper methysticum* (kava kava) which lends credence to this observation (Lebot et al., 1997; Shulgin, 1973).

The various morphological parts of plants that have been used as spices include leaf, stem bark, root bark, root, rhizome, fruit, flower, and seed. For example, leaf and fruit of *Dennettia tripetala* are used as spices in Southern parts of Nigeria (Agbakwuru et al. 1979; Oyemitan et al., 2006). Rhizome of *Curcuma* species are also used extensively as spices in India and many Asian countries (Aggarwal and Sung, 2008). Other plants with parts commonly used as spices include *Nigella sativa* (Al-Naggar et al., 2003), *Zanthoxylum capense* (Amabeoku and Kinyua, 2010), *Xylopiaa ethiopica* (Iwu, 1993), *Aframomum meleguata* (Lock et al., 1977), and *Kyllinga brevifolia* (Hellion-Ibarrola et al., 1999) among numerous examples. However, the seed and fruit are by far the most popular parts used as spices in most parts of the world. The reason could be that the seed or fruit are easily collected, stored, packaged, preserved, and transported.

The most popular *Piper* fruits are called peppercorns and are obtained from *P. nigrum* which can be preserved and presented in three different colors namely green (collected when matured just before ripening), black (when the ripe fruits are allowed to get dried), and white form obtained when green ripe fruits are treated with brine in order to remove black outer coverings (Isawumi, 1984; Jirovetz et al., 2002). *Piper nigrum* has been described as "king of spices" because of its large scale volume in international spice market (Srinivasan, 2007).

The African continent is made up of several regions falling uniquely into tropical and subtropical climates which support the growth and cultivation of several plant species. Several medicinal plants and their morphological parts obtained from the African continent have been used extensively as medicinal spices. Important families of these plants include Annonaceae, Rutaceae, Zingiberaceae, Lamiaceae, Leguminosae, Apiaceae, Piperaceae, Anacardiaceae, etc.

This chapter will focus on the medicinal spices of the genus *Piper* in the family of Piperaceae found in some regions in Africa. Discussion will cover taxonomical, distribution, ethnobotanical/ethnosocial, ethnomedicinal, biological/pharmacological, phytochemical, economic, and bioprospecting/conservation profiles of this all-important genus.

2 TAXONOMY OF THE GENUS *PIPER*

The Piperaceae family is composed of about 8 genera and 3000 species (Immelman, 2000; Smith et al., 2008; Ghosh et al., 2014). They are widespread in warm tropical and subtropical regions and are especially common in South and Central America, and central Asia particularly in India. The genus *Piper* is believed to be indigenous to India as they are found in the wild spreading from Himachal Pradesh, Arunachal Pradesh, Khasi and Jayantia hills, and adjoining regions (Srivastava et al., 2000) and mostly cultivated in the Kerala region (Parthasarathy et al., 2006). The scientific name *Piper* and the common name "pepper" were derived from the Sanskrit term "*pippali*," referring to the long pepper *Piper longum*. *Piper* is the nominate genus of the family Piperaceae. The taxonomical classification of the genus *Piper* is presented as follows:

Kingdom: Plantae (Plants)
Subkingdom: Tracheobionta (Vascular plants)
Superdivision: Spermatophyta (Seed plants)
Division: Magnoliophyta (Flowering plants)
Class: Magnoliopsida (Dicotyledons)
Subclass: Magnoliidae
Order: Piperales
Family: Piperaceae (Pepper family)
Genus: *Piper* L.

3 DIVERSITY IN THE *PIPER* SPECIES

Piper is one of the 20 most species-rich genera of flowering plants (Frodin, 2004; Jaramillo et al., 2008). The two main genera of Piperaceae are *Piper* and *Peperomia* with each containing more than a thousand species. *Peperomias* are described as small, succulent, and frequently epiphytic herbs, while pipers are woody, more diverse in habit, and include shrubs, climbing vines, and small trees (Greig, 2004). The diversity in the Piperaceae family is captured within the genera *Piper* and *Peperomia* and can easily be delineated into different geographical locations particularly in Southeast Asia. Piperaceae are less diverse in Africa with only two native species of *Piper* currently recognized namely *Piper*

Table 27.1 African *Piper* species

Number	Species	Geographical location in Africa	References
1.	*P. capense* L.f	East, West, Central and South	Walt and Breyer-Brandwik (1962)
2.	*P. guineense* Schumach Thonn.	East, Central and West	Iwu (1993)
3.	*Piper betle* L.	East	Walt and Breyer-Brandwik (1962)
4.	*Piper angustifolium* Ruiz. & Pav	East	Walt and Breyer-Brandwik (1962)
5.	*Piper clusii* C.DC.	South/Central	Walt and Breyer-Brandwik (1962)
6.	*P. nigrum* L.	East, Central and West	Walt and Breyer-Brandwik (1962); Tchoumgougnang et al. (2009)
7.	*P. umbellatum* L.	Central, West	Nwauzoma and Dawari (2013); Roersch (2010)
8.	*Piper aduncum*	West	Okunade et al. (1997)
9.	*Piper methysticum*	North	Adeneye (2014)
11.	*P. longum*	Madagascar	Manoj et al. (2004)
12.	*Piper borbonense*	South and Madagascar	Immelman (2000)

guineense and *P. capense* (Smith et al., 2008), and some of the species commonly encountered in literatures are shown in Table 27.1.

4 DISTRIBUTION OF THE GENUS *PIPER*

The largest number of *Piper* species are found in the Americas (about 700 species), with about 300 species from Southern Asia. There are smaller groups of species from the South Pacific (about 40 species) and Africa, about 15 species (Jaramillo and Manos, 2001). The American, Asian, and South Pacific groups each appear to be monophyletic, that is, organisms descended from a single ancestor or from a single taxon; the affinity of the African species is unclear. *Piper* species are classified to be pantropic, that is, occurring or distributed throughout the tropical regions of the earth (Vikash et al., 2012), commonly located in the understory (forest floor) of lowland in the tropical rain forests and sometimes in the neotropic (the major ecozones of the world, covering South America, Central America, and the Caribbean region extending to Japan and Korea where *P. kadsura* and *P. retrofractum* are reportedly found (Chaveerach et al., 2002). However, West African Pepper is a highly valued spice wherever it is found and is difficult to locate in other regions. The origins of the African species of *Piper* genus are not well established however two species are believed to be indigenous to Africa and they are *P. guineense* (Fig. 27.1) and *P. capense* (Immelman, 2000; Jaramillo and Manos, 2001).

Figure 27.1 *Photograph of P.* **guineense** *plant fresh leaves and ripened bunches of fruits.* *(Oyemitan, I.A., Kolawole, F., Oyedeji, A.O., 2014. Acute toxicity, antinociceptive and anti-inflammatory activity of the essential oil of fresh fruits of* Piper *guineense Schum & Thonn (Piperaceae) in rodents. J. Med. Plants Res. 8(40), 1191–1197).*

5 ETHNOBOTANICAL AND ETHNOSOCIAL IMPORTANCE OF THE GENUS *PIPER*

The ethnobotanical uses of the various *Piper* species found in the African continent can be broadly categorized into preservative or insecticidal, food/culinary, and flavoring/spicing agents (Tapandjou et al., 2000; Arong et al., 2011; Nwauzoma and Dawari, 2013; Oyemitan et al., 2014). Many *Piper* species depending on the dominant one in a particular community serve as means of preserving food substances and to mask the taste of food by just mixing them with grounded pepper (Saba and Tomori, 2007). The preservative purposes ascribed to pepper were attested to by recent studies (Oyedeji et al., 2005; Raju and Maridass, 2011; Udensi et al., 2012) in which the fruits or leaves extracts of different species were found to exhibit antimicrobial activities. *Piper umbellatum, P. longum, Piper cubeba, P. nigrum*, etc., are particularly used as condiments and major soup ingredient in many West African countries particularly in Southern parts of Nigeria and also in Cameroon (Iwu, 1993; Arong et al., 2011). In an undocumented report, *P. guineense* dried fruit is sometimes chewed ordinarily by elderly people in some parts of Southwest Nigeria during relaxation period or to keep the mouth refreshed. Furthermore, different pepper species have found usefulness in traditional ceremonial, rituals, and religious practices apparently due to its aromatic, spicy, and sometimes pungency nature. The use of *P. umbellatum* in rituals and medic-magic practices and various other traditional uses have also been documented (Agbor et al., 2005; Domis and Oyen, 2008; Roersch, 2010).

Traditional uses of the *Piper* species as insecticidal agents or seed protectants have been widely documented (Bernard et al., 1995). *P. guineense* and *P. nigrum* have been reportedly used as insecticides and molluskicides in several parts of Africa. Su and Horvat (1981) reported on the potentials of several African species including *P. guineense* as alternatives to conventional pesticides and it has been well articulated and supported in a review paper (Ntonifor, 2011). Kiin-Kabari et al. (2011) also demonstrated the effectiveness of some spices as preservative of smoked-dried fish and the study confirmed *P. guineense* as most potent of the three spices evaluated. The oil distilled from *P. guineense* is used in perfumery and in soap making (Besong et al., 2016). The essential oil of this species has also been used as fish protectant agent (Amusan and Okolie, 2002).

6 ETHNOMEDICINAL APPLICATIONS OF THE GENUS *PIPER*

Ethnomedicinal uses of the genus *Piper* varies from one locality to another and from one geographical area to another. However, the widespread uses of this genus in the management of several ailments have been extensively reported. In West Africa and particularly in Southern States of Nigeria the leaf or fruit of *P. guineense* has been utilized in folkloric medicine to treat respiratory disorders (Ekundayo et al., 1988), convulsion (Abila et al., 1993), infections (Olonisakin et al., 2006), fertility (Mbongue et al., 2005; Ekanem et al., 2010), pain and rheumatism (Etim et al., 2013), diabetes (Nwaichi and Igbinobaro, 2012; Oboh et al., 2013). The leaf of *P. guineense* was also used as an aphrodisiac in Cameroon (Noumi et al., 1998). The leaves of *P. umbellatum* found in the Niger-Delta region of Nigeria has been reportedly used in treating varieties of ailments including malaria, infections, gynecological problems, wound, and inflammatory disorders (Nwauzoma and Dawari, 2013). In Southwest Nigeria the species commonly used is the *P. guineense* and its ethnomedicinal uses include managing pains and related conditions, fever and as an aphrodisiac either singly or in combination with other agents (Oyemitan et al., 2014). *Piper betle* introduced to East Africa is traditionally used for cough, catarrh, and diphtheria; *P. capense* L. f. found in East and South Africa is used as a condiment, stomach, diuretic, vermifuge, heart, and kidney diseases in the Western Cape Province; *P. clusii* found in Angola is also used medicinally for varieties of diseases; *P. nigrum* is used in East Africa to treat parturient mothers against cold and as an abortifacient (Walt and Breyer-Brandwik, 1962).

7 ECONOMIC AND COMMERCIAL PROSPECT OF THE GENUS *PIPER*

Current economic activities on various species of African *Piper* genus are limited to local and national levels. There is no record to show the annual production of any of the common species found in this continent probably due to low attention to its contribution to commerce and trade when compared to cash crops grown in tropical countries that placed a great premium on cocoa, cola-nut, coffee, citrus fruits, and other high-income generating crops in the various regions. Also, large scale cultivation of these African species is absent unlike what obtains in South and Central America or India where there are large scale plantations of *P. nigrum, P. cubeba, P. betle*, etc. (Ghosh et al., 2014). It has been estimated that as at 2013, the largest producer of Black Pepper (*P. nigrum*) was Vietnam, and it accounted for about 34% of the world total of 473,000 tons (FAOSAT, 2013) and the report further indicate that other major leading producers of this pepper were Indonesia (19%), India (11%), and Brazil (9%). The black pepper constitutes an important article of commerce over many centuries and is highly valued cash crop in many tropical countries notably India, Indonesia, Vietnam, Malaysia, and Brazil (Simpson and Ogorzaly, 1995). Contrarily, there is no organized market supply chain in almost all African countries except the various local markets in each region. Prices are not fixed or stable depending mostly on the particular species, locality, seasonal availability, demand, and other market forces. In Southwest Nigeria, the demand for *P. guineense* fruit which is the most popular species found in the country is on the increase in recent times with upsurge in prices. The increase in demand can be traced to the rise in its culinary and ethnomedicinal uses among the rural and urban populace.

8 BIOPROSPECTING AND CONSERVATION STATUS OF THE GENUS *PIPER*

Bioprospecting is defined as a systematic and organized search for useful products derived from bioresources including plants, microorganisms, animals, etc., that can be developed further for commercialization and overall benefits of the society. The bioprospecting and conservation of the African *Piper* genus is almost nonexistent. Available reports from Asian and South American countries indicate institutional and governmental involvement in prospecting and conservation program of various species found in those areas (De Britto and Mahesh, 2007; Landon, 2007). Cultivation and trade in Black Pepper (*P. nigrum*) has developed to international scale in South and Central America, India, Pakistan, Malaysia, and other South Eastern Asian countries. Okafor (1990) has emphasized the imperative of efficient, economical, standardized nursery procedures, and knowledge of reliable practices for accelerated seed improvement for a large-scale production of food trees and shrubs for a conservation program or for commercial scale in Nigeria. In some West African countries, such as Nigeria and Cameroon, *P. guineense* and *P. capense* are still largely obtained from the forests which are being threatened by indiscriminate destruction of the forest reserves due to increase in land usage for diverse purposes. It is therefore suggested that a renewed attention be focused on the cultivation and preservation of the common species found in each country as this will protect these species from going into extinction and will afford the rural populace of economic benefits.

9 PHYTOCHEMICAL CONSTITUENTS OF THE GENUS *PIPER*

Several studies have been carried out to determine the various chemical composition of the *Piper* genus. The studies have covered extraction, isolation, and characterization of numerous compounds from the volatile and nonvolatile components. Various morphological parts of *Piper* species especially the fruits of the black pepper have been reported to contain up to 2.5% essential oils and 9% alkaloids among several other components (Navickene et al. 2000; Srinivasan, 2007). The first compound to be isolated from the *P. nigrum* is piperine (**1**) (Tunmann, 1918) and is responsible for the pungency of the numerous *Piper* species apart from *P. nigrum* and *P. longum* (Vasavirama and Upender, 2014). Parmar et al. (1997) in the review on the *Piper* genus phytochemistry observed that out of a total of 592 different compounds isolated from *Piper* species, 145 are alkaloids or amides, 47 are lignans, 70 are neolignans, and 89 are terpenes. However, the number of these compounds reportedly isolated from the African *Piper* species could be lower than 5% indicating that more phytochemical studies is greatly sought for on the various *Piper* species found in the Africa continent.

9.1 Nonvolatile components of *Piper* genus

Extraction with various solvents has led to isolation of several nonvolatile secondary metabolites including alkaloids, tannins, saponins, carbohydrate, lipids, sterols, flavonoids, phenols, anthraquinones, etc. (Parmar et al., 1997; Echo et al., 2012). These compounds have proven to be responsible for all the activities associated with the genus. It was observed that only about a tenth of all the plant species in the Piperaceae family has been evaluated phytochemically (Ahmad et al., 2012). Unfortunately the African *Piper* genus are the least investigated compared to species found in South and Central America and Asia especially India. The implication of this is that many useful African species of this genus are largely underreported and unknown. Most studies conducted have been on the species found in East, Central, and West Africa and they include *P. guineense*, *P. nigrum*, *P. betle*, and *P. cabense*. The paucity of data on nonvolatile compounds from African *Piper* genus can be traced to poor phytochemical technology and facilities in many African research centers compared to what obtains in other more developed regions. The commonest compounds isolated from several African *Piper* species include compound **1** and its isomers [piperidine (**2**), chavicine (**3**)], guineensimide, guineensine (**5**), pellitorine (**6**), kalecide, piplartine, etc., have been documented (Fig. 27.2). Piperine (**1**) is an alkaloid that was isolated from the fruits of *P. nigrum* by Orsted H.C. in 1819 (Handzlik et al., 2013) and it is found in almost all the *Piper* genus from around the world although to varying degree of concentrations.

9.2 Volatile components of *Piper* genus

Chemical compositions of several *Piper* species found in Africa have been conducted. Although these compounds were not isolated in pure form as will be expected

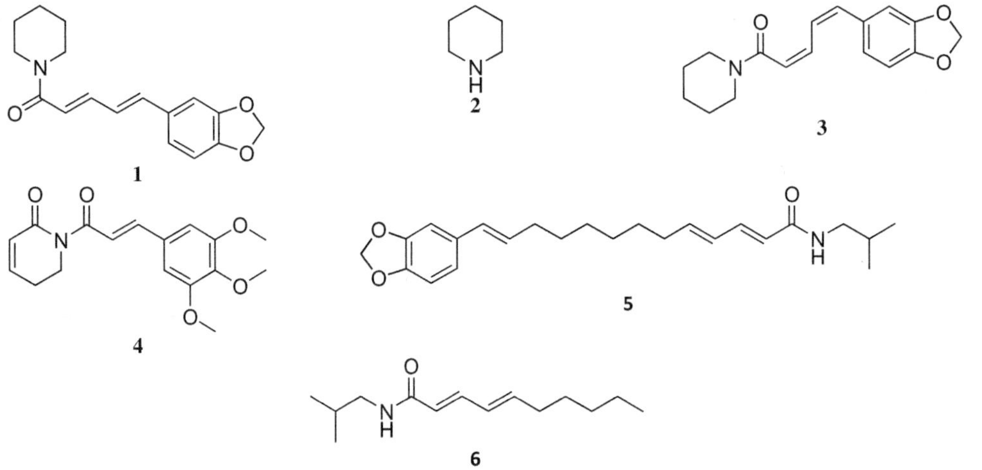

Figure 27.2 *Some nonvolatile compounds isolated from several species of the African* Piper *genus. Piperine (1); piperidine (2); chavicine (3); piplartine (4); guineensine (5); pellitorine (6).*

nevertheless they provide invaluable data on the composition of the volatile constituents of the African species compared to those from other regions of the world. For example, the results of chemical analyses of the various species across Africa provide information on the chemotype of the individual species and give insight into the potential bioactivities of the oils. Generally the commonest compounds found in African *Piper* species include: germacrene D (**13**), limonene, β-pinene (**16**), α-phellandrene, β-caryophyllene (**12**), α-pinene, *cis*-β-ocimene, β-elemene (**9**), bicyclogermacrene, α-humulene, α-humulene, β-pinene, linalool (**8**), β-sesquiphellandrene (**7**), (*Z*)-β-bisabolene (**10**), (*Z*,*Z*)-α-farnesene (**15**), ar-curcumeme, α-zingiberene (**17**), β-bisabolene, δ-cadinene, etc. Numerical analyses of these compounds in different species particularly in essential oil composition of *P. guineense* led to identification of several chemotypes. The chemotypes reported from Nigerian *P. guineense* fruit oils were β-caryophyllene/germacrene D (Onyenekwe et al., 1997), asaricin (**11**) (Ekundayo et al., 1988), α-pinene/β-pinene/germacrene B (Oyedeji et al., 2005), β-pinene/α-pinene/caryophyllene (Olonisakin et al., 2006), β-pinene/α-pinene/1,8-cineole (Oboh et al., 2013), linalool (Owolabi et al., 2013), and compound **7** (Oyemitan et al., 2015) chemotypes. From Cameroon chemotypes reported include α/β-pinene (Amvam et al., 1998), β-caryophyllene/limonene/pinene (Tchoumgougnang et al., 2009), and β-caryophyllene (Jirovetz et al., 2002) chemotypes. Some of the major components are shown in Fig. 27.3. Chemical composition of essential oils of the same species varies from one geographical locations to another besides seasonal variation, morphological, and nature (fresh/dried) of parts used, soil types, cultivation types (wild/cultivar), processing, and analytical methods (Johnson et al., 2004; Moghaddam et al., 2007; Aminzadeh et al., 2010; Chamorro et al., 2012).

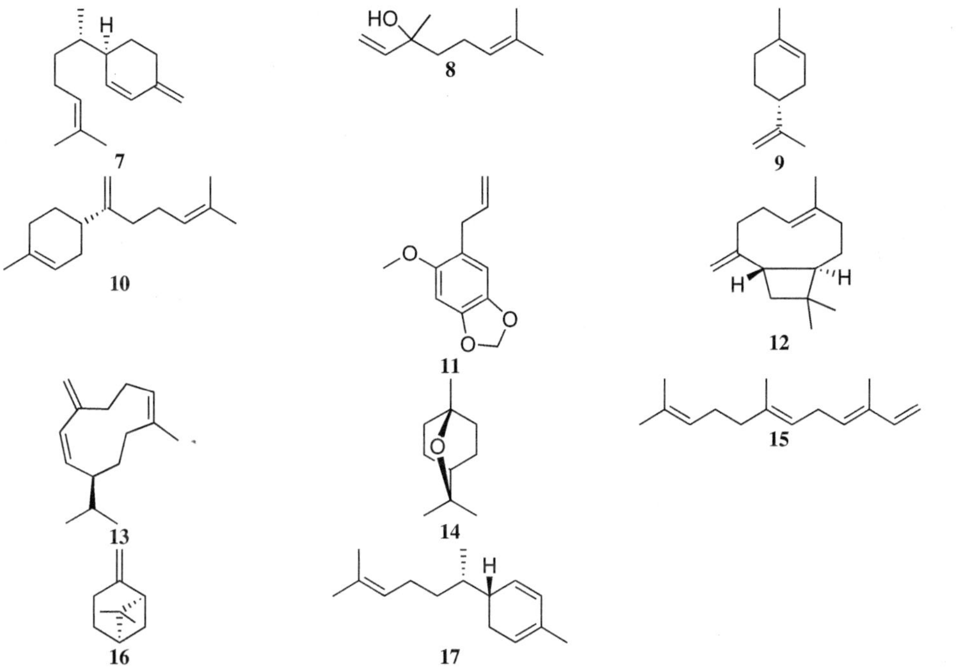

Figure 27.3 *Some major compounds identified in the* P. guineense *essential oil.* β-Sesquiphellandrene (**7**); linalool (**8**); limonene (**9**); (Z)-β-bisabolene (**10**); asaricin or sarisan (**11**); β-caryophyllene (**12**); germacrene D (**13**); eucalyptol or 1,8-cineol (**14**); (z,z)-α-farnesene (**15**); β-pinene (**16**); α-zingiberene (**17**).

10 BIOLOGICAL AND PHARMACOLOGICAL EFFECTS OF THE GENUS *PIPER*

The use of the *Piper* genus is widespread both as food, spice, or medicinal agents (Udo et al., 1999). There is scanty report on its safety profile despite its worldwide applications. However, it has been suggested that consumptions of foods prepared with excessive amounts of black pepper can cause gastrointestinal irritation and bleeding from the ulcer sites (oral communication). It was also advised that individuals with gastrointestinal diseases, such as acid–peptic disease, stomach ulcers, ulcerative colitis, and similar conditions should avoid excessive intake of this pepper. Several studies have been conducted to evaluate extracts from the African *Piper* genus but most of the reports were on the *P. guineense* maybe because it is the most widespread across this continent and particularly in West Africa countries. Some of the biological and pharmacological studies reported are discussed hereunder.

10.1 Antimicrobial activities

The extracts and essential oil of dried fruit of *P. guineense* has been demonstrated against bacterial and fungal organisms. Oyedeji et al. (2005) and Nwinyi et al. (2009) showed that essential oil of *P. guineense* was active against some bacteria, Ngane et al. (2003) and Zollo et al. (1998) reported its antifungal effect, yet its bacterial and fungal activities was demonstrated by Okigbo and Igwe (2007). Antibacterial activity of essential oil of *P. guineense* has been reported to be impressive (Olonisakin et al., 2006). Okunade et al. (1997) reported the antimicrobial activity of *P. aduncum* while Anyanwu and Nwosu (2014) also reported the antimicrobial activity of ethanolic leaf extract of *P. guineense*. These series of reports indicate that various extracts of this *Piper* genus could be exploited for its antiinfective potentials more so that the plant is used in folkloric practice to treat infections. The methanol extract of *P. capense* dried fruit obtained from Cameroon has also shown significant activity against a number of microbes with superior activity to a standard antibiotic chloramphenicol (Fankam et al., 2011). Preservative activity of *P. guineense* was tested on dried fish and the result obtained was impressive (Kiin-Kabari et al., 2011).

10.2 Antioxidant activities

It has been postulated that antioxidants play major role in the protective effect of plant and plant foods (Gey et al., 1991; Willett, 1991). Several epidemiologic reports have shown that there is a strong link between consumption of fruits and vegetables and reduction in morbidity and mortality associated with degenerative diseases. Considering the rate and manner of ingestion of the *Piper* genus either as food, spice, or medicines it is believed that it possess antioxidant potentials. Preliminary report by Ogbonna et al. (2015) indicates that essential oil of *P. guineense* displayed characteristic antiperoxide effect signifying antioxidant activity. Other studies (Etim et al., 2013; Oboh et al., 2013) also confirmed the antioxidant potentials of essential oil of *P. guineense*. These findings strongly suggest a role of antioxidative in the mechanism of action of this plant species in the ethnomedicinal treatment of several diseases, such as hypertension, pains, and diabetics. Various secondary metabolites found in this genus include saponins, flavonoids, tannins, alkaloids, and phenols are known to possess varying degree of antioxidant activity (Pal and Verma, 2013). The volatile components of *Piper* genus include the monoterpenes, sesquiterpenes, oxygenated compounds, phenols, etc. These compounds individually or synergistically displayed substantial antioxidant activities (Oboh et al., 2013).

10.3 Analgesic and antiinflammatory activities

Piper genus constitutes one major class of medicinal plants used to manage pain and inflammatory disorders in folkloric practice. *P. guineense, P. betle*, and *P. capense* are all used to treat pain, respiratory, and cough ailments in most communities they are found. Literature search on studies to validate these activities on the Africa *Piper* genus has proved unsuccessful. There are volumes of studies conducted on *Piper* genus from other regions outside Africa particularly in India as reported in a review (Ahmad et al., 2012). Recent studies conducted on the essential oil of *P. guineense* from Nigeria showed interesting analgesic and antiinflammatory potentials comparable to standard dugs (Oyemitan et al., 2014). This latest results supported the folkloric use of the African *Piper* genus in managing pains and inflammatory disorders.

10.4 Central nervous system activities

Arising from the ethnomedicinal application of the African *Piper* genus is the observation that many indications are centrally-mediated, that is, the central nervous system (CNS) is closely affected by the manifestations of the actions of various extracts of the plant species. Thus the central effects of the essential oil of *P. guineense* have been evaluated in mice. Tankam and Ito (2013) reported the sedative and anxiolytic effect of inhaled essential oil of *P. guineense* from Cameroon in mice while Oyemitan et al. (2015) demonstrated CNS depressant, hypothermic, sedative, muscle relaxant, antipsychotic, and anticonvulsant effects of intraperitoneally injected fresh fruit essential oil of the same species found in Nigeria. The later report comprehensively validated the various ethnomedicinal applications of this important species in West Africa as attested to by its popularity among the populace. Earlier study showed that aqueous extract of *P. guineense* also from Nigeria possessed anticonvulsant activity (Abila et al., 1993). Considering the importance of CNS-active drugs in managing several noninfectious diseases ravaging mankind, it is imperative that active components of this *Piper* genus be isolated and developed into useful medicines in the management of existing and upcoming diseases. Already **1** (an alkaloid) and related compounds have been isolated from several species of this *Piper* genus from several regions of the world (Parmar et al., 1997; Ghosh et al., 2014) and several studies have confirmed its various biological activities (Mujumdar et al., 1990; Reen and Singh, 1991; Vijayakumar and Nalini, 2006; Taqvi et al., 2008; Kim and Lee, 2009).

10.5 Other biological activities

Gastrointestinal (GIT) contractile effect of ethanolic extract of *P. guineense* fruit was evaluated in an isolated Guinea pig ileum and the results obtained showed that this species significantly increased GIT contraction which was blocked by cimetidine (Histamine-2 antagonist) signifying that the mediation of its GIT effect could be through H_2 receptor (Saba and Tomori, 2007). Crude extracts of *P. guineense* has been screened for antitrypanosomal activity and the results indicate that both the leaf and stem extracts of the plant exhibited significant activity against pleomorphic *Trypanosoma brucei brucei* (Abedo et al., 2013). Joshi et al. (1990) also reported the antiamoebic activity of hexane fraction of ethanolic extract of *Piper schimidtii*.

So far the highest number of studies conducted on this genus has been the insecticidal activities of the *Piper* genus on grains and other related stored food products. The first series of tests on the *Piper* genus was the insecticidal potentials of various extracts of the plant. The high content of the amide alkaloids in *P. guineense* fruit and stem has been proposed to be relevant in the protection and preservation of some selected farm products against insect pests (Adesina et al., 2003). Asawalam et al. (2007) also confirmed the efficacy of

P. guineense (Schum Thonn) seed extract against a maize weevil, *Sitophilus zeamaisi* while Lale and Alaga (2001) reported the insecticidal and larvicidal activity of essential oil of *P. guineense* against *Tribolium castaneum* (Herbst) in stored pearl millet *Pennisetum glaucum*. One of the ethnobotanical uses of the genus is the protection against insect and the application of the essential oil of *P. guineense* was shown to protect against mosquito' bites for at least 2 h (Adewoyin et al., 2006). Effect of these same species was similarly tested against maize pest *S. zeamaise* with promising results (Madubuike et al., 1990). Oben et al. (2015) also showed the potent effect of essential oil of *P. guineense* against *Sitophilus oryzae* (L.) from Cameroon. Several other studies have demonstrated the effect of various extracts of *P. guineense* against many insect crops' pests of (Ajayi and Wintola, 2006; Ukeh et al., 2008; Ntonifor et al., 2010). The significance of these reported insecticidal potentials of this *Piper* genus is in the application of natural products in preserving harvested food products with reduced adverse effects associated with synthetic chemicals (Ntonifor, 2011). *P. capense* from Cameroon has also been shown to possess insecticidal activity (Tchoumgougnang et al., 2009). Study has been conducted to confirm the insecticidal activity of the *Piper* genus from America similarly to that of Africa (Bernard et al., 1995) signifying some degree in resemblances of their chemical constituents.

11 TOXICITY PROFILE

Literature search conducted on the acute toxicity profile of the various extracts of this *Piper* genus provided little results. However the acute toxicity (LD_{50}) profiles of the essential oil obtained from the fresh fruit of *P. guineense* from Southwest Nigeria showed 693 and 1265 mg/kg for the intraperitoneal and oral routes respectively (Oyemitan et al., 2014). The implication of the results is that the essential oil of this species is moderately toxic and therefore caution is advised for its use in whatever form. It should however be noted that it is not the isolated essential oil that is being used rather it is the whole material in which the oil concentration may be less than 2%. There has been no preclinical report on the long-term (subacute, chronic, or subchronic) effect of this genus to enable categorical statement on its safety profile. Hence it is imperative that a well-designed and executed subchronic or chronic study be carried out to investigate the effect of various extracts of this genus as well articulated by Adeneye (2014) and on long term usage using at least two-animal experimental models (preferably mouse and rat).

12 CONCLUSIONS

The African medicinal *Piper* species are veritable sources of valuable spices, as well as natural sources of medicinal agents useful in managing several diseases. Research studies on this genus are at present low compared to available report on *Piper* genus from Central and South America, and Southeast Asia. The West African pepper (*P. guineense*) compared favorably with the Black pepper (*P. nigrum*) in terms of chemical composition and biological activities. There is relatively low activity in the cultivation, preservation, research, and economic activities on the African species of *Piper*. Considering the great potentials inherent in this genus effort should be made at the national, regional and international levels to coordinate a renewed effort in supporting the cultivation of the various species and protect or preserve those at the verge of extinction wherever they are found in Africa. Research institutes need funding and support to further characterize the phytochemical constituents of this genus in order to provide additional data to enhance its various applications.

REFERENCES

Abedo, J.A., Jonah, A.O., Mazadu, M.R., Abdullahi, R.S., Idris, H.Y., Shettima, F.T., Mohammed, H., Ombugadu, S., Garba, J., Daudu, M., Kugu, B.A., Abdulmalik, U., 2013. In vitro, in vivo phytochemical screening of extracts of *Piper guineense* for trypanocidal activities against *Trypanosoma brucei brucei*. Int. J. Biol. 5 (3), 120–124.

Abila, B., Richens, A., Davies, J.A., 1993. Anticonvulsant effects of extracts of the West African black pepper, *Piper guineense*. J. Ethnopharmacol. 39, 113–117.

Adeneye, A.A., 2014. Subchronic and chronic toxicities of African medicinal plants. In: Kuete, V. (Ed.), Toxicological Survey of African Medicinal Plants. Elsevier, USA, pp. 99–134.

Adesina, S.K., Adebayo, A.S., Adesina, S.K.O., Groning, R., 2003. New constituents of *Piper guineense* fruit and leaf. Pharmazie 58, 423–425.

Adewoyin, F.B., Odaibo, A.B., Adewunmi, C.O., 2006. Mosquito repellant activity of *Piper guineense* and *Xylopia aethiopica* fruits oils on *Aedes aegypti*. AJTCAM 3 (2), 79–83.

Agbakwuru, E.O.P., Osisiogu, I.U., Rucker, G., 1979. Constituents of essential oil of *Dennettia tripetala* G. Baker (Annonaceae). Niger. J. Pharm. Res. 10, 203–208.

Agbor, G.A., Oben, J.E., Ngogang, J.Y., Xinxing, C., Vinson, J.A., 2005. Antioxidant capacity of some herbs/spices from Cameroon: a comparative study of two methods. J. Agric. Food Chem. 53, 6819–6824.

Aggarwal, B.B., Sung, B., 2008. Pharmacological basis for the role of curcumin in chronic diseases: an age-old spice with modern targets. Trends Pharmcol. Sci. 30 (2), 85–94.

Ahmad, N., Fazal, H., Abbasi, B.H., Farooq, S., Ali, M., Khan, M.A., 2012. Biological role of *Piper nigrum* L. (Black pepper): a review. Asian Pac. J. Trop. Biomed. 2 (3), S1945–S1953.

Ajayi, F.A., Wintola, H.U., 2006. Suppression of the Cowpea *Bruchid callosobruhus maculatus* (F.) infesting stored Cowpea (*Vigna unguiculata* (L.) Walp.) seeds with some edible plant product powders. Pak. J. Biol. Sci. 9 (8), 1454–1459.

Al-Naggar, T.B., Gomez-Serranillos, M.P., Carretero, M.E., Villar, A.M., 2003. Neuropharmacological activity of *Nigella sativa* L. extracts. J. Neuropharmacol. 88, 63–68.

Amabeoku, G.J., Kinyua, C.G., 2010. Evaluation of the anticonvulsant activity of *Zanthoxylum capense* (Thunb.) Harv. (Rutaceae) in Mice. Int. J. Pharmcol. 6, 844–853.

Aminzadeh, M., Amiri, F., Abadi, A.E., Mahdevi, K., Fadai, Sh., 2010. Factors affecting on essential chemical composition of *Thymus kotschyanus* in Iran. World Appl. Sci. J. 8 (7), 847–856.

Amusan, A.A.S., Okorie, T.G., 2002. The use of *Piper guineense* fruit oil (PFO) as protectant of dried fish against *Dermestes maculates* (Degeer) infestation. G.J.P. Appl. Sci. 8, 197–201.

Amvam, Z.P.H., Biyiti, L., Tchoumbougnang, F., Menut, C., Lamaty, G., Bouchet, Ph., 1998. Aromatic plants of tropical central Africa. Part XXXII. Chemical composition and antifungal activity of thirteen essential oils from aromatic plants of Cameroon. Flavour Frag. J. 13, 107–114.

Anyanwu, C.U., Nwosu, G.C., 2014. Assessment of the antimicrobial activity of aqueous and ethanolic extracts of *Piper guineense* leaves. J. Med. Plant Res. 8 (10), 337–439.

Arong, G.A., Oku, E.E., Obhiokhenan, A.A., Adetunji, B.A., Mowang, D.A., 2011. Protectantability of *Xylopia aethiopica* and *Piper guineense* leaves against the cowpea Bruchid *Callosobruchus maculatus* (Fab.) (Coleoptera: Bruchidae). World J. Sci. Technol. 1 (7), 14–19.

Asawalam, E.F., Emosairue, S.O., Ekeleme, F., Wokocha, R., 2007. Efficacy of *Piper guineense* (Schum Thonn) seed extract against maize weevil, *Sitophilus zeamais* (Motschulsky) as influenced by different extraction solvents. Int. J. Pest Manag. 53 (1), 1–6.

Bernard, C.B., Krishnamurty, H.G., Chaurev, D.T., Durst, Philogène, B.J.R., Vindas, P.S., Hasbun, C., Poveda, L., San Román, L., Arnason, J.T., 1995. Insecticidal defenses of Piperaceae from the Neotropics. J. Chem. Ecol. 21 (6), 801–814.

Besong, E.E., Balogun, M.E., Djobissie, S.F.A., Mbamalu, O.S., Obimma, J.N., 2016. A Review of *Piper guineense* (African Black Pepper). IJPPR 6 (1), 368–384.

Chamorro, E.R., Zambón, S.N., Morales, W.G., Sequeira, A.F., Velasco, G.A., 2012. Study of the chemical composition of essential oils by gas chromatography, ; In: Dr., Bekir Salih, (Ed.), Gas Chromatography in Plant Science, Wine Technology, Toxicology and Some Specific Applications. InTech. Available from: http://www.intechopen.com/books/gas-chromatography-in-plant-science-winetechnology-toxicology-and-some-specific-applications/study-of-the-chemical-composition-of-essential-oils-by-gas-chromatography

Chaveerach, R., Kunitakeb, H., Nuchadomrongc, S., Sattayasaic, N., Komatsud, H., 2002. RAPD patterns as a useful tool to differentiate Thai *Piper* from morphologically alike Japanese *Piper*. Sci. Asia 28, 221–225.

De Britto, J., Mahesh, R., 2007. Exploration of Kani tribal botanical knowledge in agasthiayamalai biosphere reserve-South India. Ethnobot. Leaflets 11, 258–265.

Echo, I.A., Osuagwu, A.N., Agbor, R.B., Okpako, E.C., EkanemF B.E., 2012. Phytochemical composition of *Aframomum melegueta* and *Piper guineense* seeds. World J. Appl. Environ. Chem. 2 (1), 17–21.

Ekanem, A.P., Udoh, F.V., Oku, E.E., 2010. Effects of ethanol extracts of *Piper guineense* seeds (Schum and Thonn) on the conception of mice (*Musmusculus*). Afr. J. Pharm. Pharmacol. 4, 362–367.

Ekundayo, O., Laasko, I., Adegbola, R.M., Oguntimein, B., Sofowora, A., Hiltunen, R., 1988. Essential oil constituents of Ashanti Pepper (*Piper guineense*) fruits (Berries). J. Agric. Food Chem. 36, 880–882.

Etim, O.E., Egbuna, C.F., Odo, C.E., Udo, N.M., Awah, F.M., 2013. In-vitro antioxidant and nitric oxide scavenging activities of *Piper guineense* seeds. Glob. J. Res. Med. Plants Indigen. Med. 2 (7), 485–494.

Fankam, A.G., Kuete, V., Voukeng, I.K., Kuiate, J.R., Pages, J.M., 2011. Antibacterial activities of selected Cameroonian spices and their synergistic effects with antibiotics against multidrug-resistant phenotypes. BMC Complement. Altern. Med. 11, 104.

FAOSAT, 2013. Pepper (*Piper* spp.), Production/Crops. Food and Agriculture Organization of the United Nations: Statistical Division (FAOSTAT). Available from: http://faostat3.fao.org/browse/Q/QC/E

Frodin, D.G., 2004. History and concepts of big plant genera. Taxon 53, 753–776.

Gey, K.F., Puska, P., Jordan, P., Moser, U.K., 1991. Total antioxidant capacity of plant foods. Inverse correlation between plasma vitamin E and mortality from ischemic heart disease in cross-cultural epidemiology. Am. J. Clin. Nutr. 53, 326S–334S.

Ghosh, R., Darin, K., Nath, P., Deb, P., 2014. An overview of various *Piper* species for their biological activities. Int. J. Pharma Res. Rev. 3 (1), 67–75.

Greig, N., 2004. Introduction. In: Dyer, L.A., Palmer, A.N.D. (Eds.), Kluwer Piper: A Model Genus for Studies of Phytochemistry, Ecology and Evolution. Academic/Plenum Publishers, New York, pp. 3–4.

Handzlik, J., Matys, A., Kiec´-Kononowicz, K., 2013. Recent advances in multi-drug resistance (MDR) efflux pump inhibitors of Gram-positive bacteria *S. aureus*. Antibiotics 2 (1), 28–45.

Hellion-Ibarrola, M.C., Ibarrola, D.A., Montalbetti, Y., Heinichen, O., Ferro, E.A., 1999. Acute toxicity and general pharmacological effect on central nervous system of the crude rhizome extract of *Kyllinga brevifolia* Rottb. J. Ethnopharmacol. 66, 271–276.

Immelman, K.L., 2000. FSA contributions 15: Piperaceae. Bothalia 30 (1), 25–30, Indig. Med. 2(7), 485–494.

Isawumi, M.A., 1984. The peppery fruits of Nigeria. Niger. Field 49, 37–44.

Iwu, M.M., 1993. Handbook of African Medicinal Plants. CRC Press Taylon & Francis Group. Washington DC first ed. pp. 221–222.

Jaramillo, M.A., Manos, P.S., 2001. Phylogeny and patterns of floral diversity of the genus *Piper* (Piperaceae). Am. J. Bot. 88 (4), 706–716.

Jaramillo, M.A., Callejas, R., Davidson, C., Smith, J.F., Stevens, A.C., Tepe, E.J., 2008. A phylogeny of the tropical genus *Piper* using its and the chloroplast intron *psbJ–petA*. Syst. Bot. 33 (4), 647–660.

Jirovetz, L., Buchbauer, G., Ngassoum, M.B., Geissler, M., 2002. Aroma compound analysis of *Piper nigrum* and *Piper guineense* essential oils from Cameroon using solid-phase microextraction—gas chromatography, solid-phase microextraction—gas-chromatography mass spectrometry and olfactometry. J. Chromatogr. 976, 265–275.

Johnson, C.B., Kazantzis, A., Skoula, M., Mitteregger, U., Novak, J., 2004. Seasonal, populational and ontogenic variation in the volatile oil content and composition of individuals of *Origanum vulgare* subsp. hirtum, assessed by GC headspace analysis and by SPME sampling of individual oil glands. Phytochem. Anal. 15, 286–292.

Joshi, N., Garg, H.S., Bhakuni, D.S., 1990. Chemical constituents of *Piper schimidtii*: structure of a new Neolignan Schimiditin. J. Nat. Prod. 53 (2), 479–482.

Kiin-Kabari, D.B., Barimalaa, I.S., Achinewhu, S.C., Adeniji, T.A., 2011. Effects of extracts from three indigenous spices on the chemical stability of smoke-dried catfish (*Clariaslezera*) during storage. Afr. J. Food Agric. Nutr. Dev. 11 (6), 5335–5343.

Kim, S.H., Lee, Y.C., 2009. Piperine inhibits eosinophil infiltration and airway hyper-responsiveness by suppressing T cell activity and Th2 cytokine production in the ovalbumin-induced asthma model. J. Pharm. Pharmacol. 61, 353–359.

Lale, N.E.S., Alaga, K.A., 2001. Exploring the insecticidal, larvicidal and repellent properties of *Piper guineense* (Schum and Thonn) seed oil for the control of rust-red flour beetle *Triboliumcastaneum* (Herbst) in stored pearl millet *Pennisetumglaucum* (L.) R. Br. J. Plant Dis. Prot. 108 (3), 305–313.

Landon, A.J., 2007. Bioprospecting and Biopiracy in Latin America: The Case of MacainPerú. DigitalCommons at @University of Nebraska—Lincoln. pp. 63–73. Available from: http://digitalcommons.unl.edu/nebanthro/32

Lebot, V., Merlin, M., Lindstrom, L., 1997. Kava: The Pacific Elixir: The definitive Guide to Its Ethnobotany, History & Chemistry. New Haven, Yale University Press, Canada.

Lock, J.M., Hall, J.B., Abbiw, D.K., 1977. The cultivation of *Melegueta Pepper* (*Aframomum melegueta*) in Ghana. Econ. Bot. 31 (3), 321–330.

Madubuike, O., Nwaigbo, L.C., Orji, P.J., 1990. Protection of stored maize against *S. zeamaise* (mots) with non-toxic natural products potentials of *X. aethiopica* and *P. guineense*. Acta. Agron. Hung. 41, 131–139.

Manoj, P., Soniya, E.V., Banerjee, N.S., Ravinchandra, P., 2004. Recent studies on well-known spice, *Piper longum* Linn. Nat. Prod. Radiance 3 (4), 222–227.

Mbongue, F.G.Y., Kamtchouing, P., Essame, O.J.L., Yewah, P.M., Dimo, T., Lontsi, D., 2005. Effect of the aqueous extract of dry fruits of *Piper guineense* on the reproductive function of adult male rats. Indian J. Pharmacol. 7, 30–32.

Moghaddam, M., Omidbiagi, R., Sefidkon, F., 2007. Changes in content and chemical composition of tagetesminuta oil at various harvest times. J. Essent. Oil Res. 19, 18–20.

Mujumdar, A.M., Dhuley, J.N., Deshmukh, V.K., Raman, P.H., Thorat, S.L., Naik, S.R., 1990. Effect of piperine on pentobarbitore-induced hypnosis in rats. Indian J. Exp. Biol. 28, 486–487.

Navickene, H.M.D., Alecio, A.C., Kato, M., Bolzani, V.S., Young, M.C.M., Cavalheiro, A.J., Furlan, M., 2000. Antifungal amides from *Piper hispidum* and *Piper tuberculatum*. Phytochemistry 55, 621–626.

Ngane, A.N., Biyiti, L., Bouchet, P.H., Nkegfact, A., Zolo, P.H.A., 2003. Antifungal activity of *Piper guineense* of Cameroun. Fitoterapy 4, 464–468.

Noumi, E., Amvam, Z.P.H., Lontsi, D., 1998. Aphrodisiac plants used in Cameroon. Fitoter 69, 5–34.

Ntonifor, N.N., 2011. Potentials of tropical African spices as sources of reduced-risk pesticides. J. Entomol. 8, 16–26.

Ntonifor, N.N., Mueller-Harvey, I., Brown, R.H., 2010. Extracts of tropical African spices are active against *Plutella xylostella*. J. Food Agric. Environ. 8 (2), 498–502.

Nwaichi, E.O., Igbinobaro, O., 2012. Effects of some selected spices on some biochemical profile of Wister albino rats. Am. J. Environ. Eng. 2 (1), 8–11.

Nwauzoma, A.B., Dawari, S.L., 2013. Study on the phytochemical properties and proximate analysis of *Piper umbellatum* (L.) from Nigeria. American Journal of Research Communication 1 (7), 164–177, www.usa-journals.com.

Nwinyi, O.C., Chinedu, N.S., Ajani, O.O., Ikpo, C.O., Ogunirin, K.O., 2009. Antibacterial effects of extracts of *Ocimum gratissimum* and *Piper guineense* on *Escherichia coli* and *Staphylococcus aureus*. Afr. J. Food Sci. 3, 77–81.

Oben, E.O., McConchie, R., Phan-Thien, K., Ntonifor, N.N., 2015. Essential oil composition of different fractions of *Piper guineense* Schumach. Et Thonn. from Cameroon using gas chromatography-mass spectrometry and their insecticidal effect on *Sitophilus oryzae* (L.). Afr. J. Biotechnol. 14 (36), 2662–2671.

Oboh, G., Ademosun, A.O., Odubanjo, O.V., Akinbola, I.A., 2013. Antioxidative properties and inhibition of key enzymes relevant totype-2 diabetes and hypertension by essential oils from black pepper. Adv. Pharmacol. Sci. 2013, 6 pp.

Ogbonna, A.C., Abuajah, C.I., Hart, E.B., 2015. Preliminary evaluation of physical and chemical properties of *Piper guineense* and *Xylopia aethiopica* seed oils. Int. Food Res. J. 22 (4), 1404–1409.

Okafor, J.C., 1990. Development and selection of commercially viable cultivars from forest species for fruit. Proc. of the 12th Plenary Meeting of AETFAT, Hamburg.

Okigbo, R.N., Igwe, I.D., 2007. The antimicrobial effects of *Piper guineense* uziza and *Phyllantus amarus* ebebenizo on *Candida albicans* and *Streptococcus faecalis*. Acta Microbiologica et Immunologica Hungarica 54, 353–366.

Okunade, A.L., Hufford, C.D., Clark, A.M., Lentz, D., 1997. Antimicrobial properties of the constituents of *Piper aduncum*. Phytother. Res. 11, 142–144.

Olonisakin, A., Oladimeji, M.O., Lajide, L., 2006. Chemical composition and anti-bacterial Activity of steam distilled oil of Ashanti pepper (*Piper guineense*) fruits (Berries). Electron. J. Environ. Agric. Food Chem. 5 (5), 1531–1535.

Onyenekwe, P.C., Ogbadu, G.H., Hashimoto, S., 1997. The effect of gamma radiation on the microflora and essential oil of Ashanti pepper (*Piper guineense*) berries. Postharvest Biol. Technol. 10, 161–167.

Owolabi, M.S., Lawal, O.A., Ogunwande, I.A., Hauser, R.M., Stzer, W.N., 2013. Aroma chemical composition of *Piper guineense* Schumach. & Thonn. From Lagos, Nigeria: a new chemotype. Am. J. Essent. Oils Nat. Prod. 1 (1), 37–40.

Oyedeji, O.A., Adeniyi, B.A., Ajayi, O., König, W.A., 2005. Essential oil composition of *Piper guineense* and its antimicrobial activity. Another chemotype from Nigeria. Phytother. Res. 19, 362–364.

Oyemitan, I.A., Iwalewa, E.O., Akanmu, M.A., Asa, S.O., Olugbade, T.A., 2006. The abusive potential of habitual consumption of the fruits of *Dennettia tripetala* G. Baker (Annonaceae) among the people in Ondo Township (Nigeria). Nigerian J. Nat. Prod. Med. 10, 55–62.

Oyemitan, I.A., Kolawole, F., Oyedeji, A.O., 2014. Acute toxicity, antinociceptive and anti-inflammatory activity of the essential oil of fresh fruits of *Piper guineense* Schum & Thonn (Piperaceae) in rodents. J. Med. Plants Res. 8 (40), 1191–1197.

Oyemitan, I.A., Olayera, O.A., Alabi, A., Abass, L.A., Elusiyan, C.A., Oyedeji, A.O., Akanmu, M.A., 2015. Psychoneuropharmacological activities and chemical composition of essential oil of fresh fruits of *Piper guineense* (Piperaceae) in mice. J. Ethnopharmacol. 166, 240–249.

Pal, D., Verma, P., 2013. Flavonoids: a powerful and abundant source of antioxidants. Int. J. Pharm. Pharm. Sci. 5 (3), 95–98.

Parmar, V.S., Jain, S.C., Bisht, K.S., Jain, R., Taneja, P., Jha, A., Tyagi, O.D., Prasad, A.K., Wengel, J., Olsen, C.E., Boll, P.M., 1997. Phytochemistry of the genus *Piper*. Phytochemistry 46 (4), 591–673.

Parthasarathy, U., Saji, K.V., Jayarajan, K., Parthasarathy, V.A., 2006. Biodiversity of *Piper* in South India—application of GIS and cluster analysis. Curr. Sci. 91 (5), 652–658.

Raju, G., Maridass, M., 2011. Evaluation of antimicrobial potential of *Piper* (L.) species. Nat. Pharm. Technol. 1 (1), 19–22.

Reen, R.K., Singh, J., 1991. In vitro and in vivo inhibition of pulmonary cytochrome P450 activities by piperine, a major ingredient of *piper* species. Indian J. Exp. Biol. 29, 568–573.

Roersch, C.M.F.B., 2010. *Piper umbellatum* L.: a comparative cross-cultural analysis of its medicinal uses and an ethnopharmacological evaluation. J. Ethnopharmacol. 131, 522–537.

Saba, A.B., Tomori, O.A., 2007. The contractile effect of ethanolic extract of West African Black Pepper (*Piper guineense*) on isolated guinea pig ileum. Pak. J. Nutr. 6 (4), 366–369.

Shulgin, A.T., 1973. The narcotic pepper—The chemistry and pharmacology of *Piper methysticum* and related species. UNODC. Plate 15, 59–74. Available from: www.un.org

Simpson, B.B., Ogorzaly, M.O., 1995. Economic Botany: plants in our world. In: Simpson B.B., Ogorzaly M.O. (Eds.), second ed. McGraw-Hill Inc., New York, p. 742.

Smith, J.F., Stevens, A.C., Tepe, E.J., Davidson, C., 2008. Placing the origin of two species-rich genera in the late cretaceous with later species divergence in the tertiary: a phylogenetic, biogeographic and molecular dating analysis of *Piper* and Peperomia (Piperaceae). Plant Syst. Evol. 275, 9–30.

Srinivasan, K., 2007. Black pepper and its pungent principle piperine: A review of diverse physiological effects. Crit. Rev. Food Sci. Nutr. 47, 735–748.

Srivastava, S., Gupta, M.M., Tripathi, K.A., Kumar, S., 2000. 1,3-Benzodiaoxole-5-(2,4,8-triene-methyl nanoate) & 1,3-benzodiaoxole-5-(2,4,8-triene-isobutyl nonoate) from *Piper mullesua*. Indian J. Chem. 39B, 946–949.

Su, H.C.F., Horvat, H., 1981. Isolation, identification and insecticidal properties of *P. nigrum*. J. Agric. Food Chem. 29, 115–118.

Tankam, J.M., Ito, M., 2013. Inhalation of the essential oil of *Piper guineense* from Cameroon shows sedative and anxiolytic-like effects in mice. Biol. Pharm. Bull. 36 (10), 1608–1614.

Tapandjou, A.L., Bouda, H., Fontem, D.A., Zapfack, L., Lontsi, D., Sondengam, B.L., 2000. Local plants used for traditional stored product protection in the Menoua division of the Western highlands of Cameroun. IOBC Bull. 23, 73–77.

Taqvi, S.I., Shah, A.J., Gilani, A.H., 2008. Blood pressure lowering effects of piperine. J. Cardiovasc. Pharmacol. 52, 452–458.

Tchoumgougnang, F., Jazet-Dongmo, P.M., Sameza, M.L., Fombotioh, N., Wouatsa-Nangue, A.V., Amvam, Z.P.H., Menut, C., 2009. Comparative essential oils composition and insecticidal effect of different tissues of *Piper capense* L., *Piper guineense* Schum. Et Thonn., *Piper nigrum* L. and *Piper umbellatum* L. grown in Cameroon. Afr. J. Biotechnol. 8 (3), 424–431.

Tunmann, O., 1918. Apotheker Zeitung 33, 353 (Chemical Abstracts, 19, 2940).

Udensi, E.A., Odom, T.C., Dike, C.O., 2012. Comparative studies of ginger (*Zingiber officinale*) and West African Black Pepper (*Piper guineense*) extracts at different concentrations on the microbial quality of soymilk and kunun-zaki. Niger. Food J. 30 (2), 38–43.

Udo, F.V.I., Lot, T.Y., Braide, V.B., 1999. Effects of extracts of seed and leaf of *Piper guineense* on skeletal muscle activity in rat and frog. Phytother. Res. 13 (2), 106–110.

Ukeh, D.A., Arong, G.A., Ogban, E.I., 2008. Toxicity and oviposition deterrence of *Piper guineense* (Piperaceae) and *Monodora myristica* (Annonaceae) against *Sitophilus zeamais* (Motsch) on stored maize. J. Entomol. 5 (4), 295–299.

Vasavirama, K., Upender, M., 2014. International Journal of Pharmacy and *Pharmaceutical Sciences*, 6(4). Available from: http://www.ijppsjournal.com/Vol6Issue4/9101

Vijayakumar, R.S., Nalini, N., 2006. Piperine, an active principle from *Piper nigrum*, modulates hormonal and apo lipoprotein profiles in hyperlipidemic rats. J. Basic Clin. Physiol. Pharmacol. 17, 71–86.

Vikash, C., Tripathi, S., Verma, N.K., Singh, D.P., Chaudhary, S.K., Roshan, A., 2012. *Piper betle*: phytochemistry, traditional use and pharmacological activity—a review. Int. J. Pharm. Res. Dev. 4 (4), 216–223.

Walt, J.M., Breyer-Brandwik, M.G., 1962. The Medicinal & Poisonous Plants of Southern & Eastern Africa. Livingstone, Edingurg & London, pp. 846–847.

Willett, W.C., 1991. Micronutrients and cancer risk. Am. J. Nutr. Clin. 53, 265S–269S.

Zollo, P.H.A., Biyiti, L., Tchoumbougnang, F., Menut, C., Lamaty, G., Bouchet, Ph., 1998. Aromatic plants of tropical central Africa. Part XXXII. Chemical composition and antifungal activity of thirteen essential oils from aromatic plants of Cameroon. Flavour Frag. J. 13, 107–114.

CHAPTER 28

Thymus vulgaris

V. Kuete

1 INTRODUCTION

Thymus vulgaris is a flowering plant of the family Lamiaceae commonly known as thyme, native to Southern Europe, and has a worldwide distribution (Hosseinzadeh et al., 2015). The plant is indigenous to the Mediterranean and neighboring countries, Northern Africa, and parts of Asia. In Africa, the plant has been cultivated in Egypt, Morocco, Algeria, Tunisia, Libya (Stahl-Biskup and Sáez, 2002), Cameroon (Nkouaya Mbanjo et al., 2007), Nigeria (Kayode and Ogunleye, 2008), and South Africa (Schmitz, 2015). People have used thyme for many centuries as a flavoring agent, culinary herb, and herbal medicine (Stahl-Biskup and Venskutonis, 2012). The plant is useful as infusion to treat cough, diabetes, and cold and chest infections; and in a syrup form for digestive upset. It is also soothing for sore throat, as thyme is has antiseptic, antibiotic, and antifungal properties (Ekoh et al., 2014). *T. vulgaris* has been thought of to be astringent, anthelmintic, carminative, disinfectant, and tonic (Reddy et al., 2014). The plant has been reported incredibly useful in cases of assorted intestinal infections and infestations, such as hookworms, ascarids, Gram-positive and Gram-negative bacteria, and fungi and yeasts, as well as *Candida albicans* (Reddy et al., 2014). Its active ingredient, thymol, was reported active against enterobacteria and cocci bacteria (Reddy et al., 2014). Other properties assigned to thyme, include liver function improvement; appetite stimulant effect; treatment of cartilaginous tube, bronchial, and urinary infections; and treatment of laryngitis and inflammation (Reddy et al., 2014). Applied to the skin, thyme is reported to relieve bites and stings, neuralgy, rheumatic aches, and pains (Reddy et al., 2014). The essential oil can be used as a rub for aching joints or rheumatic pain, and can also be used in the treatment of athlete's foot (*Tinea pedis*) (Ekoh et al., 2014). In the present chapter, the synopsis of the phytochemistry and pharmacological aspects, as well as the cultivation and distribution of *T. vulgaris* (Fig. 28.1) will be provided.

2 CULTIVATION AND DISTRIBUTION OF *THYMUS VULGARIS*

T. vulgaris is an aromatic plant used for culinary and medicinal purposes almost everywhere in the world. The plant is flowering and grows up to 15–30 cm high. Thyme is a tiny perennial shrub, with stems becoming woody with age (Hosseinzadeh et al., 2015). The flowers are light violet, two-lipped, 5-mm long with a hairy glandular calyx, and

Figure 28.1 *T. vulgaris* fresh (A) and dried (B) leafy branches.

borne with leaf-like bracts in loose whorls in axillary clusters on the branchlets or in terminal oval or rounded heads (Stahl-Biskup and Venskutonis, 2012). *T. vulgaris* is best cultivated in a hot, sunny location with well-drained soil, and is generally planted in the spring. The plant can be propagated by seed, cuttings, or by dividing rooted sections. It can take deep freezes and is found growing wild on mountain highlands. Along the Riviera, it is found from sea level up to 800 m. Thyme is cultivated in most of the European countries, such as France, Spain, Italy, Bulgaria, but is also well cultivated in several African countries (Hosseinzadeh et al., 2015). *T. vulgaris* leaves are oval to rectangular in form, and somewhat fleshy aerial components are used for volatile oil production, principally by steam distillation (Hosseinzadeh et al., 2015). Thyme is grown for commercial purposes in many countries for the production of the dried leaves, thyme oil, thyme extracts, and oleoresins (Stahl-Biskup and Venskutonis, 2012).

3 CHEMISTRY OF *THYMUS VULGARIS*

The phytochemical constituents of thyme include phenolics, terpenoids, and mostly thymol, eugenol, and saponins (Ekoh et al., 2014). Thyme essential oil showed a high content of oxygenated monoterpenes (56:53%) and low contents of monoterpene hydrocarbons (28.69%), sesquiterpene hydrocarbons (5.04%), and oxygenated sesquiterpenes (1.84%) (Reddy et al., 2014).

3.1 Essential oil of *Thymus vulgaris*

The essential oil is responsible for the typical spicy aroma of thyme (Stahl-Biskup and Venskutonis, 2012). The predominant compound among the essential oil components was identified as thymol (compound **9**; 51.34%) while the amount of all other components

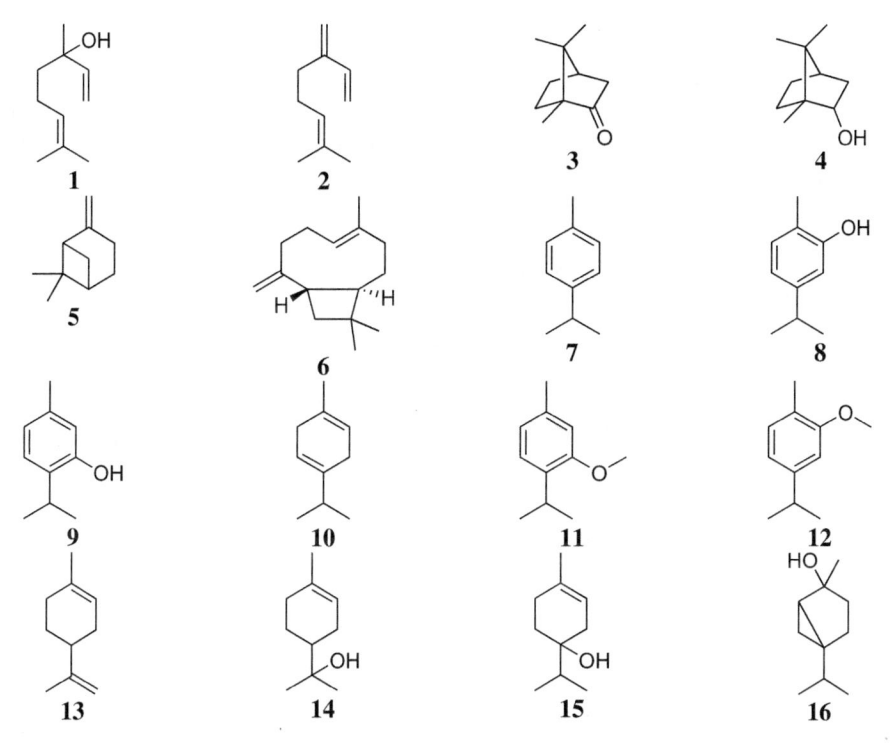

Figure 28.2 *Main terpenoids constituents of the essential of* **T. vulgaris.** Linalol (**1**), myrcene (**2**), camphor (**3**), borneol (**4**), β-pinene (**5**), β-caryophyllene (**6**), p-cymene (**7**), carvacrol (**8**), thymol (**9**), γ-terpinene (**10**), thymyl methyl ether (**11**), carvacryl methyl ether (**12**), limonene (**13**), α-terpinol (**14**), γ-terpinol (**15**), and sabinene hydrate (**16**).

was less than 19% (Reddy et al., 2014). Dried *T. vulgaris* plant contains 1–2.5% of essential oil (Stahl-Biskup and Venskutonis, 2012). The chemical structures of the major terpenoids of the essential oil are provided in Fig. 28.2 (Reddy et al., 2014). Most of the volatiles detected in thyme oil belong to the monoterpene group with compound **9**, a phenolic monoterpene, as the main representative (30–55%) (Reddy et al., 2014). These main components include linalol (**1**), myrcene (**2**), camphor (**3**), borneol (**4**), β-pinene (**5**), β-caryophyllene (**6**), p-cymene (**7**), carvacrol (**8**), thymol (**9**), γ-terpinene (**10**), thymyl methyl ether (**11**), carvacryl methyl ether (**12**), limonene (**13**), α-terpinol (**14**), γ-terpinol (**15**), and sabinene hydrate (**16**). *T. vulgaris* also contains triterpenes in the form of ursolic acid (0.94%) and oleanolic acid (0.37%) (Jager et al., 2009).

3.2 Phenolics and other constituents of *Thymus vulgaris*

A part from the main constituents of essential oil, thyme also contains phenolics, such as rosmarinic acid (**17**) that contribute to the commercial use of the herb (Kivilompolo and Hyotylainen, 2007; Stahl-Biskup and Venskutonis, 2012), caffeic acid (**18**), gentisic acid (**19**), *p*-coumaric acid, syringic acid (**20**), ferulic acid (**21**), and *p*-hydroxybenzoic acid (Proestos et al., 2005; Kivilompolo and Hyotylainen, 2007); about 25 different flavonoids (Wang et al., 1998; Stahl-Biskup and Venskutonis, 2012; Vila, 2002) among which are flavones, such as apigenin (**22**), luteolin (**23**), 6-hydroxyluteolin; and methylflavones: cirsilineol (**24**), 8-methoxycirsilineol, cirsimaritin (**25**), 5-desmethylnobiletin (**26**), 5-desmethylsinensetin (**27**), gardenin B (**28**), genkwanin (**29**), 7-methoxyluteolin, salvigenin (**30**), sideritoflavone (**31**), thymonin (**32**), thymusin (**33**), and xanthomicrol (**34**) (Stahl-Biskup and Venskutonis, 2012) (Fig. 28.3).

4 PHARMACOLOGY OF *THYMUS VULGARIS*

Thyme has been used in food and aroma industries as a flavor enhancer in a wide variety of foods, beverages, and confectionery products, as well as a food preservative, due to its antimicrobial and antioxidant properties. Apart of these effects, thyme displays a variety of biological activities.

4.1 Antiinflammatory and antinociceptive activities

Thyme extracts were reported for their antiinflammatory activities. It reduces the production and gene expression of the proinflammatory mediators, such as tumor necrosis factor alpha (TNF-α), interleukin-1 beta (IL-1β), and interleukin 6 (IL-6), and increases the expression of the antiinflammatory cytokine, interleukin-10 (IL-10) (Ocana and Reglero, 2012). Vigo et al. (2004) have demonstrated that thyme has antiinflammatory activity through significant inhibition of nitric oxide (NO) synthase mRNA expression. The hydroalcoholic extract of thyme was also shown to exhibit modulator effects on acute and chronic pain in mice (Taherian et al., 2009). Thymol (**9**), a major constituent of thyme has been reported to exhibit antiinflammatory effects in vivo and in vitro (Zhou et al., 2014). Pretreatment of ovalbumin (OVA)-induced mouse asthma with compound **9** was found to reduce the level of OVA-specific immunoglobulin E, inhibit recruitment of inflammatory cells into airway, and decrease levels of IL-4, IL-5, and IL-13 in bronchoalveolar lavage fluid (Zhou et al., 2014). These data suggested that compound **9** ameliorated airway inflammation in OVA-induced mouse asthma, possibly through inhibiting nuclear factor-κB (NF-κB) activation (Zhou et al., 2014).

4.2 Antimicrobial activities

Plants from the genus *Thymus* are important medicinal herbs, which are known to contain antimicrobial agents, and are rich in different active substances, such as compounds **7–10** (Nabavi et al., 2015). Various preparations, including essential oils, crude extracts, and constituents from *T. vulgaris* were reported for their antibacterial, antifungal, and antiviral properties.

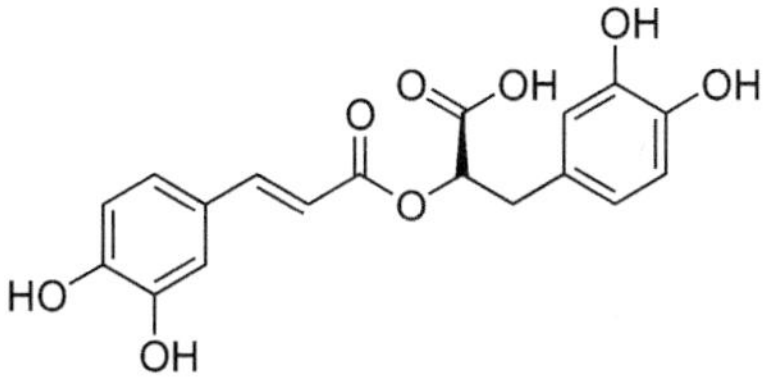

17

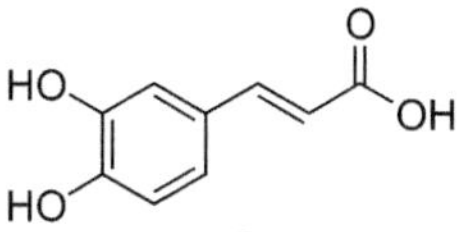

18

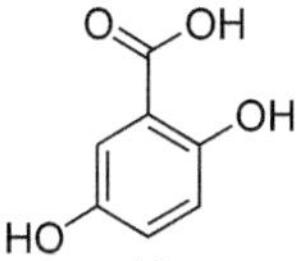

19

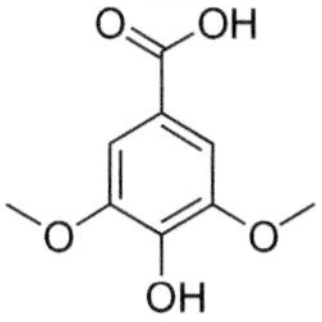

20

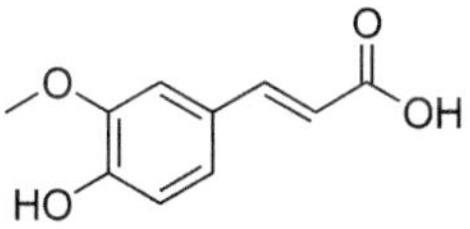

21

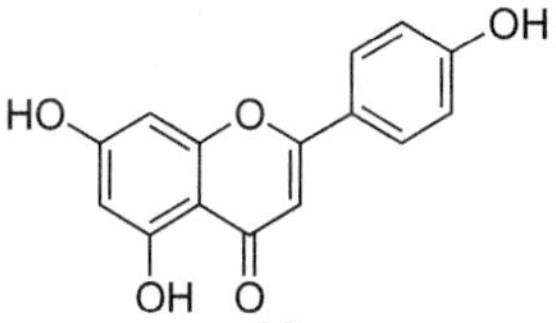

22

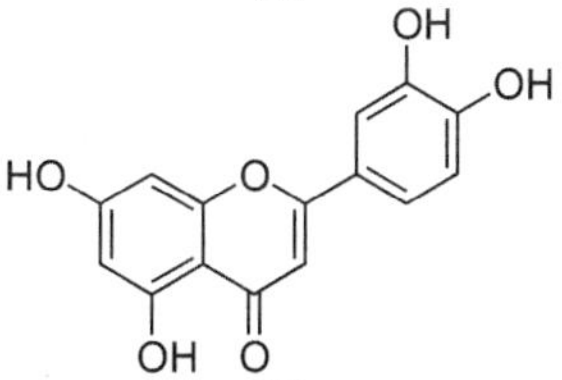

23

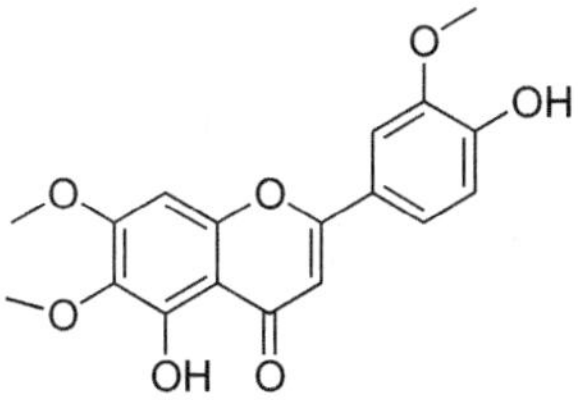

24

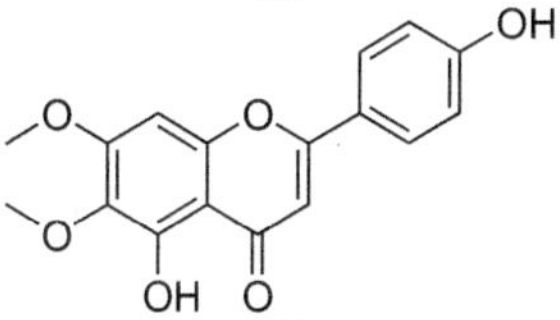

25

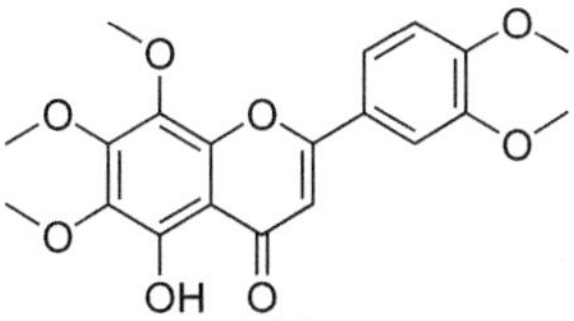

26

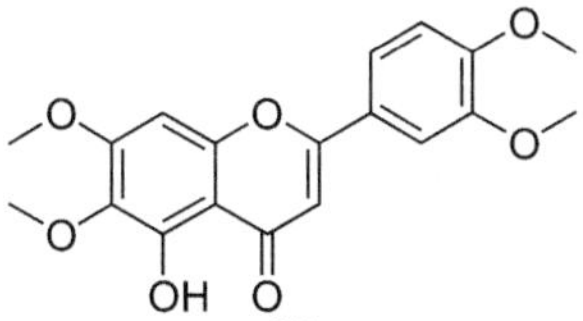

27

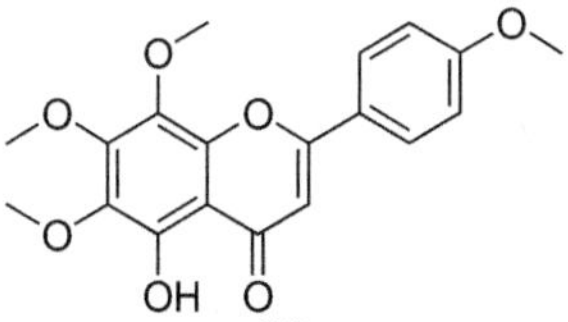

28

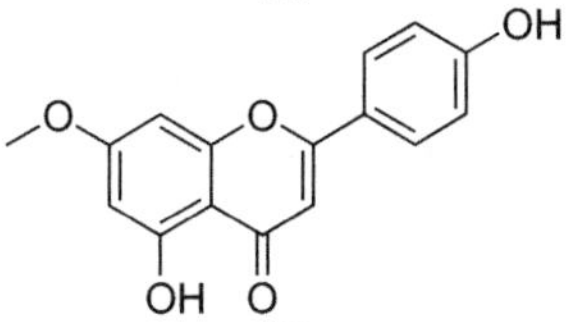

29

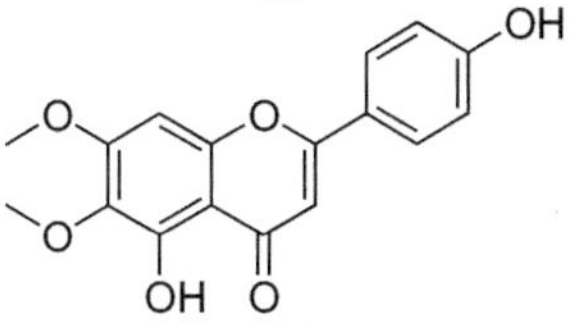

30

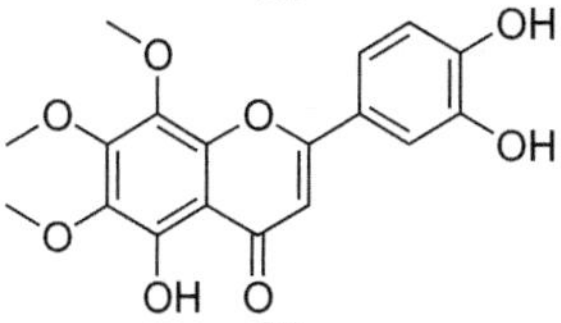

31

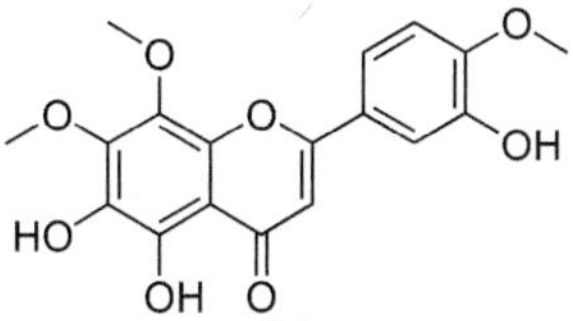

32

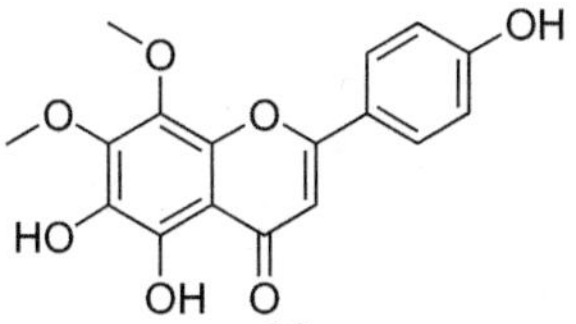

33

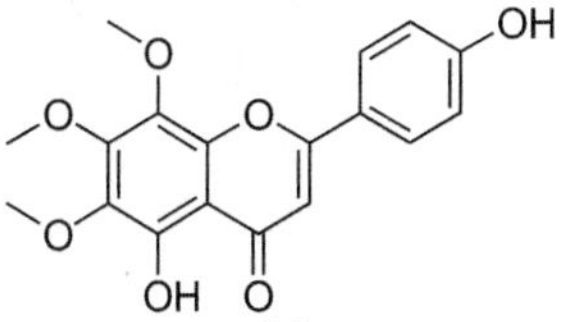

34

Figure 28.3 ***Phenolic constituents of* T. vulgaris.** Rosmarinic acid (**17**), caffeic acid (**18**), gentisic acid (**19**), syringic acid (**20**), ferulic acid (**21**), apigenin (**22**), luteolin (**23**), cirsilineol (**24**), cirsimaritin (**25**), 5-desmethylnobiletin (**26**), 5-desmethylsinensetin (**27**), gardenin B (**28**), genkwanin (**29**), salvigenin (**30**), sideritoflavone (**31**), thymonin (**32**), thymusin (**33**), and xanthomicrol (**34**).

4.2.1 Antibacterial activity

The antibacterial activity of thyme essential oil was demonstrated against a panel of 14 clinical isolates of methicillin-resistant *Staphyloccocus aureus* (MRSA) and other standard bacterial strains, such as *Bacillus cereus, Escherichia coli, Klebsiella pneumoniae* (Tohidpour et al., 2010). Tohidpour et al. (2010) reported the minimum inhibitory concentration (MIC) value of 9.25 µg/mL against *E. coli*, 55.5 µg/mL against *K. pneumoniae*, and MIC values ranged from 18.5 to 37 µg/mL against MRSA. Sienkiewicz et al. (2011) also reported a strong antibacterial activity of essential oils against 120 strains belonging to the *Staphylococcus, Enterococcus, Escherichia,* and *Pseudomonas* genera. The antibacterial activity of thyme essential oils was also reported against the Gram-positive bacteria *Streptococcus pyogenes* (Sfeir et al., 2013), as well as against *Helicobacter pylori* (Esmaeili et al., 2012). It was also shown that clinical and standard strains of *S. aureus* were highly sensitive to thyme constituents, **1, 2, 3, 5, 6, 7, 8, 9, 10, 13,** carvon, camphene, 1,4-cineol, menthone, and α-terpinene (Abu-Darwish et al., 2012; Nabavi et al., 2015). Strains of *E. coli* showed lower sensitivity to compounds **5, 8, 9, 10,** camphene, α-terpinene, 1,4-cineol, but were sensitive to **13** meanwhile *Pseudomonas aeruginosa* strains were found to be resistant to most of the tested components, with the exception of **5, 9, 10,** and α-terpinene (Abu-Darwish et al., 2012). Compounds **8** and **9** displayed antibacterial activity against some strains of verocytotoxigenic *E. coli* (Rivas et al., 2010).

4.2.2 Antifungal activity

Thyme essential oil was reported for its antifungal effects against food spoilage fungi, including *Aspergillus* species, such as *A. oryzae, A. brasiliensis,* and *A. flavus* (Dobre et al., 2011; Mandal and DebMandal, 2016), *A. flavus, A. parasiticus, A. ochracus,* and *Fusarium moniliforme* (Soliman and Badeaa, 2002; Kalemba and Kunicka, 2003; Mandal and DebMandal, 2016). Essential oils from two clonal types of *T. vulgaris* (Laval-1 and Laval-2) also displayed antifungal activities against two common storage pathogens, *Botrytis cinerea* and *Rhizopus stolonifer* (Bhaskara Reddy et al., 1998). Morphological evaluation performed by both light microscopy and scanning electron microscopy demonstrated the antifungal activity of thyme essential oils at 50 µg/mL and its fungicidal effect at 250 µg/mL against *A. flavus* (Kohiyama et al., 2015). Also, Kohiyama et al. (2015) showed that thyme essential oil was able to completely inhibit the production of both aflatoxins, B1 and B2, at 150 µg/mL, suggesting its abilities to control the growth of *A. flavus* and its production of aflatoxins. Compound **9** together with essential oil showed strong fungicidal and/or fungistatic activities against *Aspergillus, Penicillium, Cladosporium, Trichoderma, Mucor,* and *Rhizopus* species (Segvic Klaric et al., 2007). It also showed strong antifungal effects against *Rhizopus oryzae* with MICs in a range of 128–512 µg/mL for both (de Lira Mota et al., 2012).

4.2.3 Antiviral activity

The 80% ethanol extract of *T. vulgaris* had antiviral activity against Newcastle disease virus, reducing the viral potency by more than 56-fold (Rezatofighi et al., 2014). The antiviral activities of thyme were also reported against herpes simplex virus 1 and 2, as well as against acyclovir-resistant strains of herpes simplex virus 1 (Nolkemper et al., 2006).

4.3 Antioxidant activity

Thyme phenolics have excellent antioxidant activity that may be higher than the well-known butylated hydroxy toluene and α-tocopherol antioxidants (Lee and Shibamoto, 2002). Thyme essential oil was reported to have effective radical scavenging capacity, therefore acts as a natural antioxidant agent (Miladi et al., 2013). It acted as a free radical scavenger for lipid oxidation (IC_{50} value of 18.6 μg/mL), protecting lipids from oxidation during frozen and refrigerated storage (Selmi and Sadok, 2008). Thyme essential oil also possesses antioxidant activity and has a protective effect against aflatoxin toxicity (El-Nekeety et al., 2011).

4.4 Antispasmodic activity

T. vulgaris extract was shown antagonize the contraction of the *Musculus transversus tracheae*, in a reversible and concentration-dependent manner, provoked by four different spasmogens ($BaCl_2$, carbachol, histamine, and prostaglandin F_2 alpha) (Meister et al., 1999). Potent relaxant effect of thyme on guinea pig tracheal chains, comparable to that of the reference compound theophylline was also reported (Boskabady et al., 2006). Wienkotter et al. (2007) provided evidence of the effect of thyme extract on $β_2$-receptors in competition-binding experiments and relaxation experiments on rat uteri and trachea. The effects of *T. vulgaris* hydroalcoholic extract on the contractile responses of the isolated guinea pig ileum were demonstrated and it was suggested that it could affect the anticholinergic and serotoninergic pathways (Babaei et al., 2008). Using various models, Begrow et al. (2010) also demonstrated the antispasmodic effects of compound **9** on smooth muscles of trachea and ileum, as well as on ciliary activity (respiratory clearance). Thymol and compound **8** were found to contribute to the antispasmodic effect of *T. vulgaris* in several models (Stahl-Biskup and Venskutonis, 2012). Nevertheless, it was shown that compound **9** is not important for the overall antispasmodic effect, when all spasmic triggers are considered; for example, compound **9** alone had no effect on endothelin-induced trachea contraction (Engelbertz et al., 2008). van den Broucke and Lemli (1983) also showed that flavones from thyme, as well as the crude extracts were effective in test systems, which included smooth muscles of guinea pig ileum and rat vas deferens.

4.5 Other pharmacological activities

Thyme extract showed repairing effects on memory and behavioral disorders produced by scopolamine in rats, suggesting its beneficial effects in the treatment of Alzheimer's disease (Rabiei et al., 2015). The insecticidal effects of thyme essential oil were also reported. In fact, the direct toxicity of adult insects was found to be due to the inhibition of reproduction through ovicidal and larvicidal effects. The killing and repellent effects of the essential oil on the mosquito, *Culex quinquefasciatus* larvae and *Anopheles gambiae* were also reported (Stahl-Biskup and Venskutonis, 2012). The herbicidal and phytotoxic effects of thyme were also reported (Stahl-Biskup and Venskutonis, 2012).

5 PATENTS WITH *THYMUS VULGARIS*

There are some patents related to *T. vulgaris*:
1. *Patent CA2,222,563 A1* granted to Ninkov (1996), Title: Pharmaceutical compositions, based on etheric oils obtained from plants for use in the human and veterinary medical field. The invention relates to pharmaceutical compounds, which are based on the antiinflammatory properties of etheric oils selected from the group consisting of *Origanum vulgaris*, *T. vulgaris*, *Mentha piperita*, *Thymus serpilum*, *Saturea hortensis*, *Saturea montana*, *Saturea subricata*, *Carum corticum*, *Thymus zugis*, *Ocimum gratisimum*, *Moranda pungtata*, *Mosla japanoica*, and *Salvia officinalis*.
2. *Patent EP1,080,727 A1* granted to Emmanouilidis (2001), Title: Use of the extract of *T. vulgaris* for the preparation of a medicament for the treatment of ulcerative colitis and Crohn's disease. The patent demonstrates the use of the extract of *T. vulgaris* to relieve the symptoms of ulcerative colitis and Crohn's disease in a considerable percentage of patients, without any side effects.
3. *Patent WO2,012,131,732 A1* granted to Ayyathurai (2012), Title: A synergistic herbal extract composition for use in treating and preventing mastitis. The patent relates to a novel synergistic herbal extract composition comprising therapeutically effective amounts of extracts of leaves of *Abrus precatorius*, leaves of *Cleome gynandra*, roots of *Aristalochia indica*, leaves of *T. vulgaris*, and leaves of *S. officinalis* for use in treating and preventing mastitis in a subject in need thereof.

CONCLUSIONS

In the present review, reports related to the usefulness of *T. vulgaris* as potential therapeutic agents have been summarized, in addition to its common culinary use. The antiinflammatory, antibacterial, antifungal, antiviral, antioxidant, and antispasmodic properties of thyme have been highlighted. I also reviewed the phytochemical composition, cultivation, and world distribution of the plant. Some patents related to thyme were also listed and commented. In brief, thyme is an easily available source of natural antioxidants and antibiotics in food products and drugs. However, more clinical and pathological studies are needed to fully exploit the therapeutic potential of this plant.

REFERENCES

Abu-Darwish, M.S., Al-Ramamneh, E.A., Kyslychenko, V.S., Karpiuk, U.V., 2012. The antimicrobial activity of essential oils and extracts of some medicinal plants grown in Ash-shoubak region—South of Jordan. Pak. J. Pharm. Sci. 25 (1), 239–246.

Ayyathurai, K.T., 2012. A synergistic herbal extract composition for use in treating and preventing mastitis: Google Patents.

Babaei, M., Abarghoei, M.E., Ansari, R., Vafaei, A.A., Taherian, A.A., Akhavan, M.M., 2008. Antispasmodic effect of hydroalcoholic extract of *Thymus vulgaris* on the guinea-pig ileum. Nat. Prod. Res. 22 (13), 1143–1150.

Begrow, F., Engelbertz, J., Feistel, B., Lehnfeld, R., Bauer, K., Verspohl, E.J., 2010. Impact of thymol in thyme extracts on their antispasmodic action and ciliary clearance. Planta Med. 76 (4), 311–318.

Bhaskara Reddy, M.V., Angers, P., Gosselin, A., Arul, J., 1998. Characterization and use of essential oil from *Thymus vulgaris* against *Botrytis cinerea* and *Rhizopus stolonifer* in strawberry fruits. Phytochemistry 47 (8), 1515–1520.

Boskabady, M.H., Aslani, M.R., Kiani, S., 2006. Relaxant effect of *Thymus vulgaris* on guinea-pig tracheal chains and its possible mechanism(s). Phytother. Res. 20 (1), 28–33.

de Lira Mota, K.S., de Oliveira Pereira, F., de Oliveira, W.A., Lima, I.O., de Oliveira Lima, E., 2012. Antifungal activity of *Thymus vulgaris* L. essential oil and its constituent phytochemicals against *Rhizopus oryzae*: interaction with ergosterol. Molecules 17 (12), 14418–14433.

Dobre, A., Gagiu, V., Petru, N., 2011. Antimicrobial activity of essential oils against food-borne bacteria evaluated by two preliminary methods. Rom. Biotechnol. Lett. 16, 119–125.

Ekoh, S., Akubugwo, E., Chibueze Ude, V., Edwin, N., 2014. Anti-hyperglycemic and anti-hyperlipidemic effect of spices (*Thymus vulgaris, Murraya koenigii, Ocimum gratissimum* and *Piper guineense*) in alloxan-induced diabetic rats. Int. J. Biosci. 4 (2), 179–187.

El-Nekeety, A.A., Mohamed, S.R., Hathout, A.S., Hassan, N.S., Aly, S.E., Abdel-Wahhab, M.A., 2011. Antioxidant properties of *Thymus vulgaris* oil against aflatoxin-induce oxidative stress in male rats. Toxicon 57 (7–8), 984–991.

Emmanouilidis, A., 2001. Use of the extract of *Thymus vulgaris* for the preparation of a medicament for the treatment of ulcerative colitis and Crohn's disease: Google Patents.

Engelbertz, J., Schwenk, T., Kinzinger, U., Schierstedt, D., Verspohl, E.J., 2008. Thyme extract, but not thymol, inhibits endothelin-induced contractions of isolated rat trachea. Planta Med. 74 (12), 1436–1440.

Esmaeili, D., Mobarez, A.M., Tohidpour, A., 2012. Anti-*Helicobacter pylori* activities of shoya powder and essential oils of *Thymus vulgaris* and *Eucalyptus globulus*. Open Microbiol. J. 6, 65–69.

Hosseinzadeh, S., Kukhdan, A., Hosseini, A., Armand, R., 2015. The application of *Thymus vulgaris* in traditional and modern medicine: a review. Glob. J. Pharmacol. 9 (3), 260–266.

Jager, S., Trojan, H., Kopp, T., Laszczyk, M.N., Scheffler, A., 2009. Pentacyclic triterpene distribution in various plants—rich sources for a new group of multi-potent plant extracts. Molecules 14 (6), 2016–2031.

Kalemba, D., Kunicka, A., 2003. Antibacterial and antifungal properties of essential oils. Curr. Med. Chem. 10 (10), 813–829.

Kayode, J., Ogunleye, T., 2008. Checklist and status of plant species used as spices in kaduna state of Nigeria. Afr. J. Gen. Agric. 4 (1), 13–18.

Kivilompolo, M., Hyotylainen, T., 2007. Comprehensive two-dimensional liquid chromatography in analysis of Lamiaceae herbs: characterisation and quantification of antioxidant phenolic acids. J. Chromatogr. A 1145 (1–2), 155–164.

Kohiyama, C.Y., Yamamoto Ribeiro, M.M., Mossini, S.A.G., Bando, E., Bomfim, N.d.S., Nerilo, S.B., 2015. Antifungal properties and inhibitory effects upon aflatoxin production of *Thymus vulgaris* L. by *Aspergillus flavus* Link. Food Chem. 173, 1006–1010.

Lee, K.G., Shibamoto, T., 2002. Determination of antioxidant potential of volatile extracts isolated from various herbs and spices. J. Agric. Food Chem. 50 (17), 4947–4952.

Mandal, S., DebMandal, M., 2016. Chapter 94—thyme (*Thymus vulgaris* L.) oils. In: Preedy, V.R. (Ed.), Essential Oils in Food Preservation, Flavor and Safety. Academic Press, San Diego, pp. 825–834.

Meister, A., Bernhardt, G., Christoffel, V., Buschauer, A., 1999. Antispasmodic activity of *Thymus vulgaris* extract on the isolated guinea-pig trachea: discrimination between drug and ethanol effects. Planta Med. 65 (6), 512–516.

Miladi, H., Slama, R., Mili, D., Zouari, S., Bakhrouf, A., Ammar, E., 2013. Essential oil of *Thymus vulgaris* L. and *Rosmarinus officinalis* L.: gas chromatography-mass spectrometry analysis, cytotoxicity and antioxidant properties and antibacterial activities against foodborne pathogens. Nat. Sci. 5, 729–739.

Nabavi, S.M., Marchese, A., Izadi, M., Curti, V., Daglia, M., Nabavi, S.F., 2015. Plants belonging to the genus *Thymus* as antibacterial agents: from farm to pharmacy. Food Chem. 173, 339–347.

Ninkov, D., 1996. Pharmaceutical compositions, based on etheric oils obtained from plants for use in the human and veterinary medical field: Google Patents.

Nkouaya Mbanjo, E., Tchoumbougnang, M., Jazet Dongmo, P., Sameza, M., Amvam Zollo, P., Menut, C., 2007. Mosquito larvicidal activity of essential oils of *Cymbopogon citratus* and *Thymus vulgaris* grown in Cameroon. Planta Med. 73 (9), P_329.

Nolkemper, S., Reichling, J., Stintzing, F.C., Carle, R., Schnitzler, P., 2006. Antiviral effect of aqueous extracts from species of the Lamiaceae family against herpes simplex virus type 1 and type 2 in vitro. Planta Med. 72 (15), 1378–1382.

Ocana, A., Reglero, G., 2012. Effects of Thyme extract oils (from *Thymus vulgaris*, *Thymus zygis*, and *Thymus hyemalis*) on cytokine production and gene expression of oxLDL-stimulated THP-1-macrophages. J. Obes. 2012, 104706.

Proestos, C., Chorianopoulos, N., Nychas, G.J., Komaitis, M., 2005. RP-HPLC analysis of the phenolic compounds of plant extracts. investigation of their antioxidant capacity and antimicrobial activity. J. Agric. Food Chem. 53 (4), 1190–1195.

Rabiei, Z., Mokhtari, S., Asgharzade, S., Gholami, M., Rahnama, S., Rafieian-kopaei, M., 2015. Inhibitory effect of *Thymus vulgaris* extract on memory impairment induced by scopolamine in rat. Asian Pac. J. Trop. Biomed. 5 (10), 845–851.

Reddy, P., Kandisa, R., Varsha, P., Satyam, S., 2014. Review on *Thymus vulgaris* traditional uses and pharmacological properties. Med. Aromat. Plants 3, 164.

Rezatofighi, S.E., Seydabadi, A., Seyyed Nejad, S.M., 2014. Evaluating the efficacy of *Achillea millefolium* and *Thymus vulgaris* extracts against Newcastle Disease Virus *in vivo*. Jundishapur J. Microbiol. 7 (2), e9016.

Rivas, L., McDonnell, M.J., Burgess, C.M., O'Brien, M., Navarro-Villa, A., Fanning, S., 2010. Inhibition of verocytotoxigenic *Escherichia coli* in model broth and rumen systems by carvacrol and thymol. Int. J. Food Microbiol. 139 (1–2), 70–78.

Schmitz, P., 2015. *Thymus vulgaris*. Available from: http://ecoport.org/ep?Plant=2441&entityType=PLCR**&entityDisplayCategory=full

Segvic Klaric, M., Kosalec, I., Mastelic, J., Pieckova, E., Pepeljnak, S., 2007. Antifungal activity of thyme (*Thymus vulgaris* L.) essential oil and thymol against moulds from damp dwellings. Lett. Appl. Microbiol. 44 (1), 36–42.

Selmi, S., Sadok, S., 2008. The effect of natural antioxidant (*Thymus vulgaris* Linnaeus) on flesh quality of tuna *Thunnus thynnus* (Linnaeus) during chilled storage. Pan-Am. J. Aquat. Sci. 3 (1), 36–45.

Sfeir, J., Lefrancois, C., Baudoux, D., Derbre, S., Licznar, P., 2013. *In vitro* antibacterial activity of essential oils against *Streptococcus pyogenes*. Evid. Based Complement. Alternat. Med. 2013, 269161.

Sienkiewicz, M., Lysakowska, M., Ciecwierz, J., Denys, P., Kowalczyk, E., 2011. Antibacterial activity of thyme and lavender essential oils. Med. Chem. 7 (6), 674–689.

Soliman, K.M., Badeaa, R.I., 2002. Effect of oil extracted from some medicinal plants on different mycotoxigenic fungi. Food Chem. Toxicol. 40 (11), 1669–1675.

Stahl-Biskup, E., Saez, F. (Eds.), 2002. Thyme—The Genus *Thymus*. Taylor & Francis, London.

Stahl-Biskup, E., Venskutonis, R.P., 2012. 27—Thyme. In: Peter, K.V. (Ed.), Handbook of Herbs and Spices. second ed. Woodhead Publishing, Abington, Cambridge, UK, pp. 499–525.

Taherian, A.A., Babaei, M., Vafaei, A.A., Jarrahi, M., Jadidi, M., Sadeghi, H., 2009. Antinociceptive effects of hydroalcoholic extract of *Thymus vulgaris*. Pak. J. Pharm. Sci. 22 (1), 83–89.

Tohidpour, A., Sattari, M., Omidbaigi, R., Yadegar, A., Nazemi, J., 2010. Antibacterial effect of essential oils from two medicinal plants against methicillin-resistant *Staphylococcus aureus* (MRSA). Phytomedicine 17 (2), 142–145.

van den Broucke, C., Lemli, J., 1983. Spasmolytic activity of the flavonoids from *Thymus vulgaris*. Pharm. Weekbl. 5, 9–14.

Vigo, E., Cepeda, A., Gualillo, O., Perez-Fernandez, R., 2004. In-vitro anti-inflammatory effect of *Eucalyptus globulus* and *Thymus vulgaris*: nitric oxide inhibition in J774A.1 murine macrophages. J. Pharm. Pharmacol. 56 (2), 257–263.

Vila, R., 2002. Flavonoids and further polyphenols in the genus *Thymus*. In: Stahl-Biskup, E., Saez, F. (Eds.), Thyme—The Genus *Thymus*. Taylor & Francis, London, pp. 144–176.

Wang, M., Li, J., Ho, G., Peng, X., Ho, C., 1998. Isolation and identification of antioxidative flavonoid glycosides from thyme (*Thymus vulgaris* L.). J. Food Lipids 5, 313–321.

Wienkotter, N., Begrow, F., Kinzinger, U., Schierstedt, D., Verspohl, E.J., 2007. The effect of thyme extract on beta2-receptors and mucociliary clearance. Planta Med. 73 (7), 629–635.

Zhou, E., Fu, Y., Wei, Z., Yu, Y., Zhang, X., Yang, Z., 2014. Thymol attenuates allergic airway inflammation in ovalbumin (OVA)-induced mouse asthma. Fitoterapia 96, 131–137.

CHAPTER 29

Syzygium aromaticum

A.T. Mbaveng, V. Kuete

1 INTRODUCTION

Syzygium aromaticum (L.) Merr. & L.M.Perry (Syn. *Eugenia caryophyllus*) is a tree in the family Myrtaceae, native to Indonesia. The aromatic flower buds of the plant are known as cloves and are commonly used as a spice. Cloves are commercially harvested in Indonesia, India, Pakistan, Sri Lanka, as well as in African countries, such as Comoro Islands, Madagascar, Seychelles, and Tanzania. Several therapeutic uses of *S. aromaticum* have been recognized. The clove plant is used as a medicine in China and Western countries against many diseases, such as oral diseases or dental complaints (Cai and Wu, 1996; Wankhede, 2015). The plant is also used to control nausea and vomiting, cough, diarrhea, dyspepsia, flatulence, stomach distension, and gastrointestinal spasm; relieve pain; cause uterine contractions; and stimulate the nerves (Shrivastava et al., 2014). The cloves are also used in folk medicine as a diuretic, odontalgic, stomachic, tonicardiac, and condiment with carminative and stimulant effects (Pandey and Singh, 2011). Essential oil derived from this aromatic plant not only serves as a fragrance and flavor agent, but also as a dietary antioxidant expected to prevent several diseases caused by free radicals (Cai and Wu, 1996; Halliwell, 1999). It has been reported that the majority of cloves are used by kretek cigarette manufacturers in Indonesia and only about 10% for other purposes, such as folk medicine, food flavoring, food preservation, fragrance, and pharmaceuticals (Nurdjannah and Bermawie, 2012). In the present chapter, we will discuss the pharmacological potency of *S. aromaticum* with emphasis on the aromatic part, the clove (Fig. 29.1).

2 BOTANICAL ASPECT AND DISTRIBUTION OF *SYZYGIUM AROMATICUM*

S. aromaticum is an evergreen tree growing up to 8–12 m high, with large leaves and flowers grouped in terminal clusters. The flower buds initially have a pale hue, and gradually turn green. Generally, *S. aromaticum* is medium-sized, with a low crown base, and branches are semierect and numerous. The leaves are glabrous, with numerous oil glands on the lower surface. The flowers are small, in terminal cymose cluster, each peduncle bears three- or four-stalked flowers at the end, while sepals minute with triangular projection. The fruits are of typically olive shaped. The brown, dried, unopened flower buds are called cloves, a name derived from the French word *clou* meaning nail.

Figure 29.1 *Cloves of* S. aromaticum. *(Photograph by V. Kuete).*

Cloves are harvested at a 1.5–2.0 cm length. The plant is native to Indonesia and exotic to Brazil, Haiti, India, Kenya, Madagascar, Malaysia, Mauritius, Mexico, Seychelles, Sri Lanka, and Tanzania. Madagascar and Zanzibar (Tanzania) are the largest clove suppliers in Africa, with a production of approximately 20,000–27,000 tons/year while Comoros, Kenya, and Togo, together with Sri Lanka, Malaysia, China, and Grenada, supplied around 5,000–7,000 tons/year (Nurdjannah and Bermawie, 2012). Indonesia remains the largest producer in the world with a production of 70,535 tons of dried cloves in 2008 (Nurdjannah and Bermawie, 2012).

3 CHEMISTRY OF *SYZYGIUM AROMATICUM*

S. aromaticum represents one of the major vegetal sources of phenolic compounds, such as flavonoids, hydroxybenzoic acids, hydroxycinnamic acids, and hydroxyphenyl propens, as well as terpenoids (Bao et al., 2012; Kamatou et al., 2012; Cortés-Rojas et al., 2014). Eugenol (**1**) is the compound primarily responsible for the clove's aroma and constitutes 72–90% of the volatile oil of cloves (Kamatou et al., 2012). Other common constituents of essential oil include eugenyl acetate (**2**), β-caryophyllene (**3**), methyl salicylate (**4**), pinene (**5**), vanillin (**7**) (Nurdjannah and Bermawie, 2012), and α-humulene (**7**) (Jirovetz et al., 2006).

3.1 Essential oil

The variation in components and composition of the essential oil of clove depends on variety, agroecological condition, pretreatments, processing, and methods of extraction (Nurdjannah and Bermawie, 2012). It is a mixture of different compounds; with the three main active ingredients: compound **1**, **2**, and **3**. In one study, 23 constituents were identified in clove's leaf essential oil, among which were **1** (76.8%), **3** (17.4%),

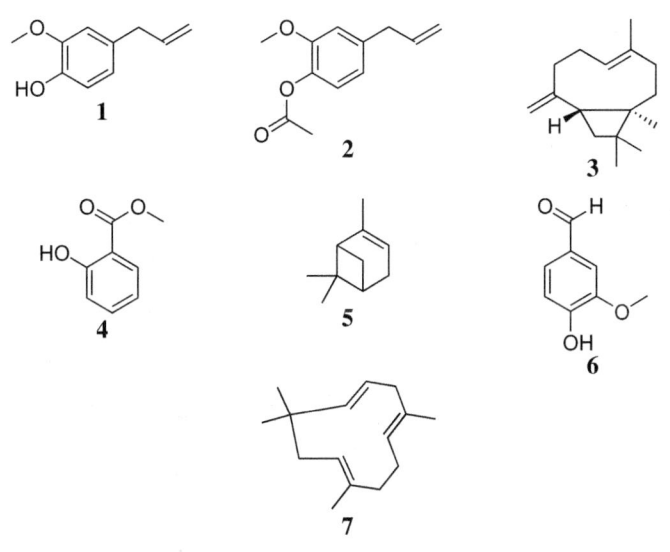

Figure 29.2 *Main consitutuents of the essential oil of* S. aromaticum. Eugenol (**1**), eugenyl acetate (**2**), caryophyllene (**3**), methyl salicylate (**4**), pinene (**5**), vanillin (**6**), and humulene (**7**).

α-humulene (**7**) (2.1%), and **2** (1.2%) were the main components (Jirovetz et al., 2006). Another study also reported compounds: **1** (89.6%), **3** (8.6%), and **2** (1.7%) (Santin et al., 2011), while Alma et al. (2007) reported **1** (87%), **2** (8%), and **3** (3.56 %) (Fig. 29.2).

3.2 Other constituents

Phenolics (Fig. 29.3), such as 7-dihydroxy-2-methylchromone 8-C-β-D-glucopyranoside, biflorin (**8**), kaempferol (**9**), rhamnocitrin (**10**), myricetin (**11**), gallic acid (**12**), ellagic acid (**13**), as well as the terpenoid, oleanolic acid (**14**) were isolated from the methanol extract of *S. aromaticum* (Cai and Wu, 1996). Eighteen hydrolyzable tannins were also isolated from an aqueous acetone extract of dried flower buds of *S. aromaticum* (Bao et al., 2012). They include aromatinin A (**15**), platycaryanin A (**16**), bicornin (**17**), syzyginin A (**18**), alunusnin A (**19**), rugosin C (**20**), tellimagrandin II (**21**), casuarictin (**22**), heterophylliin D (**23**), rugosin D (**24**), rugosin F (**25**), euprostin A (**26**), 1,2-di-O-galloyl-3-O-digalloyl-4,6-O-(S)-hexahydroxydiphenoy-β-D-glucose, alienanin B (**27**), squarrosanin A (**28**) casuarinin (**29**), syzyginin B (**30**), 1,2,3-tri-O-galloyl-β-D-glucose (**31**), and 1,2,3,6-tetra-O-galloyl-β-D-glucose (**32**) (Bao et al., 2012).

4 PHARMACOLOGY OF *SYZYGIUM AROMATICUM*

Several pharmacological properties of *S. aromaticum* have been reported including the antiseptic, antimutagenic, antiinflammatory, antioxidant, antiulcerogenic, antithrombotic, antifungal, antiviral, and antiparasitic activities.

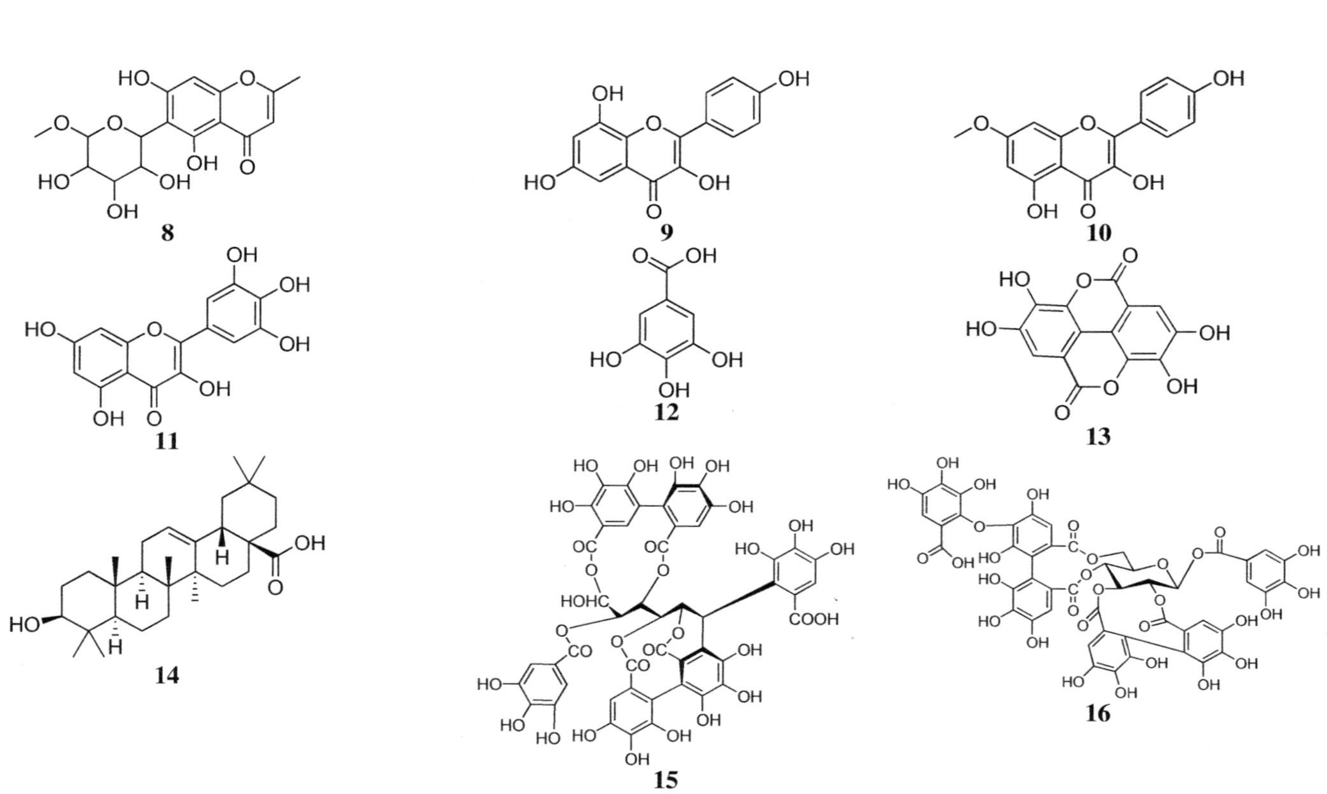

Figure 29.3 *Other constituents of S. aromaticum.* Biflorin (**8**), kaempferol (**9**), rhamnocitrin (**10**), myricetin (**11**), gallic acid (**12**), ellagic acid (**13**), oleanolic acid (**14**), aromatinin A (**15**), platycaryanin A (**16**), bicornin (**17**), syzyginin A (**18**), alunusnin A (**19**), rugosin C (**20**), tellimagrandin II (**21**), casuarictin (**22**), heterophylliin D (**23**), rugosin D (**24**), rugosin F (**25**), euprostin A (**26**), alienanin B (**27**), squarrosanin A (**28**), casuarinin (**29**), syzyginin B (**30**), 1,2,3-tri-O-galloyl-β-D-glucose (**31**), and 1,2,3,6-tetra-O-galloyl-β-D-glucose (**32**).

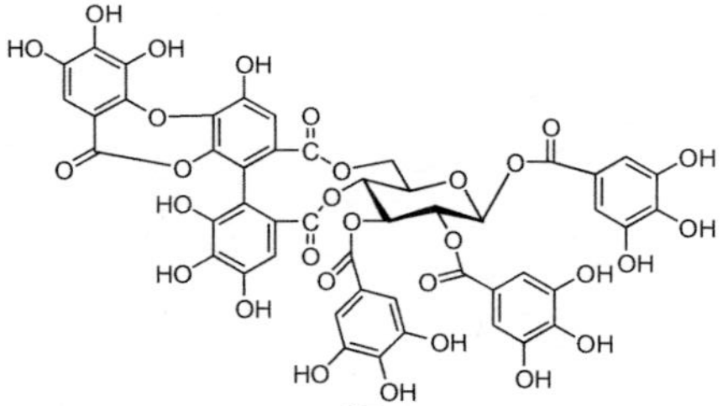

Figure 29.3 *(cont.)*

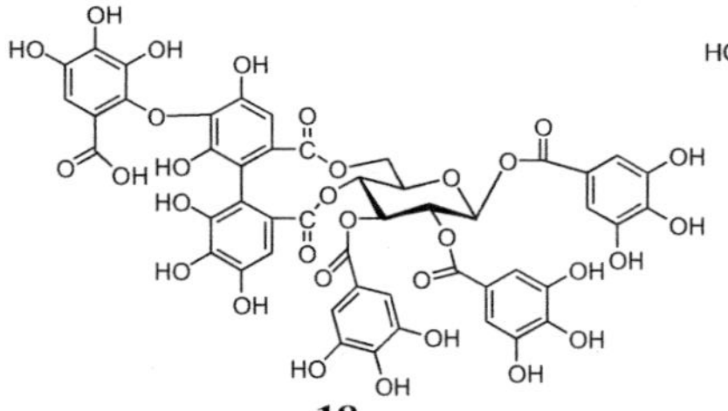

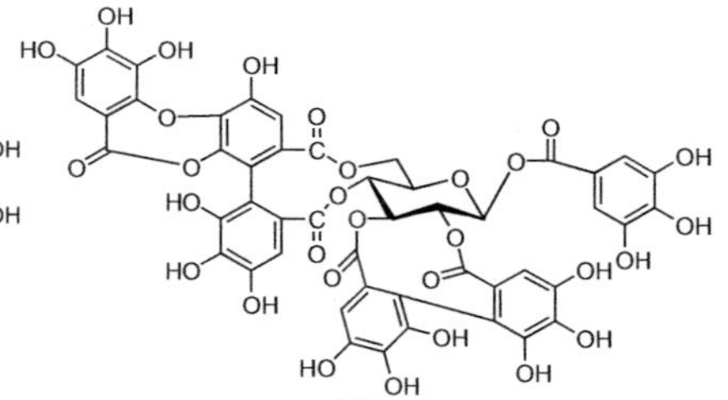

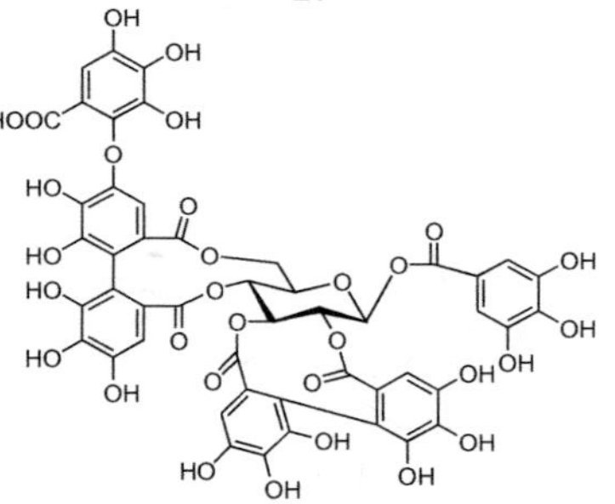

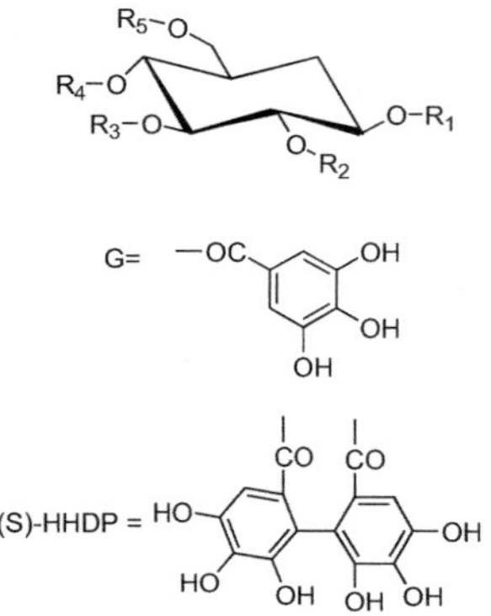

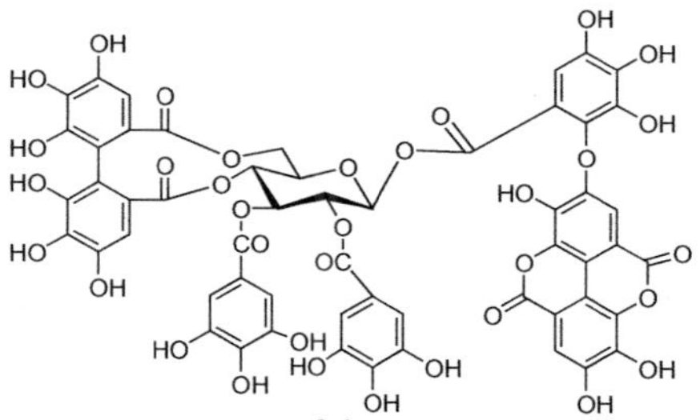

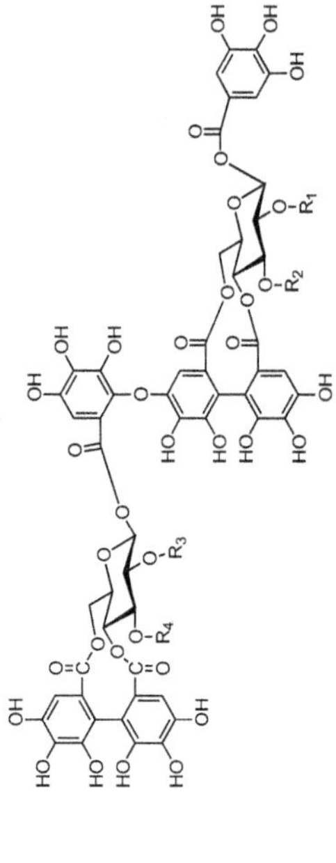

23: R1=R2=R3=R4=(S)-HHDP
24: R1=R2=R3=R4=G
25: R1=R2=(S)-HHDP; R3=R4=G

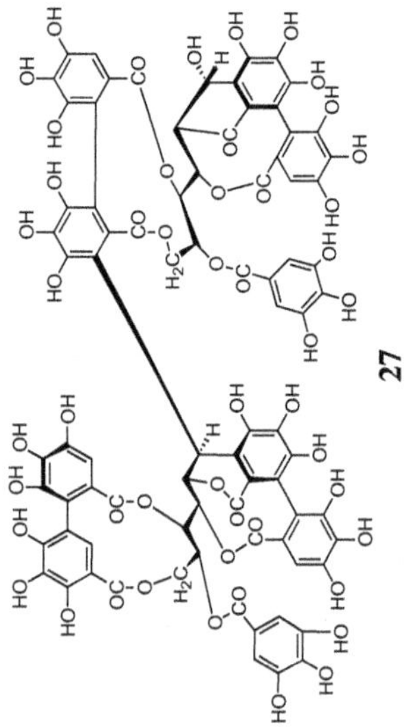

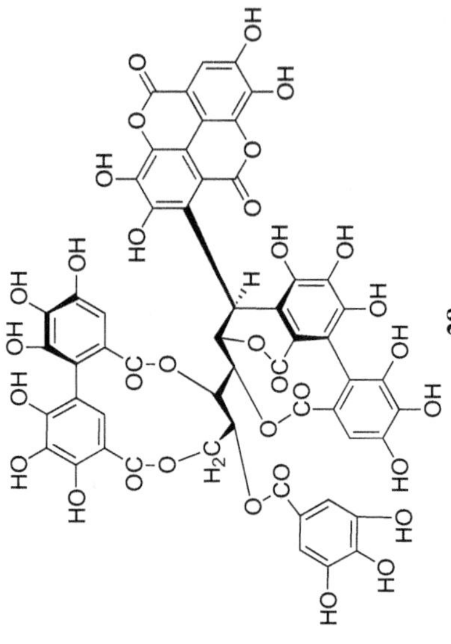

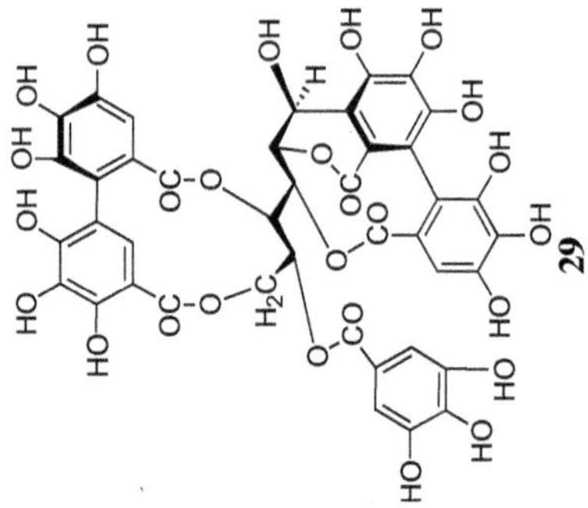

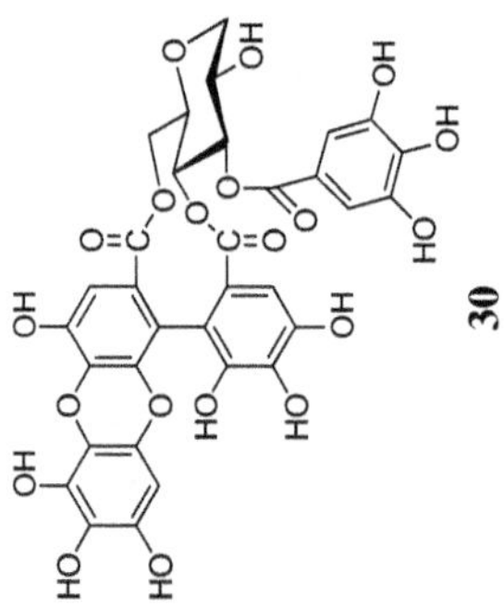

Figure 29.3 *(cont.)*

4.1 Anticancer activity

Clove essential oil displayed cytotoxic effects against cancer cell lines, as well as antimutagenic activities. Methanol extract of cloves flowers buds displayed good cytotoxic effects against colon carcinoma HCT 116 cells (IC_{50}: 31 μg/mL), breast adenocarcinoma MCF-7 cells (IC_{50}: 29.7 μg/mL), and hepatocarcinoma HepG2 cells (IC_{50}: 18.70 μg/mL) (Abd El Azim et al., 2014). The water, ethanol, and oil extracts of cloves also showed cytotoxic effects against cervical cancer HeLa cells, breast adenocarcinoma MCF-7 cells and MDA-MB-231 cells, prostate cancer DU-145 cells, and esophageal cancer TE-13 cells with limited effects on normal human peripheral blood lymphocytes (Dwivedi et al., 2011). The cytotoxicity of methanol extract from bud of clove was also reported toward melanin formation in B16 melanoma cells and **1** and **2** were identified as the active compounds, showing melanin inhibition of 60 and 40%, respectively (Arung et al., 2011). Banerjee and Das (2005) also found that aqueous infusion of cloves had a promising role in restriction of the carcinogenesis process in 9,10-dimethyl benz(a) anthracene–induced skin carcinoma in mouse. Treatment with ethyl acetate extract of clove and compound **14** isolated from the extract selectively increased protein expression of p21(WAF1/Cip1) and γ-H2AX, and downregulated expression of cell cycle–regulated proteins in colorectal cancer HT-29 cells, suggesting transcriptional regulation (Liu et al., 2014). Therefore, it was concluded that clove extract may represent a therapeutic herb for the treatment of cancers and **14** appeared to be one of the bioactive components (Liu et al., 2014). The SOS response is a global response to DNA damage in which the cell cycle is arrested and DNA repair is induced. The methanol extract from clove also had a suppressive effect of the SOS-inducing activity on the mutagen, furylfuramide, in *Salmonella typhimurium* TA1535/pSK1002 (Miyazawa and Hisama, 2001). The antimutagenic activities of clove seed extracts were also reported against two mutant bacterial strains, *S. typhimurium* TA98 and *S. typhimurium* TA100 in a ranged 34.11–79.74% (Sultana et al., 2014).

4.2 Antidiabetic activity

Several reports have documented the potential role of clove as a antidiabetic agent. It was found that clove and insulin regulate the expression of diabetes-related genes, such as phosphoenolpyruvate carboxykinase (PEPCK) and glucose 6-phosphatase (G6Pase) gene, in a similar manner (Prasad et al., 2005). Adefegha and Oboh (2012) have investigated the antidiabetic properties of free and bound phenolic extracts of clove buds against carbohydrate-hydrolyzing enzymes (alpha-amylase and alpha-glucosidase) in rat pancreas. Results revealed that both extracts inhibited alpha-amylase and alpha-glucosidase in a dose-dependent manner, highlighting their therapeutic potential on type 2 diabetes. Adefegha et al. (2014) also reported the effects of clove bud powder on biochemical parameters in a type 2 diabetes rat model, suggesting its ability to attenuate hyperglycemia in the type 2 diabetic condition. Khathi et al. (2013) found that besides improving glucose homeostasis in diabetes, oleanolic acid (**14**) and maslinic acid isolated from clove suppressed postprandial hyperglycaemia mediated in part via inhibition of carbohydrate hydrolysis and a reduction of glucose transporters in the gastrointestinal tract. *S. aromaticum* extracts have also been shown to reduce blood glucose levels in animal models via its ability to increase muscle glycolysis and mitochondria function by activating both AMP-activated protein kinase and sirtuin 1 pathways (Tu et al., 2014). Ethanol extract of flower buds of this plant significantly suppressed an increase in blood glucose level in type 2 diabetic KK-A(y) mice, indicating that it can be a potential functional food ingredient for the prevention of type 2 diabetes. The reported bioactive constituents of the extract were dehydrodieugenol and dehydrodieugenol B with potent human peroxisome proliferator–activated receptor (PPAR-γ) ligand-binding activities and to a lesser extent, compound **14** (Kuroda et al., 2012). However, **14** was mostly found to have a synergistic effect with insulin (Musabayane et al., 2010).

4.3 Antiinflammatory and antinociceptive activities

Clove and compound **1** in noncytotoxic concentrations exhibited immunomodulatory and antiinflammatory actions on cytokine production by murine macrophages (Bachiega et al., 2012). In effect, a concentration of 100 μg/well, clove inhibited (interleukin (IL)-1β, IL-6, and IL-10 production, and exerted an efficient action either before or after lipopolysaccharide challenge for all cytokines; meanwhile compound **1** did not affect IL-1β production but inhibited that of IL-6 and IL-10 (Bachiega et al., 2012). Ethanol extracts of clove flower buds were reported to have antinociceptive and antiinflammatory effects in mice and Wistar rats, as determined by acetic acid–induced abdominal contractions in mice and formalin-induced hind paw edema in Wistar rats, supporting the traditional use of the plant in painful and inflammatory conditions (Tanko et al., 2008). The local anesthetic activity of β-caryophyllene (**3**), one of the main components of clove oil was also demonstrated, as the compound was able to reduce the electrically evoked contractions of the rat phrenic hemidiaphragm drastically, in a dose-dependent manner (Ghelardini et al., 2001).

4.4 Antiinfective activities

4.4.1 Antibacterial activity

The clove methanol extract of *S. aromaticum* was shown to have good antibacterial activities against *Escherichia coli*, *Salmonella typhi*, and *Staphylococcus aureus* (Abd El Azim et al., 2014). Clove essential oil also displayed antibacterial activities, with MIC_{50} values of 1.5 μL/mL against the Gram-negative organisms: *Proteus* spp. and *Klebsiella pneumoniae*, 6.25 μl/mL against *Yersinia enterocolitica* O_9, and 25 μL/mL against *E. coli* O:157 (Al-Mariri and Safi, 2014). The antibacterial effect of the clove essential oil against *Listeria monocytogenes* ATCC19117 growth, added to bovine ground meat stored under refrigeration for 3 days, was reported and MIC was 1.56% (de Oliveira et al., 2013). Cold-pressed clove oil displayed antibacterial activity, having a drastic effect on the biosynthesis of proteins and lipids in *Bacillus subtilis* (Assiri and Hassanien, 2013). Essential oil of clove also had antibiofilm activity against the strong biofilm-forming strains of *Candida albicans* at half MIC value (Khan and Ahmad, 2012). The antibacterial activity of the ethanol extract of clove oil was also reported against clinical isolates of methicillin-resistant *S. aureus* (MRSA) with MIC values ranging from 64–512 µg/mL (Mandal et al., 2011). Quorum sensing is a system of stimuli and response correlating to population density. Many species of bacteria use quorum sensing to coordinate gene expression according to the density of their local population. Clove oil showed promising antiquorum-sensing activity in both wild and mutant strains of *Chromobacterium violaceum* CV12472 and CVO26 (Khan et al., 2009). The synergistic effects of *S. aromaticum* extract with several antibiotics, tetracycline, chloramphenicol, erythromycin, vancomycin, penicillin, oxacillin, cephalothin, ampicillin, cefixitin, cotrimoxazole, and ofloxacin were also reported against *S. aureus* (Betoni et al., 2006). Crude methanol extract of clove also exhibited preferential growth-inhibitory activity against Gram-negative anaerobic periodontal oral pathogens, *Porphyromonas gingivalis* and *Prevotella intermedia* (Cai and Wu, 1996).

4.4.2 Antifungal activity

The clove methanol extract displayed good antifungal activities against *Trichoderma* sp., *Fusarium*, *Aspergillus* sp., and *Penicillium* sp. (Abd El Azim et al., 2014). When tested on 21 isolates of oral *C. albicans*, clove extract showed a better antifungal activity than nystatin, as determined by well diffusion method (Mansourian et al., 2014). Promising antifungal effects of essential oil were also reported against isolates of *Aspergillus*, *Penicillium*, *Fusarium*, and *Scopulariopsis* species. Other fungi from plants and animals namely, *Fusarium moniliforme*, *Fusarium oxysporum*, *Aspergillus* sp., *Mucor* sp., *Trichophyton rubrum*, and *Microsporum gypseum* were found susceptible to clove oil (Rana et al., 2011). Park et al. (2007) also reported the antifungal effects of essential oil at concentrations of 0.05, 0.1, 0.15, and 0.2 mg/mL on the dermatophytes *Microsporum canis* (KCTC 6591), *Trichophyton mentagrophytes* (KCTC 6077), *T. rubrum* (KCCM 60443), *Epidermophyton floccosum* (KCCM 11667), and *Microsporum gypseum*, and identified compound **1** as the active molecule against *T. mentagrophytes* and *M. canis*. The effect of clove administered by two different routes on *C. albicans* growth, using a murine oral candidiasis model was also investigated and it was found that, when clove preparation was administered into the oral cavity of *Candida*-infected mice, their oral symptoms were improved and the number of viable *Candida* cells in the cavity was reduced. These results showed that oral intake of clove may suppress the overgrowth of *C. albicans* in the digestive tract including the oral cavity (Taguchi et al., 2005).

4.4.3 Antiprotozoal activity

The antileishmanial potential of oil from clove flower buds was reported against *Leishmania donovani* (Islamuddin et al., 2014); eugenol-rich essential oil from the plant was shown to have significant activity against *L. donovani*, with 50 % inhibitory concentration of 21 μg/mL and 15.24 μg/mL, respectively against promastigotes and intracellular amastigotes (Islamuddin et al., 2014). The leishmanicidal effect was mediated by apoptosis, as confirmed by externalization of phosphatidylserine, DNA nicking by TdT-mediated dUTP nick-end labeling (TUNEL) assay, cell cycle arrest at sub-G0/G1 phase, loss of mitochondrial membrane potential, and reactive oxygen species generation (Islamuddin et al., 2014). Ethyl acetate and methanol extracts of clove flower buds also displayed antimalarial activity against chloroquine (CQ)-sensitive (3D7) and CQ-resistant (Dd2 and INDO) strains of *Plasmodium falciparum* with lowest IC_{50} values of 13 μg/mL (for ethyl acetate) and 6.25 μg/mL (for methanol) (Bagavan et al., 2011).

4.5 Antiviral activity

Hot-water extracts of *S. aromaticum* was reported to significantly suppressed murine cytomegalovirus yields in lungs of treated mice and was suggested to be beneficial for the prophylaxis of cytomegalovirus diseases in immunocompromised patients (Yukawa et al., 1996). The strong inhibitory activity of clove extracts against HSV-1 in combination with acyclovir was also reported (Kurokawa et al., 1995). Eugeniin was purified as the anti-HSV compound from the clove extract that had anti-HSV activity in mice (Kurokawa et al., 1998). The effective concentration 50 (EC_{50}) of 5 μg/mL of eugeniin was reported against HSV-1 on Vero cells (Kurokawa et al., 1998). Eugeniin also inhibited the growth of acyclovir–phosphonoacetic acid–resistant HSV-1, thymidine kinase–deficient HSV-1, and wild HSV-2 (Kurokawa et al., 1998). Eugeniin also inhibited viral DNA and late viral protein syntheses in the infected Vero cells (Kurokawa et al., 1998). Purified HSV-1 DNA polymerase activity was inhibited by eugeniin noncompetitively with respect to dTTP (Kurokawa et al., 1998).

4.6 Antioxidant activities

The methanol extract of the flowers buds of *S. aromaticum* showed strong radicals scavenging activity against 2,2-diphenyl-1-picrylhydrazyl (DPPH) with an IC_{50} value of 44 μg/mL compared to the reference compound, vitamin C (IC_{50}: 44 μg/mL) (Abd El Azim et al., 2014). Free and bound phenolic extracts of clove buds also exhibited high antioxidant activities, as shown by their high reducing power and 2, 2-azinobis-3-ethylbenzo-thiazoline-6-sulfonate (ABTS) radical scavenging abilities, as well as inhibition of Fe^{2+}-induced lipid peroxidation in rat pancreas in vitro (Adefegha and Oboh, 2012). Clove extract also inhibited the malonaldehyde formation from horse blood plasma oxidized with Fenton's reagent (Lee and Shibamoto, 2001). The antioxidant activity of compounds **1** and **2** was found to be comparable to that of the natural antioxidant, α-tocopherol (vitamin E) (Nurdjannah and Bermawie, 2012).

4.7 Organ-protective effects

Clove extract was shown to reduce the activity of liver enzymes, such as alanine aminotransferase, aspartate aminotransferase, and alkaline phosphatase, and to elevate levels of antioxidant indices, such as glutathione, ascorbic acid, superoxide dismutase, and catalase, highlighting its hepatoprotective effects (Adefegha et al., 2014). Clove oil also exerted hepatoprotective effect against carbon tetrachloride-induced hepatotoxicity in rats, reducing the activities of aspartate aminotransferase, alanine aminotransferase, and alkaline phosphatase, as well as kidney function markers, protein, and lipid profiles (El-Hadary and Ramadan Hassanien, 2015). The effectiveness of the essential oil and compound **1** was related to their ability to stimulate the synthesis of mucus, an important gastroprotective factor (Santin et al., 2011).

4.8 Antithrombotic activity

Clove oil was reported as an inhibitor of platelet aggregation and thromboxane synthesis and may act as an antithrombotic agent. In fact, clove oil inhibited human platelet aggregation induced by arachidonic acid, platelet-activating factor, or collagen, being a more effective inhibitor for aggregation induced by arachidonic acid and platelet-activating factor (IC_{50} values of 4 and 6 µM, respectively) than collagen (IC_{50}: 132 µM) (Rasheed et al., 1984; Saeed and Gilani, 1994). The main constituent of clove oil, compound **1** also inhibited prostaglandin biosynthesis, the formation of thromboxane B2, and arachidonic acid–induced platelet aggregation in vitro (Rasheed et al., 1984).

4.9 Other pharmacological activities

Other pharmacological properties assigned to *S. aromaticum* and its constituents include the molluscicidal effects of the flower bud against the snail, *Lymnaea acuminata* (Kumar and Singh, 2006), as well as the ability of compound **1** to treat experimental arthritis (Grespan et al., 2012).

5 PATENTS WITH *SYZYGIUM AROMATICUM*

There are some patents related to *S. aromaticum*:
1. *Patent WO2,006,067,600 A2* by Katiyar et al. (2006), Title: Herbal formulations as cough lozenge. It is related to polyherbal pharmaceutical formulations in the form of lozenges, which have been found to be effective in treating and managing coughs and sore throats. The formulations comprise extracts from *Adhatoda*, *Syzygium*, and *Eucalyptus plants*.
2. *Patent WO1,994,018,994 A1* by Neiron (1994), Title: Therapeutic herbal composition. It reports a therapeutic herbal composition including *Trigonella foenum-graecum* seed, *S. aromaticum* fruit, *Allium sativum* bulb, *Cinnamonmum zyelanicum* bark, *Saussurea costus* root, and *Euphorbia lathyris* bud that have been shown effective in reducing

cholesterol and triglycerides. This herbal composition can be used to lower cholesterol and treat arthritis, blood pressure, and Alzheimer's disease. It is also effective as a bitters tonic.
3. *Patent EP0,753,305 A1* by Hozumi et al. (1997), Title: Anti-HIV composition and method for treating HIV infection with an anti-HIV agent containing crude drug. The patent provides a number of anti-HIV agents, which are effective in treating the onset of HIV and have fewer side effects. Particularly, it provides an anti-HIV composition, which includes an active agent in a therapeutically effective amount and a pharmaceutically acceptable carrier, wherein the active agent is at least one crude drug selected from the group consisting of *Alpinia officinarum* Hance, *Geum japonicum* Thunb., *Paeonia suffruticosa* Andrews, *Phellodendron amurence* Ruprecht, *Punica granatum* L., *Rhus javanica* L., *S. aromaticum* (L.) Merr. et Perry, *Terminalia arjuna* Wight et Arn., and *Terminalia chebula* Retzus.
4. *Patent WO2,005,072,253 A2* by Jeremiah (2005), Title: An improved herbal therapeutic composition for treating cardiovascular diseases. The patent reports a therapeutic herbal composition useful in lowering cholesterol and treating arthritis, blood pressure, and Alzheimer's disease. The composition includes *T. foenum-graecum* seed, *S. aromaticum* fruit, *A. sativum* bulb, *C. zyelanicum* bark, *S. costus* root, and Alfalfa, which includes an effective amount of sodium chloride to promote the digestibility and storage stability of the compositions, and has been shown effective in reducing cholesterol.

6 TOXICITY OF *SYZYGIUM AROMATICUM*

Clove is traditionally used as aphrodisiac, and its ability to enhance the sexual behavior was demonstrated on male rats and mice by Tajuddin et al. (2003, 2004). Although lower dose of the hexane extract of flower buds of *S. aromaticum* increased testosterone production, higher doses were found to cause a reduction in testosterone production thereby causing perturbation in spermatogenesis of mice (Mishra and Singh, 2008). Therefore, the flower bud used as an aphrodisiac in indigenous systems of medicine in Asian countries should be taken with caution (Mishra and Singh, 2008). Compound **1** was also reported to be toxic in relatively small quantities, for example, a dose of 5–10 mL has been found as a near-fatal dose for a 2-year-old child (Hartnoll et al., 1993).

7 CONCLUSIONS

In the present review, we brought together data related to the phytochemistry and pharmacology of clove. Though the plant is native to Indonesia, it also grows in Africa and is widely used in the majority of countries for culinary and therapeutic purposes. Herein, we have discussed the anticancer, antidiabetic, antiinflammatory, antinociceptive, antibacterial, antifungal, antiprotozoal, antioxidant, and antithrombotic properties, as well as other biological activities of the plant. Finally, *S. aromaticum* can be considered as a potential drug candidate for the management of many ailments; but more clinical and toxicological studies are needed, as some toxicity issues in animals have been documented.

REFERENCES

Abd El Azim, M., El-Mesallamy, A., El-Gerby, M., Awad, A., 2014. Anti-tumor, antioxidant and antimicrobial and the phenolic constituents of clove flower buds (*Syzygium aromaticum*). J. Microbial. Biochem. Technol. S8, 7.

Adefegha, S.A., Oboh, G., 2012. In vitro inhibition activity of polyphenol-rich extracts from *Syzygium aromaticum* (L.) Merr. & Perry (Clove) buds against carbohydrate hydrolyzing enzymes linked to type 2 diabetes and $Fe^{(2+)}$-induced lipid peroxidation in rat pancreas. Asian Pac. J. Trop. Biomed. 2 (10), 774–781.

Adefegha, S.A., Oboh, G., Adefegha, O.M., Boligon, A.A., Athayde, M.L., 2014. Antihyperglycemic, hypolipidemic, hepatoprotective and antioxidative effects of dietary clove (*Szyzgium aromaticum*) bud powder in a high-fat diet/streptozotocin-induced diabetes rat model. J. Sci. Food Agric. 94 (13), 2726–2737.

Al-Mariri, A., Safi, M., 2014. In vitro antibacterial activity of several plant extracts and oils against some Gram-negative bacteria. Iran. J. Med. Sci. 39 (1), 36–43.

Alma, M., Ertas, M., Nitz, S., Kollmannsberger, H., 2007. Chemical composition and content of essential oil from the bud of the cultivated Turkish clove (*Syzygium aromaticum* L.). Bioresources 2 (2), 265–269.

Arung, E.T., Matsubara, E., Kusuma, I.W., Sukaton, E., Shimizu, K., Kondo, R., 2011. Inhibitory components from the buds of clove (*Syzygium aromaticum*) on melanin formation in B16 melanoma cells. Fitoterapia 82 (2), 198–202.

Assiri, A.M., Hassanien, M.F., 2013. Bioactive lipids, radical scavenging potential, and antimicrobial properties of cold pressed clove (*Syzygium aromaticum*) oil. J. Med. Food. 16 (11), 1046–1056.

Bachiega, T.F., de Sousa, J.P., Bastos, J.K., Sforcin, J.M., 2012. Clove and eugenol in noncytotoxic concentrations exert immunomodulatory/anti-inflammatory action on cytokine production by murine macrophages. J. Pharm. Pharmacol. 64 (4), 610–616.

Bagavan, A., Rahuman, A.A., Kaushik, N.K., Sahal, D., 2011. In vitro antimalarial activity of medicinal plant extracts against *Plasmodium falciparum*. Parasitol. Res. 108 (1), 15–22.

Banerjee, S., Das, S., 2005. Anticarcinogenic effects of an aqueous infusion of cloves on skin carcinogenesis. Asian Pac. J. Cancer. Prev. 6 (3), 304–308.

Bao, L.-M., Eerdunbayaer, Nozaki, A., Takahashi, E., Okamoto, K., Ito, H., 2012. Hydrolysable tannins isolated from *Syzygium aromaticum*: structure of a new C-glucosidic ellagitannin and spectral features of tannins with a tergalloyl group. Heterocycles 85 (2), 365–381.

Betoni, J.E., Mantovani, R.P., Barbosa, L.N., Di Stasi, L.C., Fernandes Junior, A., 2006. Synergism between plant extract and antimicrobial drugs used on *Staphylococcus aureus* diseases. Mem. Inst. Oswaldo Cruz 101 (4), 387–390.

Cai, L., Wu, C.D., 1996. Compounds from *Syzygium aromaticum* possessing growth inhibitory activity against oral pathogens. J. Nat. Prod. 59 (10), 987–990.

Cortés-Rojas, D., de Souza, C., Oliveira, W., 2014. Clove (*Syzygium aromaticum*): a precious spice. Asian Pac. J. Trop. Biomed. 4 (2), 90–96.

de Oliveira, T.L., das Gracas Cardoso, M., de Araujo Soares, R., Ramos, E.M., Piccoli, R.H., Tebaldi, V.M., 2013. Inhibitory activity of *Syzygium aromaticum* and *Cymbopogon citratus* (DC.) Stapf. essential oils against *Listeria monocytogenes* inoculated in bovine ground meat. Braz. J. Microbiol. 44 (2), 357–365.

Dwivedi, V., Shrivastava, R., Hussain, S., Ganguly, C., Bharadwaj, M., 2011. Comparative anticancer potential of clove (*Syzygium aromaticum*)—an Indian spice—against cancer cell lines of various anatomical origin. Asian Pac. J. Cancer Prev. 12 (8), 1989–1993.

El-Hadary, A.E., Ramadan Hassanien, M.F., 2015. Hepatoprotective effect of cold-pressed *Syzygium aromaticum* oil against carbon tetrachloride (CCl_4)-induced hepatotoxicity in rats. Pharm. Biol., 1–9.

Ghelardini, C., Galeotti, N., Di Cesare Mannelli, L., Mazzanti, G., Bartolini, A., 2001. Local anaesthetic activity of beta-caryophyllene. Farmaco 56 (5–7), 387–389.

Grespan, R., Paludo, M., Lemos Hde, P., Barbosa, C.P., Bersani-Amado, C.A., Dalalio, M.M., 2012. Antiarthritic effect of eugenol on collagen-induced arthritis experimental model. Biol. Pharm. Bull. 35 (10), 1818–1820.

Halliwell, B., 1999. Free radicals in biology and medicine. Oxford University Press, Oxford.

Hartnoll, G., Moore, D., Douek, D., 1993. Near fatal ingestion of oil of cloves. Arch. Dis. Child. 69 (3), 392–393.

Hozumi, T.S., Ooyama, H.S., Shiraki, K., Kurokawa, M., Kageyama, S., Sato, H., 1997. Anti-HIV composition and method for treating HIV infection with an anti-HIV agent containing crude drug: Google Patents.

Islamuddin, M., Sahal, D., Afrin, F., 2014. Apoptosis-like death in *Leishmania donovani* promastigotes induced by eugenol-rich oil of *Syzygium aromaticum*. J. Med. Microbiol. 63 (Pt. 1), 74–85.

Jeremiah, L.C., 2005. An improved herbal therapeutic composition for treating cardiovascular diseases: Google Patents.

Jirovetz, L., Buchbauer, G., Stoilova, I., Stoyanova, A., Krastanov, A., Schmidt, E., 2006. Chemical composition and antioxidant properties of clove leaf essential oil. J. Agric. Food Chem. 54 (17), 6303–6307.

Kamatou, G.P., Vermaak, I., Viljoen, A.M., 2012. Eugenol: from the remote Maluku Islands to the international market place: a review of a remarkable and versatile molecule. Molecules 17 (6), 6953–6981.

Katiyar, C.K., Padiyar, A., Singh, R., Kumar, R., Kanaujia, A., Sharma, N.K., 2006. Herbal formulations as cough lozenge: Google Patents.

Khan, M.S., Ahmad, I., 2012. Biofilm inhibition by *Cymbopogon citratus* and *Syzygium aromaticum* essential oils in the strains of *Candida albicans*. J. Ethnopharmacol. 140 (2), 416–423.

Khan, M.S., Zahin, M., Hasan, S., Husain, F.M., Ahmad, I., 2009. Inhibition of quorum sensing regulated bacterial functions by plant essential oils with special reference to clove oil. Lett. Appl. Microbiol. 49 (3), 354–360.

Khathi, A., Serumula, M.R., Myburg, R.B., Van Heerden, F.R., Musabayane, C.T., 2013. Effects of *Syzygium aromaticum*-derived triterpenes on postprandial blood glucose in streptozotocin-induced diabetic rats following carbohydrate challenge. PLoS One 8 (11), e81632.

Kumar, P., Singh, D.K., 2006. Molluscicidal activity of *Ferula asafoetida*, *Syzygium aromaticum* and *Carum carvi* and their active components against the snail *Lymnaea acuminata*. Chemosphere 63 (9), 1568–1574.

Kuroda, M., Mimaki, Y., Ohtomo, T., Yamada, J., Nishiyama, T., Mae, T., 2012. Hypoglycemic effects of clove (*Syzygium aromaticum* flower buds) on genetically diabetic KK-Ay mice and identification of the active ingredients. J. Nat. Med. 66 (2), 394–399.

Kurokawa, M., Hozumi, T., Basnet, P., Nakano, M., Kadota, S., Namba, T., 1998. Purification and characterization of eugeniin as an anti-herpesvirus compound from *Geum japonicum* and *Syzygium aromaticum*. J. Pharmacol. Exp. Ther. 284 (2), 728–735.

Kurokawa, M., Nagasaka, K., Hirabayashi, T., Uyama, S., Sato, H., Kageyama, T., 1995. Efficacy of traditional herbal medicines in combination with acyclovir against herpes simplex virus type 1 infection in vitro and in vivo. Antiviral Res. 27 (1–2), 19–37.

Lee, K.G., Shibamoto, T., 2001. Inhibition of malonaldehyde formation from blood plasma oxidation by aroma extracts and aroma components isolated from clove and eucalyptus. Food Chem. Toxicol. 39 (12), 1199–1204.

Liu, H., Schmitz, J.C., Wei, J., Cao, S., Beumer, J.H., Strychor, S., 2014. Clove extract inhibits tumor growth and promotes cell cycle arrest and apoptosis. Oncol. Res. 21 (5), 247–259.

Mandal, S., Saha, K., Pal, N.K., 2011. In vitro antibacterial activity of three Indian spices against methicillin-resistant *Staphylococcus aureus*. Oman Med. J. 26 (5), 319–323.

Mansourian, A., Boojarpour, N., Ashnagar, S., Momen Beitolahi, J., Shamshiri, A.R., 2014. The comparative study of antifungal activity of *Syzygium aromaticum*, *Punica granatum* and nystatin on *Candida albicans*; an in vitro study. J. Mycol. Med. 24 (4), e163–e168.

Mishra, R.K., Singh, S.K., 2008. Safety assessment of *Syzygium aromaticum* flower bud (clove) extract with respect to testicular function in mice. Food Chem. Toxicol. 46 (10), 3333–3338.

Miyazawa, M., Hisama, M., 2001. Suppression of chemical mutagen-induced SOS response by alkylphenols from clove (*Syzygium aromaticum*) in the *Salmonella typhimurium* TA1535/pSK1002 umu test. J. Agric. Food Chem. 49 (8), 4019–4025.

Musabayane, C.T., Tufts, M.A., Mapanga, R.F., 2010. Synergistic antihyperglycemic effects between plant-derived oleanolic acid and insulin in streptozotocin-induced diabetic rats. Ren. Fail. 32 (7), 832–839.

Neiron, J.M., 1994. Therapeutic herbal composition: Google Patents.

Nurdjannah, N., Bermawie, N., 2012. 11—Cloves. In: Peter, K.V. (Ed.), Handbook of Herbs and Spices. second ed. Woodhead Publishing, Cambridge, UK, pp. 197–215.

Pandey, A., Singh, P., 2011. Antibacterial activity of *Syzygium aromaticum* (clove) with metal ion effect against food borne pathogens. Asian J. Plant Sci. Res. 1 (2), 69–80.

Park, M.J., Gwak, K.S., Yang, I., Choi, W.S., Jo, H.J., Chang, J.W., 2007. Antifungal activities of the essential oils in *Syzygium aromaticum* (L.) Merr. Et Perry and *Leptospermum petersonii* Bailey and their constituents against various dermatophytes. J. Microbiol. 45 (5), 460–465.

Prasad, R.C., Herzog, B., Boone, B., Sims, L., Waltner-Law, M., 2005. An extract of *Syzygium aromaticum* represses genes encoding hepatic gluconeogenic enzymes. J. Ethnopharmacol. 96 (1–2), 295–301.

Rana, I.S., Rana, A.S., Rajak, R.C., 2011. Evaluation of antifungal activity in essential oil of the *Syzygium aromaticum* (L.) by extraction, purification and analysis of its main component eugenol. Braz. J. Microbiol. 42 (4), 1269–1277.

Rasheed, A., Laekeman, G., Totte, J., Vlietinck, A.J., Herman, A.G., 1984. Eugenol and prostaglandin biosynthesis. N. Engl. J. Med. 310 (1), 50–51.

Saeed, S.A., Gilani, A.H., 1994. Antithrombotic activity of clove oil. J. Pak. Med. Assoc. 44 (5), 112–115.

Santin, J.R., Lemos, M., Klein-Junior, L.C., Machado, I.D., Costa, P., de Oliveira, A.P., 2011. Gastroprotective activity of essential oil of the *Syzygium aromaticum* and its major component eugenol in different animal models. Naunyn Schmiedebergs Arch. Pharmacol. 383 (2), 149–158.

Shrivastava, K., Sahu, S., Mishra, S., De, K., 2014. In vitro antimicrobial activity and phytochemical screening of *Syzygium aromaticum*. Asian J. Res. Pharm. Sci. 4 (1), 12–15.

Sultana, B., Anwar, F., Mushtaq, M., Aslam, M., Ijaz, S., 2014. In vitro antimutagenic, antioxidant activities and total phenolics of clove (*Syzygium aromaticum* L.) seed extracts. Pak. J. Pharm. Sci. 27 (4), 893–899.

Taguchi, Y., Ishibashi, H., Takizawa, T., Inoue, S., Yamaguchi, H., Abe, S., 2005. Protection of oral or intestinal candidiasis in mice by oral or intragastric administration of herbal food, clove (*Syzygium aromaticum*). Nihon Ishinkin Gakkai Zasshi 46 (1), 27–33.

Tajuddin, Ahmad, S., Latif, A., Qasmi, I.A., 2003. Aphrodisiac activity of 50% ethanolic extracts of *Myristica fragrans* Houtt. (nutmeg) and *Syzygium aromaticum* (L) Merr. & Perry. (clove) in male mice: a comparative study. BMC Complement. Altern. Med. 3, 6.

Tajuddin, Ahmad, S., Latif, A., Qasmi, I.A., 2004. Effect of 50% ethanolic extract of *Syzygium aromaticum* (L.) Merr. & Perry. (clove) on sexual behaviour of normal male rats. BMC Complement. Altern. Med. 4, 17.

Tanko, Y., Mohammed, A., Okasha, M.A., Umar, A.H., Magaji, R.A., 2008. Anti-nociceptive and anti-inflammatory activities of ethanol extract of *Syzygium aromaticum* flower bud in Wistar rats and mice. Afr. J. Tradit. Complement. Altern. Med. 5 (2), 209–212.

Tu, Z., Moss-Pierce, T., Ford, P., Jiang, T.A., 2014. *Syzygium aromaticum* L. (clove) extract regulates energy metabolism in myocytes. J. Med. Food. 17 (9), 1003–1010.

Wankhede, T., 2015. Evaluation of antioxidant and antimicrobial activity of the Indian clove *Syzygium aromaticum* L. Merr. & Perr. Int. Res. J. Sci. Eng. 3 (4), 166–172.

Yukawa, T.A., Kurokawa, M., Sato, H., Yoshida, Y., Kageyama, S., Hasegawa, T., 1996. Prophylactic treatment of cytomegalovirus infection with traditional herbs. Antiviral Res. 32 (2), 63–70.

CHAPTER 30

Zingiber officinale

A.T. Mbaveng, V. Kuete

1 INTRODUCTION

Z. officinale Roscoe (Zingiberaceae), commonly known as ginger, is a flowering plant with root named ginger, widely used as a spice or a folk medicine. The plant is indigenous to South China and was spread eventually to the Spice Islands, other parts of Asia, and subsequently to West Africa (https://unitproj.library.ucla.edu/biomed/spice/index.cfm, 2002). India and China are the main producers and exporters. In Africa, Nigeria is the first producer, with 156,000 tons registered in 2012 (https://en.wikipedia.org/wiki/Ginger, 2015). Ginger makes a hot, fragrant kitchen spice. Rhizomes of young ginger are often pickled in vinegar or sherry as a snack or cooked as an ingredient in many dishes. They can be steeped in boiling water to make ginger tisane, to which honey is often added; sliced orange or lemon fruit may be added (https://en.wikipedia.org/wiki/Ginger, 2015). Rhizomes of mature ginger are fibrous and nearly dry. The juice from ginger roots (Fig. 30.1) is often used as a spice in Indian recipes and is a common ingredient of Chinese, Korean, Japanese, Vietnamese, and many South Asian and African cuisines for flavoring dishes, such as seafood, meat, and vegetarian dishes. The characteristic odor and flavor of ginger is caused by a mixture of zingerone (**1**), shogaols, and gingerols, volatile oils that compose 1–3% of the weight of fresh ginger.

Z. officinale has been used since ancient times in Ayurvedic and traditional Chinese medicine to treat a wide range of ailments including common cold, fever, sore throat, pain, rheumatism, bronchitis, as a carminative and appetite stimulant, antipyretic, for digestive problems, gastrointestinal disorders, nausea and vomiting associated with motion sickness and pregnancy, and abdominal spasm (Chrubasik et al., 2005; Ali et al., 2008; Baliga et al., 2013). The plant has also been used to treat stomachaches, diarrhea, toothache, gingivitis, bronchitis, hypertension, dementia, helminthiasis, constipation, and asthmatic respiratory disorders (Chrubasik et al., 2005; Grzanna et al., 2005; Ali et al., 2008; Baliga et al., 2013). The plant is widely used as home medicine in case of dyspepsia, flatulence, abdominal discomfort, and nausea (Baliga et al., 2013). It has been recommended by herbalists for use as a carminative, diaphoretic, antispasmodic, expectorant, peripheral circulatory stimulant, and astringent (Baliga et al., 2013). Ginger is also of medicinal use in various folk systems in both traditional systems of medicines in Asia and Africa (Ali et al., 2008; Baliga et al., 2013; Chrubasik et al., 2005). In the present chapter, an overview of the chemistry and pharmacology of *Z. officinale* will be provided.

Figure 30.1 *Fresh ginger rhizome. (Photo: V. Kuete).*

2 BOTANICAL ASPECT, DISTRIBUTION, AND PRODUCTION OF *ZINGIBER OFFICINALE*

Z. officinale is a perennial, herbaceous plant that grows up to a height of about 1 m. The leaves develop from the branched rhizome and the flowers, which resemble the orchids, are inconspicuous and occur in a dense spike, consisting of several overlapping scales on an elongated stalk. Each flower has three yellowish-orange petals with an additional purplish, lip-like structure. It has been cultivated for thousands of years for medicinal purposes and as a spice (Ali et al., 2008; Chrubasik et al., 2005). *Z. officinale* has become naturalized in many countries, and now has a wide distribution throughout tropical and subtropical parts of the world. The rhizome is frequently used as a condiment for various foods and beverages (Baliga et al., 2013). It is used either fresh, dried, or as extract (mostly decoction) and globally its demand is ever increasing. The US Food and Drug Administration categorized ginger as a food additive (Chrubasik et al., 2005; Ali et al., 2008; Baliga et al., 2013). According to the 2012 statistics of the Food and Agricultural Organization of the United Nations (FAO), the world production of ginger was 2,095,056 tons, with India, having 33% (703,000 tons) of the global production, followed by China (about 20%; 425,000 tons), Nepal (about 12%; 255,208 tons), Nigeria and Thailand (each about 7%; 156,000 and 150,000, respectively), and Indonesia (about 5%; 113,851 tons) (https://en.wikipedia.org/wiki/Ginger, 2015). In Africa, other big producers of ginger as recorded by FAO in 2003 include Cameroon (about 7500 tons), Ethiopia (400 tons), Mauritius (200 tons), Kenya (150 tons), Uganda (120 tons), Madagascar (30 tons), Ghana and Zambia (0.01 ton) (Prabhakaran Nair, 2013).

3 CHEMISTRY OF *ZINGIBER OFFICINALE*

Ginger contains several compounds among which a mixture of zingerone, shogaols, gingerols and volatile oils are responsible for its characteristic odor and flavor (Fig. 30.2). Other constituents include, capsaicin, gingediol, galanolactone, gingesulfonic acid,

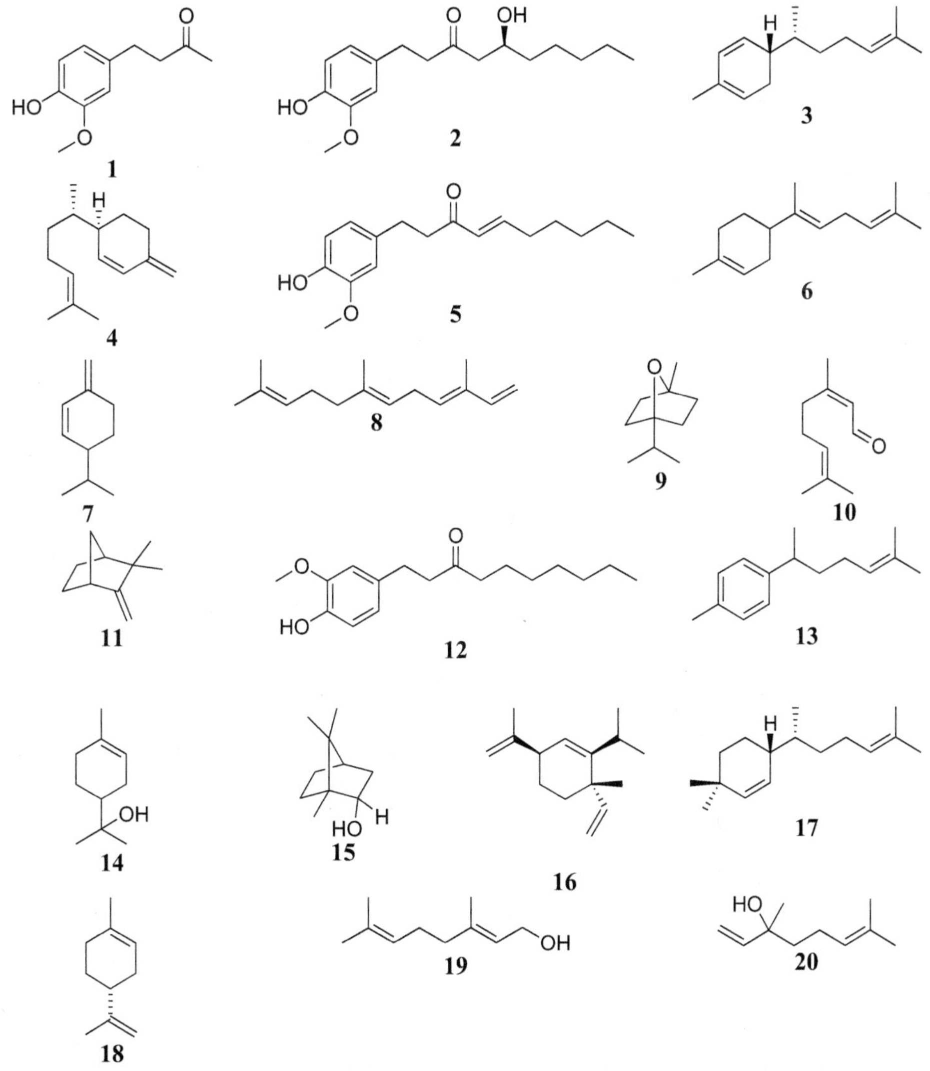

Figure 30.2 *Chemical structures of compounds from ginger.* Zingerone (**1**); gingerol (**2**); zingiberene (**3**); β-sesquiphellandrene (**4**); shogaol (**5**); bisabolene (α-form; **6**); β-phellandrene (**7**); farnesene (**8**); 1,4-cineol (**9**); citral (**10**); camphene (**11**); 6-paradol (**12**); curcumene (**13**); terpineol (α-form; **14**); borneol (**15**); β-elemene (**16**); zingiberenol (**17**); limonene (**18**); geraniol (**19**); linalool (**20**).

galactosylglycerols, gingerglycolipids, diarylheptanoids, neral, monoacyldi-vitamins, and phytosterols (Chrubasik et al., 2005; Ali et al., 2008; Baliga et al., 2013). Ginger contains up to 3% of a fragrant essential oil whose main constituents are sesquiterpenoids, with zingiberene (**3**) as the main component. Other important constituents present in the volatile oil are the mono and sesquiterpenes, camphene, β-sesquiphellandrene (**4**), β-bisabolene (**6**), α-farmesene (**8**), curcumene, cineole (**9**), citral (**10**), terpineol

(**14**), terpenes, borneol (**15**), β-elemene (**16**), zingiberenol (**17**), limonene (**18**), geraniol (**19**), zingiberol, linalool (**20**) (Chrubasik et al., 2005; Ali et al., 2008). The nonvolatile phenylpropanoid-derived compounds, particularly gingerols, shogaols (**5**), paradols (**12**), and zingerone (**1**) are responsible of the pungent taste of ginger (Baliga et al., 2013). Compound **1** is produced from gingerols (**2**) during drying or cooking processes (McGee, 2004). These compounds are responsible for the warm pungent sensation in the mouth and are also reported to account for many of the pharmacological effects of the plant (Baliga et al., 2013). The monoterpene phellandrene (**7**) as well as curcumene (**13**) have also been reported in ginger.

4 PHARMACOLOGY OF *ZINGIBER OFFICINALE*

Many scientific reports have documented the biological and pharmacological potential of ginger including antibacterial, analgesic and antiinflammatory, antidiabetic, antitumor, etc.

4.1 Ginger in the treatment and prevention of arthritis

Fouda and Berika (2009) have investigated the potential of the hydroalcoholic extract of ginger rhizomes to ameliorate inflammatory process in rat collagen-induced arthritis. They found that the plant extract in doses higher than 50 mg/kg/day intraperitoneally for 26 days can ameliorate disease incidence, joint temperature and swelling, and cartilage destruction, together with reduction of serum levels of interleukin (IL)-1β, IL-2, IL-6 and tumor necrosis factor-alpha (TNFα). Extract at the dose of 200 mg/kg/day was better than 2 mg/kg/day of indomethacin as far as biochemical parameters were concerned, making ginger a good alternative to nonsteroidal antiinflammatory drugs (NSAIDs) for patients with rheumatoid arthritis (Fouda and Berika, 2009). Animals' studies demonstrated that ginger and some of its compounds are effective in the prevention of chemically-induced arthritis (Srivastava and Mustafa, 1992). In a study carried out with 56 patients (28 with rheumatoid arthritis, 18 with osteoarthritis and 10 with muscular discomfort) who used powdered ginger against their afflictions, Srivastava and Mustafa (1992) found that among the arthritis patients, more than three-quarters experienced, to varying degrees, relief in pain and swelling. All the patients with muscular discomfort experienced relief in pain. None of the patients reported adverse effects during the period of ginger consumption which ranged from 3 months to 2.5 years (Srivastava and Mustafa, 1992). Levy et al. (2006) have also demonstrated that 6-shogaol possesses antiinflammatory and antiarthritic properties in rats. In the study, authors treated Sprague-Dawley rats with a single injection of a commercial preparation of complete Freund's adjuvant (CFA) to induce monoarthritis in the right knee over a period of 28 days. Then, during this development of arthritis, rat received a daily oral dose of 6-shogaol [6.2 mg/kg body weight (b.w.)]. As a result, in the 6-shogaol rats, significantly lower magnitudes of unsustained swelling of the knees were produced during the investigation period. They also found that 6-shogaol and indomethacin were most effective in reducing swelling of the knees on day 28 when the controls still had maximum swelling (Levy et al., 2006). It was therefore concluded that 6-shogaol reduced the inflammatory response and protected the femoral cartilage from damage produced in a CFA monoarthritic model of the knee joint of rats (Levy et al., 2006).

4.2 Analgesic and antiinflammatory activities

The analgesic and antiinflammatory effects of ginger and its constituents have been reported. Consumption of 2 g/day of ginger was reported to be capable of modestly reducing muscle pain stemming from eccentric resistance exercise and prolonged running, particularly if taken for a minimum of 5 days (Wilson, 2015). There are also studies demonstrating that ginger may accelerate recovery of maximal strength after eccentric resistance exercise and reduce the inflammatory response to cardio-respiratory exercise (Wilson, 2015). Ginger extract was found to have a potential antitolerant/antidependence property against chronic usage of morphine. In fact, it was shown that chronic morphine-injected rats displayed tolerance to the analgesic effect of morphine as well as morphine dependence and that ginger (50 and 100 mg/kg) completely prevented the development of morphine tolerance (Darvishzadeh-Mahani et al., 2012). Cyclooxygenases (COX)-1 and COX-2 are the targets of widely used NSAIDs and are essential for such physiological processes as maintenance of the gastrointestinal tract, renal function, and fever. COX-1 is expressed constitutively in all tissues, but COX-2 is induced specifically during inflammatory, degenerative, and neoplastic processes (van Breemen et al., 2011). COX-1 and COX-2 catalyze the conversion of arachidonic acid to the endoperoxide prostaglandin $(PG)H_2$ which can form a variety of prostaglandins, thromboxanes and prostacyclin through catalysis by nonrate limiting enzymes or by nonenzymatic rearrangement (van Breemen et al., 2011). Ginger rhizomes demonstrated a strong analgesic action mediated by COX-1 inhibition (van Breemen et al., 2011). Compound **2** and their derivatives showed more potent antiplatelet effects than aspirin (Nurtjahja-Tjendraputra et al., 2003). The compound 8-paradol, a natural constituent of ginger, was found to be the most potent COX-1 inhibitor and antiplatelet aggregation agent. It was suggested that the mechanism underlying arachidonic acid–induced platelet aggregation inhibition may be related to attenuation of COX-1/Tx synthase enzymatic activity (Nurtjahja-Tjendraputra et al., 2003). Intraperitoneal administration of 6-gingerol (25–50 mg/kg) produced an inhibition of acetic acid–induced writhing response and formalin-induced licking time in the late phase (Young et al., 2005). Also, 6-gingerol (50–100 mg/kg) produced an inhibition of paw edema induced by carrageenin (Young et al., 2005). Potent suppressive effect of ginger on acute and chronic inflammation, and inhibition of macrophage activation has been reported and was found to be involved in its antiinflammatory effect (Shimoda et al., 2010). Its constituents,

6-shogaol, gingerdiols, and proanthocyanidins, were identified as nitrite oxide (NO) production inhibitors (Shimoda et al., 2010). Ginger was also reported to exert an antiinflammatory effect on lungs, attenuating rat trachea hyperreactivity meanwhile COX metabolites were suggested to be involved in the process (Aimbire et al., 2007). Ojewole also reported the analgesic and antiinflammatory activities of the ethanol extract of ginger rhizomes. Using mice and rats, the investigator showed that ginger extract (50–800 mg/kg i.p.) produced dose-dependent, significant analgesic effects against thermally and chemically induced nociceptive pain in mice (Ojewole, 2006). The plant extract (50–800 mg/kg p.o.) also significantly inhibited fresh egg albumin-induced acute inflammation in rats (Ojewole, 2006). It was also demonstrated that ginger and its constituents are capable of inhibiting PGE2 production (Lantz et al., 2007). In fact, crude organic extracts of ginger were able to inhibit lipopolysaccharide (LPS)-induced PGE2 production. Extracts containing either predominantly gingerols or shogaols were both highly active at inhibiting LPS-induced PGE2 production; extracts or standards containing predominantly gingerols were able to inhibit LPS-induced COX-2 expression (Lantz et al., 2007).

4.3 Anticancer activity

Extract of *Z. officinale*, its major pungent components, 6-shogaol and 6-gingerol showed antiproliferative effects on several tumor cell lines; it was shown that ginger extract has potent anticancer activity against pancreatic cancer cells by inducing reactive oxygen species (ROS)-mediated apoptosis (Akimoto et al., 2015). In fact, ethanol extract of ginger suppressed cell cycle progression and consequently induced death of human pancreatic cancer cells, Panc-1, AsPC-1, BxPC-3, CAPAN-2, CFPAC-1, MIAPaCa-2 and SW1990, and mouse pancreatic cancer cells, Panc02 (Akimoto et al., 2015). The extract markedly increased the microtubule-associated protein light chain (LC)3-II/LC3-I ratio, decreased sequestosome 1 (SQSTM1)/p62 protein, and enhanced vacuolization of the cytoplasm in Panc-1 cells (Akimoto et al., 2015). The cytotoxic effect of the ethanolic extract of ginger was reported toward cholangiocarcinoma CL-6 cells (Plengsuriyakarn et al., 2012). Experimental studies also showed that ginger as well as 6-gingerol and 6-shogaol exerted cytotoxic effects against gastrointestinal cancer cells (Prasad and Tyagi, 2015). The anticancer activity of ginger has been attributed to its ability to modulate several signaling molecules like NF-κB, STAT3, MAPK, PI3K, ERK1/2, Akt, TNF-α, COX-2, cyclin D1, cdk, MMP-9, survivin, cIAP-1, XIAP, Bcl-2, caspases, and other cell growth regulatory proteins (Prasad and Tyagi, 2015). Qi et al. (2015) have observed that 6-shogaol (15 mg/kg) significantly inhibited colorectal tumor growth in a xenograft mouse model. The investigators showed that 6-shogaol significantly inhibited HCT-116 cells and SW-480 cells' proliferation with IC_{50} values of 7.5 and 10 μM, respectively (Qi et al., 2015). Growth of HCT-116 cells was arrested at the G2/M phase, primarily mediated by the upregulation of p53, the CDK inhibitor p21(waf1/cip1) and

GADD45α and by the downregulation of cdc2 and cdc25A (Qi et al., 2015). Using p53$^{-/-}$ and p53$^{+/+}$ HCT-116 cells, authors confirmed that p53/p21 was the main pathway that contributed to the G2/M cell cycle arrest by 6-shogaol (Qi et al., 2015). This compound also decreased cancer-induced upregulation of CC-chemokine ligand 2 in tumor-associated dendritic cells (TADCs), preventing the enhancing effects of TADCs on tumorigenesis and metastatic properties in human lung adenocarcinoma A549 cells and breast adenocarcinoma MDA-MB-231 cells (Hsu et al., 2015). Compound 6-shogaol also induced apoptosis and G2/M phase arrest in human cervical cancer HeLa cells with endoplasmic reticulum stress and mitochondrial pathway being involved (Liu et al., 2012). Gingerol was found to sensitize A549 cells to TNF-related apoptosis-inducing ligand (TRAIL)-induced apoptosis by inhibiting the autophagy flux (Nazim et al., 2015); 6-gingerol was reported to inhibit cell proliferation of human colon cancer SW-480 cells and HCT116 cells and induced apoptosis in SW-480 cells, associated with activation of caspases 8, 9, 3, and 7 and cleavage of poly ADP ribose polymerase (PARP) (Radhakrishnan et al., 2014).

4.4 Antidiabetic activity

The aqueous extract of *Z. officinale* at an oral dose of 500 mg/kg b.w. showed hypoglycemic effect in alloxan-induced diabetic rats (Abdullah et al., 2010). In a study conducted by Iranloye et al. (2011), it was found that dietary ginger had hypoglycemic effect, enhancing insulin synthesis in male rats. At a dose of 500 mg/kg b.w., raw ginger was significantly effective in lowering serum glucose, cholesterol, and triacylglycerol levels in the ginger-treated diabetic rats (Al-Amin et al., 2006). Treatment with ginger also caused a decrease in serum cholesterol, serum triglyceride (TG), and blood pressure in diabetic rats, suggesting its potential antidiabetic activity in type I diabetic rats, with 5-hydroxytryptamine receptors being involved (Akhani et al., 2004). Oral administration of ginger powder supplement was shown to improve fasting blood sugar, hemoglobin A1c, apolipoprotein B, apolipoprotein A-I, apolipoprotein B/apolipoprotein A-I and malondialdehyde (MDA) in type 2 diabetic patients, suggesting that it may have a role in alleviating the risk of some chronic complications of diabetes (Khandouzi et al., 2015). Ginger also improved insulin sensitivity and some fractions of lipid profile, and reduced PGE$_2$ in type 2 diabetic patients indicating that it can be an effective treatment for prevention of diabetes complications (Arablou et al., 2014). In a randomized clinical trial conducted on 20- to 60-year-old patients with type 2 diabetes who did not received insulin, Shidfar et al. (2015) showed that 3 months supplementation of ginger improved glycemic indices, total antioxidant capacity, and paraoxonase-1 activities.

4.5 Hepatoprotective effects

Hasan et al. (2015) have investigated the potential of ginger in the regression of liver fibrosis and its underlining mechanism of action. They found that ginger extract markedly prevented liver injury as evidenced by the decreased liver marker enzymes. Concurrent administration of ginger significantly protected against CCl_4-induced inflammation as showed by the decreased proinflammatory cytokine levels as well as the downregulation of the NF-κB/IκB and TGF-β1/Smad3 pathways in CCl_4-administered rats (Hasan et al., 2015). Using experimental rats, ginger at 200 mg/kg b.w. exhibited hepatoprotective activity in acute ethanol-induced fatty liver toxicity (Nwozo et al., 2014). The hepatoprotective effect of ginger was evidenced by decrease in the activities of serum enzymes as well as serum TG level, total cholesterol (TC) level, and hepatic MDA while it significantly restored the level of glutathione (GSH), glutathione-S-transferase (GST), and superoxide dismutase (SOD) activities (Nwozo et al., 2014). The hepatoprotective activity of the ethanolic extract of ginger rhizomes against thioacetamide-induced hepatotoxicity in rats was also reported (Abdulaziz Bardi et al., 2013). The compound 6-gingerol displayed promising hepatoprotective effect, comparable to the standard drug silymarin (Sabina et al., 2011). In effect, treatment of 6-gingerol to acetaminophen-induced hepatotoxic mice significantly lower the hepatic marker enzymes [aspartate aminotransferase (AST), alanine aminotransferase (ALT), and alkaline phosphatase (ALP)] and total bilirubin in serum (Sabina et al., 2011). In addition, 6-gingerol treatment prevented the elevation of hepatic malondialdehyde formation and the depletion of antioxidant status in the liver of acetaminophen-intoxicated mice (Sabina et al., 2011).

4.6 Nephroprotective effects

Ginger extract showed nephroprotection mediated by preventing the doxorubicin-induced decline of renal antioxidant status, and also by increasing the activity of GST in rats (Ajith et al., 2008). Ethanol extract alone and in combination with vitamin E partially ameliorated cisplatin-induced nephrotoxicity and this was mediated either by preventing the cisplatin-induced decline of renal antioxidant defense system or by their direct free radical scavenging activity (Ajith et al., 2007). Gingerols also promoted a nephroprotective effect on gentamicin-mediated nephropathy by oxidative stress, inflammatory processes, and renal dysfunction (Rodrigues et al., 2014). In effect, it was observed that consumption of ginger by Wistar rats previously treated with gentamicin resulted in the amelioration in renal function parameters and reduced lipid peroxidation and nitrosative stress, in addition to an increment in the levels of GSH and SOD activity (Rodrigues et al., 2014). Gingerols also promoted significant reductions in mRNA transcription for TNF-α, IL-2, and interferon gamma (IFN-γ) (Rodrigues et al., 2014).

4.7 Antioxidant activity

Ginger was shown to be comparatively as effective as ascorbic acid as an antioxidant agent. In effect, ginger (1% w/w) significantly lowered lipid peroxidation by maintaining the activities of the antioxidant enzymes, such as SOD, catalase, and glutathione peroxidase in rats (Ahmed et al., 2000). The blood glutathione content was significantly increased in ginger fed rats with similar effects being observed with natural antioxidant ascorbic acid upon treatment with 100 mg/kg b.w. (Ahmed et al., 2000). Compound **1** showed scavenging activity of superoxide anion (Krishnakantha and Lokesh, 1993). The scavenging effects of 6-gingerol and **1** on peroxyl radicals generated by pulse radiolysis were also reported and they were shown not to accelerate DNA damage in the bleomycin-Fe (III) system (Aeschbach et al., 1994).

4.8 Other pharmacological activities

Apart from the aforementioned activities, several other pharmacological effects of ginger and its constituents were documented.

Antiemetic effect. Ginger was shown to be effective against nausea and vomiting. However, addition of ginger to the standard antiemetic regimen has shown no advantage in reducing acute and delayed nausea and vomiting in patients with cisplatin-based regimen (Fahimi et al., 2011). The components in ginger that are responsible for the antiemetic effect are thought to be gingerols, shogaols, and galanolactone, a diterpenoid of ginger (Huang et al., 1991; Mishra et al., 2012). It was shown that oral intake of ginger improves gastroduodenal motility in the fasting state and after a standard test meal (Micklefield et al., 1999).

Molluscicidal and antischistosomal activities. Concentration-time-dependent anthelmintic activity of ginger was demonstrated in vitro on the cestodes *Raillietina cesticillus* (El-Bahy and Bazh, 2015). However, the in vivo effect of ginger on *R. cesticillus* was found to be lower than that in vitro (El-Bahy and Bazh, 2015). The constituents of ginger, gingerenone A, 6-dehydrogingerdione, 4-shogaol, 5-hydroxy-gingerdiol, and 3S,5S-[6]-gingerdiol were identified as the anthelmintic ingredients of the plant against the parasite *Hymenolepis nana* (Lin et al., 2014). The antischistosomal activity of ginger extract was also reported against *Schistosoma mansoni* but the extract was rather found to not significantly decrease (7.26%) *S. mansoni* worms in mice (Seif el-Din et al., 2014). However, Mostafa et al. (2011) rather confirmed the good antischistosomal activities against *S. mansoni* harbored in C57 mice. Gingerol and shogaol were identified as potent molluscicidal constituents of ginger against *S. mansoni* (Adewunmi et al., 1990). Ginger also exhibited a high and concentration-time dependent death rate against the nematode *Ascaridia galli* (Bazh and El-Bahy, 2013).

Cardiovascular effect. In vitro studies showed that gingerols and the related shogaols have cardio-depressant activity at low doses and cardiotonic properties at higher doses (Mishra et al., 2012). Both 6-shogaol and 6-gingerol, and the gingerdiones were found to be potent enzymatic inhibitors of prostaglandin, thromboxane, and leukotriene biosynthesis (Mishra et al., 2012).

Effect on migraine. It was reported that 500–600 mg of ginger powder administration at the onset of migraine for 3–4 days at interval of 4 h, could exert abortive and prophylactic effects in migraine headache without any side effects (Mustafa and Srivastava, 1990).

Antirheumatic properties. Sharma et al., 1994 reported that the oil of ginger was effective as an antirheumatic agent. In their study, authors induced severe arthritis in the right knee and right paw of male Sprague-Dawley rats by injecting a fine suspension of dead *Mycobacterium tuberculosis* bacilli in liquid paraffin (Sharma et al., 1994). Ginger oil (33 mg/kg b.w.), given orally for 26 days, caused a significant suppression of both paw and joint swelling, suggesting that ginger oil have potent antiinflammatory and/or antirheumatic properties (Sharma et al., 1994).

5 PATENTS WITH *ZINGIBER OFFICINALE*

1. Patent WO2003049753 A1 by Chandan et al. (2003). Title: Bioavailability enhancing activity of *Zingiber officinale* Linn and its extracts/fractions thereof. The patent relates to a bioenhancing composition containing extract and/or bioactive fraction/isolate from the plant *Z. officinale* in combination with drugs, nutrients, nutraceuticals, micronutrients and herbal drugs/products, and optionally containing piperine as an extract/active fraction obtained from *Piper nigrum*, *Piper longum* or its oleoresin as a bioavailability enhancer and its process for producing the extract or fractions from the plant source (Chandan et al., 2003).
2. Patent WO2010083967 A1 by Bombardelli (2010). Title: Compositions comprising lipophilic extracts of *Zingiber officinale* and *Echinacea angustifolia* for the prevention and treatment of gastro-oesophageal reflux and chemotherapy-induced emesis (Bombardelli, 2010).
3. Patent US20,120,015,060 A1 by Bombardelli (2012). Title: Compositions comprising a lipophilic extract of *Zingiber officinale* and an extract of *Cynara scolymus*, which are useful for the prevention and treatment of oesophageal reflux and irritable bowel syndrome (Bombardelli, 2012).
4. Patent WO2012013551 A1 by Bombardelli and Morazzoni (2012). Formulations containing extracts of *Echinacea angustifolia* and *Zingiber officinale* which are useful in reducing inflammation and peripheral pain (Bombardelli and Morazzoni, 2012).

6 CONCLUSIONS

Ginger is consumed worldwide as a spice and for its therapeutic properties. The plant has a number of chemicals responsible for its various medicinal properties, such as antiarthritis, antiinflammatory, antidiabetic, antibacterial, antifungal, anticancer effects, etc. The present chapter compiled scientific data retrieved from website, such as PubMed, ScienceDirect, Scopus, Web-of-Knowledge, Google Scholar. Herein, some patents related to ginger were also listed. In conclusion, ginger is a natural available resource that can be used in the management of several human ailments.

REFERENCES

Abdulaziz Bardi, D., Halabi, M.F., Abdullah, N.A., Rouhollahi, E., Hajrezaie, M., Abdulla, M.A., 2013. In vivo evaluation of ethanolic extract of *Zingiber officinale* rhizomes for its protective effect against liver cirrhosis. Biomed. Res. Int. 2013, 918460.

Abdullah, S., Abidin, S., Murad, N., Makpol, S., Wan Ngah, Z., Yusof, Y., 2010. Ginger extract (*Zingiber officinale*) triggers apoptosis and G0/G1 cells arrest in HCT116 and HT 29 colon cancer cell lines. Afr. J. Biochem. Res. 4 (4), 134–142.

Adewunmi, C.O., Oguntimein, B.O., Furu, P., 1990. Molluscicidal and anti-schistosomal activities of *Zingiber officinale*. Planta Med. 56 (4), 374–376.

Aeschbach, R., Loliger, J., Scott, B.C., Murcia, A., Butler, J., Halliwell, B., 1994. Antioxidant actions of thymol, carvacrol, 6-gingerol, zingerone and hydroxytyrosol. Food Chem. Toxicol. 32 (1), 31–36.

Ahmed, R.S., Seth, V., Banerjee, B.D., 2000. Influence of dietary ginger (*Zingiber officinales* Rosc) on antioxidant defense system in rat: comparison with ascorbic acid. Indian J. Exp. Biol. 38 (6), 604–606.

Aimbire, F., Penna, S.C., Rodrigues, M., Rodrigues, K.C., Lopes-Martins, R.A., Sertie, J.A., 2007. Effect of hydroalcoholic extract of *Zingiber officinalis* rhizomes on LPS-induced rat airway hyperreactivity and lung inflammation. Prostaglandins Leukot. Essent. Fatty Acids 77 (3–4), 129–138.

Ajith, T.A., Aswathy, M.S., Hema, U., 2008. Protective effect of *Zingiber officinale* roscoe against anticancer drug doxorubicin-induced acute nephrotoxicity. Food Chem. Toxicol. 46 (9), 3178–3181.

Ajith, T.A., Nivitha, V., Usha, S., 2007. *Zingiber officinale* Roscoe alone and in combination with alpha-tocopherol protect the kidney against cisplatin-induced acute renal failure. Food Chem. Toxicol. 45 (6), 921–927.

Akhani, S.P., Vishwakarma, S.L., Goyal, R.K., 2004. Anti-diabetic activity of *Zingiber officinale* in streptozotocin-induced type I diabetic rats. J. Pharm. Pharmacol. 56 (1), 101–105.

Akimoto, M., Iizuka, M., Kanematsu, R., Yoshida, M., Takenaga, K., 2015. Anticancer effect of ginger extract against pancreatic cancer cells mainly through reactive oxygen species-mediated autotic cell death. PLoS One 10 (5), e0126605.

Al-Amin, Z.M., Thomson, M., Al-Qattan, K.K., Peltonen-Shalaby, R., Ali, M., 2006. Anti-diabetic and hypolipidaemic properties of ginger (*Zingiber officinale*) in streptozotocin-induced diabetic rats. Br. J. Nutr. 96 (4), 660–666.

Ali, B.H., Blunden, G., Tanira, M.O., Nemmar, A., 2008. Some phytochemical, pharmacological and toxicological properties of ginger (*Zingiber officinale* Roscoe): a review of recent research. Food Chem. Toxicol. 46 (2), 409–420.

Arablou, T., Aryaeian, N., Valizadeh, M., Sharifi, F., Hosseini, A., Djalali, M., 2014. The effect of ginger consumption on glycemic status, lipid profile and some inflammatory markers in patients with type 2 diabetes mellitus. Int. J. Food Sci. Nutr. 65 (4), 515–520.

Baliga, M.S., Latheef, L., Haniadka, R., Fazal, F., Chacko, J., Arora, R., 2013. Chapter 41—ginger (*Zingiber officinale* Roscoe) in the treatment and prevention of arthritis. In: Preedy, R.R.W.R. (Ed.), Bioactive Food as Dietary Interventions for Arthritis and Related Inflammatory Diseases. Academic Press, San Diego, pp. 529–544.

Bazh, E.K., El-Bahy, N.M., 2013. In vitro and in vivo screening of anthelmintic activity of ginger and curcumin on *Ascaridia galli*. Parasitol. Res. 112 (11), 3679–3686.

Bombardelli, E., 2010. Compositions comprising lipophilic extracts of *Zingiber officinale* and *Echinacea angustifolia* for the prevention and treatment of gastro-oesophageal reflux and chemotherapy-induced emesis. Google Patents.

Bombardelli, E., 2012. Compositions comprising a lipophilic extract of *Zingiber officinale* and an extract of *Cynara scolymus*, which are useful for the prevention and treatment of oesophageal reflux and irritable bowel syndrome. Google Patents.

Bombardelli, E., Morazzoni, P., 2012. Formulations containing extracts of *Echinacea angustifolia* and *Zingiber officinale* which are useful in reducing inflammation and peripheral pain. Google Patents.

Chandan, K.B., Bakshi, S.K., Bedi, K.L., Tikoo, L., Gupta, A.K., Gupta, D.K., 2003. Bioavailability enhancing activity of *Zingiber officinale* Linn and its extracts/fractions thereof. Google Patents.

Chrubasik, S., Pittler, M.H., Roufogalis, B.D., 2005. *Zingiberis rhizoma*: a comprehensive review on the ginger effect and efficacy profiles. Phytomedicine 12 (9), 684–701.

Darvishzadeh-Mahani, F., Esmaeili-Mahani, S., Komeili, G., Sheibani, V., Zare, L., 2012. Ginger (*Zingiber officinale* Roscoe) prevents the development of morphine analgesic tolerance and physical dependence in rats. J. Ethnopharmacol. 141 (3), 901–907.

El-Bahy, N.M., Bazh, E.K., 2015. Anthelmintic activity of ginger, curcumin, and praziquentel against *Raillietina cesticillus* (in vitro and in vivo). Parasitol. Res. 114 (7), 2427–2434.

Fahimi, F., Khodadad, K., Amini, S., Naghibi, F., Salamzadeh, J., Baniasadi, S., 2011. Evaluating the effect of *Zingiber officinalis* on nausea and vomiting in patients receiving Cisplatin based regimens. Iran. J. Pharm. Res. 10 (2), 379–384.

Fouda, A.M., Berika, M.Y., 2009. Evaluation of the effect of hydroalcoholic extract of *Zingiber officinale* rhizomes in rat collagen-induced arthritis. Basic Clin. Pharmacol. Toxicol. 104 (3), 262–271.

Grzanna, R., Lindmark, L., Frondoza, C., 2005. Ginger—an herbal medicinal product with broad antiinflammatory actions. J. Med. Food 8, 125–132.

Hasan, I.H., El-Desouky, M.A., Hozayen, W.G., Abd El Aziz, G.M., 2015. Protective effect of *Zingiber officinale* against CCl4-induced liver fibrosis is mediated through downregulating the TGF-beta1/Smad3 and NF-x03BA; B/Ix03BA; B Pathways. Pharmacology 97 (1–2), 1–9.

Hsu, Y.L., Hung, J.Y., Tsai, Y.M., Tsai, E.M., Huang, M.S., Hou, M.F., 2015. 6-Shogaol, an active constituent of dietary ginger, impairs cancer development and lung metastasis by inhibiting the secretion of CC-chemokine ligand 2 (CCL2) in tumor-associated dendritic cells. J. Agric. Food Chem. 63 (6), 1730–1738.

https://en.wikipedia.org/wiki/Ginger, 2015. Ginger.

https://unitproj.library.ucla.edu/biomed/spice/index.cfm, 2002. Ginger.

Huang, Q.R., Iwamoto, M., Aoki, S., Tanaka, N., Tajima, K., Yamahara, J., 1991. Anti-5-hydroxytryptamine 3 effect of galanolactone, diterpenoid isolated from ginger. Chem. Pharm. Bull. 39 (2), 397–399.

Iranloye, B.O., Arikawe, A.P., Rotimi, G., Sogbade, A.O., 2011. Anti-diabetic and anti-oxidant effects of *Zingiber officinale* on alloxan-induced and insulin-resistant diabetic male rats. Niger. J. Physiol. Sci. 26 (1), 89–96.

Khandouzi, N., Shidfar, F., Rajab, A., Rahideh, T., Hosseini, P., Mir Taheri, M., 2015. The effects of ginger on fasting blood sugar, hemoglobin a1c, apolipoprotein B, apolipoprotein a-I and malondialdehyde in type 2 diabetic patients. Iran. J. Pharm. Res. 14 (1), 131–140.

Krishnakantha, T.P., Lokesh, B.R., 1993. Scavenging of superoxide anions by spice principles. Indian J. Biochem. Biophys. 30 (2), 133–134.

Lantz, R.C., Chen, G.J., Sarihan, M., Solyom, A.M., Jolad, S.D., Timmermann, B.N., 2007. The effect of extracts from ginger rhizome on inflammatory mediator production. Phytomedicine 14 (2–3), 123–128.

Levy, A.S., Simon, O., Shelly, J., Gardener, M., 2006. 6-Shogaol reduced chronic inflammatory response in the knees of rats treated with complete Freund's adjuvant. BMC Pharmacol. 6, 12.

Lin, R.J., Chen, C.Y., Lu, C.M., Ma, Y.H., Chung, L.Y., Wang, J.J., 2014. Anthelmintic constituents from ginger (*Zingiber officinale*) against *Hymenolepis nana*. Acta Trop. 140, 50–60.

Liu, Q., Peng, Y.B., Qi, L.W., Cheng, X.L., Xu, X.J., Liu, L.L., 2012. The cytotoxicity mechanism of 6-shogaol-treated Hela human cervical cancer cells revealed by label-free shotgun proteomics and bioinformatics analysis. Evid. Based Complement. Alternat. Med. 2012, 278652.

McGee, H., 2004. On Food and Cooking: The Science and Lore of the Kitchen, second ed. Scribner, New York.

Micklefield, G.H., Redeker, Y., Meister, V., Jung, O., Greving, I., May, B., 1999. Effects of ginger on gastroduodenal motility. Int. J. Clin. Pharmacol. Ther. 37 (7), 341–346.

Mishra, R.K., Kumar, A., Kumar, A., 2012. Pharmacological activity of *Zingiber officinale*. Int. J. Pharm. Chem. Sci. 1 (4), 1422–1427.

Mostafa, O.M., Eid, R.A., Adly, M.A., 2011. Anti-schistosomal activity of ginger (*Zingiber officinale*) against *Schistosoma mansoni* harbored in C57 mice. Parasitol. Res. 109 (2), 395–403.

Mustafa, T., Srivastava, K.C., 1990. Ginger (*Zingiber officinale*) in migraine headache. J. Ethnopharmacol. 29 (3), 267–273.

Nazim, U.M., Jeong, J.K., Seol, J.W., Hur, J., Eo, S.K., Lee, J.H., 2015. Inhibition of the autophagy flux by gingerol enhances TRAIL-induced tumor cell death. Oncol. Rep. 33 (5), 2331–2336.

Nurtjahja-Tjendraputra, E., Ammit, A.J., Roufogalis, B.D., Tran, V.H., Duke, C.C., 2003. Effective antiplatelet and COX-1 enzyme inhibitors from pungent constituents of ginger. Thromb. Res. 111 (4–5), 259–265.

Nwozo, S.O., Osunmadewa, D.A., Oyinloye, B.E., 2014. Anti-fatty liver effects of oils from *Zingiber officinale* and *Curcuma longa* on ethanol-induced fatty liver in rats. J. Integr. Med. 12 (1), 59–65.

Ojewole, J.A., 2006. Analgesic, antiinflammatory and hypoglycaemic effects of ethanol extract of *Zingiber officinale* (Roscoe) rhizomes (Zingiberaceae) in mice and rats. Phytother. Res. 20 (9), 764–772.

Plengsuriyakarn, T., Viyanant, V., Eursitthichai, V., Tesana, S., Chaijaroenkul, W., Itharat, A., 2012. Cytotoxicity, toxicity, and anticancer activity of *Zingiber officinale* Roscoe against cholangiocarcinoma. Asian Pac. J. Cancer Prev. 13 (9), 4597–4606.

Prabhakaran Nair, K., 2013. The Agronomy and Economy of Tumeric and Ginger. Elsevier, London, p. 446.

Prasad, S., Tyagi, A.K., 2015. Ginger and its constituents: role in prevention and treatment of gastrointestinal cancer. Gastroenterol. Res. Pract. 2015, 142979.

Qi, L.W., Zhang, Z., Zhang, C.F., Anderson, S., Liu, Q., Yuan, C.S., 2015. Anti-colon cancer effects of 6-shogaol through G_2/M cell cycle arrest by p53/p21-cdc2/cdc25A crosstalk. Am. J. Chin. Med. 43 (4), 743–756.

Radhakrishnan, E.K., Bava, S.V., Narayanan, S.S., Nath, L.R., Thulasidasan, A.K., Soniya, E.V., 2014. [6]-Gingerol induces caspase-dependent apoptosis and prevents PMA-induced proliferation in colon cancer cells by inhibiting MAPK/AP-1 signaling. PLoS One 9 (8), e104401.

Rodrigues, F.A., Prata, M.M., Oliveira, I.C., Alves, N.T., Freitas, R.E., Monteiro, H.S., 2014. Gingerol fraction from *Zingiber officinale* protects against gentamicin-induced nephrotoxicity. Antimicrob. Agents Chemother. 58 (4), 1872–1878.

Sabina, E.P., Pragasam, S.J., Kumar, S., Rasool, M., 2011. 6-gingerol, an active ingredient of ginger, protects acetaminophen-induced hepatotoxicity in mice. Zhong Xi Yi Jie He Xue Bao 9 (11), 1264–1269.

Seif el-Din, S.H., El-Lakkany, N.M., Mohamed, M.A., Hamed, M.M., Sterner, O., Botros, S.S., 2014. Potential effect of the medicinal plants *Calotropis procera* Ficus elastica and *Zingiber officinale* against *Schistosoma mansoni* in mice. Pharm. Biol. 52 (2), 144–150.

Sharma, J.N., Srivastava, K.C., Gan, E.K., 1994. Suppressive effects of eugenol and ginger oil on arthritic rats. Pharmacology 49 (5), 314–318.

Shidfar, F., Rajab, A., Rahideh, T., Khandouzi, N., Hosseini, S., Shidfar, S., 2015. The effect of ginger (*Zingiber officinale*) on glycemic markers in patients with type 2 diabetes. J. Complement. Integr. Med. 12 (2), 165–170.

Shimoda, H., Shan, S.J., Tanaka, J., Seki, A., Seo, J.W., Kasajima, N., 2010. Anti-inflammatory properties of red ginger (*Zingiber officinale* var. Rubra) extract and suppression of nitric oxide production by its constituents. J. Med. Food 13 (1), 156–162.

Srivastava, K.C., Mustafa, T., 1992. Ginger (*Zingiber officinale*) in rheumatism and musculoskeletal disorders. Med. Hypotheses 39 (4), 342–348.

van Breemen, R.B., Tao, Y., Li, W., 2011. Cyclooxygenase-2 inhibitors in ginger (*Zingiber officinale*). Fitoterapia 82 (1), 38–43.

Wilson, P.B., 2015. Ginger (*Zingiber officinale*) as an analgesic and ergogenic aid in sport: a systemic review. J. Strength Cond. Res. 29 (10), 2980–2995.

Young, H.Y., Luo, Y.L., Cheng, H.Y., Hsieh, W.C., Liao, J.C., Peng, W.H., 2005. Analgesic and anti-inflammatory activities of [6]-gingerol. J. Ethnopharmacol. 96 (1–2), 207–210.

INDEX

A

AAS. *See* Atomic absorption spectrophotometry (AAS)
Abrus precatorius, 308
ABTS. *See* 2,2-Azinobis- 3-ethylbenzo-thiazoline-6-sulfonate (ABTS)
Acalypha species, 308
Acetic acid-induced vascular permeability, 502
Acetic acid-induced writhing test, 253
6-Acetylapiin, 535
Acetylcholine, 300
Acid-peptic disease, 589
Acid-sensing ion channel (ASIC), 84
Acquired immune deficiency syndrome (AIDS), 3, 134
Acquired immunity, 78
Active component extraction
  solvents used, 213
Acute myocardial infarction (AMI)
  onion effect on, 358
Acute pain sensations, 86
Acyclovir-resistant strains, 605
Adansonia digitata, 248
  antiinflammatory or antinociceptive potential, 248–249
Adenanthera pavonina, 308
Adenovirus, 40
Afang. *See Gnetun africanum*
Aflatoxins, 604
  toxicity, 605
Aframomum alboviolaceum, 227, 277
Aframomum citratum, 226
Aframomum kayserianum, 227
Aframomum melegueta, 227, 304, 581
  antiinflammatory or antinociceptive potential, 248
Aframomum polyanthum, 227
Africa
  antimicrobial spices and vegetables
    antimicrobial mode of action, 229
      antibiotic-potentiating extracts, 229
      bacterial efflux pumps, role of, 229
    bladder cancer (BCa), 99
    breast cancer, 98
  burden of parasitic diseases on, 17
    flukes, 19
    helminthes (worms), 18
    protozoans, 18
  cancer survival rates, 271
  cancer types, 271
  cervical cancer, 97
  colorectal cancer, 99
  common cancers in, 94
  countries in Africa with operational policies, strategies, or action plans for cancer in 2013, 103
  country-specific cancer control policies in, 103, 105
  deaths occurred in, 95
  esophageal cancer, 99
  establishing and maintaining cancer control programs, 102
  estimated numbers of new cases and deaths of cancer sites 2012, 95, 96
  GLOBOCAN data for African countries, 94
  government control policies for cancer in, 100
    requirements of policies, 100–101
  Kaposi sarcoma, 100
  liver cancer (hepatocellular carcinoma), 98
  lung cancer, 98
  nasopharyngeal cancer, 100
  new cancer cases, 95
  non-Hodgkin lymphoma, 100
  other antimicrobial spices and vegetables, 228
  prostate cancer (PCa), 99
African basil. *See Ocimum gratissimum*
African biodiversity, 171
African cabbage, 286. *See also Cleome gynandra*
African elemi. *See Canarium schweinfurthii*
African medicinal spices
  antiproliferative effects, 277–286
  and vegetables
    anticancer activities, 278
    antimicrobial effects, 219–227
African Pepper, 307
African *Piper* species, 583, 587, 588

African spices and vegetables
  antiinflammatory active ingredients, 255–258
    apigenin, 257
    capsaicin, 256
    eugenol, 258
    gingerol, 258
    kaurenoic acid, 257
    luteolin, 257
    piperine, 256
    thymoquinone, 255
    xylopic acid, 257
  clinical trials, 259
  screening of antiinflammatory and antinociceptive activity
    methods used, 240–247
      in vitro methods, 247
      in vivo methods, 241–243
African spices, vegetable's extracts, and derived products
  mode of action, 290–292
    antiangiogenic effects, 292
    caspase activation, effect on, 290
    cell cycle arrest, 290
    induction of apoptosis, 290
    mitochondrial membrane potential, effect on, 291
    reactive oxygen species generation, effect on, 291
African traditional medicines (ATRM), 157
African trypanosomiasis, 56
Africa pepper, 249
  antiinflammatory action, 249
  inhibition of histamine, 249
  treatment of
    colic pain, 249
    headache, 249
    neuralgia, 249
    rheumatism, 249
Africa *Piper* genus, 590
Afzelia africana, 309
Agar dilution method, 216
Aged garlic extract (AGE), 364
Aging
  definition, 330
  mechanisms and changes associated with, 330
  medicinal spices and vegetables, 330
    Cinnamomum species, 331
    Piper nigrum, 331
    Syzygium aromaticum, 332
    Vitis vinifera, 333
    Zingiber officinale, 333
Agriculture, 60
AIDS. *See* Acquired immune deficiency syndrome (AIDS)
Ajoene, 366
Alamar Blue assay, 275
Alanine aminotransferase (ALT), 431, 633
Alkaline phosphatase (ALP), 431, 633
Alkaline reagent test, 215
Allicin, 318, 366
Alligator pepper, 248
Alliin, 366
Allium cepa, 222. *See also* Onions
Allium porrum, 222
Allium sativum, 222, 318. *See also* Garlic
  beneficial effects on MetS, 318
  origin in, 318
  other effects of, 318
Allodynia, 86
Allopathic practitioners, 156
Allyl methyl disulfide, 366
Allyl methyl tetrasulfide, 366
Allylmethyl trisulfide, 366
1-Allyl-2,3,4,5-tetra-methoxy-benzene, 532
ALP. *See* Alkaline phosphatase (ALP)
ALT. *See* Alanine aminotransferase (ALT)
Alunusnin A, 613
Alzheimer's diseases
  development of, 330
  medicinal spices and vegetables, 330
    Cinnamomum species, 331
    Piper nigrum, 331
    Syzygium aromaticum, 332
    Vitis vinifera, 333
    Zingiber officinale, 333
American Diabetes Association, 126
American trypanosomiasis, 56
Aminoglycosides, 412
AMP. *See* Ampicillin (AMP)
Amphetamine-induced stimulant effects, 522
Ampicillin (AMP), 229
Amyloid beta (1-42)-induced Alzheimer's disease rat model, 445
Anacardiaceae, 451
Anacardiaceae family, 452
Ancylostoma duodenale, 33
Andropogon ceriferus, 397

Andropogon citratus, 397
Andropogon nardus, 397
Andropogon roxburghii, 397
Anemia, 485
Anethum graveolens, 304
Angelica archangelica, 189
Anisakis simplex, 506
Annonaceae, antimicrobial effects, 219–220
Anonidium mannii, 220
Anopheles gambiae, 13, 606
Antiaging compounds
    chemical structure of, 332
Antiallergic, 452
Anti-Alzheimer's diseases compounds
    chemical structure of, 332
Antibiotic-resistant bacteria, 142
Antibone resorption, 452
Anticancer activity
    Cinnamomum cassia, 387
    Cinnamomum zeylanicum, 387
    garlic, 368
    onion, 356
Anticholinergic, 300
Anticoagulant, 425
Antidiabetic, 425
    activity, garlic, 368
    effects, onion, 357
Antidiarrhoeal, 452
Antiemetics
    antiemetic effect, determination, 303
    medicinal spices as sources of, 304
    modes of action, 300
    prevention of emesis, 299
    screening methods, 301
    treatment of emesis, 299
    vegetables as sources of, 308
    in vitro activity, 302
    in vivo activities, 303
Antiepilepsirine, in epilepsy, 339
Antiinflammatory, 74, 425
    herbal medicine, 74
    properties, 452
Antileukemic alkaloids, 277
    vinblastine, 277
    vincristine, 277
Antimicrobial activities
    garlic, 369–370
    onion, 357
Antimicrobial resistance, 142

Antimicrobial secondary metabolites, modes of
        action, 208–212
    alkaloids, 208
    phenolics, 209–212
        coumarins, 212
        quinones, 212
        tannins, 212
    terpenoids, 209
Antineoplastic, 425
Antinociceptive effect, 74
Antioxidant effects, 425, 470
    garlic, 371
    onion, 358
Antiphlogistic effect, 241
Antipyretic, 452
Antipyretic activity, 88
Antiretroviral therapy (ART), 7, 113
Antispasmodic, 452
Antithrombotic agent, 621
Antiviral, 425, 452
Anxiety, 73, 485
Apiaceae, antimicrobial effects, 220
Apigenin, 442, 535
    COX-2 gene expression, inhibition of, 257
    IL-6 activity reduction, 257
    nuclear factor-kB (NF-kB) gene expression,
        inhibition of, 257
    TNF-α reduction, 257
Apiin, 535
Apiol, 532
Apiose, 531, 535
Apium graveolens, 220
Apoptosis, 632
    reactive oxygen species (ROS)-mediated, 632
Apoptotic protease activating factor-1 (Apaf-1),
        291
Aqueous extract stimulated insulin secretion, 428
Arachidonic acid, 621
Areca catechu, 143
Aromatinin A, 613
ART. *See* Antiretroviral therapy (ART)
Artemia salina, 520
Artemisia dracunculus, 317
Artemisinin-based combination therapy (ACT), 5
Arthritis, 407
Ar-turmerone, 428
Ascariasis, 35
Ascaris lumbricoides, 33
Ascorbic acid, 535

Aspartate aminotransferase (AST), 431, 633
Aspergillus fumigatus, 226, 520
Aspirin, 88, 446
AST. *See* Aspartate aminotransferase (AST)
Asteraceae, antimicrobial effects, 221
Asthma, 485
Athlete's foot, 24
Atomic absorption spectrophotometry (AAS), 192
ATRM. *See* African traditional medicines (ATRM)
Aubergine, 306
2,2-Azinobis- 3-ethylbenzo-thiazoline-6-sulfonate (ABTS), 620
Azospirillium lipoferum, 473

B

Bacillus anthracis, 212
Bacillus cereus, 220, 472
Bacillus megaterium, 492
Bacillus pumilus, 472
Bacillus subtilis, 220, 443, 504
Bacopa monnieri, 179
Bacterial meningitis, 56
Baobab tree, 248. *See also Adansonia digitata*
   clothing, 248
   food and folk medicine source, 248
   phytochemical constituents
      adansonin, 248
      amino acids, 248
      campesterol, 248
      epicatechin, 248
      flavonoids, 248
      tocopherol, 248
      vitamin C, 248
   treatment
      bronchial asthma, 248
      dermatitis, 248
      sickle cell anemia, 248
Basil, 240, 252
   African traditional medicine, use in, 252
   antiinflammatory and antinociceptive potential, 252
   ayurveda, use in, 252
   traditional Chinese medicine, use in, 252
Benedict test, 214
Benzaldehyde, 532
Benzimidazole antihelmintic drugs, 37
Benzoil tree, 288
α−Bergamotene, 386
Bersama engleriana, 207

Beta-D-galactopyranosyl, 367
Beta-D-glucopyranosyl, 367
BHA. *See* Butylated hydroxyanisole (BHA)
BHT. *See* Butylated hydroxytoluene (BHT)
Bicornin, 613
Biflorin, 613
β−Bisabolene, 532
Bisdemethoxycurcumin, 425, 426
2,3-Bis (2-methoxy-4-nitro-5-sulfophenyl)-5-[(phenylamino) carbonyl]-2H-tetrazolium hydroxide (XTT), 272, 273
   enzymatic reduction to formazan, 273
B. koenigii. *See* Curry tree
Black cumin, 251. *See also Nigella sativa*
   active ingredients
      dithymoquinone, 251
      saponins, 251
      thymohydroquinone, 251
      thymol, 251
   asthma, treatment of, 251
   bronchitis, treatment of, 251
   eczema, treatment of, 251
   food preservative, use as, 251
   headache, treatment of, 251
   influenza, treatment of, 251
   rheumatism, treatment of, 251
   spice, use as, 251
Blackheads, 485
Black nightshade, 288. *See also Solanum nigrum*
   analgesic, use as, 288
   antitumorigenic, 288
   content
      alpha-solanine, 288
   oriental medicine, use as, 288
   sedative, use as, 288
   sudorific, use as, 288
Black pepper, 253, 331, 586. *See also Piper nigrum*
   as anticonvulsant drugs, 339
   as antiepileptic plant, 339
   antinociceptive properties, 253
   derived from, 339
   used in, traditional medicine, 331
   use in, treatment of epilepsy, 339
Blattella germanica, 507
Blood impurities, 485
Bolusanthus speciosus, 211
Bontrager test, 215
Borneol, 600

Botrytis cinerea, 225, 604
Bradykinin, 241
Brassicaceae, antimicrobial effects, 221
Brassica napus, 221
Brassica oleracea, 221, 320
Brassica rapa, 221
    isolated compounds, 286
        brassicaphenanthrene A, 286
        diarylheptanoids, 286
        6-paradol, 286
        phenanthrene derivative, 286
        trans-6-shogaol, 286
Breast adenocarcinoma MCF-7 cells, 499
Breast adenocarcinoma MDA-MB-231 cells, 426
Breast cancer MCF-7 cells, 288
Brewer's-yeast-induced pyrexia, 258
Broccoli, 335
    source of antioxidant phenolics, 336
Bronchitis, 485
Burns, 452
Butylated hydroxyanisole (BHA), 505
Butylated hydroxytoluene (BHT), 505
tert-Butyl hydroperoxide (*t*BuOOH), 410

C

δ−Cadinol, 532
Caenorhabditis elegans, 320, 333
Caffeic acid, 367
Calabash nutmeg, 305
CAM. *See* Complementary and alternative medicine (CAM)
Camphene, 532
Camphor, 386, 600
Campylobacter jejuni, 36
Canarium schweinfurthii, 379
    botanical aspects, 379
    chemistry of, 380
    description, 379
    distribution of, 379
    essential oil, 380
    pharmacology of, 381
        antidiabetic activity, 381
        antimicrobial activities, 381
        antioxidant effects, 382
Cancer, 93. *See also* Africa
    factors contributing to increased burden of, 94
    incidences of, 93
    infections leading to, 93
    risk, 93

Candida albicans, 25, 618
    occurrence of disseminated, 24
    pathogenesis of, 25
Candida utilis, 227
Candidone, 284
Canis familiaris, 301
Capsaicin, 628
    antiinflammatory properties, 256
    chemical hyperalgesia, 251
    COX activity inhibition, 251
    COX-2 inhibition, 256
    inflammatory heat, 251
    iNOS protein inhibition, 256
    rheumatoid arthritis, use in, 251
Capsaicinoids, 251
Capsicum annuum, 224, 250
Capsicum baccatum, 250
Capsicum chinense, 250
Capsicum frutescens, 224, 250
Capsicum pubescens, 250
Capsicum species
    antiinflammatory or antinociceptive potential, 250–251
    phytochemicals
        capsaicinoids, 251
        capsinoids, 251
        carotenoids, 251
        luteolin, 251
        phenolic compounds, 251
        quercetin, 251
Carbazole alkaloids, source of, 305
Cardiometabolic syndrome (CMS), 110
Cardiotonic, 452
Cardiovascular disease (CVD), 93
    effect of garlic, 370
Cardiovascular effects
    onion, 358
Cardiovascular protective, 425
3-Carene, 532
β-Carotene, 535
    assays, 410
Carotenoids, 240
Carotol, 532
Carrots
    antioxidant content, 336
    constituent, 336
Carvacrol, 600
Carvacryl methyl ether, 600
Caryophyllene, 613

β-Caryophyllene, 386, 467, 532, 600
Caryophyllene oxide, 386
Caspases, 290
Cassia italica, 212
Casuarictin, 613
Casuarinin, 613
Catarrh, 485
Catechins, antimicrobial properties, 211
Catharanthus roseus, 154, 277
Cat's whiskers, 286
CCE. *See* Counter-current extraction (CCE)
CCRF-CEM cells, 284
Cefepime (FEP), 229
Cell adhesion molecules (CAMs), 82
CEM/ADR5000 cells, 284
Centella asiatica, 179
Central nervous system (CNS), 74, 446, 591
Cephalosporins, 142
Cerebral tuberculoma, 56
CFA. *See* Complete Freund's adjuvant (CFA)
Chain-breaking antioxidants, 122
Chemokines, 77, 82
Chemotactic gradient, 302
Chest congestion, 485
Chick emesis model, 308
Chikungunya, 137
Child deaths, 3
Childhood overweight, 113
Chili, 250
   antiinflammatory potential, 250
   burning flavor, 250
Chinese cabbage, 286
Chinese hamster ovary (CHO) cells, 426
Chinese parsley. *See* Coriander
CHL. *See* Chloramphenicol (CHL)
Chloramphenicol (CHL), 229
Cholera, 139
Chronic inflammation, 409
Chronic noncommunicable diseases (CNCDs), 109
  cancers, 113
  CV disease, 113
  diabetes mellitus, 113
Chrysin, 211
Chrysoeriol, 535
Cilantro. *See* Coriander
Cinnamaldehyde, 386
Cinnamomum aromaticum, 322
Cinnamomum burmannii, 322

Cinnamomum cassia, 317, 385
  alcoholic extract, protective effects of, 392
  chemistry of, 386
  effect on, human immodeficiency virus type 1 (HIV-1), 390
  effects on protein and mRNA levels, 388
  pharmacology of, 387
    antibacterial activity, 389
    anticancer activity, 387
    antidiabetic activity, 388
    antifungal activity, 389
    antiinflammatory activity, 389
    antioxidant activity, 390
    antiviral activity, 390
    hepatoprotective activity, 391
Cinnamomum loureiroi, 322
Cinnamomum species, 322. *See also* Cinnamon
Cinnamomum verum, 317
Cinnamomum zeylanicum, 228, 322, 385
  chemistry of, 386
  essential oil, 390
  pharmacology of, 387
    antibacterial activity, 389
    anticancer activity, 387
    antidiabetic activity, 388
    antifungal activity, 389
    antiinflammatory activity, 389
    antioxidant activity, 390
    antiviral activity, 390
    hepatoprotective activity, 391
Cinnamon, 331, 385
  botanical aspects, 385
  description, 385
  treatment of cardiovascular diseases, 331
  use in treatments of, Alzheimer disease, 331
  varieties of, 331
trans-Cinnamyl acetate, 386
CIP. *See* Ciprofloxacin (CIP)
Ciprofloxacin (CIP), 229
CITES. *See* Convention on International Trade in Endangered Species (CITES)
Citral diethylacetal, 405
Citrobacter freundii, 492
Citrus aurantifolia, 226
Citrus limon, 226
Citrus paradisi, 226
Citrus sinensis, 226
Clausena anisata, 208
Cleome gynandra

antiinflammatory properties, 286
  tumor treatment, 286
Clinical and Laboratory Standards Institute (CLSI), 216
Cloves, 611
  antithrombotic activity, 621
  botanical aspect, 611
  chemistry, 612
    essential oil, 612
  distribution of, 611
  native to, 332
  organ-protective effects, 621
  patents, 621
    EP0, 753, 305 A1, 622
    WO1, 994, 018, 994 A1, 621
    WO2, 005, 072, 253 A2, 622
    WO2, 006, 067, 600 A2, 621
  pharmacology, 613
    antibacterial activity, 618
    anticancer activity, 617
    antidiabetic activity, 617
    antifungal activity, 619
    antiinflammatory activities, 618
    antinociceptive activities, 618
    antioxidant activities, 620
    antiprotozoal activity, 619
    antiviral activity, 620
  phenolics constituents, 613
  physiologic responses in aging rodents, 332
  toxicity, 622
Cloxacillin (CLX), 229
CLX. See Cloxacillin (CLX)
CM. See Cryptococcal meningitis (CM)
CMS. See Cardiometabolic syndrome (CMS)
CNCDs. See Chronic noncommunicable diseases (CNCDs)
Cnidilin, 535
Coccidioidomycosis, 30
Colon adenocarcinoma HCT116 cells, 499
Colony-stimulating factors (CSFs), 81
Common parasitic infections, in Africa, 18
Communicable diseases, 3
  in children, 3
Community empowerment, 16
Complementary and alternative medicine (CAM), 153, 171
Complete Freund's adjuvant (CFA), 630
Compression, 175
  of peripheral branches, 85

Concoctions, 173
Convention on International Trade in Endangered Species (CITES), 160
α—Copaene, 386
Copper acetate test, 214
Corchorus olitorius, 225
  chronic cystitis treatment, 287
  demulcent, use as, 287
  diuretic, use as, 287
  dysuria treatment, 287
  gonorrhea treatment, 287
Coriander, 335
  edible parts, 335
  native to, 335
  uses, 335
Corni fructus, 317
Coronary heart disease, 5
Corticosteroids, 88
Corynebacterium pseudodiphthericum, 212
Cosmosiin, 535
Coughs, 407
P-Coumaric acid, 367
Coumarin, 386
Counter-current extraction (CCE), 176
Cowpeas, 289, 336. See also *Vigna unguiculata*
  exporter, 289
  roasted seeds
    insomnia, treatment of, 289
    memory weakness, treatment of, 289
    neuritis, treatment of, 289
COX enzyme. See Cyclooxygenase (COX) enzyme
CQ-sensitive (3D7), 411
Crataegus curvisepala, 317
Crataegus laevigata, 317
Crataegus monogyna, 317
Crataegus oxyacantha, 317
Crataegus tanacetifolia, 317
Crataeva nurvala, 179
C-reactive protein (CRP), 83
Crispane, 535
Crispanone, 535
Crocus sativus, 317
Cryptococcal meningitis (CM), 139
  use of garlic extract, 369
Cryptococcus neoformans, 21
Cryptone, 532
Crysoeriol, 442
α—Cubebene, 532
Culex quinquefasciatus, 606

Culture, 10
Curcuma longa L., 321, 425. *See also* Turmeric
  aqueous extract, 430
  cardiovascular effects, 430
  chemical structures of, 427
  chemistry of, 426
  cultivation and distribution, 426
  hepatoprotective effects, 431
  methanol extract of, 430
  neuroprotective effects, 432
  patents with, 432
  patent US7220438 B2, 432
  pharmacology of, 426
    anticancer activity, 426
    antidiabetic activity, 428
    antiinflammatory activity, 429
    antimicrobial activities
      antibacterial activity, 429
      antifungal activity, 430
      antiviral activity, 430
Curcuma longa L. *See* Turmeric
Curcuma oil, 432
Curcumin, 334, 425, 426, 429
Curcuminoids, 426
  in rhizome, 425
Curcumol, chemical structures of, 338
Curry tree, 335
  native to, 335
  uses of leaves, 335
Cyanide poisoning, 522
Cyclocurcumin, 426
Cyclooxygenase-2 (COX-2), 79
Cyclooxygenase (COX) enzyme, 241, 247
Cymbopogon citratus, 397, 398
  activity, type, region, and potency, 414
  botanical aspects/distribution, 398
  chemical constituents of, 400
    flavonoids, 401
    phenolics, 400
    steroids/fatty alcohol, 400
    tannins, 401
    terpenoids, 403–407
  ethno-pharmacological applications, 400
  geographical distribution, 399
  history/distribution, 397
  nonvolatile polyphenol compounds, 400
  patents, 416
  pharmacological activities of, 407
    anticancer/antitumor activity, 413
    antidiabetic activity, 408
    anti-HIV activity, 410
    antihypertension assay, 411
    antiinflammatory, 409
    antimalarial potential, 411
    antimicrobial activity, 413
    antimutagenicity/carcinogenicity activity, 408
    antioxidant property, 410
    dermatotoxicity activity, 412
    insecticidal activity, 411
    nephrotoxicity of, 412
  root divisions, 398
  steroids and fatty alcohol and tannins, 401
Cymbopogon nardus, 397
ρ−Cymene, 532
Cynaroside, 402
Cysteine proteases, 290
Cysticercosis, 56
Cytopathic effect (CPE), 444
Cytotoxic agents, 302
Cytotoxic alkaloids and other compounds
  African medicinal spices, isolation from
    relevance to cancer cells, 285
Cytotoxicity, other African medicinal vegetables, 286–289
  Brassica rapa, 286
  Cleome gynandra, 286
  Colocasia esculenta, 287
  Corchorus olitorius, 287
  Ferula hermonis, 288
  Moringa oleifera, 288
  Solanum nigrum, 288
  Vigna unguiculata, 289
Cytotoxic phenolics
  African medicinal spices and vegetables
    effects on cancer cells, 283
Cytotoxic terpenoids
  African medicinal spices and vegetables
    effects on cancer cells, 283

D

Dalton's lymphoma cells, 426
Daucus carota, 321
DDT. *See* Dichlorodiphenyltrichloroethane (DDT)
Death rates, 3
  for children, 3
Deaths in Africa, 4
Decanal, 405, 532
Decoction, 173, 178
Demethoxy curcumin, 425, 426
Dengue, 135

Dennettia tripetala, 581
Depression, 73
Dermatitis, 412
Dermatophagoides farinae, 411
Dermatophagoides pteronyssinus, 411
5-Desmethylsinensetin, 601
Dexamethasone, 520
Diabetics, 408
Diacyl glycerol (DAG), 79
Diallyl disulfide, 366
Diallyl sulfide, 366
Diallyl tetrasulfide, 366
Diallyl trisulfide, 366
Diarrheal disease, 3, 5
Diazepam, 521
Dichlorodiphenyltrichloroethane (DDT), 188
Dichrostachys glomerata, 228
Dictyostelium chemotaxis model, 302
Dictyostelium discoideum, 302
Dietary therapy, 126
Diethylnitrosamine (DEN), 413
Dihydrocurcumin, 426
Dihydrolactucin, 439
5,7-Dihydroxy-2-(2- hydroxy-5- methoxyphenyl)-
    4*H*-chromen-4-one, 460
7-Dihydroxy-2-methylchromone 8-C-β-D-
    glucopyranoside, 613
Dill weed, 304
Dimethylbenzene, 241
3,7-Dimethyl-1,3,6-octatriene, 405, 406
Dimethyl sulfoxide (DMSO), 303
3-(4,5-Dimethylthiazol-2-yl)-5-
    (3-carboxymethoxyphenyl)-2-
    (4-sulfophenyl)-2H-tetrazolium (MTS),
    272, 274
  enzymatic reduction to formazan, 274
3-(4,5-Dimethylthiazol-2-yl)-2,5-diphenyl-2H-
    tetrazolium bromide (MTT), 272
  cell proliferation and viability analysis, 272
  colorimetric assay, 272
  cytotoxicity testing, gold standard, 272
  enzymatic reduction to formazan, 272
Dimethyl trisulfide, 366
Diosmetin 7-O-β-D-glucopyranoside, 535
1,1-Diphenyl-2-picryl-hydrazyl (DPPH), 410, 519
2,2-Diphenyl-1-picrylhydrazyl (DPPH), 441, 620
Disability, 3
Disability-adjusted life years (DALYs), 3
Dopamine, 300
Dorstenia barteri, 207

Dorstenia psilurus, 223
Dragendorff's test, 216
D2 receptor, 300
Drug-disease interaction, 125
Drumstick tree, 288
Drumstick tree. *See Moringa oleifera*

E

Ebola epidemic, in West Africa, 4
  deaths in, 4
    WHO groups, 4
Ebola virus disease (EVD), 136
Echinacea angustifolia, 636
Echinops giganteus, 220
E. coli ATCC25922, 443
Efflux pumps inhibitor (EPI), 492
Eggplant, 306
Ehrlich Ascites Carcinoma cell line, 286
Electroshock-induced hind limb tonic
    extension, 507
α−Elemene, 532
β−Elemene, 532
γ−Elemene, 532
Elemicin, 532
Elephantiasis, 407
Elevated plus-maze model, 505
ELISA. *See* Enzyme-linked immunosorbent assays
    (ELISA)
Ellagic acid, 613
EMEA. *See* European Medicines Agency (EMEA)
Emetic reflex, 302
Endothelium-independent vasorelaxation, 411
Enterobacter cloacae, 220
Enterococcus casseliflavus, 492
Enterococcus faecalis, 225
Environmental hygiene, 17
Environmental pollution, 408
Enzyme-linked immunosorbent assays (ELISA), 247
Epicatechin, 460, 461
Epilepsy, 51
  burden of, 52
  medicinal spices and vegetables, 338
    Curcuma longa, 338
    Glycyrrhiza glabra, 339
    Piper nigrum, 339
    Zingiber officinale, 339
  phytochemicals with antiepileptic properties, 340
  risk factors, 52
  signs and symptoms, 52
Epinetrum villosum, 208

Eru. *See Gnetun africanum*
Eruboside B, 367
ERY. *See* Erythromycin (ERY)
Erysiphe graminis, 430
Erythromycin (ERY), 229
Escherichia coli, 413, 472, 504
E-selectin, 82
Essential oils, comparative compositional analysis of, 407
Ethinyl estradiol, 522
Ethiopia, 165
   health sector strategy, 165
   medicinal plants, 162
Ethiopian pepper, 249. *See also Xylopia aethiopica*
Eudesma-7, 404
Eugeniin, 620
Eugenol, 386, 532, 613
   antiinflammatory and antinociceptive properties, 258
   arachidonic acid metabolism, inhibition of, 258
   prostaglandins synthesis, inhibition of, 258
Eugenyl acetate, 613
Euprostin A, 613
European Medicines Agency (EMEA), 198
EVD. *See* Ebola virus disease (EVD)
Experimental inflammatory methods, 241–242
   carrageenan-induced paw edema, 241
   cotton pellet granuloma test, 242
   egg albumin-induced paw edema testin rodents, 241
   xylene-induced ear edema assay, 241
Experimental models of nociceptive pain, 242–243
   acetic acid-induced writhing test, 242
   formalin-induced paw edema, 243
   hot plate latency tests, 243
   tail-flick test, 243
Ex vivo iron-induced cell death, 519

F

Fabaceae, 339
Fagara leprieurii, 226
FAO. *See* Food and Agricultural Organization of the United Nations (FAO)
Fatoua pilosa, 212
Fatty alcohols, as hexaconsanol, 400
FDA. *See* US Food and Drug Administration (FDA)
Fehling test, 214
Fenchone, 403
FEP. *See* Cefepime (FEP)
Ferric chloride test, 215
Ferric reducing ability of plasma (FRAP), 410
Ferric-reducing antioxidant power (FRAP) assay, 470
Ferulic acid, 367, 601
Fibrous finer roots, 452
Ficus cordata, 207
Ficus ovata, 207
Ficus polita, 207
Flavonoids
   anticonvulsant effects, 340
   from *C. citratus*, 402
   inhibitory mechanisms of, 343
   sources in
      Brassica vegetables, 340
      C. sativum, 340
      lettuce, 340
Fluoroquinolones, 142
Foam test, 215
Food and Agricultural Organization of the United Nations (FAO), 628
Food carcinogen, 426
Food tree. *See Adenanthera pavonina*
Forced swim test (FST), 501
Fraxinus xanthoxyloides, 226
Froth test, 219
Fruits, 329
FST. *See* Forced swim test (FST)
Fungal infection, 21, 24
   aspergillosis *(Aspergillus fumigatus)*, 26
   Athlete's foot, 24
   Candida albicans, pathogenesis of, 25
   candidiasis, 24
      clinical manifestations of, 25
   coccidioidomycosis (valley fever), 30
   Cryptococcus neoformans infections, 28
   epidemiology of, 21
   fungal pneumonia, 30
   histoplasmosis, 31
   opportunistic infections, 27
   paracoccidioidomycosis (PCM), 29
   prevention and control, 31
   Sporothrix schenckii, 26, 27
   sporotrichosis, 26
   supersaturated potassium iodide (SSKI), 27
   Tinea capitis, 23
   Tinea corporis, 23
   Tinea unguium, 24
   types of, 22
Fusarium oxysporum, 520

G

GABAergic transmission, 88
Galangin, 211
Galanolactone, 628
Gallic acid, 613
Gallic acid equivalent (GAE), 443
Gamma-amino butyric acid (GABA), 446
Gardenin B, 601
Garlic, 333
    antiinflammatory properties, 240
    antioxidant effects, 371
    cardiovascular effects, 370
        lipid lowering effects, 370
        reduction of cholesterol level, 370
        vascular protection effect of, 370
    chemistry of, 365
        phenolics and other compounds, 367
        phytochemicals in, 367
        saponins, 366
        sulfur-containing compounds from, 365, 366
    cloves of, 364, 365
    commercial use of oil, 364
    culinary purpose, 333
    cultivation, 364
    distribution of, 364
    effect of consumption, 372
    effects in Alzheimer disease, 333
    medicinal purpose, 333
    nematodes, 364
    origin, 363
    pharmacology of, 368
        anticancer activity, 368
        antidiabetic activity, 368
        antifungal potency of, 369
        antimicrobial activity, 369
        antiviral activity, 370
    products of, 364
    publication records from pubmed database, 364
    toxicity of, 373
    white rot disease, 364
Gas chromatography-mass spectrometry (GC-MS) analysis, 406
Gelatinase matrix metalloproteinases, 520
Gelatin test, 215
Gene Xpert (GXP), 10
Genkwanin, 601
Genus *Thymus*, 307
Geranial, 403
Geranic acid, 404

Geranium nepalense, 317
Geranyl acetate, 403
Gingediol, 628
Ginger, 333, 627
    anticonvulsant effects, 339
    as antiepileptic plant, 339
    antiglycating properties, 333
    botanical aspect, 628
    chemistry, 628
    compounds, 628
        capsaicin, 628
        chemical structures, 629
        galanolactone, 628
        gingediol, 628
        gingerols, 628
        gingesulfonic acid, 628
        shogaols, 628
        volatile oils, 628
        zingerone, 628
    distribution, 628
    herbaceous plant, 628
    patents, 636
        US20, 120, 015, 060 A1, 636
        WO2003049753 A1, 636
        WO2010083967 A1, 636
        WO2012013551 A1, 636
    pharmacology, 630
        analgesic and antiinflammatory activities, 631
        anticancer activity, 632
        antidiabetic activity, 633
        antiemetic effect, 635
        antioxidant activity, 634
        antirheumatic properties, 636
        cardiovascular effect, 635
        effect on migraine, 635
        hepatoprotective effects, 633
        molluscicidal and antischistosomal activities, 635
        nephroprotective effects, 634
        in treatment and prevention of arthritis, 630
    potent suppressive effect, 631
    production, 628
    use, 333
Gingerol, 628
    antiinflammatory activity, 258
    inflammatory mediators production, inhibition of, 258
Gingesulfonic acid, 628

Glioblastoma U87MG cells, 284
Global Burden of Disease (GBD) study, 4
Glucomoringin (4(α-L-rhamnosyloxy)-benzyl glucosinolate)
  antiinflammatory activity of, 490
D-Glucose, 535
Glucose 6-phosphatase (G6Pase) gene, 617
Glutamate, 339
  excitatory transmitter, 339
Glutathione (GSH), 121, 442, 488
Glutathione reductase (GR), 124
Glutathione-S-transferase (GST), 441, 633
Glycosides, 248
Glycyrrhiza glabra. *See* Liquorice
Glycyrrhizin, 211
Gnetaceae
  antimicrobial effects, 221
Gnetum africanum, 221, 336
Gnetun africanum, 286, 329
G6Pase gene. *See* Glucose 6-phosphatase (G6Pase) gene
GR. *See* Glutathione reductase (GR)
Grains of paradise, 248
Griseofulvin, 24
GSH. *See* Glutathione (GSH)
GST. *See* Glutathione-S-transferase (GST)
Guinea grains, 248
Guinea pepper, 248
Guinea worm disease, 17

H

HACCP. *See* Hazard analysis and critical control point (HACCP)
Hager's test, 216
Haloperidol-induced catalepsy (HIC), 501
Hamburg parsley, 528
Hansenula anomala, 227
Hansenula mrakii, 221
Harpagophytum procumbens, 154
Hazard analysis and critical control point (HACCP), 192
HDL. *See* High density lipoprotein (HDL)
Headache, 407
HeLa cells, 499
HeLa cervix adenocarcinoma, 286
Helichrysum aureonitens, 211
Helminths, 16
Hemoglobin, 428
Hepatic gluconeogenesis, 490

Hepatic marker enzymes, 633
  alanine aminotransferase (ALT), 633
  alkaline phosphatase (ALP), 633
  aspartate aminotransferase (AST), 633
Hepatitis C virus (HCV), 430
Hepatocellular carcinoma (HepG2) cells, 287
Hepatoprotective, 425
HepG2 cell line, 489, 492
HepG2 hepatocarcinoma cells, 487
Heptadeca-1,9*(Z)*-diene-4, 6-diyn-3, 8-diol, 532
Heptadeca-1,9*(Z)*-diene-4, 6-diyn-3-ol, 532
1-Heptyl-(1-phenyl)-3-xylose benzoate, 464
Herbal drugs, 191
  parameters for quality control, 191
    analytical methods, 194
    ash content, 191
    foreign matter, 191
    heavy metals, 192
    labeling of herbal products, 196
    microbial contaminants and aflatoxins, 192
    microscopic evaluation, 191
    pesticide residues, 193
    radioactive contamination, 193
    validation, 195
Herbal ingredients, contaminants, 187
  ash values, 187
  foreign organic matter, 187
  fumigants, 188
  microbial contamination, 187
  pesticides, 188
  radioactive contamination, 188
  toxic metals, 188
Herbal medicine, 154
  policy for trade, 161
  regulation in Africa, 163
Hermaphrodite, 453
Herpes simplex virus 1, 605
Heterophylliin D, 613
Hexanal, 532
cis-Hex-3-en-1-ol, 532
Hibiscus sabdariffa, 154
HIC. *See* Haloperidol-induced catalepsy (HIC)
High density lipoprotein (HDL), 109, 428
Highly active antiretroviral therapy (HAART), 54
High performance liquid chromatography (HPLC), 180, 439
Himalayan plants, 162
Histamine, 239, 241, 249, 300
Histoplasmosis, 31

HIV. *See* Human immunodeficiency virus (HIV)
HIV-related cryptococcal meningitis, 29
H460 large cell lung cancer, 499
Holy basil. *See Ocimum sanctum*
   COX-1 inhibitory compounds
      cirsilineol, 253
      eugenol, 253
      isothymonin, 253
      isothymusin, 253
      rosmarinic acid, 253
Homeostasis, 83
Hoodia gordonii, 317
Horizontal-wire tests, 521
Horseradish tree, 288
HPLC. *See* High performance liquid chromatography (HPLC)
5-HT. *See* 5-hydroxytryptamine (5-HT) *See* Serotonin (5-HT)
Human cytomegalovirus (HCMV) infection
  in vitro antiviral activity of garlic extract, 370
Human cytomegalovirus (HCMV) strain AD-169 (ATCC Ref.VR538), 444
Human esophageal cancer cells (SNO), 487
Human immunodeficiency virus infection and acquired immune deficiency syndrome (HIV/AIDS), 3, 133, 207, 410
  in Africa, 6
    control, 6
      challenges, 7
      prevention, 6
      treatment, 7
      WHO policies, 7
    epidemiology, 6
Humoral immunity, 79
Humulene, 613
Hydatidosis, 56
Hydrodistillation (HD), 467
Hydrodistilled essential oil, 407
Hydrolysable tannins
  dry weight (DW) of, 401
Hydroxycinnamic acids, 212
5-Hydroxy-3-(4-hydroxylphenyl) pyrano [3, 2-g] chromene-4(8H)-one, 461
Hydroxymangiferonic acid, 457
5-Hydroxytryptamine (5-HT), 239, 241
Hyperalgesia, 86
Hyperglycemia-induced oxidative stress, 125
Hyperglycemia toxicity, 122
Hyperinsulinemia, 408

Hypertension, 5
Hypotensive, 452

I

Iatrogenic phenytoin toxicity, 507
ICP. *See* Inductively coupled plasma (ICP)
IDF. *See* International Diabetes Federation (IDF)
IFN−α. *See* Interferon-α (IFN−α)
ILs. *See* Interleukins (ILs)
Immunoblotting, 247
Immunomodulation, 452
Immunostimulant activity, 425
Imperata cylindrica, 228
Imperatorin, 535
Indole alkaloid N, α-L-rhamnopyranosyl vincosamide (VR), 486
Inducible nitric oxide synthase (iNOS), 82, 409
Inductively coupled plasma (ICP), 192
Infectious diseases, 6, 17, 133
  acquired immune deficiency syndrome, 134
  chikungunya, 137
  cholera, 139
  cryptococcal meningitis, 139
  dengue, 135
  ebola virus disease (EVD), 136
  HIV/AIDS in Africa, 6
  human immune-deficiency virus, 134
  malaria, 140
  management using African biodiversity, 143
  schistosomiasis, 141
Infectious organisms, 93
Inflammation, 73
  symptoms of, 73
Inflammatory disorders, 74, 407
  conventional drugs used for the management, 87
  epidemiology of, 74
  gender and age in Africa, 75
  medicinal plants used for the management, 88
  prevention, 75–76
Inflammatory pain control policy
  in the African context, 76
Inflammatory response, 77
  biology and physiology of inflammation, 77–78
  innate and acquired immunity, 78
  interactions of cellular and humoral immunity as defense against invaders, 79
  proinflammatory compounds
    acute-phase proteins, 83
    acute-phase response, 83

Inflammatory response (*cont.*)
   regulated by nuclear factor kappa B and their physiological effects, 80
      cell adhesion molecules (CAMs), 82
      chemokines, 82
      colony-stimulating factors (CSFs), 81
      enzymes, 82
      interferon, 81
      interleukins, 81
      tumor necrosis factor, 80–81
   triggers of immune response and inflammatory, 80
Influenza, 407
Infusions, 174, 178
Initiator caspases, 290
Injury, 85
Injury response, 86
Innate immunity, 78
Insulin receptor (IR), 501
Integral quality control (IQC), 199
Interferon, 82
Interferon-α (IFN$-\alpha$), 239
Interleukin (IL)-6, 490
Interleukins (ILs), 77, 81, 239, 252, 261
International Agency for Research on Cancer (IARC) projects, 93
International Diabetes Federation (IDF), 110, 315
Intestinal parasitism, 35
In vitro emesis model, 302
In vivo emesis models, 301
   leopard models, 301
   ranid frog models, 301
Iodine deficiency, 59
IQC. *See* Integral quality control (IQC)
IR. *See* Insulin receptor (IR)
Irvingia africana, 207
Ischemia-reperfusion (I/R), 430
Isomangiferin, 460, 461
Isomangiferolic acid, 457
Isopimpinellin, 535
Isorhamnetin, 535
Isothiocyanates, from *M. oleifera* leaves, 486
Itraconazole, 27

J

Japonicin A, 440
Jew's mallow, 287. *See also Corchorus olitorius*
JIS. *See* Joint Interim Statement (JIS)
Joint Interim Statement (JIS), 111

Juices, 176
Jurkat human T-cell leukemia cell, 277

K

Kaempferol, 401, 442, 460, 535, 613
Kaempferol 3-O-β-D-glucopyranoside, 535
Kaempferol 3-O-glycoside, 463
Kauranes, 249
Kaurenoic acid, 249
   antiinflammatory activity, 257
   histamine production, inhibition of, 257
   phospholipase A2, inhibition of, 257
Kedde's test, 216
Klebsiella pneumoniae, 142, 413, 492
   clinical isolate of, 443
Kok. *See Gnetun africanum*
Koko. *See Gnetun africanum*
Kwashiorkor, 61
Kyllinga brevifolia, 581

L

Lactate dehydrogenase (LDH), 276, 431, 519
Lactones, 209
Lactuca dregeana, 439
Lactuca saligna L., 439
Lactuca sativa, 221, 439
Lactuca sativa, 437
   antioxidant activities of, 442
   butterhead lettuce, 437
   chemical composition, 438–441
   chemical structures of, 440
   clinical trials, 447
   high performance liquid chromatography (HPLC), 439
   history, 438
   methanol extract, chemical investigation of, 441
   pharmacological activities, 446
      anticancer activity, 445
      antimicrobial activities, 443–444
      antioxidant effects, 441–443
      hypnoptic effects, 446
      neurodegenerative disorders, effects, 444–445
   urinary tract infections, 437
   vitamin C, 437
   world/African distribution, 438, 439
Lactucasativoside A, 440
Lactuca virosa L., 439
Lamiaceae
   antimicrobial effects, 222
Lasioderma serricorne, 507

Latex saps
 antifungal effect of, 444
LC-MS/MS. *See* Liquid chromatography tandem mass spectrometry (LC-MS/MS)
LDL. *See* Low-density lipoprotein (LDL)
Lead acetate test, 215
Learning and memory
 central cholinergic pathways, role of, 505
 diazepam-induced impairment, 505
 scopolamine-induced impairment, 505
Leaves, of plant, 453
Leishmania donovani, 619
Lemongrass oil, 416
Leprosy, 5, 407
Lettuce
 antioxidant effects, 337
 source of flavonoids, 337
Lettuce. *See Lactuca sativa*
Leukemia, 488
L-glutamyl-S-alkyl-l-cysteine, 366
Licochalcones, 211
Licorice. *See* Liquorice
Liebermann-Burchard test, 214
Liliaceae
 antimicrobial effects, 222
Limonene, 381, 532, 600
Linalol, 600
Lipid peroxyl radical, 122
Lipopolysaccharide (LPS), 502
 induced RAW264.7 macrophages, 490
 stimulated J774A.1 murine monocyte macrophage cells, 445
Lipoxygenase (LOX), 247
Liquid chromatography tandem mass spectrometry (LC-MS/MS), 521
Liquorice, 339
 antioxidative effects, 339
 antispasmodic herbs, 339
 use in, prevention and treatment of seizure-induced brain injury, 339
Listeria monocytogenes, 228
Liver, clinical examination of, 488
Liver enzymes, 621
 alanine aminotransferase, 621
 alkaline phosphatase, 621
 aspartate aminotransferase, 621
Low birth-weight (LBW) baby, 58
Low-density lipoprotein (LDL), 111
 oxidation, 410
LOX. *See* Lipoxygenase (LOX)
Lucigenin-enhanced chemiluminescence, 519
Lutein, 535
Luteolin, 442, 535, 601
 adhesion molecules, inhibition of, 257
 antiinflammatory activity, 257
 interleukin-1β, inhibition of, 257
Luteolin 6-O-glucosides, 401, 402
Lymphatic filariasis, 17
Lymphocytes, 88
Lymphokines, interferon (IFN), 77
Lyophilization, 180

M

Mace
 Indonesian folk medicine, 497
 rheumatism, use in, 497
Maceration, 178
Maclurin-3-C-(2-O-galloyl)-β-D-glucoside, iriflophenone-3-C-(2-O-galloyl)-β-D-glucoside, 465
Macrophages, 82, 88
Madagascar periwinkle, 277
Major facilitator superfamily (MFS), 229
Malaria, 3, 5, 13, 140, 407, 452
 cases of malaria globally, 13
 challenge, 16
 endemic malaria distribution, on the African Continent, 14
 life cycle, 13
 malarial deaths, 13
 overview of the burden of malaria in Africa, 13
 prevention and control in Africa, 15
 vulnerable population, 15
 World Malaria Report 2015, 13
Malnutrition, 57
 causes of, 59
  health and infectious diseases, 60
  water and sanitation, 60
 consequences of, 61
  economic, 62
  health, 61
  kwashiorkor, 61
  marasmus, 61
  social, 61
 global response, 63
  global initiatives, 63
  sustainable development Goal 3, 63

Malnutrition (cont.)
   treatments for, 62
      artificial nutritional support, 63
      NICE guidelines, 62
   in women and children, 58–59
Malondialdehyde (MDA), 633
   in blood and brain, 442
   hemin-induced, formation, 410
   MB-231 breast adenocarcinoma, 487
Mangifera indica L., 451, 456
   acute toxicity of seed kernel, 477
   alcoholic extract, 476
   antiallergic activity, 476
   antiinflammatory, 476
   antimicrobial activity, 472
      antibacterial activity, 472–473
      antifungal activity, 473
      antiparasitic activity, 473
      antiprotozoal, 473
      antiviral activity, 474
   antioxidant activity, 470–472
   antiplasmodial activity, 474
   antitumor activity, 474
      antidiabetic activity, 475
   botanical description, 452–453
   carbohydrates, 467–469
   classification and cultivars of, 455
   clinical trials, 476–477
   essential oil, 469
   immunomodulatory effect, 476
   methanol and ethyl acetate extracts, 472
   origin/distribution, 454
   pharmacological activity of, 469
   pharmacological properties, 476
   phenolic derivatives, 464–465, 467
   phytochemicals distribution, 471
   phytochemical study, 451
   phytochemistry of, 455
      polyphenolic/phenolic compounds, 460
         flavonoids, 460–463
         tannins, 463–464
      saponins, 456
      steroids, 456
      triterpenoids, 456
   structures of alkaloids, 470
   tannins isolated, 463
   toxicity status of, 477
   volatile components, 467
   volatile compounds, structures of, 468

Mangiferin, anti-HIV-1 activity of, 474
Mangiferolic acid, 457
Mango. See *Mangifera indica* L.
Mangocoumarin, 467
Mango seed kernel, 472
Mangostin, 467
MAPK. See Mitogen-activated protein kinase (MAPK)
Maracuja, 515
Marasmus, 61
Marburg hemorrhagic fever, 207
Mass spectrometry (MS), 186
Mayer's test, 216
MCF-7 breast cancer cells, 445
MDA. See Malondialdehyde (MDA)
MDCK multicystic dysplastic kidney cells, 499
MDR. See Multidrug-resistant (MDR)
Measles deaths, 5
Medical doctors, 154
Medical system, in Africa, 154
Medicinal herbs, 121
Medicinal plants, 144, 154, 172
   in drug development, 160
   ethnomedicinal uses, against infectious diseases in Africa, 144
   extraction method, parameters for, 179
      authentication of plant material, 179
      drying conditions for plant material, 179
      grinding methods, 179
      nature of constituents, 179
      part of plant material, 179
   extracts, modes of preparation, 176
      counter-current extraction, 176
      decoction, 178
      digestion, 178
      infusion, 178
      maceration, 178
      percolation, 177
      plant tissue homogenization, 176
      serial exhaustive extraction, 178
      sonication, 177
      soxhlet extraction, 178
   factors affecting in use of preparations, 172
      drug interactions, 173
      inadequate standardization, 172
      ineffective in acute medical care, 172
      lack of quality specifications, 172
      lack of scientific data, 172
   government support on research, 159

in health care programs, 154
incentives to collectors and farmers for sustainable production, 159
informal trade, 162
modes of preparation, 173
  compresses, 175
  concoctions, 173
  decoctions, 173
  infusions, 174
  juices, 176
  pills, 174
  poultice, 175
  powder, 174
  syrup, 175
  tablets, 175
  tinctures, 175
in pharmacological research, 160
policies
  for conservation, 157
  regarding export, 160
regulation of exploitation, 160
regulation of exportation, 160
standardization of preparation
  in Africa, 180
  in other parts of world, 180
Melegueta. *See Aframomum melegueta*
Melegueta pepper, 248. *See also Aframomum melegueta*
Mentha longifolia, 304
ρ−1,3,8−Menthatriene, 532
Metabolic syndrome (MetS), 109
  in Africa, 109
    burden, 113
    complications, 113
    medicinal herbs, framework to develop public health policy, 121
    prevalence, 111
    public health policies, systematic review, 117
  definition, 315
  diabetes mellitus, risk of, 315
  effects of medicinal spices, 318
  management, 315
  obesity, risk factor for, 317
  potential of phytochemicals against, 317
  prevalence of, 315
  risk of cardiovascular disease, 315
  in vitro screening methods of phytochemicals and, 316
Methanol, antiviral activities of, 444
Methanol/chloroform extract, 446

Methanol extracts, 303
Methicillin-resistant *S. aureus* (MRSA), 142, 618
5′-Methoxycurcumin, 426
7-Methoxyluteolin, 601
5-Methoxypsoralen, 535
8-Methoxypsoralen, 535
Methyl allyl thiosulfinate, 366
1-methyl-4-(methylethenyl)-2,3-dioxabicyclo [2.2.2] oct-5-ene, 535
4-methyl-7-(methylethenyl)-3, 8- dioxatricyclo [5.1.0] octane, 535
Methyl salicylate, 613
MetS. *See* Metabolic syndrome (MetS)
Meyer method, 520
MFS. *See* Major facilitator superfamily (MFS)
MIC. *See* Minimal inhibitory concentration (MIC)
Micrococcus luteus, 504
Microsporum canis, 412
Microsporum gypseum, 226
Microwave-assisted hydrodistillation (MAHD), 404
Millennium development goals (MDGs), 8
Minimal inhibitory concentration (MIC), 208, 217, 218, 429, 443
  minimum inhibitory concentration 50 (MIC_{50}), 428
Mitochondrial membrane integrity, 291
Mitogen-activated protein kinase (MAPK), 432, 502
M. oleifera against multidrug-resistant (MDR), 492
Molisch test, 214
Momordica charantia, 317
Mondia whitei, 228
Monodora myristica, 286, 305
Monoembryonic seeds, 453
Moraceae, antimicrobial effects, 223
Moringa, 288. *See also Moringa oleifera*
  antibacterial activities, 288
  antidiabetic property, 288
  antiepileptic property, 288
  antifungal activities, 288
  antipyretic property, 288
  antiulcer property, 288
  cholesterol lowering property, 288
  circulatory stimulants, 288
  content
    beta-sitosterol, 288
    caffeoylquinic acid, 288
    kaempferol, 288
    quercetin, 288
    zeatin, 288

Moringa. *See also Moringa oleifera*
Moringaceae, 485
Moringa fruit extract, 490
Moringa oil, 485
Moringa oleifera, 485
   antidiabetic effect of, 490
   anti-herpes simplex virus type 1 (HSV-1) effect, 492
   antiinflammatory effect, 490
   chemical structures, 487
   cultivation and distribution of, 485
   cytotoxicity of, 487
   essential oil, 486
   glucose-lowering effects, 489
   isoproterenol (ISO)-induced cardiac toxicity, 493
   methanol extract, 486
   nitric oxide (NO) production, 490
   nonvolatile constituents, 486
   pharmacology of, 487
      anticancer activity, 487–489
      antidiabetic activity, 489–490
      antiinflammatory and analgesic activities, 490
      antimicrobial activities, 492
      pharmacological effects, 493
Morphine consumption, 73
Mortality rates, 3, 28
   rose for women, 3
MRSA. *See* Methicillin-resistant *S. aureus* (MRSA)
MS. *See* Mass spectrometry (MS)
MTS. *See* 3-(4,5-dimethylthiazol-2-yl)-5-(3-carboxymethoxyphenyl)-2-(4-sulfophenyl)-2H-tetrazolium) (MTS)
MTT. *See* 3-(4,5-dimethylthiazol-2-yl)-2,5-diphenyl-2H-tetrazolium bromide (MTT)
Mucor mucedo, 227
Mucosal infections, 22
Multidrug-resistant (MDR), 11, 212
Murraya koenigii, 305. *See also* Curry tree
m-Xylene, 532
Mycobacterium tuberculosis, 211
Mycoses (fungi infection), 21
Mycosis, 54
Myrcene, 403, 532, 600
Myricetin, 613
Myrislignan, 503
Myristica fragrans, 305
   botanical aspects, 497
      fruit, 497
      leaves, 497
      seed, 497
   chemistry, 498
   compounds
      chemical structures, 500
   distribution, 497
   patents, 508
   pharmacology, 499–507
      analgesic effect, 502–504
      anticancer activity, 499
      antidepressant effect, 501
      antidiabetic effect, 501
      antiinflammatory effect, 502–504
      antimicrobial activities, 504
      antiobesity effect, 501
      antioxidant effects, 505
      hepatoprotective effects, 505
      memory enhancing effects, 505
      other effects, 506–507
   producers, 497
   source of
      mace, 497
      nutmeg, 497
   toxicity, 507–508
Myristicin, 532

N

NAA. *See* Neutron activation analysis (NAA)
NAC. *See n*-acetylcysteine (NAC)
N-acetylcysteine (NAC), 121
Nalta jute, 287. *See also Corchorus olitorius*
Naphtalene, 405
Nardostachys grandiflora, 162
Nardostachys jatamansi, 162
National Cancer Institute (NCI), 274
National Cholesterol Education Program (NCEP), 315
National Cholesterol Education Program Adult Treatment Panel (NCEP-ATP III), 111
Nauplii, 520
Nausea, 299
   physiological mechanisms, 300
NCDs. *See* Noncommunicable diseases (NCDs)
NCEP-ATP III. *See* National Cholesterol Education Program Adult Treatment Panel (NCEP-ATP III)
NCI. *See* National Cancer Institute (NCI)
Necator americanus, 33

Neglected tropical diseases (NTDs), 17, 32, 133
   epidemiology and burden of disease, 34
      clinical features, 35
      early larval migration, 35
     intestinal parasitism, 35
      ascariasis, 35
      hookworm infection, 36–39
      trachoma, 39
        clinical features, 40
        detection of *Chlamydia trachomatis* infection, 42
        differential diagnosis, 40
        epidemiology, 41
        trachoma control, 43
        transmission of infection, 41
      trichuriasis, 36
   schistosomiasis in Africa, 32
   soil-transmitted helminthes, 33
Neisseria meningitidis, 56
Neointermediol, 404
Neonatal asphyxia, 52
Neonatal blindness, 142
Neonatal death rate, 5
Neoxanthin, 535
Neral, 403
Nerolidol, 381
Nerve root, injury, 85
Neurocysticercosis, 56
Neurological infections, 54
   age-adjusted death rates for the most common cancers in males and females in Africa, 2008, 46
   age-adjusted incidence rates for the most common cancers in males and females in Africa, 2008, 45
   Alzheimer and other dementias, 57
   bacterial infections, 55
   bacterial meningitis, 56
   estimated numbers of new cases and deaths for leading cancer sites in Africa, 44
   neurological diseases in the HIV-infected individual, 55
   number of undernourished and prevalence (%) of undernourishment, 55
   parasitic diseases, 56
      cysticercosis, 56
      malaria, 56
   Parkinson's disease, 57
Neuropathic pain, 86

Neutral red uptake assay, 276
   biomedical application, 276
   chemical structure, 276
   cytotoxicity test, 276
   environmental application, 276
Neutron activation analysis (NAA), 192
Newborn conditions, 3
Newcastle disease virus, 605
Nickel metal ($NiCl_2$), 408
Nicotinamide adenine dinucleotide phosphate (NADPH)-dependent cellular oxidoreductase enzymes, 272
Niemann-Pick disease, 47
Nigella sativa
   antiinflammatory or antinociceptive potential, 251
Nigella sativa, 581
Nigeria Natural Medicine Development Agency (NNMDA), 160
Nigerian *Piper guineense* fruit oils, 587
Nigeria possessed anticonvulsant activity, 591
Nitric oxide (NO), 602
NK cells, 88
N-methyl-*N*-nitrosurea (MNU), 408
NMR. *See* Nuclear magnetic resonance (NMR)
NNMDA. *See* Nigeria Natural Medicine Development Agency (NNMDA)
NO. *See* Nitric oxide (NO)
Nociception
   activities, 86
   pathophysiology of, 84
   pathway, 84
   in uninjured skin, 85
Nociceptive disorders, 74
   conventional drugs used for the management, 87
   disorders, 88
   epidemiology of, 74
   gender and age in Africa, 75
   medicinal plants used for the management, 88
   prevention, 75–76
Nociceptors, 85
Nonanal, 532
Noncommunicable diseases (NCDs), 3, 5, 43, 110
   cancers in Africa, 43
   causes and risk factors of neurological disorders, 47
   consequences of neurological disorders, 48
   in East African countries, 119
   epilepsy, 51
   lifestyle diseases, in sub-Saharan Africa, 4

Noncommunicable diseases (NCDs) (cont.)
  metabolic syndrome, 109
  neurological disorders, 43
  in North African countries, 119
  Parkinson's disease, 49
  in Southern African countries, 119
  stroke, 52
  in West African countries, 118
Nonsteroidal antiinflammatory drugs (NSAIDs), 88, 409, 630
Nonvolatile polyphenol compounds, 400
NOR. See Norfloxacin (NOR)
Norfloxacin (NOR), 229
Nrf2 transcription factor, 489
NSAIDs. See Nonsteroidal antiinflammatory drugs (NSAIDs)
NTDs. See Neglected tropical diseases (NTDs)
Ntoumou. See Gnetun africanum
Nuclear factor-kappa B (NF-κB), 502
Nuclear factor kappa (NF-κ)-light-chain-enhancer, 490
Nuclear magnetic resonance (NMR), 195
Nucleic acid amplification tests, 42
Nutmeg, 305. See also *Myristica fragrans*
  acetylcholinesterase-inhibiting activity, 505
  anticonvulsant activity, 507
  antihelmintic effects, 506
  anxiolytic effects, 506
  aphrodisiac activity, 506
  content
    camphene, 498
    elemicin, 498
    γ−terpinene, 498
    γ−terpinol, 498
    lignans, 498
    malabaricone B, 498
    myristin, 498
    nectandrin B, 498
    polyphenol, 498
    sabinene, 498
    saponins, 498
    saucernetindiol, 498
    volatile oil, 498
  hallucinogenic properties, 497
  hepatoprotective activity, 505
    plasma aminotranferase activities, assessed by, 505
  kidney stone disolution, role in, 497
  medicinal use, 497
  platelet antiaggregatory ability, 507
  toxic effects
    myristicin oil, role of, 507
Nutrition transition, 115

O

Obesity, 112
  classification, 341
  definition, 341
  dietary phytochemicals, 341
  spices and vegetables, potential effects on, 341
cis-β−Ocimene, 532
trans-β−Ocimene, 532
Ocimum basilicum, 222, 252
Ocimum gratissimum, 252
Ocimum sanctum, 252
Ocimum species, 305
  antiinflammatory compounds
    antinociceptive potential, 252, 252–253
    sieboldogenin, 253
Ocimum tenuiflorum, 252
η−Octanol, 381
Octyl acetate, 381
Okazi. See Gnetun africanum
Olax subscorpioidea, 228
Olitoriside, 287
Onchocerciasis, 17
Onions, 353
  bulbs of, 354
  chemistry of, 354
    essential oil, 354
    phenolics, 355
  cultivation, 353
  distribution, 353
  effective against gram-positive bacteria, 357
  level of antioxidant, 336
  pharmacology of, 356
    anticancer activity, 356
    antidiabetic effects, 357
    antiinflammatory activity, 359
    antimicrobial activities, 357
    antioxidant effects, 358
    cardiovascular effects, 358
    use in blood-brain barrier (BBB), 359
  phenolic compounds in, 356
  sources of flavonoids, 336
  use of oil, in fungal growth, 357
  volatile constituents in onion oil, 355
OPLC. See Overpressured layer chromatography (OPLC)

2″-O-rhamnosyl isoorientin, 401
12-O-tetradecanoylphorbol-13-acetate-derived promotion, 413
Other African spices and vegetables
    antiinflammatory or antinociceptive potential, 244, 254
Overpressured layer chromatography (OPLC), 195
Overweight, 112
Oxidative medicine, 125
Oxypeucedanin hydrate, 535

P

Pain, 73
Painful syndromes, 86
Pain, possible ways of classifying, 85
Pain processing pathway, 84
    biology and physiology of nociceptors, 84
Paracoccidiocomycosis infection, 30
Parasitic infections, 16
    burden of parasitic diseases on Africa, 17
    control and preventive measures, 20
    diagnoses, 20
    in humans, 16
    vulnerable population, 20
Parkinson's disease, 49
    mortality, 51
    prevalence, incidence, and distribution, 49
    risk factors for, 50
    signs and symptoms, 51
Parsley. See *Petroselinum crispum*
Passifloraceae, 513
Passiflora edulis, 247
Passion fruit. See *Passiflora edulis*
    analgesic effect, 515
    antidiarrheal effect, 515
    antispasmodic property, 515
    botanical description, 514
        flower, 514
        fruit, 514
        leaves, 514
        seed, 514
    clinical trials, 522
    compounds
        chemical structures, 517
    congestive heart failure, effect on, 522
    distribution, 515
    emetic effect, 515
    flower drink, 515
    pharmacovigilance data, 522–523
    phytochemistry, 516–518
        alkaloids, 518
            harmaline, 518
            harmalol, 518
            harmane, 518
            harmine, 518
        flavonoids, 516–518
            isoorientin, 516
            isoschaftoside, 516
            isovitexin, 516
            lucenin, 518
            luteolin, 518
            luteolin-6-C-chinovoside, 516
            orientin, 516
            schaftoside, 516
        glycosides, 518
            luteolin-6-C-chinovoside, 518
            luteolin-6-C-fucoside, 518
            passibiflorin, 518
            passicapsin, 518
        miscellaneous and phytoconstituents, 518
        terpenoids, 518
            α−terpeneol, 518
            furaneol, 518
            terpene alcohols linalool, 518
    propagation, 515
    safety profile, 522–523
    sedative activities, 515
    traditional or ethnomedicinal uses, 515
    tranquillizer activities, 515
    in vitro and in vivo pharmacological studies, 519–521
        antianxiety activity, 521
        antifungal activity, 520
        antihypertensive activity, 521
        antiinflammatory activity, 520
        antioxidant activity, 519
        antitumor activity, 520
        cytotoxic activity, 520
Passive-avoidance apparatus, 505
Patents
    EP2, 689, 806 A1, 508
    garlic, 373
    onion, 359
    Petroselinum crispum, 542
    WO2, 008, 096, 998 A1, 512
Pathogen-associated molecular patterns (PAMPs), 80
Pausinystalia johimbe, 161
Pausinystalia yohimbe, 154

PCR. *See* Polymerase chain reaction (PCR)
p-Cymene, 381
Pedaliaceae
  antimicrobial effects, 223
Penicillium crustosum, 227
Penta-O-galloylglucose, 463
Pentobarbitone sleeping time (PST), 501
2-Pentylfuran, 532
PEPCK. *See* Phosphoenolpyruvate carboxykinase (PEPCK)
Pepper, 121
  African medicinal spice, 121
  antioxidant potential of soup for oxidative medicine, 121
  as medicinal spice, 127
  oxidative medicine value, 122
    adoptability in policy formulation, 125
    blood flow/viscosity, maintenance of, 124
    diabetes management, 125
    as therapy, 126
    prevents lipid peroxidation, 127
Pepper soup, 121
  antioxidant potential for oxidative medicine, 121
  nutritional value, 121
Percolation, 177
Periplocaceae, 228
Peroxisome proliferator-activated receptor (PPAR)-gamma activation, 428
Petroselinic acid, 532
Petroselinum crispum, 220, 527
  clinical trial, 541
  ethnomedicinal uses, 529
  flavonoids, 531
  geographical distribution, 529
  leaves and aerial parts of, 536
  origin, 527
  oxypeucedanin, 531
  pharmacological properties, 534, 537
    analgesic activity, 539
    anticancer activity, 540
    antimicrobial activity, 534
    antioxidant activity, 534
    cardiovascular activity, 538
    gastrointestinal activity, 539
    genitourinary activity, 539
    immunomodulating activity, 539
    neuroprotective effect, 541
    nutraceutical intervention, 540
    spasmolytic activity, 540
  plant description, 528
  pytochemical constituents, 530
    essential oil, 530
  toxicity of, 541
  varieties, 528
Petroside, 535
P-glycoprotein activation, 464
Phagocytes, 79
Pharmacopoeia, 160
PHC. *See* Primary health care (PHC)
α−Phellandrene, 381
α−Phellandrene, 532
β−Phellandrene, 381
β−Phellandrene, 532
Phemacoccus solenopsis, 411
Phenazine methosulfate (PMS), 274
Phenolic antioxidants, 464
Phenolic compounds, 88
Phenolics, from *Allium sativum*, 367
Phenylacetaldehyde, 532
Phenylalanine arginine β-naphthylamide (PAβN), 229, 492
4-Phenyl gallate, 464
6-Phenyl-*n*-hexyl gallate, 464
Phosphoenolpyruvate carboxykinase (PEPCK), 617
Phytochemicals
  anticancer activities
    in vitro screening, 272–276
      MTS assays, 274
      MTT assay, 272
      neutral red uptake assay, 276
      resazurin assay, 275
      sulforhodamine B assays, 274
      XTT assays, 273
  anticancer potential, 277
  antimicrobial activities, in vitro screening methods, 213–219
  antimicrobial activity interpretation, cutoff points, 218–219
  biological activity screenings, extraction solvent selection, 213
  chemical structures of, 319
  microorganisms, mode of action, 208
Phytolacca dodecandra, 154
Phytomedicines, 160
  harmonization, 167
  in international market, 164
Phytophthora infestans, 430
Pichia membranifaciens, 221

Pills, 174
Pinene, 613
α−Pinene, 381
α−Pinene, 532
β−Pinene, 381, 600
β−Pinene, 532
Pinoresinol, 559
Piperaceae, 331
   antimicrobial effects, 223–224
Piper, African medicinal spices, 581
   biological/pharmacological effects, 589
      analgesic/antiinflammatory activities, 590
      antimicrobial activities, 589
      antioxidant activities, 590
      central nervous system activities, 591
      gastrointestinal (GIT), 591
   bioprospecting and conservation status, 586
   distribution of, 583
   diversity in, 582
   economic activities, 585
   economic and commercial prospect, 585
   ethnobotanical/ethnosocial importance, 584–585
   ethnomedicinal applications of, 585
   ethnomedicinal uses, 585
   gastrointestinal (GIT) contractile effect of ethanolic extract, 591
   phytochemical constituents, 587
      nonvolatile components of, 587
      volatile components of, 587
   taxonomy of, 582
   toxicity profile, 592
Piper capense, 223
Piper capensis, 253
Piper flaviflorum, 253
Piper guineense, 224, 253
   essential oil, 589
   photograph of, 584
Piperine
   antiinflammatory properties, 256
   chemical structures of, 338
   edema reduction in submucosa, 256
   LPS-induced endotoxin shock inhibition, 256
   mononuclear infiltration alleviation, 256
   prostaglandin E2 production, inhibition of, 256
   synovial hyperplasia alleviation, 256
Piperitenone oxide, 304
Piper kadsura, 254
Piper longum, 582

Piper miniatum
   antiinflammatory activity, 253
Piper nigrum, 223, 253, 306. *See also* Black pepper
Piper nigrum, 581
Piper species
   antiinflammatory and antinociceptive activities, 253–254
   phyto-constituents
      alkaloids, 254
      amides, 254
      futoquinol, 254
      kawapyrones, 254
      lignan, 254
      neolignans, 254
      piperolides, 254
      propenylphenol, 254
Piper vicosanum, 253
   edema formation reduction, 253
   leukocyte migration inhibition, 253
Plant antimicrobials, chemical structures, 210
Plant-derived substances, 172
Plant extracts
   phytochemical screening, 214–216
      alkaloids detection, 216
      anthraquinone glycosides detection, 215
         bound anthraquinone, 215
         free anthraquinone, 215
      carbohydrates detection, 214
      cardiac glycosides detection, 216
      diterpenes detection, 214
      flavonoids detection, 215
      phenols detection, 215
      saponins detection, 215
      tannins detection, 215
      terpenoids and steroids tests, 214
Plant tissue homogenization, 176
Plasmodium falciparum, 13
Plasmodium vivax, 13
Platycaryanin A, 613
Poaceae family, *Cymbopogon citratus*, 397
Policymakers, 4
Poly [ADP-ribose] polymerase 1 (PARP-1), 487
Polyembryonic seeds, 453
Polygonum lapathifolium, 309
Polymerase chain reaction (PCR), 10, 247
Polyoxyethylene sorbitan monooleate, 303
Potato
   anthocyanidin content, 337
   phenolics in, 337

Poultice, 175
Powder, 174
Preexposure antiretroviral prophylaxis (PrEP), 6
Premature death, 3
Primary health care (PHC), 16, 155
Procaspase-9, 291
Procaspases-3, 287
Proinflammatory cytokines, 86
1-Propenyl allyl thiosulfinate, 366
Protein-energy malnutrition, 3
Protein tyrosine phosphatase 1B (PTP1B), 501
Proteus vulgaris, 472
Proto eruboside B, 367
Providencia stuartii, 220
Provisional tolerable weekly intake (PTWI), 192
Pseudomonas aeruginosa, 212, 413, 472, 504
  ATCC 9027, 443
Pseudomonas cichorii, 439
Pseudomonas fluorescens, 492
Pseudomonas syringae pv. *phaseolicola*, 443
Psoralen, 535
PST. *See* Pentobarbitone sleeping time (PST)
PTP1B. *See* Protein tyrosine phosphatase 1B (PTP1B)
PTWI. *See* Provisional tolerable weekly intake (PTWI)
P2X receptor sites, 84

Q

QTLC. *See* Quantitative TLC (QTLC)
Quality-adjusted life years (QALYs), 3
Quality control, of medicinal preparations, 188
  approaches in herbal medicine, 196
    fingerprint, 197
    integral quality control, 199
    metabolomics, 198
  herbal drugs, challenges and factors affecting, 189
    adulteration and contamination of botanicals, 190
    plant material identity, 189
    quality control parameters of herbal drugs, 191
    variations in botanicals, 189
Quantitative TLC (QTLC), 195
Quercetin, 401, 402, 442, 535
  antimicrobial properties, 211
Quercetin 3-O-gallate, 463

Quercetin-3-O-glucoside, 489
Quercetin-3-O-glucuronide, 441
Quercetin-3-O-rhamnoside, 463
Quercetin-3-O-rutinoside, 463
Quercetin-3-sulfate, 460

R

Ramineae, 228
Raphanus sativus, 221
RDI. *See* Recommended dietary indication (RDI)
Reactive oxygen species (ROS), 410, 470, 632
  production, 426
Recommended dietary indication (RDI), 123
Red pepper. *See Capsicum baccatum*
  antiinflammatory activity, 250
  neutrophil migration inhibition, 250
Resazurin assay, 275
  advantages, 275
  disadvantages, 275
  enzymatic reduction to resorubin, 275
Reserpine reversal test (RRT), 501
Resistance/nodulation/cell division (RND) family, 229
Resistant pathogens, 142
Rhamnocitrin, 613
Rheumatoid arthritis, 86
  medicinal spices and vegetables, 340
    Indian ginseng, 341
    poison gooseberry, 341
    winter cherry, 341
    W. somnifera, 341
Rhizoctonia solani, 225, 430
Rhizomes, 425, 627
  of *Curcuma* species, 581
  oils, 404
Rhizopus chinensis, 227
Rhizopus stolonifer, 604
River blindness, 5
RND family. *See* Resistance/nodulation/cell division (RND) family
Road injuries, 3
ROS. *See* Reactive oxygen species (ROS)
RRT. *See* Reserpine reversal test (RRT)
Rubi fructus, 317
Rugosin C, 613
Rugosin D, 613
Rugosin F, 613
Rutaceae, antimicrobial effects, 226
Rutin, chemical structures of, 338

S

Sabinene, 381, 532
Sabinene hydrate, 600
Saccharomyces cerevisiae, 221
SAFE strategy, 43
Salicis radicis, 317
Salkowski's test, 214
S-allyl cysteine, 366
S-allyl mercaptocysteine, 366
Salmonella typhi, 221, 504
Salvigenin, 601
Saponins, from *Allium sativum*, 366
Scalds, 452
SCC. *See* Spectral correlative chromatograms (SCC)
Schistosoma guineensis, 32
Schistosoma haematobium, 32
Schistosoma intercalatum, 32
Schistosoma japonicum, 32
Schistosoma mansoni, 32
Schistosoma mekongi, 32
Schistosomiasis, 141
    control programs, 32
Schitosomiasis, 56
Schizosaccharomyces pombe, 227
Sclerotinia libertiana, 227
Scopulariopsis brevicaulis, 226
Seizures, clinical manifestations of, 51
Selected African spices
    antiinflammatory/antinociceptive potential, 247–254
Selina-6-en-4-ol, 404
Septicemia, 142
Serial exhaustive extraction, 178
Serotonin (5-HT), 241
Serotonin antagonists, 301
Sesame, 549
    antiinflammatory effects of, 552
    biological activities of solvent extracts from different parts, 561
    chemical properties of, 551
    chemical structures of common bioactive compounds isolated from, 559
    countries of consumption, 550
    cultivation of, 572
    description, 549
    ethnomedicinal value, 549
    fatty acid composition, 552, 553
    gastronomic uses, 550
    growth, 574
    industrial raw material, 572
    isolated compounds from different parts and biological activities, 554
    mineral composition of, 552
    nutritional and medicinal uses, 572
    oil content of, 552
    origin of, 551
    pharmacological activities, 553
    pharmacological evaluation of plant extracts, 559
    phytochemical studies, 552
    production, 572
    safety and toxicology, 572
    seeds, 550
    top twenty producers of sesame seeds for year 2013, 573
    use in traditional medicine, 572
    use of fruits, 549
    wound-healing activity, 560
    yield of, 574
Sesamin, 559
Sesamol, 559
Sesamolin, 559
Sesamum indicum, 223. *See also* Sesame
Sesamum radiatum, 223
SFDA. *See* State Food and Drug Administration (SFDA) of China
Shingles, 86
Shogaols, 628
SH-SY5Y neuroblastoma cells, 432
Sideritoflavone, 601
Sinapic acid, 367
Single nucleotide polymorphisms (SNP) technique, 29
Siphonochilus aethiopicus, 162
β-Sitosterol, 400
Skin infections, 485
Small multidrug resistance (SMR) family, 229
SMR family. *See* Small multidrug resistance (SMR) family
SOD. *See* Superoxide dismutase (SOD)
Soils ranging, 398
Soil-transmitted helminthes, 33
    infections, 34
Solanaceae, antimicrobial effects, 224–225
Solanum aethiopicum, 309
Solanum khasianum, 208
Solanum lycopersicum, 224
Solanum melongena, 224

Solanum nigrum, 288
Solanum species, 306
Solanum tuberosum, 224
Sonication, 177
Sophoraflavone G, 211
Soxhlet extraction, 178
Spectral correlative chromatograms (SCC), 195
Spectrophotometer, 272
Spectrophotometric microtiter well plate reader, 275
Spices
  antioxidant effects of, 334
  chemical structures of, antioxidant compounds, 337
  conventional classification of, 329
  definition, 329
  oxidative stress, 334
  role in in human fertility, 344
  tree, 249
Spiderwisp, 286
Spleen, clinical examination of, 488
Spontaneous standardization processes, 180
Sporothrix infections, 27
Sporothrix schenckii, 26
Sporotrichosis, 26, 27
Squarrosanin A, 613
SRB. *See* Sulforhodamine B assay (SRB)
SSA. *See* Sub-Saharan Africa (SSA)
Standardization of medicinal plant, 180
  necessity, 180
  steps involved in preparations, 182
    ash values, 182
    chromatographic examination, 183
    crude fiber, 182
    extractive values, 182
    foreign organic matter, 182
    macro and microscopic examination, 182
    moisture content, 182
    qualitative chemical evaluation, 183
    quantitative chemical evaluation, 183
    toxicological studies, 183
  techniques for preparations, 183
    adulteration, 185
    analytical methods, 184
    biological evaluation, 184
    chemical evaluation, 183
    chromatography, 184
    contaminants of herbal ingredients, 187
    control of starting material, 185
    good agricultural/manufacturing practices, 187
    identity and purity, 186
    microscopic evaluation, 183
    physical evaluation, 183
    purity determination, 184
    quantitative analysis, 184
    substitution, 185
Staphylococcus aureus, 40, 504
Staphylococcus aureus, 492
Staphylococcus aureus ATCC 6538P, 443
Staphylococcus epidermidis, 224
State Food and Drug Administration (SFDA) of China, 198
Stems include myricetin, 486
Steroids, 426
Stevens-Johnson syndrome, 40
Stigmasterol, 457
Stinging nettle, 191
Streptococcus faecalis, 220
Streptococcus pneumoniae, 56
Streptozotocin-induced diabetic rats, 428, 501
Stroke, 52
  burden of, 53
  risk factors, 53
  signs and symptoms, 53
Strophanthin, 287
Sub-Saharan Africa (SSA), 109
Succinic semialdehyde dehydrogenase (SSADH), 446
Sulforhodamine B assay (SRB), 272, 274
  chemical structure, 275
Sulfur-containing compounds from *Allium sativum*, 365
Superoxide dismutase (SOD), 633
  in blood and brain, 442
Supersaturated potassium iodide (SSKI), 27
Susceptibility testing methods, 216–217
  bioautography, 217
  diffusion methods, 217
  microbroth dilution method, 216
Sutherlandia frutescens, 143
Swertiajaponin, 401
Swertifrancheside, 211
Synthetic drugs, 143
  antimalarial, 143
  antithelmintic, 143
  arecolin, 143
  bacillary dysentery, 143
  hemsleyadin, 143
  quinidine, 143
  quinine, 143

Syrup, 175
Sysmex KX-21N, 488
Syzyginin A, 613
Syzyginin B, 613
Syzygium aromaticum, 228, 307, 322, 611. *See also* Cloves

T

Tablets, 175
Tail-flick test, variants, 243
Tail suspension test (TST), 501
Taraxerol, 457
Taraxerone, 457
Taro, 287. *See* Colocasia esculenta
    contents
        carbohydrate, 287
        dietary fiber, 287
        minerals, 287
        protein, 287
        sodium, 287
        sugars, 287
    1,8-dinitropyrene adsorbtion, role in, 287
TC. *See* Total cholesterol (TC)
TCAM (Traditional Complementary and Alternative Medicine) systems, 155
    definition, 157
    practitioners, 156
    safety and efficacy, 157
    usage in developing countries, 156
TdT-mediated dUTP nick-end labeling (TUNEL) assay, 619
Telfairia occidentalis, 228
Tellimagrandin II, 613
Terpenoids, 240, 456
    from *C. citratus*, 403–405
    structures of, 457
α-Terpinene, 532
γ-Terpinene, 381, 532, 600
α-Terpineol, 381, 532
Terpineone, 404
α-Terpinol, 600
γ-Terpinol, 600
α-Terpinolene, 532
TET. *See* Tetracycline (TET).
Tetracycline (TET), 229
TG. *See* Triglyceride (TG)
Thai basil. *See Ocimum basilicum*
Thermolabile drugs, 178
Thin layer chromatography (TLC), 180, 217

Thiocarbamate, 486
Throat infection, 452
α-Thujene, 381, 532
Thyme, 599
    antioxidant activity, 605
    antispasmodic activity, 605
    chemistry, 600
        essential oil, 600
        phenolics constituents, 601
    cultivation, 599
    distribution, 599
    patents, 606
        CA2, 222, 563 A1, 606
        EP1, 080, 727 A1, 606
        WO2, 012, 131, 732 A1, 606
    pharmacology, 602
        antibacterial activities, 604
        antifungal activities, 604
        antiinflammatory activities, 602
        antimicrobial activities, 602
        antinociceptive activities, 602
        antiviral activities, 605
Thymol, 600
Thymonin, 601
Thymoquinone
    antiinflammatory effect, 255
    IFN-γ production, 255
    mRNA expression inhibition, 255
    NO production reduction, 255
Thymusin, 601
Thymus vulgaris, 323, 599. *See also* Thyme
Thymyl methyl ether, 600
Tiliaceae, antimicrobial effects, 225
Tinctures, 175
Tinea capitis, 23
Tinea corporis, 23
Tinea unguium, 24
TLC. *See* Thin layer chromatography (TLC)
TM. *See* Traditional medicine (TM)
TNFα. *See* Tumor necrosis factor-alpha (TNFα)
α-Tocopherol, 620
Toenails, 24
Toluene, 532
2-(ρ-Tolyl) propan-2-ol, 532
Torulopsis etchellsii, 221
Total cholesterol (TC), 633
Total phenolic content (TPC), 443
Total radical-trapping antioxidant potential (TRAP), 447

Toxoplamosis, 56
Toxoplasmolisis, 17
TPA-induced mouse ear edema model, 253
Trachoma, 39
   clinical features, 40
   detection of *Chlamydia trachomatis* infection, 42
   differential diagnosis, 40
   epidemiology, 41
   trachoma control, 43
   transmission of infection, 41
Traditional doctors, 154
Traditional health care systems, 154
Traditional medicine (TM), 153, 171
   for national health, 155
   resource base, 155
   for traditional knowledge, 155
Transient ischemic attack (TIA), 53
Transient receptor potential vanilloid (TRPV), 84
Treatment policies, 5
Triacontanol, 400
Tribolium castaneum, 591
Trichoderma harzianum, 520
Trichophyton mentagrophytes, 226
Trichuriasis, 36
Tridesmostemon omphalocarpoides, 207
Triglyceride (TG), 114, 412
Trigonella foenum-graecum, 317
Tripetroselinin, 532
Triphenyl tetrazolium chloride (TTC), 493
Triterpenoids, from C. citratus, 406
Triumfetta cordifolia, 225
Trolox equivalent antioxidant capacity (TEAC) assay, 505
Trypanosoma brucei brucei, 591
TST. *See* Tail suspension test (TST)
Tuberculosis (TB), 5
   diagnosis, 10–11
   epidemiology, 9–10
   overview in Africa, 8
   treatment, 11
   tuberculosis program, 11
      DOTS and TB/HIV collaborative activities, 11
      WHO response, 11
Tumor necrosis factor (TNF), 77, 80
Tumor necrosis factor-α (TNF-α), 239, 241, 255–257, 490, 502, 630
   induced neutrophil adherence, 409

Tumor necrosis factor-beta (TNF-β), 81
Tumor necrosis factor-induced nuclear factor (NF-κB) activation, 429
Tumor suppressor genes, 291
TUNEL assay. *See* TdT-mediated dUTP nick-end labeling (TUNEL) assay
Turmeric, 334, 338
   antimicrobial property, 338
   as heal sores, 338
   pharmacological potential, 338
   treatment of Epilepsy, 338
   use in, treatment of stomach and liver ailments, 338
Turnip, 286. *See also Brassica rapa*
   root parsley, 528
Type 2 diabetes, effect of onion peel extracts, 357

U

U937 leukemia cells, 284
U87MGΔ*EGFR* cells, 284
Undernutrition, in early childhood, 62
Uric acid (URIC), 412
US Food and Drug Administration (FDA), 195

V

Vaccinium angustifolium, 317
Valeriana jatamansi, 162
Valeric ester, 404
valley fever. *See* Coccidioidomycosis
Vanillic acid, 367
Vanillin, 613
Vasculopathy, 125
Vegetable
   antioxidant effects, 334
   chemical structures of, antioxidant compounds, 337
   definition, 329
   as dietary fiber, 330
   health-promoting effects, 330
   oxidative stress, 334
   role in in human fertility, 344
Vernodalol, 209
Vernolide, 209
Vernonia amygdalina, 209
Vibrio cholerae, 211
Vigna unguiculata. *See* Cowpea
Vimang, analgesic/antiinflammatory effects of, 476

2-Vinyl-4-*H*-1,3-dithiin, 366
3-Vinyl-4-*H*-1,2-dithiin, 366
Violaxanthin, 535
Vismia rubescens, 212
Vitamins A/D, 485
Vitis vinifera, 317
Volatile oils, 628
Vomiting, 299
    physiological mechanisms, 300

W

Wagner's test, 216
Wakana syndrome, 35
Warburgia salutaris, 162
Western medicine, 163
WHO. *See* World Health Organization (WHO)
Wild mint, 304
World Food Programme (WFP), 57
World Health Organization (WHO), 4, 110, 171, 271
Writhing, definition, 242

X

Xanthine oxidase model system, 493
Xanthomicrol, 601
XTT. *See* 2,3-Bis (2-methoxy-4-nitro-5-sulfophenyl)-5-[(phenylamino) carbonyl]-2H-tetrazolium hydroxide (XTT)
XTT-microculture tetrazolium assay, 273
ρ−Xylene, 532
Xylopia aethiopia, 307
Xylopia aethiopica, 220, 249, 581
    antiinflammatory or antinociceptive potential, 249–250
Xylopia parviflora, 286
Xylopic acid, 249

Y

Years lost due to disease (YLDs), 3

Z

Zingerone, 628
Zingiberaceae, 248, 338, 425
    antimicrobial effects, 226–227
Zingiber officinale, 227, 292, 307, 323, 628. *See also* Ginger; ginger
    mechanism of action, 323
    medicinal values of, 323
Z-ocimene, 405